Matrix Computations

Johns Hopkins Studies in the Mathematical Sciences
in association with the Department of Mathematical Sciences
The Johns Hopkins University

Matrix Computations

THIRD EDITION

Gene H. Golub
Department of Computer Science
Stanford University

Charles F. Van Loan
Department of Computer Science
Cornell University

The Johns Hopkins University Press
Baltimore and London

©1983, 1989, 1996 The Johns Hopkins University Press
All rights reserved. Published 1996
Printed in the United States of America on acid-free paper
05 04 03 02 01 00 99 98 97 96   5 4 3 2 1

First edition 1983
Second edition 1989
Third Edition 1996

The Johns Hopkins University Press
2715 North Charles Street
Baltimore, Maryland 21218-4319
The Johns Hopkins Press Ltd., London

Library of Congress Cataloging-in-Publication Data will be found
at the end of this book.

A catalog record for this book is available from the British Library.

ISBN 0-8018-5413-X
ISBN 0-8018-5414-8   (pbk.)

Contents

4   Special Linear Systems   133

5   Orthogonalization and Least Squares   206

6   Parallel Matrix Computations   275

7   The Unsymmetric Eigenvalue Problem   308

8   The Symmetric Eigenvalue Problem   391

Preface to the Third Edition

The field of matrix computations continues to grow and mature. In the *Third Edition* we have added over 300 new references and 100 new problems. The LINPACK and EISPACK citations have been replaced with appropriate pointers to LAPACK with key codes tabulated at the *beginning* of appropriate chapters.

In the *First Edition* and *Second Edition* we identified a small number of global references: Wilkinson (1965), Forsythe and Moler (1967), Stewart (1973), Hanson and Lawson (1974) and Parlett (1980). These volumes are as important as ever to the research landscape, but there are some magnificent new textbooks and monographs on the scene. See *The Literature* section that follows.

We continue as before with the practice of giving references at the end of each section and a master bibliography at the end of the book.

The earlier editions suffered from a large number of typographical errors and we are obliged to the dozens of readers who have brought these to our attention. Many corrections and clarifications have been made.

Here are some specific highlights of the new edition. Chapter 1 (Matrix Multiplication Problems) and Chapter 6 (Parallel Matrix Computations) have been completely rewritten with less formality. We think that this facilitates the building of intuition for high performance computing and draws a better line between algorithm and implementation on the printed page.

In Chapter 2 (Matrix Analysis) we expanded the treatment of CS decomposition and included a proof. The overview of floating point arithmetic has been brought up to date. In Chapter 4 (Special Linear Systems) we embellished the Toeplitz section with connections to circulant matrices and the fast Fourier transform. A subsection on equilibrium systems has been included in our treatment of indefinite systems.

A more accurate rendition of the modified Gram-Schmidt process is offered in Chapter 5 (Orthogonalization and Least Squares). Chapter 8 (The Symmetric Eigenproblem) has been extensively rewritten and rearranged so as to minimize its dependence upon Chapter 7 (The Unsymmetric Eigenproblem). Indeed, the coupling between these two chapters is now so minimal that it is possible to read either one first.

In Chapter 9 (Lanczos Methods) we have expanded the discussion of the unsymmetric Lanczos process and the Arnoldi iteration. The "unsymmetric component" of Chapter 10 (Iterative Methods for Linear Systems) has likewise been broadened with a whole new section devoted to various Krylov space methods designed to handle the sparse unsymmetric linear system problem.

In §12.5 (Updating Orthogonal Decompositions) we included a new subsection on ULV updating. Toeplitz matrix eigenproblems and orthogonal matrix eigenproblems are discussed in §12.6.

Both of us look forward to continuing the dialog with our readers. As we said in the Preface to the *Second Edition*, "It has been a pleasure to deal with such an interested and friendly readership."

Many individuals made valuable *Third Edition* suggestions, but Greg Ammar, Mike Heath, Nick Trefethen, and Steve Vavasis deserve special thanks.

Finally, we would like to acknowledge the support of Cindy Robinson at Cornell. A dedicated assistant makes a big difference.

Software

LAPACK

Many of the algorithms in this book are implemented in the software package LAPACK:

E. Anderson, Z. Bai, C. Bischof, J. Demmel, J. Dongarra, J. DuCroz, A. Greenbaum, S. Hammarling, A. McKenney, S. Ostrouchov, and D. Sorensen (1995). *LAPACK Users' Guide, Release 2.0, 2nd ed.*, SIAM Publications, Philadelphia.

Pointers to some of the more important routines in this package are given at the *beginning* of selected chapters:

Chapter 1.	Level-1, Level-2, Level-3 BLAS
Chapter 3.	General Linear Systems
Chapter 4.	Positive Definite and Band Systems
Chapter 5.	Orthogonalization and Least Squares Problems
Chapter 7.	The Unsymmetric Eigenvalue Problem
Chapter 8.	The Symmetric Eigenvalue Problem

Our LAPACK references are spare in detail but rich enough to "get you started." Thus, when we say that _TRSV can be used to solve a triangular system $Ax = b$, we leave it to you to discover through the LAPACK manual that A can be either upper or lower triangular and that the transposed system $A^T x = b$ can be handled as well. Moreover, the underscore is a placeholder whose mission is to designate type (single, double, complex, etc).

LAPACK stands on the shoulders of two other packages that are milestones in the history of software development. EISPACK was developed in the early 1970s and is dedicated to solving symmetric, unsymmetric, and generalized eigenproblems:

B.T. Smith, J.M. Boyle, Y. Ikebe, V.C. Klema, and C.B. Moler (1970). *Matrix Eigensystem Routines: EISPACK Guide, 2nd ed.*, Lecture Notes in Computer Science, Volume 6, Springer-Verlag, New York.

B.S. Garbow, J.M. Boyle, J.J. Dongarra, and C.B. Moler (1972). *Matrix Eigensystem Routines: EISPACK Guide Extension,* Lecture Notes in Computer Science, Volume 51, Springer-Verlag, New York.

LINPACK was developed in the late 1970s for linear equations and least squares problems:

EISPACK and LINPACK have their roots in sequence of papers that feature Algol implementations of some of the key matrix factorizations. These papers are collected in

J.H. Wilkinson and C. Reinsch, eds. (1971). *Handbook for Automatic Computation, Vol. 2, Linear Algebra,* Springer-Verlag, New York.

NETLIB

A wide range of software including LAPACK, EISPACK, and LINPACK is available electronically via Netlib:

World Wide Web: `http://www.netlib.org/index.html`
Anonymous ftp: `ftp://ftp.netlib.org`

Via email, send a one-line message:

```
mail netlib@ornl.gov
send index
```

to get started.

MATLAB®

Complementing LAPACK and defining a very popular matrix computation environment is MATLAB:

MATLAB *User's Guide,* The MathWorks Inc., Natick, Massachusetts.

M. Marcus (1993). *Matrices and* MATLAB: *A Tutorial,* Prentice Hall, Upper Saddle River, NJ.

R. Pratap (1995). *Getting Started with* MATLAB, Saunders College Publishing, Fort Worth, TX.

Many of the problems in *Matrix Computations* are best posed to students as MATLAB problems. We make extensive use of MATLAB notation in the presentation of algorithms.

Selected References

Each section in the book concludes with an annotated list of references. A master bibliography is given at the end of the text.

Useful books that collectively cover the field, are cited below. Chapter titles are included if appropriate but do not infer too much from the level of detail because one author's chapter may be another's subsection. The citations are classified as follows:

Pre-1970 Classics. Early volumes that set the stage.
Introductory (General). Suitable for the undergraduate classroom.
Advanced (General). Best for practitioners and graduate students.
Analytical. For the supporting mathematics.
Linear Equation Problems. $Ax = b$.
Linear Fitting Problems. $Ax \approx b$.
Eigenvalue Problems. $Ax = \lambda x$.
High Performance. Parallel/vector issues.
Edited Volumes. Useful, thematic collections.

Within each group the entries are specified in chronological order.

Pre-1970 Classics

V.N. Faddeeva (1959). *Computational Methods of Linear Algebra*, Dover, New York.

Basic Material from Linear Algebra. Systems of Linear Equations. The Proper Numbers and Proper Vectors of a Matrix.

E. Bodewig (1959). *Matrix Calculus*, North Holland, Amsterdam.

Matrix Calculus. Direct Methods for Linear Equations. Indirect Methods for Linear Equations. Inversion of Matrices. Geodetic Matrices. Eigenproblems.

R.S. Varga (1962). *Matrix Iterative Analysis*, Prentice-Hall, Englewood Cliffs, NJ.

Matrix Properties and Concepts. Nonnegative Matrices. Basic Iterative Methods and Comparison Theorems. Successive Overrelaxation Iterative Methods. Semi-Iterative Methods. Derivation and Solution of Elliptic Difference Equations. Alternating Direction Implicit Iterative Methods. Matrix Methods for Parabolic Partial Differential Equations. Estimation of Acceleration Parameters.

J.H. Wilkinson (1963). *Rounding Errors in Algebraic Processes*, Prentice-Hall, Englewood Cliffs, NJ.

The Fundamental Arithmetic Operations. Computations Involving Polynomials. Matrix Computations.

A.S. Householder (1964). *Theory of Matrices in Numerical Analysis*, Blaisdell, New York. Reprinted in 1974 by Dover, New York.

Some Basic Identities and Inequalities. Norms, Bounds, and Convergence. Localization Theorems and Other Inequalities. The Solution of Linear Systems: Methods of Successive Approximation. Direct Methods of Inversion. Proper Values and Vectors: Normalization and Reduction of the Matrix. Proper Values and Vectors: Successive Approximation.

L. Fox (1964). *An Introduction to Numerical Linear Algebra*, Oxford University Press, Oxford, England.

Introduction, Matrix Algebra. Elimination Methods of Gauss, Jordan, and Aitken. Compact Elimination Methods of Doolittle, Crout, Banachiewicz, and Cholesky. Orthogonalization Methods. Condition, Accuracy, and Precision. Comparison of Methods, Measure of Work. Iterative and Gradient Methods. Iterative methods for Latent Roots and Vectors. Transformation Methods for Latent Roots and Vectors. Notes on Error Analysis for Latent Roots and Vectors.

J.H. Wilkinson (1965). *The Algebraic Eigenvalue Problem,* Clarendon Press, Oxford, England.

Theoretical Background. Perturbation Theory. Error Analysis. Solution of Linear Algebraic Equations. Hermitian Matrices. Reduction of a General Matrix to Condensed Form. Eigenvalues of Matrices of Condensed Forms. The LR and QR Algorithms. Iterative Methods.

G.E. Forsythe and C. Moler (1967). *Computer Solution of Linear Algebraic Systems,* Prentice-Hall, Englewood Cliffs, NJ.

Reader's Background and Purpose of Book. Vector and Matrix Norms. Diagonal Form of a Matrix Under Orthogonal Equivalence. Proof of Diagonal Form Theorem. Types of Computational Problems in Linear Algebra. Types of Matrices encountered in Practical Problems. Sources of Computational Problems of Linear Algebra. Condition of a Linear System. Gaussian Elimination and LU Decomposition. Need for Interchanging Rows. Scaling Equations and Unknowns. The Crout and Doolittle Variants. Iterative Improvement. Computing the Determinant. Nearly Singular Matrices. Algol 60 Program. Fortran, Extended Algol, and PL/I Programs. Matrix Inversion. An Example: Hilbert Matrices. Floating Point Round-Off Analysis. Rounding Error in Gaussian Elimination. Convergence of Iterative Improvement. Positive Definite Matrices; Band Matrices. Iterative Methods for Solving Linear Systems. Nonlinear Systems of Equations.

Introductory (General)

A.R. Gourlay and G.A. Watson (1973). *Computational Methods for Matrix Eigenproblems*, John Wiley & Sons, New York.

Introduction. Background Theory. Reductions and Transformations. Methods for the Dominant Eigenvalue. Methods for the Subdominant Eigenvalue. Inverse Iteration. Jacobi's Methods. Givens and Householder's Methods. Eigensystem of a Symmetric Tridiagonal Matrix. The LR and QR Algorithms. Extensions of Jacobi's Method. Extension of Givens' and Householder's Methods. QR Algorithm for Hessenberg Matrices. generalized Eigenvalue Problems. Available Implementations.

G.W. Stewart (1973). *Introduction to Matrix Computations*, Academic Press, New York.

Preliminaries. Practicalities. The Direct Solution of Linear Systems. Norms, Limits, and Condition Numbers. The Linear Least Squares Problem. Eigenvalues and Eigenvectors. The QR Algorithm.

R.J. Goult, R.F. Hoskins, J.A. Milner and M.J. Pratt (1974). *Computational Methods in Linear Algebra*, John Wiley and Sons, New York.

Eigenvalues and Eigenvectors. Error Analysis. The Solution of Linear Equations by Elimination and Decomposition Methods. The Solution of Linear Systems of Equations by Iterative Methods. Errors in the Solution Sets of Equations. Computation of Eigenvalues and Eigenvectors. Errors in Eigenvalues and Eigenvectors. Appendix – A Survey of Essential Results from Linear Algebra.

T.F. Coleman and C.F. Van Loan (1988). *Handbook for Matrix Computations*, SIAM Publications, Philadelphia, PA.

Fortran 77, The Basic Linear Algebra Subprograms, Linpack, MATLAB.

W.W. Hager (1988). *Applied Numerical Linear Algebra*, Prentice-Hall, Englewood Cliffs, NJ.

Introduction. Elimination Schemes. Conditioning. Nonlinear Systems. Least Squares. Eigenproblems. Iterative Methods.

P.G. Ciarlet (1989). *Introduction to Numerical Linear Algebra and Optimisation*, Cambridge University Press.

A Summary of Results on Matrices. General Results in the Numerical Analysis of Matrices. Sources of Problems in the Numerical Analysis of Matrices. Direct Methods for the Solution of Linear Systems. Iterative Methods for the Solution of Linear Systems. Methods for the Calculation of Eigenvalues and Eigenvectors. A Review of Differential Calculus. Some Applications. General Results on Optimization. Some Algorithms. Introduction to Nonlinear Programming. Linear Programming.

D.S. Watkins (1991). *Fundamentals of Matrix Computations*, John Wiley and Sons, New York.

Gaussian Elimination and Its Variants. Sensitivity of Linear Systems; Effects of Roundoff Errors. Orthogonal Matrices and the Least-Squares Problem. Eigenvalues and Eigenvectors I. Eigenvalues and Eigenvectors II. Other Methods for the Symmetric Eigenvalue Problem. The Singular Value Decomposition.

P. Gill, W. Murray, and M.H. Wright (1991). *Numerical Linear Algebra and Optimization, Vol. 1*, Addison-Wesley, Reading, MA.

Introduction. Linear Algebra Background. Computation and Condition. Linear Equations. Compatible Systems. Linear Least Squares. Linear Constraints I: Linear Programming. The Simplex Method.

A. Jennings and J.J. McKeowen (1992). *Matrix Computation (2nd ed)*, John Wiley and Sons, New York.

Basic Algebraic and Numerical Concepts. Some Matrix Problems. Computer Implementation. Elimination Methods for Linear Equations. Sparse Matrix Elimination. Some Matrix Eigenvalue Problems. Transformation Methods for Eigenvalue Problems. Sturm Sequence Methods. Vector Iterative Methods for Partial Eigensolution. Orthogonalization and Re-Solution Techniques for Linear Equations. Iterative Methods for Linear Equations. Non-linear Equations. Parallel and Vector Computing.

B.N. Datta (1995). *Numerical Linear Algebra and Applications*. Brooks/Cole Publishing Company, Pacific Grove, California.

Review of Required Linear Algebra Concepts. Floating Point Numbers and Errors in Computations. Stability of Algorithms and Conditioning of Problems. Numerically Effective Algorithms and Mathematical Software. Some Useful Transformations in Numerical Linear Algebra and Their Applications. Numerical Matrix Eigenvalue Problems. The Generalized Eigenvalue Problem. The Singular Value Decomposition. A Taste of Roundoff Error Analysis.

M.T. Heath (1997). *Scientific Computing: An Introductory Survey*, McGraw-Hill, New York.

Scientific Computing. Systems of Linear Equations. Linear Least Squares. Eigenvalues and Singular Values. Nonlinear Equations. Optimization. Interpolation. Numerical Integration and Differentiation. Initial Value Problems for ODEs. Boundary Value Problems for ODEs. Partial Differential Equations. Fast Fourier Transform. Random Numbers and Simulation.

C.F. Van Loan (1997). *Introduction to Scientific Computing: A Matrix-Vector Approach Using Matlab*, Prentice Hall, Upper Saddle River, NJ.

Power Tools of the Trade. Polynomial Interpolation. Piecewise Polynomial Interpolation. Numerical Integration. Matrix Computations. Linear Systems. The QR and Cholesky Factorizations. Nonlinear Equations and Optimization. The Initial Value Problem.

Advanced (General)

N.J. Higham (1996). *Accuracy and Stability of Numerical Algorithms*, SIAM Publications, Philadelphia, PA.

Principles of Finite Precision Computation. Floating Point Arithmetic. Basics. Summation. Polynomials. Norms. Perturbation Theory for Linear Systems. Triangular Systems. LU Factorization and Linear Equations. Cholesky Factorization. Iterative Refinement. Block LU Factorization. Matrix Inversion. Condition Number Estimation. The Sylvester Equation. Stationary Iterative Methods. Matrix Powers. QR Factorization. The Least Squares Problem. Underdetermined Systems. Vandermonde Systems. Fast Matrix Multiplication. The Fast Fourier Transform and Applications. Automatic Error Analysis. Software Issues in Floating Point Arithmetic. A Gallery of Test Matrices.

J.W. Demmel (1996). *Numerical Linear Algebra*, SIAM Publications, Philadelphia, PA.

> Introduction. Linear Equation Solving. Linear Least Squares Problems. Nonsymmetric Eigenvalue Problems. The Symmetric Eigenproblem and Singular Value Decomposition. Iterative Methods for Linear Systems and Eigenvalue Problems. Iterative Algorithms for Eigenvalue Problems.

L.N. Trefethen and D. Bau III (1997). *Numerical Linear Algebra*, SIAM Publications, Philadelphia, PA.

> Matrix-Vector Multiplication. Orthogonal Vectors and Matrices. Norms. The Singular Value Decomposition. More on the SVD. Projectors. QR Factorization. Gram-Schmidt Orthogonalization. MATLAB. Householder Triangularization. Least-Squares Problems. Conditioning and Condition Numbers. Floating Point Arithmetic. Stability. More on Stability. Stability of Householder Triangularization. Stability of Back Substitution. Conditioning of Least-Squares Problems. Stability of Least-Squares Algorithms. Gaussian Elimination. Pivoting. Stability of Gaussian Elimination. Cholesky Factorization. Eigenvalue Problems. Overview of Eigenvalue Algorithms. Reduction to Hessenberg/Tridiagonal Form. Rayleigh Quotient, Inverse Iteration. QR Algorithm Without Shifts. QR Algorithm With Shifts. Other Eigenvalue Algorithms. Computing the SVD. Overview of Iterative Methods. The Arnoldi Iteration. How Arnoldi Locates Eigenvalues. GMRES. The Lanczos Iteration. Orthogonal Polynomials and Gauss Quadrature. Conjugate Gradients. Biorthogonalization Methods. Preconditioning. The Definition of Numerical Analysis.

Analytical

F.R. Gantmacher (1959). *The Theory of Matrices Vol. 1*, Chelsea, New York.

> Matrices and Operations on Matrices. The Algorithm of Gauss and Some of its Applications. Linear Operators in an n-dimensional Vector Space. The Characteristic Polynomial and the Minimum Polynomial of a Matrix. Functions of Matrices, Equivalent Transformations of Polynomial Matrices, Analytic Theory of Elementary Divisors. The Structure of a Linear Operator in an n-dimensional Space. Matrix Equations. Linear Operators in a Unitary Space. Quadratic and Hermitian Forms.

F.R. Gantmacher (1959). *The Theory of Matrices Vol. 2*, Chelsea, New York.

> Complex Symmetric, Skew-Symmetric, and Orthogonal Matrices. Singular Pencils of Matrices. Matrices with Nonnegative Elements. Application of the Theory of Matrices to the Investigation of Systems of Linear Differential Equations. The Problem of Routh-Hurwitz and Related Questions.

A. Berman and R.J. Plemmons (1979). *Nonnegative Matrices in the Mathematical Sciences*, Academic Press, New York. Reprinted with additions in 1994 by SIAM Publications, Philadelphia, PA.

> Matrices Which Leave a Cone Invariant. Nonnegative Matrices. Semigroups of Nonnegative Matrices. Symmetric Nonnegative Matrices. Generalized Inverse-Positivity. M-Matrices. Iterative Methods for Linear Systems. Finite Markov Chains. Input-Output Analysis in Economics. The Linear Complementarity Problem.

G.W. Stewart and J. Sun (1990). *Matrix Perturbation Theory*, Academic Press, San Diego.

Preliminaries. Norms and Metrics. Linear Systems and Least Squares Problems. The Perturbation of Eigenvalues. Invariant Subspaces. Generalized Eigenvalue Problems.

R. Horn and C. Johnson (1985). *Matrix Analysis*, Cambridge University Press, New York.

Review and Miscellanea. Eigenvalues, Eigenvectors, and Similarity. Unitary Equivalence and Normal Matrices. Canonical Forms. Hermitian and Symmetric Matrices. Norms for Vectors and Matrices. Location and Perturbation of Eigenvalues. Positive Definite Matrices.

R. Horn and C. Johnson (1991). *Topics in Matrix Analysis*, Cambridge University Press, New York.

The Field of Values. Stable Matrices and Inertia. Singular Value Inequalities. Matrix Equations and the Kronecker Product. The Hadamard Product. Matrices and Functions.

Linear Equation Problems

D.M. Young (1971). *Iterative Solution of Large Linear Systems*, Academic Press, New York.

Introduction. Matrix Preliminaries. Linear Stationary Iterative Methods. Convergence of the Basic Iterative Methods. Eigenvalues of the SOR Method for Consistently Ordered Matrices. Determination of the Optimum Relaxation Parameter. Norms of the SOR Method. The Modified SOR Method: Fixed Parameters. Nonstationary Linear Iterative Methods. The Modified SOR Method: Variable Parameters. Semi-Iterative Methods. Extensions of the SOR Theory; Stieltjes Matrices. Generalized Consistently Ordered Matrices. Group Iterative Methods. Symmetric SOR Method and Related Methods. Second Degree Methods. Alternating Direction Implicit Methods. Selection of an Iterative Method.

L.A. Hageman and D.M. Young (1981). *Applied Iterative Methods*, Academic Press, New York.

Background on Linear Algebra and Related Topics. Background on Basic Iterative Methods. Polynomial Acceleration. Chebyshev Acceleration. An Adaptive Chebyshev Procedure Using Special Norms. Adaptive Chebyshev Acceleration. Conjugate Gradient Acceleration. Special Methods for Red/Black Partitionings. Adaptive Procedures for Successive Overrelaxation Method. The Use of Iterative Methods in the Solution of Partial Differential Equations. Case Studies. The Nonsymmetrizable Case.

A. George and J. W-H. Liu (1981). *Computer Solution of Large Sparse Positive Definite Systems.* Prentice-Hall Inc., Englewood Cliffs, New Jersey.

Introduction. Fundamentals. Some Graph Theory Notation and Its Use in the Study of Sparse Symmetric Matrices. BAnd and Envelope Methods. General Sparse Methods. Quotient Tree Methods for Finite Element and Finite Difference Problems. One-Way Dissection Methods for Finite Element Problems. Nested Dissection Methods. Numerical Experiments.

S. Pissanetsky (1984). *Sparse Matrix Technology,* Academic Press, New York.

Fundamentals. Linear Algebraic Equations. Numerical Errors in Gaussian Elimination. Ordering for Gauss Elimination: Symmetric Matrices. Ordering for Gauss Elimination: General Matrices. Sparse Eigenanalysis. Sparse Matrix Algebra. Connectivity and Nodal Assembly. General Purpose Algorithms.

I.S. Duff, A.M. Erisman, and J.K. Reid (1986). *Direct Methods for Sparse Matrices,* Oxford University Press, New York.

Introduction. Sparse Matrices:Storage Schemes and Simple Operations. Gaussian Elimination for Dense Matrices: The Algebraic Problem. Gaussian Elimination for Dense Matrices: Numerical Considerations. Gaussian Elimination for Sparse Matrices: An Introduction. Reduction to Block Triangular Form. Local Pivotal Strategies for Sparse Matrices. Ordering Sparse Matrices to Special Forms. Implementing Gaussian Elimination: Analyse with Numerical Values. Implementing Gaussian Elimination with Symbolic Analyse. Partitioning, Matrix Modification, and Tearing. Other Sparsity-Oriented Issues.

R. Barrett, M. Berry, T.F. Chan, J. Demmel, J. Donato, J. Dongarra, V. Eijkhout, R. Pozo, C. Romine, H. van der Vorst (1993). *Templates for the Solution of Linear Systems: Building Blocks for Iterative Methods,* SIAM Publications, Philadelphia, PA.

Introduction. Why Use Templates? What Methods are Covered? Iterative Methods. Stationary Methods. Nonstationary Iterative Methods. Survey of Recent Krylov Methods. Jacobi, Incomplete, SSOR, and Polynomial Preconditioners. Complex Systems. Stopping Criteria. Data Structures. Parallelism. The Lanczos Connection. Block Iterative Methods. Reduced System Preconditioning. Domain Decomposition Methods. Multigrid Methods. Row Projection Methods.

W. Hackbusch (1994). *Iterative Solution of Large Sparse Systems of Equations,* Springer-Verlag, New York.

Introduction. Recapitulation of Linear Algebra. Iterative Methods. Methods of Jacobi and Gauss-Seidel and SOR Iteration in the Positive Definite Case. Analysis in the 2-Cyclic Case. Analysis for M-Matrices. Semi-Iterative Methods. Transformations, Secondary Iterations, Incomplete Triangular Decompositions. Conjugate Gradient Methods. Multi-Grid Methods. Domain Decomposition Methods.

O. Axelsson (1994). *Iterative Solution Methods*, Cambridge University Press.

> Direct Solution Methods. Theory of Matrix Eigenvalues. Positive Definite Matrices, Schur Complements, and Generalized Eigenvalue Probems. Reducible and Irreducible Matrices and the Perron-Frobenious Theory for Nonnegative Matrices. Basic Iterative Methods and Their Rates of Convergence. M-Matrices, Convergent Splittings, and the SOR Method. Incomplete Factorization Preconditioning Methods. Approximate Matrix Inverses and Corresponding Preconditioning Methods. Block Diagonal and Schur Complement Preconditionngs. Estimates of Eigenvalues and Condition Numbers for Preconditioned Matrices. Conjugate Gradient and Lanczos-Type Methods. Generalized Conjugate Gradient Methods. The Rate of Convergence of the Conjugate Gradient Method.

Y. Saad (1996). *Iterative Methods for Sparse Linear Systems*, PWS Publishing Co., Boston.

> Background in Linear Algebra. Discretization of PDEs. Sparse Matrices. Basic Iterative Methods. Projection Methods. Krylov Subspace Methods – Part I. Krylov Subspace Methods – Part II. Methods Related to the Normal Equations. Preconditioned Iterations. Preconditioning Techniques. Parallel Implementations. Parallel Preconditioners. Domain Decomposition Methods.

Linear Fitting Problems

C.L. Lawson and R.J. Hanson (1974). *Solving Least Squares Problems*, Prentice-Hall, Englewood Cliffs, NJ. Reprinted with a detailed "new developments" appendix in 1996 by SIAM Publications, Philadelphia, PA.

> Introduction. Analysis of the Least Squares Problem. Orthogonal Decomposition by Certain Elementary Transformations. Orthogonal Decomposition by Singular Value Decomposition. Perturbation Theorems for Singular Values. Bounds for the Condition Number of a Triangular Matrix. The Pseudoinverse. Perturbation Bounds for the Pseudoinverse. Perturbation Bounds for the Solution of Problem LS. Numerical Computations Using Elementary Orthogonal Transformations. Computing the Solution for the Overdetermined or Exactly Determined Full Rank Problem. Computation of the Covariance Matrix of the Solution Parameters. Computing the Solution for the Underdetermined Full Rank Problem. Computing the Solution for Problem LS with Possibly Deficient Pseudorank. Analysis of Computing Errors for Householder Transformations. Analysis of Computing Errors for the Problem LS. Analysis of Computing Errors for the Problem LS Using Mixed Precision Arithmetic. Computation of the Singular Value Decomposition and the Solution of Problem LS. Other Methods for Least Squares Problems. Linear Least Squares with Linear Equality Constraints Using a Basis of the Null Space. Linear Least Squares with Linear Equality Constraints by Direct Elimination. Linear Least Squares with Linear Equality Constraints by Weighting. Linear least Squares with Linear Inequality Constraints. Modifying a QR Decomposition to Add or Remove Column Vectors. Practical Analysis of Least Squares Problems. Examples of Some Methods of Analyzing a Least Squares Problem. Modifying a QR Decomposition to Add or Remove Row Vectors with Application to Sequential Processing of Problems Having a Large or Banded Coefficient Matrix.

R.W. Farebrother (1987). *Linear Least Squares Computations*, Marcel Dekker, New York.

> The Gauss and Gauss–Jordan Methods. Matrix Analysis of Gauss's Method: The Cholesky and Doolittle Decompositions. The Linear Algebraic Model: The Method of Averages and the Method of Least Squares. The Cauchy-Bienayme, Laplace, and Schmidt Procedures. Householder Procedures. Givens Procedures. Updating the QU Decomposition. Pseudorandom Numbers. The Standard Linear Model. Condition Numbers. Instrumental Variable Estimators. Generalized Least Squares Estimation. Iterative Solutions of Linear and Nonlinear Least Squares Problems. Canonical Expressions for the Least Squares Estimators and Test Statistics. Traditional Expressions for the Least Squares Updating Formulas and test Statistics. Least Squares Estimation Subject to Linear Constraints.

S. Van Huffel and J. Vandewalle (1991). *The Total Least Squares Problem: Computational Aspects and Analysis*, SIAM Publications, Philadelphia, PA.

> Introduction. Basic Principles of the Total Least Squares Problem. Extensions of the Basic Total Least Squares Problem. Direct Speed Improvement of the Total Least Squares Computations. Iterative Speed Improvement for Solving Slowly Varying Total Least Squares Problems. Algebraic Connections Between Total Least Squares and Least Squares Problems. Sensitivity Analysis of Total Least Squares and Least Squares Problems in the Presence of Errors in All Data. Statistical Properties of the Total Least Squares Problem. Algebraic Connections Between Total Least Squares Estimation and Classical Linear Regression in Multicollinearity Problems. Conclusions.

Å. Björck (1996). *Numerical Methods for Least Squares Problems*, SIAM Publications, Philadelphia, PA.

> Mathematical and Statistical Properties of Least Squares Solutions. Basic Numerical Methods. Modified Least Squares Problems. Generalized Least Squares Problems. Constrained Least Squares Problems. Direct Methods for Sparse Least Squares Problems. Iterative Methods for Least Squares Problems. Least Squares with Special Bases. Nonlinear Least Squares Problems.

Eigenvalue Problems

B.N. Parlett (1980). *The Symmetric Eigenvalue Problem,* Prentice-Hall, Englewood Cliffs, NJ.

> Basic Facts about Self-Adjoint Matrices. Tasks, Obstacles, and Aids. Counting Eigenvalues. Simple Vector Iterations. Deflation. Useful Orthogonal Matrices. Tridiagonal Form. The QL and QR Algorithms. Jacobi Methods. Eigenvalue Bounds. Approximation from a Subspace. Krylov Subspaces. Lanczos Algorithms. Subspace Iteration. The General Linear Eigenvalue Problem.

J. Cullum and R.A. Willoughby (1985a). *Lanczos Algorithms for Large Symmetric Eigenvalue Computations, Vol. I Theory*, Birkhaüser, Boston.

> Preliminaries: Notation and Definitions. Real Symmetric Problems. Lanczos Procedures, Real Symmetric Problems. Tridiagonal Matrices. Lanczos Procedures with No Reorthogonalization for Symmetric Problems. Real Rectangular Matrices. Non-Defective Complex Symmetric Matrices. Block Lanczos Procedures, Real Symmetric Matrices.

J. Cullum and R.A. Willoughby (1985b). *Lanczos Algorithms for Large Symmetric Eigenvalue Computations, Vol. II Programs,* Birkhaüser, Boston.

> Lanczos Procedures. Real Symmetric Matrices. Hermitian Matrices. Factored Inverses of Real Symmetric Matrices. Real Symmetric Generalized Problems. Real Rectangular Problems. Nondefective Complex Symmetric Matrices. Real Symmetric Matrices, Block Lanczos Code. Factored Inverses, Real Symmetric Matrices, Block Lanczos Code.

Y. Saad (1992). *Numerical Methods for Large Eigenvalue Problems: Theory and Algorithms,* John Wiley and Sons, New York.

> Background in Matrix Theory and Linear Algebra. Perturbation Theory and Error Analysis. The Tools of Spectral Approximation. Subspace Iteration. Krylov Subspace Methods. Acceleration Techniques and Hybrid Methods. Preconditioning Techniques. Non-Standard Eigenvalue Problems. Origins of Matrix Eigenvalue Problems.

F. Chatelin (1993). *Eigenvalues of Matrices,* John Wiley and Sons, New York.

> Supplements from Linear Algebra. Elements of Spectral Theory. Why Compute Eigenvalues. Error Analysis. Foundations of Methods for Computing Eigenvalues. Numerical Methods for Large Matrices. Chebyshev's Iterative Methods.

High Performance

W. Schönauer (1987). *Scientific Computing on Vector Computers,* North Holland, Amsterdam.

> Introduction. The First Commercially Significant Vector Computer. The Arithmetic Performance of the First Commercially Significant Vector Computer. Hockney's $n^{1/2}$ and Timing Formulae. Fortran and Autovectorization. Behavior of Programs. Some Basic Algorithms, Recurrences. Matrix Operations. Systems of Linear Equations with Full Matrices. Tridiagonal Linear Systems. The Iterative Solution of Linear Equations. Special Applications. The Fujitsu VPs and Other Japanese Vector Computers. The Cray-2. The IBM VF and Other Vector Processors. The Convex C1.

R.W. Hockney and C.R. Jesshope (1988). *Parallel Computers 2,* Adam Hilger, Bristol and Philadelphia.

> Introduction. Pipelined Computers. Processor Arrays. Parallel Languages. Parallel Algorithms. Future Developments.

J.J. Modi (1988).*Parallel Algorithms and Matrix Computation,* Oxford University Press, Oxford.

> General Principles of Parallel Computing. Parallel Techniques and Algorithms. Parallel Sorting Algorithms. Solution of a System of Linear Algebraic Equations. The Symmetric Eigenvalue Problem: Jacobi's Method. QR Factorization. Singular Value Decomposition and Related Problems.

J. Ortega (1988). *Introduction to Parallel and Vector Solution of Linear Systems*, Plenum Press, New York.

Introduction. Direct Methods for Linear Equations. Iterative Methods for Linear Equations.

J. Dongarra, I. Duff, D. Sorensen, and H. van der Vorst (1990). *Solving Linear Systems on Vector and Shared Memory Computers*, SIAM Publications, Philadelphia, PA.

Vector and Parallel Processing. Overview of Current High-Performance Computers. Implementation Details and Overhead. Performance Analysis, Modeling, and Measurements. Building Blocksin Linear Algebra. Direct Solution of Sparse Linear Systems. Iterative Solution of Sparse Linear Systems.

Y. Robert (1990). *The Impact of Vector and Parallel Architectures on the Gaussian Elimination Algorithm*, Halsted Press, New York.

Introduction. Vector and Parallel Architectures. Vector Multiprocessor Computing. Hypercube Computing. Systolic Computing. Task Graph Scheduling. Analysis of Distributed Algorithms. Design Methodologies.

G.H. Golub and J.M. Ortega (1993). *Scientific Computing: An Introduction with Parallel Computing*, Academic Press, Boston.

The World of Scientific Computing. Linear Algebra. Parallel and Vector Computing. Polynomial Approximation. Continuous Problems Solved Discretely. Direct Solution of Linear Equations. Parallel Direct Methods. Iterative Methods. Conjugate Gradient-Type Methods.

Edited Volumes

D.J. Rose and R. A. Willoughby, eds. (1972). *Sparse Matrices and Their Applications*, Plenum Press, New York, 1972

J.R. Bunch and D.J. Rose, eds. (1976). *Sparse Matrix Computations*, Academic Press, New York.

I.S. Duff and G.W. Stewart, eds. (1979). *Sparse Matrix Proceedings, 1978*, SIAM Publications, Philadelphia, PA.

I.S. Duff, ed. (1981). *Sparse Matrices and Their Uses*, Academic Press, New York.

Å. Björck, R.J. Plemmons, and H. Schneider, eds. (1981). *Large-Scale Matrix Problems*, North-Holland, New York.

G. Rodrigue, ed. (1982). *Parallel Computation*, Academic Press, New York.

B. Kågström and A. Ruhe, eds. (1983). *Matrix Pencils*, Proc. Pite Havs-bad, 1982, Lecture Notes in Mathematics 973, Springer-Verlag, New York and Berlin.

J. Cullum and R.A. Willoughby, eds. (1986). *Large Scale Eigenvalue Problems*, North-Holland, Amsterdam.

A. Wouk, ed. (1986). *New Computing Environments: Parallel, Vector, and Systolic, SIAM Publications*, Philadelphia, PA.

M.T. Heath, ed. (1986). *Proceedings of First SIAM Conference on Hypercube Multiprocessors*, SIAM Publications, Philadelphia, PA.

M.T. Heath, ed. (1987). *Hypercube Multiprocessors*, SIAM Publications, Philadelphia, PA.

G. Fox, ed. (1988). *The Third Conference on Hypercube Concurrent Computers and Applications, Vol. II – Applications*, ACM Press, New York.

M.H. Schultz, ed. (1988). *Numerical Algorithms for Modern Parallel Computer Architectures*, IMA Volumes in Mathematics and Its Applications, Number 13, Springer-Verlag, Berlin.

E.F. Deprettere, ed. (1988). *SVD and Signal Processing*. Elsevier, Amsterdam.

B.N. Datta, C.R. Johnson. M.A. Kaashoek, R. Plemmons, and E.D. Sontag, eds. (1988), *Linear Algebra in Signals, Systems, and Control*, SIAM Publications, Philadelphia, PA.

J. Dongarra, I. Duff, P. Gaffney, and S. McKee, eds. (1989), *Vector and Parallel Computing*, Ellis Horwood, Chichester, England.

O. Axelsson, ed. (1989). "Preconditioned Conjugate Gradient Methods," *BIT 29:4*.

K. Gallivan, M. Heath, E. Ng, J. Ortega, B. Peyton, R. Plemmons, C. Romine, A. Sameh, and B. Voigt (1990), *Parallel Algorithms for Matrix Computations*, SIAM Publications, Philadelphia, PA.

G.H. Golub and P. Van Dooren, eds. (1991). *Numerical Linear Algebra, Digital Signal Processing, and Parallel Algorithms*. Springer-Verlag, Berlin.

R. Vaccaro, ed. (1991). *SVD and Signal Processing II: Algorithms, Analysis, and Applications*. Elsevier, Amsterdam.

R. Beauwens and P. de Groen, eds. (1992). *Iterative Methods in Linear Algebra*, Elsevier (North-Holland), Amsterdam.

R.J. Plemmons and C.D. Meyer, eds. (1993). *Linear Algebra, Markov Chains, and Queuing Models*, Springer-Verlag, New York.

M.S. Moonen, G.H. Golub, and B.L.R. de Moor, eds. (1993). *Linear Algebra for Large Scale and Real-Time Applications*, Kluwer, Dordrecht, The Netherlands.

J.D. Brown, M.T. Chu, D.C. Ellison, and R.J. Plemmons, eds. (1994). *Proceedings of the Cornelius Lanczos International Centenary Conference*, SIAM Publications, Philadelphia, PA.

R.V. Patel, A.J. Laub, and P.M. Van Dooren, eds. (1994). *Numerical Linear Algebra Techniques for Systems and Control*, IEEE Press, Piscataway, New Jersey.

J. Lewis, ed. (1994). *Proceedings of the Fifth SIAM Conference on Applied Linear Algebra*, SIAM Publications, Philadelphia, PA.

A. Bojanczyk and G. Cybenko, eds. (1995). *Linear Algebra for Signal Processing*, IMA Volumes in Mathematics and Its Applications, Springer-Verlag, New York.

M. Moonen and B. De Moor, eds. (1995). *SVD and Signal Processing III: Algorithms, Analysis, and Applications*. Elsevier, Amsterdam.

Matrix Computations

Chapter 1

Matrix Multiplication Problems

The proper study of matrix computations begins with the study of the matrix-matrix multiplication problem. Although this problem is simple mathematically it is very rich from the computational point of view. We begin in §1.1 by looking at the several ways that the matrix multiplication problem can be organized. The "language" of partitioned matrices is established and used to characterize several linear algebraic "levels" of computation.

If a matrix has structure, then it is usually possible to exploit it. For example, a symmetric matrix can be stored in half the space as a general matrix. A matrix-vector product that involves a matrix with many zero entries may require much less time to execute than a full matrix times a vector. These matters are discussed in §1.2.

In §1.3 block matrix notation is established. A block matrix is a matrix with matrix entries. This concept is very important from the standpoint of both theory and practice. On the theoretical side, block matrix notation allows us to prove important matrix factorizations very succinctly. These factorizations are the cornerstone of numerical linear algebra. From the computational point of view, block algorithms are important because they are rich in matrix multiplication, the operation of choice for many new high performance computer architectures.

These new architectures require the algorithm designer to pay as much attention to memory traffic as to the actual amount of arithmetic. This aspect of scientific computation is illustrated in §1.4 where the critical issues of vector pipeline computing are discussed: stride, vector length, the number of vector loads and stores, and the level of vector re-use.

Before You Begin

It is important to be familiar with the MATLAB language. See the texts by Pratap(1995) and Van Loan (1996). A richer introduction to high performance matrix computations is given in Dongarra, Duff, Sorensen, and Duff (1991). This chapter's LAPACK connections include

LAPACK: Some General Operations		
_SCAL	$x \leftarrow \alpha x$	Vector scale
DOT	$\mu \leftarrow x^T y$	Dot product
_AXPY	$y \leftarrow \alpha x + y$	Saxpy
_GEMV	$y \leftarrow \alpha A x + \beta y$	Matrix-vector multiplication
_GER	$A \leftarrow A + \alpha x y^T$	Rank-1 update
_GEMM	$C \leftarrow \alpha A B + \beta C$	Matrix multiplication

LAPACK: Some Symmetric Operations		
_SYMV	$y \leftarrow \alpha A x + \beta y$	Matrix-vector multiplication
_SPMV	$y \leftarrow \alpha A x + \beta y$	Matrix-vector multiplication (Packed)
_SYR	$A \leftarrow \alpha x x^T + A$	Rank-1 update
_SYR2	$A \leftarrow \alpha x y^T + \alpha y x^T + A$	Rank-2 update
_SYRK	$C \leftarrow \alpha A A^T + \beta C$	Rank-k update
_SYR2K	$C \leftarrow \alpha A B^T + \alpha B A^T + \beta C$	Rank-$2k$ update
_SYMM	$C = \alpha A B + \beta C$ or $(\alpha B A + \beta C)$	Symmetric/General Product

LAPACK: Some Band/Triangular Operations		
_GBMV	$y \leftarrow \alpha A x + \beta y$	General Band
_SBMV	$y \leftarrow \alpha A x + \beta y$	Symmetric Band
_TBMV	$x \leftarrow \alpha A x$	Triangular
_TPMV	$x \leftarrow \alpha A x$	Triangular Packed
_TRMM	$B \leftarrow \alpha A B$ (or BA)	Triangular/General Product

1.1   Basic Algorithms and Notation

Matrix computations are built upon a hierarchy of linear algebraic operations. Dot products involve the scalar operations of addition and multiplication. Matrix-vector multiplication is made up of dot products. Matrix-matrix multiplication amounts to a collection of matrix-vector products. All of these operations can be described in algorithmic form or in the language of linear algebra. Our primary objective in this section is to show

how these two styles of expression complement each another. Along the way we pick up notation and acquaint the reader with the kind of thinking that underpins the matrix computation area. The discussion revolves around the matrix multiplication problem, a computation that can be organized in several ways.

1.1.1 Matrix Notation

Let $\mathbb{R}$ denote the set of real numbers. We denote the vector space of all m-by-n real matrices by $\mathbb{R}^{m \times n}$:

$$A \in \mathbb{R}^{m \times n} \iff A = (a_{ij}) = \begin{bmatrix} a_{11} & \cdots & a_{1n} \\ \vdots & & \vdots \\ a_{m1} & \cdots & a_{mn} \end{bmatrix} \quad a_{ij} \in \mathbb{R}.$$

If a capital letter is used to denote a matrix (e.g. A, B, Δ), then the corresponding lower case letter with subscript ij refers to the (i,j) entry (e.g., a_{ij}, b_{ij}, δ_{ij}). As appropriate, we also use the notation $[A]_{ij}$ and $A(i,j)$ to designate the matrix elements.

1.1.2 Matrix Operations

Basic matrix operations include *transposition* $(\mathbb{R}^{m \times n} \to \mathbb{R}^{n \times m})$,

$$C = A^T \quad \Longrightarrow \quad c_{ij} = a_{ji},$$

addition $(\mathbb{R}^{m \times n} \times \mathbb{R}^{m \times n} \to \mathbb{R}^{m \times n})$,

$$C = A + B \quad \Longrightarrow \quad c_{ij} = a_{ij} + b_{ij},$$

scalar-matrix multiplication, $(\mathbb{R} \times \mathbb{R}^{m \times n} \to \mathbb{R}^{m \times n})$,

$$C = \alpha A \quad \Longrightarrow \quad c_{ij} = \alpha a_{ij},$$

and *matrix-matrix multiplication* $(\mathbb{R}^{m \times p} \times \mathbb{R}^{p \times n} \to \mathbb{R}^{m \times n})$,

$$C = AB \quad \Longrightarrow \quad c_{ij} = \sum_{k=1}^{r} a_{ik} b_{kj}.$$

These are the building blocks of matrix computations.

1.1.3   Vector Notation

Let $\mathbb{R}^n$ denote the vector space of real n-vectors:

$$x \in \mathbb{R}^n \quad \Longleftrightarrow \quad x = \begin{bmatrix} x_1 \\ \vdots \\ x_n \end{bmatrix} \quad x_i \in \mathbb{R}.$$

We refer to x_i as the ith component of x. Depending upon context, the alternative notations $[x]_i$ and $x(i)$ are sometimes used.

Notice that we are identifying $\mathbb{R}^n$ with $\mathbb{R}^{n \times 1}$ and so the members of $\mathbb{R}^n$ are *column* vectors. On the other hand, the elements of $\mathbb{R}^{1 \times n}$ are *row* vectors:

$$x \in \mathbb{R}^{1 \times n} \quad \Longleftrightarrow \quad x = (x_1, \ldots, x_n).$$

If x is a column vector, then $y = x^T$ is a row vector.

1.1.4   Vector Operations

Assume $a \in \mathbb{R}$, $x \in \mathbb{R}^n$, and $y \in \mathbb{R}^n$. Basic vector operations include *scalar-vector multiplication*,

$$z = ax \quad \Longrightarrow \quad z_i = ax_i,$$

vector addition,

$$z = x + y \quad \Longrightarrow \quad z_i = x_i + y_i,$$

the *dot product* (or *inner product*),

$$c = x^T y \quad \Longrightarrow \quad c = \sum_{i=1}^{n} x_i y_i,$$

and *vector multiply* (or the *Hadamard product*)

$$z = x.*y \quad \Longrightarrow \quad z_i = x_i y_i.$$

Another very important operation which we write in "update form" is the *saxpy*:

$$y = ax + y \quad \Longrightarrow \quad y_i = ax_i + y_i$$

Here, the symbol "=" is being used to denote assignment, not mathematical equality. The vector y is being updated. The name "saxpy" is used in LAPACK, a software package that implements many of the algorithms in this book. One can think of "saxpy" as a mnemonic for "scalar a x plus y."

1.1.5 The Computation of Dot Products and Saxpys

We have chosen to express algorithms in a stylized version of the MATLAB language. MATLAB is a powerful interactive system that is ideal for matrix computation work. We gradually introduce our stylized MATLAB notation in this chapter beginning with an algorithm for computing dot products.

Algorithm 1.1.1 (Dot Product) If $x, y \in \mathbb{R}^n$, then this algorithm computes their dot product $c = x^T y$.

$$c = 0$$
$$\textbf{for } i = 1{:}n$$
$$\qquad c = c + x(i)y(i)$$
$$\textbf{end}$$

The dot product of two n-vectors involves n multiplications and n additions. It is an "$O(n)$" operation, meaning that the amount of work is linear in the dimension. The saxpy computation is also an $O(n)$ operation, but it returns a vector instead of a scalar.

Algorithm 1.1.2 (Saxpy) If $x, y \in \mathbb{R}^n$ and $a \in \mathbb{R}$, then this algorithm overwrites y with $ax + y$.

$$\textbf{for } i = 1{:}n$$
$$\qquad y(i) = ax(i) + y(i)$$
$$\textbf{end}$$

It must be stressed that the algorithms in this book are encapsulations of critical computational ideas and not "production codes."

1.1.6 Matrix-Vector Multiplication and the Gaxpy

Suppose $A \in \mathbb{R}^{m \times n}$ and that we wish to compute the update

$$y = Ax + y$$

where $x \in \mathbb{R}^n$ and $y \in \mathbb{R}^m$ are given. This generalized saxpy operation is referred to as a *gaxpy*. A standard way that this computation proceeds is to update the components one at a time:

$$y_i = \sum_{j=1}^{n} a_{ij} x_j + y_i \qquad i = 1{:}m.$$

This gives the following algorithm.

Algorithm 1.1.3 (Gaxpy: Row Version) If $A \in \mathbb{R}^{m \times n}$, $x \in \mathbb{R}^n$, and $y \in \mathbb{R}^m$, then this algorithm overwrites y with $Ax + y$.

```
for i = 1:m
    for j = 1:n
        y(i) = A(i,j)x(j) + y(i)
    end
end
```

An alternative algorithm results if we regard Ax as a linear combination of A's columns, e.g.,

$$\begin{bmatrix} 1 & 2 \\ 3 & 4 \\ 5 & 6 \end{bmatrix}\begin{bmatrix} 7 \\ 8 \end{bmatrix} = \begin{bmatrix} 1\cdot7+2\cdot8 \\ 3\cdot7+4\cdot8 \\ 5\cdot7+6\cdot8 \end{bmatrix} = 7\begin{bmatrix} 1 \\ 3 \\ 5 \end{bmatrix} + 8\begin{bmatrix} 2 \\ 4 \\ 6 \end{bmatrix} = \begin{bmatrix} 23 \\ 53 \\ 83 \end{bmatrix}.$$

Algorithm 1.1.4 (Gaxpy: Column Version) If $A \in \mathbb{R}^{m\times n}$, $x \in \mathbb{R}^n$, and $y \in \mathbb{R}^m$, then this algorithm overwrites y with $Ax + y$.

```
for j = 1:n
    for i = 1:m
        y(i) = A(i,j)x(j) + y(i)
    end
end
```

Note that the inner loop in either gaxpy algorithm carries out a saxpy operation. The column version was derived by rethinking what matrix-vector multiplication "means" at the vector level, but it could also have been obtained simply by interchanging the order of the loops in the row version. In matrix computations, it is important to relate loop interchanges to the underlying linear algebra.

1.1.7   Partitioning a Matrix into Rows and Columns

Algorithms 1.1.3 and 1.1.4 access the data in A by row and by column respectively. To highlight these orientations more clearly we introduce the language of *partitioned matrices*.

From the row point of view, a matrix is a stack of row vectors:

$$A \in \mathbb{R}^{m\times n} \quad \Longleftrightarrow \quad A = \begin{bmatrix} r_1^T \\ \vdots \\ r_m^T \end{bmatrix} \quad r_k \in \mathbb{R}^n. \qquad (1.1.1)$$

This is called a *row partition* of A. Thus, if we row partition

$$\begin{bmatrix} 1 & 2 \\ 3 & 4 \\ 5 & 6 \end{bmatrix},$$

then we are choosing to think of A as a collection of rows with

$$r_1^T = [\,1 \quad 2\,], \qquad r_2^T = [\,3 \quad 4\,], \quad \text{and} \quad r_3^T = [\,5 \quad 6\,].$$

With the row partitioning (1.1.1) Algorithm 1.1.3 can be expressed as follows:

> **for** $i = 1{:}m$
> $\qquad y_i = r_i^T x + y(i)$
> **end**

Alternatively, a matrix is a collection of column vectors:

$$A \in \mathbb{R}^{m \times n} \quad \Longleftrightarrow \quad A = [\,c_1, \ldots, c_n\,], \quad c_k \in \mathbb{R}^m . \qquad (1.1.2)$$

We refer to this as a *column partition* of A. In the 3-by-2 example above, we thus would set c_1 and c_2 to be the first and second columns of A respectively:

$$c_1 = \begin{bmatrix} 1 \\ 3 \\ 5 \end{bmatrix} \qquad c_2 = \begin{bmatrix} 2 \\ 4 \\ 6 \end{bmatrix} .$$

With (1.1.2) we see that Algorithm 1.1.4 is a saxpy procedure that accesses A by columns:

> **for** $j = 1{:}n$
> $\qquad y = x_j c_j + y$
> **end**

In this context appreciate y as a running vector sum that undergoes repeated saxpy updates.

1.1.8 The Colon Notation

A handy way to specify a column or row of a matrix is with the "colon" notation. If $A \in \mathbb{R}^{m \times n}$, then $A(k,:)$ designates the kth row, i.e.,

$$A(k,:) = [a_{k1}, \ldots, a_{kn}] .$$

The kth column is specified by

$$A(:,k) = \begin{bmatrix} a_{1k} \\ \vdots \\ a_{mk} \end{bmatrix} .$$

With these conventions we can rewrite Algorithms 1.1.3 and 1.1.4 as

> **for** $i = 1{:}m$
> $\qquad y(i) = A(i,:)x + y(i)$
> **end**

and

> **for** $j = 1{:}n$
> $y = x(j)A(:,j) + y$
> **end**

respectively. With the colon notation we are able to suppress iteration details. This frees us to think at the vector level and focus on larger computational issues.

1.1.9 The Outer Product Update

As a preliminary application of the colon notation, we use it to understand the *outer product update*

$$A = A + xy^T, \qquad A \in \mathbb{R}^{m \times n}, \ x \in \mathbb{R}^m, \ y \in \mathbb{R}^n .$$

The outer product operation xy^T "looks funny" but is perfectly legal, e.g.,

$$\begin{bmatrix} 1 \\ 2 \\ 3 \end{bmatrix} \begin{bmatrix} 4 & 5 \end{bmatrix} = \begin{bmatrix} 4 & 5 \\ 8 & 10 \\ 12 & 15 \end{bmatrix} .$$

This is because xy^T is the product of two "skinny" matrices and the number of columns in the left matrix x equals the number of rows in the right matrix y^T. The entries in the outer product update are prescribed by

> **for** $i = 1{:}m$
>   **for** $j = 1{:}n$
> $a_{ij} = a_{ij} + x_i y_j$
>   **end**
> **end**

The mission of the j loop is to add a multiple of y^T to the i-th row of A, i.e.,

> **for** $i = 1{:}m$
> $A(i,:) = A(i,:) + x(i)y^T$
> **end**

On the other hand, if we make the i-loop the inner loop, then its task is to add a multiple of x to the jth column of A:

> **for** $j = 1{:}n$
> $A(:,j) = A(:,j) + y(j)x$
> **end**

Note that both outer product algorithms amount to a set of saxpy updates.

1.1.10 Matrix-Matrix Multiplication

Consider the 2-by-2 matrix-matrix multiplication AB. In the dot product formulation each entry is computed as a dot product:

$$\begin{bmatrix} 1 & 2 \\ 3 & 4 \end{bmatrix} \begin{bmatrix} 5 & 6 \\ 7 & 8 \end{bmatrix} = \begin{bmatrix} 1 \cdot 5 + 2 \cdot 7 & 1 \cdot 6 + 2 \cdot 8 \\ 3 \cdot 5 + 4 \cdot 7 & 3 \cdot 6 + 4 \cdot 8 \end{bmatrix}.$$

In the saxpy version each column in the product is regarded as a linear combination of columns of A:

$$\begin{bmatrix} 1 & 2 \\ 3 & 4 \end{bmatrix} \begin{bmatrix} 5 & 6 \\ 7 & 8 \end{bmatrix} = \begin{bmatrix} 5 \begin{bmatrix} 1 \\ 3 \end{bmatrix} + 7 \begin{bmatrix} 2 \\ 4 \end{bmatrix}, & 6 \begin{bmatrix} 1 \\ 3 \end{bmatrix} + 8 \begin{bmatrix} 2 \\ 4 \end{bmatrix} \end{bmatrix}.$$

Finally, in the outer product version, the result is regarded as the sum of outer products:

$$\begin{bmatrix} 1 & 2 \\ 3 & 4 \end{bmatrix} \begin{bmatrix} 5 & 6 \\ 7 & 8 \end{bmatrix} = \begin{bmatrix} 1 \\ 3 \end{bmatrix} \begin{bmatrix} 5 & 6 \end{bmatrix} + \begin{bmatrix} 2 \\ 4 \end{bmatrix} \begin{bmatrix} 7 & 8 \end{bmatrix}.$$

Although equivalent mathematically, it turns out that these versions of matrix multiplication can have very different levels of performance because of their memory traffic properties. This matter is pursued in §1.4. For now, it is worth detailing the above three approaches to matrix multiplication because it gives us a chance to review notation and to practice thinking at different linear algebraic levels.

1.1.11 Scalar-Level Specifications

To fix the discussion we focus on the following matrix multiplication update:

$$C = AB + C \qquad A \in \mathbb{R}^{m \times p}, \ B \in \mathbb{R}^{p \times n}, \ C \in \mathbb{R}^{m \times n}.$$

The starting point is the familiar triply-nested loop algorithm:

Algorithm 1.1.5 (Matrix Multiplication: ijk Variant) If $A \in \mathbb{R}^{m \times p}$, $B \in \mathbb{R}^{p \times n}$, and $C \in \mathbb{R}^{m \times n}$ are given, then this algorithm overwrites C with $AB + C$.

> **for** $i = 1{:}m$
> > **for** $j = 1{:}n$
> > > **for** $k = 1{:}p$
> > > > $C(i,j) = A(i,k)B(k,j) + C(i,j)$
> > >
> > > **end**
> >
> > **end**
>
> **end**

This is the "ijk variant" because we identify the rows of C (and A) with i, the columns of C (and B) with j, and the summation index with k.

We consider the update $C = AB + C$ instead of just $C = AB$ for two reasons. We do not have to bother with $C = 0$ initializations and updates of the form $C = AB + C$ arise more frequently in practice.

The three loops in the matrix multiplication update can be arbitrarily ordered giving $3! = 6$ variations. Thus,

> **for** $j = 1:n$
>   **for** $k = 1:p$
>   **for** $i = 1:m$
> $C(i,j) = A(i,k)B(k,j) + C(i,j)$
>   **end**
>   **end**
> **end**

is the jki variant. Each of the six possibilities (ijk, jik, ikj, jki, kij, kji) features an inner loop operation (dot product or saxpy) and has its own pattern of data flow. For example, in the ijk variant, the inner loop oversees a dot product that requires access to a row of A and a column of B. The jki variant involves a saxpy that requires access to a column of C and a column of A. These attributes are summarized in Table 1.1.1 along with an interpretation of what is going on when the middle and inner loop are considered together. Each variant involves the same amount of floating

Loop Order	Inner Loop	Middle Loop	Inner Loop Data Access
ijk	dot	vector × matrix	A by row, B by column
jik	dot	matrix × vector	A by row, B by column
ikj	saxpy	row gaxpy	B by row, C by row
jki	saxpy	column gaxpy	A by column, C by column
kij	saxpy	row outer product	B by row, C by row
kji	saxpy	column outer product	A by column, C by column

TABLE 1.1.1. *Matrix Multiplication: Loop Orderings and Properties*

point arithmetic, but accesses the A, B, and C data differently.

1.1.12   A Dot Product Formulation

The usual matrix multiplication procedure regards AB as an array of dot products to be computed one at a time in left-to-right, top-to-bottom order.

This is the idea behind Algorithm 1.1.5. Using the colon notation we can highlight this dot-product formulation:

Algorithm 1.1.6 (Matrix Multiplication: Dot Product Version)
If $A \in \mathbb{R}^{m \times p}$, $B \in \mathbb{R}^{p \times n}$, and $C \in \mathbb{R}^{m \times n}$ are given, then this algorithm overwrites C with $AB + C$.

> **for** $i = 1{:}m$
> > **for** $j = 1{:}n$
> > > $C(i,j) = A(i,:)B(:,j) + C(i,j)$
> >
> > **end**
>
> **end**

In the language of partitioned matrices, if

$$A = \begin{bmatrix} a_1^T \\ \vdots \\ a_m^T \end{bmatrix} \qquad a_k \in \mathbb{R}^p$$

and

$$B = [\, b_1, \ldots, b_n \,] \qquad b_k \in \mathbb{R}^p$$

then Algorithm 1.1.6 has this interpretation:

> **for** $i = 1{:}m$
> > **for** $j = 1{:}n$
> > > $c_{ij} = a_i^T b_j + c_{ij}$
> >
> > **end**
>
> **end**

Note that the "mission" of the j-loop is to compute the ith row of the update. To emphasize this we could write

> **for** $i = 1{:}m$
> > $c_i^T = a_i^T B + c_i^T$
>
> **end**

where

$$C = \begin{bmatrix} c_1^T \\ \vdots \\ c_m^T \end{bmatrix}$$

is a row partitioning of C. To say the same thing with the colon notation we write

> **for** $i = 1{:}m$
> > $C(i,:) = A(i,:)B + C(i,:)$
>
> **end**

Either way we see that the inner two loops of the ijk variant define a row-oriented gaxpy operation.

1.1.13   A Saxpy Formulation

Suppose A and C are column-partitioned as follows

$$A = [\, a_1, \ldots, a_p \,] \qquad a_j \in \mathbb{R}^m$$

$$C = [\, c_1, \ldots, c_n \,] \qquad c_j \in \mathbb{R}^m.$$

By comparing jth columns in $C = AB + C$ we see that

$$c_j = \sum_{k=1}^{p} b_{kj} a_k + c_j, \qquad j = 1{:}n.$$

These vector sums can be put together with a sequence of saxpy updates.

Algorithm 1.1.7 (Matrix Multiplication: Saxpy Version) If the matrices $A \in \mathbb{R}^{m \times p}$, $B \in \mathbb{R}^{p \times n}$, and $C \in \mathbb{R}^{m \times n}$ are given, then this algorithm overwrites C with $AB + C$.

> **for** $j = 1{:}n$
> > **for** $k = 1{:}p$
> > > $C(:,j) = A(:,k)B(k,j) + C(:,j)$
> >
> > **end**
>
> **end**

Note that the k-loop oversees a gaxpy operation:

> **for** $j = 1{:}n$
> > $C(:,j) = AB(:,j) + C(:,j)$
>
> **end**

1.1.14   An Outer Product Formulation

Consider the kij variant of Algorithm 1.1.5:

> **for** $k = 1{:}p$
> > **for** $j = 1{:}n$
> > > **for** $i = 1{:}m$
> > > > $C(i,j) = A(i,k)B(k,j) + C(i,j)$
> > >
> > > **end**
> >
> > **end**
>
> **end**

The inner two loops oversee the outer product update

$$C = a_k b_k^T + C$$

where

$$A = [\, a_1, \ldots, a_p \,] \quad \text{and} \quad B = \begin{bmatrix} b_1^T \\ \vdots \\ b_p^T \end{bmatrix} \tag{1.1.3}$$

with $a_k \in \mathbb{R}^m$ and $b_k \in \mathbb{R}^n$. We therefore obtain

Algorithm 1.1.8 (Matrix Multiplication: Outer Product Version)
If $A \in \mathbb{R}^{m \times p}$, $B \in \mathbb{R}^{p \times n}$, and $C \in \mathbb{R}^{m \times n}$ are given, then this algorithm overwrites C with $AB + C$.

> **for** $k = 1{:}p$
> $\quad C = A(:,k)B(k,:) + C$
> **end**

This implementation revolves around the fact that AB is the sum of p outer products.

1.1.15   The Notion of "Level"

The dot product and saxpy operations are examples of "level-1" operations. Level-1 operations involve an amount of data and an amount of arithmetic that is linear in the dimension of the operation. An m-by-n outer product update or gaxpy operation involves a quadratic amount of data $(O(mn))$ and a quadratic amount of work $(O(mn))$. They are examples of "level-2" operations.

The matrix update $C = AB + C$ is a "level-3" operation. Level-3 operations involve a quadratic amount of data and a cubic amount of work. If A, B, and C are n-by-n matrices, then $C = AB + C$ involves $O(n^2)$ matrix entries and $O(n^3)$ arithmetic operations.

The design of matrix algorithms that are rich in high-level linear algebra operations is a recurring theme in the book. For example, a high performance linear equation solver may require a level-3 organization of Gaussian elimination. This requires some algorithmic rethinking because that method is usually specified in level-1 terms, e.g., "multiply row 1 by a scalar and add the result to row 2."

1.1.16   A Note on Matrix Equations

In striving to understand matrix multiplication via outer products, we essentially established the matrix equation

$$AB = \sum_{k=1}^{p} a_k b_k^T$$

where the a_k and b_k are defined by the partitionings in (1.1.3).

Numerous matrix equations are developed in subsequent chapters. Sometimes they are established algorithmically like the above outer product expansion and other times they are proved at the ij-component level. As an example of the latter, we prove an important result that characterizes transposes of products.

Theorem 1.1.1 *If $A \in \mathbb{R}^{m \times p}$ and $B \in \mathbb{R}^{p \times n}$, then $(AB)^T = B^T A^T$.*

Proof. If $C = (AB)^T$, then

$$c_{ij} = [(AB)^T]_{ij} = [AB]_{ji} = \sum_{k=1}^{p} a_{jk} b_{ki} \, .$$

On the other hand, if $D = B^T A^T$, then

$$d_{ij} = [B^T A^T]_{ij} = \sum_{k=1}^{p} [B^T]_{ik} [A^T]_{kj} = \sum_{k=1}^{p} b_{ki} a_{jk}.$$

Since $c_{ij} = d_{ij}$ for all i and j, it follows that $C = D$. $\square$

Scalar-level proofs such as this one are usually not very insightful. However, they are sometimes the only way to proceed.

1.1.17 Complex Matrices

From time to time computations that involve complex matrices are discussed. The vector space of m-by-n complex matrices is designated by $\mathbb{C}^{m \times n}$. The scaling, addition, and multiplication of complex matrices corresponds exactly to the real case. However, transposition becomes *conjugate transposition*:

$$C = A^H \quad \Longrightarrow \quad c_{ij} = \bar{a}_{ji} \, .$$

The vector space of complex n-vectors is designated by $\mathbb{C}^n$. The dot product of complex n-vectors x and y is prescribed by

$$s = x^H y = \sum_{i=1}^{n} \bar{x}_i y_i \, .$$

Finally, if $A = B + iC \in \mathbb{C}^{m \times n}$, then we designate the real and imaginary parts of A by $\text{Re}(A) = B$ and $\text{Im}(A) = C$ respectively.

Problems

P1.1.1 Suppose $A \in \mathbb{R}^{n \times n}$ and $x \in \mathbb{R}^r$ are given. Give a saxpy algorithm for computing the first column of $M = (A - x_1 I) \cdots (A - x_r I)$.

P1.1.2 In the conventional 2-by-2 matrix multiplication $C = AB$, there are eight multiplications: $a_{11}b_{11}$, $a_{11}b_{12}$, $a_{21}b_{11}$, $a_{21}b_{12}$, $a_{12}b_{21}$, $a_{12}b_{22}$, $a_{22}b_{21}$ and $a_{22}b_{22}$. Make a table that indicates the order that these multiplications are performed for the ijk, jik, kij, ikj, jki, and kji matrix multiply algorithms.

P1.1.3 Give an algorithm for computing $C = (xy^T)^k$ where x and y are n-vectors.

P1.1.4 Specify an algorithm for computing $(XY^T)^k$ where $X, Y \in \mathbb{R}^{n \times 2}$.

P1.1.5 Formulate an outer product algorithm for the update $C = AB^T + C$ where $A \in \mathbb{R}^{m \times r}$, $B \in \mathbb{R}^{n \times r}$, and $C \in \mathbb{R}^{m \times n}$.

P1.1.6 Suppose we have real n-by-n matrices C, D, E, and F. Show how to compute real n-by-n matrices A and B with just three real n-by-n matrix multiplications so that $(A + iB) = (C + iD)(E + iF)$. Hint: Compute $W = (C + D)(E - F)$.

Notes and References for Sec. 1.1

It must be stressed that the development of quality software from any of our "semi-formal" algorithmic presentations is a long and arduous task. Even the implementation of the level-1,2, and 3 BLAS require care:

C.L. Lawson, R.J. Hanson, , D.R. Kincaid, and F.T. Krogh (1979). "Basic Linear Algebra Subprograms for FORTRAN Usage," *ACM Trans. Math. Soft.* 5, 308–323.

C.L. Lawson, R.J. Hanson, D.R. Kincaid, and F.T. Krogh (1979). "Algorithm 539, Basic Linear Algebra Subprograms for FORTRAN Usage," *ACM Trans. Math. Soft.* 5, 324–325.

J.J. Dongarra, J. Du Croz, S. Hammarling, and R.J. Hanson (1988). "An Extended Set of Fortran Basic Linear Algebra Subprograms," *ACM Trans. Math. Soft. 14,* 1–17.

J.J. Dongarra, J. Du Croz, S. Hammarling, and R.J. Hanson (1988). "Algorithm 656 An Extended Set of Fortran Basic Linear Algebra Subprograms: Model Implementation and Test Programs," *ACM Trans. Math. Soft. 14,* 18–32.

J.J. Dongarra, J. Du Croz, I.S. Duff, and S.J. Hammarling (1990). "A Set of Level 3 Basic Linear Algebra Subprograms," *ACM Trans. Math. Soft. 16,* 1–17.

J.J. Dongarra, J. Du Croz, I.S. Duff, and S.J. Hammarling (1990). "Algorithm 679. A Set of Level 3 Basic Linear Algebra Subprograms: Model Implementation and Test Programs," *ACM Trans. Math. Soft. 16,* 18–28.

Other BLAS references include

B. Kågström, P. Ling, and C. Van Loan (1991). "High-Performance Level-3 BLAS: Sample Routines for Double Precision Real Data," in *High Performance Computing II*, M. Durand and F. El Dabaghi (eds), North-Holland, 269–281.

B. Kågström, P. Ling, and C. Van Loan (1995). "GEMM-Based Level-3 BLAS: High-Performance Model Implementations and Performance Evaluation Benchmark," in *Parallel Programming and Applications*, P. Fritzon and L. Finmo (eds), ISO Press, 184–188.

For an appreciation of the subtleties associated with software development we recommend

J.R. Rice (1981). *Matrix Computations and Mathematical Software*, Academic Press, New York.

and a browse through the LAPACK manual.

1.2   Exploiting Structure

The efficiency of a given matrix algorithm depends on many things. Most obvious and what we treat in this section is the amount of required arithmetic and storage. We continue to use matrix-vector and matrix-matrix multiplication as a vehicle for introducing the key ideas. As examples of exploitable structure we have chosen the properties of bandedness and symmetry. Band matrices have many zero entries and so it is no surprise that band matrix manipulation allows for many arithmetic and storage shortcuts. Arithmetic complexity and data structures are discussed in this context.

Symmetric matrices provide another set of examples that can be used to illustrate structure exploitation. Symmetric linear systems and eigenvalue problems have a very prominent role to play in matrix computations and so it is important to be familiar with their manipulation.

1.2.1   Band Matrices and the x-0 Notation

We say that $A \in \mathbb{R}^{m \times n}$ has *lower bandwidth* p if $a_{ij} = 0$ whenever $i > j + p$ and *upper bandwidth* q if $j > i + q$ implies $a_{ij} = 0$. Here is an example of an 8-by-5 matrix that has lower bandwidth 1 and upper bandwidth 2:

$$
\begin{bmatrix}
\times & \times & \times & 0 & 0 \\
\times & \times & \times & \times & 0 \\
0 & \times & \times & \times & \times \\
0 & 0 & \times & \times & \times \\
0 & 0 & 0 & \times & \times \\
0 & 0 & 0 & 0 & \times \\
0 & 0 & 0 & 0 & 0 \\
0 & 0 & 0 & 0 & 0
\end{bmatrix} .
$$

The $\times$'s designates arbitrary nonzero entries. This notation is handy to indicate the zero-nonzero structure of a matrix and we use it extensively. Band structures that occur frequently are tabulated in Table 1.2.1.

1.2.2   Diagonal Matrix Manipulation

Matrices with upper and lower bandwidth zero are *diagonal*. If $D \in \mathbb{R}^{m \times n}$ is diagonal, then

$$ D = \text{diag}(d_1, \ldots, d_q), \quad q = \min\{m, n\} \quad \Longleftrightarrow \quad d_i = d_{ii} $$

If D is diagonal and A is a matrix, then DA is a *row scaling* of A and AD is a *column scaling* of A.

Type of Matrix	Lower Bandwidth	Upper Bandwidth
diagonal	0	0
upper triangular	0	$n-1$
lower triangular	$m-1$	0
tridiagonal	1	1
upper bidiagonal	0	1
lower bidiagonal	1	0
upper Hessenberg	1	$n-1$
lower Hessenberg	$m-1$	1

TABLE 1.2.1. *Band Terminology for m-by-n Matrices*

1.2.3 Triangular Matrix Multiplication

To introduce band matrix "thinking" we look at the matrix multiplication problem $C = AB$ when A and B are both n-by-n and upper triangular. The 3-by-3 case is illuminating:

$$
C = \begin{bmatrix} a_{11}b_{11} & a_{11}b_{12} + a_{12}b_{22} & a_{11}b_{13} + a_{12}b_{23} + a_{13}b_{33} \\ 0 & a_{22}b_{22} & a_{22}b_{23} + a_{23}b_{33} \\ 0 & 0 & a_{33}b_{33} \end{bmatrix}.
$$

It suggests that the product is upper triangular and that its upper triangular entries are the result of abbreviated inner products. Indeed, since $a_{ik}b_{kj} = 0$ whenever $k < i$ or $j < k$ we see that

$$
c_{ij} = \sum_{k=i}^{j} a_{ik}b_{kj}
$$

and so we obtain:

Algorithm 1.2.1 (Triangular Matrix Multiplication) If $A, B \in \mathbb{R}^{n \times n}$ are upper triangular, then this algorithm computes $C = AB$.

```
C = 0
for i = 1:n
    for j = i:n
        for k = i:j
            C(i,j) = A(i,k)B(k,j) + C(i,j)
        end
    end
end
```

To quantify the savings in this algorithm we need some tools for measuring the amount of work.

1.2.4 Flops

Obviously, upper triangular matrix multiplication involves less arithmetic than when the matrices are full. One way to quantify this is with the notion of a *flop*. A flop[1] is a floating point operation. A dot product or saxpy operation of length n involves $2n$ flops because there are n multiplications and n adds in either of these vector operations.

The gaxpy $y = Ax + y$ where $A \in \mathbb{R}^{m \times n}$ involves $2mn$ flops as does an m-by-n outer product update of the form $A = A + xy^T$.

The matrix multiply update $C = AB + C$ where $A \in \mathbb{R}^{m \times p}$, $B \in \mathbb{R}^{p \times n}$, and $C \in \mathbb{R}^{m \times n}$ involves $2mnp$ flops.

Flop counts are usually obtained by summing the amount of arithmetic associated with the most deeply nested statements in an algorithm. For matrix-matrix multiplication, this is the statement,

$$C(i,j) = A(i,k)B(k,j) + C(i,j)$$

which involves two flops and is executed mnp times as a simple loop accounting indicates. Hence the conclusion that general matrix multiplication requires $2mnp$ flops.

Now let us investigate the amount of work involved in Algorithm 1.2.1. Note that c_{ij}, $(i \leq j)$ requires $2(j - i + 1)$ flops. Using the heuristics

$$\sum_{p=1}^{q} p = \frac{q(q+1)}{2} \approx \frac{q^2}{2}$$

and

$$\sum_{p=1}^{q} p^2 = \frac{q^3}{3} + \frac{q^2}{2} + \frac{q}{6} \approx \frac{q^3}{3}$$

we find that triangular matrix multiplication requires one-sixth the number of flops as full matrix multiplication:

$$\sum_{i=1}^{n} \sum_{j=i}^{n} 2(j - i + 1) = \sum_{i=1}^{n} \sum_{j=1}^{n-i+1} 2j \approx \sum_{i=1}^{n} \frac{2(n - i + 1)^2}{2} = \sum_{i=1}^{n} i^2 \approx \frac{n^3}{3}.$$

We throw away the low order terms since their inclusion does not contribute to what the flop count "says." For example, an exact flop count of Algorithm 1.2.1 reveals that precisely $n^3/3 + n^2 + 2n/3$ flops are involved. For

---

[1]In the first edition of this book we defined a flop to be the amount of work associated with an operation of the form $a_{ij} = a_{ij} + a_{ik}a_{kj}$, i.e., a floating point add, a floating point multiply, and some subscripting. Thus, an "old flop" involves two "new flops." In defining a flop to be a single floating point operation we are opting for a more precise measure of arithmetic complexity.

large n (the typical situation of interest) we see that the exact flop count offers no insight beyond the $n^3/3$ approximation.

Flop counting is a necessarily crude approach to the measuring of program efficiency since it ignores subscripting, memory traffic, and the countless other overheads associated with program execution. We must not infer too much from a comparison of flops counts. We cannot conclude, for example, that triangular matrix multiplication is six times faster than square matrix multiplication. Flop counting is just a "quick and dirty" accounting method that captures only one of the several dimensions of the efficiency issue.

1.2.5 The Colon Notation–Again

The dot product that the k-loop performs in Algorithm 1.2.1 can be succinctly stated if we extend the colon notation introduced in §1.1.8. Suppose $A \in \mathbb{R}^{m \times n}$ and the integers p, q, and r satisfy $1 \leq p \leq q \leq n$ and $1 \leq r \leq m$. We then define

$$A(r, p{:}q) = [\, a_{rp}, \ldots, a_{rq} \,] \in \mathbb{R}^{1 \times (q-p+1)} .$$

Likewise, if $1 \leq p \leq q \leq m$ and $1 \leq c \leq n$, then

$$A(p{:}q, c) = \begin{bmatrix} a_{pc} \\ \vdots \\ a_{qc} \end{bmatrix} \in \mathbb{R}^{q-p+1}.$$

With this notation we can rewrite Algorithm 1.2.1 as

```
C(1:n, 1:n) = 0
for i = 1:n
    for j = i:n
        C(i, j) = A(i, i:j)B(i:j, j) + C(i, j)
    end
end
```

We mention one additional feature of the colon notation. Negative increments are allowed. Thus, if x and y are n-vectors, then $s = x^T y(n{:}{-}1{:}1)$ is the summation

$$s = \sum_{i=1}^{n} x_i y_{n-i+1} \,.$$

1.2.6 Band Storage

Suppose $A \in \mathbb{R}^{n \times n}$ has lower bandwidth p and upper bandwidth q and assume that p and q are much smaller than n. Such a matrix can be stored in a $(p+q+1)$-by-n array $A.band$ with the convention that

$$a_{ij} = A.band(i - j + q + 1, j) \tag{1.2.1}$$

for all (i, j) that fall inside the band. Thus, if

$$
A = \begin{bmatrix}
a_{11} & a_{12} & a_{13} & 0 & 0 & 0 \\
a_{21} & a_{22} & a_{23} & a_{24} & 0 & 0 \\
0 & a_{32} & a_{33} & a_{34} & a_{35} & 0 \\
0 & 0 & a_{43} & a_{44} & a_{45} & a_{46} \\
0 & 0 & 0 & a_{54} & a_{55} & a_{56} \\
0 & 0 & 0 & 0 & a_{65} & a_{66}
\end{bmatrix},
$$

then

$$
A.band = \begin{bmatrix}
0 & 0 & a_{13} & a_{24} & a_{35} & a_{46} \\
0 & a_{12} & a_{23} & a_{34} & a_{45} & a_{56} \\
a_{11} & a_{22} & a_{33} & a_{44} & a_{55} & a_{66} \\
a_{21} & a_{32} & a_{43} & a_{54} & a_{65} & 0
\end{bmatrix}.
$$

Here, the "0" entries are unused. With this data structure, our column-oriented gaxpy algorithm transforms to the following:

Algorithm 1.2.2 (Band Gaxpy)   Suppose $A \in \mathbb{R}^{n \times n}$ has lower bandwidth p and upper bandwidth q and is stored in the $A.band$ format (1.2.1). If $x, y \in \mathbb{R}^n$, then this algorithm overwrites y with $Ax + y$.

> **for** $j = 1{:}n$
> $\qquad y_{top} = \max(1, j - q)$
> $\qquad y_{bot} = \min(n, j + p)$
> $\qquad a_{top} = \max(1, q + 2 - j)$
> $\qquad a_{bot} = a_{top} + y_{bot} - y_{top}$
> $\qquad y(y_{top}{:}y_{bot}) = x(j)A.band(a_{top}{:}a_{bot}, j) \ + \ y(y_{top}{:}y_{bot})$
> **end**

Notice that by storing A by column in $A.band$, we obtain a saxpy, column access procedure. Indeed, Algorithm 1.2.2 is obtained from Algorithm 1.1.4 by recognizing that each saxpy involves a vector with a small number of nonzeros. Integer arithmetic is used to identify the location of these nonzeros. As a result of this careful zero/nonzero analysis, the algorithm involves just $2n(p + q + 1)$ flops with the assumption that p and q are much smaller than n.

1.2.7   Symmetry

We say that $A \in \mathbb{R}^{n \times n}$ is *symmetric* if $A^T = A$. Thus,

$$
A = \begin{bmatrix}
1 & 2 & 3 \\
2 & 4 & 5 \\
3 & 5 & 6
\end{bmatrix}
$$

is symmetric. Storage requirements can be halved if we just store the lower triangle of elements, e.g., $A.vec = \begin{bmatrix} 1 & 2 & 3 & 4 & 5 & 6 \end{bmatrix}$. In general, with this data structure we agree to store the a_{ij} as follows:

$$a_{ij} = A.vec((j-1)n - j(j-1)/2 + i) \qquad (i \geq j) \qquad (1.2.2)$$

Let us look at the column-oriented gaxpy operation with the matrix A represented in $A.vec$.

Algorithm 1.2.3 (Symmetric Storage Gaxpy) Suppose $A \in \mathbb{R}^{n \times n}$ is symmetric and stored in the $A.vec$ style (1.2.2). If $x, y \in \mathbb{R}^n$, then this algorithm overwrites y with $Ax + y$.

> **for** $j = 1{:}n$
>   **for** $i = 1{:}j-1$
> $y(i) = A.vec((i-1)n - i(i-1)/2 + j)x(j) + y(i)$
>   **end**
>   **for** $i = j{:}n$
> $y(i) = A.vec((j-1)n - j(j-1)/2 + i)x(j) + y(i)$
>   **end**
> **end**

This algorithm requires the same $2n^2$ flops that an ordinary gaxpy requires. Notice that the halving of the storage requirement is purchased with some awkward subscripting.

1.2.8 Store by Diagonal

Symmetric matrices can also be stored by diagonal. If

$$A = \begin{bmatrix} 1 & 2 & 3 \\ 2 & 4 & 5 \\ 3 & 5 & 6 \end{bmatrix},$$

then in a store-by-diagonal scheme we represent A with the vector

$$A.diag = \begin{bmatrix} 1 & 4 & 6 & 2 & 5 & 3 \end{bmatrix}.$$

In general, if $i \geq j$, then

$$a_{i+k,i} = A.diag(i + nk - k(k-1)/2) \qquad (k \geq 0) \qquad (1.2.3)$$

Some notation simplifies the discussion of how to use this data structure in a matrix-vector multiplication.

If $A \in \mathbb{R}^{m \times n}$, then let $D(A, k) \in \mathbb{R}^{m \times n}$ designate the kth diagonal of A as follows:

$$[D(A,k)]_{ij} = \begin{cases} a_{ij} & j = i + k, \quad 1 \leq i \leq m, \quad 1 \leq j \leq n \\ 0 & \text{otherwise.} \end{cases}$$

Thus,

$$
A \;=\; \begin{bmatrix} 1 & 2 & 3 \\ 2 & 4 & 5 \\ 3 & 5 & 6 \end{bmatrix} = \underbrace{\begin{bmatrix} 0 & 0 & 3 \\ 0 & 0 & 0 \\ 0 & 0 & 0 \end{bmatrix}}_{D(A,2)} + \underbrace{\begin{bmatrix} 0 & 2 & 0 \\ 0 & 0 & 5 \\ 0 & 0 & 0 \end{bmatrix}}_{D(A,1)}
$$

$$
+ \underbrace{\begin{bmatrix} 1 & 0 & 0 \\ 0 & 4 & 0 \\ 0 & 0 & 6 \end{bmatrix}}_{D(A,0)} + \underbrace{\begin{bmatrix} 0 & 0 & 0 \\ 2 & 0 & 0 \\ 0 & 5 & 0 \end{bmatrix}}_{D(A,-1)} + \underbrace{\begin{bmatrix} 0 & 0 & 0 \\ 0 & 0 & 0 \\ 3 & 0 & 0 \end{bmatrix}}_{D(A,-2)}.
$$

Returning to our store-by-diagonal data structure, we see that the nonzero parts of $D(A,0)$, $D(A,1),\ldots,$ $D(A,n-1)$ are sequentially stored in the $A.diag$ scheme (1.2.3). The gaxpy $y = Ax + y$ can then be organized as follows:

$$
y \;=\; D(A,0)x \;+\; \sum_{k=1}^{n-1}(D(A,k) + D(A,k)^T)x \;+\; y\,.
$$

Working out the details we obtain the following algorithm.

Algorithm 1.2.4 (Store-By-Diagonal Gaxpy) Suppose $A \in \mathbb{R}^{n \times n}$ is symmetric and stored in the $A.diag$ style (1.2.3). If $x, y \in \mathbb{R}^n$, then this algorithm overwrites y with $Ax + y$.

> **for** $i = 1{:}n$
>> $y(i) = A.diag(i)x(i) + y(i)$
>
> **end**
> **for** $k = 1{:}n-1$
>> $t = nk - k(k-1)/2$
>> $\{y = D(A,k)x + y\}$
>> **for** $i = 1{:}n-k$
>>> $y(i) = A.diag(i+t)x(i+k) + y(i)$
>>
>> **end**
>> $\{y = D(A,k)^T x + y\}$
>> **for** $i = 1{:}n-k$
>>> $y(i+k) = A.diag(i+t)x(i) + y(i+k)$
>>
>> **end**
>
> **end**

Note that the inner loops oversee vector multiplications:

$$
y(1{:}n-k) = A.diag(t+1{:}t+n-k). * x(k+1{:}n) + y(1{:}n-k)
$$
$$
y(k+1{:}n) = A.diag(t+1{:}t+n-k). * x(1{:}n-k) + y(k+1{:}n)
$$

1.2.9   A Note on Overwriting and Workspaces

An undercurrent in the above discussion has been the economical use of
storage. Overwriting input data is another way to control the amount of
memory that a matrix computation requires. Consider the n-by-n matrix
multiplication problem $C = AB$ with the proviso that the "input matrix"
B is to be overwritten by the "output matrix" C . We cannot simply
transform

$$C(1{:}n, 1{:}n) = 0$$
$$\textbf{for } j = 1{:}n$$
$$\quad \textbf{for } k = 1{:}n$$
$$\quad\quad C(:,j) = C(:,j) + A(:,k)B(k,j)$$
$$\quad \textbf{end}$$
$$\textbf{end}$$

to

$$\textbf{for } j = 1{:}n$$
$$\quad \textbf{for } k = 1{:}n$$
$$\quad\quad B(:,j) = B(:,j) + A(:,k)B(k,j)$$
$$\quad \textbf{end}$$
$$\textbf{end}$$

because $B(:,j)$ is needed throughout the entire k-loop. A linear *workspace*
is needed to hold the jth column of the product until it is "safe" to overwrite
$B(:,j)$:

$$\textbf{for } j = 1{:}n$$
$$\quad w(1{:}n) = 0$$
$$\quad \textbf{for } k = 1{:}n$$
$$\quad\quad w(:) = w(:) + A(:,k)B(k,j)$$
$$\quad \textbf{end}$$
$$\quad B(:,j) = w(:)$$
$$\textbf{end}$$

A linear workspace overhead is usually not important in a matrix computation that has a 2-dimensional array of the same order.

Problems

P1.2.1   Give an algorithm that overwrites A with A^2 where $A \in \mathbf{R}^{n \times n}$ is (a) upper
triangular and (b) square. Strive for a minimum workspace in each case.

P1.2.2   Suppose $A \in \mathbf{R}^{n \times n}$ is upper Hessenberg and that scalars $\lambda_1, \ldots, \lambda_r$ are given.
Give a saxpy algorithm for computing the first column of $M = (A - \lambda_1 I) \cdots (A - \lambda_r I)$.

P1.2.3   Give a column saxpy algorithm for the n-by-n matrix multiplication problem

$C = AB$ where A is upper triangular and B is lower triangular.

P1.2.4   Extend Algorithm 1.2.2 so that it can handle rectangular band matrices. Be sure to describe the underlying data structure.

P1.2.5 $A \in \mathbf{R}^{n \times n}$ is *Hermitian* if $A^H = A$. If $A = B + iC$, then it is easy to show that $B^T = B$ and $C^T = -C$. Suppose we represent A in an array $A.herm$ with the property that $A.herm(i, j)$ houses b_{ij} if $i \geq j$ and c_{ij} if $j > i$. Using this data structure write a matrix-vector multiply function that computes $\mathrm{Re}(z)$ and $\mathrm{Im}(z)$ from $\mathrm{Re}(x)$ and $\mathrm{Im}(x)$ so that $z = Ax$.

P1.2.6   Suppose $X \in \mathbf{R}^{n \times p}$ and $A \in \mathbf{R}^{n \times n}$, with A symmetric and stored by diagonal. Give an algorithm that computes $Y = X^T A X$ and stores the result by diagonal. Use separate arrays for A and Y.

P1.2.7   Suppose $a \in \mathbf{R}^n$ is given and that $A \in \mathbf{R}^{n \times n}$ has the property that $a_{ij} = a_{|i-j|+1}$. Give an algorithm that overwrites y with $Ax + y$ where $x, y \in \mathbf{R}^n$ are given.

P1.2.8   Suppose $a \in \mathbf{R}^n$ is given and that $A \in \mathbf{R}^{n \times n}$ has the property that $a_{ij} = a_{((i+j-1) \bmod n)+1}$. Give an algorithm that overwrites y with $Ax + y$ where $x, y \in \mathbf{R}^n$ are given.

P1.2.9   Develop a compact store-by-diagonal scheme for unsymmetric band matrices and write the corresponding gaxpy algorithm.

P1.2.10   Suppose p and q are n-vectors and that $A = (a_{ij})$ is defined by $a_{ij} = a_{ji} = p_i q_j$ for $1 \leq i \leq j \leq n$. How many flops are required to compute $y = Ax$ where $x \in \mathbf{R}^n$ is given?

Notes and References for Sec. 1.2

Consult the LAPACK manual for a discussion about appropriate data structures when symmetry and/or bandedness is present. See also

N. Madsen, G. Roderigue, and J. Karush (1976). "Matrix Multiplication by Diagonals on a Vector Parallel Processor," *Infomation Processing Letters 5*, 41-45.

1.3   Block Matrices and Algorithms

Having a facility with block matrix notation is crucial in matrix computations because it simplifies the derivation of many central algorithms. Moreover, "block algorithms" are increasingly important in high performance computing. By a block algorithm we essentially mean an algorithm that is rich in matrix-matrix multiplication. Algorithms of this type turn out to be more efficient in many computing environments than those that are organized at a lower linear algebraic level.

1.3.1   Block Matrix Notation

Column and row partitionings are special cases of matrix blocking. In general we can partition both the rows and columns of an m-by-n matrix

A to obtain

$$A = \begin{bmatrix} A_{11} & \dots & A_{1r} \\ \vdots & & \vdots \\ A_{q1} & \cdots & A_{qr} \end{bmatrix} \begin{matrix} m_1 \\ \\ m_q \end{matrix}$$

$$\begin{matrix} n_1 & & n_r \end{matrix}$$

where $m_1 + \cdots + m_q = m$, $n_1 + \cdots + n_r = n$, and $A_{\alpha\beta}$ designates the (α, β) block or submatrix. With this notation, block $A_{\alpha\beta}$ has dimension m_α-by-n_β and we say that $A = (A_{\alpha\beta})$ is a q-by-r block matrix.

1.3.2 Block Matrix Manipulation

Block matrices combine just like matrices with scalar entries as long as certain dimension requirements are met. For example, if

$$B = \begin{bmatrix} B_{11} & \dots & B_{1r} \\ \vdots & & \vdots \\ B_{q1} & \cdots & B_{qr} \end{bmatrix} \begin{matrix} m_1 \\ \\ m_q \end{matrix} \quad ,$$

$$\begin{matrix} n_1 & & n_r \end{matrix}$$

then we say that B is partitioned *conformably* with the matrix A above. The sum $C = A + B$ can also be regarded as a q-by-r block matrix:

$$C = \begin{bmatrix} C_{11} & \cdots & C_{1r} \\ \vdots & & \vdots \\ C_{q1} & \cdots & C_{qr} \end{bmatrix} = \begin{bmatrix} A_{11} + B_{11} & \cdots & A_{1r} + B_{1r} \\ \vdots & & \vdots \\ A_{q1} + B_{q1} & \cdots & A_{qr} + B_{qr} \end{bmatrix}.$$

The multiplication of block matrices is a little trickier. We start with a pair of lemmas.

Lemma 1.3.1 *If $A \in \mathbb{R}^{m \times p}$, $B \in \mathbb{R}^{p \times n}$,*

$$A = \begin{bmatrix} A_1 \\ \vdots \\ A_q \end{bmatrix} \begin{matrix} m_1 \\ \\ m_q \end{matrix} \qquad B = \begin{bmatrix} B_1 & ,\dots, & B_r \end{bmatrix} \quad ,$$

$$\begin{matrix} n_1 & & n_r \end{matrix}$$

then

$$AB = C = \begin{bmatrix} C_{11} & \dots & C_{1r} \\ \vdots & & \vdots \\ C_{q1} & \cdots & C_{qr} \end{bmatrix} \begin{matrix} m_1 \\ \\ m_q \end{matrix}$$

$$\begin{matrix} n_1 & & n_r \end{matrix}$$

where $C_{\alpha\beta} = A_\alpha B_\beta$ for $\alpha = 1{:}q$ and $\beta = 1{:}r$.

Proof. First we relate scalar entries in block $C_{\alpha\beta}$ to scalar entries in C. For $1 \le \alpha \le q$, $1 \le \beta \le r$, $1 \le i \le m_\alpha$, and $1 \le j \le n_\beta$ we have

$$[C_{\alpha\beta}]_{ij} = c_{\lambda+i,\mu+j}$$

where

$$\begin{aligned} \lambda &= m_1 + \cdots + m_{\alpha-1} \\ \mu &= n_1 + \cdots + n_{\beta-1}. \end{aligned}$$

But

$$c_{\lambda+i,\mu+j} = \sum_{k=1}^{p} a_{\lambda+i,k} b_{k,\mu+j} = \sum_{k=1}^{p} [A_\alpha]_{ik} [B_\beta]_{kj} = [A_\alpha B_\beta]_{ij}.$$

Thus, $C_{\alpha\beta} = A_\alpha B_\beta$. $\square$

Lemma 1.3.2 *If* $A \in \mathbb{R}^{m \times p}$, $B \in \mathbb{R}^{p \times n}$,

$$A = \begin{bmatrix} A_1 & ,\ldots, & A_s \\ p_1 & & p_s \end{bmatrix} \qquad , and \qquad B = \begin{bmatrix} B_1 \\ \vdots \\ B_s \end{bmatrix} \begin{matrix} p_1 \\ \\ p_s \end{matrix} \quad ,$$

then

$$AB = C = \sum_{\gamma=1}^{s} A_\gamma B_\gamma.$$

Proof. We set $s = 2$ and leave the general s case to the reader. (See P1.3.6.) For $1 \le i \le m$ and $1 \le j \le n$ we have

$$\begin{aligned} c_{ij} &= \sum_{k=1}^{p} a_{ik} b_{kj} = \sum_{k=1}^{p_1} a_{ik} b_{kj} + \sum_{k=p_1+1}^{p_1+p_2} a_{ik} b_{kj} \\ &= [A_1 B_1]_{ij} + [A_2 B_2]_{ij} = [A_1 B_1 + A_2 B_2]_{ij}. \end{aligned}$$

Thus, $C = A_1 B_1 + A_2 B_2$. $\square$

For general block matrix multiplication we have the following result:

Theorem 1.3.3 *If*

$$A = \begin{bmatrix} A_{11} & \ldots & A_{1s} \\ \vdots & & \vdots \\ A_{q1} & \cdots & A_{qs} \end{bmatrix} \begin{matrix} m_1 \\ \\ m_q \end{matrix} \quad , \qquad B = \begin{bmatrix} B_{11} & \ldots & B_{1r} \\ \vdots & & \vdots \\ B_{s1} & \cdots & B_{sr} \end{bmatrix} \begin{matrix} p_1 \\ \\ p_s \end{matrix} \quad ,$$
$$\quad\quad p_1 \quad\quad p_s \qquad\qquad\qquad\quad n_1 \quad\quad n_r$$

and we partition the product $C = AB$ as follows,

$$C = \begin{bmatrix} C_{11} & \dots & C_{1r} \\ \vdots & & \vdots \\ C_{q1} & \cdots & C_{qr} \end{bmatrix} \begin{matrix} m_1 \\ \\ m_q \end{matrix} \quad ,$$
$$\begin{matrix} n_1 & & n_r \end{matrix}$$

then

$$C_{\alpha\beta} = \sum_{\gamma=1}^{s} A_{\alpha\gamma} B_{\gamma\beta} \qquad \alpha = 1{:}q, \quad \beta = 1{:}r.$$

Proof. See P1.3.7. $\square$

A very important special case arises if we set $s = 2$, $r = 1$, and $n_1 = 1$:

$$\begin{bmatrix} A_{11} & A_{12} \\ A_{21} & A_{22} \end{bmatrix} \begin{bmatrix} x_1 \\ x_2 \end{bmatrix} = \begin{bmatrix} A_{11}x_1 + A_{12}x_2 \\ A_{21}x_1 + A_{22}x_2 \end{bmatrix}.$$

This partitioned matrix-vector product is used over and over again in subsequent chapters.

1.3.3 Submatrix Designation

As with "ordinary" matrix multiplication, block matrix multiplication can be organized in several ways. To specify the computations precisely, we need some notation.

Suppose $A \in \mathbb{R}^{m \times n}$ and that $i = (i_1, \dots, i_r)$ and $j = (j_1, \dots, j_c)$ are integer vectors with the property that

$$\begin{aligned} i_1, \dots, i_r &\in \{1, 2, \dots, m\} \\ j_1, \dots, j_c &\in \{1, 2, \dots, n\}. \end{aligned}$$

We let $A(i, j)$ denote the r-by-c submatrix

$$A(i, j) = \begin{bmatrix} A(i_1, j_1) & \cdots & A(i_1, j_c) \\ \vdots & & \vdots \\ A(i_r, j_1) & \cdots & A(i_r, j_c) \end{bmatrix}.$$

If the entries in the subscript vectors i and j are contiguous, then the "colon" notation can be used to define $A(i, j)$ in terms of the scalar entries in A. In particular, if $1 \le i_1 \le i_2 \le m$ and $1 \le j_1 \le j_2 \le n$, then $A(i_1{:}i_2, j_1{:}j_2)$ is the submatrix obtained by extracting rows i_1 through i_2 and columns j_1 through j_2, e.g,

$$A(3{:}5, 1{:}2) = \begin{bmatrix} a_{31} & a_{32} \\ a_{41} & a_{42} \\ a_{51} & a_{52} \end{bmatrix}.$$

While on the subject of submatrices, recall from §1.1.8 that if i and j are scalars, then $A(i,:)$ designates the ith row of A and $A(:,j)$ designates the jth column of A.

1.3.4   Block Matrix Times Vector

An important situation covered by Theorem 1.3.3 is the case of a block matrix times vector. Let us consider the details of the gaxpy $y = Ax + y$ where $A \in \mathbb{R}^{m \times n}$, $x \in \mathbb{R}^n$, $y \in \mathbb{R}^m$, and

$$A = \begin{bmatrix} A_1 \\ \vdots \\ A_q \end{bmatrix} \begin{matrix} m_1 \\ \\ m_q \end{matrix} \qquad y = \begin{bmatrix} y_1 \\ \vdots \\ y_q \end{bmatrix} \begin{matrix} m_1 \\ \\ m_q \end{matrix}.$$

We refer to A_i as the ith block row. If $m.vec = (m_1, \ldots, m_q)$ is the vector of block row "heights", then from

$$\begin{bmatrix} y_1 \\ \vdots \\ y_q \end{bmatrix} = \begin{bmatrix} A_1 \\ \vdots \\ A_q \end{bmatrix} x + \begin{bmatrix} y_1 \\ \vdots \\ y_q \end{bmatrix}$$

we obtain

$$
\begin{aligned}
&last = 0 \\
&\textbf{for } i = 1{:}q \\
&\qquad first = last + 1 \\
&\qquad last = first + m.vec(i) - 1 \\
&\qquad y(first{:}last) = A(first{:}last,:)x + y(first{:}last) \\
&\textbf{end}
\end{aligned}
\qquad (1.3.1)
$$

Each time through the loop an "ordinary" gaxpy is performed so Algorithms 1.1.3 and 1.1.4 apply.

Another way to block the gaxpy computation is to partition A and x as follows:

$$A = \begin{matrix} [\ A_1 \ , \ldots, \ A_r \ ] \\ n_1 n_r \end{matrix} \qquad x = \begin{bmatrix} x_1 \\ \vdots \\ x_r \end{bmatrix} \begin{matrix} n_1 \\ \\ n_r \end{matrix}.$$

In this case we refer to A_j as the jth block column of A. If $n.vec = (n_1, \ldots, n_r)$ is the vector of block column widths, then from

$$y = [\ A_1 \ , \ldots, \ A_r\ ] \begin{bmatrix} x_1 \\ \vdots \\ x_r \end{bmatrix} + y = \sum_{j=1}^{r} A_j x_j + y$$

we obtain

$$last = 0$$
$$\textbf{for } j = 1\text{:}r$$
$$\quad first = last + 1$$
$$\quad last = first + n.vec(j) - 1 \qquad\qquad (1.3.2)$$
$$\quad y = A(:, first\text{:}last)x(first\text{:}last) + y$$
$$\textbf{end}$$

Again, the gaxpy's performed each time through the loop can be carried out with Algorithm 1.1.3 or 1.1.4.

1.3.5   Block Matrix Multiplication

Just as ordinary, scalar-level matrix multiplication can be arranged in several possible ways, so can the multiplication of block matrices. Different blockings for A, B, and C can set the stage for block versions of the dot product, saxpy, and outer product algorithms of §1.1. To illustrate this with a minimum of subscript clutter, we assume that these three matrices are all n-by-n and that $n = N\ell$ where N and ℓ are positive integers.

If $A = (A_{\alpha\beta})$, $B = (B_{\alpha\beta})$, and $C = (C_{\alpha\beta})$ are N-by-N block matrices with ℓ-by-ℓ blocks, then from Theorem 1.3.3

$$C_{\alpha\beta} = \sum_{\gamma=1}^{N} A_{\alpha\gamma} B_{\gamma\beta} + C_{\alpha\beta} \qquad \alpha = 1\text{:}N, \quad \beta = 1\text{:}N.$$

If we organize a matrix multiplication procedure around this summation, then we obtain a block analog of Algorithm 1.1.5:

$$\textbf{for } \alpha = 1\text{:}N$$
$$\quad i = (\alpha - 1)\ell + 1\text{:}\alpha\ell$$
$$\quad \textbf{for } \beta = 1\text{:}N$$
$$\quad\quad j = (\beta - 1)\ell + 1\text{:}\beta\ell \qquad\qquad (1.3.3)$$
$$\quad\quad \textbf{for } \gamma = 1\text{:}N$$
$$\quad\quad\quad k = (\gamma - 1)\ell + 1\text{:}\gamma\ell$$
$$\quad\quad\quad C(i, j) = A(i, k)B(k, j) + C(i, j)$$
$$\quad\quad \textbf{end}$$
$$\quad \textbf{end}$$
$$\textbf{end}$$

Note that if $\ell = 1$, then $\alpha \equiv i$, $\beta \equiv j$, and $\gamma \equiv k$ and we revert to Algorithm 1.1.5.

To obtain a block saxpy matrix multiply, we write $C = AB + C$ as

$$[\, C_1, \ldots, C_N \,] = [\, A_1, \ldots, A_N \,] \begin{bmatrix} B_{11} & \cdots & B_{1N} \\ \vdots & \ddots & \vdots \\ B_{N1} & \cdots & B_{NN} \end{bmatrix} + [\, C_1, \ldots, C_N \,]$$

where $A_\alpha, C_\alpha \in \mathbb{R}^{n \times \ell}$, and $B_{\alpha\beta} \in \mathbb{R}^{\ell \times \ell}$. From this we obtain

$$\textbf{for } \beta = 1:N$$
$$\quad j = (\beta - 1)\ell + 1:\beta\ell$$
$$\quad \textbf{for } \alpha = 1:N$$
$$\qquad i = (\alpha - 1)\ell + 1:\alpha\ell \qquad\qquad\qquad (1.3.4)$$
$$\qquad C(:,j) = A(:,i)B(i,j) + C(:,j)$$
$$\quad \textbf{end}$$
$$\textbf{end}$$

This is the block version of Algorithm 1.1.7.

A *block outer product* scheme results if we work with the blockings

$$A = [\, A_1, \ldots, A_N\,] \qquad B = \begin{bmatrix} B_1^T \\ \vdots \\ B_N^T \end{bmatrix}$$

where $A_\gamma, B_\gamma \in \mathbb{R}^{n \times \ell}$. From Lemma 1.3.2 we have

$$C = \sum_{\gamma=1}^{N} A_\gamma B_\gamma^T + C$$

and so

$$\textbf{for } \gamma = 1:N$$
$$\quad k = (\gamma - 1)\ell + 1:\gamma\ell$$
$$\quad C = A(:,k)B(k,:) + C \qquad\qquad\qquad (1.3.5)$$
$$\textbf{end}$$

This is the block version of Algorithm 1.1.8.

1.3.6   Complex Matrix Multiplication

Consider the complex matrix multiplication update

$$C_1 + iC_2 = (A_1 + iA_2)(B_1 + iB_2) + (C_1 + iC_2)$$

where all the matrices are real and $i^2 = -1$. Comparing the real and imaginary parts we find

$$\begin{aligned} C_1 &= A_1 B_1 - A_2 B_2 + C_1 \\ C_2 &= A_1 B_2 + A_2 B_1 + C_2 \end{aligned}$$

and this can be expressed as follows:

$$\begin{bmatrix} C_1 \\ C_2 \end{bmatrix} = \begin{bmatrix} A_1 & -A_2 \\ A_2 & A_1 \end{bmatrix} \begin{bmatrix} B_1 \\ B_2 \end{bmatrix} + \begin{bmatrix} C_1 \\ C_2 \end{bmatrix}.$$

This suggests how real matrix software might be applied to solve complex matrix problems. The only snag is that the explicit formation of

$$\tilde{A} = \begin{bmatrix} A_1 & -A_2 \\ A_2 & A_1 \end{bmatrix}$$

requires the "double storage" of the matrices A_1 and A_2.

1.3.7   A Divide and Conquer Matrix Multiplication

We conclude this section with a completely different approach to the matrix-matrix multiplication problem. The starting point in the discussion is the 2-by-2 block matrix multiplication

$$\begin{bmatrix} C_{11} & C_{12} \\ C_{21} & C_{22} \end{bmatrix} = \begin{bmatrix} A_{11} & A_{12} \\ A_{21} & A_{22} \end{bmatrix} \begin{bmatrix} B_{11} & B_{12} \\ B_{21} & B_{22} \end{bmatrix}$$

where each block is square. In the ordinary algorithm, $C_{ij} = A_{i1}B_{1j} + A_{i2}B_{2j}$. There are 8 multiplies and 4 adds. Strassen (1969) has shown how to compute C with just 7 multiplies and 18 adds:

$$
\begin{aligned}
P_1 &= (A_{11} + A_{22})(B_{11} + B_{22}) \\
P_2 &= (A_{21} + A_{22})B_{11} \\
P_3 &= A_{11}(B_{12} - B_{22}) \\
P_4 &= A_{22}(B_{21} - B_{11}) \\
P_5 &= (A_{11} + A_{12})B_{22} \\
P_6 &= (A_{21} - A_{11})(B_{11} + B_{12}) \\
P_7 &= (A_{12} - A_{22})(B_{21} + B_{22}) \\
C_{11} &= P_1 + P_4 - P_5 + P_7 \\
C_{12} &= P_3 + P_5 \\
C_{21} &= P_2 + P_4 \\
C_{22} &= P_1 + P_3 - P_2 + P_6
\end{aligned}
$$

These equations are easily confirmed by substitution. Suppose $n = 2m$ so that the blocks are m-by-m. Counting adds and multiplies in the computation $C = AB$ we find that conventional matrix multiplication involves $(2m)^3$ multiplies and $(2m)^3 - (2m)^2$ adds. In contrast, if Strassen's algorithm is applied *with conventional multiplication at the block level*, then $7m^3$ multiplies and $7m^3 + 11m^2$ adds are required. If $m \gg 1$, then the Strassen method involves about 7/8ths the arithmetic of the fully conventional algorithm.

Now recognize that we can recur on the Strassen idea. In particular, we can apply the Strassen algorithm to each of the half-sized block multiplications associated with the P_i. Thus, if the original A and B are n-by-n and $n = 2^q$, then we can repeatedly apply the Strassen multiplication algorithm. At the bottom "level," the blocks are 1-by-1. Of course, there is no need to recur down to the $n = 1$ level. When the block size gets sufficiently small, $(n \leq n_{min})$, it may be sensible to use conventional matrix multiplication when finding the P_i. Here is the overall procedure:

Algorithm 1.3.1 (Strassen Multiplication) Suppose $n = 2^q$ and that $A \in \mathbb{R}^{n \times n}$ and $B \in \mathbb{R}^{n \times n}$. If $n_{min} = 2^d$ with $d \leq q$, then this algorithm computes $C = AB$ by applying Strassen procedure recursively $q - d$ times.

> **function:** $C = \mathbf{strass}(A, B, n, n_{min})$
>   **if** $n \leq n_{min}$
> $C = AB$
>   **else**
> $m = n/2; u = 1{:}m; v = m+1{:}n;$
> $P_1 = \mathbf{strass}(A(u, u) + A(v, v), B(u, u) + B(v, v), m, n_{min})$
> $P_2 = \mathbf{strass}(A(v, u) + A(v, v), B(u, u), m, n_{min})$
> $P_3 = \mathbf{strass}(A(u, u), B(u, v) - B(v, v), m, n_{min})$
> $P_4 = \mathbf{strass}(A(v, v), B(v, u) - B(u, u), m, n_{min})$
> $P_5 = \mathbf{strass}(A(u, u) + A(u, v), B(v, v), m, n_{min})$
> $P_6 = \mathbf{strass}(A(v, u) - A(u, u), B(u, u) + B(u, v), m, n_{min})$
> $P_7 = \mathbf{strass}(A(u, v) - A(v, v), B(v, u) + B(v, v), m, n_{min})$
> $C(u, u) = P_1 + P_4 - P_5 + P_7$
> $C(u, v) = P_3 + P_5$
> $C(v, u) = P_2 + P_4$
> $C(v, v) = P_1 + P_3 - P_2 + P_6$
>   **end**

Unlike any of our previous algorithms **strass** is recursive, meaning that it calls itself. Divide and conquer algorithms are often best described in this manner. We have presented this algorithm in the style of a MATLAB function so that the recursive calls can be stated with precision.

The amount of arithmetic associated with **strass** is a complicated function of n and n_{min}. If $n_{min} \gg 1$, then it suffices to count multiplications as the number of additions is roughly the same. If we just count the multiplications, then it suffices to examine the deepest level of the recursion as that is where all the multiplications occur. In **strass** there are $q - d$ subdivisions and thus, 7^{q-d} conventional matrix-matrix multiplications to perform. These multiplications have size n_{min} and thus **strass** involves about $s = (2^d)^3 7^{q-d}$ multiplications compared to $c = (2^q)^3$, the number of multiplications in the conventional approach. Notice that

$$\frac{s}{c} = \left(\frac{2^d}{2^q}\right)^3 7^{q-d} = \left(\frac{7}{8}\right)^{q-d}.$$

If $d = 0$, i.e., we recur on down to the 1-by-1 level, then

$$s = \left(\frac{7}{8}\right)^q c = 7^q = n^{\log_2 7} \approx n^{2.807} .$$

Thus, asymptotically, the number of multiplications in the Strassen procedure is $O(n^{2.807})$. However, the number of additions (relative to the number
of multiplications) becomes significant as n_{min} gets small.

Example 1.3.1 If $n = 1024$ and $n_{min} = 64$, then **strass** involves $(7/8)^{10-6} \approx .6$ the
arithmetic of the conventional algorithm.

Problems

P1.3.1 Generalize (1.3.3) so that it can handle the variable block-size problem covered
by Theorem 1.3.3.

P1.3.2 Generalize (1.3.4) and (1.3.5) so that they can handle the variable block-size
case.

P1.3.3 Adapt **strass** so that it can handle square matrix multiplication of any order.
Hint: If the "current" A has odd dimension, append a zero row and column.

P1.3.4 Prove that if

$$A = \begin{bmatrix} A_{11} & \cdots & A_{1r} \\ \vdots & \ddots & \vdots \\ A_{q1} & \cdots & A_{qr} \end{bmatrix}$$

is a blocking of the matrix A, then

$$A^T = \begin{bmatrix} A_{11}^T & \cdots & A_{q1}^T \\ \vdots & \ddots & \vdots \\ A_{1r}^T & \cdots & A_{qr}^T \end{bmatrix}.$$

P1.3.5 Suppose n is even and define the following function from $\mathbf{R}^n$ to $\mathbf{R}$:

$$f(x) = x(1{:}2{:}n)^T x(2{:}n) = \sum_{i=1}^{n/2} x_{2i-1} x_{2i}$$

(a) Show that if $x, y \in \mathbf{R}^n$ then

$$x^T y = \sum_{i=1}^{n/2} (x_{2i-1} + y_{2i})(x_{2i} + y_{2i-1}) - f(x) - f(y)$$

(b) Now consider the n-by-n matrix multiplication $C = AB$. Give an algorithm for
computing this product that requires $n^3/2$ multiplies once f is applied to the rows of A
and the columns of B. See Winograd (1968) for details.

P1.3.6 Prove Lemma 1.3.2 for general s. Hint. Set

$$\rho_\gamma = p_1 + \cdots + p_{\gamma-1} \qquad \gamma = 1{:}s+1$$

and show that

$$c_{ij} = \sum_{\gamma=1}^{s} \sum_{k=\rho_\gamma+1}^{\rho_{\mu+1}} a_{ik} b_{kj}.$$

P1.3.7 Use Lemmas 1.3.1 and 1.3.2 to prove Theorem 1.3.3. In particular, set

$$
A_\gamma = \begin{bmatrix} A_{1\gamma} \\ \vdots \\ A_{q\gamma} \end{bmatrix} \quad \text{and} \quad B_\gamma = \begin{bmatrix} B_{\gamma 1} & \cdots & B_{\gamma r} \end{bmatrix}
$$

and note from Lemma 1.3.2 that

$$
C = \sum_{\gamma=1}^{s} A_\gamma B_\gamma .
$$

Now analyze each $A_\gamma B_\gamma$ with the help of Lemma 1.3.1.

Notes and References for Sec. 1.3

For quite some time fast methods for matrix multiplication have attracted a lot of attention within computer science. See

S. Winograd (1968). "A New Algorithm for Inner Product," *IEEE Trans. Comp. C-17*, 693–694.
V. Strassen (1969). "Gaussian Elimination is Not Optimal," *Numer. Math. 13*, 354–356.
V. Pan (1984). "How Can We Speed Up Matrix Multiplication?," *SIAM Review 26*, 393–416.

Many of these methods have dubious practical value. However, with the publication of

D. Bailey (1988). "Extra High Speed Matrix Multiplication on the Cray-2," *SIAM J. Sci. and Stat. Comp. 9*, 603–607.

it is clear that the blanket dismissal of these fast procedures is unwise. The "stability" of the Strassen algorithm is discussed in §2.4.10. See also

N.J. Higham (1990). "Exploiting Fast Matrix Multiplication within the Level 3 BLAS," *ACM Trans. Math. Soft. 16*, 352–368.
C.C. Douglas, M. Heroux, G. Slishman, and R.M. Smith (1994). "GEMMW: A Portable Level 3 BLAS Winograd Variant of Strassen's Matrix-Matrix Multiply Algorithm," *J. Comput. Phys. 110*, 1–10.

1.4   Vectorization and Re-Use Issues

The matrix manipulations discussed in this book are mostly built upon dot products and saxpy operations. *Vector pipeline computers* are able to perform vector operations such as these very fast because of special hardware that is able to exploit the fact that a vector operation is a very regular sequence of scalar operations. Whether or not high performance is extracted from such a computer depends upon the length of the vector operands and a number of other factors that pertain to the movement of data such as vector stride, the number of vector loads and stores, and the level of data re-use. Our goal is to build a useful awareness of these issues. We are *not* trying to build a comprehensive model of vector pipeline computing that might be used to predict performance. We simply want to identify the kind of *thinking* that goes into the design of an effective vector pipeline code. We do not mention any particular machine. The literature is filled with case studies.

1.4.1 Pipelining Arithmetic Operations

The primary reason why vector computers are fast has to do with *pipelining*. The concept of pipelining is best understood by making an analogy to assembly line production. Suppose the assembly of an individual automobile requires one minute at each of sixty workstations along an assembly line. If the line is well staffed and able to initiate the assembly of a new car every minute, then 1000 cars can be produced from scratch in about 1000 + 60 = 1060 minutes. For a work order of this size the line has an effective "vector speed" of 1000/1060 automobiles per minute. On the other hand, if the assembly line is understaffed and a new assembly can be initiated just once an hour, then 1000 hours are required to produce 1000 cars. In this case the line has an effective "scalar speed" of 1/60th automobile per minute.

So it is with a pipelined vector operation such as the vector add $z = x+y$. The scalar operations $z_i = x_i + y_i$ are the cars. The number of elements is the size of the work order. If the start-to-finish time required for each z_i is τ, then a pipelined, length n vector add could be completed in time much less than $n\tau$. This gives vector speed. Without the pipelining, the vector computation would proceed at a scalar rate and would approximately require time $n\tau$ for completion.

Let us see how a sequence of floating point operations can be pipelined. Floating point operations usually require several cycles to complete. For example, a 3-cycle addition of two scalars x and y may proceed as in FIG.1.4.1. To visualize the operation, continue with the above metaphor

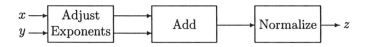

FIG. 1.4.1 *A 3-Cycle Adder*

and think of the addition unit as an assembly line with three "work stations". The input scalars x and y proceed along the assembly line spending one cycle at each of three stations. The sum z emerges after three cycles.

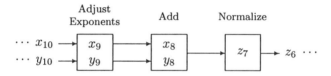

FIG. 1.4.2 *Pipelined Addition*

Note that when a single, "free standing" addition is performed, only one of the three stations is active during the computation.

Now consider a vector addition $z = x + y$. With pipelining, the x and y vectors are streamed through the addition unit. Once the pipeline is filled and steady state reached, a z_i is produced every cycle. In FIG.1.4.2 we depict what the pipeline might look like once this steady state is achieved. In this case, vector speed is about three times scalar speed because the time for an individual add is three cycles.

1.4.2 Vector Operations

A vector pipeline computer comes with a repertoire of *vector instructions*, such as vector add, vector multiply, vector scale, dot product, and saxpy. We assume for clarity that these operations take place in *vector registers*. Vectors travel between the registers and memory by means of *vector load* and *vector store* instructions.

An important attribute of a vector processor is the length of its vector registers which we designate by v_L. A length-n vector operation must be broken down into subvector operations of length v_L or less. Here is how such a partitioning might be managed in the case of a vector addition $z = x + y$ where x and y are n-vectors:

$$first = 1$$
while $first \leq n$
     $last = \min\{n, first + v_L - 1\}$
     Vector load $x(first:last)$.
     Vector load $y(first:last)$.
     Vector add: $z(first:last) = x(first:last) + y(first:last)$.
     Vector store $z(first:last)$.
     $first = last + 1$
end

A reasonable compiler for a vector computer would automatically generate these vector instructions from a programmer specified $z = x + y$ command.

1.4.3   The Vector Length Issue

Suppose the pipeline for the vector operation op takes τ_{op} cycles to "set up." Assume that one component of the result is obtained per cycle once the pipeline is filled. The time required to perform an n-dimensional op is then given by

$$T_{op}(n) = (\tau_{op} + n)\mu \qquad n \le v_L$$

where μ is the cycle time and v_L is the length of the vector hardware.

If the vectors to be combined are longer than the vector hardware length, then as we have seen the overall vector operation must be broken down into hardware-manageable chunks. Thus, if

$$n = n_1 v_L + n_0 \qquad 0 \le n_0 < v_L \, ,$$

then we assume that

$$T_{op}(n) = \left\{ \begin{array}{ll} n_1(\tau_{op} + v_L)\mu & n_0 = 0 \\ (n_1(\tau_{op} + v_L) + \tau_{op} + n_0)\mu & n_0 \ne 0 \end{array} \right.$$

specifies the overall time required to perform a length-n op. This simplifies to

$$T_{op}(n) = (n + \tau_{op}\mathrm{ceil}(n/v_L))\,\mu$$

where $\mathrm{ceil}(\alpha)$ is the smallest integer such that $\alpha \le \mathrm{ceil}(\alpha)$. If ρ flops per component are involved, then the effective rate of computation for general n is given by

$$R_{op}(n) = \frac{\rho n}{T_{op}(n)} = \frac{\rho}{\mu} \frac{1}{1 + \frac{\tau_{op}}{n} \mathrm{ceil}\left(\frac{n}{v_L}\right)} .$$

(If μ is in seconds, then R_{op} is in flops per second.) The asymptotic rate of performance is given by

$$\lim_{n \to \infty} R_{op}(n) = \frac{1}{1 + \frac{\tau_{op}}{v_L}} \frac{\rho}{\mu} .$$

As a way of assessing how serious the start-up overhead is for a vector operation, Hockney and Jesshope (1988) define the quantity $n_{1/2}$ to be the smallest n for which half of peak performance is achieved, i.e.,

$$\frac{\rho n_{1/2}}{T_{op}(n_{1/2})} = \frac{1}{2} \frac{\rho}{\mu} .$$

Machines that have big $n_{1/2}$ factors do not perform well on short vector operations.

Let us see what the above performance model says about the design of the matrix multiply update $C = AB + C$ where $A \in \mathbb{R}^{m \times p}$, $B \in \mathbb{R}^{p \times n}$, and $C \in \mathbb{R}^{m \times n}$. Recall from §1.1.11 that there are six possible versions of the conventional algorithm and they correspond to the six possible loop orderings of

```
for i = 1:m
    for j = 1:n
        for k = 1:p
            C(i, j) = A(i, k)B(k, j) + C(i, j)
        end
    end
end
```

This is the ijk variant and its innermost loop oversees a length-p dot product. Thus, our performance model predicts that

$$T_{ijk} = mnp + mn \cdot \text{ceil}(p/v_L)\tau_{dot}$$

cycles are required. A similar analysis for each of the other variants leads to the following table:

Variant	Cycles
ijk	$mnp + mn \cdot \tau_{dot}(p/v_L)$
jik	$mnp + mn \cdot \tau_{dot}(p/v_L)$
ikj	$mnp + mp \cdot \tau_{sax}(n/v_L)$
jki	$mnp + np \cdot \tau_{sax}(m/v_L)$
kij	$mnp + mp \cdot \tau_{sax}(n/v_L)$
kji	$mnp + np \cdot \tau_{sax}(m/v_L)$

We make a few observations based upon some elementary integer arithmetic manipulation. Assume that τ_{sax} and τ_{dot} are roughly equal. If m, n, and p are all less than v_L, then the most efficient variants will have the longest inner loops. If m, n, and p are much bigger than v_L, then the distinction between the six options is small.

1.4.4   The Stride Issue

The "layout" of a vector operand in memory often has a bearing on execution speed. The key factor is stride. The *stride* of a stored floating point vector is the distance (in logical memory locations) between the vector's components. Accessing a row in a two-dimensional Fortran array is not a unit stride operation because arrays are stored by column. In C, it is just the opposite as matrices are stored by row. Nonunit stride vector operations may interfere with the pipelining capability of a computer degrading performance.

To clarify the stride issue we consider how the six variants of matrix multiplication "pull up" data from the A, B, and C matrices in the inner loop. This is where the vector calculation occurs (dot product or saxpy) and there are three possibilities:

jki or kji: **for** $i = 1{:}m$
$$C(i,j) = C(i,j) + A(i,k)B(k,j)$$
                                   end

ikj or kij: **for** $j = 1{:}n$
$$C(i,j) = C(i,j) + A(i,k)B(k,j)$$
                                   end

ijk or jik: **for** $k = 1{:}p$
$$C(i,j) = C(i,j) + A(i,k)B(k,j)$$
                                   end

Here is a table that specifies the A, B, and C strides associated with each of these possibilities:

Variant	A Stride	B Stride	C Stride
jki or kji	Unit	0	Unit
ikj or kij	0	Non-Unit	Non-Unit
ijk or jik	Non-Unit	Unit	0

Storage in column-major order is assumed. A stride of zero means that only a single array element is accessed in the inner loop. From the stride point of view, it is clear that we should favor the jki and kji variants. This may not coincide with a preference that is based on vector length considerations. Dilemmas of this type are typical in high performance computing. One goal (maximize vector length) can conflict with another (impose unit stride).

Sometimes a vector stride/vector length conflict can s be resolved through the intelligent choice of data structures. Consider the gaxpy $y = Ax + y$ where $A \in \mathbb{R}^{n \times n}$ is symmetric. Assume that $n \leq v_L$ for simplicity. If A is stored conventionally and Algorithm 1.1.4 is used, then the central computation entails n, unit stride saxpy's each having length n:

> **for** $j = 1{:}n$
> $\quad y = A(:,j)x(j) + y$
> **end**

Our simple execution model tells us that

$$T_1 = n(\tau_{sax} + n)$$

cycles are required.

In §1.2.7 we introduced the lower triangular storage scheme for symmetric matrices and obtained this version of the gaxpy:

```
for j = 1:n
    for i = 1:j − 1
        y(i) = A.vec((i − 1)n − i(i − 1)/2 + j)x(j) + y(i)
    end
    for i = j:n
        y(i) = A.vec((j − 1)n − j(j − 1)/2 + i)x(j) + y(i)
    end
end
```

Notice that the first i-loop does *not* define a unit stride saxpy. If we assume that a length n, nonunit stride saxpy is equivalent to n unit-length saxpys (a worst case scenario), then this implementation involves

$$T_2 = n \left(\frac{n}{2} \tau_{sax} + n \right)$$

cycles.

In §1.2.8 we developed the store-by-diagonal version:

```
for i = 1:n
    y(i) = A.diag(i)x(i) + y(i)
end
for k = 1:n − 1
    t = nk − k(k − 1)/2
    {y = D(A, k)x + y}
    for i = 1:n − k
        y(i) = A.diag(i + t)x(i + k) + y(i)
    end
    {y = D(A, k)ᵀx + y}
    for i = 1:n − k
        y(i + k) = A.diag(i + t)x(i) + y(i + k)
    end
end
```

In this case both inner loops define a unit stride vector multiply (vm) and our model of execution predicts

$$T_3 = n \left(2\tau_{vm} + n \right)$$

cycles.

The example shows how the choice of data structure can effect the stride attributes of an algorithm. Store by diagonal seems attractive because it represents the matrix compactly and has unit stride. However, a careful which-is-best analysis would depend upon the values of τ_{sax} and τ_{vm} and the precise penalties for nonunit stride computation and excess storage. The complexity of the situation would call for careful benchmarking.

1.4.5   Thinking About Data Motion

Another important attribute of a matrix algorithm concerns the actual volume of data that has to be moved around during execution. Matrices sit in memory but the computations that involve their entries take place in functional units. The control of memory traffic is crucial to performance in many computers. To continue with the factory metaphor used at the beginning of this section: *Can we keep the superfast arithmetic units busy with enough deliveries of matrix data and can we ship the results back to memory fast enough to avoid backlog?* FIG.1.4.3 depicts the typical situation in an advanced uniprocessor environment. Details vary from machine

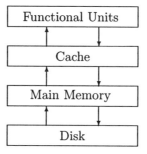

FIG. 1.4.3 *Memory Hierarchy*

to machine, but two "axioms;; prevail:

- Each level in the hierarchy has a limited capacity and for economic reasons this capacity is usually smaller as we ascend the hierarchy.

- There is a cost, sometimes relatively great, associated with the moving of data between two levels in the hierarchy.

The design of an efficient matrix algorithm requires careful thinking about the flow of data in between the various levels of storage. The vector touch and data re-use issues are important in this regard.

1.4.6   The Vector Touch Issue

In many advanced computers, data is moved around in chunks, e.g., vectors. The time required to read or write a vector to memory is comparable to the time required to engage the vector in a dot product or saxpy. Thus, the number of *vector touches* associated with a matrix code is a very important statistic. By a "vector touch" we mean either a vector load or store.

Let's count the number of vector touches associated with an m-by-n outer product. Assume that $m = m_1 v_L$ and $n = n_1 v_L$ where v_L is the vector hardware length. (See §1.4.3.) In this environment, the outer product update $A = A + xy^T$ would be arranged as follows:

    for $\alpha = 1{:}m_1$
        $i = (\alpha - 1)v_L + 1{:}\alpha v_L$
        for $\beta = 1{:}n_1$
            $j = (\beta - 1)v_L + 1{:}\beta v_L$
            $A(i, j) = A(i, j) + x(i)y(j)^T$
        end
    end

Each column of the submatrix $A(i, j)$ must be loaded, updated, and then stored. Not forgetting to account for the vector touches associated with x and y we see that approximately

$$\sum_{\alpha=1}^{m_1} \left(1 + \sum_{\beta=1}^{n_1} (1 + 2v_L) \right) \approx 2m_1 n$$

vector touches are required. (Low order terms do not contribute to the analysis.)

Now consider the gaxpy update $y = Ax + y$ where $y \in \mathbb{R}^m$, $x \in \mathbb{R}^n$ and $A \in \mathbb{R}^{m \times n}$. Breaking this computation down into segments of length v_L gives

    for $\alpha = 1{:}m_1$
        $i = (\alpha - 1)v_L + 1{:}\alpha v_L$
        for $\beta = 1{:}n_1$
            $j = (\beta - 1)v_L + 1{:}\beta v_L$
            $y(i) = y(i) + A(i, j)x(j)$
        end
    end

Again, each column of submatrix $A(i, j)$ must be read but the only writing to memory involves subvectors of y. Thus, the number of vector touches for an m-by-n gaxpy is

$$\sum_{\alpha=1}^{m_1} \left(2 + \sum_{\beta=1}^{n_1} (1 + v_L) \right) \approx m_1 n .$$

This is half the number required by an identically-sized the outer product. Thus, if a computation can be arranged in terms of either outer products or gaxpys, then the former is preferable from the vector touch standpoint.

1.4.7 Blocking and Re-Use

A *cache* is a small high-speed memory situated in between the functional units and main memory. See Fig.1.4.3. Cache utilization colors performance because it has a direct bearing upon how data flows in between the functional units and the lower levels of memory.

To illustrate this we consider the computation of the matrix multiply update $C = AB + C$ where $A, B, C \in \mathbb{R}^{n \times n}$ reside in main memory[2]. All data must pass through the cache on its way to the functional units where the floating point computations are carried out. If the cache is small and n is big, then the update must be broken down into smaller parts so that the cache can "gracefully" process the flow of data.

One strategy is to block the B and C matrices,

$$B = [\ \underset{\ell}{B_1}\ , \ldots, \underset{\ell}{B_N}\ ] \qquad C = [\ \underset{\ell}{C_1}\ , \ldots, \underset{\ell}{C_N}\ ]$$

where we assume that $n = \ell N$. From the expansion

$$C_\alpha = AB_\alpha + C_\alpha = \sum_{k=1}^{n} A(:,k)B_\alpha(k,:) + C_\alpha$$

we obtain the following computational framework:

> **for** $\alpha = 1{:}N$
>   Load B_α and C_α into cache.
>   **for** $k = 1{:}n$
>   Load $A(:,k)$ into cache and update C_α:
> $C_\alpha = A(:,k)B_\alpha(k,:) + C_\alpha$
>   **end**
>   Store C_α in main memory.
> **end**

Note that if M is the cache size measured in floating point words, then we must have

$$2n\ell + n \le M. \tag{1.4.1}$$

Let Γ_1 be the number of floating point numbers that flow (in either direction) between cache and main memory. Note that every entry in B is loaded into cache once, every entry in C is loaded into cache once and stored back in main memory once, and every entry in A is loaded into cache $N = n/\ell$ times. It follows that

$$\Gamma_1 = 3n^2 + \frac{n^3}{\ell}.$$

---

[2]The discussion which follows would also apply if the matrices were on a disk and needed to be brought into main memory.

In the interest of keeping data motion to a minimum, we choose ℓ to be as large as possible subject to the constraint (1.4.1). We therefore set

$$\ell \approx \frac{1}{2}\left(\frac{M}{n} - 1\right)$$

obtaining

$$\Gamma_1 \approx 3n^2 + \frac{2n^4}{M-n}.$$

(We use "$\approx$" to emphasize the approximate nature of our analysis.) If cache is large enough to house the entire B and C matrices with room left over for a column of A, then $\ell = n$ and $\Gamma_1 = 4n^2$. At the other extreme, if we can just fit three columns in cache, then $\ell = 1$ and $\Gamma_1 \approx n^3$.

Now let us regard $A = (A_{\alpha\beta})$, $B = (B_{\alpha\beta})$, and $C = (C_{\alpha\beta})$ as N-by-N block matrices with uniform block size $\ell = n/N$. With this blocking the computation of

$$C_{\alpha\beta} = \sum_{\gamma=1}^{N} A_{\alpha\gamma}B_{\gamma\beta} \qquad \alpha = 1{:}N, \ \beta = 1{:}N$$

can be arranged as follows:

> **for** $\alpha = 1{:}N$
> > **for** $\beta = 1{:}N$
> > > Load $C_{\alpha\beta}$ into cache.
> > > **for** $\gamma = 1{:}N$
> > > > Load $A_{\alpha\gamma}$ and $B_{\gamma\beta}$ into cache.
> > > > $C_{\alpha\beta} = C_{\alpha\beta} + A_{\alpha\gamma}B_{\gamma\beta}$
> > > **end**
> > > Store $C_{\alpha\beta}$ in main memory.
> > **end**
> **end**

In this case the main memory/cache traffic sums to

$$\Gamma_2 = 2n^2 + \frac{2n^3}{\ell}$$

because each entry in A and B is loaded $N = n/\ell$ times and each entry in C is loaded once and stored once. We can minimize this by choosing ℓ to be as large as possible subject to the constraint that three blocks fit in cache, i.e.,

$$3\ell^2 \leq M$$

Setting $\ell \approx \sqrt{M/3}$ gives

$$\Gamma_2 \approx 2n^2 + 2n^3\sqrt{\frac{3}{M}}.$$

A manipulation shows that

$$\frac{\Gamma_1}{\Gamma_2} \approx \frac{3n^2 + \frac{2n^4}{M-n}}{2n^2 + 2n^3\sqrt{\frac{3}{M}}} \geq \frac{3 + 2\frac{n^2}{M}}{2 + 2\sqrt{3}\sqrt{\frac{n^2}{M}}}.$$

The key quantity here is n^2/M, the ratio of matrix size (in floating point words) to cache size. As this ratio grows the we find that

$$\frac{\Gamma_1}{\Gamma_2} \approx \frac{n}{\sqrt{3M}}$$

showing that the second blocking strategy is superior from the standpoint of data motion to and from the cache. The fundamental conclusion to be reached from all of this is that *blocking effects data motion.*

1.4.8 Block Matrix Data Structures

We conclude this section with a discussion about block data structures. A programming language that supports two-dimensional arrays must have a convention for storing such a structure in memory. For example, Fortran stores two-dimensional arrays in *column major order*. This means that the entries within a column are contiguous in memory. Thus, if 24 storage locations are allocated for $A \in \mathbb{R}^{4 \times 6}$, then in traditional store-by-column format the matrix entries are "lined up" in memory as depicted in FIG. 1.4.4. In other words, if $A \in \mathbb{R}^{m \times n}$ is stored in $v(1:mn)$, then we identify

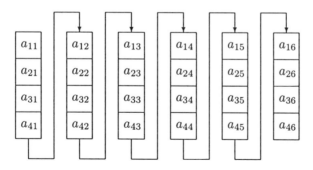

FIG. 1.4.4 *Store by Column (4-by-6 case)*

$A(i,j)$ with $v((j-1)m + i)$. For algorithms that access matrix data by column this is a good arrangement since the column entries are contiguous in memory.

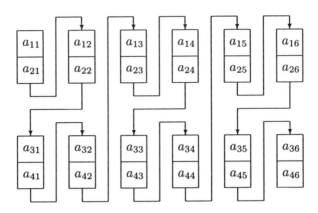

FIG. 1.4.5 *Store-by-Blocks (4-by-6 case with 2-by-2 Blocks)*

In certain block matrix algorithms it is sometimes useful to store matrices by blocks rather than by column. Suppose, for example, that the matrix A above is a 2-by-3 block matrix with 2-by-2 blocks. In a store-by-column block scheme with store-by-column within each block, the 24 entries are arranged in memory as shown in FIG. 1.4.5. This data structure can be attractive for block algorithms because the entries within a given block are contiguous in memory.

Problems

P1.4.1   Consider the matrix product $D = ABC$ where $A \in \mathbf{R}^{m \times r}$, $B \in \mathbf{R}^{r \times n}$ and $C \in \mathbf{R}^{n \times q}$. Assume that all the matrices are stored by column and that the time required to execute a unit-stride saxpy operation of length k is of the form $t(k) = (L+k)\mu$ where L is a constant and μ is the cycle time. Based on this model, when is it more economical to compute D as $D = (AB)C$ instead of as $D = A(BC)$? Assume that all matrix multiplies are done using the jki, (gaxpy) algorithm.

P1.4.2   What is the total time spent in jki variant on the saxpy operations assuming that all the matrices are stored by column and that the time required to execute a unit-stride saxpy operation of length k is of the form $t(k) = (L+k)\mu$ where L is a constant and μ is the cycle time? Specialize the algorithm so that it efficiently handles the case when A and B are n-by-n and upper triangular. Does it follow that the triangular implementation is six times faster as the flop count suggests?

P1.4.3   Give an algorithm for computing $C = A^T BA$ where A and B are n–by–n and B is symmetric. Arrays should be accessed in unit stride fashion within all innermost loops.

P1.4.4   Suppose $A \in \mathbf{R}^{m \times n}$ is stored by column in $A.col(1:mn)$. Assume that $m = \ell_1 M$

and $n = \ell_2 N$ and that we regard A as an M-by-N block matrix with ℓ_1-by-ℓ_2 blocks. Given i, j, α, and β that satisfy $1 \leq i \leq \ell_1$, $1 \leq j \leq \ell_2$, $1 \leq \alpha \leq M$, and $1 \leq \beta \leq N$ determine k so that $A.col(k)$ houses the (i, j) entry of $A_{\alpha\beta}$. Give an algorithm that overwrites $A.col$ with A stored by block as in Figure 1.4.5. How big of a work array is required?

Notes and References for Sec. 1.4

Two excellent expositions about vector computation are

J.J. Dongarra, F.G. Gustavson, and A. Karp (1984). "Implementing Linear Algebra Algorithms for Dense Matrices on a Vector Pipeline Machine," *SIAM Review 26*, 91-112.
J.M. Ortega and R.G. Voigt (1985). "Solution of Partial Differential Equations on Vector and Parallel Computers," *SIAM Review 27*, 149-240.

A very detailed look at matrix computations in hierarchical memory systems can be found in

K. Gallivan, W. Jalby, U. Meier, and A.H. Sameh (1988). "Impact of Hierarchical Memory Systems on Linear Algebra Algorithm Design," *Int'l J. Supercomputer Applic.* 2, 12-48.

See also

W. Schönauer (1987). *Scientific Computing on Vector Computers*, North Holland, Amsterdam.
R.W. Hockney and C.R. Jesshope (1988). *Parallel Computers 2*, Adam Hilger, Bristol and Philadelphia.

where various models of vector processor performance are set forth. Papers on the practical aspects of vector computing include

J.J. Dongarra and A. Hinds (1979). "Unrolling Loops in Fortran," *Software Practice and Experience 9*, 219-229.
J.J. Dongarra and S. Eisenstat (1984). "Squeezing the Most Out of an Algorithm in Cray Fortran," *ACM Trans. Math. Soft. 10*, 221-230.
B.L. Buzbee (1986) "A Strategy for Vectorization," *Parallel Computing 3*, 187-192.
K. Gallivan, W. Jalby, and U. Meier (1987). "The Use of BLAS3 in Linear Algebra on a Parallel Processor with a Hierarchical Memory," *SIAM J. Sci. and Stat. Comp. 8*, 1079-1084.
J.J. Dongarra and D. Walker (1995). "Software Libraries for Linear Algebra Computations on High Performance Computers," *SIAM Review 37*, 151–180.

Chapter 2

Matrix Analysis

The analysis and derivation of algorithms in the matrix computation area requires a facility with certain aspects of linear algebra. Some of the basics are reviewed in §2.1. Norms and their manipulation are covered in §2.2 and §2.3. In §2.4 we develop a model of finite precision arithmetic and then use it in a typical roundoff analysis.

The next two sections deal with orthogonality, which has a prominent role to play in matrix computations. The singular value decomposition and the CS decomposition are a pair of orthogonal reductions that provide critical insight into the important notions of rank and distance between subspaces. In §2.7 we examine how the solution of a linear system $Ax = b$ changes if A and b are perturbed. The important concept of matrix condition is introduced.

Before You Begin

References that complement this chapter include Forsythe and Moler (1967), Stewart (1973), Stewart and Sun (1990), and Higham (1996).

2.1   Basic Ideas from Linear Algebra

This section is a quick review of linear algebra. Readers who wish a more detailed coverage should consult the references at the end of the section.

2.1.1 Independence, Subspace, Basis, and Dimension

A set of vectors $\{a_1, \ldots, a_n\}$ in $\mathbb{R}^m$ is *linearly independent* if $\sum_{j=1}^{n} \alpha_j a_j = 0$ implies $\alpha(1{:}n) = 0$. Otherwise, a nontrivial combination of the a_i is zero and $\{a_1, \ldots, a_n\}$ is said to be *linearly dependent* .

A *subspace* of $\mathbb{R}^m$ is a subset that is also a vector space. Given a collection of vectors $a_1, \ldots, a_n \in \mathbb{R}^m$, the set of all linear combinations of these vectors is a subspace referred to as the *span* of $\{a_1, \ldots, a_n\}$:

$$\text{span}\{a_1, \ldots, a_n\} \;=\; \Big\{ \sum_{j=1}^{n} \beta_j a_j : \beta_j \in \mathbb{R} \Big\}.$$

If $\{a_1, \ldots, a_n\}$ is independent and $b \in \text{span}\{a_1, \ldots, a_n\}$, then b is a unique linear combination of the a_j.

If $S_1, \ldots, S_k$ are subspaces of $\mathbb{R}^m$, then their sum is the subspace defined by $S = \{\, a_1 + a_2 + \cdots + a_k : a_i \in S_i, \; i = 1{:}k \,\}$. S is said to be a *direct sum* if each $v \in S$ has a unique representation $v = a_1 + \cdots + a_k$ with $a_i \in S_i$. In this case we write $S = S_1 \oplus \cdots \oplus S_k$. The intersection of the S_i is also a subspace, $S = S_1 \cap S_2 \cap \cdots \cap S_k$.

The subset $\{a_{i_1}, \ldots, a_{i_k}\}$ is a *maximal linearly independent subset* of $\{a_1, \ldots, a_n\}$ if it is linearly independent and is not properly contained in any linearly independent subset of $\{a_1, \ldots, a_n\}$. If $\{a_{i_1}, \ldots, a_{i_k}\}$ is maximal, then $\text{span}\{a_1, \ldots, a_n\} \;=\; \text{span}\{a_{i_1}, \ldots, a_{i_k}\}$ and $\{a_{i_1}, \ldots, a_{i_k}\}$ is a *basis* for $\text{span}\{a_1, \ldots, a_n\}$. If $S \subseteq \mathbb{R}^m$ is a subspace, then it is possible to find independent basic vectors $a_1, \ldots, a_k \in S$ such that $S = \text{span}\{a_1, \ldots, a_k\}$. All bases for a subspace S have the same number of elements. This number is the *dimension* and is denoted by $\dim(S)$.

2.1.2 Range, Null Space, and Rank

There are two important subspaces associated with an m-by-n matrix A. The *range* of A is defined by

$$\text{ran}(A) = \{y \in \mathbb{R}^m : y = Ax \text{ for some } x \in \mathbb{R}^n\},$$

and the *null space* of A is defined by

$$\text{null}(A) = \{x \in \mathbb{R}^n : Ax = 0\}.$$

If $A = [\, a_1, \ldots, a_n \,]$ is a column partitioning, then

$$\text{ran}(A) = \text{span}\{a_1, \ldots, a_n\}.$$

The *rank* of a matrix A is defined by

$$\text{rank}(A) = \dim(\text{ran}(A)).$$

It can be shown that $\text{rank}(A) = \text{rank}(A^T)$. We say that $A \in \mathbb{R}^{m \times n}$ is *rank deficient* if $\text{rank}(A) < \min\{m, n\}$. If $A \in \mathbb{R}^{m \times n}$, then

$$\dim(\text{null}(A)) + \text{rank}(A) = n.$$

2.1.3   Matrix Inverse

The n-by-n *identity matrix* I_n is defined by the column partitioning

$$I_n = [\, e_1, \ldots, e_n \,]$$

where e_k is the kth "canonical" vector:

$$e_k = (\, \underbrace{0, \ldots, 0}_{k-1}, 1, \underbrace{0, \ldots, 0}_{n-k} \,)^T.$$

The canonical vectors arise frequently in matrix analysis and if their dimension is ever ambiguous, we use superscripts, i.e., $e_k^{(n)} \in \mathbb{R}^n$.

If A and X are in $\mathbb{R}^{n \times n}$ and satisfy $AX = I$, then X is the *inverse* of A and is denoted by A^{-1}. If A^{-1} exists, then A is said to be *nonsingular*. Otherwise, we say A is *singular*.

Several matrix inverse properties have an important role to play in matrix computations. The inverse of a product is the reverse product of the inverses:

$$(AB)^{-1} = B^{-1}A^{-1}. \tag{2.1.1}$$

The transpose of the inverse is the inverse of the transpose:

$$(A^{-1})^T = (A^T)^{-1} \equiv A^{-T}. \tag{2.1.2}$$

The identity

$$B^{-1} = A^{-1} - B^{-1}(B - A)A^{-1} \tag{2.1.3}$$

shows how the inverse changes if the matrix changes.

The *Sherman-Morrison-Woodbury formula* gives a convenient expression for the inverse of $(A + UV^T)$ where $A \in \mathbb{R}^{n \times n}$ and U and V are n-by-k:

$$(A + UV^T)^{-1} = A^{-1} - A^{-1}U(I + V^T A^{-1}U)^{-1}V^T A^{-1}. \tag{2.1.4}$$

A rank k correction to a matrix results in a rank k correction of the inverse. In (2.1.4) we assume that both A and $(I + V^T A^{-1}U)$ are nonsingular.

Any of these facts can be verified by just showing that the "proposed" inverse does the job. For example, here is how to confirm (2.1.3):

$$B\left(A^{-1} - B^{-1}(B - A)A^{-1}\right) = BA^{-1} - (B - A)A^{-1} = I.$$

2.1.4   The Determinant

If $A = (a) \in \mathbb{R}^{1 \times 1}$, then its *determinant* is given by $\det(A) = a$. The determinant of $A \in \mathbb{R}^{n \times n}$ is defined in terms of order $n - 1$ determinants:

$$\det(A) = \sum_{j=1}^{n} (-1)^{j+1} a_{1j} \det(A_{1j}).$$

Here, A_{1j} is an $(n-1)$-by-$(n-1)$ matrix obtained by deleting the first row and jth column of A. Useful properties of the determinant include

$$
\begin{array}{lll}
\det(AB) & = & \det(A)\det(B) \quad A, B \in \mathbb{R}^{n \times n} \\
\det(A^T) & = & \det(A) \qquad\qquad A \in \mathbb{R}^{n \times n} \\
\det(cA) & = & c^n \det(A) \qquad\; c \in \mathbb{R}, A \in \mathbb{R}^{n \times n} \\
\det(A) \neq 0 & \Leftrightarrow & A \text{ is nonsingular} \quad A \in \mathbb{R}^{n \times n}
\end{array}
$$

2.1.5   Differentiation

Suppose α is a scalar and that $A(\alpha)$ is an m-by-n matrix with entries $a_{ij}(\alpha)$. If $a_{ij}(\alpha)$ is a differentiable function of α for all i and j, then by $\dot{A}(\alpha)$ we mean the matrix

$$
\dot{A}(\alpha) = \frac{d}{d\alpha} A(\alpha) = \left(\frac{d}{d\alpha} a_{ij}(\alpha) \right) = (\dot{a}_{ij}(\alpha)) .
$$

The differentiation of a parameterized matrix turns out to be a handy way to examine the sensitivity of various matrix problems.

Problems

P2.1.1   Show that if $A \in \mathbb{R}^{m \times n}$ has rank p, then there exists an $X \in \mathbb{R}^{m \times p}$ and a $Y \in \mathbb{R}^{n \times p}$ such that $A = XY^T$, where $\text{rank}(X) = \text{rank}(Y) = p$.

P2.1.2   Suppose $A(\alpha) \in \mathbb{R}^{m \times r}$ and $B(\alpha) \in \mathbb{R}^{r \times n}$ are matrices whose entries are differentiable functions of the scalar α. Show

$$
\frac{d}{d\alpha}[A(\alpha)B(\alpha)] = \left[\frac{d}{d\alpha} A(\alpha) \right] B(\alpha) + A(\alpha) \left[\frac{d}{d\alpha} B(\alpha) \right] .
$$

P2.1.3   Suppose $A(\alpha) \in \mathbb{R}^{n \times n}$ has entries that are differentiable functions of the scalar α. Assuming $A(\alpha)$ is always nonsingular, show

$$
\frac{d}{d\alpha} \left[A(\alpha)^{-1} \right] = -A(\alpha)^{-1} \left[\frac{d}{d\alpha} A(\alpha) \right] A(\alpha)^{-1}.
$$

P2.1.4   Suppose $A \in \mathbb{R}^{n \times n}$, $b \in \mathbb{R}^n$ and that $\phi(x) = \frac{1}{2} x^T A x - x^T b$. Show that the gradient of ϕ is given by $\nabla\phi(x) = \frac{1}{2}(A^T + A)x - b$.

P2.1.5   Assume that both A and $A + uv^T$ are nonsingular where $A \in \mathbb{R}^{n \times n}$ and $u, v \in \mathbb{R}$. Show that if x solves $(A + uv^T)x = b$, then it also solves a perturbed right hand side problem of the form $Ax = b + \alpha u$. Give an expression for α in terms of A, u, and v.

Notes and References for Sec. 2.1

There are many introductory linear algebra texts. Among them, the following are particularly useful:

P.R. Halmos (1958). *Finite Dimensional Vector Spaces*, 2nd ed., Van Nostrand-Reinhold, Princeton.

S.J. Leon (1980). *Linear Algebra with Applications*. Macmillan, New York.

G. Strang (1993). *Introduction to Linear Algebra*, Wellesley-Cambridge Press, Wellesley MA.

D. Lay (1994). *Linear Algebra and Its Applications*, Addison-Wesley, Reading, MA.

C. Meyer (1997). *A Course in Applied Linear Algebra*, SIAM Publications, Philadelphia, PA.

More advanced treatments include Gantmacher (1959), Horn and Johnson (1985, 1991), and

A.S. Householder (1964). *The Theory of Matrices in Numerical Analysis*, Ginn (Blaisdell), Boston.

M. Marcus and H. Minc (1964). *A Survey of Matrix Theory and Matrix Inequalities*, Allyn and Bacon, Boston.

J.N. Franklin (1968). *Matrix Theory* Prentice Hall, Englewood Cliffs, NJ.

R. Bellman (1970). *Introduction to Matrix Analysis, Second Edition*, McGraw-Hill, New York.

P. Lancaster and M. Tismenetsky (1985). *The Theory of Matrices, Second Edition*, Academic Press, New York.

J.M. Ortega (1987). *Matrix Theory: A Second Course*, Plenum Press, New York.

2.2 Vector Norms

Norms serve the same purpose on vector spaces that absolute value does on the real line: they furnish a measure of distance. More precisely, $\mathbb{R}^n$ together with a norm on $\mathbb{R}^n$ defines a metric space. Therefore, we have the familiar notions of neighborhood, open sets, convergence, and continuity when working with vectors and vector-valued functions.

2.2.1 Definitions

A *vector norm* on $\mathbb{R}^n$ is a function $f : \mathbb{R}^n \to \mathbb{R}$ that satisfies the following properties:

$$
\begin{aligned}
f(x) &\geq 0 & x \in \mathbb{R}^n, \quad (f(x) = 0 \text{ iff } x = 0) \\
f(x + y) &\leq f(x) + f(y) & x, y \in \mathbb{R}^n \\
f(\alpha x) &= |\alpha| f(x) & \alpha \in \mathbb{R}, x \in \mathbb{R}^n
\end{aligned}
$$

We denote such a function with a double bar notation: $f(x) = \| \, x \, \|$. Subscripts on the double bar are used to distinguish between various norms.

A useful class of vector norms are the *p-norms* defined by

$$
\| \, x \, \|_p = (|x_1|^p + \cdots + |x_n|^p)^{\frac{1}{p}} \qquad p \geq 1 . \tag{2.2.1}
$$

Of these the 1, 2, and ∞ norms are the most important:

$$
\begin{aligned}
\| \, x \, \|_1 &= |x_1| + \cdots + |x_n| \\
\| \, x \, \|_2 &= \left(|x_1|^2 + \cdots + |x_n|^2\right)^{\frac{1}{2}} = \left(x^T x\right)^{\frac{1}{2}} \\
\| \, x \, \|_\infty &= \max_{1 \leq i \leq n} |x_i|
\end{aligned}
$$

A *unit vector* with respect to the norm $\| \cdot \|$ is a vector x that satisfies $\| x \| = 1$.

2.2.2 Some Vector Norm Properties

A classic result concerning p-norms is the *Holder inequality*:

$$|x^T y| \leq \| x \|_p \| y \|_q \qquad \frac{1}{p} + \frac{1}{q} = 1. \qquad (2.2.2)$$

A very important special case of this is the *Cauchy-Schwartz inequality*:

$$|x^T y| \leq \| x \|_2 \| y \|_2. \qquad (2.2.3)$$

All norms on $\mathbb{R}^n$ are *equivalent* , i.e., if $\| \cdot \|_\alpha$ and $\| \cdot \|_\beta$ are norms on $\mathbb{R}^n$, then there exist positive constants, c_1 and c_2 such that

$$c_1 \| x \|_\alpha \leq \| x \|_\beta \leq c_2 \| x \|_\alpha \qquad (2.2.4)$$

for all $x \in \mathbb{R}^n$. For example, if $x \in \mathbb{R}^n$, then

$$\| x \|_2 \leq \| x \|_1 \leq \sqrt{n} \| x \|_2 \qquad (2.2.5)$$
$$\| x \|_\infty \leq \| x \|_2 \leq \sqrt{n} \| x \|_\infty \qquad (2.2.6)$$
$$\| x \|_\infty \leq \| x \|_1 \leq n \| x \|_\infty. \qquad (2.2.7)$$

2.2.3 Absolute and Relative Error

Suppose $\hat{x} \in \mathbb{R}^n$ is an approximation to $x \in \mathbb{R}^n$. For a given vector norm $\| \cdot \|$ we say that

$$\epsilon_{abs} = \| \hat{x} - x \|$$

is the *absolute error* in $\hat{x}$. If $x \neq 0$, then

$$\epsilon_{rel} = \frac{\| \hat{x} - x \|}{\| x \|}$$

prescribes the *relative error* in $\hat{x}$. Relative error in the ∞-norm can be translated into a statement about the number of correct significant digits in $\hat{x}$. In particular, if

$$\frac{\| \hat{x} - x \|_\infty}{\| x \|_\infty} \approx 10^{-p},$$

then the largest component of $\hat{x}$ has approximately p correct significant digits.

Example 2.2.1 If $x = (1.234 \ .05674)^T$ and $\hat{x} = (1.235 \ .05128)^T$, then $\| \hat{x} - x \|_\infty / \| x \|_\infty \approx .0043 \approx 10^{-3}$. Note than $\hat{x}_1$ has about three significant digits that are correct while only one significant digit in $\hat{x}_2$ is correct.

2.2.4   Convergence

We say that a sequence $\{x^{(k)}\}$ of n-vectors *converges* to x if

$$\lim_{k\to\infty} \| x^{(k)} - x \| = 0 .$$

Note that because of (2.2.4), convergence in the α-norm implies convergence in the β-norm and vice versa.

Problems

P2.2.1 Show that if $x \in \mathbf{R}^n$, then $\lim_{p\to\infty} \| x \|_p = \| x \|_\infty$.

P2.2.2 Prove the Cauchy-Schwartz inequality (2.2.3) by considering the inequality $0 \le (ax + by)^T (ax + by)$ for suitable scalars a and b .

P2.2.3 Verify that $\| \cdot \|_1, \| \cdot \|_2$, and $\| \cdot \|_\infty$ are vector norms.

P2.2.4 Verify (2.2.5)-(2.2.7). When is equality achieved in each result?

P2.2.5 Show that in $\mathbf{R}^n$, $x^{(i)} \to x$ if and only if $x_k^{(i)} \to x_k$ for $k = 1{:}n$.

P2.2.6 Show that any vector norm on $\mathbf{R}^n$ is uniformly continuous by verifying the inequality $|\, \| x \| - \| y \| \,| \le \| x - y \|$.

P2.2.7 Let $\| \cdot \|$ be a vector norm on $\mathbf{R}^m$ and assume $A \in \mathbf{R}^{m \times n}$. Show that if $\text{rank}(A) = n$, then $\| x \|_A = \| Ax \|$ is a vector norm on $\mathbf{R}^n$.

P2.2.8 Let x and y be in $\mathbf{R}^n$ and define $\psi{:}\mathbf{R} \to \mathbf{R}$ by $\psi(\alpha) = \| x - \alpha y \|_2$. Show that ψ is minimized when $\alpha = x^T y / y^T y$.

P2.2.9 (a) Verify that $\| x \|_p = (|x_1|^p + \cdots + |x_n|^p)^{\frac{1}{p}}$ is a vector norm on $\mathbf{C}^n$. (b) Show that if $x \in \mathbf{C}^n$ then $\| x \|_p \le c \left(\| \,\text{Re}(x) \,\|_p + \| \,\text{Im}(x) \,\|_p \right)$. (c) Find a constant c_n such that $c_n \left(\| \,\text{Re}(x) \,\|_2 + \| \,\text{Im}(x) \,\|_2 \right) \le \| x \|_2$ for all $x \in \mathbf{C}^n$.

P2.2.10 Prove or disprove:

$$v \in \mathbf{R}^n \ \Rightarrow\ \| v \|_1 \| v \|_\infty \le \frac{1 + \sqrt{n}}{2} \| v \|_2.$$

Notes and References for Sec. 2.2

Although a vector norm is "just" a generalization of the absolute value concept, there are some noteworthy subtleties:

J.D. Pryce (1984). "A New Measure of Relative Error for Vectors," *SIAM J. Num. Anal.* **21**, 202-21.

2.3   Matrix Norms

The analysis of matrix algorithms frequently requires use of matrix norms. For example, the quality of a linear system solver may be poor if the matrix of coefficients is "nearly singular." To quantify the notion of near-singularity we need a measure of distance on the space of matrices. Matrix norms provide that measure.

2.3.1 Definitions

Since $\mathbb{R}^{m \times n}$ is isomorphic to $\mathbb{R}^{mn}$, the definition of a matrix norm should be equivalent to the definition of a vector norm. In particular, $f{:}\mathbb{R}^{m \times n} \to \mathbb{R}$ is a matrix norm if the following three properties hold:

$$
\begin{array}{lll}
f(A) \geq 0 & A \in \mathbb{R}^{m \times n}, & (f(A) = 0 \text{ iff } A = 0) \\
f(A + B) \leq f(A) + f(B) & A, B \in \mathbb{R}^{m \times n}, & \\
f(\alpha A) = |\alpha| f(A) & \alpha \in \mathbb{R}, A \in \mathbb{R}^{m \times n}. &
\end{array}
$$

As with vector norms, we use a double bar notation with subscripts to designate matrix norms, i.e., $\| A \| = f(A)$.

The most frequently used matrix norms in numerical linear algebra are the Frobenius norm,

$$
\| A \|_F = \sqrt{\sum_{i=1}^{m} \sum_{j=1}^{n} |a_{ij}|^2} \tag{2.3.1}
$$

and the p-norms

$$
\| A \|_p = \sup_{x \neq 0} \frac{\| Ax \|_p}{\| x \|_p}. \tag{2.3.2}
$$

Note that the matrix p-norms are defined in terms of the vector p-norms that we discussed in the previous section. The verification that (2.3.1) and (2.3.2) are matrix norms is left as an exercise. It is clear that $\| A \|_p$ is the p-norm of the largest vector obtained by applying A to a unit p-norm vector:

$$
\| A \|_p = \sup_{x \neq 0} \left\| A \left(\frac{x}{\| x \|_p} \right) \right\|_p = \max_{\| x \|_p = 1} \| Ax \|_p.
$$

It is important to understand that (2.3.1) and (2.3.2) define families of norms—the 2-norm on $\mathbb{R}^{3 \times 2}$ is a different function from the 2-norm on $\mathbb{R}^{5 \times 6}$. Thus, the easily verified inequality

$$
\| AB \|_p \leq \| A \|_p \| B \|_p \qquad A \in \mathbb{R}^{m \times n}, B \in \mathbb{R}^{n \times q} \tag{2.3.3}
$$

is really an observation about the relationship between three different norms. Formally, we say that norms f_1, f_2, and f_3 on $\mathbb{R}^{m \times q}$, $\mathbb{R}^{m \times n}$, and $\mathbb{R}^{n \times q}$ are *mutually consistent* if for all $A \in \mathbb{R}^{m \times n}$ and $B \in \mathbb{R}^{n \times q}$ we have $f_1(AB) \leq f_2(A) f_3(B)$.

Not all matrix norms satisfy the submultiplicative property

$$
\| AB \| \leq \| A \| \| B \|. \tag{2.3.4}
$$

For example, if $\| A \|_{\Delta} = \max |a_{ij}|$ and

$$A = B = \begin{bmatrix} 1 & 1 \\ 1 & 1 \end{bmatrix},$$

then $\| AB \|_{\Delta} > \| A \|_{\Delta} \| B \|_{\Delta}$. For the most part we work with norms that satisfy (2.3.4).

The p-norms have the important property that for every $A \in \mathbb{R}^{m \times n}$ and $x \in \mathbb{R}^n$ we have $\| Ax \|_p \leq \| A \|_p \| x \|_p$. More generally, for any vector norm $\| \cdot \|_{\alpha}$ on $\mathbb{R}^n$ and $\| \cdot \|_{\beta}$ on $\mathbb{R}^m$ we have $\| Ax \|_{\beta} \leq \| A \|_{\alpha,\beta} \| x \|_{\alpha}$ where $\| A \|_{\alpha,\beta}$ is a matrix norm defined by

$$\| A \|_{\alpha,\beta} = \sup_{x \neq 0} \frac{\| Ax \|_{\beta}}{\| x \|_{\alpha}}. \tag{2.3.5}$$

We say that $\| \cdot \|_{\alpha,\beta}$ is *subordinate* to the vector norms $\| \cdot \|_{\alpha}$ and $\| \cdot \|_{\beta}$. Since the set $\{ x \in \mathbb{R}^n : \| x \|_{\alpha} = 1 \}$ is compact and $\| \cdot \|_{\beta}$ is continuous, it follows that

$$\| A \|_{\alpha,\beta} = \max_{\| x \|_{\alpha} = 1} \| Ax \|_{\beta} = \| Ax^* \|_{\beta} \tag{2.3.6}$$

for some $x^* \in \mathbb{R}^n$ having unit α-norm.

2.3.2   Some Matrix Norm Properties

The Frobenius and p-norms (especially $p = 1, 2, \infty$) satisfy certain inequalities that are frequently used in the analysis of matrix computations. For $A \in \mathbb{R}^{m \times n}$ we have

$$\| A \|_2 \leq \| A \|_F \leq \sqrt{n} \| A \|_2 \tag{2.3.7}$$

$$\max_{i,j} |a_{ij}| \leq \| A \|_2 \leq \sqrt{mn} \max_{i,j} |a_{ij}| \tag{2.3.8}$$

$$\| A \|_1 = \max_{1 \leq j \leq n} \sum_{i=1}^{m} |a_{ij}| \tag{2.3.9}$$

$$\| A \|_{\infty} = \max_{1 \leq i \leq m} \sum_{j=1}^{n} |a_{ij}| \tag{2.3.10}$$

$$\frac{1}{\sqrt{n}} \| A \|_{\infty} \leq \| A \|_2 \leq \sqrt{m} \| A \|_{\infty} \tag{2.3.11}$$

$$\frac{1}{\sqrt{m}} \| A \|_1 \leq \| A \|_2 \leq \sqrt{n} \| A \|_1 \tag{2.3.12}$$

If $A \in \mathbb{R}^{m \times n}$, $1 \le i_1 \le i_2 \le m$, and $1 \le j_1 \le j_2 \le n$, then

$$\| A(i_1{:}i_2, j_1{:}j_2) \|_p \le \| A \|_p \qquad (2.3.13)$$

The proofs of these relations are not hard and are left as exercises.

A sequence $\{A^{(k)}\} \in \mathbb{R}^{m \times n}$ *converges* if $\lim_{k \to \infty} \| A^{(k)} - A \| = 0$. Choice of norm is irrelevant since all norms on $\mathbb{R}^{m \times n}$ are equivalent.

2.3.3 The Matrix 2-Norm

A nice feature of the matrix 1-norm and the matrix ∞-norm is that they are easily computed from (2.3.9) and (2.3.10). A characterization of the 2-norm is considerably more complicated.

Theorem 2.3.1 *If $A \in \mathbb{R}^{m \times n}$, then there exists a unit 2-norm n-vector z such that $A^T A z = \mu^2 z$ where $\mu = \| A \|_2$.*

Proof. Suppose $z \in \mathbb{R}^n$ is a unit vector such that $\| Az \|_2 = \| A \|_2$. Since z maximizes the function

$$g(x) = \frac{1}{2} \frac{\| Ax \|_2^2}{\| x \|_2^2} = \frac{1}{2} \frac{x^T A^T A x}{x^T x}$$

it follows that it satisfies $\nabla g(z) = 0$ where ∇g is the gradient of g. But a tedious differentiation shows that for $i = 1{:}n$

$$\frac{\partial g(z)}{\partial z_i} = \left[(z^T z) \sum_{j=1}^{n} (A^T A)_{ij} z_j - (z^T A^T A z) z_i \right] \Big/ (z^T z)^2 .$$

In vector notation this says $A^T A z = (z^T A^T A z) z$. The theorem follows by setting $\mu = \| Az \|_2$. $\square$

The theorem implies that $\| A \|_2^2$ is a zero of the polynomial $p(\lambda) = \det(A^T A - \lambda I)$. In particular, the 2-norm of A is the square root of the largest eigenvalue of $A^T A$. We have much more to say about eigenvalues in Chapters 7 and 8. For now, we merely observe that 2-norm computation is iterative and decidedly more complicated than the computation of the matrix 1-norm or ∞-norm. Fortunately, if the object is to obtain an order-of-magnitude estimate of $\| A \|_2$, then (2.3.7), (2.3.11), or (2.3.12) can be used.

As another example of "norm analysis," here is a handy result for 2-norm estimation.

Corollary 2.3.2 *If $A \in \mathbb{R}^{m \times n}$, then $\| A \|_2 \le \sqrt{\| A \|_1 \| A \|_\infty}$.*

Proof. If $z \ne 0$ is such that $A^T A z = \mu^2 z$ with $\mu = \| A \|_2$, then $\mu^2 \| z \|_1 = \| A^T A z \|_1 \le \| A^T \|_1 \| A \|_1 \| z \|_1 = \| A \|_\infty \| A \|_1 \| z \|_1$. $\square$

2.3.4   Perturbations and the Inverse

We frequently use norms to quantify the effect of perturbations or to prove that a sequence of matrices converges to a specified limit. As an illustration of these norm applications, let us quantify the change in A^{-1} as a function of change in A.

Lemma 2.3.3 *If $F \in \mathbb{R}^{n \times n}$ and $\| F \|_p < 1$, then $I - F$ is nonsingular and*

$$(I - F)^{-1} = \sum_{k=0}^{\infty} F^k$$

with

$$\| (I - F)^{-1} \|_p \le \frac{1}{1 - \| F \|_p}.$$

Proof. Suppose $I - F$ is singular. It follows that $(I - F)x = 0$ for some nonzero x. But then $\| x \|_p = \| Fx \|_p$ implies $\| F \|_p \ge 1$, a contradiction. Thus, $I - F$ is nonsingular. To obtain an expression for its inverse consider the identity

$$\left(\sum_{k=0}^{N} F^k \right) (I - F) = I - F^{N+1}.$$

Since $\| F \|_p < 1$ it follows that $\lim_{k \to \infty} F^k = 0$ because $\| F^k \|_p \le \| F \|_p^k$. Thus,

$$\left(\lim_{N \to \infty} \sum_{k=0}^{N} F^k \right) (I - F) = I.$$

It follows that $(I - F)^{-1} = \lim_{N \to \infty} \sum_{k=0}^{N} F^k$. From this it is easy to show that

$$\| (I - F)^{-1} \|_p \le \sum_{k=0}^{\infty} \| F \|_p^k = \frac{1}{1 - \| F \|_p}. \quad \square$$

Note that $\| (I - F)^{-1} - I \|_p \le \| F \|_p / (1 - \| F \|_p)$ as a consequence of the lemma. Thus, if $\epsilon \ll 1$, then $O(\epsilon)$ perturbations in I induce $O(\epsilon)$ perturbations in the inverse. We next extend this result to general matrices.

Theorem 2.3.4 *If A is nonsingular and $r \equiv \| A^{-1}E \|_p < 1$, then $A + E$ is nonsingular and $\| (A + E)^{-1} - A^{-1} \|_p \le \| E \|_p \| A^{-1} \|_p^2 / (1 - r)$.*

Proof. Since A is nonsingular $A + E = A(I - F)$ where $F = -A^{-1}E$. Since $\| F \|_p = r < 1$ it follows from Lemma 2.3.3 that $I - F$ is nonsingular and $\| (I - F)^{-1} \|_p < 1/(1 - r)$. Now $(A + E)^{-1} = (I - F)^{-1}A^{-1}$ and so

$$\| (A + E)^{-1} \|_p \le \frac{\| A^{-1} \|_p}{1 - r}.$$

Equation (2.1.3) says that $(A + E)^{-1} - A^{-1} = -A^{-1}E(A + E)^{-1}$ and so by taking norms we find

$$\| (A + E)^{-1} - A^{-1} \|_p \leq \| A^{-1} \|_p \| E \|_p \| (A + E)^{-1} \|_p$$

$$\leq \frac{\| A^{-1} \|_p^2 \| E \|_p}{1 - r}. \quad \square$$

Problems

P2.3.1 Show $\| AB \|_p \leq \| A \|_p \| B \|_p$ where $1 \leq p \leq \infty$.

P2.3.2 Let B be any submatrix of A. Show that $\| B \|_p \leq \| A \|_p$.

P2.3.3 Show that if $D = \mathrm{diag}(\mu_1, \ldots, \mu_k) \in \mathbb{R}^{m \times n}$ with $k = \min\{m, n\}$, then $\| D \|_p = \max |\mu_i|$.

P2.3.4 Verify (2.3.7) and (2.3.8).

P2.3.5 Verify (2.3.9) and (2.3.10).

P2.3.6 Verify (2.3.11) and (2.3.12).

P2.3.7 Verify (2.3.13).

P2.3.8 Show that if $0 \neq s \in \mathbb{R}^n$ and $E \in \mathbb{R}^{n \times n}$, then

$$\left\| E \left(I - \frac{ss^T}{s^T s} \right) \right\|_F^2 = \| E \|_F^2 - \frac{\| Es \|_2^2}{s^T s}.$$

P2.3.9 Suppose $u \in \mathbb{R}^m$ and $v \in \mathbb{R}^n$. Show that if $E = uv^T$ then $\| E \|_F = \| E \|_2 = \| u \|_2 \| v \|_2$ and that $\| E \|_\infty \leq \| u \|_\infty \| v \|_1$.

P2.3.10 Suppose $A \in \mathbb{R}^{m \times n}$, $y \in \mathbb{R}^m$, and $0 \neq s \in \mathbb{R}^n$. Show that $E = (y - As)s^T / s^T s$ has the smallest 2-norm of all m-by-n matrices E that satisfy $(A + E)s = y$.

Notes and References for Sec. 2.3

For deeper issues concerning matrix/vector norms, see

F.L. Bauer and C.T. Fike (1960). "Norms and Exclusion Theorems," *Numer. Math. 2,* 137-44.

L. Mirsky (1960). "Symmetric Gauge Functions and Unitarily Invariant Norms," *Quart. J. Math. 11,* 50-59.

A.S. Householder (1964). *The Theory of Matrices in Numerical Analysis* , Dover Publications, New York.

N.J. Higham (1992). "Estimating the Matrix p-Norm," *Numer. Math. 62,* 539–556.

2.4 Finite Precision Matrix Computations

In part, rounding errors are what makes the matrix computation area so nontrivial and interesting. In this section we set up a model of floating point arithmetic and then use it to develop error bounds for floating point dot products, saxpy's, matrix-vector products and matrix-matrix products. For

a more comprehensive treatment than what we offer, see Higham (1996) or
Wilkinson (1965). The coverage in Forsythe and Moler (1967) and Stewart
(1973) is also excellent.

2.4.1 The Floating Point Numbers

When calculations are performed on a computer, each arithmetic operation is generally affected by *roundoff error*. This error arises because the
machine hardware can only represent a subset of the real numbers. We
denote this subset by **F** and refer to its elements as *floating point numbers.*
Following conventions set forth in Forsythe, Malcolm, and Moler (1977, pp.
10-29), the floating point number system on a particular computer is characterized by four integers: the *base* β, the *precision* t, and the *exponent
range* $[L, U]$. In particular, **F** consists of all numbers f of the form

$$f = \pm .d_1 d_2 \ldots d_t \times \beta^e \qquad 0 \le d_i < \beta, \quad d_1 \neq 0, \quad L \le e \le U$$

together with zero. Notice that for a nonzero $f \in \mathbf{F}$ we have $m \le |f| \le M$
where

$$m = \beta^{L-1} \quad \text{and} \quad M = \beta^U (1 - \beta^{-t}). \tag{2.4.1}$$

As an example, if $\beta = 2$, $t = 3$, $L = 0$, and $U = 2$, then the non-negative
elements of **F** are represented by hash marks on the axis displayed in FIG.
2.4.1. Notice that the floating point numbers are not equally spaced. A

$$-2 \qquad -1 \ -.5 \quad 0 \quad .5 \quad 1 \qquad 2$$

FIGURE 2.4.1 *Sample Floating Point Number System*

typical value for (β, t, L, U) might be (2, 56, -64, 64).

2.4.2 A Model of Floating Point Arithmetic

To make general pronouncements about the effect of rounding errors on a
given algorithm, it is necessary to have a model of computer arithmetic on
F. To this end define the set G by

$$G = \{ \, x \in \mathbb{R} : m \le |x| \le M \, \} \cup \{0\} \tag{2.4.2}$$

and the operator $fl\colon G \to \mathbf{F}$ by

$$fl(x) = \left\{ \begin{array}{l} \text{nearest } c \in \mathbf{F} \text{ to } x \text{ with ties handled} \\ \text{by rounding away from zero.} \end{array} \right\}$$

The fl operator can be shown to satisfy

$$fl(x) = x(1 + \epsilon) \qquad |\epsilon| \leq \mathbf{u} \qquad (2.4.3)$$

where $\mathbf{u}$ is the *unit roundoff* defined by

$$\mathbf{u} = \frac{1}{2}\beta^{1-t}. \qquad (2.4.4)$$

Let a and b be any two floating point numbers and let "op" denote any of the four arithmetic operations $+, -, \times, \div$. If a op $b \in G$, then in our model of floating point arithmetic we assume that the computed version of $(a$ op $b)$ is given by $fl(a$ op $b)$. It follows that $fl(a$ op $b) = (a$ op $b)(1 + \epsilon)$ with $|\epsilon| \leq \mathbf{u}$. Thus,

$$\frac{|fl(a \text{ op } b) - (a \text{ op } b)|}{|a \text{ op } b|} \leq \mathbf{u} \qquad a \text{ op } b \neq 0 \qquad (2.4.5)$$

showing that there is small relative error associated with individual arithmetic operations[1]. It is important to realize, however, that this is not necessarily the case when a sequence of operations is involved.

Example 2.4.1 If $\beta = 10$, $t = 3$ floating point arithmetic is used, then it can be shown that $fl[fl(10^{-4} + 1) - 1] = 0$ implying a relative error of 1. On the other hand the exact answer is given by $fl[fl(10^{-4} + fl(1 - 1)] = 10^{-4}$. Floating point arithmetic is not always associative.

If a op $b \notin G$, then an *arithmetic exception* occurs. *Overflow* and *underflow* results whenever $|a$ op $b| > M$ or $0 < |a$ op $b| < m$ respectively. The handling of these and other exceptions is hardware/system dependent.

2.4.3 Cancellation

Another important aspect of finite precision arithmetic is the phenomenon of *catastrophic cancellation*. Roughly speaking, this term refers to the extreme loss of correct significant digits when small numbers are additively computed from large numbers. A well-known example taken from Forsythe, Malcolm and Moler (1977, pp. 14-16) is the computation of e^{-a} via Taylor series with $a > 0$. The roundoff error associated with this method is

---

[1]There are important examples of machines whose additive floating point operations satisfy $fl(a \pm b) = (1 + \epsilon_1)a \pm (1 + \epsilon_2)b$ where $|\epsilon_1|, |\epsilon_2| \leq \mathbf{u}$. In such an environment, the inequality $|fl(a \pm b) - (a \pm b)| \leq \mathbf{u}|a \pm b|$ need not hold.

approximately **u** times the largest partial sum. For large a, this error can actually be larger than the exact exponential and there will be no correct digits in the answer no matter how many terms in the series are summed. On the other hand, if enough terms in the Taylor series for e^a are added and the result reciprocated, then an estimate of e^{-a} to full precision is attained.

2.4.4   The Absolute Value Notation

Before we proceed with the roundoff analysis of some basic matrix calculations, we acquire some useful notation. Suppose $A \in \mathbb{R}^{m \times n}$ and that we wish to quantify the errors associated with its floating point representation. Denoting the stored version of A by $fl(A)$, we see that

$$[fl(A)]_{ij} = fl(a_{ij}) = a_{ij}(1 + \epsilon_{ij}) \qquad |\epsilon_{ij}| \leq \mathbf{u} \qquad (2.4.6)$$

for all i and j. A better way to say the same thing results if we adopt two conventions. If A and B are in $\mathbb{R}^{m \times n}$, then

$$B = |A| \quad \Rightarrow \quad b_{ij} = |a_{ij}|, \ i = 1{:}m, \ j = 1{:}n$$

$$B \leq A \quad \Rightarrow \quad b_{ij} \leq a_{ij}, \ i = 1{:}m, \ j = 1{:}n \ .$$

With this notation we see that (2.4.6) has the form

$$|fl(A) - A| \leq \mathbf{u}|A| \ .$$

A relation such as this can be easily turned into a norm inequality, e.g., $\| fl(A) - A \|_1 \leq \mathbf{u}\| A \|_1$. However, when quantifying the rounding errors in a matrix manipulation, the absolute value notation can be a lot more informative because it provides a comment on each (i, j) entry.

2.4.5   Roundoff in Dot Products

We begin our study of finite precision matrix computations by considering the rounding errors that result in the standard dot product algorithm:

$$
\begin{aligned}
&s = 0 \\
&\mathbf{for}\ k = 1{:}n \\
&\qquad s = s + x_k y_k \\
&\mathbf{end}
\end{aligned}
\qquad (2.4.7)
$$

Here, x and y are n-by-1 floating point vectors.

In trying to quantify the rounding errors in this algorithm, we are immediately confronted with a notational problem: the distinction between computed and exact quantities. When the underlying computations are clear, we shall use the $fl(\cdot)$ operator to signify computed quantities.

Thus, $fl(x^Ty)$ denotes the computed output of (2.4.7). Let us bound $|fl(x^Ty) - x^Ty|$. If

$$s_p = fl\left(\sum_{k=1}^{p} x_k y_k\right),$$

then $s_1 = x_1 y_1(1 + \delta_1)$ with $|\delta_1| \le \mathbf{u}$ and for $p = 2{:}n$

$$\begin{aligned} s_p &= fl(s_{p-1} + fl(x_p y_p)) \\ &= (s_{p-1} + x_p y_p(1 + \delta_p))\,(1 + \epsilon_p) \qquad |\delta_p|, |\epsilon_p| \le \mathbf{u}. \quad (2.4.8) \end{aligned}$$

A little algebra shows that

$$fl(x^Ty) = s_n = \sum_{k=1}^{n} x_k y_k(1 + \gamma_k)$$

where

$$(1 + \gamma_k) = (1 + \delta_k)\prod_{j=k}^{n}(1 + \epsilon_j)$$

with the convention that $\epsilon_1 = 0$. Thus,

$$|fl(x^Ty) - x^Ty| \quad \le \quad \sum_{k=1}^{n} |x_k y_k||\gamma_k|. \qquad (2.4.9)$$

To proceed further, we must bound the quantities $|\gamma_k|$ in terms of $\mathbf{u}$. The following result is useful for this purpose.

Lemma 2.4.1 *If* $(1+\alpha) = \displaystyle\prod_{k=1}^{n}(1+\alpha_k)$ *where* $|\alpha_k| \le \mathbf{u}$ *and* $n\mathbf{u} \le .01$, *then* $|\alpha| \le 1.01 n\mathbf{u}$.

Proof. See Higham (1996, p. 75). □

Applying this result to (2.4.9) under the "reasonable" assumption $n\mathbf{u} \le .01$ gives

$$|fl(x^Ty) - x^Ty| \le 1.01 n\mathbf{u}|x|^T|y|. \qquad (2.4.10)$$

Notice that if $|x^Ty| \ll |x|^T|y|$, then the relative error in $fl(x^Ty)$ may not be small.

2.4.6 Alternative Ways to Quantify Roundoff Error

An easier but less rigorous way of bounding α in Lemma 2.4.1 is to say $|\alpha| \le n\mathbf{u} + O(\mathbf{u}^2)$. With this convention we have

$$|fl(x^Ty) - x^Ty| \le n\mathbf{u}|x|^T|y| + O(\mathbf{u}^2). \qquad (2.4.11)$$

Other ways of expressing the same result include

$$|fl(x^T y) - x^T y| \le \phi(n)\mathbf{u}|x|^T |y| \qquad (2.4.12)$$

and

$$|fl(x^T y) - x^T y| \le cn\mathbf{u}|x|^T |y|, \qquad (2.4.13)$$

where in (2.4.12) $\phi(n)$ is a "modest" function of n and in (2.4.13) c is a constant of order unity.

We shall not express a preference for any of the error bounding styles shown in (2.4.10)-(2.4.13). This spares us the necessity of translating the roundoff results that appear in the literature into a fixed format. Moreover, paying overly close attention to the details of an error bound is inconsistent with the "philosophy" of roundoff analysis. As Wilkinson (1971, p. 567) says,

> There is still a tendency to attach too much importance to the precise error bounds obtained by an à priori error analysis. In my opinion, the bound itself is usually the least important part of it. The main object of such an analysis is to expose the potential instabilities, if any, of an algorithm so that hopefully from the insight thus obtained one might be led to improved algorithms. Usually the bound itself is weaker than it might have been because of the necessity of restricting the mass of detail to a reasonable level and because of the limitations imposed by expressing the errors in terms of matrix norms. À priori bounds are not, in general, quantities that should be used in practice. Practical error bounds should usually be determined by some form of à posteriori error analysis, since this takes full advantage of the statistical distribution of rounding errors and of any special features, such as sparseness, in the matrix.

It is important to keep these perspectives in mind.

2.4.7   Dot Product Accumulation

Some computers have provision for accumulating dot products in *double precision*. This means that if x and y are floating point vectors with length t mantissas, then the running sum s in (2.4.7) is built up in a register with a $2t$ digit mantissa. Since the multiplication of two t-digit floating point numbers can be stored exactly in a double precision variable, it is only when s is written to single precision memory that any roundoff occurs. In this situation one can usually assert that a computed dot product has good *relative* error, i.e., $fl(x^T y) = x^T y(1 + \delta)$ where $|\delta| \approx \mathbf{u}$. Thus, the ability to accumulate dot products is very appealing.

2.4.8 Roundoff in Other Basic Matrix Computations

It is easy to show that if A and B are floating point matrices and α is a floating point number, then

$$fl(\alpha A) = \alpha A + E \qquad |E| \leq \mathbf{u}|\alpha A| \qquad (2.4.14)$$

and

$$fl(A + B) = (A + B) + E \qquad |E| \leq \mathbf{u}|A + B|. \qquad (2.4.15)$$

As a consequence of these two results, it is easy to verify that computed saxpy's and outer product updates satisfy

$$fl(\alpha x + y) = \alpha x + y + z \qquad |z| \leq \mathbf{u}\,(2|\alpha x| + |y|) + O(\mathbf{u}^2) \qquad (2.4.16)$$

$$fl(C + uv^T) = C + uv^T + E \qquad |E| \leq \mathbf{u}\,(|C| + 2|uv^T|) + O(\mathbf{u}^2). \qquad (2.4.17)$$

Using (2.4.10) it is easy to show that a dot product based multiplication of two floating point matrices A and B satisfies

$$fl(AB) = AB + E \qquad |E| \leq n\mathbf{u}|A||B| + O(\mathbf{u}^2). \qquad (2.4.18)$$

The same result applies if a gaxpy or outer product based procedure is used. Notice that matrix multiplication does not necessarily give small relative error since $|AB|$ may be much smaller than $|A||B|$, e.g.,

$$\begin{bmatrix} 1 & 1 \\ 0 & 0 \end{bmatrix} \begin{bmatrix} 1 & 0 \\ -.99 & 0 \end{bmatrix} = \begin{bmatrix} .01 & 0 \\ 0 & 0 \end{bmatrix}.$$

It is easy to obtain norm bounds from the roundoff results developed thus far. If we look at the 1-norm error in floating point matrix multiplication, then it is easy to show from (2.4.18) that

$$\| fl(AB) - AB \|_1 \leq n\mathbf{u}\| A \|_1 \| B \|_1 + O(\mathbf{u}^2). \qquad (2.4.19)$$

2.4.9 Forward and Backward Error Analyses

Each roundoff bound given above is the consequence of a *forward error analysis*. An alternative style of characterizing the roundoff errors in an algorithm is accomplished through a technique known as *backward error analysis*. Here, the rounding errors are related to the data of the problem rather than to its solution. By way of illustration, consider the $n = 2$ version of triangular matrix multiplication. It can be shown that:

$$fl(AB) = \begin{bmatrix} a_{11}b_{11}(1 + \epsilon_1) & (a_{11}b_{12}(1 + \epsilon_2) + a_{12}b_{22}(1 + \epsilon_3))(1 + \epsilon_4) \\ 0 & a_{22}b_{22}(1 + \epsilon_5) \end{bmatrix}$$

where $|\epsilon_i| \leq u$, for $i = 1{:}5$. However, if we define

$$\hat{A} = \left[\begin{array}{cc} a_{11} & a_{12}(1+\epsilon_3)(1+\epsilon_4) \\ \\ 0 & a_{22}(1+\epsilon_5) \end{array} \right]$$

and

$$\hat{B} = \left[\begin{array}{cc} b_{11}(1+\epsilon_1) & b_{12}(1+\epsilon_2)(1+\epsilon_4) \\ \\ 0 & b_{22} \end{array} \right],$$

then it is easily verified that $fl(AB) = \hat{A}\hat{B}$. Moreover,

$$\hat{A} = A + E \qquad |E| \leq 2\mathbf{u}|A| + O(\mathbf{u}^2)$$

$$\hat{B} = B + F \qquad |F| \leq 2\mathbf{u}|B| + O(\mathbf{u}^2).$$

In other words, the computed product is the exact product of slightly perturbed A and B.

2.4.10   Error in Strassen Multiplication

In §1.3.8 we outlined an unconventional matrix multiplication procedure due to Strassen (1969). It is instructive to compare the effect of roundoff in this method with the effect of roundoff in any of the conventional matrix multiplication methods of §1.1.

It can be shown that the Strassen approach (Algorithm 1.3.1) produces a $\hat{C} = fl(AB)$ that satisfies an inequality of the form (2.4.19). This is perfectly satisfactory in many applications. However, the $\hat{C}$ that Strassen's method produces does *not* always satisfy an inequality of the form (2.4.18). To see this, suppose

$$A = B = \left[\begin{array}{cc} .99 & .0010 \\ .0010 & .99 \end{array} \right]$$

and that we execute Algorithm 1.3.1 using 2-digit floating point arithmetic. Among other things, the following quantities are computed:

$$\begin{array}{rcl} \hat{P}_3 & = & fl(.99(.001 - .99)) = -.98 \\ \hat{P}_5 & = & fl((.99 + .001).99) = .98 \\ \hat{c}_{12} & = & fl(\hat{P}_3 + \hat{P}_5) = 0.0 \end{array}$$

Now in exact arithmetic $c_{12} = 2(.001)(.99) = .00198$ and thus Algorithm 1.3.1 produces a $\hat{c}_{12}$ with no correct significant digits. The Strassen approach gets into trouble in this example because small off-diagonal entries are combined with large diagonal entries. Note that in conventional matrix multiplication neither b_{12} and b_{22} nor a_{11} and a_{12} are summed. Thus the contribution of the small off-diagonal elements is not lost. Indeed, for the above A and B a conventional matrix multiply gives $\hat{c}_{12} = .0020$.

Failure to produce a componentwise accurate $\hat{C}$ can be a serious shortcoming in some applications. For example, in Markov processes the a_{ij}, b_{ij}, and c_{ij} are transition probabilities and are therefore nonnegative. It may be critical to compute c_{ij} accurately if it reflects a particularly important probability in the modeled phenomena. Note that if $A \geq 0$ and $B \geq 0$, then conventional matrix multiplication produces a product $\hat{C}$ that has small componentwise relative error:

$$|\hat{C} - C| \leq n\mathbf{u}|A|\,|B| + O(\mathbf{u}^2) = n\mathbf{u}|C| + O(\mathbf{u}^2).$$

This follows from (2.4.18). Because we cannot say the same for the Strassen approach, we conclude that Algorithm 1.3.1 is not attractive for *certain* nonnegative matrix multiplication problems *if* relatively accurate $\hat{c}_{ij}$ are required.

Extrapolating from this discussion we reach two fairly obvious but important conclusions:

- Different methods for computing the same quantity can produce substantially different results.

- Whether or not an algorithm produces satisfactory results depends upon the type of problem solved and the goals of the user.

These observations are clarified in subsequent chapters and are intimately related to the concepts of algorithm stability and problem condition.

Problems

P2.4.1 Show that if (2.4.7) is applied with $y = x$, then $fl(x^T x) = x^T x(1 + \alpha)$ where $|\alpha| \leq n\mathbf{u} + O(\mathbf{u}^2)$.

P2.4.2 Prove (2.4.3).

P2.4.3 Show that if $E \in \mathbb{R}^{m \times n}$ with $m \geq n$, then $\|\,|E|\,\|_2 \leq \sqrt{n}\|\,E\,\|_2$. This result is useful when deriving norm bounds from absolute value bounds.

P2.4.4 Assume the existence of a square root function satisfying $fl(\sqrt{x}) = \sqrt{x}(1 + \epsilon)$ with $|\epsilon| \leq \mathbf{u}$. Give an algorithm for computing $\|\,x\,\|_2$ and bound the rounding errors.

P2.4.5 Suppose A and B are n-by-n upper triangular floating point matrices. If $\hat{C} = fl(AB)$ is computed using one of the conventional §1.1 algorithms, does it follow that $\hat{C} = \hat{A}\hat{B}$ where $\hat{A}$ and $\hat{B}$ are close to A and B?

P2.4.6 Suppose A and B are n-by-n floating point matrices and that A is nonsingular with $\|\,|A^{-1}|\,|A|\,\|_\infty = \tau$. Show that if $\hat{C} = fl(AB)$ is obtained using any of the algorithms in §1.1, then there exists a $\hat{B}$ so $\hat{C} = A\hat{B}$ and $\|\,\hat{B} - B\,\|_\infty \leq n\mathbf{u}\tau\|\,B\,\|_\infty + O(\mathbf{u}^2)$.

P2.4.7 Prove (2.4.18).

Notes and References for Sec. 2.4

For a general introduction to the effects of roundoff error, we recommend

J.H. Wilkinson (1963). *Rounding Errors in Algebraic Processes*, Prentice-Hall, Engle-
    wood Cliffs, NJ.
J.H. Wilkinson (1971). "Modern Error Analysis," *SIAM Review 13*, 548–68.
D. Kahaner, C.B. Moler, and S. Nash (1988). *Numerical Methods and Software*, Prentice-
    Hall, Englewood Cliffs, NJ.
F. Chaitin-Chatelin and V. Fraysed (1996). *Lectures on Finite Precision Computations*,
    SIAM Publications, Philadelphia.

More recent developments in error analysis involve interval analysis, the building of statistical models of roundoff error, and the automating of the analysis itself:

T.E. Hull and J.R. Swensen (1966). "Tests of Probabilistic Models for Propagation of
    Roundoff Errors," *Comm. ACM. 9*, 108–13.
J. Larson and A. Sameh (1978). "Efficient Calculation of the Effects of Roundoff Errors,"
    ACM Trans. Math. Soft. 4, 228–36.
W. Miller and D. Spooner (1978). "Software for Roundoff Analysis, II," *ACM Trans.
    Math. Soft. 4*, 369–90.
J.M. Yohe (1979). "Software for Interval Arithmetic: A Reasonable Portable Package,"
    ACM Trans. Math. Soft. 5, 50–63.

Anyone engaged in serious software development needs a thorough understanding of
floating point arithmetic. A good way to begin acquiring knowledge in this direction is
to read about the IEEE floating point standard in

D. Goldberg (1991). "What Every Computer Scientist Should Know About Floating
    Point Arithmetic," *ACM Surveys 23*, 5–48.

See also

R.P. Brent (1978). "A Fortran Multiple Precision Arithmetic Package," *ACM Trans.
    Math. Soft. 4*, 57–70.
R.P. Brent (1978). "Algorithm 524 MP, a Fortran Multiple Precision Arithmetic Pack-
    age," *ACM Trans. Math. Soft. 4*, 71–81.
J.W. Demmel (1984). "Underflow and the Reliability of Numerical Software," *SIAM J.
    Sci. and Stat. Comp. 5*, 887–919.
U.W. Kulisch and W.L. Miranker (1986). "The Arithmetic of the Digital Computer,"
    SIAM Review 28, 1–40.
W.J. Cody (1988). "ALGORITHM 665 MACHAR: A Subroutine to Dynamically De-
    termine Machine Parameters," *ACM Trans. Math. Soft. 14*, 303–311.
D.H. Bailey, H.D. Simon, J. T. Barton, M.J. Fouts (1989). "Floating Point Arithmetic
    in Future Supercomputers," *Int'l J. Supercomputing Appl. 3*, 86–90.
D.H. Bailey (1993). "Algorithm 719: Multiprecision Translation and Execution of FOR-
    TRAN Programs," *ACM Trans. Math. Soft. 19*, 288–319.

The subtleties associated with the development of high-quality software, even for "simple" problems, are immense. A good example is the design of a subroutine to compute
2-norms

J.M. Blue (1978). "A Portable FORTRAN Program to Find the Euclidean Norm of a
    Vector," *ACM Trans. Math. Soft. 4*, 15–23.

For an analysis of the Strassen algorithm and other "fast" linear algebra procedures see

R.P. Brent (1970). "Error Analysis of Algorithms for Matrix Multiplication and Triangular Decomposition Using Winograd's Identity," *Numer. Math. 16*, 145–156.

W. Miller (1975). "Computational Complexity and Numerical Stability," *SIAM J. Computing 4*, 97–107.

N.J. Higham (1992). "Stability of a Method for Multiplying Complex Matrices with Three Real Matrix Multiplications," *SIAM J. Matrix Anal. Appl. 13*, 681–687.

J.W. Demmel and N.J. Higham (1992). "Stability of Block Algorithms with Fast Level-3 BLAS," *ACM Trans. Math. Soft. 18*, 274–291.

2.5 Orthogonality and the SVD

Orthogonality has a very prominent role to play in matrix computations. After establishing a few definitions we prove the extremely useful singular value decomposition (SVD). Among other things, the SVD enables us to intelligently handle the matrix rank problem. The concept of rank, though perfectly clear in the exact arithmetic context, is tricky in the presence of roundoff error and fuzzy data. With the SVD we can introduce the practical notion of numerical rank.

2.5.1 Orthogonality

A set of vectors $\{x_1, \ldots, x_p\}$ in $\mathbb{R}^m$ is *orthogonal* if $x_i^T x_j = 0$ whenever $i \neq j$ and *orthonormal* if $x_i^T x_j = \delta_{ij}$. Intuitively, orthogonal vectors are maximally independent for they point in totally different directions.

A collection of subspaces $S_1, \ldots, S_p$ in $\mathbb{R}^m$ is *mutually orthogonal* if $x^T y = 0$ whenever $x \in S_i$ and $y \in S_j$ for $i \neq j$. The *orthogonal complement* of a subspace $S \subseteq \mathbb{R}^m$ is defined by

$$S^\perp = \{y \in \mathbb{R}^m : y^T x = 0 \text{ for all } x \in S\}$$

and it is not hard to show that $\mathrm{ran}(A)^\perp = \mathrm{null}(A^T)$. The vectors $v_1, \ldots, v_k$ form an *orthonormal* basis for a subspace $S \subseteq \mathbb{R}^m$ if they are orthonormal and span S.

A matrix $Q \in \mathbb{R}^{m \times m}$ is said to be *orthogonal* if $Q^T Q = I$. If $Q = [\, q_1, \ldots, q_m \,]$ is orthogonal, then the q_i form an orthonormal basis for $\mathbb{R}^m$. It is always possible to extend such a basis to a full orthonormal basis $\{v_1, \ldots, v_m\}$ for $\mathbb{R}^m$:

Theorem 2.5.1 *If $V_1 \in \mathbb{R}^{n \times r}$ has orthonormal columns, then there exists $V_2 \in \mathbb{R}^{n \times (n-r)}$ such that*

$$V = [\, V_1 \; V_2 \,]$$

is orthogonal. Note that $\mathrm{ran}(V_1)^\perp = \mathrm{ran}(V_2)$.

Proof. This is a standard result from introductory linear algebra. It is also a corollary of the QR factorization that we present in §5.2. $\square$

2.5.2 Norms and Orthogonal Transformations

The 2-norm is invariant under orthogonal transformation, for if $Q^T Q = I$, then $\| Qx \|_2^2 = x^T Q^T Q x = x^T x = \| x \|_2^2$. The matrix 2-norm and the Frobenius norm are also invariant with respect to orthogonal transformations. In particular, it is easy to show that for all orthogonal Q and Z of appropriate dimensions we have

$$\| QAZ \|_F = \| A \|_F \tag{2.5.1}$$

and

$$\| QAZ \|_2 = \| A \|_2 . \tag{2.5.2}$$

2.5.3 The Singular Value Decomposition

The theory of norms developed in the previous two sections can be used to prove the extremely useful singular value decomposition.

Theorem 2.5.2 (Singular Value Decomposition (SVD)) *If A is a real m-by-n matrix, then there exist orthogonal matrices*

$$U = [\, u_1, \ldots, u_m \,] \in \mathbb{R}^{m \times m} \quad and \quad V = [\, v_1, \ldots, v_n \,] \in \mathbb{R}^{n \times n}$$

such that

$$U^T A V = \mathrm{diag}(\sigma_1, \ldots, \sigma_p) \in \mathbb{R}^{m \times n} \qquad p = \min\{m, n\}$$

where $\sigma_1 \geq \sigma_2 \geq \ldots \geq \sigma_p \geq 0$.

Proof. Let $x \in \mathbb{R}^n$ and $y \in \mathbb{R}^m$ be unit 2-norm vectors that satisfy $Ax = \sigma y$ with $\sigma = \| A \|_2$. From Theorem 2.5.1 there exist $V_2 \in \mathbb{R}^{n \times (n-1)}$ and $U_2 \in \mathbb{R}^{m \times (m-1)}$ so $V = [\, x \; V_2 \,] \in \mathbb{R}^{n \times n}$ and $U = [\, y \; U_2 \,] \in \mathbb{R}^{m \times m}$ are orthogonal. It is not hard to show that $U^T A V$ has the following structure:

$$U^T A V = \begin{bmatrix} \sigma & w^T \\ 0 & B \end{bmatrix} \equiv A_1.$$

Since

$$\left\| A_1 \left(\begin{bmatrix} \sigma \\ w \end{bmatrix} \right) \right\|_2^2 \geq (\sigma^2 + w^T w)^2$$

we have $\| A_1 \|_2^2 \geq (\sigma^2 + w^T w)$. But $\sigma^2 = \| A \|_2^2 = \| A_1 \|_2^2$, and so we must have $w = 0$. An obvious induction argument completes the proof of the theorem. $\square$

The σ_i are the *singular values* of A and the vectors u_i and v_i are the *i*th *left singular vector* and the *i*th *right singular vector* respectively. It is easy to verify by comparing columns in the equations $AV = U\Sigma$ and $A^T U = V\Sigma^T$ that

$$\left. \begin{array}{rcl} Av_i & = & \sigma_i u_i \\ A^T u_i & = & \sigma_i v_i \end{array} \right\} \quad i = 1{:}\min\{m, n\}$$

It is convenient to have the following notation for designating singular values:

$$\begin{array}{rcl} \sigma_i(A) & = & \text{the } i\text{th largest singular value of } A, \\ \sigma_{max}(A) & = & \text{the largest singular value of } A, \\ \sigma_{min}(A) & = & \text{the smallest singular value of } A. \end{array}$$

The singular values of a matrix A are precisely the lengths of the semi-axes of the hyperellipsoid E defined by $E = \{\, Ax : \| x \|_2 = 1 \,\}$.

Example 2.5.1

$$A = \left[\begin{array}{cc} .96 & 1.72 \\ 2.28 & .96 \end{array} \right] = U\Sigma V^T = \left[\begin{array}{cc} .6 & -.8 \\ .8 & .6 \end{array} \right] \left[\begin{array}{cc} 3 & 0 \\ 0 & 1 \end{array} \right] \left[\begin{array}{cc} .8 & .6 \\ .6 & -.8 \end{array} \right]^T .$$

The SVD reveals a great deal about the structure of a matrix. If the SVD of A is given by Theorem 2.5.2, and we define r by

$$\sigma_1 \geq \cdots \geq \sigma_r > \sigma_{r+1} = \cdots = \sigma_p = 0,$$

then

$$\begin{array}{rcll} \text{rank}(A) & = & r & (2.5.3) \\ \text{null}(A) & = & \text{span}\{v_{r+1}, \ldots, v_n\} & (2.5.4) \\ \text{ran}(A) & = & \text{span}\{u_1, \ldots, u_r\}\,, & (2.5.5) \end{array}$$

and we have the *SVD expansion*

$$A = \sum_{i=1}^{r} \sigma_i u_i v_i^T . \qquad (2.5.6)$$

Various 2-norm and Frobenius norm properties have connections to the SVD. If $A \in \mathbb{R}^{m \times n}$, then

$$\begin{array}{rcll} \| A \|_F^2 & = & \sigma_1^2 + \cdots + \sigma_p^2 \qquad p = \min\{m, n\} & (2.5.7) \\ \| A \|_2 & = & \sigma_1 & (2.5.8) \end{array}$$

$$\min_{x \neq 0} \frac{\| Ax \|_2}{\| x \|_2} = \sigma_n \qquad (m \geq n)\,. \qquad (2.5.9)$$

2.5.4   The Thin SVD

If $A = U\Sigma V^T \in \mathbb{R}^{m \times n}$ is the SVD of A and $m \geq n$, then

$$A = U_1 \Sigma_1 V^T$$

where

$$U_1 \;=\; U(:, 1{:}n) \;=\; [\, u_1, \ldots, u_n \,] \in \mathbb{R}^{m \times n}$$

and

$$\Sigma_1 \;=\; \Sigma(1{:}n, 1{:}n) \;=\; \mathrm{diag}(\sigma_1, \ldots, \sigma_n) \in \mathbb{R}^{n \times n}.$$

We refer to this much-used, trimmed down version of the SVD as the *thin SVD*.

2.5.5   Rank Deficiency and the SVD

One of the most valuable aspects of the SVD is that it enables us to deal sensibly with the concept of matrix rank. Numerous theorems in linear algebra have the form "if such-and-such a matrix has full rank, then such-and-such a property holds." While neat and aesthetic, results of this flavor do not help us address the numerical difficulties frequently encountered in situations where near rank deficiency prevails. Rounding errors and fuzzy data make rank determination a nontrivial exercise. Indeed, for some small ϵ we may be interested in the ϵ-rank of a matrix which we define by

$$\mathrm{rank}(A, \epsilon) = \min_{\|A - B\|_2 \leq \epsilon} \mathrm{rank}(B).$$

Thus, if A is obtained in a laboratory with each a_{ij} correct to within $\pm.001$, then it might make sense to look at $\mathrm{rank}(A, .001)$. Along the same lines, if A is an m-by-n floating point matrix then it is reasonable to regard A as *numerically rank deficient* if $\mathrm{rank}(A, \epsilon) < \min\{m, n\}$ with $\epsilon = \mathbf{u}\| A \|_2$.

Numerical rank deficiency and ϵ-rank are nicely characterized in terms of the SVD because the singular values indicate how near a given matrix is to a matrix of lower rank.

Theorem 2.5.3 *Let the SVD of $A \in \mathbb{R}^{m \times n}$ be given by Theorem 2.5.2. If $k < r = \mathrm{rank}(A)$ and*

$$A_k = \sum_{i=1}^{k} \sigma_i u_i v_i^T, \tag{2.5.10}$$

then

$$\min_{\mathrm{rank}(B)=k} \| A - B \|_2 \;=\; \| A - A_k \|_2 \;=\; \sigma_{k+1}. \tag{2.5.11}$$

Proof. Since $U^T A_k V = \text{diag}(\sigma_1, \ldots, \sigma_k, 0, \ldots, 0)$ it follows that $\text{rank}(A_k) = k$ and that $U^T (A - A_k) V = \text{diag}(0, \ldots, 0, \sigma_{k+1}, \ldots, \sigma_p)$ and so $\| A - A_k \|_2 = \sigma_{k+1}$.

Now suppose $\text{rank}(B) = k$ for some $B \in \mathbb{R}^{m \times n}$. It follows that we can find orthonormal vectors $x_1, \ldots, x_{n-k}$ so $\text{null}(B) = \text{span}\{x_1, \ldots, x_{n-k}\}$. A dimension argument shows that

$$\text{span}\{x_1, \ldots, x_{n-k}\} \cap \text{span}\{v_1, \ldots, v_{k+1}\} \neq \{0\}.$$

Let z be a unit 2-norm vector in this intersection. Since $Bz = 0$ and

$$Az = \sum_{i=1}^{k+1} \sigma_i (v_i^T z) u_i$$

we have

$$\| A - B \|_2^2 \geq \| (A - B)z \|_2^2 = \| Az \|_2^2 = \sum_{i=1}^{k+1} \sigma_i^2 (v_i^T z)^2 \geq \sigma_{k+1}^2$$

completing the proof of the theorem. $\square$

Theorem 2.5.3 says that the smallest singular value of A is the 2-norm distance of A to the set of all rank-deficient matrices. It also follows that the set of full rank matrices in $\mathbb{R}^{m \times n}$ is both open and dense.

Finally, if $r_\epsilon = \text{rank}(A, \epsilon)$, then

$$\sigma_1 \geq \cdots \geq \sigma_{r_\epsilon} > \epsilon \geq \sigma_{r_\epsilon+1} \geq \cdots \geq \sigma_p \qquad p = \min\{m, n\}.$$

We have more to say about the numerical rank issue in §5.5 and §12.2.

2.5.6 Unitary Matrices

Over the complex field the unitary matrices correspond to the orthogonal matrices. In particular, $Q \in \mathbb{C}^{n \times n}$ is *unitary* if $Q^H Q = Q Q^H = I_n$. Unitary matrices preserve 2-norm. The SVD of a complex matrix involves unitary matrices. If $A \in \mathbb{C}^{m \times n}$, then there exist unitary matrices $U \in \mathbb{C}^{m \times m}$ and $V \in \mathbb{C}^{n \times n}$ such that

$$U^H AV = \text{diag}(\sigma_1, \ldots, \sigma_p) \in \mathbb{R}^{m \times n} \qquad p = \min\{m, n\}$$

where $\sigma_1 \geq \sigma_2 \geq \ldots \geq \sigma_p \geq 0$.

Problems

P2.5.1 Show that if S is real and $S^T = -S$, then $I - S$ is nonsingular and the matrix $(I - S)^{-1}(I + S)$ is orthogonal. This is known as the *Cayley transform* of S.

P2.5.2 Show that a triangular orthogonal matrix is diagonal.

P2.5.3 Show that if $Q = Q_1 + iQ_2$ is unitary with $Q_1, Q_2 \in \mathbf{R}^{n \times n}$, then the $2n$-by-$2n$ real matrix

$$Z = \begin{bmatrix} Q_1 & -Q_2 \\ Q_2 & Q_1 \end{bmatrix}$$

is orthogonal.

P2.5.4 Establish properties (2.5.3)-(2.5.9).

P2.5.5 Prove that

$$\sigma_{max}(A) = \max_{y \in \mathbf{R}^m, x \in \mathbf{R}^n} \frac{y^T A x}{\| x \|_2 \| y \|_2}$$

P2.5.6 For the 2-by-2 matrix $A = \begin{bmatrix} w & x \\ y & z \end{bmatrix}$, derive expressions for $\sigma_{max}(A)$ and $\sigma_{min}(A)$ that are functions of w, x, y, and z.

P2.5.7 Show that any matrix in $\mathbf{R}^{m \times n}$ is the limit of a sequence of full rank matrices.

P2.5.8 Show that if $A \in \mathbf{R}^{m \times n}$ has rank n, then $\| A(A^T A)^{-1} A^T \|_2 = 1$.

P2.5.9 What is the nearest rank-one matrix to $A = \begin{bmatrix} 1 & M \\ 0 & 1 \end{bmatrix}$ in the Frobenius norm?

P2.5.10 Show that if $A \in \mathbf{R}^{m \times n}$ then $\| A \|_F \leq \sqrt{\operatorname{rank}(A)} \, \| A \|_2$, thereby sharpening (2.3.7).

Notes and References for Sec. 2.5

Forsythe and Moler (1967) offer a good account of the SVD's role in the analysis of the $Ax = b$ problem. Their proof of the decomposition is more traditional than ours in that it makes use of the eigenvalue theory for symmetric matrices. Historical SVD references include

E. Beltrami (1873). "Sulle Funzioni Bilineari," *Gionale di Mathematiche 11*, 98–106.

C. Eckart and G. Young (1939). "A Principal Axis Transformation for Non-Hermitian Matrices," *Bull. Amer. Math. Soc. 45*, 118–21.

G.W. Stewart (1993). "On the Early History of the Singular Value Decomposition," *SIAM Review 35*, 551–566.

One of the most significant developments in scientific computation has been the increased use of the SVD in application areas that require the intelligent handling of matrix rank. The range of applications is impressive. One of the most interesting is

C.B. Moler and D. Morrison (1983). "Singular Value Analysis of Cryptograms," *Amer. Math. Monthly 90*, 78–87.

For generalizations of the SVD to infinite dimensional Hilbert space, see

I.C. Gohberg and M.G. Krein (1969). *Introduction to the Theory of Linear Non-Self Adjoint Operators* , Amer. Math. Soc., Providence, R.I.

F. Smithies (1970). *Integral Equations*, Cambridge University Press, Cambridge.

Reducing the rank of a matrix as in Theorem 2.5.3 when the perturbing matrix is constrained is discussed in

J.W. Demmel (1987). "The smallest perturbation of a submatrix which lowers the rank and constrained total least squares problems, *SIAM J. Numer. Anal. 24*, 199–206.

G.H. Golub, A. Hoffman, and G.W. Stewart (1988). "A Generalization of the Eckart-Young-Mirsky Approximation Theorem." *Lin. Alg. and Its Applic. 88/89*, 317–328.
G.A. Watson (1988). "The Smallest Perturbation of a Submatrix which Lowers the Rank of the Matrix," *IMA J. Numer. Anal. 8*, 295–304.

2.6 Projections and the CS Decomposition

If the object of a computation is to compute a matrix or a vector, then norms are useful for assessing the accuracy of the answer or for measuring progress during an iteration. If the object of a computation is to compute a subspace, then to make similar comments we need to be able to quantify the distance between two subspaces. Orthogonal projections are critical in this regard. After the elementary concepts are established we discuss the CS decomposition. This is an SVD-like decomposition that is handy when having to compare a pair of subspaces. We begin with the notion of an orthogonal projection.

2.6.1 Orthogonal Projections

Let $S \subseteq \mathbb{R}^n$ be a subspace. $P \in \mathbb{R}^{n \times n}$ is the *orthogonal projection* onto S if $\text{ran}(P) = S$, $P^2 = P$, and $P^T = P$. From this definition it is easy to show that if $x \in \mathbb{R}^n$, then $Px \in S$ and $(I - P)x \in S^\perp$.

If P_1 and P_2 are each orthogonal projections, then for any $z \in \mathbb{R}^n$ we have

$$\| (P_1 - P_2)z \|_2^2 = (P_1 z)^T (I - P_2)z + (P_2 z)^T (I - P_1)z.$$

If $\text{ran}(P_1) = \text{ran}(P_2) = S$, then the right-hand side of this expression is zero showing that the orthogonal projection for a subspace is unique. If the columns of $V = [\, v_1, \ldots, v_k \,]$ are an orthonormal basis for a subspace S, then it is easy to show that $P = VV^T$ is the unique orthogonal projection onto S. Note that if $v \in \mathbb{R}^n$, then $P = vv^T/v^Tv$ is the orthogonal projection onto $S = \text{span}\{v\}$.

2.6.2 SVD-Related Projections

There are several important orthogonal projections associated with the singular value decomposition. Suppose $A = U\Sigma V^T \in \mathbb{R}^{m \times n}$ is the SVD of A and that $r = \text{rank}(A)$. If we have the U and V partitionings

$$U = [\, \underset{r}{U_r} \quad \underset{m-r}{\tilde{U}_r} \,] \qquad V = [\, \underset{r}{V_r} \quad \underset{n-r}{\tilde{V}_r} \,]$$

then

$$
\begin{aligned}
V_r V_r^T &= \text{projection on to null}(A)^\perp = \text{ran}(A^T) \\
\tilde{V}_r \tilde{V}_r^T &= \text{projection on to null}(A) \\
U_r U_r^T &= \text{projection on to ran}(A) \\
\tilde{U}_r \tilde{U}_r^T &= \text{projection on to ran}(A)^\perp = \text{null}(A^T)
\end{aligned}
$$

2.6.3   Distance Between Subspaces

The one-to-one correspondence between subspaces and orthogonal projections enables us to devise a notion of distance between subspaces. Suppose S_1 and S_2 are subspaces of $\mathbb{R}^n$ and that $\dim(S_1) = \dim(S_2)$. We define the *distance* between these two spaces by

$$\text{dist}(S_1, S_2) \;=\; \| \, P_1 - P_2 \, \|_2 \qquad\qquad (2.6.1)$$

where P_i is the orthogonal projection onto S_i. The distance between a pair of subspaces can be characterized in terms of the blocks of a certain orthogonal matrix.

Theorem 2.6.1 *Suppose*

$$W = [\, W_1 \quad W_2 \,] \qquad\qquad Z = [\, Z_1 \quad Z_2 \,]$$
$$k \quad\;\; n-k \qquad\qquad\qquad k \quad\;\; n-k$$

are n-by-n orthogonal matrices. If $S_1 = \text{ran}(W_1)$ and $S_2 = \text{ran}(Z_1)$, then

$$\text{dist}(S_1, S_2) \;=\; \| \, W_1^T Z_2 \, \|_2 \;=\; \| \, Z_1^T W_2 \, \|_2.$$

Proof.

$$\text{dist}(S_1, S_2) \;=\; \| \, W_1 W_1^T - Z_1 Z_1^T \, \|_2 \;=\; \| \, W^T (W_1 W_1^T - Z_1 Z_1^T) Z \, \|_2$$

$$=\; \left\| \begin{bmatrix} 0 & W_1^T Z_2 \\ -W_2^T Z_1 & 0 \end{bmatrix} \right\|_2 .$$

Note that the matrices $W_2^T Z_1$ and $W_1^T Z_2$ are submatrices of the orthogonal matrix

$$Q = \begin{bmatrix} Q_{11} & Q_{12} \\ Q_{21} & Q_{22} \end{bmatrix} \equiv \begin{bmatrix} W_1^T Z_1 & W_1^T Z_2 \\ W_2^T Z_1 & W_2^T Z_2 \end{bmatrix} = W^T Z.$$

Our goal is to show that $\| \, Q_{21} \, \|_2 \;=\; \| \, Q_{12} \, \|_2$. Since Q is orthogonal it follows from

$$Q \begin{bmatrix} x \\ 0 \end{bmatrix} = \begin{bmatrix} Q_{11} x \\ Q_{21} x \end{bmatrix}$$

that

$$1 = \| \, Q_{11} x \, \|_2^2 \;+\; \| \, Q_{21} x \, \|_2^2$$

for all unit 2-norm $x \in \mathbb{R}^k$. Thus,

$$\| \, Q_{21} \, \|_2^2 \;=\; \max_{\| x \|_2 = 1} \| \, Q_{21} x \, \|_2^2 = 1 - \min_{\| x \|_2 = 1} \| \, Q_{11} x \, \|_2^2$$

$$=\; 1 - \sigma_{min}(Q_{11})^2.$$

Analogously, by working with Q^T (which is also orthogonal) it is possible to show that

$$\| Q_{12}^T \|_2^2 = 1 - \sigma_{min}(Q_{11}^T)^2.$$

and therefore

$$\| Q_{12} \|_2^2 = 1 - \sigma_{min}(Q_{11})^2.$$

Thus, $\| Q_{21} \|_2 = \| Q_{12} \|_2$. $\square$

Note that if S_1 and S_2 are subspaces in $\mathbb{R}^n$ with the same dimension, then

$$0 \leq \text{dist}(S_1, S_2) \leq 1.$$

The distance is zero if $S_1 = S_2$ and one if $S_1 \cap S_2^\perp \neq \{0\}$.

A more refined analysis of the blocks of the Q matrix above sheds more light on the difference between a pair of subspaces. This requires a special SVD-like decomposition for orthogonal matrices.

2.6.4 The CS Decomposition

The blocks of an orthogonal matrix partitioned into 2-by-2 form have highly related SVDs. This is the gist of the *CS decomposition*. We prove a very useful special case first.

Theorem 2.6.2 (The CS Decomposition (Thin Version)) *Consider the matrix*

$$Q = \begin{bmatrix} Q_1 \\ Q_2 \end{bmatrix} \qquad Q_1 \in \mathbb{R}^{m_1 \times n}, \ Q_2 \in \mathbb{R}^{m_2 \times n}$$

where $m_1 \geq n$ and $m_2 \geq n$. If the columns of Q are orthonormal, then there exist orthogonal matrices $U_1 \in \mathbb{R}^{m_1 \times m_1}$, $U_2 \in \mathbb{R}^{m_2 \times m_2}$, and $V_1 \in \mathbb{R}^{n \times n}$ such that

$$\begin{bmatrix} U_1 & 0 \\ 0 & U_2 \end{bmatrix}^T \begin{bmatrix} Q_1 \\ Q_2 \end{bmatrix} V_1 = \begin{bmatrix} C \\ S \end{bmatrix}$$

where

$$
\begin{aligned}
C &= \text{diag}(\cos(\theta_1), \ldots, \cos(\theta_n)), \\
S &= \text{diag}(\sin(\theta_1), \ldots, \sin(\theta_n)),
\end{aligned}
$$

and

$$0 \leq \theta_1 \leq \theta_2 \leq \cdots \leq \theta_n \leq \frac{\Pi}{2}.$$

Proof. Since $\| Q_{11} \|_2 \leq \| Q \|_2 = 1$, the singular values of Q_{11} are all in the interval $[0, 1]$. Let

$$U_1^T Q_1 V_1 = C = \text{diag}(c_1, \ldots, c_n) = \begin{bmatrix} I_t & 0 \\ 0 & \Sigma \\ t & n-t \end{bmatrix} \begin{matrix} t \\ m_1 - t \end{matrix}$$

be the SVD of Q_1 where we assume

$$1 = c_1 = \cdots = c_t > c_{t+1} \geq \cdots \geq c_n \geq 0.$$

To complete the proof of the theorem we must construct the orthogonal matrix U_2. If

$$Q_2 V_1 = [\; W_1 \quad W_2 \;] \atop t \qquad n-t \qquad ,$$

then

$$\left[\begin{array}{cc} U_1 & 0 \\ 0 & I_{m_2} \end{array}\right]^T \left[\begin{array}{c} Q_1 \\ Q_2 \end{array}\right] V_1 = \left[\begin{array}{cc} I_t & 0 \\ 0 & \Sigma \\ W_1 & W_2 \end{array}\right].$$

Since the columns of this matrix have unit 2-norm, $W_1 = 0$. The columns of W_2 are nonzero and mutually orthogonal because

$$W_2^T W_2 = I_{n-t} - \Sigma^T \Sigma \equiv \operatorname{diag}(1 - c_{t+1}^2, \ldots, 1 - c_n^2)$$

is nonsingular. If $s_k = \sqrt{1 - c_k^2}$ for $k = 1{:}n$, then the columns of

$$Z = W_2 \operatorname{diag}(1/s_{t+1}, \ldots, 1/s_n)$$

are orthonormal. By Theorem 2.5.1 there exists an orthogonal matrix $U_2 \in \mathbb{R}^{m_2 \times m_2}$ with $U_2(:, t+1{:}n) = Z$. It is easy to verify that

$$U_2^T Q_2 V_1 = \operatorname{diag}(s_1, \ldots, s_n) \equiv S.$$

Since $c_k^2 + s_k^2 = 1$ for $k = 1{:}n$, it follows that these quantities are the required cosines and sines. $\square$

Using the same sort of techniques it is possible to prove the following more general version of the decomposition:

Theorem 2.6.3 (CS Decomposition (General Version)) *If*

$$Q = \left[\begin{array}{c|c} Q_{11} & Q_{12} \\ \hline Q_{21} & Q_{22} \end{array}\right]$$

is a 2-by-2 (arbitrary) partitioning of an n-by-n orthogonal matrix, then there exist orthogonal

$$U = \left[\begin{array}{c|c} U_1 & 0 \\ \hline 0 & U_2 \end{array}\right] \qquad \text{and} \qquad V = \left[\begin{array}{c|c} V_1 & 0 \\ \hline 0 & V_2 \end{array}\right]$$

such that

$$U^T Q V = \left[\begin{array}{ccc|ccc} I & 0 & 0 & 0 & 0 & 0 \\ 0 & C & 0 & 0 & S & 0 \\ 0 & 0 & 0 & 0 & 0 & I \\ \hline 0 & 0 & 0 & I & 0 & 0 \\ 0 & S & 0 & 0 & -C & 0 \\ 0 & 0 & I & 0 & 0 & 0 \end{array}\right]$$

where $C = \text{diag}(c_1, \ldots, c_p)$ and $S = \text{diag}(s_1, \ldots, s_p)$ are square diagonal matrices with $0 < c_i, s_i < 1$.

Proof. See Paige and Saunders (1981) for details. We have suppressed the dimensions of the zero submatrices, some of which may be empty. □

The essential message of the decomposition is that the SVDs of the Q_{ij} are highly related.

Example 2.6.1 The matrix

$$
Q = \left[
\begin{array}{cc|ccc}
-0.7576 & 0.3697 & 0.3838 & 0.2126 & -0.3112 \\
-0.4077 & -0.1552 & -0.1129 & 0.2676 & 0.8517 \\
-0.0488 & 0.7240 & -0.6730 & -0.1301 & 0.0602 \\
\hline
-0.2287 & 0.0088 & 0.2235 & -0.9235 & 0.2120 \\
0.4530 & 0.5612 & 0.5806 & 0.1162 & 0.3595
\end{array}
\right]
$$

is orthogonal and with the indicated partitioning can be reduced to

$$
U^T Q V = \left[
\begin{array}{cc|ccc}
0.9837 & 0.0000 & 0.1800 & 0.0000 & 0.0000 \\
0.0000 & 0.6781 & 0.0000 & 0.7349 & 0.0000 \\
0.0000 & 0.0000 & 0.0000 & 0.0000 & 1.0000 \\
\hline
0.1800 & 0.0000 & -0.9837 & 0.0000 & 0.0000 \\
0.0000 & 0.7349 & 0.0000 & -0.6781 & 0.0000
\end{array}
\right]
$$

The angles associated with the cosines and sines turn out to be very important in a number of applications. See §12.4.

Problems

P2.6.1 Show that if P is an orthogonal projection, then $Q = I - 2P$ is orthogonal.

P2.6.2 What are the singular values of an orthogonal projection?

P2.6.3 Suppose $S_1 = \text{span}\{x\}$ and $S_2 = \text{span}\{y\}$, where x and y are unit 2-norm vectors in $\mathbf{R}^2$. Working only with the definition of $\text{dist}(\cdot, \cdot)$, show that $\text{dist}(S_1, S_2) = \sqrt{1 - (x^T y)^2}$ verifying that the distance between S_1 and S_2 equals the sine of the angle between x and y.

Notes and References for Sec. 2.6

The following papers discuss various aspects of the CS decomposition:

C. Davis and W. Kahan (1970). "The Rotation of Eigenvectors by a Perturbation III," *SIAM J. Num. Anal. 7*, 1–46.

G.W. Stewart (1977). "On the Perturbation of Pseudo-Inverses, Projections and Linear Least Squares Problems," *SIAM Review 19*, 634–662.

C.C. Paige and M. Saunders (1981). "Toward a Generalized Singular Value Decomposition," *SIAM J. Num. Anal. 18*, 398–405.

C.C. Paige and M. Wei (1994). "History and Generality of the CS Decomposition," *Lin. Alg. and Its Applic. 208/209*, 303–326.

See §8.7 for some computational details.

For a deeper geometrical understanding of the CS decomposition and the notion of distance between subspaces, see

T.A. Arias, A. Edelman, and S. Smith (1996). "Conjugate Gradient and Newton's Method on the Grassman and Stiefel Manifolds," to appear in *SIAM J. Matrix Anal. Appl.*

2.7   The Sensitivity of Square Systems

We now use some of the tools developed in previous sections to analyze the linear system problem $Ax = b$ where $A \in \mathbb{R}^{n \times n}$ is nonsingular and $b \in \mathbb{R}^n$. Our aim is to examine how perturbations in A and b affect the solution x. A much more detailed treatment may be found in Higham (1996).

2.7.1   An SVD Analysis

If

$$A = \sum_{i=1}^{n} \sigma_i u_i v_i^T = U\Sigma V^T$$

is the SVD of A, then

$$x = A^{-1}b = (U\Sigma V^T)^{-1}b = \sum_{i=1}^{n} \frac{u_i^T b}{\sigma_i} v_i. \qquad (2.7.1)$$

This expansion shows that small changes in A or b can induce relatively large changes in x if σ_n is small.

It should come as no surprise that the magnitude of σ_n should have a bearing on the sensitivity of the $Ax = b$ problem when we recall from Theorem 2.5.3 that σ_n is the distance from A to the set of singular matrices. As the matrix of coefficients approaches this set, it is intuitively clear that the solution x should be increasingly sensitive to perturbations.

2.7.2   Condition

A precise measure of linear system sensitivity can be obtained by considering the parameterized system

$$(A + \epsilon F)x(\epsilon) = b + \epsilon f \qquad x(0) = x$$

where $F \in \mathbb{R}^{n \times n}$ and $f \in \mathbb{R}^n$. If A is nonsingular, then it is clear that $x(\epsilon)$ is differentiable in a neighborhood of zero. Moreover, $\dot{x}(0) = A^{-1}(f - Fx)$ and thus, the Taylor series expansion for $x(\epsilon)$ has the form

$$x(\epsilon) = x + \epsilon \dot{x}(0) + O(\epsilon^2).$$

Using any vector norm and consistent matrix norm we obtain

$$\frac{\|x(\epsilon) - x\|}{\|x\|} \le |\epsilon| \, \|A^{-1}\| \left\{ \frac{\|f\|}{\|x\|} + \|F\| \right\} + O(\epsilon^2). \qquad (2.7.2)$$

For square matrices A define the *condition number* $\kappa(A)$ by

$$\kappa(A) = \|A\| \, \|A^{-1}\| \qquad (2.7.3)$$

with the convention that $\kappa(A) = \infty$ for singular A. Using the inequality $\|b\| \le \|A\| \, \|x\|$ it follows from (2.7.2) that

$$\frac{\|x(\epsilon) - x\|}{\|x\|} \le \kappa(A)(\rho_A + \rho_b) + O(\epsilon^2) \qquad (2.7.4)$$

where

$$\rho_A = |\epsilon| \frac{\|F\|}{\|A\|} \quad \text{and} \quad \rho_b = |\epsilon| \frac{\|f\|}{\|b\|}$$

represent the relative errors in A and b, respectively. Thus, the relative error in x can be $\kappa(A)$ times the relative error in A and b. In this sense, the condition number $\kappa(A)$ quantifies the sensitivity of the $Ax = b$ problem.

Note that $\kappa(\cdot)$ depends on the underlying norm and subscripts are used accordingly, e.g.,

$$\kappa_2(A) = \|A\|_2 \|A^{-1}\|_2 = \frac{\sigma_1(A)}{\sigma_n(A)}. \qquad (2.7.5)$$

Thus, the 2-norm condition of a matrix A measures the elongation of the hyperellipsoid $\{Ax : \|x\|_2 = 1\}$.

We mention two other characterizations of the condition number. For p-norm condition numbers, we have

$$\frac{1}{\kappa_p(A)} = \min_{A+\Delta A \text{ singular}} \frac{\|\Delta A\|_p}{\|A\|_p}. \qquad (2.7.6)$$

This result may be found in Kahan (1966) and shows that $\kappa_p(A)$ measures the relative p-norm distance from A to the set of singular matrices.

For any norm, we also have

$$\kappa(A) = \lim_{\epsilon \to 0} \sup_{\|\Delta A\| \le \epsilon \|A\|} \frac{\|(A + \Delta A)^{-1} - A^{-1}\|}{\epsilon} \cdot \frac{1}{\|A^{-1}\|} . \qquad (2.7.7)$$

This imposing result merely says that the condition number is a normalized Frechet derivative of the map $A \to A^{-1}$. Further details may be found in Rice (1966b). Recall that we were initially led to $\kappa(A)$ through differentiation.

If $\kappa(A)$ is large, then A is said to be an *ill-conditioned* matrix. Note that this is a norm-dependent property[2]. However, any two condition numbers $\kappa_\alpha(\cdot)$ and $\kappa_\beta(\cdot)$ on $\mathbb{R}^{n \times n}$ are equivalent in that constants c_1 and c_2 can be found for which

$$c_1 \kappa_\alpha(A) \; \leq \; \kappa_\beta(A) \; \leq \; c_2 \kappa_\alpha(A) \qquad A \in \mathbb{R}^{n \times n} \,.$$

For example, on $\mathbb{R}^{n \times n}$ we have

$$\frac{1}{n} \kappa_2(A) \; \leq \; \kappa_1(A) \; \leq \; n \kappa_2(A)$$

$$\frac{1}{n} \kappa_\infty(A) \; \leq \; \kappa_2(A) \; \leq \; n \kappa_\infty(A) \tag{2.7.8}$$

$$\frac{1}{n^2} \kappa_1(A) \; \leq \; \kappa_\infty(A) \leq n^2 \kappa_1(A) \,.$$

Thus, if a matrix is ill-conditioned in the α-norm, it is ill-conditioned in the β-norm modulo the constants c_1 and c_2 above.

For any of the p-norms, we have $\kappa_p(A) \geq 1$. Matrices with small condition numbers are said to be *well-conditioned* . In the 2-norm, orthogonal matrices are perfectly conditioned in that $\kappa_2(Q) = 1$ if Q is orthogonal.

2.7.3   Determinants and Nearness to Singularity

It is natural to consider how well determinant size measures ill-conditioning. If $\det(A) = 0$ is equivalent to singularity, is $\det(A) \approx 0$ equivalent to near singularity? Unfortunately, there is little correlation between $\det(A)$ and the condition of $Ax = b$. For example, the matrix B_n defined by

$$B_n \; = \; \begin{bmatrix} 1 & -1 & \cdots & -1 \\ 0 & 1 & \cdots & -1 \\ \vdots & \vdots & \ddots & \vdots \\ 0 & 0 & \cdots & 1 \end{bmatrix} \in \mathbb{R}^{n \times n} \tag{2.7.9}$$

has determinant 1, but $\kappa_\infty(B_n) = n2^{n-1}$. On the other hand, a very well conditioned matrix can have a very small determinant. For example,

$$D_n = \text{diag}(10^{-1}, \ldots, 10^{-1}) \in \mathbb{R}^{n \times n}$$

satisfies $\kappa_p(D_n) = 1$ although $\det(D_n) = 10^{-n}$.

2.7.4   A Rigorous Norm Bound

Recall that the derivation of (2.7.4) was valuable because it highlighted the connection between $\kappa(A)$ and the rate of change of $x(\epsilon)$ at $\epsilon = 0$. However,

---

[2]It also depends upon the definition of "large." The matter is pursued in §3.5

it is a little unsatisfying because it is contingent on ϵ being "small enough" and because it sheds no light on the size of the $O(\epsilon^2)$ term. In this and the next subsection we develop some additional $Ax = b$ perturbation theorems that are completely rigorous.

We first establish a useful lemma that indicates in terms of $\kappa(A)$ when we can expect a perturbed system to be nonsingular.

Lemma 2.7.1 *Suppose*

$$Ax = b \qquad\qquad A \in \mathbb{R}^{n \times n},\ 0 \neq b \in \mathbb{R}^n$$

$$(A + \Delta A)y = b + \Delta b \qquad \Delta A \in \mathbb{R}^{n \times n},\ \Delta b \in \mathbb{R}^n$$

with $\| \Delta A \| \leq \epsilon \| A \|$ *and* $\| \Delta b \| \leq \epsilon \| b \|$. *If* $\epsilon \kappa(A) = r < 1$, *then* $A + \Delta A$ *is nonsingular and*

$$\frac{\| y \|}{\| x \|} \leq \frac{1 + r}{1 - r}.$$

Proof. Since $\| A^{-1} \Delta A \| \leq \epsilon \| A^{-1} \| \| A \| = r < 1$ it follows from Theorem 2.3.4 that $(A + \Delta A)$ is nonsingular. Using Lemma 2.3.3 and the equality $(I + A^{-1}\Delta A)y = x + A^{-1}\Delta b$ we find

$$\| y \| \leq \| (I + A^{-1}\Delta A)^{-1} \| \left(\| x \| + \epsilon \| A^{-1} \| \| b \| \right)$$

$$\leq \frac{1}{1 - r} \left(\| x \| + \epsilon \| A^{-1} \| \| b \| \right) = \frac{1}{1 - r} \left(\| x \| + r\frac{\| b \|}{\| A \|} \right).$$

Since $\| b \| = \| Ax \| \leq \| A \| \| x \|$ it follows that

$$\| y \| \leq \frac{1}{1 - r} \left(\| x \| + r\| x \| \right). \ \square$$

We are now set to establish a rigorous $Ax = b$ perturbation bound.

Theorem 2.7.2 *If the conditions of Lemma 2.7.1 hold, then*

$$\frac{\| y - x \|}{\| x \|} \leq \frac{2\epsilon}{1 - r}\kappa(A) \qquad\qquad (2.7.10)$$

Proof. Since

$$y - x = A^{-1}\Delta b - A^{-1}\Delta A y \qquad\qquad (2.7.11)$$

we have $\| y - x \| \leq \epsilon\| A^{-1} \| \| b \| + \epsilon \| A^{-1} \| \| A \| \| y \|$ and so

$$\frac{\| y - x \|}{\| x \|} \leq \epsilon \kappa(A)\frac{\| b \|}{\| A \| \| x \|} + \epsilon \kappa(A)\frac{\| y \|}{\| x \|}$$

$$\leq \epsilon\kappa(A) \left(1 + \frac{1 + r}{1 - r} \right) = \frac{2\epsilon}{1 - r}\kappa(A). \ \square$$

Example 2.7.1 The $Ax = b$ problem

$$\begin{bmatrix} 1 & 0 \\ 0 & 10^{-6} \end{bmatrix} \begin{bmatrix} x_1 \\ x_2 \end{bmatrix} = \begin{bmatrix} 1 \\ 10^{-6} \end{bmatrix}$$

has solution $x = (\,1,\,1\,)^T$ and condition $\kappa_\infty(A) = 10^6$. If $\Delta b = (\,10^{-6},\,0\,)^T$, $\Delta A = 0$, and $(A + \Delta A)y = b + \Delta b$, then $y = (\,1 + 10^{-6},\,1\,)^T$ and the inequality (2.7.10) says

$$10^{-6} = \frac{\|\,x - y\,\|_\infty}{\|\,x\,\|_\infty} \ll \frac{\|\,\Delta b\,\|_\infty}{\|\,b\,\|_\infty}\kappa_\infty(A) = 10^{-6}10^6 = 1.$$

Thus, the upper bound in (2.7.10) can be a gross overestimate of the error induced by the perturbation. On the other hand, if $\Delta b = (0,\,10^{-6})^T$, $\Delta A = 0$, and $(A + \Delta A)y = b + \Delta b$, then this inequality says

$$\frac{10^0}{10^0} \le 2 \times 10^{-6}10^6.$$

Thus, there are perturbations for which the bound in (2.7.10) is essentially attained.

2.7.5   Some Rigorous Componentwise Bounds

We conclude this section by showing that a more refined perturbation theory is possible if componentwise perturbation bounds are in effect and if we make use of the absolute value notation.

Theorem 2.7.3 *Suppose*

$$Ax = b \qquad A \in \mathbb{R}^{n \times n},\ 0 \ne b \in \mathbb{R}^n$$

$$(A + \Delta A)y = b + \Delta b \quad \Delta A \in \mathbb{R}^{n \times n},\ \Delta b \in \mathbb{R}^n$$

and that $|\Delta A| \le \epsilon|A|$ *and* $|\Delta b| \le \epsilon|b|$. *If* $\delta\kappa_\infty(A) = r < 1$, *then* $(A + \Delta A)$ *is nonsingular and*

$$\frac{\|\,y - x\,\|_\infty}{\|\,x\,\|_\infty} \le \frac{2\delta}{1 - r}\|\,|A^{-1}|\,|A|\,\|_\infty.$$

Proof. Since $\|\,\Delta A\,\|_\infty \le \epsilon\|\,A\,\|_\infty$ and $\|\,\Delta b\,\|_\infty \le \epsilon\|\,b\,\|_\infty$ the conditions of Lemma 2.7.1 are satisfied in the infinity norm. This implies that $A + \Delta A$ is nonsingular and

$$\frac{\|\,y\,\|_\infty}{\|\,x\,\|_\infty} \le \frac{1 + r}{1 - r}.$$

Now using (2.7.11) we find

$$|y - x| \le |A^{-1}|\,|\Delta b| + |A^{-1}|\,|\Delta A|\,|y|$$

$$\le \epsilon|A^{-1}|\,|b| + \epsilon|A^{-1}|\,|A|\,|y| \le \epsilon|A^{-1}|\,|A|\,(|x| + |y|).$$

If we take norms, then

$$\|\,y - x\,\|_\infty \le \epsilon\|\,|A^{-1}|\,|A|\,\|_\infty \left(\|\,x\,\|_\infty + \frac{1 + r}{1 - r}\|\,x\,\|_\infty\right).$$

The theorem follows upon division by $\| x \|_\infty$. $\square$

We refer to the quantity $\| \, |A^{-1}| \, |A| \, \|_\infty$ as the *Skeel condition number*. It has been effectively used in the analysis of several important linear system computations. See §3.5.

Lastly, we report on the results of Oettli and Prager (1964) that indicate when an approximate solution $\hat{x} \in \mathbb{R}^n$ to the n-by-n system $Ax = b$ satisfies a perturbed system with prescribed structure. In particular, suppose $E \in \mathbb{R}^{n \times n}$ and $f \in \mathbb{R}^n$ are given and have nonnegative entries. We seek $\Delta A \in \mathbb{R}^{n \times n}$, $\Delta b \in \mathbb{R}^n$, and $\omega \geq 0$ such that

$$(A + \Delta A)\hat{x} = b + \Delta b \qquad |\Delta A| \leq \omega E, \; |\Delta b| \leq \omega f. \qquad (2.7.12)$$

Note that by properly choosing E and f the perturbed system can take on certain qualities. For example, if $E = |A|$ and $f = |b|$ and ω is small, then $\hat{x}$ satisfies a nearby system in the componentwise sense. Oettli and Prager (1964) show that for a given A, b, $\hat{x}$, E, and f the smallest ω possible in (2.7.12) is given by

$$\omega_{min} = \max_{1 \leq i \leq n} \frac{|A\hat{x} - b|_i}{(E|\hat{x}| + f)_i}.$$

If $A\hat{x} = b$ then $\omega_{min} = 0$. On the other hand, if $\omega_{min} = \infty$, then $\hat{x}$ does not satisfy any system of the prescribed perturbation structure.

Problems

P2.7.1 Show that if $\| I \| \geq 1$, then $\kappa(A) \geq 1$.

P2.7.2 Show that for a given norm, $\kappa(AB) \leq \kappa(A)\kappa(B)$ and that $\kappa(\alpha A) = \kappa(A)$ for all nonzero α.

P2.7.3 Relate the 2-norm condition of $X \in \mathbf{R}^{m \times n}$ ($m \geq n$) to the 2-norm condition of the matrices

$$B = \left[\begin{array}{cc} I_m & X \\ 0 & I_n \end{array} \right]$$

and

$$C = \left[\begin{array}{c} X \\ I_n \end{array} \right].$$

Notes and References for Sec. 2.7

The condition concept is thoroughly investigated in

J. Rice (1966). "A Theory of Condition," *SIAM J. Num. Anal.* 3, 287–310.
W. Kahan (1966). "Numerical Linear Algebra," *Canadian Math. Bull.* 9, 757–801.

References for componentwise perturbation theory include

W. Oettli and W. Prager (1964). "Compatibility of Approximate Solutions of Linear Equations with Given Error Bounds for Coefficients and Right Hand Sides," *Numer. Math. 6*, 405–409.

J.E. Cope and B.W. Rust (1979). "Bounds on solutions of systems with accurate data," *SIAM J. Num. Anal. 16*, 950–63.

R.D. Skeel (1979). "Scaling for numerical stability in Gaussian Elimination," *J. ACM 26*, 494–526.

J.W. Demmel (1992). "The Componentwise Distance to the Nearest Singular Matrix," *SIAM J. Matrix Anal. Appl. 13*, 10–19.

D.J. Higham and N.J. Higham (1992). "Componentwise Perturbation Theory for Linear Systems with Multiple Right-Hand Sides," *Lin. Alg. and Its Applic. 174*, 111–129.

N.J. Higham (1994). "A Survey of Componentwise Perturbation Theory in Numerical Linear Algebra," in *Mathematics of Computation 1943-1993: A Half Century of Computational Mathematics*, W. Gautschi (ed.), Volume 48 of *Proceedings of Symposia in Applied Mathematics*, American Mathematical Society, Providence, Rhode Island.

S. Chandrasekaren and I.C.F. Ipsen (1995). "On the Sensitivity of Solution Components in Linear Systems of Equations," *SIAM J. Matrix Anal. Appl. 16*, 93–112.

The reciprocal of the condition number measures how near a given $Ax = b$ problem is to singularity. The importance of knowing how near a given problem is to a difficult or insoluble problem has come to be appreciated in many computational settings. See

A. Laub(1985). "Numerical Linear Algebra Aspects of Control Design Computations," *IEEE Trans. Auto. Cont. AC-30*, 97–108.

J. L. Barlow (1986). "On the Smallest Positive Singular Value of an M-Matrix with Applications to Ergodic Markov Chains," *SIAM J. Alg. and Disc. Struct. 7*, 414–424.

J.W. Demmel (1987). "On the Distance to the Nearest Ill-Posed Problem," *Numer. Math. 51*, 251–289.

J.W. Demmel (1988). "The Probability that a Numerical Analysis Problem is Difficult," *Math. Comp. 50*, 449–480.

N.J. Higham (1989). "Matrix Nearness Problems and Applications," in *Applications of Matrix Theory*, M.J.C. Gover and S. Barnett (eds), Oxford University Press, Oxford UK, 1–27.

Chapter 3

General Linear Systems

The problem of solving a linear system $Ax = b$ is central in scientific computation. In this chapter we focus on the method of Gaussian elimination, the algorithm of choice when A is square, dense, and unstructured. When A does not fall into this category, then the algorithms of Chapters 4, 5, and 10 are of interest. Some parallel $Ax = b$ solvers are discussed in Chapter 6.

We motivate the method of Gaussian elimination in §3.1 by discussing the ease with which triangular systems can be solved. The conversion of a general system to triangular form via Gauss transformations is then presented in §3.2 where the "language" of matrix factorizations is introduced. Unfortunately, the derived method behaves very poorly on a nontrivial class of problems. Our error analysis in §3.3 pinpoints the difficulty and motivates §3.4, where the concept of pivoting is introduced. In the final section we comment upon the important practical issues associated with scaling, iterative improvement, and condition estimation.

Before You Begin

Chapter 1, §§2.1-2.5, and §2.7 are assumed. Complementary references include Forsythe and Moler (1967), Stewart (1973), Hager (1988), Watkins

(1991), Ciarlet (1992), Datta (1995), Higham (1996), Trefethen and Bau (1996), and Demmel (1996). Some MATLABfunctions important to this chapter are `lu`, `cond`, `rcond`, and the "backslash" operator "\". LAPACK connections include

LAPACK: Triangular Systems	
_TRSV	Solves $Ax = b$
_TRSM	Solves $AX = B$
_TRCON	Condition estimate
_TRRFS	Solve $AX = B$, $A^T X = B$ with error bounds
_TRTRS	Solve $AX = B$, $A^T X = B$
_TRTRI	A^{-1}

LAPACK: General Linear Systems	
_GESV	Solve $AX = B$
_GECON	Condition estimate via $PA = LU$
_GERFS	Improve $AX = B$, $A^T X = B$, $A^H X = B$ solutions with error bounds
_GESVX	Solve $AX = B$, $A^T X = B$, $A^H X = B$ with condition estimate
_GETRF	$PA = LU$
_GETRS	Solve $AX = B$, $A^T X = B$, $A^H X = B$ via $PA = LU$
_GETRI	A^{-1}
_GEEQU	Equilibration

3.1   Triangular Systems

Traditional factorization methods for linear systems involve the conversion of the given square system to a triangular system that has the same solution. This section is about the solution of triangular systems.

3.1.1   Forward Substitution

Consider the following 2-by-2 lower triangular system:

$$\left[\begin{array}{cc} \ell_{11} & 0 \\ \ell_{21} & \ell_{22} \end{array} \right] \left[\begin{array}{c} x_1 \\ x_2 \end{array} \right] = \left[\begin{array}{c} b_1 \\ b_2 \end{array} \right].$$

If $\ell_{11}\ell_{22} \neq 0$, then the unknowns can be determined sequentially:

$$\begin{aligned} x_1 &= b_1/\ell_{11} \\ x_2 &= (b_2 - \ell_{21}x_1)/\ell_{22}. \end{aligned}$$

This is the 2-by-2 version of an algorithm known as *forward substitution*. The general procedure is obtained by solving the ith equation in $Lx = b$ for x_i:

$$x_i = \left(b_i - \sum_{j=1}^{i-1} \ell_{ij}x_j \right) \Big/ \ell_{ii}$$

If this is evaluated for $i = 1{:}n$, then a complete specification of x is obtained. Note that at the ith stage the dot product of $L(i, 1{:}i - 1)$ and $x(1{:}i - 1)$ is required. Since b_i only is involved in the formula for x_i, the former may be overwritten by the latter:

Algorithm 3.1.1 (Forward Substitution: Row Version) If $L \in \mathbb{R}^{n \times n}$ is lower triangular and $b \in \mathbb{R}^n$, then this algorithm overwrites b with the solution to $Lx = b$. L is assumed to be nonsingular.

$b(1) = b(1)/L(1,1)$
for $i = 2{:}n$
$\quad b(i) = (b(i) - L(i, 1{:}i - 1)b(1{:}i - 1))/L(i, i)$
end

This algorithm requires n^2 flops. Note that L is accessed by row. The computed solution $\hat{x}$ satisfies:

$$(L + F)\hat{x} = b \qquad |F| \leq nu|L| + O(\mathbf{u}^2) \qquad (3.1.1)$$

For a proof, see Higham (1996). It says that the computed solution exactly satisfies a slightly perturbed system. Moreover, each entry in the perturbing matrix F is small relative to the corresponding element of L.

3.1.2 Back Substitution

The analogous algorithm for upper triangular systems $Ux = b$ is called *back-substitution*. The recipe for x_i is prescribed by

$$x_i = \left(b_i - \sum_{j=i+1}^{n} u_{ij}x_j\right) / u_{ii}$$

and once again b_i can be overwritten by x_i.

Algorithm 3.1.2 (Back Substitution: Row Version) If $U \in \mathbb{R}^{n \times n}$ is upper triangular and $b \in \mathbb{R}^n$, then the following algorithm overwrites b with the solution to $Ux = b$. U is assumed to be nonsingular.

$b(n) = b(n)/U(n,n)$
for $i = n - 1{:} -1{:}1$
$\quad b(i) = (b(i) - U(i, i + 1{:}n)b(i + 1{:}n))/U(i, i)$
end

This algorithm requires n^2 flops and accesses U by row. The computed solution $\hat{x}$ obtained by the algorithm can be shown to satisfy

$$(U + F)\hat{x} = b \qquad |F| \leq nu|U| + O(\mathbf{u}^2). \qquad (3.1.2)$$

3.1.3   Column Oriented Versions

Column oriented versions of the above procedures can be obtained by reversing loop orders. To understand what this means from the algebraic
point of view, consider forward substitution. Once x_1 is resolved, it can
be removed from equations 2 through n and we proceed with the reduced
system $L(2{:}n, 2{:}n)x(2{:}n) = b(2{:}n) - x(1)L(2{:}n, 1)$. We then compute x_2 and
remove it from equations 3 through n, etc. Thus, if this approach is applied
to

$$\begin{bmatrix} 2 & 0 & 0 \\ 1 & 5 & 0 \\ 7 & 9 & 8 \end{bmatrix} \begin{bmatrix} x_1 \\ x_2 \\ x_3 \end{bmatrix} = \begin{bmatrix} 6 \\ 2 \\ 5 \end{bmatrix}$$

we find $x_1 = 3$ and then deal with the 2-by-2 system

$$\begin{bmatrix} 5 & 0 \\ 9 & 8 \end{bmatrix} \begin{bmatrix} x_2 \\ x_3 \end{bmatrix} = \begin{bmatrix} 2 \\ 5 \end{bmatrix} - 3 \begin{bmatrix} 1 \\ 7 \end{bmatrix} = \begin{bmatrix} -1 \\ -16 \end{bmatrix}.$$

Here is the complete procedure with overwriting.

Algorithm 3.1.3 (Forward Substitution: Column Version) If $L \in \mathbb{R}^{n \times n}$
is lower triangular and $b \in \mathbb{R}^n$, then this algorithm overwrites b with the
solution to $Lx = b$. L is assumed to be nonsingular.

> **for** $j = 1{:}n - 1$
> $\quad b(j) = b(j)/L(j,j)$
> $\quad b(j + 1{:}n) = b(j + 1{:}n) - b(j)L(j + 1{:}n, j)$
> **end**
> $b(n) = b(n)/L(n, n)$

It is also possible to obtain a column-oriented saxpy procedure for back-
substitution.

Algorithm 3.1.4 (Back Substitution: Column Version) If $U \in \mathbb{R}^{n \times n}$
is upper triangular and $b \in \mathbb{R}^n$, then this algorithm overwrites b with the
solution to $Ux = b$. U is assumed to be nonsingular.

> **for** $j = n{:} - 1{:}2$
> $\quad b(j) = b(j)/U(j,j)$
> $\quad b(1{:}j - 1) = b(1{:}j - 1) - b(j)U(1{:}j - 1, j)$
> **end**
> $b(1) = b(1)/U(1, 1)$

Note that the dominant operation in both Algorithms 3.1.3 and 3.1.4 is
the saxpy operation. The roundoff behavior of these saxpy implementations
is essentially the same as for the dot product versions.

The accuracy of a computed solution to a triangular system is often
surprisingly good. See Higham (1996).

3.1.4   Multiple Right Hand Sides

Consider the problem of computing a solution $X \in \mathbb{R}^{n \times q}$ to $LX = B$ where $L \in \mathbb{R}^{n \times n}$ is lower triangular and $B \in \mathbb{R}^{n \times q}$. This is the *multiple right hand side* forward substitution problem. We show that such a problem can be solved by a block algorithm that is rich in matrix multiplication assuming that q and n are large enough. This turns out to be important in subsequent sections where various block factorization schemes are discussed. We mention that although we are considering here just the lower triangular problem, everything we say applies to the upper triangular case as well.

To develop a block forward substitution algorithm we partition the equation $LX = B$ as follows:

$$
\begin{bmatrix}
L_{11} & 0 & \cdots & 0 \\
L_{21} & L_{22} & \cdots & 0 \\
\vdots & \vdots & \ddots & \vdots \\
L_{N1} & L_{N2} & \cdots & L_{NN}
\end{bmatrix}
\begin{bmatrix}
X_1 \\ X_2 \\ \vdots \\ X_N
\end{bmatrix}
=
\begin{bmatrix}
B_1 \\ B_2 \\ \vdots \\ B_N
\end{bmatrix}. \tag{3.1.3}
$$

Assume that the diagonal blocks are square. Paralleling the development of Algorithm 3.1.3, we solve the system $L_{11}X_1 = B_1$ for X_1 and then remove X_1 from block equations 2 through N:

$$
\begin{bmatrix}
L_{22} & 0 & \cdots & 0 \\
L_{32} & L_{33} & \cdots & 0 \\
\vdots & \vdots & \ddots & \vdots \\
L_{N2} & L_{N3} & \cdots & L_{NN}
\end{bmatrix}
\begin{bmatrix}
X_2 \\ X_3 \\ \vdots \\ X_N
\end{bmatrix}
=
\begin{bmatrix}
B_2 - L_{21}X_1 \\
B_3 - L_{31}X_1 \\
\vdots \\
B_N - L_{N1}X_1
\end{bmatrix}.
$$

Continuing in this way we obtain the following block saxpy forward elimination scheme:

$$
\begin{aligned}
&\textbf{for } j = 1{:}N \\
&\qquad \text{Solve } L_{jj}X_j = B_j \\
&\qquad \textbf{for } i = j + 1{:}N \\
&\qquad\qquad B_i = B_i - L_{ij}X_j \\
&\qquad \textbf{end} \\
&\textbf{end}
\end{aligned} \tag{3.1.4}
$$

Notice that the i-loop oversees a single block saxpy update of the form

$$
\begin{bmatrix}
B_{j+1} \\ \vdots \\ B_N
\end{bmatrix}
=
\begin{bmatrix}
B_{j+1} \\ \vdots \\ B_N
\end{bmatrix}
-
\begin{bmatrix}
L_{j+1,j} \\ \vdots \\ L_{N,j}
\end{bmatrix}
X_j .
$$

For this to be handled as a matrix multiplication in a given architecture it is clear that the blocking in (3.1.3) must give sufficiently "big" X_j. Let us assume that this is the case if each X_j has at least r rows. This can be accomplished if $N = \text{ceil}(n/r)$ and $X_1, \ldots, X_{N-1} \in \mathbb{R}^{r \times q}$ and $X_N \in \mathbb{R}^{(n-(N-1)r) \times q}$.

3.1.5   The Level-3 Fraction

It is handy to adopt a measure that quantifies the amount of matrix multi-plication in a given algorithm. To this end we define the *level-3 fraction* of an algorithm to be the fraction of flops that occur in the context of matrix multiplication. We call such flops *level-3 flops*.

Let us determine the level-3 fraction for (3.1.4) with the simplifying assumption that $n = rN$. (The same conclusions hold with the unequal blocking described above.) Because there are N applications of r-by-r forward elimination (the level-2 portion of the computation) and n^2 flops overall, the level-3 fraction is approximately given by

$$1 - \frac{Nr^2}{n^2} = 1 - \frac{1}{N}$$

Thus, for large N almost all flops are level-3 flops and it makes sense to choose N as large as possible subject to the constraint that the underlying architecture can achieve a high level of performance when processing block saxpy's of width at least $r = n/N$.

3.1.6   Non-square Triangular System Solving

The problem of solving nonsquare, m-by-n triangular systems deserves some mention. Consider first the lower triangular case when $m \geq n$, i.e.,

$$\begin{bmatrix} L_{11} \\ L_{21} \end{bmatrix} x = \begin{bmatrix} b_1 \\ b_2 \end{bmatrix} \qquad \begin{matrix} L_{11} \in \mathbb{R}^{n \times n} & b_1 \in \mathbb{R}^n \\ L_{21} \in \mathbb{R}^{(m-n) \times n} & b_2 \in \mathbb{R}^{m-n} \end{matrix}$$

Assume that L_{11} is lower triangular, and nonsingular. If we apply forward elimination to $L_{11}x = b_1$ then x solves the system provided $L_{21}(L_{11}^{-1}b_1) = b_2$. Otherwise, there is no solution to the overall system. In such a case least squares minimization may be appropriate. See Chapter 5.

Now consider the lower triangular system $Lx = b$ when the number of columns n exceeds the number of rows m. In this case apply forward substitution to the square system $L(1{:}m, 1{:}m)x(1{:}m, 1{:}m) = b$ and prescribe an arbitrary value for $x(m + 1{:}n)$. See §5.7 for additional comments on systems that have more unknowns than equations.

The handling of nonsquare upper triangular systems is similar. Details are left to the reader.

3.1.7   Unit Triangular Systems

A *unit* triangular matrix is a triangular matrix with ones on the diagonal. Many of the triangular matrix computations that follow have this added bit of structure. It clearly poses no difficulty in the above procedures.

3.1.8 The Algebra of Triangular Matrices

For future reference we list a few properties about products and inverses of triangular and unit triangular matrices.

- The inverse of an upper (lower) triangular matrix is upper (lower) triangular.

- The product of two upper (lower) triangular matrices is upper (lower) triangular.

- The inverse of a unit upper (lower) triangular matrix is unit upper (lower) triangular.

- The product of two unit upper (lower) triangular matrices is unit upper (lower) triangular.

Problems

P3.1.1 Give an algorithm for computing a nonzero $z \in \mathbf{R}^n$ such that $Uz = 0$ where $U \in \mathbf{R}^{n \times n}$ is upper triangular with $u_{nn} = 0$ and $u_{11} \cdots u_{n-1,n-1} \neq 0$.

P3.1.2 Discuss how the determinant of a square triangular matrix could be computed with minimum risk of overflow and underflow.

P3.1.3 Rewrite Algorithm 3.1.4 given that U is stored by column in a length $n(n+1)/2$ array u.vec.

P3.1.4 Write a detailed version of (3.1.4). Do not assume that N divides n.

P3.1.5 Prove all the facts about triangular matrices that are listed in §3.1.8.

P3.1.6 Suppose $S, T \in \mathbf{R}^{n \times n}$ are upper triangular and that $(ST - \lambda I)x = b$ is a nonsingular system. Give an $O(n^2)$ algorithm for computing x. Note that the explicit formation of $ST - \lambda I$ requires $O(n^3)$ flops. Hint. Suppose

$$
S_+ = \begin{bmatrix} \sigma & u^T \\ 0 & S_c \end{bmatrix}, \quad T_+ = \begin{bmatrix} \tau & v^T \\ 0 & T_c \end{bmatrix}, \quad b_+ = \begin{bmatrix} \beta \\ b_c \end{bmatrix}
$$

where $S_+ = S(k-1{:}n, k-1{:}n)$, $T_+ = T(k-1{:}n, k-1{:}n)$, $b_+ = b(k-1{:}n)$, and $\sigma, \tau, \beta \in \mathbf{R}$. Show that if we have a vector x_c such that

$$(S_c T_c - \lambda I)x_c = b_c$$

and $w_c = T_c x_c$ is available, then

$$
x_+ = \begin{bmatrix} \gamma \\ x_c \end{bmatrix} \qquad \gamma = \frac{\beta - \sigma v^T x_c - u^T w_c}{\sigma \tau - \lambda}
$$

solves $(S_+ T_+ - \lambda I)x_+ = b_+$. Observe that x_+ and $w_+ = T_+ x_+$ each require $O(n - k)$ flops.

P3.1.7 Suppose the matrices $R_1, \ldots, R_p \in \mathbf{R}^{n \times n}$ are all upper triangular. Give an $O(pn^2)$ algorithm for solving the system $(R_1 \cdots R_p - \lambda I)x = b$ assuming that the matrix of coefficients is nonsingular. Hint. Generalize the solution to the previous problem.

Notes and References for Sec. 3.1

The accuracy of triangular system solvers is analyzed in

N.J. Higham (1989). "The Accuracy of Solutions to Triangular Systems," *SIAM J. Num. Anal.* **26**, 1252–1265.

3.2   The LU Factorization

As we have just seen, triangular systems are "easy" to solve. The idea behind Gaussian elimination is to convert a given system $Ax = b$ to an equivalent triangular system. The conversion is achieved by taking appropriate linear combinations of the equations. For example, in the system

$$
\begin{aligned}
3x_1 + 5x_2 &= 9 \\
6x_1 + 7x_2 &= 4
\end{aligned}
$$

if we multiply the first equation by 2 and subtract it from the second we obtain

$$
\begin{aligned}
3x_1 + 5x_2 &= 9 \\
-3x_2 &= -14
\end{aligned}
$$

This is $n = 2$ Gaussian elimination. Our objective in this section is to give a complete specification of this central procedure and to describe what it does in the language of matrix factorizations. This means showing that the algorithm computes a unit lower triangular matrix L and an upper triangular matrix U so that $A = LU$, e.g.,

$$
\begin{bmatrix} 3 & 5 \\ 6 & 7 \end{bmatrix} = \begin{bmatrix} 1 & 0 \\ 2 & 1 \end{bmatrix} \begin{bmatrix} 3 & 5 \\ 0 & -3 \end{bmatrix} .
$$

The solution to the original $Ax = b$ problem is then found by a two step triangular solve process:

$$
Ly = b, \quad Ux = y \quad \Longrightarrow \quad Ax = LUx = Ly = b .
$$

The LU factorization is a "high-level" algebraic description of Gaussian elimination. Expressing the outcome of a matrix algorithm in the "language" of matrix factorizations is a worthwhile activity. It facilitates generalization and highlights connections between algorithms that may appear very different at the scalar level.

3.2.1   Gauss Transformations

To obtain a factorization description of Gaussian elimination we need a matrix description of the zeroing process. At the $n = 2$ level if $x_1 \neq 0$ and $\tau = x_2/x_1$, then

$$
\begin{bmatrix} 1 & 0 \\ -\tau & 1 \end{bmatrix} \begin{bmatrix} x_1 \\ x_2 \end{bmatrix} = \begin{bmatrix} x_1 \\ 0 \end{bmatrix}
$$

More generally, suppose $x \in \mathbb{R}^n$ with $x_k \neq 0$. If

$$
\tau^T = (\; \underbrace{0,\ldots,0}_{k}, \tau_{k+1}, \ldots, \tau_n) \qquad \tau_i = \frac{x_i}{x_k} \qquad i = k + 1{:}n
$$

and we define

$$M_k = I - \tau e_k^T, \tag{3.2.1}$$

then

$$
M_k x =
\begin{bmatrix}
1 & \cdots & 0 & 0 & \cdots & 0 \\
\vdots & \ddots & \vdots & \vdots & & \vdots \\
0 & & 1 & 0 & & 0 \\
0 & & -\tau_{k+1} & 1 & & 0 \\
\vdots & \vdots & \vdots & \vdots & \ddots & \vdots \\
0 & \cdots & -\tau_n & 0 & \cdots & 1
\end{bmatrix}
\begin{bmatrix}
x_1 \\ \vdots \\ x_k \\ x_{k+1} \\ \vdots \\ x_n
\end{bmatrix}
=
\begin{bmatrix}
x_1 \\ \vdots \\ x_k \\ 0 \\ \vdots \\ 0
\end{bmatrix}.
$$

In general, a matrix of the form $M_k = I - \tau e_k^T \in \mathbb{R}^{n \times n}$ is a *Gauss trans-formation* if the first k components of $\tau \in \mathbb{R}^n$ are zero. Such a matrix is unit lower triangular. The components of $\tau(k+1:n)$ are called *multipliers*. The vector τ is called the *Gauss vector*.

3.2.2 Applying Gauss Transformations

Multiplication by a Gauss transformation is particularly simple. If $C \in \mathbb{R}^{n \times r}$ and $M_k = I - \tau e_k^T$ is a Gauss transform, then

$$M_k C = (I - \tau e_k^T)C = C - \tau(e_k^T C) = C - \tau C(k,:).$$

is an outer product update. Since $\tau(1:k) = 0$ only $C(k+1:n,:)$ is affected and the update $C = M_k C$ can be computed row-by-row as follows:

> **for** $i = k+1:n$
> $\quad C(i,:) = C(i,:) - \tau_i C(k,:)$
> **end**

This computation requires $2(n-1)r$ flops.

Example 3.2.1

$$
C =
\begin{bmatrix}
1 & 4 & 7 \\
2 & 5 & 8 \\
3 & 6 & 10
\end{bmatrix},
\tau =
\begin{bmatrix}
0 \\ 1 \\ -1
\end{bmatrix}
\Rightarrow (I - \tau e_1^T)C =
\begin{bmatrix}
1 & 4 & 7 \\
1 & 1 & 1 \\
4 & 10 & 17
\end{bmatrix}.
$$

3.2.3 Roundoff Properties of Gauss Transforms

If $\hat{\tau}$ is the computed version of an exact Gauss vector τ, then it is easy to verify that

$$\hat{\tau} = \tau + e \qquad |e| \le \mathbf{u}|\tau|.$$

If $\hat{\tau}$ is used in a Gauss transform update and $fl((I - \hat{\tau}e_k^T)C)$ denotes the computed result, then

$$fl\left((I - \hat{\tau}e_k^T)C\right) = (I - \tau e_k^T)C + E,$$

where

$$|E| \leq 3\mathbf{u}\left(|C| + |\tau||C(k,:)|\right) + O(\mathbf{u}^2).$$

Clearly, if τ has large components, then the errors in the update may be large in comparison to $|C|$. For this reason, care must be exercised when Gauss transformations are employed, a matter that is pursued in §3.4.

3.2.4   Upper Triangularizing

Assume that $A \in \mathbb{R}^{n \times n}$. Gauss transformations $M_1, \ldots, M_{n-1}$ can usually be found such that $M_{n-1} \cdots M_2 M_1 A = U$ is upper triangular. To see this we first look at the $n = 3$ case. Suppose

$$A = \begin{bmatrix} 1 & 4 & 7 \\ 2 & 5 & 8 \\ 3 & 6 & 10 \end{bmatrix}.$$

If

$$M_1 = \begin{bmatrix} 1 & 0 & 0 \\ -2 & 1 & 0 \\ -3 & 0 & 1 \end{bmatrix},$$

then

$$M_1 A = \begin{bmatrix} 1 & 4 & 7 \\ 0 & -3 & -6 \\ 0 & -6 & -11 \end{bmatrix}.$$

Likewise,

$$M_2 = \begin{bmatrix} 1 & 0 & 0 \\ 0 & 1 & 0 \\ 0 & -2 & 1 \end{bmatrix} \quad \Rightarrow \quad M_2(M_1 A) = \begin{bmatrix} 1 & 4 & 7 \\ 0 & -3 & -6 \\ 0 & 0 & 1 \end{bmatrix}.$$

Extrapolating from this example observe that during the kth step

- We are confronted with a matrix $A^{(k-1)} = M_{k-1} \cdots M_1 A$ that is upper triangular in columns 1 to $k - 1$.

- The multipliers in M_k are based on $A^{(k-1)}(k+1{:}n, k)$. In particular, we need $a_{kk}^{(k-1)} \neq 0$ to proceed.

Noting that complete upper triangularization is achieved after $n - 1$ steps we therefore obtain

$$k = 1$$
$$\textbf{while } (A(k,k) \neq 0) \ \& \ (k \leq n-1)$$
$$\tau(k+1{:}n) = A(k+1{:}n,k)/A(k,k) \qquad\qquad (3.2.2)$$
$$A(k+1{:}n,:) = A(k+1{:}n,:) - \tau(k+1{:}n)A(k,:)$$
$$k = k+1$$
$$\textbf{end}$$

The entry $A(k,k)$ must be checked to avoid a zero divide. These quantities are referred to as the *pivots* and their relative magnitude turns out to be critically important.

3.2.5   The LU Factorization

In matrix language, if (3.2.2) terminates with $k = n$, then it computes Gauss transforms $M_1, \ldots, M_{n-1}$ such that $M_{n-1} \cdots M_1 A = U$ is upper triangular. It is easy to check that if $M_k = I - \tau^{(k)} e_k^T$, then its inverse is prescribed by $M_k^{-1} = I + \tau^{(k)} e_k^T$ and so

$$A = LU \qquad\qquad (3.2.3)$$

where

$$L = M_1^{-1} \cdots M_{n-1}^{-1}. \qquad\qquad (3.2.4)$$

It is clear that L is a unit lower triangular matrix because each M_k^{-1} is unit lower triangular. The factorization (3.2.3) is called the *LU factorization* of A.

As suggested by the need to check for zero pivots in (3.2.2), the LU factorization need not exist. For example, it is impossible to find l_{ij} and u_{ij} so

$$\begin{bmatrix} 1 & 2 & 3 \\ 2 & 4 & 7 \\ 3 & 5 & 3 \end{bmatrix} = \begin{bmatrix} 1 & 0 & 0 \\ \ell_{21} & 1 & 0 \\ \ell_{31} & \ell_{32} & 1 \end{bmatrix} \begin{bmatrix} u_{11} & u_{12} & u_{13} \\ 0 & u_{22} & u_{23} \\ 0 & 0 & u_{33} \end{bmatrix}.$$

To see this equate entries and observe that we must have $u_{11} = 1$, $u_{12} = 2$, $\ell_{21} = 2$, $u_{22} = 0$, and $\ell_{31} = 3$. But when we then look at the (3,2) entry we obtain the contradictory equation $5 = \ell_{31}u_{12} + \ell_{32}u_{22} = 6$.

As we now show, a zero pivot in (3.2.2) can be identified with a singular leading principal submatrix.

Theorem 3.2.1 $A \in \mathbb{R}^{n \times n}$ has an LU factorization if $\det(A(1{:}k, 1{:}k)) \neq 0$ for $k = 1{:}n-1$. If the LU factorization exists and A is nonsingular, then the LU factorization is unique and $\det(A) = u_{11} \cdots u_{nn}$.

Proof. Suppose $k-1$ steps in (3.2.2) have been executed. At the beginning of step k the matrix A has been overwritten by $M_{k-1} \cdots M_1 A = A^{(k-1)}$. Note that $a_{kk}^{(k-1)}$ is the kth pivot. Since the Gauss transformations are unit lower triangular it follows by looking at the leading k-by-k portion of this equation that $\det(A(1{:}k,1{:}k)) = a_{11}^{(k-1)}\cdots a_{kk}^{(k-1)}$. Thus, if $A(1{:}k,1{:}k)$ is nonsingular then the kth pivot is nonzero.

As for uniqueness, if $A = L_1 U_1$ and $A = L_2 U_2$ are two LU factorizations of a nonsingular A, then $L_2^{-1} L_1 = U_2 U_1^{-1}$. Since $L_2^{-1} L_1$ is unit lower triangular and $U_2 U_1^{-1}$ is upper triangular, it follows that both of these matrices must equal the identity. Hence, $L_1 = L_2$ and $U_1 = U_2$.

Finally, if $A = LU$ then $\det(A) = \det(LU) = \det(L)\det(U) = \det(U) = u_{11}\cdots u_{nn}$. $\square$

3.2.6   Some Practical Details

From the practical point of view there are several improvements that can be made to (3.2.2). First, because zeros have already been introduced in columns 1 through $k-1$, the Gauss transform update need only be applied to columns k through n. Of course, we need not even apply the kth Gauss transform to $A(:,k)$ since we know the result. So the efficient thing to do is simply to update $A(k+1{:}n, k+1{:}n)$. Another worthwhile observation is that the multipliers associated with M_k can be stored in the locations that they zero, i.e., $A(k+1{:}n, k)$. With these changes we obtain the following version of (3.2.2):

Algorithm 3.2.1 (Outer Product Gaussian Elimination) Suppose $A \in \mathbb{R}^{n \times n}$ has the property that $A(1{:}k, 1{:}k)$ is nonsingular for $k = 1{:}n-1$. This algorithm computes the factorization $M_{n-1}\cdots M_1 A = U$ where U is upper triangular and each M_k is a Gauss transform. U is stored in the upper triangle of A. The multipliers associated with M_k are stored in $A(k+1{:}n, k)$, i.e., $A(k+1{:}n, k) = -M_k(k+1{:}n, k)$.

> for $k = 1{:}n-1$
> $rows = k+1{:}n$
> $A(rows, k) = A(rows, k)/A(k, k)$
> $A(rows, rows) = A(rows, rows) - A(rows, k)A(k, rows)$
> end

This algorithm involves $2n^3/3$ flops and it is one of several formulations of *Gaussian Elimination*. Note that each pass through the k-loop involves an outer product.

3.2.7   Where is L?

Algorithm 3.2.3 represents L in terms of the multipliers. In particular, if $\tau^{(k)}$ is the vector of multipliers associated with M_k then upon termination, $A(k+1{:}n, k) = \tau^{(k)}$. One of the more happy "coincidences" in matrix computations is that if $L = M_1^{-1} \cdots M_{n-1}^{-1}$, then $L(k+1:n, k) = \tau^{(k)}$. This follows from a careful look at the product that defines L. Indeed,

$$L = \left(I + \tau^{(1)} e_1^T\right) \cdots \left(I + \tau^{(n-1)} e_{n-1}^T\right) = I + \sum_{k=1}^{n-1} \tau^{(k)} e_k^T .$$

Since $A(k+1:n, k)$ houses the kth vector of multipliers $\tau^{(k)}$, it follows that $A(i, k)$ houses ℓ_{ik} for all $i > k$.

3.2.8   Solving a Linear System

Once A has been factored via Algorithm 3.2.1, then L and U are represented in the array A. We can then solve the system $Ax = b$ via the triangular systems $Ly = b$ and $Ux = y$ by using the methods of §3.1.

Example 3.2.2 If Algorithm 3.2.1 is applied to

$$A = \begin{bmatrix} 1 & 4 & 7 \\ 2 & 5 & 8 \\ 3 & 6 & 10 \end{bmatrix} = \begin{bmatrix} 1 & 0 & 0 \\ 2 & 1 & 0 \\ 3 & 2 & 1 \end{bmatrix} \begin{bmatrix} 1 & 4 & 7 \\ 0 & -3 & -6 \\ 0 & 0 & 1 \end{bmatrix} ,$$

then upon completion,

$$A = \begin{bmatrix} 1 & 4 & 7 \\ 2 & -3 & -6 \\ 3 & 2 & 1 \end{bmatrix} .$$

If $b = (1, 1, 1)^T$, then $y = (1, -1, 0)^T$ solves $Ly = b$ and $x = (-1/3, 1/3, 0)^T$ solves $Ux = y$.

3.2.9   Other Versions

Gaussian elimination, like matrix multiplication, is a triple-loop procedure that can be arranged in several ways. Algorithm 3.2.1 corresponds to the "kij" version of Gaussian elimination if we compute the outer product update row-by-row:

$$
\begin{aligned}
&\textbf{for } k = 1{:}n - 1 \\
&\qquad A(k+1{:}n, k) = A(k+1{:}n, k)/A(k, k) \\
&\qquad \textbf{for } i = k + 1{:}n \\
&\qquad\qquad \textbf{for } j = k + 1{:}n \\
&\qquad\qquad\qquad A(i, j) = A(i, j) - A(i, k)A(k, j) \\
&\qquad\qquad \textbf{end} \\
&\qquad \textbf{end} \\
&\textbf{end}
\end{aligned}
$$

There are five other versions: kji, ikj, ijk, jik, and jki. The last of these results in an implementation that features a sequence of gaxpy's and forward eliminations. In this formulation, the Gauss transformations are not immediately applied to A as they are in the outer product version. Instead, their application is delayed. The original $A(:,j)$ is untouched until step j. At that point in the algorithm $A(:,j)$ is overwritten by $M_{j-1}\cdots M_1 A(:,j)$. The jth Gauss transformation is then computed.

To be precise, suppose $1 \leq j \leq n-1$ and assume that $L(:,1{:}j-1)$ and $U(1{:}j-1,1{:}j-1)$ are known. This means that the first $j-1$ columns of L and U are available. To get the jth columns of L and U we equate jth columns in the equation $A = LU$: $A(:,j) = LU(:,j)$. From this we conclude that

$$A(1{:}j-1,j) = L(1{:}j-1,1{:}j-1)U(1{:}j-1,j)$$

and

$$A(j{:}n,j) = \sum_{k=1}^{j} L(j{:}n,k)U(k,j).$$

The first equation is a lower triangular system that can be solved for the vector $U(1{:}j-1,j)$. Once this is accomplished, the second equation can be rearranged to produce recipes for $U(j,j)$ and $L(j+1{:}n,j)$. Indeed, if we set

$$\begin{aligned} v(j{:}n) &= A(j{:}n,j) - \sum_{k=1}^{j-1} L(j{:}n,k)U(k,j) \\ &= A(j{:}n,j) - L(j{:}n,1{:}j-1)U(1{:}j-1,j), \end{aligned}$$

then $L(j+1{:}n,j) = v(j+1{:}n)/v(j)$ and $U(j,j) = v(j)$. Thus, $L(j+1{:}n,j)$ is a scaled gaxpy and we obtain

$L = I;\ U = 0$
for $j = 1{:}n$
    if $j = 1$
        $v(j{:}n) = A(j{:}n,j)$
    else
        Solve $L(1{:}j-1,1{:}j-1)z = A(1{:}j-1,j)$ for z (3.2.5)
        and set $U(1{:}j-1,j) = z$.
        $v(j{:}n) = A(j{:}n,j) - L(j{:}n,1{:}j-1)z$
    end
    if $j < n$
        $L(j+1{:}n,j) = v(j+1{:}n)/v(j)$
    end
    $U(j,j) = v(j)$
end

This arrangement of Gaussian elimination is rich in forward eliminations and gaxpy operations and, like Algorithm 3.2.1, requires $2n^3/3$ flops.

3.2.10   Block LU

It is possible to organize Gaussian elimination so that matrix multiplication becomes the dominant operation. The key to the derivation of this block procedure is to partition $A \in \mathbb{R}^{n \times n}$ as follows

$$
A = \begin{bmatrix} A_{11} & A_{12} \\ A_{21} & A_{22} \end{bmatrix} \begin{matrix} r \\ n-r \end{matrix}
$$
$$
\phantom{A = \begin{bmatrix}} r \quad n-r
$$

where r is a blocking parameter. Suppose we compute the LU factorization $L_{11}U_{11} = A_{11}$ and then solve the multiple right hand side triangular systems $L_{11}U_{12} = A_{12}$ and $L_{21}U_{11} = A_{21}$ for U_{12} and L_{21} respectively. It follows that

$$
\begin{bmatrix} A_{11} & A_{12} \\ A_{21} & A_{22} \end{bmatrix} = \begin{bmatrix} L_{11} & 0 \\ L_{21} & I_{n-r} \end{bmatrix} \begin{bmatrix} I_r & 0 \\ 0 & \tilde{A} \end{bmatrix} \begin{bmatrix} U_{11} & U_{12} \\ 0 & I_{n-r} \end{bmatrix}
$$

where $\tilde{A} = A_{22} - L_{21}U_{12}$. The matrix $\tilde{A}$ is the *Schur complement* of A_{11} with respect to A. Note that if $\tilde{A} = L_{22}U_{22}$ is the LU factorization of $\tilde{A}$, then

$$
\begin{bmatrix} A_{11} & A_{12} \\ A_{21} & A_{22} \end{bmatrix} = \begin{bmatrix} L_{11} & 0 \\ L_{21} & L_{22} \end{bmatrix} \begin{bmatrix} I_r & 0 \\ 0 & \tilde{A} \end{bmatrix} \begin{bmatrix} U_{11} & U_{12} \\ 0 & U_{22} \end{bmatrix}
$$

is the LU factorization of A. Thus, after L_{11}, L_{21}, U_{11} and U_{22}, are computed, we repeat the process on the level-3 updated (2,2) block $\tilde{A}$.

Algorithm   3.2.2 (Block Outer Product LU)   Suppose $A \in \mathbb{R}^{n \times n}$ and that $\det(A(1:k, 1:k)$ is nonzero for $k = 1:n-1$. Assume that r satisfies $1 \le r \le n$. The following algorithm computes $A = LU$ via rank r updates. Upon completion, $A(i,j)$ is overwritten with $L(i,j)$ for $i > j$ and $A(i,j)$ is overwritten with $U(i,j)$ if $j \ge i$.

> $\lambda = 1$
> **while** $\lambda \le n$
> > $\mu = \min(n, \lambda + r - 1)$
> > Use Algorithm 3.2.1 to overwrite $A(\lambda:\mu, \lambda:\mu)$
> > > with its LU factors $\tilde{L}$ and $\tilde{U}$.
> >
> > Solve $\tilde{L}Z = A(\lambda:\mu, \mu+1:n)$ for Z and overwrite
> > > $A(\lambda:\mu, \mu+1:n)$ with Z.
> >
> > Solve $W\tilde{U} = A(\mu+1:n, \lambda:\mu)$ for W and overwrite
> > > $A(\mu+1:n, \lambda:\mu)$ with W.
> >
> > $A(\mu+1:n, \mu+1:n) = A(\mu+1:n, \mu+1:n) - WZ$
> > $\lambda = \mu + 1$
>
> **end**

This algorithm involves $2n^3/3$ flops.

Recalling the discussion in §3.1.5, let us consider the level-3 fraction for this procedure assuming that r is large enough so that the underlying computer is able to compute the matrix multiply update $A(\mu + 1{:}n, \mu + 1{:}n) = A(\mu + 1{:}n, \mu + 1{:}n) - WZ$ at "level-3 speed." Assume for clarity that $n = rN$. The only flops that are not level-3 flops occur in the context of the r-by-r LU factorizations $A(\lambda{:}\mu, \lambda{:}\mu) = \tilde{L}\tilde{U}$. Since there are N such systems solved in the overall computation, we see that the level-3 fraction is given by

$$1 - \frac{N(2r^3/3)}{2n^3/3} = 1 - \frac{1}{N^2}.$$

Thus, for large N almost all arithmetic takes place in the context of matrix multiplication. As we have mentioned, this ensures high performance on a wide range of computing environments.

3.2.11   The LU Factorization of a Rectangular Matrix

The LU factorization of a rectangular matrix $A \in \mathbb{R}^{m \times n}$ can also be performed. The $m > n$ case is illustrated by

$$\begin{bmatrix} 1 & 2 \\ 3 & 4 \\ 5 & 6 \end{bmatrix} = \begin{bmatrix} 1 & 0 \\ 3 & 1 \\ 5 & 2 \end{bmatrix} \begin{bmatrix} 1 & 2 \\ 0 & -2 \end{bmatrix}$$

while

$$\begin{bmatrix} 1 & 2 & 3 \\ 4 & 5 & 6 \end{bmatrix} = \begin{bmatrix} 1 & 0 \\ 4 & 1 \end{bmatrix} \begin{bmatrix} 1 & 2 & 3 \\ 0 & -3 & -6 \end{bmatrix}$$

depicts the $m < n$ situation. The LU factorization of $A \in \mathbb{R}^{m \times n}$ is guaranteed to exist if $A(1{:}k, 1{:}k)$ is nonsingular for $k = 1{:}\min(m, n)$.

The square LU factorization algorithms above need only minor modification to handle the rectangular case. For example, to handle the $m > n$ case we modify Algorithm 3.2.1 as follows:

> **for** $k = 1{:}n$
> > $rows = k + 1{:}m$
> > $A(rows, k) = A(rows, k)/A(k, k)$
> > **if** $k < n$
> > > $cols = k + 1{:}n$
> > > $A(rows, cols) = A(rows, cols) - A(rows, k)A(k, cols)$
> >
> > **end**
>
> **end**

This algorithm requires $mn^2 - n^3/3$ flops.

3.2.12   A Note on Failure

As we know, Gaussian elimination fails unless the first $n - 1$ principal submatrices are nonsingular. This rules out some very simple matrices, e.g.,

$$A = \begin{bmatrix} 0 & 1 \\ 1 & 0 \end{bmatrix}.$$

While A has perfect 2-norm condition, it fails to have an LU factorization because it has a singular leading principal submatrix.

Clearly, modifications are necessary if Gaussian elimination is to be effectively used in general linear system solving. The error analysis in the following section suggests the needed modifications.

Problems

P3.2.1 Suppose the entries of $A(\epsilon) \in \mathbb{R}^{n \times n}$ are continuously differentiable functions of the scalar ϵ. Assume that $A \equiv A(0)$ and all its principal submatrices are nonsingular. Show that for sufficiently small ϵ, the matrix $A(\epsilon)$ has an LU factorization $A(\epsilon) = L(\epsilon)U(\epsilon)$ and that $L(\epsilon)$ and $U(\epsilon)$ are both continuously differentiable.

P3.2.2 Suppose we partition $A \in \mathbb{R}^{n \times n}$

$$A = \begin{bmatrix} A_{11} & A_{12} \\ A_{21} & A_{22} \end{bmatrix}$$

where A_{11} is r-by-r. Assume that A_{11} is nonsingular. The matrix $S = A_{22} - A_{21}A_{11}^{-1}A_{12}$ is called the *Schur complement* of A_{11} in A. Show that if A_{11} has an LU factorization, then after r steps of Algorithm 3.2.1, $A(r + 1{:}n, r + 1{:}n)$ houses S. How could S be obtained after r steps of (3.2.5)?

P3.2.3 Suppose $A \in \mathbb{R}^{n \times n}$ has an LU factorization. Show how $Ax = b$ can be solved without storing the multipliers by computing the LU factorization of the n-by-$(n + 1)$ matrix $[A\ b]$.

P3.2.4 Describe a variant of Gaussian elimination that introduces zeros into the columns of A in the order, $n{:}-1{:}2$ and which produces the factorization $A = UL$ where U is unit upper triangular and L is lower triangular.

P3.2.5 Matrices in $\mathbb{R}^{n \times n}$ of the form $N(y, k) = I - ye_k^T$ where $y \in \mathbb{R}^n$ are said to be *Gauss-Jordan transformations*. (a) Give a formula for $N(y, k)^{-1}$ assuming it exists. (b) Given $x \in \mathbb{R}^n$, under what conditions can y be found so $N(y, k)x = e_k$? (c) Give an algorithm using Gauss-Jordan transformations that overwrites A with A^{-1}. What conditions on A ensure the success of your algorithm?

P3.2.6 Extend (3.2.5) so that it can also handle the case when A has more rows than columns.

P3.2.7 Show how A can be overwritten with L and U in (3.2.5). Organize the three loops so that unit stride access prevails.

P3.2.8 Develop a version of Gaussian elimination in which the innermost of the three loops oversees a dot product.

Notes and References for Sec. 3.2

Schur complements (P3.2.2) arise in many applications. For a survey of both practical and theoretical interest, see

R.W. Cottle (1974). "Manifestations of the Schur Complement," *Lin. Alg. and Its Applic. 8*, 189-211.

Schur complements are known as "Gauss transforms" in some application areas. The use of Gauss-Jordan transformations (P3.2.5) is detailed in Fox (1964). See also

T. Dekker and W. Hoffman (1989). "Rehabilitation of the Gauss-Jordan Algorithm," *Numer. Math. 54*, 591-599.

As we mentioned, inner product versions of Gaussian elimination have been known and used for some time. The names of Crout and Doolittle are associated with these ijk techniques. They were popular during the days of desk calculators because there are far fewer intermediate results than in Gaussian elimination. These methods still have attraction because they can be implemented with accumulated inner products. For remarks along these lines see Fox (1964) as well as Stewart (1973, pp. 131–39). See also:

G.E. Forsythe (1960). "Crout with Pivoting," *Comm. ACM 3*, 507-8.
W.M. McKeeman (1962). "Crout with Equilibration and Iteration," *Comm. ACM. 5*, 553-55.

Loop orderings and block issues in LU computations are discussed in

J.J. Dongarra, F.G. Gustavson, and A. Karp (1984). "Implementing Linear Algebra Algorithms for Dense Matrices on a Vector Pipeline Machine," *SIAM Review 26*, 91–112.
J.M. Ortega (1988). "The ijk Forms of Factorization Methods I: Vector Computers," *Parallel Computers 7*, 135–147.
D.H. Bailey, K.Lee, and H.D. Simon (1991). "Using Strassen's Algorithm to Accelerate the Solution of Linear Systems," *J. Supercomputing 4*, 357–371.
J.W. Demmel, N.J. Higham, and R.S. Schreiber (1995). "Stability of Block LU Factorization," *Numer. Lin. Alg. with Applic. 2*, 173–190.

3.3   Roundoff Analysis of Gaussian Elimination

We now assess the effect of rounding errors when the algorithms in the previous two sections are used to solve the linear system $Ax = b$. A much more detailed treatment of roundoff error in Gaussian elimination is given in Higham (1996).

Before we proceed with the analysis, it is useful to consider the nearly ideal situation in which no roundoff occurs during the entire solution process except when A and b are stored. Thus, if $fl(b) = b+e$ and the stored matrix $fl(A) = A + E$ is nonsingular, then we are assuming that the computed solution $\hat{x}$ satisfies

$$(A + E)\hat{x} = (b + e) \qquad \| E \|_\infty \leq \mathbf{u} \| A \|_\infty, \quad \| e \|_\infty \leq \mathbf{u} \| b \|_\infty . \qquad (3.3.1)$$

That is, $\hat{x}$ solves a "nearby" system exactly. Moreover, if $\mathbf{u}\kappa_\infty(A) \leq \frac{1}{2}$ (say), then by using Theorem 2.7.2, it can be shown that

$$\frac{\|\, x - \hat{x}\, \|_\infty}{\|\, x\, \|_\infty} \leq 4\mathbf{u}\kappa_\infty(A)\,. \tag{3.3.2}$$

The bounds (3.3.1) and (3.3.2) are "best possible" norm bounds. No general ∞-norm error analysis of a linear equation solver that requires the storage of A and b can render sharper bounds. As a consequence, we cannot justifiably criticize an algorithm for returning an inaccurate $\hat{x}$ if A is ill-conditioned relative to the machine precision, e.g., $\mathbf{u}\kappa_\infty(A) \approx 1$.

3.3.1 Errors in the LU Factorization

Let us see how the error bounds for Gaussian elimination compare with the ideal bounds above. We work with the infinity norm for convenience and focus our attention on Algorithm 3.2.3, the outer product version. The error bounds that we derive also apply to Algorithm 3.2.4, the gaxpy formulation.

Our first task is to quantify the roundoff errors associated with the computed triangular factors.

Theorem 3.3.1 *Assume that A is an n-by-n matrix of floating point numbers. If no zero pivots are encountered during the execution of Algorithm 3.2.3, then the computed triangular matrices $\hat{L}$ and $\hat{U}$ satisfy*

$$\hat{L}\hat{U} \;=\; A + H \tag{3.3.3}$$

$$|H| \;\leq\; 3(n-1)\mathbf{u}\left(|A| + |\hat{L}||\hat{U}|\right) + O(\mathbf{u}^2)\,. \tag{3.3.4}$$

Proof. The proof is by induction on n. The theorem obviously holds for $n = 1$. Assume it holds for all $(n-1)$-by-$(n-1)$ floating point matrices. If

$$A \;=\; \begin{array}{c} \\ \left[\begin{array}{cc} \alpha & w^T \\ v & B \end{array}\right] \\ \begin{array}{cc} 1 & n-1 \end{array} \end{array} \begin{array}{c} 1 \\ n-1 \end{array}$$

then $\hat{z} = fl(v/\alpha)$ and $\hat{A}_1 = fl(B - \hat{z}w^T)$ are computed in the first step of the algorithm. We therefore have

$$\hat{z} \;=\; \frac{1}{\alpha}v + f \qquad |f| \leq \mathbf{u}\frac{|v|}{|\alpha|} \tag{3.3.5}$$

and

$$\hat{A}_1 \;=\; B - \hat{z}w^T + F \qquad |F| \leq 2\mathbf{u}\left(|B| + |\hat{z}||w|^T\right) + O(\mathbf{u}^2)\,. \tag{3.3.6}$$

The algorithm now proceeds to calculate the LU factorization of $\hat{A}_1$. By induction, we compute approximate factors $\hat{L}_1$ and $\hat{U}_1$ for $\hat{A}_1$ that satisfy

$$\hat{L}_1\hat{U}_1 \; = \; \hat{A}_1 + H_1 \tag{3.3.7}$$

$$|H_1| \; \leq \; 3(n-2)\mathbf{u}\left(|\hat{A}_1| + |\hat{L}_1||\hat{U}_1|\right) + O(\mathbf{u}^2). \tag{3.3.8}$$

Thus,

$$\begin{aligned}
\hat{L}\hat{U} \; &\equiv \; \begin{bmatrix} 1 & 0 \\ \hat{z} & \hat{L}_1 \end{bmatrix}\begin{bmatrix} \alpha & w^T \\ 0 & \hat{U}_1 \end{bmatrix} \\
&= \; A + \begin{bmatrix} 0 & 0 \\ \alpha f & H_1 + F \end{bmatrix} \equiv A + H.
\end{aligned}$$

From (3.3.6) it follows that

$$|\hat{A}_1| \; \leq \; (1+2\mathbf{u})\left(|B| + |\hat{z}||w|^T\right) + O(\mathbf{u}^2),$$

and therefore by using (3.3.7) and (3.3.8) we have

$$|H_1 + F| \; \leq \; 3(n-1)\mathbf{u}\left(|B| + |\hat{z}||w|^T + |\hat{L}_1||\hat{U}_1|\right) \quad + \quad O(\mathbf{u}^2).$$

Since $|\alpha f| \leq \mathbf{u}|v|$ it is easy to verify that

$$|H| \; \leq \; 3(n-1)\mathbf{u}\left\{\begin{bmatrix} |\alpha| & |w|^T \\ |v| & |B| \end{bmatrix} + \begin{bmatrix} 1 & 0 \\ |\hat{z}| & |\hat{L}_1| \end{bmatrix}\begin{bmatrix} |\alpha| & |w|^T \\ 0 & |\hat{U}_1| \end{bmatrix}\right\} + O(\mathbf{u}^2)$$

thereby proving the theorem. □

We mention that if A is m-by-n, then the theorem applies with n in (3.3.4) replaced by the smaller of n and m.

3.3.2   Triangular Solving with Inexact Triangles

We next examine the effect of roundoff error when $\hat{L}$ and $\hat{U}$ are used by the triangular system solvers of §3.1.

Theorem 3.3.2 *Let $\hat{L}$ and $\hat{U}$ be the computed LU factors of the n-by-n floating point matrix A obtained by either Algorithm 3.2.3 or 3.2.4. Suppose the methods of §3.1 are used to produce the computed solution $\hat{y}$ to $\hat{L}y = b$ and the computed solution $\hat{x}$ to $\hat{U}x = \hat{y}$. Then $(A + E)\hat{x} = b$ with*

$$|E| \leq n\mathbf{u}\left(3|A| + 5|\hat{L}||\hat{U}|\right) + O(\mathbf{u}^2). \tag{3.3.9}$$

Proof. From (3.1.1) and (3.1.2) we have

$$(\hat{L} + F)\hat{y} = b \qquad |F| \leq n\mathbf{u}|\hat{L}| + O(\mathbf{u}^2)$$
$$(\hat{U} + G)\hat{x} = \hat{y} \qquad |G| \leq n\mathbf{u}|\hat{U}| + O(\mathbf{u}^2)$$

and thus

$$(\hat{L} + F)(\hat{U} + G)\hat{x} = (\hat{L}\hat{U} + F\hat{U} + \hat{L}G + FG)\hat{x} = b.$$

From Theorem 3.3.1

$$\hat{L}\hat{U} = A + H,$$

with $|H| \leq 3(n-1)\mathbf{u}(|A| + |\hat{L}||\hat{U}|) + O(\mathbf{u}^2)$, and so by defining

$$E = H + F\hat{U} + \hat{L}G + FG$$

we find $(A + E)\hat{x} = b$. Moreover,

$$\begin{aligned} |E| &\leq |H| + |F|\,|\hat{U}| + |\hat{L}|\,|G| + O(\mathbf{u}^2) \\ &\leq 3n\mathbf{u}\left(|A| + |\hat{L}||\hat{U}|\right) + 2n\mathbf{u}\left(|\hat{L}||\hat{U}|\right) + O(\mathbf{u}^2). \quad \square \end{aligned}$$

Were it not for the possibility of a large $|\hat{L}||\hat{U}|$ term, (3.3.9) would compare favorably with the ideal bound in (3.3.1). (The factor n is of no consequence, cf. the Wilkinson quotation in §2.4.6.) Such a possibility exists, for there is nothing in Gaussian elimination to rule out the appearance of small pivots. If a small pivot is encountered, then we can expect large numbers to be present in $\hat{L}$ and $\hat{U}$.

We stress that small pivots are not necessarily due to ill-conditioning as the example $A = \begin{bmatrix} \epsilon & 1 \\ 1 & 0 \end{bmatrix}$ bears out. Thus, Gaussian elimination can give arbitrarily poor results, even for well-conditioned problems. The method is unstable.

In order to repair this shortcoming of the algorithm, it is necessary to introduce row and/or column interchanges during the elimination process with the intention of keeping the numbers that arise during the calculation suitably bounded. This idea is pursued in the next section.

Example 3.3.1 Suppose $\beta = 10, t = 3$, floating point arithmetic is used to solve:

$$\begin{bmatrix} .001 & 1.00 \\ 1.00 & 2.00 \end{bmatrix}\begin{bmatrix} x_1 \\ x_2 \end{bmatrix} = \begin{bmatrix} 1.00 \\ 3.00 \end{bmatrix}.$$

Applying Gaussian elimination we get

$$\hat{L} = \begin{bmatrix} 1 & 0 \\ 1000 & 1 \end{bmatrix} \qquad \hat{U} = \begin{bmatrix} .001 & 1 \\ 0 & -1000 \end{bmatrix}$$

and a calculation shows

$$\hat{L}\hat{U} = \begin{bmatrix} .001 & 1 \\ 1 & 2 \end{bmatrix} + \begin{bmatrix} 0 & 0 \\ 0 & -2 \end{bmatrix} \equiv A + H.$$

Moreover, $6 \begin{bmatrix} 10^{-6} & .001 \\ 10^{-3} & 1.0001 \end{bmatrix}$ is the bounding matrix in (3.3.4), not a severe overesti-mate of $|H|$. If we go on to solve the problem using the triangular system solvers of §3.1, then using the same precision arithmetic we obtain a computed solution $\hat{x} = (0,1)^T$. This is in contrast to the exact solution $x = (1.002..., .998...)^T$.

Problems

P3.3.1 Show that if we drop the assumption that A is a floating point matrix in Theorem 3.3.1, then (3.3.4) holds with the coefficient "3"replaced by "4."

P3.3.2 Suppose A is an n-by-n matrix and that $\hat{L}$ and $\hat{U}$ are produced by Algorithm 3.2.1. (a) How many flops are required to compute $\| \, |\hat{L}| \, |\hat{U}| \, \|_\infty$? (b) Show $fl(|\hat{L}||\hat{U}|) \leq (1+2n\mathbf{u})|\hat{L}||\hat{U}| + O(\mathbf{u}^2)$.

P3.3.3 Suppose $x = A^{-1}b$. Show that if $e = x - \hat{x}$ (the error) and $r = b - A\hat{x}$ (the residual), then

$$\frac{\| \, r \, \|}{\| \, A \, \|} \leq \| \, e \, \| \leq \| \, A^{-1} \, \| \, \| \, r \, \|.$$

Assume consistency between the matrix and vector norm.

P3.3.4 Using 2-digit, base 10, floating point arithmetic, compute the LU factorization of

$$A = \begin{bmatrix} 7 & 6 \\ 9 & 8 \end{bmatrix}.$$

For this example, what is the matrix H in (3.3.3)?

Notes and References for Sec. 3.3

The original roundoff analysis of Gaussian elimination appears in

J.H. Wilkinson (1961). "Error Analysis of Direct Methods of Matrix Inversion," *J. ACM* *8*, 281-330.

Various improvements in the bounds and simplifications in the analysis have occurred over the years. See

B.A. Chartres and J.C. Geuder (1967). "Computable Error Bounds for Direct Solution of Linear Equations," *J. ACM 14*, 63–71.

J.K. Reid (1971). "A Note on the Stability of Gaussian Elimination," *J. Inst. Math. Applic. 8*, 374–75.

C.C. Paige (1973). "An Error Analysis of a Method for Solving Matrix Equations," *Math. Comp. 27*, 355–59.

C. de Boor and A. Pinkus (1977). "A Backward Error Analysis for Totally Positive Linear Systems," *Numer. Math. 27*, 485-90.

H.H. Robertson (1977). "The Accuracy of Error Estimates for Systems of Linear Algebraic Equations," *J. Inst. Math. Applic. 20*, 409–14.

J.J. Du Croz and N.J. Higham (1992). "Stability of Methods for Matrix Inversion," *IMA J. Num. Anal. 12*, 1–19.

3.4 Pivoting

The analysis in the previous section shows that we must take steps to ensure that no large entries appear in the computed triangular factors $\hat{L}$ and $\hat{U}$. The example

$$A = \begin{bmatrix} .0001 & 1 \\ 1 & 1 \end{bmatrix} = \begin{bmatrix} 1 & 0 \\ 10,000 & 1 \end{bmatrix} \begin{bmatrix} .0001 & 1 \\ 0 & -9999 \end{bmatrix} = LU$$

correctly identifies the source of the difficulty: relatively small pivots. A way out of this difficulty is to interchange rows. In our example, if P is the permutation

$$P = \begin{bmatrix} 0 & 1 \\ 1 & 0 \end{bmatrix}$$

then

$$PA = \begin{bmatrix} 1 & 1 \\ .0001 & 1 \end{bmatrix} = \begin{bmatrix} 1 & 0 \\ .0001 & 1 \end{bmatrix} \begin{bmatrix} 1 & 1 \\ 0 & .9999 \end{bmatrix} = LU .$$

Now the triangular factors are comprised of acceptably small elements.

In this section we show how to determine a permuted version of A that has a reasonably stable LU factorization. There are several ways to do this and they each correspond to a different pivoting strategy. We focus on partial pivoting and complete pivoting. The efficient implementation of these strategies and their properties are discussed. We begin with a discussion of permutation matrix manipulation.

3.4.1 Permutation Matrices

The stabilizations of Gaussian elimination that are developed in this section involve data movements such as the interchange of two matrix rows. In keeping with our desire to describe all computations in "matrix terms," it is necessary to acquire a familiarity with *permutation matrices*. A permutation matrix is just the identity with its rows re-ordered, e.g.,

$$P = \begin{bmatrix} 0 & 0 & 0 & 1 \\ 1 & 0 & 0 & 0 \\ 0 & 0 & 1 & 0 \\ 0 & 1 & 0 & 0 \end{bmatrix} .$$

An n-by-n permutation matrix should never be explicitly stored. It is much more efficient to represent a general permutation matrix P with an integer n-vector p. One way to do this is to let $p(k)$ be the column index of the sole "1" in P's kth row. Thus, $p = [4\,1\,3\,2]$ is the appropriate encoding of the above P. It is also possible to encode P on the basis of where the "1" occurs in each column, e.g., $p = [2\,4\,3\,1]$.

If P is a permutation and A is a matrix, then PA is a row permuted version of A and AP is a column permuted version of A. Permutation matrices are orthogonal and so if P is a permutation, then $P^{-1} = P^T$. A product of permutation matrices is a permutation matrix.

In this section we are particularly interested in *interchange permutations*. These are permutations obtained by merely swapping two rows in the identity, e.g.,

$$
E \; = \; \begin{bmatrix} 0 & 0 & 0 & 1 \\ 0 & 1 & 0 & 0 \\ 0 & 0 & 1 & 0 \\ 1 & 0 & 0 & 0 \end{bmatrix}.
$$

Interchange permutations can be used to describe row and column swapping. With the above 4-by-4 example, EA is A with rows 1 and 4 interchanged. Likewise, AE is A with columns 1 and 4 swapped.

If $P = E_n \cdots E_1$ and each E_k is the identity with rows k and $p(k)$ interchanged, then $p(1{:}n)$ is a useful vector encoding of P. Indeed, $x \in \mathbb{R}^n$ can be overwritten by Px as follows:

> **for** $k = 1{:}n$
> $\qquad x(k) \leftrightarrow x(p(k))$
> **end**

Here, the "$\leftrightarrow$" notation means "swap contents." Since each E_k is symmetric and $P^T = E_1 \cdots E_n$, the representation can also be used to overwrite x with $P^T x$:

> **for** $k = n{:} - 1{:}1$
> $\qquad x(k) \leftrightarrow x(p(k))$
> **end**

It should be noted that no floating point arithmetic is involved in a permutation operation. However, permutation matrix operations often involve the irregular movement of data and can represent a significant computational overhead.

3.4.2   Partial Pivoting: The Basic Idea

We show how interchange permutations can be used in LU computations to guarantee that no multiplier is greater than one in absolute value. Suppose

$$
A \; = \; \begin{bmatrix} 3 & 17 & 10 \\ 2 & 4 & -2 \\ 6 & 18 & -12 \end{bmatrix}.
$$

To get the smallest possible multipliers in the first Gauss transform using row interchanges we need a_{11} to be the largest entry in the first column. Thus, if E_1 is the interchange permutation

$$E_1 = \begin{bmatrix} 0 & 0 & 1 \\ 0 & 1 & 0 \\ 1 & 0 & 0 \end{bmatrix}$$

then

$$E_1 A = \begin{bmatrix} 6 & 18 & -12 \\ 2 & 4 & -2 \\ 3 & 17 & 10 \end{bmatrix}$$

and

$$M_1 = \begin{bmatrix} 1 & 0 & 0 \\ -1/3 & 1 & 0 \\ -1/2 & 0 & 1 \end{bmatrix} \implies M_1 E_1 A = \begin{bmatrix} 6 & 18 & -12 \\ 0 & -2 & 2 \\ 0 & 8 & 16 \end{bmatrix}.$$

Now to get the smallest possible multiplier in M_2 we need to swap rows 2 and 3. Thus, if

$$E_2 = \begin{bmatrix} 1 & 0 & 0 \\ 0 & 0 & 1 \\ 0 & 1 & 0 \end{bmatrix} \quad \text{and} \quad M_2 = \begin{bmatrix} 1 & 0 & 0 \\ 0 & 1 & 0 \\ 0 & 1/4 & 1 \end{bmatrix}$$

then

$$M_2 E_2 M_1 E_1 A = \begin{bmatrix} 6 & 18 & -12 \\ 0 & 8 & 16 \\ 0 & 0 & 6 \end{bmatrix}.$$

The example illustrates the basic idea behind the row interchanges. In general we have:

> **for** $k = 1{:}n - 1$
>> Determine an interchange matrix E_k with $E_k(1{:}k, 1{:}k) = I_k$
>> such that if z is the kth column of $E_k A$, then
>> $|z(k)| = \| z(k{:}n) \|_\infty$.
>> $A = E_k A$
>> Determine the Gauss transform M_k such that if v is the
>> kth column of $M_k A$, then $v(k + 1{:}n) = 0$.
>> $A = M_k A$
> **end**

This particular row interchange strategy is called *partial pivoting*. Upon completion we emerge with $M_{n-1} E_{n-1} \cdots M_1 E_1 A = U$, an upper triangular matrix.

As a consequence of the partial pivoting, no multiplier is larger than one in absolute value. This is because

$$|(E_k M_{k-1} \cdots M_1 E_1 A)_{kk}| = \max_{k \le i \le n} |(E_k M_{k-1} \cdots M_1 E_1 A)_{ik}|$$

for $k = 1{:}n-1$. Thus, partial pivoting effectively guards against arbitrarily large multipliers.

3.4.3   Partial Pivoting Details

We are now set to detail the overall Gaussian Elimination with partial pivoting algorithm.

Algorithm 3.4.1 (Gauss Elimination with Partial Pivoting)   If $A \in \mathbb{R}^{n \times n}$, then this algorithm computes Gauss transforms $M_1, \cdots M_{n-1}$ and interchange permutations $E_1, \cdots E_{n-1}$ such that $M_{n-1} E_{n-1} \cdots M_1 E_1 A = U$ is upper triangular. No multiplier is bigger than 1 in absolute value. $A(1{:}k, k)$ is overwritten by $U(1{:}k, k)$, $k = 1{:}n$. $A(k+1{:}n, k)$ is overwritten by $-M_k(k+1{:}n, k)$, $k = 1{:}n-1$. The integer vector $p(1{:}n-1)$ defines the interchange permutations. In particular, E_k interchanges rows k and $p(k), k = 1{:}n-1$.

> **for** $k = 1{:}n-1$
>   Determine μ with $k \le \mu \le n$ so $|A(\mu, k)| = \| A(k{:}n, k) \|_\infty$
> $A(k, k{:}n) \leftrightarrow A(\mu, k{:}n)$
> $p(k) = \mu$
>   **if** $A(k, k) \ne 0$
> $rows = k+1{:}n$
> $A(rows, k) = A(rows, k)/A(k, k)$
> $A(rows, rows) = A(rows, rows) - A(rows, k)A(k, rows)$
>   **end**
> **end**

Note that if $\| A(k{:}n, k) \|_\infty = 0$ in step k, then in exact arithmetic the first k columns of A are linearly dependent. In contrast to Algorithm 3.2.1, this poses no difficulty. We merely skip over the zero pivot.

The overhead associated with partial pivoting is minimal from the standpoint of floating point arithmetic as there are only $O(n^2)$ comparisons associated with the search for the pivots. The overall algorithm involves $2n^3/3$ flops.

To solve the linear system $Ax = b$ after invoking Algorithm 3.4.1 we

- Compute $y = M_{n-1} E_{n-1} \cdots M_1 E_1 b$.

- Solve the upper triangular system $Ux = y$.

All the information necessary to do this is contained in the array A and the pivot vector p. Indeed, the calculation

> **for** $k = 1{:}n - 1$
> $\quad b(k) \leftrightarrow b(p(k))$
> $\quad b(k + 1{:}n) = b(k + 1{:}n) - b(k)A(k + 1{:}n, k)$
> **end**

overwrites b with $M_{n-1}E_{n-1} \cdots M_1 E_1$.

Example 3.4.1 If Algorithm 3.4.1 is applied to

$$A = \begin{bmatrix} 3 & 17 & 10 \\ 2 & 4 & -2 \\ 6 & 18 & -12 \end{bmatrix},$$

then upon exit

$$A = \begin{bmatrix} 6 & 18 & -12 \\ 1/3 & 8 & 16 \\ 1/2 & -1/4 & 6 \end{bmatrix}$$

and $p = [3 , 3]$. These two quantities encode all the information associated with the reduction:

$$\begin{bmatrix} 1 & 0 & 0 \\ 0 & 1 & 0 \\ 0 & 1/4 & 1 \end{bmatrix}\begin{bmatrix} 1 & 0 & 0 \\ 0 & 0 & 1 \\ 0 & 1 & 0 \end{bmatrix}\begin{bmatrix} 1 & 0 & 0 \\ -1/3 & 1 & 0 \\ -1/2 & 0 & 1 \end{bmatrix}\begin{bmatrix} 0 & 0 & 1 \\ 0 & 1 & 0 \\ 1 & 0 & 0 \end{bmatrix}A = \begin{bmatrix} 6 & 18 & -12 \\ 0 & 8 & 16 \\ 0 & 0 & 6 \end{bmatrix}.$$

3.4.4 Where is L?

Gaussian elimination with partial pivoting computes the LU factorization of a row permuted version of A. The proof is a messy subscripting argument.

Theorem 3.4.1 *If Gaussian elimination with partial pivoting is used to compute the upper triangularization*

$$M_{n-1}E_{n-1} \cdots M_1 E_1 A = U \tag{3.4.1}$$

via Algorithm 3.4.1, then

$$PA = LU$$

where $P = E_{n-1} \cdots E_1$ and L is a unit lower triangular matrix with $|\ell_{ij}| \leq 1$. The kth column of L below the diagonal is a permuted version of the kth Gauss vector. In particular, if $M_k = I - \tau^{(k)}e_k^T$, then $L(k + 1{:}n, k) = g(k + 1{:}n)$ where $g = E_{n-1} \cdots E_{k+1}\tau^{(k)}$.

Proof. A manipulation of (3.4.1) reveals that $\tilde{M}_{n-1} \cdots \tilde{M}_1 PA = U$ where $\tilde{M}_{n-1} = M_{n-1}$ and

$$\tilde{M}_k = E_{n-1} \cdots E_{k+1}M_k E_{k+1} \cdots E_{n-1} \qquad k \leq n - 2 .$$

Since each E_j is an interchange permutation involving row j and a row μ with $\mu \geq j$ we have $E_j(1{:}j - 1, 1{:}j - 1) = I_{j-1}$. It follows that each $\tilde{M}_k$ is a Gauss transform with Gauss vector $\tilde{\tau}^{(k)} = E_{n-1} \cdots E_{k+1} \tau^{(k)}$. $\square$

As a consequence of the theorem, it is easy to see how to change Algorithm 3.4.1 so that upon completion, $A(i, j)$ houses $L(i, j)$ for all $i > j$. We merely apply each E_k to all the previously computed Gauss vectors. This is accomplished by changing the line "$A(k, k{:}n) \leftrightarrow A(\mu, k{:}n)$" in Algorithm 3.4.1 to "$A(k, 1{:}n) \leftrightarrow A(\mu, 1{:}n)$."

Example 3.4.2 The factorization $PA = LU$ of the matrix in Example 3.4.1 is given by

$$
\begin{bmatrix} 0 & 0 & 1 \\ 1 & 0 & 0 \\ 0 & 1 & 0 \end{bmatrix}
\begin{bmatrix} 3 & 17 & 10 \\ 2 & 4 & -2 \\ 6 & 18 & -12 \end{bmatrix}
=
\begin{bmatrix} 1 & 0 & 0 \\ 1/2 & 1 & 0 \\ 1/3 & -1/4 & 1 \end{bmatrix}
\begin{bmatrix} 6 & 18 & -12 \\ 0 & 8 & 16 \\ 0 & 0 & 6 \end{bmatrix}.
$$

3.4.5   The Gaxpy Version

In §3.2 we developed outer product and gaxpy schemes for computing the LU factorization. Having just incorporated pivoting in the outer product version, it is natural to do the same with the gaxpy approach. Recall from (3.2.5) the general structure of the gaxpy LU process:

$L = I$
$U = 0$
for $j = 1{:}n$
    if $j = 1$
        $v(j{:}n) = A(j{:}n, j)$
    else
        Solve $L(1{:}j - 1, 1{:}j - 1)z = A(1{:}j - 1, j)$ for z
           and set $U(1{:}j - 1, j) = z$.
        $v(j{:}n) = A(j{:}n, j) - L(j{:}n, 1{:}j - 1)z$
    end
    if $j < n$
        $L(j + 1{:}n, j) = v(j + 1{:}n)/v(j)$
    end
    $U(j, j) = v(j)$
end

With partial pivoting we search $|v(j{:}n)|$ for its maximal element and proceed accordingly. Assuming A is nonsingular so no zero pivots are encountered we obtain

$$L = I$$
$$U = 0$$
for $j = 1{:}n$
    if $j = 1$
        $v(j{:}n) = A(j{:}n, j)$
    else
        Solve $L(1{:}j-1, 1{:}j-1)z = A(1{:}j-1, j)$
            for z and set $U(1{:}j-1, j) = z$.
        $v(j{:}n) = A(j{:}n, j) - L(j{:}n, 1{:}j-1)z$
    end   (3.4.2)
    if $j < n$
        Determine μ with $k \le \mu \le n$ so $|v(\mu)| = \| v(j{:}n) \|_\infty$.
        $p(j) = \mu$
        $v(j) \leftrightarrow v(\mu)$
        $L(j+1{:}n, j) = v(j+1{:}n)/v(j)$
        if $j > 1$
            $L(j, 1{:}j-1) \leftrightarrow L(\mu, 1{:}j-1)$
        end
    end
    $U(j, j) = v(j)$
end

In this implementation, we emerge with the factorization $PA = LU$ where $P = E_{n-1} \cdots E_1$ where E_k is obtained by interchanging rows k and $p(k)$ of the n-by-n identity. As with Algorithm 3.4.1, this procedure requires $2n^3/3$ flops and $O(n^2)$ comparisons.

3.4.6 Error Analysis

We now examine the stability that is obtained with partial pivoting. This requires an accounting of the rounding errors that are sustained during elimination and during the triangular system solving. Bearing in mind that there are no rounding errors associated with permutation, it is not hard to show using Theorem 3.3.2 that the computed solution $\hat{x}$ satisfies $(A + E)\hat{x} = b$ where

$$|E| \le n\mathbf{u}\left(3|A| + 5\hat{P}^T|\hat{L}||\hat{U}|\right) + O(\mathbf{u}^2). \qquad (3.4.3)$$

Here we are assuming that $\hat{P}$, $\hat{L}$, and $\hat{U}$ are the computed analogs of P, L, and U as produced by the above algorithms. Pivoting implies that the elements of $\hat{L}$ are bounded by one. Thus $\| \hat{L} \|_\infty \le n$ and we obtain the bound

$$\| E \|_\infty \le n\mathbf{u}\left(3\| A \|_\infty + 5n\| \hat{U} \|_\infty\right) + O(\mathbf{u}^2). \qquad (3.4.4)$$

The problem now is to bound $\| \hat{U} \|_\infty$. Define the *growth factor* ρ by

$$\rho = \max_{i,j,k} \frac{|\hat{a}_{ij}^{(k)}|}{\| A \|_\infty} \tag{3.4.5}$$

where $\hat{A}^{(k)}$ is the computed version of the matrix $A^{(k)} = M_k E_k \cdots M_1 E_1 A$. It follows that

$$\| E \|_\infty \leq 8n^3 \rho \| A \|_\infty u + O(u^2). \tag{3.4.6}$$

Whether or not this compares favorably with the ideal bound (3.3.1) hinges upon the size of the growth factor of ρ. (The factor n^3 is not an operating factor in practice and may be ignored in this discussion.) The growth factor measures how large the numbers become during the process of elimination. In practice, ρ is usually of order 10 but it can also be as large as 2^{n-1}. Despite this, most numerical analysts regard the occurrence of serious element growth in Gaussian elimination with partial pivoting as highly unlikely in practice. The method can be used with confidence.

Example 3.4.3 If Gaussian elimination with partial pivoting is applied to the problem

$$\begin{bmatrix} .001 & 1.00 \\ 1.00 & 2.00 \end{bmatrix} \begin{bmatrix} x_1 \\ x_2 \end{bmatrix} = \begin{bmatrix} 1.00 \\ 3.00 \end{bmatrix}$$

with $\beta = 10, t = 3$, floating point arithmetic, then

$$P = \begin{bmatrix} 0 & 1 \\ 1 & 0 \end{bmatrix}, \quad \hat{L} = \begin{bmatrix} 1.00 & 0 \\ .001 & 1.00 \end{bmatrix}, \quad \hat{U} = \begin{bmatrix} 1.00 & 2.00 \\ 0 & 1.00 \end{bmatrix}$$

and $\hat{x} = (1.00, .996)^T$. Compare with Example 3.3.1.

Example 3.4.2 If $A \in \mathbb{R}^{n \times n}$ is defined by

$$a_{ij} = \begin{cases} 1 & \text{if } i = j \text{ or } j = n \\ -1 & \text{if } i > j \\ 0 & \text{otherwise} \end{cases}$$

then A has an LU factorization with $|\ell_{ij}| \leq 1$ and $u_{nn} = 2^{n-1}$.

3.4.7   Block Gaussian Elimination

Gaussian Elimination with partial pivoting can be organized so that it is rich in level-3 operations. We detail a block outer product procedure but block gaxpy and block dot product formulations are also possible. See Dayde and Duff (1988).

Assume $A \in \mathbb{R}^{n \times n}$ and for clarity that $n = rN$. Partition A as follows:

$$A = \begin{bmatrix} A_{11} & A_{12} \\ A_{21} & A_{22} \end{bmatrix} \begin{matrix} r \\ n - r \end{matrix} \quad .$$
$$\begin{matrix} r & n - r \end{matrix}$$

The first step in the block reduction is typical and proceeds as follows:

- Use scalar Gaussian elimination with partial pivoting (e.g. a rectangular version of Algorithm 3.4.1) to compute permutation $P_1 \in \mathbb{R}^{n \times n}$, unit lower triangular $L_{11} \in \mathbb{R}^{r \times r}$ and upper triangular $U_{11} \in \mathbb{R}^{r \times r}$ so

$$P_1 \begin{bmatrix} A_{11} \\ A_{21} \end{bmatrix} = \begin{bmatrix} L_{11} \\ L_{21} \end{bmatrix} U_{11} .$$

- Apply the P_1 across the rest of A:

$$\begin{bmatrix} \tilde{A}_{12} \\ \tilde{A}_{22} \end{bmatrix} = P_1 \begin{bmatrix} A_{12} \\ A_{22} \end{bmatrix} .$$

- Solve the lower triangular multiple right hand side problem

$$L_{11} U_{12} = \tilde{A}_{12} .$$

- Perform the level-3 update

$$\tilde{A} = \tilde{A}_{22} - L_{21} U_{12} .$$

With these computations we obtain the factorization

$$P_1 A = \begin{bmatrix} L_{11} & 0 \\ L_{21} & I_{n-r} \end{bmatrix} \begin{bmatrix} I_r & 0 \\ 0 & \tilde{A} \end{bmatrix} \begin{bmatrix} U_{11} & U_{12} \\ 0 & I_{n-r} \end{bmatrix} .$$

The process is then repeated on the first r columns of $\tilde{A}$.

In general, during step k ($1 \le k \le N - 1$) of the block algorithm we apply scalar Gaussian elimination to a matrix of size $(n - (k-1)r)$-by-r. An r-by-$(n - kr)$ multiple right hand side system is solved and a level 3 update of size $(n - kr)$-by-$(n - kr)$ is performed. The level 3 fraction for the overall process is approximately given by $1 - 3/(2N)$. Thus, for large N the procedure is rich in matrix multiplication.

3.4.8 Complete Pivoting

Another pivot strategy called *complete pivoting* has the property that the associated growth factor bound is considerably smaller than 2^{n-1}. Recall that in partial pivoting, the kth pivot is determined by scanning the current subcolumn $A(k{:}n, k)$. In complete pivoting, the largest entry in the current submatrix $A(k{:}n, k{:}n)$ is permuted into the (k, k) position. Thus, we compute the upper triangularization $M_{n-1} E_{n-1} \cdots M_1 E_1 A F_1 \cdots F_{n-1} = U$ with the property that in step k we are confronted with the matrix

$$A^{(k-1)} = M_{k-1} E_{k-1} \cdots M_1 E_1 A F_1 \cdots F_{k-1}$$

and determine interchange permutations E_k and F_k such that

$$\left|\left(E_k A^{(k-1)} F_k\right)_{kk}\right| = \max_{k \le i,j \le n} \left|\left(E_k A^{(k-1)} F_k\right)_{ij}\right|.$$

We have the analog of Theorem 3.4.1

Theorem 3.4.2 *If Gaussian elimination with complete pivoting is used to compute the upper triangularization*

$$M_{n-1}E_{n-1}\cdots M_1 E_1 A F_1 \cdots F_{n-1} = U \qquad (3.4.7)$$

then
$$PAQ = LU$$

where $P = E_{n-1}\cdots E_1$, $Q = F_1 \cdots F_{n-1}$ and L is a unit lower triangular matrix with $|\ell_{ij}| \le 1$. The kth column of L below the diagonal is a permuted version of the kth Gauss vector. In particular, if $M_k = I - \tau^{(k)}e_k^T$ then $L(k+1:n,k) = g(k+1:n)$ where $g = E_{n-1}\cdots E_{k+1}\tau^{(k)}$.

Proof. The proof is similar to the proof of Theorem 3.4.1. Details are left to the reader. $\square$

Here is Gaussian elimination with complete pivoting in detail:

Algorithm 3.4.2 (Gaussian Elimination with Complete Pivoting)
This algorithm computes the complete pivoting factorization $PAQ = LU$ where L is unit lower triangular and U is upper triangular. $P = E_{n-1}\cdots E_1$ and $Q = F_1 \cdots F_{n-1}$ are products of interchange permutations. $A(1:k,k)$ is overwritten by $U(1:k,k), k = 1:n$. $A(k+1:n,k)$ is overwritten by $L(k+1:n,k), k = 1:n-1$. E_k interchanges rows k and $p(k)$. F_k interchanges columns k and $q(k)$.

```
for k = 1:n − 1
    Determine μ with k ≤ μ ≤ n and λ with k ≤ λ ≤ n so
        |A(μ, λ)| = max{ |A(i, j)| : i = k:n, j = k:n }
    A(k, 1:n) ↔ A(μ, 1:n)
    A(1:n, k) ↔ A(1:n, λ)
    p(k) = μ
    q(k) = λ
    if A(k, k) ≠ 0
        rows = k + 1:n
        A(rows, k) = A(rows, k)/A(k, k)
        A(rows, rows) = A(rows, rows) − A(rows, k)A(k, rows)
    end
end
```

This algorithm requires $2n^3/3$ flops and $O(n^3)$ comparisons. Unlike partial pivoting, complete pivoting involves a significant overhead because of the two-dimensional search at each stage.

3.4.9 Comments on Complete Pivoting

Suppose $\text{rank}(A) = r < n$. It follows that at the beginning of step $r + 1$, $A(r+1{:}n, r+1{:}n) = 0$. This implies that $E_k = F_k = M_k = I$ for $k = r+1{:}n$ and so the algorithm can be terminated after step r with the following factorization in hand:

$$ PAQ = LU = \begin{bmatrix} L_{11} & 0 \\ L_{21} & I_{n-r} \end{bmatrix} \begin{bmatrix} U_{11} & U_{12} \\ 0 & 0 \end{bmatrix}. $$

Here L_{11} and U_{11} are r-by-r and L_{21} and U_{12}^T are $(n-r)$-by-r. Thus, Gaussian elimination with complete pivoting can in principle be used to determine the rank of a matrix. Yet roundoff errors make the probability of encountering an exactly zero pivot remote. In practice one would have to "declare" A to have rank k if the pivot element in step $k+1$ was sufficiently small. The numerical rank determination problem is discussed in detail in §5.4.

Wilkinson (1961) has shown that in exact arithmetic the elements of the matrix $A^{(k)} = M_k E_k \cdots M_1 E_1 A F_1 \cdots F_k$ satisfy

$$ |a_{ij}^{(k)}| \leq k^{1/2}(2 \cdot 3^{1/2} \cdots k^{1/(k-1)})^{1/2} \max|a_{ij}|. \tag{3.4.8} $$

The upper bound is a rather slow-growing function of k. This fact coupled with vast empirical evidence suggesting that ρ is always modestly sized (e.g, $\rho = 10$) permit us to conclude that *Gaussian elimination with complete pivoting is stable.* The method solves a nearby linear system $(A + E)\hat{x} = b$ exactly in the sense of (3.3.1). However, there appears to be no practical justification for choosing complete pivoting over partial pivoting except in cases where rank determination is an issue.

Example 3.4.6 If Gaussian elimination with complete pivoting is applied to the problem

$$ \begin{bmatrix} .001 & 1.00 \\ 1.00 & 2.00 \end{bmatrix} \begin{bmatrix} x_1 \\ x_2 \end{bmatrix} = \begin{bmatrix} 1.00 \\ 3.00 \end{bmatrix} $$

with $\beta = 10, t = 3$, floating arithmetic, then

$$ P = \begin{bmatrix} 0 & 1 \\ 1 & 0 \end{bmatrix}, \quad Q = \begin{bmatrix} 0 & 1 \\ 1 & 0 \end{bmatrix}, \quad \hat{L} = \begin{bmatrix} 1.00 & 0.00 \\ .500 & 1.00 \end{bmatrix}, \quad \hat{U} = \begin{bmatrix} 2.00 & 1.00 \\ 0.00 & .499 \end{bmatrix} $$

and $\hat{x} = [1.00, \ 1.00]^T$. Compare with Examples 3.3.1 and 3.4.3.

3.4.10 The Avoidance of Pivoting

For certain classes of matrices it is not necessary to pivot. It is important to identify such classes because pivoting usually degrades performance. To illustrate the kind of analysis required to prove that pivoting can be safely avoided, we consider the case of diagonally dominant matrices. We say that $A \in \mathbb{R}^{n \times n}$ is *strictly diagonally dominant* if

$$|a_{ii}| > \sum_{\substack{j=1 \\ j \neq i}}^{n} |a_{ij}| \qquad i = 1{:}n .$$

The following theorem shows how this property can ensure a nice, no-pivoting LU factorization.

Theorem 3.4.3 *If A^T is strictly diagonally dominant, then A has an LU factorization and $|l_{ij}| \leq 1$. In other words, if Algorithm 3.4.1 is applied, then $P = I$.*

Proof. Partition A as follows

$$A = \begin{bmatrix} \alpha & w^T \\ v & C \end{bmatrix}$$

where α is 1-by-1 and note that after one step of the outer product LU process we have the factorization

$$\begin{bmatrix} \alpha & w^T \\ v & C \end{bmatrix} = \begin{bmatrix} 1 & 0 \\ v/\alpha & I \end{bmatrix} \begin{bmatrix} 1 & 0 \\ 0 & C - vw^T/\alpha \end{bmatrix} \begin{bmatrix} \alpha & w^T \\ 0 & I \end{bmatrix} .$$

The theorem follows by induction on n if we can show that the transpose of $B = C - vw^T/\alpha$ is strictly diagonally dominant. This is because we may then assume that B has an LU factorization $B = L_1 U_1$ and that implies

$$A = \begin{bmatrix} 1 & 0 \\ v/\alpha & L_1 \end{bmatrix} \begin{bmatrix} \alpha & w^T \\ 0 & U_1 \end{bmatrix} \equiv LU .$$

But the proof that B^T is strictly diagonally dominant is straight forward. From the definitions we have

$$\sum_{\substack{i=1 \\ i \neq j}}^{n-1} |b_{ij}| = \sum_{\substack{i=1 \\ i \neq j}}^{n-1} |c_{ij} - v_i w_j/\alpha| \leq \sum_{\substack{i=1 \\ i \neq j}}^{n-1} |c_{ij}| + \frac{|w_j|}{|\alpha|} \sum_{\substack{i=1 \\ i \neq j}}^{n-1} |v_i|$$

$$\leq (|c_{jj}| - |w_j|) + \frac{|w_j|}{|\alpha|}(|\alpha| - |v_j|)$$

$$\leq \left| c_{jj} - \frac{w_j v_j}{\alpha} \right| = |b_{jj}| . \square$$

3.4.11   Some Applications

We conclude with some examples that illustrate how to think in terms of matrix factorizations when confronted with various linear equation situations.

Suppose A is nonsingular and n-by-n and that B is n-by-p. Consider the problem of finding X (n-by-p) so $AX = B$, i.e., the *multiple right hand side problem*. If $X = [\,x_1, \ldots, x_p\,]$ and $B = [\,b_1, \ldots, b_p\,]$ are column partitions, then

$$
\begin{aligned}
&\text{Compute } PA = LU. \\
&\textbf{for } k = 1{:}p \\
&\qquad \text{Solve } Ly = Pb_k \\
&\qquad \text{Solve } Ux_k = y \\
&\textbf{end}
\end{aligned}
\qquad (3.4.9)
$$

Note that A is factored just once. If $B = I_n$ then we emerge with a computed A^{-1} .

As another example of getting the LU factorization "outside the loop," suppose we want to solve the linear system $A^k x = b$ where $A \in \mathbb{R}^{n \times n}$, $b \in \mathbb{R}^n$, and k is a positive integer. One approach is to compute $C = A^k$ and then solve $Cx = b$. However, the matrix multiplications can be avoided altogether:

$$
\begin{aligned}
&\text{Compute } PA = LU \\
&\textbf{for } j = 1{:}k \\
&\qquad \text{Overwrite } b \text{ with the solution to } Ly = Pb. \\
&\qquad \text{Overwrite } b \text{ with the solution to } Ux = b. \\
&\textbf{end}
\end{aligned}
\qquad (3.4.10)
$$

As a final example we show how to avoid the pitfall of explicit inverse computation. Suppose we are given $A \in \mathbb{R}^{n \times n}$, $d \in \mathbb{R}^n$, and $c \in \mathbb{R}^n$ and that we want to compute $s = c^T A^{-1} d$. One approach is to compute $X = A^{-1}$ as suggested above and then compute $s = c^T X d$. A more economical procedure is to compute $PA = LU$ and then solve the triangular systems $Ly = Pd$ and $Ux = y$. It follows that $s = c^T x$. The point of this example is to stress that when a matrix inverse is encountered in a formula, we must think in terms of solving equations rather than in terms of explicit inverse formation.

Problems

P3.4.1 Let $A = LU$ be the LU factorization of n-by-n A with $|\ell_{ij}| \leq 1$. Let a_i^T and u_i^T denote the ith rows of A and U, respectively. Verify the equation

$$
u_i^T = a_i^T - \sum_{j=1}^{i-1} \ell_{ij} u_j^T.
$$

and use it to show that $\| U \|_\infty \leq 2^{n-1} \| A \|_\infty$. (Hint: Take norms and use induction.)

P3.4.2 Show that if $PAQ = $ LU is obtained via Gaussian elimination with complete pivoting, then no element of $U(i, i{:}n)$ is larger in absolute value than $|u_{ii}|$.

P3.4.3 Suppose $A \in \mathbb{R}^{n \times n}$ has an LU factorization and that L and U are known. Give an algorithm which can compute the (i, j) entry of A^{-1} in approximately $(n-j)^2 + (n-i)^2$ flops.

P3.4.4 Suppose $\hat{X}$ is the computed inverse obtained via (3.4.9). Give an upper bound for $\| A\hat{X} - I \|_F$.

P3.4.5 Prove Theorem 3.4.2.

P3.4.6 Extend Algorithm 3.4.3 so that it can factor an arbitrary rectangular matrix.

P3.4.7 Write a detailed version of the block elimination algorithm outlined in §3.4.7.

Notes and References for Sec. 3.4

An Algol version of Algorithm 3.4.1 is given in

H.J. Bowdler, R.S. Martin, G. Peters, and J.H. Wilkinson (1966). "Solution of Real and Complex Systems of Linear Equations," *Numer. Math. 8*, 217–34. See also Wilkinson and Reinsch (1971, 93–110).

The conjecture that $|a_{ij}^{(k)}| \leq n \max|a_{ij}|$ when complete pivoting is used has been proven in the real $n = 4$ case in

C.W. Cryer (1968). "Pivot Size in Gaussian Elimination," *Numer. Math. 12*, 335–45.

Other papers concerned with element growth and pivoting include

J.K. Reid (1971). "A Note on the Stability of Gaussian Elimination," *J.Inst. Math. Applics. 8*, 374–75.

P.A. Businger (1971). "Monitoring the Numerical Stability of Gaussian Elimination," *Numer. Math. 16*, 360–61.

A.M. Cohen (1974). "A Note on Pivot Size in Gaussian Elimination," *Lin. Alg. and Its Applic. 8*, 361–68.

A.M. Erisman and J.K. Reid (1974). "Monitoring the Stability of the Triangular Factorization of a Sparse Matrix," *Numer. Math. 22*, 183–86.

J. Day and B. Peterson (1988). "Growth in Gaussian Elimination," *Amer. Math. Monthly 95*, 489–513.

N.J. Higham and D.J. Higham (1989). "Large Growth Factors in Gaussian Elimination with Pivoting," *SIAM J. Matrix Anal. Appl. 10*, 155–164.

L.N. Trefethen and R.S. Schreiber (1990). "Average-Case Stability of Gaussian Elimination," *SIAM J. Matrix Anal. Appl. 11*, 335–360.

N. Gould (1991). "On Growth in Gaussian Elimination with Complete Pivoting," *SIAM J. Matrix Anal. Appl. 12*, 354–361.

A. Edelman (1992). "The Complete Pivoting Conjecture for Gaussian Elimination is False," *The Mathematica Journal 2*, 58–61.

S.J. Wright (1993). "A Collection of Problems for Which Gaussian Elimination with Partial Pivoting is Unstable," *SIAM J. Sci. and Stat. Comp. 14*, 231–238.

L.V. Foster (1994). "Gaussian Elimination with Partial Pivoting Can Fail in Practice," *SIAM J. Matrix Anal. Appl. 15*, 1354–1362.

A. Edelman and W. Mascarenhas (1995). "On the Complete Pivoting Conjecture for a Hadamard Matrix of Order 12," *Linear and Multilinear Algebra 38*, 181–185.

The designers of sparse Gaussian elimination codes are interested in the topic of element growth because multipliers greater than unity are sometimes tolerated for the sake of minimizing fill-in. See

I.S. Duff, A.M. Erisman, and J.K. Reid (1986). *Direct Methods for Sparse Matrices*, Oxford University Press.

The connection between small pivots and near singularity is reviewed in

T.F. Chan (1985). "On the Existence and Computation of LU Factorizations with small pivots," *Math. Comp. 42*, 535–548.

A pivot strategy that we did not discuss is *pairwise pivoting*. In this approach, 2-by-2 Gauss transformations are used to zero the lower triangular portion of A. The technique is appealing in certain multiprocessor environments because only adjacent rows are combined in each step. See

D. Sorensen (1985). "Analysis of Pairwise Pivoting in Gaussian Elimination," *IEEE Trans. on Computers C-34*, 274–278.

As a sample paper in which a class of matrices is identified that require no pivoting, see

S. Serbin (1980). "On Factoring a Class of Complex Symmetric Matrices Without Pivoting," *Math. Comp. 35*, 1231-1234.

Just as there are six "conventional" versions of scalar Gaussian elimination, there are also six conventional block formulations of Gaussian elimination. For a discussion of these procedures and their implementation see

K. Gallivan, W. Jalby, U. Meier, and A.H. Sameh (1988). "Impact of Hierarchical Memory Systems on Linear Algebra Algorithm Design," *Int'l J. Supercomputer Applic. 2*, 12–48.

3.5 Improving and Estimating Accuracy

Suppose Gaussian elimination with partial pivoting is used to solve the n-by-n system $Ax = b$. Assume t-digit, base β floating point arithmetic is used. Equation (3.4.6) essentially says that if the growth factor is modest then the computed solution $\hat{x}$ satisfies

$$(A + E)\hat{x} = b, \qquad \| E \|_\infty \approx \mathbf{u} \| A \|_\infty, \quad \mathbf{u} = \frac{1}{2}\beta^{-t}. \qquad (3.5.1)$$

In this section we explore the practical ramifications of this result. We begin by stressing the distinction that should be made between residual size and accuracy. This is followed by a discussion of scaling, iterative improvement, and condition estimation. See Higham (1996) for a more detailed treatment of these topics.

We make two notational remarks at the outset. The infinity norm is used throughout since it is very handy in roundoff error analysis and in practical error estimation. Second, whenever we refer to "Gaussian elimination" in this section we really mean Gaussian elimination with some stabilizing pivot strategy such as partial pivoting.

3.5.1   Residual Size Versus Accuracy

The *residual* of a computed solution $\hat{x}$ to the linear system $Ax = b$ is the vector $b - A\hat{x}$. A small residual means that $A\hat{x}$ effectively "predicts" the right hand side b. From (3.5.1) we have $\| b - A\hat{x} \|_\infty \approx \mathbf{u} \| A \|_\infty \| \hat{x} \|_\infty$ and so we obtain

Heuristic I. Gaussian elimination produces a solution $\hat{x}$ with a relatively small residual.

Small residuals do not imply high accuracy. Combining (3.3.2) and (3.5.1), we see that

$$\frac{\| \hat{x} - x \|_\infty}{\| x \|_\infty} \approx \mathbf{u} \kappa_\infty(A) . \tag{3.5.2}$$

This justifies a second guiding principle.

Heuristic II. If the unit roundoff and condition satisfy $\mathbf{u} \approx 10^{-d}$ and $\kappa_\infty(A) \approx 10^q$, then Gaussian elimination produces a solution $\hat{x}$ that has about $d - q$ correct decimal digits.

If $\mathbf{u}\kappa_\infty(A)$ is large, then we say that A is ill-conditioned with respect to the machine precision.

As an illustration of the Heuristics I and II, consider the system

$$\begin{bmatrix} .986 & .579 \\ .409 & .237 \end{bmatrix} \begin{bmatrix} x_1 \\ x_2 \end{bmatrix} = \begin{bmatrix} .235 \\ .107 \end{bmatrix}$$

in which $\kappa_\infty(A) \approx 700$ and $x = (2, -3)^T$. Here is what we find for various machine precisions:

β	t	$\hat{x}_1$	$\hat{x}_2$	$\dfrac{\| \hat{x} - x \|_\infty}{\| x \|_\infty}$	$\dfrac{\| b - A\hat{x} \|_\infty}{\| A \|_\infty \| \hat{x} \|_\infty}$
10	3	2.11	-3.17	$5 \cdot 10^{-2}$	$2.0 \cdot 10^{-3}$
10	4	1.986	-2.975	$8 \cdot 10^{-3}$	$1.5 \cdot 10^{-4}$
10	5	2.0019	-3.0032	$1 \cdot 10^{-3}$	$2.1 \cdot 10^{-6}$
10	6	2.00025	-3.00094	$3 \cdot 10^{-4}$	$4.2 \cdot 10^{-7}$

Whether or not one is content with the computed solution $\hat{x}$ depends on the requirements of the underlying source problem. In many applications accuracy is not important but small residuals are. In such a situation, the $\hat{x}$ produced by Gaussian elimination is probably adequate. On the other hand, if the number of correct digits in $\hat{x}$ is an issue then the situation is more complicated and the discussion in the remainder of this section is relevant.

3.5.2   Scaling

Let β be the machine base and define the diagonal matrices D_1 and D_2 by

$$\begin{aligned} D_1 &= \operatorname{diag}(\beta^{r_1} \ldots \beta^{r_n}) \\ D_2 &= \operatorname{diag}(\beta^{c_1} \ldots \beta^{c_n}). \end{aligned}$$

The solution to the n-by-n linear system $Ax = b$ can be found by solving the *scaled system* $(D_1^{-1}AD_2)y = D_1^{-1}b$ using Gaussian elimination and then setting $x = D_2 y$. The scalings of A, b, and y require only $O(n^2)$ flops and may be accomplished without roundoff. Note that D_1 scales equations and D_2 scales unknowns.

It follows from Heuristic II that if $\hat{x}$ and $\hat{y}$ are the computed versions of x and y, then

$$\frac{\| D_2^{-1}(\hat{x} - x) \|_\infty}{\| D_2^{-1}x \|_\infty} = \frac{\| \hat{y} - y \|_\infty}{\| y \|_\infty} \approx \mathbf{u}\kappa_\infty(D_1^{-1}AD_2). \qquad (3.5.3)$$

Thus, if $\kappa_\infty(D_1^{-1}AD_2)$ can be made considerably smaller than $\kappa_\infty(A)$, then we might expect a correspondingly more accurate $\hat{x}$, provided errors are measured in the "D_2" norm defined by $\| z \|_{D_2} = \| D_2^{-1}z \|_\infty$. This is the objective of scaling. Note that it encompasses two issues: the condition of the scaled problem and the appropriateness of appraising error in the D_2-norm.

An interesting but very difficult mathematical problem concerns the exact minimization of $\kappa_p(D_1^{-1}AD_2)$ for general diagonal D_i and various p. What results there are in this direction are not very practical. This is hardly discouraging, however, when we recall that (3.5.3) is heuristic and it makes little sense to minimize exactly a heuristic bound. What we seek is a fast, approximate method for improving the quality of the computed solution $\hat{x}$.

One technique of this variety is *simple row scaling*. In this scheme D_2 is the identity and D_1 is chosen so that each row in $D_1^{-1}A$ has approximately the same ∞-norm. Row scaling reduces the likelihood of adding a very small number to a very large number during elimination—an event that can greatly diminish accuracy.

Slightly more complicated than simple row scaling is *row-column equilibration*. Here, the object is to choose D_1 and D_2 so that the ∞-norm of each row and column of $D_1^{-1}AD_2$ belongs to the interval $[1/\beta, 1]$ where β is the base of the floating point system. For work along these lines see McKeeman (1962).

It cannot be stressed too much that simple row scaling and row-column equilibration do not "solve" the scaling problem. Indeed, either technique can render a worse $\hat{x}$ than if no scaling whatever is used. The ramifications of this point are thoroughly discussed in Forsythe and Moler (1967, chapter 11). The basic recommendation is that the scaling of equations and unknowns must proceed on a problem-by-problem basis. General scaling strategies are unreliable. It is best to scale (if at all) on the basis of what the source problem proclaims about the significance of each a_{ij}. Measurement units and data error may have to be considered.

Example 3.5.1 (Forsythe and Moler (1967, pp. 34, 40]) . If

$$\begin{bmatrix} 10 & 100,000 \\ 1 & 1 \end{bmatrix} \begin{bmatrix} x_1 \\ x_2 \end{bmatrix} = \begin{bmatrix} 100,000 \\ 2 \end{bmatrix}$$

and the equivalent row-scaled problem

$$\begin{bmatrix} .0001 & 1 \\ 1 & 1 \end{bmatrix} \begin{bmatrix} x_1 \\ x_2 \end{bmatrix} = \begin{bmatrix} 1 \\ 2 \end{bmatrix}$$

are each solved using $\beta = 10, t = 3$ arithmetic, then solutions $\hat{x} = (0.00, \ 1.00)^T$ and $\hat{x} = (1.00, \ 1.00)^T$ are respectively computed. Note that $x = (1.0001\ldots, \ .9999\ldots)^T$ is the exact solution.

3.5.3   Iterative Improvement

Suppose $Ax = b$ has been solved via the partial pivoting factorization $PA = LU$ and that we wish to improve the accuracy of the computed solution $\hat{x}$. If we execute

$$\begin{aligned} & r = b - A\hat{x} \\ & \text{Solve } Ly = Pr. \\ & \text{Solve } Uz = y. \\ & x_{new} = \hat{x} + z \end{aligned} \qquad (3.5.4)$$

then in exact arithmetic $Ax_{new} = A\hat{x} + Az = (b-r)+r = b$. Unfortunately, the naive floating point execution of these formulae renders an x_{new} that is no more accurate than $\hat{x}$. This is to be expected since $\hat{r} = fl(b - A\hat{x})$ has few, if any, correct significant digits. (Recall Heuristic I.) Consequently, $\hat{z} = fl(A^{-1}r) \approx A^{-1} \cdot$ noise $\approx$ noise is a very poor correction *from the standpoint of improving the accuracy of* $\hat{x}$. However, Skeel (1980) has done an error analysis that indicates when (3.5.4) gives an improved x_{new} *from the standpoint of backwards error*. In particular, if the quantity

$$\tau = \left(\| \, |A| \, |A^{-1}| \, \|_\infty \right) \left(\max_i \ (|A||x|)_i \, / \min_i \ (|A||x|)_i \right)$$

is not too big, then (3.5.4) produces an x_{new} such that $(A + E)x_{new} = b$ for very small E. Of course, if Gaussian elimination with partial pivoting is used then the computed $\hat{x}$ already solves a nearby system. However, this may not be the case for some of the pivot strategies that are used to preserve sparsity. In this situation, the *fixed precision iterative improvement* step (3.5.4) can be very worthwhile and cheap. See Arioli, Demmel, and Duff (1988).

For (3.5.4) to produce a more accurate x, it is necessary to compute the residual $b - A\hat{x}$ with extended precision floating point arithmetic. Typically, this means that if t-digit arithmetic is used to compute $PA = LU$, x, y, and z, then $2t$-digit arithmetic is used to form $b - A\hat{x}$, i.e., double precision. The process can be iterated. In particular, once we have computed $PA = LU$ and initialize $x = 0$, we repeat the following:

$$
\begin{aligned}
&r = b - Ax \text{ (Double Precision)} \\
&\text{Solve } Ly = Pr \text{ for } y. \\
&\text{Solve } Uz = y \text{ for } z. \\
&x = x + z
\end{aligned}
\tag{3.5.5}
$$

We refer to this process as *mixed precision iterative improvement*. The original A must be used in the double precision computation of r. The basic result concerning the performance of (3.5.5) is summarized in the following heuristic:

Heuristic III. If the machine precision $\mathbf{u}$ and condition satisfy $\mathbf{u} = 10^{-d}$ and $\kappa_\infty(A) \approx 10^q$, then after k executions of (3.5.5), x has approximately $\min(d, k(d - q))$ correct digits.

Roughly speaking, if $\mathbf{u}\kappa_\infty(A) \leq 1$, then iterative improvement can ultimately produce a solution that is correct to full (single) precision. Note that the process is relatively cheap. Each improvement costs $O(n^2)$, to be compared with the original $O(n^3)$ investment in the factorization $PA = LU$. Of course, no improvement may result if A is badly enough conditioned with respect to the machine precision.

The primary drawback of mixed precision iterative improvement is that its implementation is somewhat machine-dependent. This discourages its use in software that is intended for wide distribution. The need for retaining an original copy of A is another aggravation associated with the method.

On the other hand, mixed precision iterative improvement is usually very easy to implement on a given machine that has provision for the accumulation of inner products, i.e., provision for the double precision calculation of inner products between the rows of A and x. In a short mantissa computing environment the presence of an iterative improvement routine can significantly widen the class of solvable $Ax = b$ problems.

Example 3.5.2 If (3.5.5) is applied to the system
$$
\begin{bmatrix} .986 & .579 \\ .409 & .237 \end{bmatrix} \begin{bmatrix} x_1 \\ x_2 \end{bmatrix} = \begin{bmatrix} .235 \\ .107 \end{bmatrix}
$$
and $\beta = 10$ and $t = 3$, then iterative improvement produces the following sequence of computed solutions:
$$
\hat{x} = \begin{bmatrix} 2.11 \\ -3.17 \end{bmatrix}, \begin{bmatrix} 1.99 \\ -2.99 \end{bmatrix}, \begin{bmatrix} 2.00 \\ -3.00 \end{bmatrix}, \ldots
$$

The exact solution is $x = [2, \ -3]^T$.

3.5.4   Condition Estimation

Suppose that we have solved $Ax = b$ via $PA = LU$ and that we now wish to ascertain the number of correct digits in the computed solution $\hat{x}$. It follows from Heuristic II that in order to do this we need an estimate of the condition $\kappa_\infty(A) = \| A \|_\infty \| A^{-1} \|_\infty$. Computing $\| A \|_\infty$ poses no problem as we merely use the formula

$$\| A \|_\infty \ = \ \max_{1 \le i \le n} \sum_{j=1}^{n} |a_{ij}|.$$

The challenge is with respect to the factor $\| A^{-1} \|_\infty$. Conceivably, we could estimate this quantity by $\| \hat{X} \|_\infty$, where $\hat{X} = [\, \hat{x}_1, \ldots, \hat{x}_n \,]$ and $\hat{x}_i$ is the computed solution to $Ax_i = e_i$. (See §3.4.9.) The trouble with this approach is its expense: $\hat{\kappa}_\infty = \| A \|_\infty \| \hat{X} \|_\infty$ costs about three times as much as $\hat{x}$.

The central problem of *condition estimation* is how to estimate the condition number in $O(n^2)$ flops assuming the availability of $PA = LU$ or some other factorizations that are presented in subsequent chapters. An approach described in Forsythe and Moler (SLE, p. 51) is based on iterative improvement and the heuristic $u\kappa_\infty(A) \approx \| z \|_\infty / \| x \|_\infty$ where z is the first correction of x in (3.5.5). While the resulting condition estimator is $O(n^2)$, it suffers from the shortcoming of iterative improvement, namely, machine dependency.

Cline, Moler, Stewart, and Wilkinson (1979) have proposed a very successful approach to the condition estimation problem without this flaw. It is based on exploitation of the implication

$$Ay = d \quad \Longrightarrow \quad \| A^{-1} \|_\infty \ge \| y \|_\infty / \| d \|_\infty.$$

The idea behind their estimator is to choose d so that the solution y is large in norm and then set

$$\hat{\kappa}_\infty \ = \ \| A \|_\infty \| y \|_\infty / \| d \|_\infty.$$

The success of this method hinges on how close the ratio $\| y \|_\infty / \| d \|_\infty$ is to its maximum value $\| A^{-1} \|_\infty$.

Consider the case when $A = T$ is upper triangular. The relation between d and y is completely specified by the following column version of back substitution:

$p(1{:}n) = 0$

> **for** $k = n: -1:1$
>   Choose $d(k)$.
> $y(k) = (d(k) - p(k))/T(k,k)$ (3.5.6)
> $p(1{:}k - 1) = p(1{:}k - 1) + y(k)T(1{:}k - 1, k)$
> **end**

Normally, we use this algorithm to solve a *given* triangular system $Ty = d$. Now, however, we are free to pick the right-hand side d subject to the "constraint" that y is large relative to d.

One way to encourage growth in y is to choose $d(k)$ from the set $\{-1, +1\}$ so as to maximize $y(k)$. If $p(k) \geq 0$, then set $d(k) = -1$. If $p(k) < 0$, then set $d(k) = +1$. In other words, (3.5.6) is invoked with $d(k) = -\text{sign}(p(k))$. Since d is then a vector of the form $d(1{:}n) = (\pm 1, \ldots, \pm 1)^T$, we obtain the estimator $\hat{\kappa}_\infty = \| T \|_\infty \| y \|_\infty$.

A more reliable estimator results if $d(k) \in \{-1, +1\}$ is chosen so as to encourage growth both in $y(k)$ and the updated running sum given by $p(1{:}k - 1, k) + T(1{:}k - 1, k)y(k)$. In particular, at step k we compute

$$y(k)^+ = (1 - p(k))/T(k,k)$$

$$s(k)^+ = |y(k)^+| + \| p(1{:}k - 1) + T(1{:}k - 1, k)y(k)^+ \|_1$$

$$y(k)^- = (-1 - p(k))/T(k,k)$$

$$s(k)^- = |y(k)^-| + \| p(1{:}k - 1) + T(1{:}k - 1, k)y(k)^- \|_1$$

and set

$$y(k) = \begin{cases} y(k)^+ & \text{if } s(k)^+ \geq s(k)^- \\ y(k)^- & \text{if } s(k)^+ < s(k)^- \end{cases}$$

This gives

Algorithm 3.5.1 (Condition Estimator) Let $T \in \mathbb{R}^{n \times n}$ be a nonsingular upper triangular matrix. This algorithm computes unit ∞-norm y and a scalar κ so $\| Ty \|_\infty \approx 1/\| T^{-1} \|_\infty$ and $\kappa \approx \kappa_\infty(T)$

> $p(1{:}n) = 0$
> **for** $k = n: -1:1$
> $y(k)^+ = (1 - p(k))/T(k,k)$
> $y(k)^- = (-1 - p(k))/T(k,k)$
> $p(k)^+ = p(1{:}k - 1) + T(1{:}k - 1, k)y(k)^+$
> $p(k)^- = p(1{:}k - 1) + T(1{:}k - 1, k)y(k)^-$

$$\textbf{if } |y(k)^+| + \| p(k)^+ \|_1 \ge |y(k)^-| + \| p(k)^- \|_1$$
$$y(k) = y(k)^+$$
$$p(1{:}k - 1) = p(k)^+$$
$$\textbf{else}$$
$$y(k) = y(k)^-$$
$$p(1{:}k - 1) = p(k)^-$$
$$\textbf{end}$$
$$\textbf{end}$$
$$\kappa = \| y \|_\infty \| T \|_\infty;$$
$$y = y/\| y \|_\infty$$

The algorithm involves several times the work of ordinary back substitution.

We are now in a position to describe a procedure for estimating the condition of a square nonsingular matrix A whose $PA = LU$ factorization we know:

- Apply the lower triangular version of Algorithm 3.5.1 to U^T and obtain a large norm solution to $U^T y = d$.

- Solve the triangular systems $L^T r = y$, $Lw = Pr$, and $Uz = w$.

- $\hat{\kappa}_\infty = \| A \|_\infty \| z \|_\infty / \| r \|_\infty$.

Note that $\| z \|_\infty \le \| A^{-1} \|_\infty \| r \|_\infty$. The method is based on several heuristics. First, if A is ill-conditioned and $PA = LU$, then it is usually the case that U is correspondingly ill-conditioned. The lower triangle L tends to be fairly well-conditioned. Thus, it is more profitable to apply the condition estimator to U than to L. The vector r, because it solves $A^T P^T r = d$, tends to be rich in the direction of the left singular vector associated with $\sigma_{min}(A)$. Righthand sides with this property render large solutions to the problem $Az = r$.

In practice, it is found that the condition estimation technique that we have outlined produces good order-of-magnitude estimates of the actual condition number.

Problems

P3.5.1 Show by example that there may be more than one way to equilibrate a matrix.

P3.5.2 Using $\beta = 10, t = 2$ arithmetic, solve

$$\begin{bmatrix} 11 & 15 \\ 5 & 7 \end{bmatrix} \begin{bmatrix} x_1 \\ x_2 \end{bmatrix} = \begin{bmatrix} 7 \\ 3 \end{bmatrix}$$

using Gaussian elimination with partial pivoting. Do one step of iterative improvement using $t = 4$ arithmetic to compute the residual. (Do not forget to round the computed residual to two digits.)

P3.5.3 Suppose $P(A + E) = \hat{L}\hat{U}$, where P is a permutation, $\hat{L}$ is lower triangular with $|\hat{\ell}_{ij}| \le 1$, and $\hat{U}$ is upper triangular. Show that $\hat{\kappa}_\infty(A) \ge \| A \|_\infty / (\| E \|_\infty + \mu)$ where

$\mu = \min |\hat{u}_{ii}|$. Conclude that if a small pivot is encountered when Gaussian elimination with pivoting is applied to A, then A is ill-conditioned. The converse is not true. (Let $A = B_n$).

P3.5.4 (Kahan 1966) The system $Ax = b$ where

$$A = \begin{bmatrix} 2 & -1 & 1 \\ -1 & 10^{-10} & 10^{-10} \\ 1 & 10^{-10} & 10^{-10} \end{bmatrix} \qquad b = \begin{bmatrix} 2(1+10^{-10}) \\ -10^{-10} \\ 10^{-10} \end{bmatrix}$$

has solution $x = (10^{-10} \ -1 \ 1)^T$. (a) Show that if $(A + E)y = b$ and $|E| \le 10^{-8}|A|$, then $|x - y| \le 10^{-7}|x|$. That is, small relative changes in A's entries do not induce large changes in x even though $\kappa_\infty(A) = 10^{10}$. (b) Define $D = \text{diag}(10^{-5}, 10^5, 10^5)$. Show $\kappa_\infty(DAD) \le 5$. (c) Explain what is going on in terms of Theorem 2.7.3.

P3.5.5 Consider the matrix:

$$T = \begin{bmatrix} 1 & 0 & M & -M \\ 0 & 1 & -M & M \\ 0 & 0 & 1 & 0 \\ 0 & 0 & 0 & 1 \end{bmatrix} \qquad M \in \mathbf{R}.$$

What estimate of $\kappa_\infty(T)$ is produced when (3.5.6) is applied with $d(k) = -\text{sign}(p(k))$? What estimate does Algorithm 3.5.1 produce? What is the true $\kappa_\infty(T)$?

P3.5.6 What does Algorithm 3.5.1 produce when applied to the matrix B_n given in (2.7.9)?

Notes and References for Sec. 3.5

The following papers are concerned with the scaling of $Ax = b$ problems:

F.L. Bauer (1963). "Optimally Scaled Matrices," *Numer. Math. 5*, 73–87.

P.A. Businger (1968). "Matrices Which Can be Optimally Scaled," *Numer. Math. 12*, 346–48.

A. van der Sluis (1969). "Condition Numbers and Equilibration Matrices," *Numer. Math. 14*, 14–23.

A. van der Sluis (1970). "Condition, Equilibration, and Pivoting in Linear Algebraic Systems," *Numer. Math. 15*, 74–86.

C. McCarthy and G. Strang (1973). "Optimal Conditioning of Matrices," *SIAM J. Num. Anal. 10*, 370–88.

T. Fenner and G. Loizou (1974). "Some New Bounds on the Condition Numbers of Optimally Scaled Matrices," *J. ACM 21*, 514–24.

G.H. Golub and J.M. Varah (1974). "On a Characterization of the Best L_2-Scaling of a Matrix," *SIAM J. Num. Anal. 11*, 472–79.

R. Skeel (1979). "Scaling for Numerical Stability in Gaussian Elimination," *J. ACM. 26*, 494–526.

R. Skeel (1981). "Effect of Equilibration on Residual Size for Partial Pivoting," *SIAM J. Num. Anal. 18*, 449–55.

Part of the difficulty in scaling concerns the selection of a norm in which to measure errors. An interesting discussion of this frequently overlooked point appears in

W. Kahan (1966). "Numerical Linear Algebra," *Canadian Math. Bull. 9*, 757–801.

For a rigorous analysis of iterative improvement and related matters, see

M. Jankowski and M. Wozniakowski (1977). "Iterative Refinement Implies Numerical Stability," *BIT 17*, 303–311.

C.B. Moler (1967). "Iterative Refinement in Floating Point," *J. ACM 14*, 316-71.

R.D. Skeel (1980). "Iterative Refinement Implies Numerical Stability for Gaussian Elimination," *Math. Comp. 35*, 817–832.

G.W. Stewart (1981). "On the Implicit Deflation of Nearly Singular Systems of Linear Equations," *SIAM J. Sci. and Stat. Comp. 2*, 136-140.

The condition estimator that we described is given in

A.K. Cline, C.B. Moler, G.W. Stewart, and J.H. Wilkinson (1979). "An Estimate for the Condition Number of a Matrix," *SIAM J. Num. Anal.* 16, 368-75.

Other references concerned with the condition estimation problem include

C.G. Broyden (1973). "Some Condition Number Bounds for the Gaussian Elimination Process," *J. Inst. Math. Applic. 12*, 273–86.

F. Lemeire (1973). "Bounds for Condition Numbers of Triangular Value of a Matrix," *Lin. Alg. and Its Applic. 11*, 1–2.

R.S. Varga (1976). "On Diagonal Dominance Arguments for Bounding $\| A^{-1} \|_\infty$," *Lin. Alg. and Its Applic. 14*, 211–17.

G.W. Stewart (1980). "The Efficient Generation of Random Orthogonal Matrices with an Application to Condition Estimators," *SIAM J. Num. Anal. 17*, 403–9.

D.P. O'Leary (1980). "Estimating Matrix Condition Numbers," *SIAM J. Sci. Stat. Comp. 1*, 205–9.

R.G. Grimes and J.G. Lewis (1981). "Condition Number Estimation for Sparse Matrices," *SIAM J. Sci. and Stat. Comp. 2*, 384–88.

A.K. Cline, A.R. Conn, and C. Van Loan (1982). "Generalizing the LINPACK Condition Estimator," in *Numerical Analysis* , ed., J.P. Hennart, Lecture Notes in Mathematics no. 909, Springer-Verlag, New York.

A.K. Cline and R.K. Rew (1983). "A Set of Counter examples to Three Condition Number Estimators," *SIAM J. Sci. and Stat. Comp. 4*, 602–611.

W. Hager (1984). "Condition Estimates," *SIAM J. Sci. and Stat. Comp. 5*, 311–316.

N.J. Higham (1987). "A Survey of Condition Number Estimation for Triangular Matrices," *SIAM Review 29*, 575–596.

N.J. Higham (1988). "Fortran Codes for Estimating the One-norm of a Real or Complex Matrix, with Applications to Condition Estimation," *ACM Trans. Math. Soft. 14*, 381–396.

N.J. Higham (1988). "FORTRAN Codes for Estimating the One-Norm of a Real or Complex Matrix with Applications to Condition Estimation (Algorithm 674)," *ACM Trans. Math. Soft. 14*, 381–396.

C.H. Bischof (1990). "Incremental Condition Estimation," *SIAM J. Matrix Anal. Appl. 11*, 644–659.

C.H. Bischof (1990). "Incremental Condition Estimation for Sparse Matrices," *SIAM J. Matrix Anal. Appl. 11*, 312–322.

G. Auchmuty (1991). "A Posteriori Error Estimates for Linear Equations," *Numer. Math. 61*, 1–6.

N.J. Higham (1993). "Optimization by Direct Search in Matrix Computations," *SIAM J. Matrix Anal. Appl. 14*, 317–333.

D.J. Higham (1995). "Condition Numbers and Their Condition Numbers," *Lin. Alg. and Its Applic. 214*, 193–213.

Chapter 4

Special Linear Systems

It is a basic tenet of numerical analysis that structure should be exploited whenever solving a problem. In numerical linear algebra, this translates into an expectation that algorithms for general matrix problems can be streamlined in the presence of such properties as symmetry, definiteness, and sparsity. This is the central theme of the current chapter, where our principal aim is to devise special algorithms for computing special variants of the LU factorization.

We begin by pointing out the connection between the triangular factors L and U when A is symmetric. This is achieved by examining the LDM^T factorization in §4.1. We then turn our attention to the important case when A is both symmetric and positive definite, deriving the stable Cholesky factorization in §4.2. Unsymmetric positive definite systems are also investigated in this section. In §4.3, banded versions of Gaussian elimination and other factorization methods are discussed. We then examine the interesting situation when A is symmetric but indefinite. Our treatment of this problem in §4.4 highlights the numerical analyst's ambivalence towards pivoting. We love pivoting for the stability it induces but despise it for the structure that it can destroy. Fortunately, there is a happy resolution to this conflict in the symmetric indefinite problem.

Any block banded matrix is also banded and so the methods of §4.3 are applicable. Yet, there are occasions when it pays not to adopt this point of view. To illustrate this we consider the important case of block tridiagonal systems in §4.5. Other block systems are discussed as well.

In the final two sections we examine some very interesting $O(n^2)$ algorithms that can be used to solve Vandermonde and Toeplitz systems.

Before You Begin

Chapter 1, §§2.1-2.5, and §2.7, and Chapter 3 are assumed. Within this chapter there are the following dependencies:

$$§4.5$$
$$\uparrow$$
$$§4.1 \quad \rightarrow \quad §4.2 \quad \rightarrow \quad §4.3 \quad \rightarrow \quad §4.4$$
$$\downarrow$$
$$§4.6 \quad \rightarrow \quad §4.7$$

Complementary references include George and Liu (1981), Gill, Murray, and Wright (1991), Higham (1996), Trefethen and Bau (1996), and Demmel (1996). Some MATLAB functions important to this chapter: chol, tril, triu, vander, toeplitz, fft. LAPACK connections include

LAPACK: General Band Matrices	
_GBSV	Solve $AX = B$
_CGBCON	Condition estimator
_GBRFS	Improve $AX = B$, $A^T X = B$, $A^H X = B$ solutions with error bounds
_GBSVX	Solve $AX = B$, $A^T X = B$, $A^H X = B$ with condition estimate
_GBTRF	$PA = LU$
_GBTRS	Solve $AX = B$, $A^T X = B$, $A^H X = B$ via $PA = LU$
_GBEQU	Equilibration

LAPACK: General Tridiagonal Matrices	
_GTSV	Solve $AX = B$
_GTCON	Condition estimator
_GTRFS	Improve $AX = B$, $A^T X = B$, $A^H X = B$ solutions with error bounds
_GTSVX	Solve $AX = B$, $A^T X = B$, $A^H X = B$ with condition estimate
_GTTRF	$PA = LU$
_GTTRS	Solve $AX = B$, $A^T X = B$, $A^H X = B$ via $PA = LU$

LAPACK: Full Symmetric Positive Definite	
_POSV	Solve $AX = B$
_POCON	Condition estimate via $PA = LU$
_PORFS	Improve $AX = B$ solutions with error bounds
_POSVX	Solve $AX = B$ with condition estimate
_POTRF	$A = GG^T$
_POTRS	Solve $AX = B$ via $A = GG^T$
_POTRI	A^{-1}
_POEQU	Equilibration

LAPACK: Banded Symmetric Positive Definite	
_PBSV	Solve $AX = B$
_PBCON	Condition estimate via $A = GG^T$
_PBRFS	Improve $AX = B$ solutions with error bounds
_PBSVX	Solve $AX = B$ with condition estimate
_PBTRF	$A = GG^T$
_PBTRS	Solve $AX = B$ via $A = GG^T$

LAPACK: Tridiagonal Symmetric Positive Definite	
_PTSV	Solve $AX = B$
_PTCON	Condition estimate via $A = LDL^T$
_PTRFS	Improve $AX = B$ solutions with error bounds
_PTSVX	Solve $AX = B$ with condition estimate
_PTTRF	$A = LDL^T$
_PTTRS	Solve $AX = B$ via $A = LDL^T$

LAPACK: Full Symmetric Indefinite	
_SYSV	Solve $AX = B$
_SYCON	Condition estimate via $PAP^T = LDL^T$
_SYRFS	Improve $AX = B$ solutions with error bounds
_SYSVX	Solve $AX = B$ with condition estimate
_SYTRF	$PAP^T = LDL^T$
_SYTRS	Solve $AX = B$ via $PAP^T = LDL^T$
_SYTRI	A^{-1}

LAPACK: Triangular Banded Matrices	
_TBCON	Condition estimate
_TBRFS	Improve $AX = B$, $A^T X = B$ solutions with error bounds
_TBTRS	Solve $AX = B$, $A^T X = B$

4.1 The LDM^T and LDL^T Factorizations

We want to develop a structure-exploiting method for solving symmetric $Ax = b$ problems. To do this we establish a variant of the LU factorization in which A is factored into a three-matrix product LDM^T where D is diagonal and L and M are unit lower triangular. Once this factorization is obtained, the solution to $Ax = b$ may be found in $O(n^2)$ flops by solving $Ly = b$ (forward elimination), $Dz = y$, and $M^T x = z$ (back substitution). The reason for developing the LDM^T factorization is to set the stage for the symmetric case for if $A = A^T$ then $L = M$ and the work associated with the factorization is half of that required by Gaussian elimination. The issue of pivoting is taken up in subsequent sections.

4.1.1 The LDM^T Factorization

Our first result connects the LDM^T factorization with the LU factorization.

Theorem 4.1.1 *If all the leading principal submatrices of $A \in \mathbb{R}^{n \times n}$ are nonsingular, then there exist unique unit lower triangular matrices L and M and a unique diagonal matrix $D = \text{diag}(d_1, \ldots, d_n)$ such that $A = LDM^T$.*

Proof. By Theorem 3.2.1 we know that A has an LU factorization $A = LU$. Set $D = \text{diag}(d_1, \ldots, d_n)$ with $d_i = u_{ii}$ for $i = 1{:}n$. Notice that D is non-singular and that $M^T = D^{-1}U$ is unit upper triangular. Thus, $A = LU = LD(D^{-1}U) = LDM^T$. Uniqueness follows from the uniqueness of the LU factorization as described in Theorem 3.2.1. $\square$

The proof shows that the LDM^T factorization can be found by using Gaussian elimination to compute $A = LU$ and then determining D and M from the equation $U = DM^T$. However, an interesting alternative algorithm can be derived by computing L, D, and M directly.

Assume that we know the first $j - 1$ columns of L, diagonal entries $d_1, \ldots, d_{j-1}$ of D, and the first $j - 1$ rows of M for some j with $1 \le j \le n$. To develop recipes for $L(j + 1{:}n, j)$, $M(j, 1{:}j - 1)$, and d_j we equate jth columns in the equation $A = LDM^T$. In particular,

$$A(1{:}n, j) = Lv \qquad\qquad (4.1.1)$$

where $v = DM^T e_j$. The "top" half of (4.1.1) defines $v(1{:}j)$ as the solution of a known lower triangular system:

$$L(1{:}j, 1{:}j)v(1{:}j) = A(1{:}j, j) .$$

Once we know v then we compute

$$
\begin{aligned}
d(j) &= v(j) \\
M(j, i) &= v(i)/d(i) \qquad i = 1{:}j - 1.
\end{aligned}
$$

The "bottom" half of (4.1.1) says $L(j+1{:}n, 1{:}j)v(1{:}j) = A(j+1{:}n, j)$ which can be rearranged to obtain a recipe for the jth column of L:

$$L(j + 1{:}n, j)v(j) = A(j + 1{:}n, j) - L(j + 1{:}n, 1{:}j - 1)v(1{:}j - 1) .$$

Thus, $L(j + 1{:}n, j)$ is a scaled gaxpy operation and overall we obtain

> for $j = 1{:}n$
>   Solve $L(1{:}j, 1{:}j)v(1{:}j) = A(1{:}j, j)$ for $v(1{:}j)$.
>   for $i = 1{:}j - 1$
> $M(j, i) = v(i)/d(i)$ (4.1.2)
>   end
> $d(j) = v(j)$
> $L(j + 1{:}n, j) =$
> $(A(j + 1{:}n, j) - L(j + 1{:}n, 1{:}j - 1)v(1{:}j - 1)) / v(j)$
> end

As with the LU factorization, it is possible to overwrite A with the L, D, and M factors. If the column version of forward elimination is used to solve for $v(1{:}j)$ then we obtain the following procedure:

Algorithm 4.1.1 (LDM^T) If $A \in \mathbb{R}^{n \times n}$ has an LU factorization then this algorithm computes unit lower triangular matrices L and M and a diagonal matrix $D = \text{diag}(d_1, \ldots, d_n)$ such that $A = LDM^T$. The entry a_{ij} is overwritten with ℓ_{ij} if $i > j$, with d_i if $i = j$, and with m_{ji} if $i < j$.

> for $j = 1{:}n$
>> { Solve $L(1{:}j, 1{:}j)v(1{:}j) = A(1{:}j, j)$. }
>> $v(1{:}j) = A(1{:}j, j)$
>> for $k = 1{:}j - 1$
>>> $v(k+1{:}j) = v(k+1{:}j) - v(k)A(k+1{:}j, k)$
>> end
>> { Compute $M(j, 1{:}j - 1)$ and store in $A(1{:}j - 1, j)$. }
>> for $i = 1{:}j - 1$
>>> $A(i, j) = v(i)/A(i, i)$
>> end
>> { Store $d(j)$ in $A(j, j)$. }
>> $A(j, j) = v(j)$
>> { Compute $L(j+1{:}n, j)$ and store in $A(j+1{:}n, j)$ }
>> for $k = 1{:}j - 1$
>>> $A(j+1{:}n) = A(j+1{:}n, j) - v(k)A(j+1{:}n, k)$
>> end
>> $A(j+1{:}n, j) = A(j+1{:}n, j)/v(j)$
> end

This algorithm involves the same amount of work as the LU factorization, about $2n^3/3$ flops.

The computed solution $\hat{x}$ to $Ax = b$ obtained via Algorithm 4.1.1 and the usual triangular system solvers of §3.1 can be shown to satisfy a perturbed system $(A + E)\hat{x} = b$, where

$$|E| \leq n\mathbf{u}\left(3|A| + 5|\hat{L}||\hat{D}||\hat{M}^T|\right) + O(\mathbf{u}^2) \qquad (4.1.3)$$

and $\hat{L}$, $\hat{D}$, and $\hat{M}$ are the computed versions of L, D, and M, respectively.

As in the case of the LU factorization considered in the previous chapter, the upper bound in (4.1.3) is without limit unless some form of pivoting is done. Hence, for Algorithm 4.1.1 to be a practical procedure, it must be modified so as to compute a factorization of the form $PA = LDM^T$, where P is a permutation matrix chosen so that the entries in L satisfy $|\ell_{ij}| \leq 1$. The details of this are not pursued here since they are straightforward and since our main object for introducing the LDM^T factorization is to motivate special methods for symmetric systems.

Example 4.1.1

$$A = \begin{bmatrix} 10 & 10 & 20 \\ 20 & 25 & 40 \\ 30 & 50 & 61 \end{bmatrix} = \begin{bmatrix} 1 & 0 & 0 \\ 2 & 1 & 0 \\ 3 & 4 & 1 \end{bmatrix} \begin{bmatrix} 10 & 0 & 0 \\ 0 & 5 & 0 \\ 0 & 0 & 1 \end{bmatrix} \begin{bmatrix} 1 & 1 & 2 \\ 0 & 1 & 0 \\ 0 & 0 & 1 \end{bmatrix}$$

and upon completion, Algorithm 4.1.1 overwrites A as follows:

$$A = \begin{bmatrix} 10 & 1 & 2 \\ 2 & 5 & 0 \\ 3 & 4 & 1 \end{bmatrix}.$$

4.1.2   Symmetry and the LDL^T Factorization

There is redundancy in the LDM^T factorization if A is symmetric.

Theorem 4.1.2 *If $A = LDM^T$ is the LDM^T factorization of a nonsingular symmetric matrix A, then $L = M$.*

Proof. The matrix $M^{-1}AM^{-T} = M^{-1}LD$ is both symmetric and lower triangular and therefore diagonal. Since D is nonsingular, this implies that $M^{-1}L$ is also diagonal. But $M^{-1}L$ is unit lower triangular and so $M^{-1}L = I$.□

In view of this result, it is possible to halve the work in Algorithm 4.1.1 when it is applied to a symmetric matrix. In the jth step we already know $M(j, 1{:}j - 1)$ since $M = L$ and we presume knowledge of L's first $j - 1$ columns. Recall that in the jth step of (4.1.2) the vector $v(1{:}j)$ is defined by the first j components of $DM^T e_j$. Since $M = L$, this says that

$$v(1{:}j) = \begin{bmatrix} d(1)L(j,1) \\ \vdots \\ d(j-1)L(j,j-1) \\ d(j) \end{bmatrix}.$$

Hence, the vector $v(1{:}j - 1)$ can be obtained by a simple scaling of L's jth row. The formula $v(j) = A(j,j) - L(j, 1{:}j - 1)v(1{:}j - 1)$ can be derived from the jth equation in $L(1{:}j, 1{:}j)v = A(1{:}j, j)$ rendering

> **for** $j = 1{:}n$
>   **for** $i = 1{:}j - 1$
> $v(i) = L(j,i)d(i)$
>   **end**
> $v(j) = A(j,j) - L(j, 1{:}j - 1)v(1{:}j - 1)$
> $d(j) = v(j)$
> $L(j + 1{:}n, j) =$
> $(A(j + 1{:}n, j) - L(j + 1{:}n, 1{:}j - 1)v(1{:}j - 1))/v(j)$
> **end**

With overwriting this becomes

Algorithm 4.1.2 (LDL^T) If $A \in \mathbb{R}^{n \times n}$ is symmetric and has an LU factorization then this algorithm computes a unit lower triangular matrix L and a diagonal matrix $D = \text{diag}(d_1, \ldots, d_n)$ so $A = LDL^T$. The entry a_{ij} is overwritten with ℓ_{ij} if $i > j$ and with d_i if $i = j$.

 for $j = 1{:}n$
  { Compute $v(1{:}j)$. }
  for $i = 1{:}j - 1$
   $v(i) = A(j, i)A(i, i)$
  end
  $v(j) = A(j, j) - A(j, 1{:}j - 1)v(1{:}j - 1)$
  { Store $d(j)$ and compute $L(j + 1{:}n, j)$. }
  $A(j, j) = v(j)$
  $A(j + 1{:}n, j) =$
   $(A(j + 1{:}n, j) - A(j + 1{:}n, 1{:}j - 1)v(1{:}j - 1))/v(j)$
 end

This algorithm requires $n^3/3$ flops, about half the number of flops involved in Gaussian elimination.

 In the next section, we show that if A is both symmetric and positive definite, then Algorithm 4.1.2 not only runs to completion, but is extremely stable. If A is symmetric but not positive definite, then pivoting may be necessary and the methods of §4.4 are relevant.

Example 4.1.2

$$A = \begin{bmatrix} 10 & 20 & 30 \\ 20 & 45 & 80 \\ 30 & 80 & 171 \end{bmatrix} = \begin{bmatrix} 1 & 0 & 0 \\ 2 & 1 & 0 \\ 3 & 4 & 1 \end{bmatrix} \begin{bmatrix} 10 & 0 & 0 \\ 0 & 5 & 0 \\ 0 & 0 & 1 \end{bmatrix} \begin{bmatrix} 1 & 2 & 3 \\ 0 & 1 & 4 \\ 0 & 0 & 1 \end{bmatrix}$$

and so if Algorithm 4.1.2 is applied, A is overwritten by

$$A = \begin{bmatrix} 10 & 20 & 30 \\ 2 & 5 & 80 \\ 3 & 4 & 1 \end{bmatrix}.$$

Problems

P4.1.1 Show that the LDM^T factorization of a nonsingular A is unique if it exists.

P4.1.2 Modify Algorithm 4.1.1 so that it computes a factorization of the form $PA = LDM^T$, where L and M are both unit lower triangular, D is diagonal, and P is a permutation that is chosen so $|\ell_{ij}| \leq 1$.

P4.1.3 Suppose the n-by-n symmetric matrix $A = (a_{ij})$ is stored in a vector c as follows: $c = (a_{11}, a_{21}, \ldots, a_{n1}, a_{22}, \ldots, a_{n2}, \ldots, a_{nn})$. Rewrite Algorithm 4.1.2 with A stored in this fashion. Get as much indexing outside the inner loops as possible.

P4.1.4 Rewrite Algorithm 4.1.2 for A stored by diagonal. See §1.2.8.

Notes and References for Sec. 4.1

Algorithm 4.1.1 is related to the methods of Crout and Doolittle in that outer product updates are avoided. See Chapter 4 of Fox (1964) or Stewart (1973,131–149). An Algol procedure may be found in

H.J. Bowdler, R.S. Martin, G. Peters, and J.H. Wilkinson (1966), "Solution of Real and Complex Systems of Linear Equations," *Numer. Math. 8*, 217–234.

See also

G.E. Forsythe (1960). "Crout with Pivoting," *Comm. ACM 3*, 507–08.
W.M. McKeeman (1962). "Crout with Equilibration and Iteration," *Comm. ACM 5*, 553–55.

Just as algorithms can be tailored to exploit structure, so can error analysis and perturbation theory:

M. Arioli, J. Demmel, and I. Duff (1989). "Solving Sparse Linear Systems with Sparse Backward Error," *SIAM J. Matrix Anal. Appl. 10*, 165–190.
J.R. Bunch, J.W. Demmel, and C.F. Van Loan (1989). "The Strong Stability of Algorithms for Solving Symmetric Linear Systems," *SIAM J. Matrix Anal. Appl. 10*, 494–499.
A. Barrlund (1991). "Perturbation Bounds for the LDL^T and LU Decompositions," *BIT 31*, 358–363.
D.J. Higham and N.J. Higham (1992). "Backward Error and Condition of Structured Linear Systems," *SIAM J. Matrix Anal. Appl. 13*, 162–175.

4.2   Positive Definite Systems

A matrix $A \in \mathbb{R}^{n \times n}$ is *positive definite* if $x^T A x > 0$ for all nonzero $x \in \mathbb{R}^n$. Positive definite systems constitute one of the most important classes of special $Ax = b$ problems. Consider the 2-by-2 symmetric case. If

$$A = \begin{bmatrix} a_{11} & a_{12} \\ a_{21} & a_{22} \end{bmatrix}$$

is positive definite then

$$
\begin{array}{llll}
x = (1, 0)^T & \Rightarrow & x^T A x = a_{11} > 0 \\
x = (0, 1)^T & \Rightarrow & x^T A x = a_{22} > 0 \\
x = (1, 1)^T & \Rightarrow & x^T A x = a_{11} + 2a_{12} + a_{22} > 0 \\
x = (1, -1)^T & \Rightarrow & x^T A x = a_{11} - 2a_{12} + a_{22} > 0.
\end{array}
$$

The last two equations imply $|a_{12}| \le (a_{11} + a_{22})/2$. From these results we see that the largest entry in A is on the diagonal and that it is positive. This turns out to be true in general. A symmetric positive definite matrix has a "weighty" diagonal. The mass on the diagonal is not blatantly obvious as in the case of diagonal dominance but it has the same effect in that it precludes the need for pivoting. See §3.4.10.

We begin with a few comments about the property of positive definiteness and what it implies in the unsymmetric case with respect to pivoting. We then focus on the efficient organization of the Cholesky procedure which can be used to safely factor a symmetric positive definite A. Gaxpy, outer product, and block versions are developed. The section concludes with a few comments about the semidefinite case.

4.2.1 Positive Definiteness

Suppose $A \in \mathbb{R}^{n \times n}$ is positive definite. It is obvious that a positive definite matrix is nonsingular for otherwise we could find a nonzero x so $x^T A x = 0$. However, much more is implied by the positivity of the *quadratic form* $x^T A x$ as the following results show.

Theorem 4.2.1 *If $A \in \mathbb{R}^{n \times n}$ is positive definite and $X \in \mathbb{R}^{n \times k}$ has rank k, then $B = X^T A X \in \mathbb{R}^{k \times k}$ is also positive definite.*

Proof. If $z \in \mathbb{R}^k$ satisfies $0 \geq z^T B z = (Xz)^T A(Xz)$ then $Xz = 0$. But since X has full column rank, this implies that $z = 0$. $\square$

Corollary 4.2.2 *If A is positive definite then all its principal submatrices are positive definite. In particular, all the diagonal entries are positive.*

Proof. If $v \in \mathbb{R}^k$ is an integer vector with $1 \leq v_1 < \cdots < v_k \leq n$, then $X = I_n(:, v)$ is a rank k matrix made up columns $v_1, \ldots, v_k$ of the identity. It follows from Theorem 4.2.1 that $A(v, v) = X^T A X$ is positive definite. $\square$

Corollary 4.2.3 *If A is positive definite then the factorization $A = LDM^T$ exists and $D = \operatorname{diag}(d_1, \ldots, d_n)$ has positive diagonal entries.*

Proof. From Corollary 4.2.2, it follows that the submatrices $A(1:k, 1:k)$ are nonsingular for $k = 1:n$ and so from Theorem 4.1.1 the factorization $A = LDM^T$ exists. If we apply Theorem 4.2.1 with $X = L^{-T}$ then $B = DM^T L^{-T} = L^{-1} A L^{-T}$ is positive definite. Since $M^T L^{-T}$ is unit upper triangular, B and D have the same diagonal and it must be positive. $\square$

There are several typical situations that give rise to positive definite matrices in practice:

- The quadratic form is an energy function whose positivity is guaranteed from physical principles.

- The matrix A equals a *cross-product* $X^T X$ where X has full column rank. (Positive definiteness follows by setting $A = I_n$ in Theorem 4.2.1.)

- Both A and A^T are diagonally dominant and each a_{ii} is positive.

4.2.2   Unsymmetric Positive Definite Systems

The mere existence of an LDM^T factorization does not mean that its computation is advisable because the resulting factors may have unacceptably large elements. For example, if $\epsilon > 0$ then the matrix

$$A = \begin{bmatrix} \epsilon & m \\ -m & \epsilon \end{bmatrix} = \begin{bmatrix} 1 & 0 \\ -m/\epsilon & 1 \end{bmatrix} \begin{bmatrix} \epsilon & 0 \\ 0 & \epsilon + m^2/\epsilon \end{bmatrix} \begin{bmatrix} 1 & m/\epsilon \\ 0 & 1 \end{bmatrix}$$

is positive definite. But if $m/\epsilon \gg 1$, then pivoting is recommended.

The following result suggests when to expect element growth in the LDM^T factorization of a positive definite matrix.

Theorem 4.2.4 *Let $A \in \mathbb{R}^{n \times n}$ be positive definite and set $T = (A + A^T)/2$ and $S = (A - A^T)/2$. If $A = LDM^T$, then*

$$\| \, |L||D||M^T| \, \|_F \le n \left(\| \, T \, \|_2 + \| \, ST^{-1}S \, \|_2 \right) \qquad (4.2.1)$$

Proof. See Golub and Van Loan (1979). □

The theorem suggests when it is safe not to pivot. Assume that the computed factors $\hat{L}$, $\hat{D}$, and $\hat{M}$ satisfy:

$$\| \, |\hat{L}||\hat{D}||\hat{M}^T| \, \|_F \le c\| \, |L||D||M^T| \, \|_F, \qquad (4.2.2)$$

where c is a constant of modest size. It follows from (4.2.1) and the analysis in §3.3 that if these factors are used to compute a solution to $Ax = b$, then the computed solution $\hat{x}$ satisfies $(A + E)\hat{x} = b$ with

$$\| \, E \, \|_F \le \mathbf{u} \left(3n\| \, A \, \|_F + 5cn^2 \left(\| \, T \, \|_2 + \| \, ST^{-1}S \, \|_2 \right) \right) + O(\mathbf{u}^2). \quad (4.2.3)$$

It is easy to show that $\| \, T \, \|_2 \le \| \, A \, \|_2$, and so it follows that if

$$\Omega \;=\; \frac{\| \, ST^{-1}S \, \|_2}{\| \, A \, \|_2} \qquad (4.2.4)$$

is not too large then it is safe not to pivot. In other words, the norm of the skew part S has to be modest relative to the condition of the symmetric part T. Sometimes it is possible to estimate Ω in an application. This is trivially the case when A is symmetric for then $\Omega = 0$.

4.2.3   Symmetric Positive Definite Systems

When we apply the above results to a symmetric positive definite system we know that the factorization $A = LDL^T$ exists and moreover is stable to compute. However, in this situation another factorization is available.

Theorem 4.2.5 (Cholesky Factorization) *If $A \in \mathbb{R}^{n \times n}$ is symmetric positive definite, then there exists a unique lower triangular $G \in \mathbb{R}^{n \times n}$ with positive diagonal entries such that $A = GG^T$.*

Proof. From Theorem 4.1.2, there exists a unit lower triangular L and a diagonal $D = \mathrm{diag}(d_1, \ldots, d_n)$ such that $A = LDL^T$. Since the d_k are positive, the matrix $G = L\,\mathrm{diag}(\sqrt{d_1}, \ldots, \sqrt{d_n})$ is real lower triangular with positive diagonal entries. It also satisfies $A = GG^T$. Uniqueness follows from the uniqueness of the LDL^T factorization. $\square$

The factorization $A = GG^T$ is known as the *Cholesky factorization* and G is referred to as the *Cholesky triangle*. Note that if we compute the Cholesky factorization and solve the triangular systems $Gy = b$ and $G^Tx = y$, then $b = Gy = G(G^Tx) = (GG^T)x = Ax$.

Our proof of the Cholesky factorization in Theorem 4.2.5 is constructive. However, more effective methods for computing the Cholesky triangle can be derived by manipulating the equation $A = GG^T$. This can be done in several ways as we show in the next few subsections.

Example 4.2.1 The matrix

$$\begin{bmatrix} 2 & -2 \\ -2 & 5 \end{bmatrix} = \begin{bmatrix} 1 & 0 \\ -1 & 1 \end{bmatrix} \begin{bmatrix} 2 & 0 \\ 0 & 3 \end{bmatrix} \begin{bmatrix} 1 & -1 \\ 0 & 1 \end{bmatrix} = \begin{bmatrix} \sqrt{2} & 0 \\ -\sqrt{2} & \sqrt{3} \end{bmatrix} \begin{bmatrix} \sqrt{2} & -\sqrt{2} \\ 0 & \sqrt{3} \end{bmatrix}$$

is positive definite.

4.2.4 Gaxpy Cholesky

We first derive an implementation of Cholesky that is rich in the gaxpy operation. If we compare jth columns in the equation $A = GG^T$ then we obtain

$$A(:,j) = \sum_{k=1}^{j} G(j,k)G(:,k).$$

This says that

$$G(j,j)G(:,j) = A(:,j) - \sum_{k=1}^{j-1} G(j,k)G(:,k) \equiv v. \quad (4.2.5)$$

If we know the first $j - 1$ columns of G, then v is computable. It follows by equating components in (4.2.5) that

$$G(j{:}n, j) = v(j{:}n)/\sqrt{v(j)}.$$

This is a scaled gaxpy operation and so we obtain the following gaxpy-based method for computing the Cholesky factorization:

$$\textbf{for } j = 1{:}n$$
$$\quad v(j{:}n) = A(j{:}n, j)$$
$$\quad \textbf{for } k = 1{:}j - 1$$
$$\quad\quad v(j{:}n) = v(j{:}n) - G(j,k)G(j{:}n,k)$$
$$\quad \textbf{end}$$
$$\quad G(j{:}n, j) = v(j{:}n)/\sqrt{v(j)}$$
$$\textbf{end}$$

It is possible to arrange the computations so that G overwrites the lower triangle of A.

Algorithm 4.2.1 (Cholesky: Gaxpy Version) Given a symmetric positive definite $A \in \mathbb{R}^{n \times n}$, the following algorithm computes a lower triangular $G \in \mathbb{R}^{n \times n}$ such that $A = GG^T$. For all $i \geq j$, $G(i,j)$ overwrites $A(i,j)$.

$$\textbf{for } j = 1{:}n$$
$$\quad \textbf{if } j > 1$$
$$\quad\quad A(j{:}n, j) = A(j{:}n, j) - A(j{:}n, 1{:}j - 1)A(j, 1{:}j - 1)^T$$
$$\quad \textbf{end}$$
$$\quad A(j{:}n, j) = A(j{:}n, j)/\sqrt{A(j,j)}$$
$$\textbf{end}$$

This algorithm requires $n^3/3$ flops.

4.2.5   Outer Product Cholesky

An alternative Cholesky procedure based on outer product (rank-1) updates can be derived from the partitioning

$$A = \begin{bmatrix} \alpha & v^T \\ v & B \end{bmatrix} = \begin{bmatrix} \beta & 0 \\ v/\beta & I_{n-1} \end{bmatrix} \begin{bmatrix} 1 & 0 \\ 0 & B - vv^T/\alpha \end{bmatrix} \begin{bmatrix} \beta & v^T/\beta \\ 0 & I_{n-1} \end{bmatrix}.$$
$$(4.2.6)$$

Here, $\beta = \sqrt{\alpha}$ and we know that $\alpha > 0$ because A is positive definite. Note that $B - vv^T/\alpha$ is positive definite because it is a principal submatrix of $X^T A X$ where

$$X = \begin{bmatrix} 1 & -v^T/\alpha \\ 0 & I_{n-1} \end{bmatrix}.$$

If we have the Cholesky factorization $G_1 G_1^T = B - vv^T/\alpha$, then from (4.2.6) it follows that $A = GG^T$ with

$$G = \begin{bmatrix} \beta & 0 \\ v/\beta & G_1 \end{bmatrix}.$$

Thus, the Cholesky factorization can be obtained through the repeated application of (4.2.6), much in the the the style of kji Gaussian elimination.

Algorithm 4.2.2 (Cholesky: Outer product Version) Given a symmetric positive definite $A \in \mathbb{R}^{n \times n}$, the following algorithm computes a lower triangular $G \in \mathbb{R}^{n \times n}$ such that $A = GG^T$. For all $i \geq j$, $G(i, j)$ overwrites $A(i, j)$.

> for $k = 1{:}n$
> $\quad A(k, k) = \sqrt{A(k, k)}$
> $\quad A(k + 1{:}n, k) = A(k + 1{:}n, k)/A(k, k)$
> $\quad$ for $j = k + 1{:}n$
> $\quad\quad A(j{:}n, j) = A(j{:}n, j) - A(j{:}n, k)A(j, k)$
> $\quad$ end
> end

This algorithm involves $n^3/3$ flops. Note that the j-loop computes the lower triangular part of the outer product update

$$A(k + 1{:}n, k + 1{:}n) = A(k + 1{:}n, k + 1{:}n) - A(k + 1{:}n, k)A(k + 1{:}n, k)^T.$$

Recalling our discussion in §1.4.8 about gaxpy versus outer product updates, it is easy to show that Algorithm 4.2.1 involves fewer vector touches than Algorithm 4.2.2 by a factor of two.

4.2.6 Block Dot Product Cholesky

Suppose $A \in \mathbb{R}^{n \times n}$ is symmetric positive definite. Regard $A = (A_{ij})$ and its Cholesky factor $G = (G_{ij})$ as N-by-N block matrices with square diagonal blocks. By equating (i, j) blocks in the equation $A = GG^T$ with $i \geq j$ it follows that

$$A_{ij} = \sum_{k=1}^{j} G_{ik} G_{jk}^T.$$

Defining

$$S = A_{ij} - \sum_{k=1}^{j-1} G_{ik} G_{jk}^T$$

we see that $G_{jj} G_{jj}^T = S$ if $i = j$ and that $G_{ij} G_{jj}^T = S$ if $i > j$. Properly sequenced, these equations can be arranged to compute all the G_{ij}:

Algorithm 4.2.3 (Cholesky: Block Dot Product Version) Given a symmetric positive definite $A \in \mathbb{R}^{n \times n}$, the following algorithm computes a lower triangular $G \in \mathbb{R}^{n \times n}$ such that $A = GG^T$. The lower triangular part of A is overwritten by the lower triangular part of G. A is regarded as an N-by-N block matrix with square diagonal blocks.

for $j = 1{:}N$
    for $i = j{:}N$
$$S = A_{ij} - \sum_{k=1}^{j-1} G_{ik} G_{jk}^T$$
        if $i = j$
            Compute Cholesky factorization $S = G_{jj} G_{jj}^T$.
        else
            Solve $G_{ij} G_{jj}^T = S$ for G_{ij}
        end
        Overwrite A_{ij} with G_{ij}.
    end
end

The overall process involves $n^3/3$ flops like the other Cholesky procedures that we have developed. The procedure is rich in matrix multiplication assuming a suitable blocking of the matrix A. For example, if $n = rN$ and each A_{ij} is r-by-r, then the level-3 fraction is approximately $1 - (1/N^2)$.

Algorithm 4.2.3 is incomplete in the sense that we have not specified how the products $G_{ik} G_{jk}$ are formed or how the r-by-r Cholesky factorizations $S = G_{jj} G_{jj}^T$ are computed. These important details would have to be worked out carefully in order to extract high performance.

Another block procedure can be derived from the gaxpy Cholesky algorithm. After r steps of Algorithm 4.2.1 we know the matrices $G_{11} \in \mathbb{R}^{r \times r}$ and $G_{21} \in \mathbb{R}^{(n-r) \times r}$ in

$$\begin{bmatrix} A_{11} & A_{12} \\ A_{21} & A_{22} \end{bmatrix} = \begin{bmatrix} G_{11} & 0 \\ G_{21} & I_{n-r} \end{bmatrix} \begin{bmatrix} I_r & 0 \\ 0 & \tilde{A} \end{bmatrix} \begin{bmatrix} G_{11} & 0 \\ G_{21} & I_{n-r} \end{bmatrix}^T .$$

We then perform r more steps of gaxpy Cholesky not on A but on the reduced matrix $\tilde{A} = A_{22} - G_{21} G_{21}^T$ which we *explicitly* form exploiting symmetry. Continuing in this way we obtain a block Cholesky algorithm whose kth step involves r gaxpy Cholesky steps on a matrix of order $n - (k-1)r$ followed a level-3 computation having order $n - kr$. The level-3 fraction is approximately equal to $1 - 3/(2N)$ if $n \approx rN$.

4.2.7   Stability of the Cholesky Process

In exact arithmetic, we know that a symmetric positive definite matrix has a Cholesky factorization. Conversely, if the Cholesky process runs to completion with strictly positive square roots, then A is positive definite. Thus, to find out if a matrix A is positive definite, we merely try to compute its Cholesky factorization using any of the methods given above.

The situation in the context of roundoff error is more interesting. The numerical stability of the Cholesky algorithm roughly follows from the inequality

$$g_{ij}^2 \leq \sum_{k=1}^{i} g_{ik}^2 = a_{ii}.$$

This shows that the entries in the Cholesky triangle are nicely bounded. The same conclusion can be reached from the equation $\| G \|_2^2 = \| A \|_2$.

The roundoff errors associated with the Cholesky factorization have been extensively studied in a classical paper by Wilkinson (1968). Using the results in this paper, it can be shown that if $\hat{x}$ is the computed solution to $Ax = b$, obtained via any of our Cholesky procedures then $\hat{x}$ solves the perturbed system $(A + E)\hat{x} = b$ where $\| E \|_2 \leq c_n u \| A \|_2$ and c_n is a small constant depending upon n. Moreover, Wilkinson shows that if $q_n u \kappa_2(A) \leq 1$ where q_n is another small constant, then the Cholesky process runs to completion, i.e, no square roots of negative numbers arise.

Example 4.2.2 If Algorithm 4.2.2 is applied to the positive definite matrix

$$A = \begin{bmatrix} 100 & 15 & .01 \\ 15 & 2.3 & .01 \\ .01 & .01 & 1.00 \end{bmatrix}$$

and $\beta = 10$, $t = 2$, rounded arithmetic used, then $\hat{g}_{11} = 10$, $\hat{g}_{21} = 1.5$, $\hat{g}_{31} = .001$ and $\hat{g}_{22} = 0.00$. The algorithm then breaks down trying to compute g_{32}.

4.2.8   The Semidefinite Case

A matrix is said to be *positive semidefinite* if $x^T A x \geq 0$ for all vectors x. Symmetric positive semidefinite (*sps*) matrices are important and we briefly discuss some Cholesky-like manipulations that can be used to solve various *sps* problems. Results about the diagonal entries in an *sps* matrix are needed first.

Theorem 4.2.6 *If $A \in \mathbb{R}^{n \times n}$ is symmetric positive semidefinite, then*

$$|a_{ij}| \leq (a_{ii} + a_{jj})/2 \tag{4.2.7}$$

$$|a_{ij}| \leq \sqrt{a_{ii}a_{jj}} \quad (i \neq j) \tag{4.2.8}$$

$$\max_{i,j} |a_{ij}| = \max_{i} a_{ii} \tag{4.2.9}$$

$$a_{ii} = 0 \quad \Rightarrow \quad A(i,:) = 0, \ A(:,i) = 0 \tag{4.2.10}$$

Proof. If $x = e_i + e_j$ then $0 \leq x^T A x = a_{ii} + a_{jj} + 2a_{ij}$ while $x = e_i - e_j$ implies $0 \leq x^T A x = a_{ii} + a_{jj} - 2a_{ij}$. Inequality (4.2.7) follows from these two results. Equation (4.2.9) is an easy consequence of (4.2.7).

To prove (4.2.8) assume without loss of generality that $i = 1$ and $j = 2$ and consider the inequality

$$0 \leq \begin{bmatrix} x \\ 1 \end{bmatrix}^T \begin{bmatrix} a_{11} & a_{12} \\ a_{21} & a_{22} \end{bmatrix} \begin{bmatrix} x \\ 1 \end{bmatrix} = a_{11}x^2 + 2a_{12}x + a_{22}$$

which holds since $A(1{:}2, 1{:}2)$ is also semidefinite. This is a quadratic equation in x and for the inequality to hold, the discriminant $4a_{12}^2 - 4a_{11}a_{22}$ must be negative. Implication (4.2.10) follows from (4.2.8).   □

Consider what happens when outer product Cholesky is applied to an *sps* matrix. If a zero $A(k, k)$ is encountered then from (4.2.10) $A(k{:}n, k)$ is zero and there is "nothing to do" and we obtain

> **for** $k = 1{:}n$
>   **if** $A(k, k) > 0$
> $A(k, k) = \sqrt{A(k, k)}$
> $A(k + 1{:}n, k) = A(k + 1{:}n, k)/A(k, k)$
>   **for** $j = k + 1{:}n$
> $A(j{:}n, j) = A(j{:}n, j) - A(j{:}n, k)A(j, k)$ (4.2.11)
>   **end**
>   **end**
> **end**

Thus, a simple change makes Algorithm 4.2.2 applicable to the semidefinite case. However, in practice rounding errors preclude the generation of exact zeros and it may be preferable to incorporate pivoting.

4.2.9 Symmetric Pivoting

To preserve symmetry in a symmetric A we only consider data reorderings of the form PAP^T where P is a permutation. Row permutations $(A \leftarrow PA)$ or column permutations $(A \leftarrow AP)$ alone destroy symmetry. An update of the form

$$A \leftarrow PAP^T$$

is called a *symmetric permutation* of A. Note that such an operation *does not* move off-diagonal elements to the diagonal. The diagonal of PAP^T is a reordering of the diagonal of A.

Suppose at the beginning of the kth step in (4.2.11) we symmetrically permute the largest diagonal entry of $A(k{:}n, k{:}n)$ into the lead position. If that largest diagonal entry is zero then $A(k{:}n, k{:}n) = 0$ by virtue of (4.2.10). In this way we can compute the factorization $PAP^T = GG^T$ where $G \in \mathbb{R}^{n \times (k-1)}$ is lower triangular.

Algorithm 4.2.4 Suppose $A \in \mathbb{R}^{n \times n}$ is symmetric positive semidefinite and that $\text{rank}(A) = r$. The following algorithm computes a permutation P, the index r, and an n-by-r lower triangular matrix G such that $PAP^T = GG^T$. The lower triangular part of $A(:, 1{:}r)$ is overwritten by the lower triangular part of G. $P = P_r \cdots P_1$ where P_k is the identity with rows k and $piv(k)$ interchanged.

$r = 0$
for $k = 1{:}n$
       Find q $(k \le q \le n)$ so $A(q,q) = \max\{A(k,k), .., A(n,n)\}$
       if $A(q,q) > 0$
           $r = r + 1$
           $piv(k) = q$
           $A(k,:) \leftrightarrow A(q,:)$
           $A(:,k) \leftrightarrow A(:,q)$
           $A(k,k) = \sqrt{A(k,k)}$
           $A(k+1{:}n, k) = A(k+1{:}n, k)/A(k,k)$
           for $j = k+1{:}n$
                $A(j{:}n, j) = A(j{:}n, j) - A(j{:}n, k)A(j,k)$
           end
       end
end

In practice, a tolerance is used to detect small $A(k,k)$. However, the situation is quite tricky and the reader should consult Higham (1989). In addition, §5.5 has a discussion of tolerances in the rank detection problem. Finally, we remark that a truly efficient implementation of Algorithm 4.2.4 would only access the lower triangular portion of A.

4.2.10   The Polar Decomposition and Square Root

Let $A = U_1 \Sigma_1 V^T$ be the thin SVD of $A \in \mathbb{R}^{m \times n}$ where $m \ge n$. Note that

$$A = (U_1 V^T)(V \Sigma_1 V^T) \equiv ZP \qquad (4.2.12)$$

where $Z = U_1 V^T$ and $P = V \Sigma_1 V^T$. Z has orthonormal columns and P is symmetric positive semidefinite because

$$x^T P x = (V^T x)^T \Sigma_1 (V^T x) = \sum_{k=1}^{n} \sigma_k y_k^2 \ge 0$$

where $y = V^T x$. The decomposition (4.2.12) is called the *polar decomposition* because it is analogous to the complex number factorization $z = e^{i \cdot arg(z)}|z|$. See §12.4.1 for further discussion.

Another important decomposition is the *matrix square root*. Suppose $A \in \mathbb{R}^{n \times n}$ is symmetric positive semidefinite and that $A = GG^T$ is its Cholesky factorization. If $G = U \Sigma V^T$ is G's SVD and $X = U \Sigma U^T$, then X is symmetric positive semidefinite and

$$A = GG^T = (U \Sigma V^T)(U \Sigma V^T)^T = U \Sigma^2 U^T = (U \Sigma U^T)(U \Sigma U^T) = X^2.$$

Thus, X is a *square root* of A. It can be shown (most easily with eigenvalue theory) that a symmetric positive semidefinite matrix has a unique symmetric positive semidefinite square root.

Problems

P4.2.1 Suppose that $H = A + iB$ is Hermitian and positive definite with $A, B \in \mathbf{R}^{n \times n}$. This means that $x^H H x > 0$ whenever $x \neq 0$. (a) Show that

$$C = \left[\begin{array}{cc} A & -B \\ B & A \end{array} \right]$$

is symmetric and positive definite. (b) Formulate an algorithm for solving $(A+iB)(x+iy) = (b + ic)$, where b, c, x, and y are in $\mathbf{R}^n$. It should involve $8n^3/3$ flops. How much storage is required?

P4.2.2 Suppose $A \in \mathbf{R}^{n \times n}$ is symmetric and positive definite. Give an algorithm for computing an upper triangular matrix $R \in \mathbf{R}^{n \times n}$ such that $A = RR^T$.

P4.2.3 Let $A \in \mathbf{R}^{n \times n}$ be positive definite and set $T = (A+A^T)/2$ and $S = (A-A^T)/2$. (a) Show that $\| A^{-1} \|_2 \leq \| T^{-1} \|_2$ and $x^T A^{-1} x \leq x^T T^{-1} x$ for all $x \in \mathbf{R}^n$. (b) Show that if $A = LDM^T$, then $d_k \geq 1/\| T^{-1} \|_2$ for $k = 1{:}n$

P4.2.4 Find a 2-by-2 real matrix A with the property that $x^T A x > 0$ for all real nonzero 2-vectors but which is not positive definite when regarded as a member of $\mathbf{C}^{2 \times 2}$.

P4.2.5 Suppose $A \in \mathbf{R}^{n \times n}$ has a positive diagonal. Show that if both A and A^T are strictly diagonally dominant, then A is positive definite.

P4.2.6 Show that the function $f(x) = (x^T A x)/2$ is a vector norm on $\mathbf{R}^n$ if and only if A is positive definite.

P4.2.7 Modify Algorithm 4.2.1 so that if the square root of a negative number is encountered, then the algorithm finds a unit vector x so $x^T A x < 0$ and terminates.

P4.2.8 The numerical range $W(A)$ of a complex matrix A is defined to be the set $W(A) = \{x^H A x : x^H x = 1\}$. Show that if $0 \notin W(A)$, then A has an LU factorization.

P4.2.9 Formulate an $m < n$ version of the polar decomposition for $A \in \mathbf{R}^{m \times n}$.

P4.2.10 Suppose $A = I + uu^T$ where $A \in \mathbf{R}^{n \times n}$ and $\| u \|_2 = 1$. Give explicit formulae for the diagonal and subdiagonal of A's Cholesky factor.

P4.2.11 Suppose $A \in \mathbf{R}^{n \times n}$ is symmetric positive definite and that its Cholesky factor is available. Let $e_k = I_n(:, k)$. For $1 \leq i < j \leq n$, let α_{ij} be the smallest real that makes $A + \alpha(e_i e_j^T + e_j e_i^T)$ singular. Likewise, let α_{ii} be the smallest real that makes $(A + \alpha e_i e_i^T)$ singular. Show how to compute these quantities using the Sherman-Morrison-Woodbury formula. How many flops are required to find all the α_{ij}?

Notes and References for Sec. 4.2

The definiteness of the quadratic form $x^T A x$ can frequently be established by considering the mathematics of the underlying problem. For example, the discretization of certain partial differential operators gives rise to provably positive definite matrices. Aspects of the unsymmetric positive definite problem are discussed in

A. Buckley (1974). "A Note on Matrices $A = I + H$, H Skew-Symmetric," *Z. Angew. Math. Mech. 54*, 125–26.

A. Buckley (1977). "On the Solution of Certain Skew-Symmetric Linear Systems," *SIAM J. Num. Anal. 14*, 566–70.

G.H. Golub and C. Van Loan (1979). "Unsymmetric Positive Definite Linear Systems," *Lin. Alg. and Its Applic. 28*, 85–98.

R. Mathias (1992). "Matrices with Positive Definite Hermitian Part: Inequalities and Linear Systems," *SIAM J. Matrix Anal. Appl. 13*, 640–654.

Symmetric positive definite systems constitute the most important class of special $Ax = b$ problems. Algol programs for these problems are given in

R.S. Martin, G. Peters, and J.H. Wilkinson (1965). "Symmetric Decomposition of a Positive Definite Matrix," *Numer. Math. 7*, 362-83.

R.S. Martin, G. Peters, and J.H. Wilkinson (1966). "Iterative Refinement of the Solution of a Positive Definite System of Equations," *Numer. Math. 8*, 203–16.

F.L. Bauer and C. Reinsch (1971). "Inversion of Positive Definite Matrices by the Gauss-Jordan Method," in *Handbook for Automatic Computation Vol. 2, Linear Algebra*, J.H. Wilkinson and C. Reinsch, eds. Springer-Verlag, New York, 45–49.

The roundoff errors associated with the method are analyzed in

J.H. Wilkinson (1968). "À Priori Error Analysis of Algebraic Processes," *Proc. International Congress Math.* (Moscow: Izdat. Mir, 1968), pp. 629–39.

J. Meinguet (1983). "Refined Error Analyses of Cholesky Factorization," *SIAM J. Numer. Anal. 20*, 1243–1250.

A. Kielbasinski (1987). "A Note on Rounding Error Analysis of Cholesky Factorization," *Lin. Alg. and Its Applic. 88/89*, 487–494.

N.J. Higham (1990). "Analysis of the Cholesky Decomposition of a Semidefinite Matrix," in *Reliable Numerical Computation*, M.G. Cox and S.J. Hammarling (eds), Oxford University Press, Oxford, UK, 161–185.

R. Carter (1991). "Y-MP Floating Point and Cholesky Factorization," *Int'l J. High Speed Computing 3*, 215–222.

J-Guang Sun (1992). "Rounding Error and Perturbation Bounds for the Cholesky and LDL^T Factorizations," *Lin. Alg. and Its Applic. 173*, 77–97.

The question of how the Cholesky triangle G changes when $A = GG^T$ is perturbed is analyzed in

G.W. Stewart (1977b). "Perturbation Bounds for the QR Factorization of a Matrix," *SIAM J. Num. Anal. 14*, 509-18.

Z. Dramăc, M. Omladič, and K. Veselič (1994). "On the Perturbation of the Cholesky Factorization," *SIAM J. Matrix Anal. Appl. 15*,1319–1332.

Nearness/sensitivity issues associated with positive semi-definiteness and the polar decomposition are presented in

N.J. Higham (1988). "Computing a Nearest Symmetric Positive Semidefinite Matrix," *Lin. Alg. and Its Applic. 103*, 103–118.

R. Mathias (1993). "Perturbation Bounds for the Polar Decomposition," *SIAM J. Matrix Anal. Appl. 14*, 588–597.

R-C. Li (1995). "New Perturbation Bounds for the Unitary Polar Factor," *SIAM J. Matrix Anal. Appl. 16*, 327–332.

Computationally-oriented references for the polar decomposition and the square root are given in §8.6 and §11.2 respectively.

4.3   Banded Systems

In many applications that involve linear systems, the matrix of coefficients is *banded*. This is the case whenever the equations can be ordered so that each unknown x_i appears in only a few equations in a "neighborhood" of the ith equation. Formally, we say that $A = (a_{ij})$ has *upper bandwidth q* if $a_{ij} = 0$ whenever $j > i + q$ and *lower bandwidth p* if $a_{ij} = 0$ whenever $i > j + p$. Substantial economies can be realized when solving banded systems because the triangular factors in LU, GG^T, LDM^T, etc., are also banded.

Before proceeding the reader is advised to review §1.2 where several aspects of band matrix manipulation are discussed.

4.3.1   Band LU Factorization

Our first result shows that if A is banded and $A = LU$ then $L(U)$ inherits the lower (upper) bandwidth of A.

Theorem 4.3.1 *Suppose $A \in \mathbb{R}^{n \times n}$ has an LU factorization $A = LU$. If A has upper bandwidth q and lower bandwidth p, then U has upper bandwidth q and L has lower bandwidth p.*

Proof. The proof is by induction on n. From (3.2.6) we have the factorization

$$A = \begin{bmatrix} \alpha & w^T \\ v & B \end{bmatrix} = \begin{bmatrix} 1 & 0 \\ v/\alpha & I_{n-1} \end{bmatrix} \begin{bmatrix} 1 & 0 \\ 0 & B - vw^T/\alpha \end{bmatrix} \begin{bmatrix} \alpha & w^T \\ 0 & I_{n-1} \end{bmatrix}.$$

It is clear that $B - vw^T/\alpha$ has upper bandwidth q and lower bandwidth p because only the first q components of w and the first p components of v are nonzero. Let $L_1 U_1$ be the LU factorization of this matrix. Using the induction hypothesis and the sparsity of w and v, it follows that

$$L = \begin{bmatrix} 1 & 0 \\ v/\alpha & L_1 \end{bmatrix} \quad \text{and} \quad U = \begin{bmatrix} \alpha & w^T \\ 0 & U_1 \end{bmatrix}$$

have the desired bandwidth properties and satisfy $A = LU$. $\square$

The specialization of Gaussian elimination to banded matrices having an LU factorization is straightforward.

Algorithm 4.3.1 (Band Gaussian Elimination: Outer Product Version) Given $A \in \mathbb{R}^{n \times n}$ with upper bandwidth q and lower bandwidth p, the following algorithm computes the factorization $A = LU$, assuming it exists. $A(i,j)$ is overwritten by $L(i,j)$ if $i > j$ and by $U(i,j)$ otherwise.

```
for k = 1:n − 1
    for i = k + 1:min(k + p, n)
        A(i, k) = A(i, k)/A(k, k)
    end
    for j = k + 1:min(k + q, n)
        for i = k + 1:min(k + p, n)
            A(i, j) = A(i, j) − A(i, k)A(k, j)
        end
    end
end
```

If $n \gg p$ and $n \gg q$ then this algorithm involves about $2npq$ flops. Band versions of Algorithm 4.1.1 (LDM^T) and all the Cholesky procedures also exist, but we leave their formulation to the exercises.

4.3.2 Band Triangular System Solving

Analogous savings can also be made when solving banded triangular systems.

Algorithm 4.3.2 (Band Forward Substitution: Column Version) Let $L \in \mathbb{R}^{n \times n}$ be a unit lower triangular matrix having lower bandwidth p. Given $b \in \mathbb{R}^n$, the following algorithm overwrites b with the solution to $Lx = b$.

```
for j = 1:n
    for i = j + 1:min(j + p, n)
        b(i) = b(i) − L(i, j)b(j)
    end
end
```

If $n \gg p$ then this algorithm requires about $2np$ flops.

Algorithm 4.3.3 (Band Back-Substitution: Column Version) Let $U \in \mathbb{R}^{n \times n}$ be a nonsingular upper triangular matrix having upper bandwidth q. Given $b \in \mathbb{R}^n$, the following algorithm overwrites b with the solution to $Ux = b$.

```
for j = n: − 1:1
    b(j) = b(j)/U(j, j)
    for i = max(1, j − q):j − 1
        b(i) = b(i) − U(i, j)b(j)
    end
end
```

If $n \gg q$ then this algorithm requires about $2nq$ flops.

4.3.3   Band Gaussian Elimination with Pivoting

Gaussian elimination with partial pivoting can also be specialized to exploit band structure in A. If, however, $PA = LU$, then the band properties of L and U are not quite so simple. For example, if A is tridiagonal and the first two rows are interchanged at the very first step of the algorithm, then u_{13} is nonzero. Consequently, row interchanges expand bandwidth. Precisely how the band enlarges is the subject of the following theorem.

Theorem 4.3.2 *Suppose $A \in \mathbb{R}^{n \times n}$ is nonsingular and has upper and lower bandwidths q and p, respectively. If Gaussian elimination with partial pivoting is used to compute Gauss transformations*

$$M_j = I - \alpha^{(j)} e_j^T \qquad j = 1{:}n - 1$$

and permutations $P_1, \ldots, P_{n-1}$ such that $M_{n-1} P_{n-1} \cdots M_1 P_1 A = U$ is upper triangular, then U has upper bandwidth $p + q$ and $\alpha_i^{(j)} = 0$ whenever $i \le j$ or $i > j + p$.

Proof. Let $PA = LU$ be the factorization computed by Gaussian elimination with partial pivoting and recall that $P = P_{n-1} \cdots P_1$. Write $P^T = [\, e_{s_1}, \ldots, e_{s_n} \,]$, where $\{s_1, \ldots, s_n\}$ is a permutation of $\{1, 2, \ldots, n\}$. If $s_i > i + p$ then it follows that the leading i-by-i principal submatrix of PA is singular, since $(PA)_{ij} = a_{s_i,j}$ for $j = 1{:}s_i - p - 1$ and $s_i - p - 1 \ge i$. This implies that U and A are singular, a contradiction. Thus, $s_i \le i + p$ for $i = 1{:}n$ and therefore, PA has upper bandwidth $p + q$. It follows from Theorem 4.3.1 that U has upper bandwidth $p + q$.

The assertion about the $\alpha^{(j)}$ can be verified by observing that M_j need only zero elements $(j + 1, j), \ldots, (j + p, j)$ of the partially reduced matrix $P_j M_{j-1} P_{j-1} \cdots_1 P_1 A$. $\square$

Thus, pivoting destroys band structure in the sense that U becomes "wider" than A's upper triangle, while nothing at all can be said about the bandwidth of L. However, since the jth column of L is a permutation of the jth Gauss vector α_j, it follows that L has at most $p + 1$ nonzero elements per column.

4.3.4   Hessenberg LU

As an example of an unsymmetric band matrix computation, we show how Gaussian elimination with partial pivoting can be applied to factor an upper Hessenberg matrix H. (Recall that if H is upper Hessenberg then $h_{ij} = 0$, $i > j + 1$). After $k - 1$ steps of Gaussian elimination with partial pivoting we are left with an upper Hessenberg matrix of the form:

$$
\begin{bmatrix}
\times & \times & \times & \times & \times \\
0 & \times & \times & \times & \times \\
0 & 0 & \times & \times & \times \\
0 & 0 & \times & \times & \times \\
0 & 0 & 0 & \times & \times
\end{bmatrix}
\qquad k = 3, n = 5
$$

By virtue of the special structure of this matrix, we see that the next permutation, P_3, is either the identity or the identity with rows 3 and 4 interchanged. Moreover, the next Gauss transformation M_k has a single nonzero multiplier in the $(k+1, k)$ position. This illustrates the kth step of the following algorithm.

Algorithm 4.3.4 (Hessenberg LU)   Given an upper Hessenberg matrix $H \in \mathbb{R}^{n \times n}$, the following algorithm computes the upper triangular matrix $M_{n-1}P_{n-1} \cdots M_1 P_1 H = U$ where each P_k is a permutation and each M_k is a Gauss transformation whose entries are bounded by unity. $H(i, k)$ is overwritten with $U(i, k)$ if $i \le k$ and by $(M_k)_{k+1,k}$ if $i = k+1$. An integer vector $piv(1{:}n-1)$ encodes the permutations. If $P_k = I$, then $piv(k) = 0$. If P_k interchanges rows k and $k+1$, then $piv(k) = 1$.

> **for** $k = 1{:}n-1$
> > **if** $|H(k, k)| < |H(k+1, k)|$
> > > $piv(k) = 1$; $H(k, k{:}n) \leftrightarrow H(k+1, k{:}n)$
> >
> > **else**
> > > $piv(k) = 0$
> >
> > **end**
> > **if** $H(k, k) \ne 0$
> > > $t = -H(k+1, k)/H(k, k)$
> > > **for** $j = k+1{:}n$
> > > > $H(k+1, j) = H(k+1, j) + tH(k, j)$
> > >
> > > **end**
> > > $H(k+1, k) = t$
> >
> > **end**
>
> **end**

This algorithm requires n^2 flops.

4.3.5   Band Cholesky

The rest of this section is devoted to banded $Ax = b$ problems where the matrix A is also symmetric positive definite. The fact that pivoting is unnecessary for such matrices leads to some very compact, elegant algorithms. In particular, it follows from Theorem 4.3.1 that if $A = GG^T$ is the Cholesky factorization of A, then G has the same lower bandwidth as A.

This leads to the following banded version of Algorithm 4.2.1, gaxpy-based Cholesky

Algorithm 4.3.5 (Band Cholesky: Gaxpy Version) Given a symmetric positive definite $A \in \mathbb{R}^{n \times n}$ with bandwidth p, the following algorithm computes a lower triangular matrix G with lower bandwidth p such that $A = GG^T$. For all $i \geq j$, $G(i,j)$ overwrites $A(i,j)$.

> **for** $j = 1{:}n$
> >   **for** $k = \max(1, j - p){:}j - 1$
> > > $\lambda = \min(k + p, n)$
> > > $A(j{:}\lambda, j) = A(j{:}\lambda, j) - A(j, k)A(j{:}\lambda, k)$
> >
> >   **end**
> > $\lambda = \min(j + p, n)$
> > $A(j{:}\lambda, j) = A(j{:}\lambda, j) / \sqrt{A(j,j)}$
>
> **end**

If $n \gg p$ then this algorithm requires about $n(p^2 + 3p)$ flops and n square roots. Of course, in a serious implementation an appropriate data structure for A should be used. For example, if we just store the nonzero lower triangular part, then a $(p + 1)$-by-n array would suffice. (See §1.2.6)

If our band Cholesky procedure is coupled with appropriate band triangular solve routines then approximately $np^2 + 7np + 2n$ flops and n square roots are required to solve $Ax = b$. For small p it follows that the square roots represent a significant portion of the computation and it is preferable to use the LDL^T approach. Indeed, a careful flop count of the steps $A = LDL^T$, $Ly = b$, $Dz = y$, and $L^T x = z$ reveals that $np^2 + 8np + n$ flops and no square roots are needed.

4.3.6   Tridiagonal System Solving

As a sample narrow band LDL^T solution procedure, we look at the case of symmetric positive definite tridiagonal systems. Setting

$$
L = \begin{bmatrix}
1 & & & \cdots & & 0 \\
e_1 & 1 & & & & \vdots \\
& & \ddots & \ddots & & \\
\vdots & & & \ddots & & \\
0 & \cdots & & & e_{n-1} & 1
\end{bmatrix}
$$

and $D = \text{diag}(d_1, \ldots, d_n)$ we deduce from the equation $A = LDL^T$ that:

$$
\begin{array}{rcll}
a_{11} & = & d_1 & \\
a_{k,k-1} & = & e_{k-1}d_{k-1} & k = 2{:}n \\
a_{kk} & = & d_k + e_{k-1}^2 d_{k-1} = d_k + e_{k-1}a_{k,k-1} & k = 2{:}n
\end{array}
$$

Thus, the d_i and e_i can be resolved as follows:

$$
\begin{aligned}
& d_1 = a_{11} \\
& \textbf{for } k = 2{:}n \\
& \qquad e_{k-1} = a_{k,k-1}/d_{k-1}; \; d_k = a_{kk} - e_{k-1}a_{k,k-1} \\
& \textbf{end}
\end{aligned}
$$

To obtain the solution to $Ax = b$ we solve $Ly = b$, $Dz = y$, and $L^T x = z$. With overwriting we obtain

Algorithm 4.3.6 (Symmetric, Tridiagonal, Positive Definite System Solver) Given an n-by-n symmetric, tridiagonal, positive definite matrix A and $b \in \mathbb{R}^n$, the following algorithm overwrites b with the solution to $Ax = b$. It is assumed that the diagonal of A is stored in $d(1{:}n)$ and the superdiagonal in $e(1{:}n-1)$.

$$
\begin{aligned}
& \textbf{for } k = 2{:}n \\
& \qquad t = e(k-1); \; e(k-1) = t/d(k-1); \; d(k) = d(k) - te(k-1) \\
& \textbf{end} \\
& \textbf{for } k = 2{:}n \\
& \qquad b(k) = b(k) - e(k-1)b(k-1) \\
& \textbf{end} \\
& b(n) = b(n)/d(n) \\
& \textbf{for } k = n - 1{:} - 1{:}1 \\
& \qquad b(k) = b(k)/d(k) - e(k)b(k+1) \\
& \textbf{end}
\end{aligned}
$$

This algorithm requires $8n$ flops.

4.3.7 Vectorization Issues

The tridiagonal example brings up a sore point: narrow band problems and vector/pipeline architectures do not mix well. The narrow band implies short vectors. However, it is sometimes the case that large, independent sets of such problems must be solved at the same time. Let us look at how such a computation should be arranged in light of the issues raised in §1.4.

For simplicity, assume that we must solve the n-by-n unit lower bidiagonal systems

$$
A^{(k)}x^{(k)} = b^{(k)} \qquad k = 1{:}m
$$

and that $m \gg n$. Suppose we have arrays $E(1{:}n - 1, 1{:}m)$ and $B(1{:}n, 1{:}m)$ with the property that $E(1{:}n - 1, k)$ houses the subdiagonal of $A^{(k)}$ and $B(1{:}n, k)$ houses the kth right hand side $b^{(k)}$. We can overwrite $b^{(k)}$ with the solution $x^{(k)}$ as follows:

> **for** $k = 1{:}m$
>> **for** $i = 2{:}n$
>>> $B(i, k) = B(i, k) - E(i - 1, k)B(i - 1, k)$
>> **end**
> **end**

The problem with this algorithm, which sequentially solves each bidiagonal system in turn, is that the inner loop does not vectorize. This is because of the dependence of $B(i, k)$ on $B(i - 1, k)$. If we interchange the k and i loops we get

> **for** $i = 2{:}n$
>> **for** $k = 1{:}m$
>>> $B(i, k) = B(i, k) - E(i - 1, k)B(i - 1, k)$ $\hspace{2cm}$ (4.3.1)
>> **end**
> **end**

Now the inner loop vectorizes well as it involves a vector multiply and a vector add. Unfortunately, (4.3.1) is not a unit stride procedure. However, this problem is easily rectified if we store the subdiagonals and right-hand-sides by row. That is, we use the arrays $E(1{:}m, 1{:}n-1)$ and $B(1{:}m, 1{:}n-1)$ and store the subdiagonal of $A^{(k)}$ in $E(k, 1{:}n - 1)$ and $b^{(k)^T}$ in $B(k, 1{:}n)$. The computation (4.3.1) then transforms to

> **for** $i = 2{:}n$
>> **for** $k = 1{:}m$
>>> $B(k, i) = B(k, i) - E(k, i - 1)B(k, i - 1)$
>> **end**
> **end**

illustrating once again the effect of data structure on performance.

4.3.8   Band Matrix Data Structures

The above algorithms are written as if the matrix A is conventionally stored in an n-by-n array. In practice, a band linear equation solver would be organized around a data structure that takes advantage of the many zeroes in A. Recall from §1.2.6 that if A has lower bandwidth p and upper bandwidth q it can be represented in a $(p + q + 1)$-by-n array $A.band$ where band entry a_{ij} is stored in $A.band(i - j + q + 1, j)$. In this arrangement, the nonzero portion of A's jth column is housed in the jth column of $A.band$. Another possible band matrix data structure that we discussed in §1.2.8

involves storing A by diagonal in a 1-dimensional array $A.diag$. Regardless of the data structure adopted, the design of a matrix computation with a band storage arrangement requires care in order to minimize subscripting overheads.

Problems

P4.3.1 Derive a banded LDM^T procedure similar to Algorithm 4.3.1.

P4.3.2 Show how the output of Algorithm 4.3.4 can be used to solve the upper Hessenberg system $Hx = b$.

P4.3.3 Give an algorithm for solving an unsymmetric tridiagonal system $Ax = b$ that uses Gaussian elimination with partial pivoting. It should require only four n-vectors of floating point storage for the factorization.

P4.3.4 For $C \in \mathbf{R}^{n \times n}$ define the *profile indices* $m(C, i) = \min\{j: c_{ij} \neq 0\}$, where $i = 1{:}n$. Show that if $A = GG^T$ is the Cholesky factorization of A, then $m(A, i) = m(G, i)$ for $i = 1{:}n$. (We say that G has the same *profile* as A.)

P4.3.5 Suppose $A \in \mathbf{R}^{n \times n}$ is symmetric positive definite with profile indices $m_i = m(A, i)$ where $i = 1{:}n$. Assume that A is stored in a one-dimensional array v as follows: $v = (a_{11}, a_{2,m_2}, \ldots, a_{22}, a_{3,m_3}, \ldots, a_{33}, \ldots, a_{n,m_n}, \ldots, a_{nn})$. Write an algorithm that overwrites v with the corresponding entries of the Cholesky factor G and then uses this factorization to solve $Ax = b$. How many flops are required?

P4.3.6 For $C \in \mathbf{R}^{n \times n}$ define $p(C, i) = \max\{j: c_{ij} \neq 0\}$. Suppose that $A \in \mathbf{R}^{n \times n}$ has an LU factorization $A = LU$ and that:
$$\begin{matrix} m(A, 1) & \leq & m(A, 2) & \leq & \cdots & \leq & m(A, n) \\ p(A, 1) & \leq & p(A, 2) & \leq & \cdots & \leq & p(A, n) \end{matrix}$$
Show that $m(A, i) = m(L, i)$ and $p(A, i) = p(U, i)$ for $i = 1{:}n$. Recall the definition of $m(A, i)$ from P4.3.4.

P4.3.7 Develop a gaxpy version of Algorithm 4.3.1.

P4.3.8 Develop a unit stride, vectorizable algorithm for solving the symmetric positive definite tridiagonal systems $A^{(k)}x^{(k)} = b^{(k)}$. Assume that the diagonals, superdiagonals, and right hand sides are stored by row in arrays D, E, and B and that $b^{(k)}$ is overwritten with $x^{(k)}$.

P4.3.9 Develop a version of Algorithm 4.3.1 in which A is stored by diagonal.

P4.3.10 Give an example of a 3-by-3 symmetric positive definite matrix whose tridiagonal part is not positive definite.

P4.3.11 Consider the $Ax = b$ problem where

$$A = \begin{bmatrix} 2 & -1 & 0 & \cdots & 0 & -1 \\ -1 & 2 & -1 & \ddots & \vdots & 0 \\ 0 & -1 & 2 & \ddots & \ddots & \vdots \\ \vdots & \ddots & \ddots & \ddots & \ddots & 0 \\ 0 & \cdots & \cdots & \ddots & 2 & -1 \\ -1 & 0 & \cdots & 0 & -1 & 2 \end{bmatrix}$$

This kind of matrix arises in boundary value problems with periodic boundary conditions. (a) Show A is singular. (b) Give conditions that b must satisfy for there to exist a solution and specify an algorithm for solving it. (c). Assume that n is even and consider the permutation

$$P = [\, e_1 \ e_n \ e_2 \ e_{n-1} \ e_3 \ \cdots \,]$$

where e_k is the kth column of I_n. Describe the transformed system $P^T AP(P^T x) = P^T b$ and show how to solve it. Assume that there is a solution and ignore pivoting.

Notes and References for Sec. 4.3

The literature concerned with banded systems is immense. Some representative papers include

R.S. Martin and J.H. Wilkinson (1965). "Symmetric Decomposition of Positive Definite Band Matrices," *Numer. Math. 7*, 355–61.
R. S. Martin and J.H. Wilkinson (1967). "Solution of Symmetric and Unsymmetric Band Equations and the Calculation of Eigenvalues of Band Matrices," *Numer. Math. 9*, 279–301.
E.L. Allgower (1973). "Exact Inverses of Certain Band Matrices," *Numer. Math. 21*, 279–84.
Z. Bohte (1975). "Bounds for Rounding Errors in the Gaussian Elimination for Band Systems," *J. Inst. Math. Applic. 16*, 133–42.
I.S. Duff (1977). "A Survey of Sparse Matrix Research," *Proc. IEEE 65*, 500–535.
N.J. Higham (1990). "Bounding the Error in Gaussian Elimination for Tridiagonal Systems," *SIAM J. Matrix Anal. Appl. 11*, 521–530.

A topic of considerable interest in the area of banded matrices deals with methods for reducing the width of the band. See

E. Cuthill (1972). "Several Strategies for Reducing the Bandwidth of Matrices," in *Sparse Matrices and Their Applications*, ed. D.J. Rose and R.A. Willoughby, Plenum Press, New York.
N.E. Gibbs, W.G. Poole, Jr., and P.K. Stockmeyer (1976). "An Algorithm for Reducing the Bandwidth and Profile of a Sparse Matrix," *SIAM J. Num. Anal. 13*, 236–50.
N.E. Gibbs, W.G. Poole, Jr., and P.K. Stockmeyer (1976). "A Comparison of Several Bandwidth and Profile Reduction Algorithms," *ACM Trans. Math. Soft. 2*, 322–30.

As we mentioned, tridiagonal systems arise with particular frequency. Thus, it is not surprising that a great deal of attention has been focused on special methods for this class of banded problems.

C. Fischer and R.A. Usmani (1969). "Properties of Some Tridiagonal Matrices and Their Application to Boundary Value Problems," *SIAM J. Num. Anal. 6*, 127–42.
D.J. Rose (1969). "An Algorithm for Solving a Special Class of Tridiagonal Systems of Linear Equations," *Comm. ACM 12*, 234–36.
H.S. Stone (1973). "An Efficient Parallel Algorithm for the Solution of a Tridiagonal Linear System of Equations,"*J. ACM 20*, 27–38.
M.A. Malcolm and J. Palmer (1974). "A Fast Method for Solving a Class of Tridiagonal Systems of Linear Equations," *Comm. ACM 17*, 14–17.
J. Lambiotte and R.G. Voigt (1975). "The Solution of Tridiagonal Linear Systems of the CDC-STAR 100 Computer," *ACM Trans. Math. Soft. 1*, 308–29.
H.S. Stone (1975). "Parallel Tridiagonal Equation Solvers," *ACM Trans. Math. Soft.1*, 289–307.
D. Kershaw(1982). "Solution of Single Tridiagonal Linear Systems and Vectorization of the ICCG Algorithm on the Cray-1," in G. Roderigue (ed), *Parallel Computation*, Academic Press, NY, 1982.
N.J. Higham (1986). "Efficient Algorithms for computing the condition number of a tridiagonal matrix," *SIAM J. Sci. and Stat. Comp. 7*, 150–165.

Chapter 4 of George and Liu (1981) contains a nice survey of band methods for positive definite systems.

4.4   Symmetric Indefinite Systems

A symmetric matrix whose quadratic form $x^T A x$ takes on both positive and negative values is called *indefinite*. Although an indefinite A may have an LDL^T factorization, the entries in the factors can have arbitrary magnitude:

$$\begin{bmatrix} \epsilon & 1 \\ 1 & 0 \end{bmatrix} = \begin{bmatrix} 1 & 0 \\ 1/\epsilon & 1 \end{bmatrix} \begin{bmatrix} \epsilon & 0 \\ 0 & -1/\epsilon \end{bmatrix} \begin{bmatrix} 1 & 0 \\ 1/\epsilon & 1 \end{bmatrix}^T.$$

Of course, any of the pivot strategies in §3.4 could be invoked. However, they destroy symmetry and with it, the chance for a "Cholesky speed" indefinite system solver. Symmetric pivoting, i.e., data reshufflings of the form $A \leftarrow PAP^T$, must be used as we discussed in §4.2.9. Unfortunately, symmetric pivoting does not always stabilize the LDL^T computation. If ϵ_1 *and* ϵ_2 are small then regardless of P, the matrix

$$\tilde{A} = P \begin{bmatrix} \epsilon_1 & 1 \\ 1 & \epsilon_2 \end{bmatrix} P^T$$

has small diagonal entries and large numbers surface in the factorization. With symmetric pivoting, the pivots are always selected from the diagonal and trouble results if these numbers are small relative to what must be zeroed off the diagonal. Thus, LDL^T with symmetric pivoting cannot be recommended as a reliable approach to symmetric indefinite system solving. It seems that the challenge is to involve the off-diagonal entries in the pivoting process while at the same time maintaining symmetry.

In this section we discuss two ways to do this. The first method is due to Aasen(1971) and it computes the factorization

$$PAP^T = LTL^T \tag{4.4.1}$$

where $L = (\ell_{ij})$ is unit lower triangular and T is tridiagonal. P is a permutation chosen such that $|\ell_{ij}| \le 1$. In contrast, the *diagonal pivoting method* due to Bunch and Parlett (1971) computes a permutation P such that

$$PAP^T = LDL^T \tag{4.4.2}$$

where D is a direct sum of 1-by-1 and 2-by-2 pivot blocks. Again, P is chosen so that the entries in the unit lower triangular L satisfy $|\ell_{ij}| \le 1$. Both factorizations involve $n^3/3$ flops and once computed, can be used to solve $Ax = b$ with $O(n^2)$ work:

$$PAP^T = LTL^T, Lz = Pb, Tw = z, L^T y = w, x = Py \quad \Rightarrow \quad Ax = b$$

$$PAP^T = LDL^T, Lz = Pb, Dw = z, L^T y = w, x = Py \quad \Rightarrow \quad Ax = b$$

The only thing "new" to discuss in these solution procedures are the $Tw = z$ and $Dw = z$ systems.

In Aasen's method, the symmetric indefinite tridiagonal system $Tw = z$ is solved in $O(n)$ time using band Gaussian elimination with pivoting. Note that there is no serious price to pay for the disregard of symmetry at this level since the overall process is $O(n^3)$.

In the diagonal pivoting approach, the $Dw = z$ system amounts to a set of 1-by-1 and 2-by-2 symmetric indefinite systems. The 2-by-2 problems can be handled via Gaussian elimination with pivoting. Again, there is no harm in disregarding symmetry during this $O(n)$ phase of the calculation.

Thus, the central issue in this section is the efficient computation of the factorizations (4.4.1) and (4.4.2).

4.4.1   The Parlett-Reid Algorithm

Parlett and Reid (1970) show how to compute (4.4.1) using Gauss transforms. Their algorithm is sufficiently illustrated by displaying the $k = 2$ step for the case $n = 5$. At the beginning of this step the matrix A has been transformed to

$$A^{(1)} = M_1 P_1 A P_1^T M_1^T = \begin{bmatrix} \alpha_1 & \beta_1 & 0 & 0 & 0 \\ \beta_1 & \alpha_2 & v_3 & v_4 & v_5 \\ 0 & v_3 & \times & \times & \times \\ 0 & v_4 & \times & \times & \times \\ 0 & v_5 & \times & \times & \times \end{bmatrix}$$

where P_1 is a permutation chosen so that the entries in the Gauss transformation M_1 are bounded by unity in modulus. Scanning the vector $(v_3 \; v_4 \; v_5)^T$ for its largest entry, we now determine a 3-by-3 permutation $\tilde{P}_2$ such that

$$\tilde{P}_2 \begin{bmatrix} v_3 \\ v_4 \\ v_5 \end{bmatrix} = \begin{bmatrix} \tilde{v}_3 \\ \tilde{v}_4 \\ \tilde{v}_5 \end{bmatrix} \qquad \Rightarrow \qquad |\tilde{v}_3| = \max\{|\tilde{v}_3|, |\tilde{v}_4|, |\tilde{v}_5|\} \; .$$

If this maximal element is zero, we set $M_2 = P_2 = I$ and proceed to the next step. Otherwise, we set $P_2 = \text{diag}(I_2, \tilde{P}_2)$ and $M_2 = I - \alpha^{(2)} e_3^T$ with

$$\alpha^{(2)} = \begin{pmatrix} 0 & 0 & 0 & \tilde{v}_4/\tilde{v}_3 & \tilde{v}_5/\tilde{v}_3 \end{pmatrix}^T$$

and observe that

$$A^{(2)} = M_2 P_2 A^{(1)} P_2^T M_2^T = \begin{bmatrix} \alpha_1 & \beta_1 & 0 & 0 & 0 \\ \beta_1 & \alpha_2 & \tilde{v}_3 & 0 & 0 \\ 0 & \tilde{v}_3 & \times & \times & \times \\ 0 & 0 & \times & \times & \times \\ 0 & 0 & \times & \times & \times \end{bmatrix} \; .$$

In general, the process continues for $n-2$ steps leaving us with a tridiagonal matrix

$$T = A^{(n-2)} = (M_{n-2}P_{n-2} \cdots M_1 P_1) A (M_{n-2}P_{n-2} \cdots M_1 P_1)^T \; .$$

It can be shown that (4.4.1) holds with $P = P_{n-2} \cdots P_1$ and

$$L = (M_{n-2}P_{n-2} \cdots M_1 P_1 P^T)^{-1}.$$

Analysis of L reveals that its first column is e_1 and that its subdiagonal entries in column k with $k > 1$ are "made up" of the multipliers in M_{k-1}.

The efficient implementation of the Parlett-Reid method requires care when computing the update

$$A^{(k)} = M_k(P_k A^{(k-1)} P_k^T)M_k^T. \qquad (4.4.3)$$

To see what is involved with a minimum of notation, suppose $B = B^T$ has order $n - k$ and that we wish to form: $B_+ = (I - we_1^T)B(I - we_1^T)^T$ where $w \in \mathbb{R}^{n-k}$ and e_1 is the first column of I_{n-k}. Such a calculation is at the heart of (4.4.3). If we set

$$u = Be_1 - \frac{b_{11}}{2}w,$$

then the lower half of the symmetric matrix $B_+ = B - wu^T - uw^T$ can be formed in $2(n - k)^2$ flops. Summing this quantity as k ranges from 1 to $n - 2$ indicates that the Parlett-Reid procedure requires $2n^3/3$ flops—twice what we would like.

Example 4.4.1 If the Parlett-Reid algorithm is applied to

$$A = \begin{bmatrix} 0 & 1 & 2 & 3 \\ 1 & 2 & 2 & 2 \\ 2 & 2 & 3 & 3 \\ 3 & 2 & 3 & 4 \end{bmatrix}$$

then

$$\begin{aligned}
P_1 &= [\, e_1\ e_4\ e_3\ e_2\,] \\
M_1 &= I_4 - (0,\, 0,\, 2/3,\, 1/3,\,)^T e_2^T \\
P_2 &= [\, e_1\ e_2\ e_4\ e_3\,] \\
M_2 &= I_4 - (0,\, 0,\, 0,\, 1/2)^T e_3^T
\end{aligned}$$

and $PAP^T = LTL^T$, where $P = [\, e_1,\, e_3,\, e_4,\, e_2\,]$,

$$L = \begin{bmatrix} 1 & 0 & 0 & 0 \\ 0 & 1 & 0 & 0 \\ 0 & 1/3 & 1 & 0 \\ 0 & 2/3 & 1/2 & 1 \end{bmatrix} \quad \text{and} \quad T = \begin{bmatrix} 0 & 3 & 0 & 0 \\ 3 & 4 & 2/3 & 0 \\ 0 & 2/3 & 10/9 & 0 \\ 0 & 0 & 0 & 1/2 \end{bmatrix}.$$

4.4.2   The Method of Aasen

An $n^3/3$ approach to computing (4.4.1) due to Aasen (1971) can be derived by reconsidering some of the computations in the Parlett-Reid approach.

We need a notation for the tridiagonal T:

$$T = \begin{bmatrix} \alpha_1 & \beta_1 & & \cdots & & 0 \\ \beta_1 & \alpha_2 & \ddots & & & \vdots \\ & \ddots & \ddots & \ddots & & \\ \vdots & & \ddots & \ddots & \beta_{n-1} & \\ 0 & \cdots & & & \beta_{n-1} & \alpha_n \end{bmatrix}.$$

For clarity, we temporarily ignore pivoting and assume that the factorization $A = LTL^T$ exists where L is unit lower triangular with $L(:,1) = e_1$. Aasen's method is organized as follows:

$$\begin{aligned} &\textbf{for } j = 1{:}n \\ &\qquad \text{Compute } h(1{:}j) \text{ where } h = TL^T e_j = H e_j. \\ &\qquad \text{Compute } \alpha(j). \\ &\qquad \textbf{if } j \le n - 1 \\ &\qquad\qquad \text{Compute } \beta(j) \\ &\qquad \textbf{end} \\ &\qquad \textbf{if } j \le n - 2 \\ &\qquad\qquad \text{Compute } L(j + 2{:}n, j + 1). \\ &\qquad \textbf{end} \\ &\textbf{end} \end{aligned} \qquad (4.4.4)$$

Thus, the mission of the jth Aasen step is to compute the jth column of T and the $(j + 1)$-st column of L. The algorithm exploits the fact that the matrix $H = TL^T$ is upper Hessenberg. As can be deduced from (4.4.4), the computation of $\alpha(j)$, $\beta(j)$, and $L(j + 2{:}n, j + 1)$ hinges upon the vector $h(1{:}j) = H(1{:}j, j)$. Let us see why.

Consider the jth column of the equation $A = LH$:

$$A(:, j) = L(:, 1{:}j + 1)h(1{:}j + 1) . \qquad (4.4.5)$$

This says that $A(:, j)$ is a linear combination of the first $j + 1$ columns of L. In particular,

$$A(j + 1{:}n, j) = L(j + 1{:}n, 1{:}j)h(1{:}j) + L(j + 1{:}n, j + 1)h(j + 1) .$$

It follows that if we compute

$$v(j + 1{:}n) = A(j + 1{:}n, j) - L(j + 1{:}n, 1{:}j)h(1{:}j) ,$$

then

$$L(j + 1{:}n, j + 1)h(j + 1) = v(j + 1{:}n) . \qquad (4.4.6)$$

Thus, $L(j + 2{:}n, j + 1)$ is a scaling of $v(j + 2{:}n)$. Since L is unit lower triangular we have from (4.4.6) that

$$v(j + 1) = h(j + 1)$$

and so from that same equation we obtain the following recipe for the $(j + 1)$-st column of L:

$$L(j + 2{:}n, j + 1) = v(j + 2{:}n)/v(j + 1) .$$

Note that $L(j + 2{:}n, j + 1)$ is a scaled gaxpy.

We next develop formulae for $\alpha(j)$ and $\beta(j)$. Compare the (j, j) and $(j + 1, j)$ entries in the equation $H = TL^T$. With the convention $\beta(0) = 0$ we find that $h(j) = \beta(j - 1)L(j, j - 1) + \alpha(j)$ and $h(j + 1) = v(j + 1)$ and so

$$\alpha(j) \quad = \quad h(j) - \beta(j - 1)L(j, j - 1)$$

$$\beta(j) \quad = \quad v(j + 1) .$$

With these recipes we can completely describe the Aasen procedure:

> for $j = 1{:}n$
> > Compute $h(1{:}j)$ where $h = TL^T e_j$.
> > if $j = 1 \vee j = 2$
> > > $\alpha(j) = h(j)$
> >
> > else
> > > $\alpha(j) = h(j) - \beta(j - 1)L(j, j - 1)$
> >
> > end
> > if $j \le n - 1$ (4.4.7)
> > > $v(j + 1{:}n) = A(j + 1{:}n, j) - L(j + 1{:}n, 1{:}j)h(1{:}j)$
> > > $\beta(j) = v(j + 1)$
> >
> > end
> > if $j \le n - 2$
> > > $L(j + 2{:}n, j + 1) = v(j + 2{:}n)/v(j + 1)$
> >
> > end
>
> end

To complete the description we must detail the computation of $h(1{:}j)$. From (4.4.5) it follows that

$$A(1{:}j, j) = L(1{:}j, 1{:}j)h(1{:}j) . \tag{4.4.8}$$

This lower triangular system can be solved for $h(1{:}j)$ since we know the first j columns of L. However, a much more efficient way to compute $H(1{:}j, j)$

is obtained by exploiting the jth column of the equation $H = TL^T$. In particular, with the convention that $\beta(0)L(j,0) = 0$ we have

$$h(k) = \beta(k-1)L(j, k-1) + \alpha(k)L(j, k) + \beta(k)L(j, k+1) .$$

for $k = 1{:}j$. These are working formulae except in the case $k = j$ because we have not yet computed $\alpha(j)$ and $\beta(j)$. However, once $h(1{:}j-1)$ is known we can obtain $h(j)$ from the last row of the triangular system (4.4.8), i.e.,

$$h(j) = A(j,j) - \sum_{k=1}^{j-1} L(j, k)h(k) .$$

Collecting results and using a work array $\ell(1{:}n)$ for $L(j, 1{:}j)$ we see that the computation of $h(1{:}j)$ in (4.4.7) can be organized as follows:

$$
\begin{aligned}
&\textbf{if } j = 1 \\
&\qquad h(1) = A(1, 1) \\
&\textbf{elseif } j = 2 \\
&\qquad h(1) = \beta(1); \ h(2) = A(2, 2) \qquad\qquad\qquad (4.4.9)\\
&\textbf{else} \\
&\qquad \ell(0) = 0; \ \ell(1) = 0; \ \ell(2{:}j-1) = L(j, 2{:}j-1); \ \ell(j) = 1 \\
&\qquad h(j) = A(j, j) \\
&\qquad \textbf{for } k = 1{:}j-1 \\
&\qquad\qquad h(k) = \beta(k-1)\ell(k-1) + \alpha(k)\ell(k) + \beta(k)\ell(k+1) \\
&\qquad\qquad h(j) = h(j) - \ell(k)h(k) \\
&\qquad \textbf{end} \\
&\textbf{end}
\end{aligned}
$$

Note that with this $O(j)$ method for computing $h(1{:}j)$, the gaxpy calculation of $v(j+1{:}n)$ is the dominant operation in (4.4.7). During the jth step this gaxpy involves about $2j(n-j)$ flops. Summing this for $j = 1{:}n$ shows that Aasen's method requires $n^3/3$ flops. Thus, the Aasen and Cholesky algorithms entail the same amount of arithmetic.

4.4.3   Pivoting in Aasen's Method

As it now stands, the columns of L are scalings of the v-vectors in (4.4.7). If any of these scalings are large, i.e., if any of the $v(j+1)$'s are small, then we are in trouble. To circumvent this problem we need only permute the largest component of $v(j+1{:}n)$ to the top position. Of course, this permutation must be suitably applied to the unreduced portion of A and the previously computed portion of L.

Algorithm 4.4.1 (Aasen's Method)   If $A \in \mathbb{R}^{n \times n}$ is symmetric then the following algorithm computes a permutation P, a unit lower triangular

L, and a tridiagonal T such that $PAP^T = LTL^T$ with $|L(i,j)| \le 1$. The permutation P is encoded in an integer vector piv. In particular, $P = P_1 \cdots P_{n-2}$ where P_j is the identity with rows $piv(j)$ and $j+1$ interchanged. The diagonal and subdiagonal of T are stored in $\alpha(1{:}n)$ and $\beta(1{:}n-1)$, respectively. Only the subdiagonal portion of $L(2{:}n, 2{:}n)$ is computed.

> **for** $j = 1{:}n$
>   Compute $h(1{:}j)$ via (4.4.9).
>   **if** $j = 1 \vee j = 2$
> $\alpha(j) = h(j)$
>   **else**
> $\alpha(j) = h(j) - \beta(j-1)L(j, j-1)$
>   **end**
>   **if** $j \le n-1$
> $v(j+1{:}n) = A(j+1{:}n, j) - L(j+1{:}n, 1{:}j)h(1{:}j)$
>   Find q so $|v(q)| = \| v(j+1{:}n) \|_\infty$ with $j+1 \le q \le n$.
> $piv(j) = q$; $v(j+1) \leftrightarrow v(q)$; $L(j+1, 2{:}j) \leftrightarrow L(q, 2{:}j)$
> $A(j+1, j+1{:}n) \leftrightarrow A(q, j+1{:}n)$
> $A(j+1{:}n, j+1) \leftrightarrow A(j+1{:}n, q)$
> $\beta(j) = v(j+1)$
>   **end**
>   **if** $j \le n-2$
> $L(j+2{:}n, j+1) = v(j+2{:}n)$
>   **if** $v(j+1) \ne 0$
> $L(j+2{:}n, j+1) = L(j+2{:}n, j+1)/v(j+1)$
>   **end**
>   **end**
> **end**

Aasen's method is stable in the same sense that Gaussian elimination with partial pivoting is stable. That is, the exact factorization of a matrix near A is obtained provided $\| \hat{T} \|_2 / \| A \|_2 \approx 1$, where $\hat{T}$ is the computed version of the tridiagonal matrix T. In general, this is almost always the case.

In a practical implementation of the Aasen algorithm, the lower triangular portion of A would be overwritten with L and T. Here is $n = 5$ case:

$$
A \leftarrow
\begin{bmatrix}
\alpha_1 & & & & \\
\beta_1 & \alpha_2 & & & \\
\ell_{32} & \beta_2 & \alpha_3 & & \\
\ell_{42} & \ell_{43} & \beta_3 & \alpha_4 & \\
\ell_{52} & \ell_{53} & \ell_{54} & \beta_4 & \alpha_5
\end{bmatrix}
$$

Notice that the columns of L are shifted left in this arrangement.

4.4.4   Diagonal Pivoting Methods

We next describe the computation of the block LDL^T factorization (4.4.2). We follow the discussion in Bunch and Parlett (1971). Suppose

$$P_1 A P_1^T = \begin{bmatrix} E & C^T \\ C & B \end{bmatrix} \begin{matrix} s \\ n-s \end{matrix}$$
$$\phantom{P_1 A P_1^T = \begin{bmatrix}} s \quad n-s$$

where P_1 is a permutation matrix and $s = 1$ or 2. If A is nonzero, then it is always possible to choose these quantities so that E is nonsingular thereby enabling us to write

$$P_1 A P_1^T = \begin{bmatrix} I_s & 0 \\ CE^{-1} & I_{n-s} \end{bmatrix} \begin{bmatrix} E & 0 \\ 0 & B - CE^{-1}C^T \end{bmatrix} \begin{bmatrix} I_s & E^{-1}C^T \\ 0 & I_{n-s} \end{bmatrix}$$

For the sake of stability, the s-by-s "pivot" E should be chosen so that the entries in

$$\tilde{A} = (\tilde{a}_{ij}) \equiv B - CE^{-1}C^T \tag{4.4.10}$$

are suitably bounded. To this end, let $\alpha \in (0,1)$ be given and define the size measures

$$\mu_0 = \max_{i,j} |a_{ij}|$$

$$\mu_1 = \max_{i} |a_{ii}|.$$

The Bunch-Parlett pivot strategy is as follows:

> if $\mu_1 \geq \alpha\mu_0$
> $\quad s = 1$
> $\quad$ Choose P_1 so $|e_{11}| = \mu_1$.
> else
> $\quad s = 2$
> $\quad$ Choose P_1 so $|e_{21}| = \mu_0$.
> end

It is easy to verify from (4.4.10) that if $s = 1$ then

$$|\tilde{a}_{ij}| \leq (1 + \alpha^{-1})\mu_0 \tag{4.4.11}$$

while $s = 2$ implies

$$|\tilde{a}_{ij}| \leq \frac{3 - \alpha}{1 - \alpha}\mu_0 . \tag{4.4.12}$$

By equating $(1+\alpha^{-1})^2$, the growth factor associated with two $s = 1$ steps, and $(3-\alpha)/(1-\alpha)$, the corresponding $s = 2$ factor, Bunch and Parlett conclude that $\alpha = (1 + \sqrt{17})/8$ is optimum from the standpoint of minimizing the bound on element growth.

The reductions outlined above are then repeated on the $n - s$ order symmetric matrix $\tilde{A}$. A simple induction argument establishes that the factorization (4.4.2) exists and that $n^3/3$ flops are required if the work associated with pivot determination is ignored.

4.4.5   Stability and Efficiency

Diagonal pivoting with the above strategy is shown by Bunch (1971) to be as stable as Gaussian elimination with complete pivoting. Unfortunately, the overall process requires between $n^3/12$ and $n^3/6$ comparisons, since μ_0 involves a two-dimensional search at each stage of the reduction. The actual number of comparisons depends on the total number of 2-by-2 pivots but in general the Bunch-Parlett method for computing (4.4.2) is considerably slower than the technique of Aasen. See Barwell and George(1976).

This is not the case with the diagonal pivoting method of Bunch and Kaufman (1977). In their scheme, it is only necessary to scan two columns at each stage of the reduction. The strategy is fully illustrated by considering the very first step in the reduction:

$$\alpha = (1 + \sqrt{17})/8; \; \lambda = |a_{r1}| = \max\{|a_{21}|, \ldots, |a_{n1}|\}$$

> if $\lambda > 0$
>> if $|a_{11}| \geq \alpha\lambda$
>>> $s = 1; P_1 = I$
>> else
>>> $\sigma = |a_{pr}| = \max\{|a_{1r}|, \ldots, |a_{r-1,r}|, |a_{r+1,r}|, \ldots, |a_{nr}|\}$
>>> if $\sigma|a_{11}| \geq \alpha\lambda^2$
>>>> $s = 1, P_1 = I$
>>> elseif $|a_{rr}| \geq \alpha\sigma$
>>>> $s = 1$ and choose P_1 so $(P_1^T A P_1)_{11} = a_{rr}$.
>>> else
>>>> $s = 2$ and choose P_1 so $(P_1^T A P_1)_{21} = a_{r1}$.
>>> end
>> end
> end

Overall, the Bunch-Kaufman algorithm requires $n^3/3$ flops, $O(n^2)$ comparisons, and, like all the methods of this section, $n^2/2$ storage.

Example 4.4.2 If the Bunch-Kaufman algorithm is applied to

$$A = \begin{bmatrix} 1 & 10 & 20 \\ 10 & 1 & 30 \\ 20 & 30 & 1 \end{bmatrix}$$

then in the first step $\lambda = 20$, $r = 3$, $\sigma = 30$, and $p = 2$. The permutation $P = [\, e_3 \; e_2 \; e_1 \,]$
is applied giving

$$PAP^T = \begin{bmatrix} 1 & 30 & 20 \\ 30 & 1 & 10 \\ 20 & 10 & 1 \end{bmatrix}.$$

A 2-by-2 pivot is then used to produce the reduction

$$PAP^T = \begin{bmatrix} 1 & 0 & 0 \\ 0 & 1 & 0 \\ \dfrac{280}{899} & \dfrac{170}{399} & 1 \end{bmatrix} \begin{bmatrix} 1 & 30 & 0 \\ 30 & 1 & 0 \\ 0 & 0 & \dfrac{-10601}{899} \end{bmatrix} \begin{bmatrix} 1 & 0 & 0 \\ 0 & 1 & 0 \\ \dfrac{280}{899} & \dfrac{170}{399} & 1 \end{bmatrix}^T$$

4.4.6   A Note on Equilibrium Systems

A very important class of symmetric indefinite matrices have the form

$$A = \begin{bmatrix} C & B \\ B^T & 0 \end{bmatrix} \begin{matrix} n \\ p \end{matrix} \qquad (4.4.13)$$
$$\begin{matrix} n & p \end{matrix}$$

where C is symmetric positive definite and B has full column rank. These
conditions ensure that A is nonsingular.

Of course, the methods of this section apply to A. However, they do not
exploit its structure because the pivot strategies "wipe out" the zero (2,2)
block. On the other hand, here is a tempting approach that does exploit
A's block structure:

(a) Compute the Cholesky factorization of C, $C = GG^T$.

(b) Solve $GK = B$ for $K \in \mathbb{R}^{n \times p}$.

(c) Compute the Cholesky factorization of $K^T K = B^T C^{-1} B$, $HH^T = K^T K$.

From this it follows that

$$A = \begin{bmatrix} G & 0 \\ K^T & H \end{bmatrix} \begin{bmatrix} G^T & K \\ 0 & -H^T \end{bmatrix}.$$

In principle, this triangular factorization can be used to solve the *equilibrium system*

$$\begin{bmatrix} C & B \\ B^T & 0 \end{bmatrix} \begin{bmatrix} x \\ y \end{bmatrix} = \begin{bmatrix} f \\ g \end{bmatrix}. \qquad (4.4.14)$$

However, it is clear by considering steps (b) and (c) above that the accuracy of the computed solution depends upon $\kappa(C)$ and this quantity may be much greater than $\kappa(A)$. The situation has been carefully analyzed and various structure-exploiting algorithms have been proposed. A brief review of the literature is given at the end of the section.

But before we close it is interesting to consider a special case of (4.4.14) that clarifies what it means for an algorithm to be stable and illustrates how perturbation analysis can structure the search for better methods. In several important applications, $g = 0$, C is diagonal, and the solution subvector y is of primary importance. A manipulation of (4.4.14) shows that this vector is specified by

$$y = (B^T C^{-1} B)^{-1} B^T C^{-1} f. \qquad (4.4.15)$$

Looking at this we are again led to believe that $\kappa(C)$ should have a bearing on the accuracy of the computed y. However, it can be shown that

$$\| (B^T C^{-1} B)^{-1} B^T C^{-1} \| \leq \psi_B \qquad (4.4.16)$$

where the upper bound ψ_B is independent of C, a result that (correctly) suggests that y is *not* sensitive to perturbations in C. A stable method for computing this vector should respect this, meaning that the accuracy of the computed y should be independent of C. Vavasis (1994) has developed a method with this property. It involves the careful assembly of a matrix $V \in \mathbb{R}^{n \times (n-p)}$ whose columns are a basis for the nullspace of $B^T C^{-1}$. The n-by-n linear system

$$[\, B \,, V\,] \begin{bmatrix} y \\ q \end{bmatrix} = f$$

is then solved implying $f = By + Vq$. Thus, $B^T C^{-1} f = B^T C^{-1} By$ and (4.4.15) holds.

Problems

P4.4.1   Show that if all the 1-by-1 and 2-by-2 principal submatrices of an n-by-n symmetric matrix A are singular, then A is zero.

P4.4.2   Show that no 2-by-2 pivots can arise in the Bunch-Kaufman algorithm if A is positive definite.

P4.4.3   Arrange Algorithm 4.4.1 so that only the lower triangular portion of A is referenced and so that $\alpha(j)$ overwrites $A(j, j)$ for $j = 1{:}n$, $\beta(j)$ overwrites $A(j+1, j)$ for $j = 1{:}n - 1$, and $L(i, j)$ overwrites $A(i, j - 1)$ for $j = 2{:}n - 1$ and $i = j + 1{:}n$.

P4.4.4   Suppose $A \in \mathbb{R}^{n \times n}$ is nonsingular, symmetric, and strictly diagonally dominant. Give an algorithm that computes the factorization

$$\Pi A \Pi^T = \begin{bmatrix} R & 0 \\ S & -M \end{bmatrix} \begin{bmatrix} R^T & S^T \\ 0 & M^T \end{bmatrix}$$

where $R \in \mathbb{R}^{k \times k}$ and $M \in \mathbb{R}^{(n-k) \times (n-k)}$ are lower triangular and nonsingular and Π is a permutation.

P4.4.5 Show that if

$$A = \begin{bmatrix} A_{11} & A_{12} \\ A_{21} & -A_{22} \end{bmatrix} \begin{matrix} n \\ p \end{matrix}$$
$$\quad\;\; n \quad\;\; p$$

is symmetric with A_{11} and A_{22} positive definite, then it has an LDL^T factorization with
the property that

$$D = \begin{bmatrix} D_1 & 0 \\ 0 & -D_2 \end{bmatrix}$$

where $D_1 \in \mathbf{R}^{n \times n}$ and $D_2 \in \mathbf{R}^{p \times p}$ have positive diagonal entries.

P4.4.6 Prove (4.4.11) and (4.4.12).

P4.4.7 Show that $-(B^T C^{-1} B)^{-1}$ is the (2,2) block of A^{-1} where A is given by (4.4.13).

P4.4.8 The point of this problem is to consider a special case of (4.4.15). Define the
matrix

$$M(\alpha) = (B^T C^{-1} B)^{-1} B^T C^{-1}$$

where

$$C = (I_n + \alpha e_k e_k^T) \qquad \alpha > -1.$$

and $e_k = I_n(:, k)$. (Note that C is just the identity with α added to the (k, k) entry.)
Assume that $B \in \mathbf{R}^{n \times p}$ has rank p and show that

$$M(\alpha) = (B^T B)^{-1} B^T \left(I_n - \frac{\alpha}{1 + \alpha w^T w} e_k w^T \right)$$

where $w = (I_n - B(B^T B)^{-1} B^T) e_k$. Show that if $\| w \|_2 = 0$ or $\| w \|_2 = 1$, then
$\| M(\alpha) \|_2 = 1/\sigma_{min}(B)$. Show that if $0 < \| w \|_2 < 1$, then

$$\| M(\alpha) \|_2 \leq \max \left\{ \frac{1}{1 - \| w \|_2}, 1 + \frac{1}{\| w \|_2} \right\} \Big/ \sigma_{min}(B).$$

Thus, $\| M(\alpha) \|_2$ has an α-independent upper bound.

Notes and References for Sec. 4.4

The basic references for computing (4.4.1) are

J.O. Aasen (1971). "On the Reduction of a Symmetric Matrix to Tridiagonal Form,"
     BIT 11, 233–42.
B.N. Parlett and J.K. Reid (1970). "On the Solution of a System of Linear Equations
     Whose Matrix is Symmetric but not Definite," *BIT 10*, 386–97.

The diagonal pivoting literature includes

J.R. Bunch and B.N. Parlett (1971). "Direct Methods for Solving Symmetric Indefinite
     Systems of Linear Equations," *SIAM J. Num. Anal. 8*, 639–55.
J.R. Bunch (1971). "Analysis of the Diagonal Pivoting Method," *SIAM J. Num. Anal.
     8*, 656–680.
J.R. Bunch (1974). "Partial Pivoting Strategies for Symmetric Matrices," *SIAM J. Num.
     Anal. 11*, 521–528.
J.R. Bunch, L. Kaufman, and B.N. Parlett (1976). "Decomposition of a Symmetric
     Matrix," *Numer. Math. 27*, 95–109.
J.R. Bunch and L. Kaufman (1977). "Some Stable Methods for Calculating Inertia and
     Solving Symmetric Linear Systems," *Math. Comp. 31*, 162–79.
I.S. Duff, N.I.M. Gould, J.K. Reid, J.A. Scott, and K. Turner (1991). "The Factorization
     of Sparse Indefinite Matrices," *IMA J. Num. Anal. 11*, 181–204.

M.T. Jones and M.L. Patrick (1993). "Bunch-Kaufman Factorization for Real Symmetric Indefinite Banded Matrices," *SIAM J. Matrix Anal. Appl.* *14*, 553–559.

Because "future" columns must be scanned in the pivoting process, it is awkward (but possible) to obtain a gaxpy-rich diagonal pivoting algorithm. On the other hand, Aasen's method is naturally rich in gaxpy's. Block versions of both procedures are possible. LA-PACK uses the diagonal pivoting method. Various performance issues are discussed in

V. Barwell and J.A. George (1976). "A Comparison of Algorithms for Solving Symmetric Indefinite Systems of Linear Equations," *ACM Trans. Math. Soft.* *2*, 242–51.
M.T. Jones and M.L. Patrick (1994). "Factoring Symmetric Indefinite Matrices on High-Performance Architectures," *SIAM J. Matrix Anal. Appl.* *15*, 273–283.

Another idea for a cheap pivoting strategy utilizes error bounds based on more liberal interchange criteria, an idea borrowed from some work done in the area of sparse elimination methods. See

R. Fletcher (1976). "Factorizing Symmetric Indefinite Matrices," *Lin. Alg. and Its Applic.* *14*, 257–72.

Before using any symmetric $Ax = b$ solver, it may be advisable to equilibrate A. An $O(n^2)$ algorithm for accomplishing this task is given in

J.R. Bunch (1971). "Equilibration of Symmetric Matrices in the Max-Norm," *J. ACM* *18*, 566–72.

Analogues of the symmetric indefinite solvers that we have presented exist for skew-symmetric systems. See

J.R. Bunch (1982). "A Note on the Stable Decomposition of Skew Symmetric Matrices," *Math. Comp.* *158*, 475–480.

The equilibrium system literature is scattered among the several application areas where it has an important role to play. Nice overviews with pointers to this literature include

G. Strang (1988). "A Framework for Equilibrium Equations," *SIAM Review 30*, 283–297.
S.A. Vavasis (1994). "Stable Numerical Algorithms for Equilibrium Systems," *SIAM J. Matrix Anal. Appl.* *15*, 1108–1131.

Other papers include

C.C. Paige (1979). "Fast Numerically Stable Computations for Generalized Linear Least Squares Problems," *SIAM J. Num. Anal.* *16*, 165–71.
Å. Björck and I.S. Duff (1980). "A Direct Method for the Solution of Sparse Linear Least Squares Problems," *Lin. Alg. and Its Applic.* *34*, 43–67.
Å. Björck (1992). "Pivoting and Stability in the Augmented System Method," *Proceedings of the 14th Dundee Conference*, D.F. Griffiths and G.A. Watson (eds), Longman Scientific and Technical, Essex, U.K.
P.D. Hough and S.A. Vavasis (1996). "Complete Orthogonal Decomposition for Weighted Least Squares," *SIAM J. Matrix Anal. Appl.*, to appear.

Some of these papers make use of the QR factorization and other least squares ideas that are discussed in the next chapter and §12.1.

Problems with structure abound in matrix computations and perturbation theory has a key role to play in the search for stable, efficient algorithms. For equilibrium systems, there are several results like (4.4.15) that underpin the most effective algorithms. See

A. Forsgren (1995). "On Linear Least-Squares Problems with Diagonally Dominant Weight Matrices," Technical Report TRITA-MAT-1995-OS2, Department of Mathematics, Royal Institute of Technology, S-100 44, Stockholm, Sweden.

and the included references. A discussion of (4.4.15) may be found in

G.W. Stewart (1989). "On Scaled Projections and Pseudoinverses," *Lin. Alg. and Its Applic. 112*, 189–193.
D.P. O'Leary (1990). "On Bounds for Scaled Projections and Pseudoinverses," *Lin. Alg. and Its Applic. 132*, 115–117.
M.J. Todd (1990). "A Dantzig-Wolfe-like Variant of Karmarker's Interior-Point Linear Programming Algorithm," *Operations Research 38*, 1006–1018.

4.5 Block Systems

In many application areas the matrices that arise have exploitable block structure. As a case study we have chosen to discuss block tridiagonal systems of the form

$$
\begin{bmatrix}
D_1 & F_1 & & \cdots & & 0 \\
E_1 & D_2 & \ddots & & & \vdots \\
 & \ddots & \ddots & \ddots & & \\
\vdots & & \ddots & \ddots & & F_{n-1} \\
0 & \cdots & & & E_{n-1} & D_n
\end{bmatrix}
\begin{bmatrix} x_1 \\ x_2 \\ \vdots \\ \\ x_n \end{bmatrix}
=
\begin{bmatrix} b_1 \\ b_2 \\ \vdots \\ \\ b_n \end{bmatrix}
\tag{4.5.1}
$$

Here we assume that all blocks are q-by-q and that the x_i and b_i are in $\mathbb{R}^q$. In this section we discuss both a block LU approach to this problem as well as a divide and conquer scheme known as *cyclic reduction*. Kronecker product systems are briefly mentioned.

4.5.1 Block Tridiagonal LU Factorization

We begin by considering a block LU factorization for the matrix in (4.5.1). Define the block tridiagonal matrices A_k by

$$
A_k =
\begin{bmatrix}
D_1 & F_1 & & \cdots & & 0 \\
E_1 & D_2 & \ddots & & & \vdots \\
 & \ddots & \ddots & \ddots & & \\
\vdots & & \ddots & \ddots & & F_{k-1} \\
0 & \cdots & & & E_{k-1} & D_k
\end{bmatrix}
\qquad k = 1{:}n .
\tag{4.5.2}
$$

Comparing blocks in

$$
A_n = \begin{bmatrix} I & & \cdots & 0 \\ L_1 & I & & \vdots \\ & \ddots & \ddots & \\ \vdots & & \ddots & \\ 0 & \cdots & L_{n-1} & I \end{bmatrix} \begin{bmatrix} U_1 & F_1 & \cdots & 0 \\ 0 & U_2 & \ddots & \vdots \\ & \ddots & \ddots & \ddots \\ \vdots & & \ddots & \ddots & F_{n-1} \\ 0 & \cdots & & 0 & U_n \end{bmatrix} \qquad (4.5.3)
$$

we formally obtain the following algorithm for the L_i and U_i:

> $U_1 = D_1$
> for $i = 2:n$
>   Solve $L_{i-1}U_{i-1} = E_{i-1}$ for L_{i-1}.   (4.5.4)
> $U_i = D_i - L_{i-1}F_{i-1}$
> end

The procedure is defined so long as the U_i are nonsingular. This is assured, for example, if the matrices $A_1, \ldots, A_n$ are nonsingular.

Having computed the factorization (4.5.3), the vector x in (4.5.1) can be obtained via block forward and back substitution:

> $y_1 = b_1$
> for $i = 2:n$
> $y_i = b_i - L_{i-1}y_{i-1}$
> end   (4.5.5)
> Solve $U_n x_n = y_n$ for x_n.
> for $i = n - 1: -1:1$
>   Solve $U_i x_i = y_i - F_i x_{i+1}$ for x_i.
> end

To carry out both (4.5.4) and (4.5.5), each U_i must be factored since linear systems involving these submatrices are solved. This could be done using Gaussian elimination with pivoting. However, this does not guarantee the stability of the overall process. To see this just consider the case when the block size q is unity.

4.5.2   Block Diagonal Dominance

In order to obtain satisfactory bounds on the L_i and U_i it is necessary to make additional assumptions about the underlying block matrix. For example, if for $i = 1:n$ we have the block diagonal dominance relations

$$
\| D_i^{-1} \|_1 (\| F_{i-1} \|_1 + \| E_i \|_1) < 1 \qquad E_n \equiv F_0 \equiv 0 \qquad (4.5.6)
$$

then the factorization (4.5.3) exists and it is possible to show that the L_i and U_i satisfy the inequalities

$$\| L_i \|_1 \; \le \; 1 \tag{4.5.7}$$
$$\| U_i \|_1 \; \le \; \| A_n \|_1 \tag{4.5.8}$$

4.5.3   Block Versus Band Solving

At this point it is reasonable to ask why we do not simply regard the matrix A in (4.5.1) as a qn-by-qn matrix having scalar entries and bandwidth $2q - 1$. Band Gaussian elimination as described in §4.3 could be applied. The effectiveness of this course of action depends on such things as the dimensions of the blocks and the sparsity patterns within each block.

To illustrate this in a very simple setting, suppose that we wish to solve

$$\begin{bmatrix} D_1 & F_1 \\ E_1 & D_2 \end{bmatrix} \begin{bmatrix} x_1 \\ x_2 \end{bmatrix} = \begin{bmatrix} b_1 \\ b_2 \end{bmatrix} \tag{4.5.9}$$

where D_1 and D_2 are diagonal and F_1 and E_1 are tridiagonal. Assume that each of these blocks is n-by-n and that it is "safe" to solve (4.5.9) via (4.5.3) and (4.5.5). Note that

$$\begin{aligned} U_1 &= D_1 & \text{(diagonal)} \\ L_1 &= E_1 U_1^{-1} & \text{(tridiagonal)} \\ U_2 &= D_2 - L_1 F_1 & \text{(pentadiagonal)} \\ y_1 &= b_1 \\ y_2 &= b_2 - E_1(D_1^{-1}y_1) \\ U_2 x_2 &= y_2 \\ D_1 x_1 &= y_1 - F_1 x_2. \end{aligned}$$

Consequently, some very simple n-by-n calculations with the original banded blocks renders the solution.

On the other hand, the naive application of band Gaussian elimination to the system (4.5.9) would entail a great deal of unnecessary work and storage as the system has bandwidth $n + 1$. However, we mention that by permuting the rows and columns of the system via the permutation

$$P = [e_1, e_{n+1}, e_2, \ldots, e_n, e_{2n}] \tag{4.5.10}$$

we find (in the $n = 5$ case) that

$$PAP^T = \begin{bmatrix} \times & \times & 0 & \times & 0 & 0 & 0 & 0 & 0 & 0 \\ \times & \times & \times & 0 & 0 & 0 & 0 & 0 & 0 & 0 \\ 0 & \times & \times & \times & 0 & \times & 0 & 0 & 0 & 0 \\ \times & 0 & \times & \times & \times & 0 & 0 & 0 & 0 & 0 \\ 0 & 0 & 0 & \times & \times & \times & 0 & \times & 0 & 0 \\ 0 & 0 & \times & 0 & \times & \times & \times & 0 & 0 & 0 \\ 0 & 0 & 0 & 0 & 0 & \times & \times & \times & 0 & \times \\ 0 & 0 & 0 & 0 & \times & 0 & \times & \times & \times & 0 \\ 0 & 0 & 0 & 0 & 0 & 0 & 0 & \times & \times & \times \\ 0 & 0 & 0 & 0 & 0 & 0 & \times & 0 & \times & \times \end{bmatrix}$$

This matrix has upper and lower bandwidth equal to three and so a very reasonable solution procedure results by applying band Gaussian elimination to this permuted version of A.

The subject of bandwidth-reducing permutations is important. See George and Liu (1981, Chapter 4). We also refer to the reader to Varah (1972) and George (1974) for further details concerning the solution of block tridiagonal systems.

4.5.4   Block Cyclic Reduction

We next describe the method of *block cyclic reduction* that can be used to solve some important special instances of the block tridiagonal system (4.5.1). For simplicity, we assume that A has the form

$$A = \begin{bmatrix} D & F & & \cdots & 0 \\ F & D & \ddots & & \vdots \\ & \ddots & \ddots & \ddots & \\ \vdots & & \ddots & \ddots & F \\ 0 & \cdots & & F & D \end{bmatrix} \in \mathbb{R}^{nq \times nq} \qquad (4.5.11)$$

where F and D are q-by-q matrices that satisfy $DF = FD$. We also assume that $n = 2^k - 1$. These conditions hold in certain important applications such as the discretization of Poisson's equation on a rectangle. In that situation,

$$D = \begin{bmatrix} 4 & -1 & & \cdots & 0 \\ -1 & 4 & \ddots & & \vdots \\ & \ddots & \ddots & \ddots & \\ \vdots & & \ddots & \ddots & -1 \\ 0 & \cdots & & -1 & 4 \end{bmatrix} \qquad (4.5.12)$$

and $F = -I_q$. The integer n is determined by the size of the mesh and can often be chosen to be of the form $n = 2^k - 1$. (Sweet (1977) shows how to proceed when the dimension is not of this form.)

The basic idea behind cyclic reduction is to halve the dimension of the problem on hand repeatedly until we are left with a single q-by-q system for the unknown subvector $x_{2^{k-1}}$. This system is then solved by standard means. The previously eliminated x_i are found by a back-substitution process.

The general procedure is adequately motivated by considering the case $n = 7$:

$$
\begin{aligned}
b_1 &= Dx_1 + Fx_2 \\
b_2 &= Fx_1 + Dx_2 + Fx_3 \\
b_3 &= Fx_2 + Dx_3 + Fx_4 \\
b_4 &= Fx_3 + Dx_4 + Fx_5 \\
b_5 &= Fx_4 + Dx_5 + Fx_6 \\
b_6 &= Fx_5 + Dx_6 + Fx_7 \\
b_7 &= Fx_6 + Dx_7
\end{aligned}
$$

$$(4.5.13)$$

For $i = 2$, 4, and 6 we multiply equations $i - 1$, i, and $i + 1$ by F, $-D$, and F, respectively, and add the resulting equations to obtain

$$
\begin{aligned}
(2F^2 - D^2)x_2 + F^2x_4 &= F(b_1 + b_3) - Db_2 \\
F^2x_2 + (2F^2 - D^2)x_4 + F^2x_6 &= F(b_3 + b_5) - Db_4 \\
F^2x_4 + (2F^2 - D^2)x_6 &= F(b_5 + b_7) - Db_6
\end{aligned}
$$

Thus, with this tactic we have removed the odd-indexed x_i and are left with a reduced block tridiagonal system of the form

$$
\begin{aligned}
D^{(1)}x_2 + F^{(1)}x_4 \phantom{+ F^{(1)}x_6} &= b_2^{(1)} \\
F^{(1)}x_2 + D^{(1)}x_4 + F^{(1)}x_6 &= b_4^{(1)} \\
F^{(1)}x_4 + D^{(1)}x_6 &= b_6^{(1)}
\end{aligned}
$$

where $D^{(1)} = 2F^2 - D^2$ and $F^{(1)} = F^2$ commute. Applying the same elimination strategy as above, we multiply these three equations respectively by $F^{(1)}$, $-D^{(1)}$, and $F^{(1)}$. When these transformed equations are added together, we obtain the single equation

$$
\left(2[F^{(1)}]^2 - D^{(1)2}\right)x_4 = F^{(1)}\left(b_2^{(1)} + b_6^{(1)}\right) - D^{(1)}b_4^{(1)}
$$

which we write as

$$
D^{(2)}x_4 = b^{(2)}.
$$

This completes the cyclic reduction. We now solve this (small) q-by-q system for x_4. The vectors x_2 and x_6 are then found by solving the systems

$$
\begin{aligned}
D^{(1)}x_2 &= b_2^{(1)} - F^{(1)}x_4 \\
D^{(1)}x_6 &= b_6^{(1)} - F^{(1)}x_4
\end{aligned}
$$

Finally, we use the first, third, fifth, and seventh equations in (4.5.13) to compute x_1, x_3, x_5, and x_7, respectively.

For general n of the form $n = 2^k - 1$ we set $D^{(0)} = D$, $F^{(0)} = F$, $b^{(0)} = b$ and compute:

$$
\begin{aligned}
&\textbf{for } p = 1{:}k - 1 \\
&\qquad F^{(p)} = [F^{(p-1)}]^2 \\
&\qquad D^{(p)} = 2F^{(p)} - [D^{(p-1)}]^2 \\
&\qquad r = 2^p \\
&\qquad \textbf{for } j = 1{:}2^{k-p} - 1 \\
&\qquad\qquad b_{jr}^{(p)} = F^{(p-1)}\left(b_{jr-r/2}^{(p-1)} + b_{jr+r/2}^{(p-1)}\right) - D^{(p-1)}b_{jr}^{(p-1)} \\
&\qquad \textbf{end} \\
&\textbf{end}
\end{aligned}
\qquad (4.5.14)
$$

The x_i are then computed as follows:

$$
\begin{aligned}
&\text{Solve } D^{(k-1)}x_{2^{k-1}} = b_1^{(k-1)} \text{ for } x_{2^{k-1}}. \\
&\textbf{for } p = k - 2{:}-1{:}0 \\
&\qquad r = 2^p \\
&\qquad \textbf{for } j = 1{:}2^{k-p-1} \\
&\qquad\qquad \textbf{if } j = 1 \\
&\qquad\qquad\qquad c = b_{(2j-1)r}^{(p)} - F^{(p)}x_{2jr} \\
&\qquad\qquad \textbf{elseif } j = 2^{k-p+1} \\
&\qquad\qquad\qquad c = b_{(2j-1)r}^{(p)} - F^{(p)}x_{(2j-2)r} \\
&\qquad\qquad \textbf{else} \\
&\qquad\qquad\qquad c = b_{(2j-1)r}^{(p)} - F^{(p)}\left(x_{2jr} + x_{(2j-2)r}\right) \\
&\qquad\qquad \textbf{end} \\
&\qquad\qquad \text{Solve } D^{(p)}x_{(2j-1)r} = c \text{ for } x_{(2j-1)r} \\
&\qquad \textbf{end} \\
&\textbf{end}
\end{aligned}
\qquad (4.5.15)
$$

The amount of work required to perform these recursions depends greatly upon the sparsity of the $D^{(p)}$ and $F^{(p)}$. In the worse case when these matrices are full, the overall flop count has order $\log(n)q^3$. Care must be exercised in order to ensure stability during the reduction. For further details, see Buneman (1969).

Example 4.5.1 Suppose $q = 1$, $D = (4)$, and $F = (-1)$ in (4.5.14) and that we wish to solve:

$$
\begin{bmatrix}
4 & -1 & 0 & 0 & 0 & 0 & 0 \\
-1 & 4 & -1 & 0 & 0 & 0 & 0 \\
0 & -1 & 4 & -1 & 0 & 0 & 0 \\
0 & 0 & -1 & 4 & -1 & 0 & 0 \\
0 & 0 & 0 & -1 & 4 & -1 & 0 \\
0 & 0 & 0 & 0 & -1 & 4 & -1 \\
0 & 0 & 0 & 0 & 0 & -1 & 4
\end{bmatrix}
\begin{bmatrix}
x_1 \\ x_2 \\ x_3 \\ x_4 \\ x_5 \\ x_6 \\ x_7
\end{bmatrix}
=
\begin{bmatrix}
2 \\ 4 \\ 6 \\ 8 \\ 10 \\ 12 \\ 22
\end{bmatrix}
$$

By executing (4.5.15) we obtain the reduced systems:

$$
\begin{bmatrix} -14 & 1 & 0 \\ 1 & -14 & 1 \\ 0 & 1 & -14 \end{bmatrix} \begin{bmatrix} x_2 \\ x_4 \\ x_6 \end{bmatrix} = \begin{bmatrix} -24 \\ -48 \\ -80 \end{bmatrix} \qquad p = 1
$$

and

$$
\begin{bmatrix} -194 \end{bmatrix} = \begin{bmatrix} x_4 \end{bmatrix} \begin{bmatrix} -776 \end{bmatrix} \qquad p = 2
$$

The x_i are then determined via (4.5.16):

$$
\begin{array}{llll}
p = 2: & x_4 = 4 & & \\
p = 1: & x_2 = 2 & x_6 = 6 & \\
p = 0: & x_1 = 1 & x_3 = 3 & x_5 = 5 \qquad x_7 = 7
\end{array}
$$

Cyclic reduction is an example of a divide and conquer algorithm. Other divide and conquer procedures are discussed in §1.3.8 and §8.6.

4.5.5 Kronecker Product Systems

If $B \in \mathbb{R}^{m \times n}$ and $C \in \mathbb{R}^{p \times q}$, then their *Kronecker product* is given by

$$
A = B \otimes C = \begin{bmatrix} b_{11}C & b_{12}C & \cdots & b_{1n}C \\ b_{21}C & b_{22}C & \cdots & b_{2n}C \\ \vdots & \vdots & \ddots & \vdots \\ b_{m1}C & b_{m2}C & \cdots & b_{mn}C \end{bmatrix}.
$$

Thus, A is an m-by-n block matrix whose (i, j) block is $b_{ij}C$. Kronecker products arise in conjunction with various mesh discretizations and throughout signal processing. Some of the more important properties that the Kronecker product satisfies include

$$
\begin{align}
(A \otimes B)(C \otimes D) &= = AC \otimes BD & (4.5.16) \\
(A \otimes B)^T &= A^T \otimes B^T & (4.5.17) \\
(A \otimes B)^{-1} &= A^{-1} \otimes B^{-1} & (4.5.18)
\end{align}
$$

where it is assumed that all the factor operations are defined.

Related to the Kronecker product is the "vec" operation:

$$
X \in \mathbb{R}^{m \times n} \qquad \Leftrightarrow \qquad \text{vec}(X) = \begin{bmatrix} X(:,1) \\ \vdots \\ X(:,n) \end{bmatrix} \in \mathbb{R}^{mn}.
$$

Thus, the vec of a matrix amounts to a "stacking" of its columns. It can be shown that

$$
Y = CXB^T \qquad \Leftrightarrow \qquad \text{vec}(Y) = (B \otimes C)\text{vec}(X). \qquad (4.5.19)
$$

It follows that solving a Kronecker product system,

$$(B \otimes C)x = d$$

is equivalent to solving the matrix equation $CXB^T = D$ for X where $x = \text{vec}(X)$ and $d = \text{vec}(D)$. This has efficiency ramifications. To illustrate, suppose $B, C \in \mathbb{R}^{n \times n}$ are symmetric positive definite. If $A = B \otimes C$ is treated as a general matrix and factored in order to solve for x, then $O(n^6)$ flops are required since $B \otimes C \in \mathbb{R}^{n^2 \times n^2}$. On the other hand, the solution approach

1. Compute the Cholesky factorizations $B = GG^T$ and $C = HH^T$.

2. Solve $BZ = D^T$ for Z using G.

3. Solve $CX = Z^T$ for X using H.

4. $x = \text{vec}(X)$.

involves $O(n^3)$ flops. Note that

$$B \otimes C = GG^T \otimes HH^T = (G \otimes H)(G \otimes H)^T$$

is the Cholesky factorization of $B \otimes C$ because the Kronecker product of a pair of lower triangular matrices is lower triangular. Thus, the above four-step solution approach is a structure-exploiting, Cholesky method applied to $B \otimes C$.

We mention that if B is sparse, then $B \otimes C$ has the same sparsity at the block level. For example, if B is tridiagonal, then $B \otimes C$ is block tridiagonal.

Problems

P4.5.1 Show that a block diagonally dominant matrix is nonsingular.

P4.5.2 Verify that (4.5.6) implies (4.5.7) and (4.5.8).

P4.5.3 Suppose block cyclic reduction is applied with D given by (4.5.12) and $F = -I_q$. What can you say about the band structure of the matrices $F^{(p)}$ and $D^{(p)}$ that arise?

P4.5.4 Suppose $A \in \mathbb{R}^{n \times n}$ is nonsingular and that we have solutions to the linear systems $Az = b$ and $Ay = g$ where $b, g \in \mathbb{R}^n$ are given. Show how to solve the system

$$\begin{bmatrix} A & g \\ h^T & \alpha \end{bmatrix} \begin{bmatrix} x \\ \mu \end{bmatrix} = \begin{bmatrix} b \\ \beta \end{bmatrix}$$

in $O(n)$ flops where $\alpha, \beta \in \mathbb{R}$ and $h \in \mathbb{R}^n$ are given and the matrix of coefficients A_+ is nonsingular. The advisability of going for such a quick solution is a complicated issue that depends upon the condition numbers of A and A_+ and other factors.

P4.5.5 Verify (4.5.16)-(4.5.19).

P4.5.7 Show how to construct the SVD of $B \otimes C$ from the SVDs of B and C.

P4.5.8 If A, B, and C are matrices, then it can be shown that $(A \otimes B) \otimes C = A \otimes (B \otimes C)$

and so we just write $A \otimes B \otimes C$ for this matrix. Show how to solve the linear system $(A \otimes B \otimes C)x = d$ assuming that A, B, and C are symmetric positive definite.

Notes and References for Sec. 4.5

The following papers provide insight into the various nuances of block matrix computations:

J.M. Varah (1972). "On the Solution of Block-Tridiagonal Systems Arising from Certain Finite-Difference Equations," *Math. Comp. 26*, 859–68.

J.A. George (1974). "On Block Elimination for Sparse Linear Systems," *SIAM J. Num. Anal. 11*, 585–603.

R. Fourer (1984). "Staircase Matrices and Systems," *SIAM Review 26*, 1-71.

M.L. Merriam (1985). "On the Factorization of Block Tridiagonals With Storage Constraints," *SIAM J. Sci. and Stat. Comp. 6*, 182-192.

The property of block diagonal dominance and its various implications is the central theme in

D.G. Feingold and R.S. Varga (1962). "Block Diagonally Dominant Matrices and Generalizations of the Gershgorin Circle Theorem," *Pacific J. Math. 12*, 1241–50.

Early methods that involve the idea of cyclic reduction are described in

R.W. Hockney (1965). "A Fast Direct Solution of Poisson's Equation Using Fourier Analysis, " *J. ACM 12*, 95-113.

B.L. Buzbee, G.H. Golub, and C.W. Nielson (1970). "On Direct Methods for Solving Poisson's Equations," *SIAM J. Num. Anal. 7*, 627-56.

The accumulation of the right-hand side must be done with great care, for otherwise there would be a significant loss of accuracy. A stable way of doing this is described in

O. Buneman (1969). "A Compact Non-Iterative Poisson Solver," Report 294, Stanford University Institute for Plasma Research, Stanford, California.

Other literature concerned with cyclic reduction includes

F.W. Dorr (1970). "The Direct Solution of the Discrete Poisson Equation on a Rectangle," *SIAM Review 12*, 248–63.

B.L. Buzbee, F.W. Dorr, J.A. George, and G.H. Golub (1971). "The Direct Solution of the Discrete Poisson Equation on Irregular Regions," *SIAM J. Num. Anal. 8*, 722–36.

F.W. Dorr (1973). "The Direct Solution of the Discrete Poisson Equation in $O(n^2)$ Operations," *SIAM Review 15*, 412–415.

P. Concus and G.H. Golub (1973). "Use of Fast Direct Methods for the Efficient Numerical Solution of Nonseparable Elliptic Equations," *SIAM J. Num. Anal. 10*, 1103–20.

B.L. Buzbee and F.W. Dorr (1974). "The Direct Solution of the Biharmonic Equation on Rectangular Regions and the Poisson Equation on Irregular Regions," *SIAM J. Num. Anal. 11*, 753–63.

D. Heller (1976). "Some Aspects of the Cyclic Reduction Algorithm for Block Tridiagonal Linear Systems," *SIAM J. Num. Anal. 13*, 484–96.

Various generalizations and extensions to cyclic reduction have been proposed:

P.N. Swarztrauber and R.A. Sweet (1973). "The Direct Solution of the Discrete Poisson Equation on a Disk," *SIAM J. Num. Anal. 10*, 900–907.

R.A. Sweet (1974). "A Generalized Cyclic Reduction Algorithm," *SIAM J. Num. Anal.*
   11, 506–20.
M.A. Diamond and D.L.V. Ferreira (1976). "On a Cyclic Reduction Method for the
   Solution of Poisson's Equation," *SIAM J. Num. Anal. 13*, 54–70.
R.A. Sweet (1977). "A Cyclic Reduction Algorithm for Solving Block Tridiagonal Sys-
   tems of Arbitrary Dimension," *SIAM J. Num. Anal. 14*, 706–20.
P.N. Swarztrauber and R. Sweet (1989). "Vector and Parallel Methods for the Direct
   Solution of Poisson's Equation," *J. Comp. Appl. Math. 27*, 241–263.
S. Bondeli and W. Gander (1994). "Cyclic Reduction for Special Tridiagonal Systems,"
   SIAM J. Matrix Anal. Appl. 15, 321–330.

For certain matrices that arise in conjunction with elliptic partial differential equations,
block elimination corresponds to rather natural operations on the underlying mesh. A
classical example of this is the method of nested dissection described in

A. George (1973). "Nested Dissection of a Regular Finite Element Mesh," *SIAM J.
   Num. Anal. 10*, 345–63.

We also mention the following general survey:

J.R. Bunch (1976). "Block Methods for Solving Sparse Linear Systems," in *Sparse
   Matrix Computations*, J.R. Bunch and D.J. Rose (eds), Academic Press, New York.

Bordered linear systems as presented in P4.5.4 are discussed in

W. Govaerts and J.D. Pryce (1990). "Block Elimination with One Iterative Refinement
   Solves Bordered Linear Systems Accurately," *BIT 30*, 490–507.
W. Govaerts (1991). "Stable Solvers and Block Elimination for Bordered Systems,"
   SIAM J. Matrix Anal. Appl. 12, 469–483.
W. Govaerts and J.D. Pryce (1993). "Mixed Block Elimination for Linear Systems with
   Wider Borders," *IMA J. Num. Anal. 13*, 161–180.

Kronecker product references include

H.C. Andrews and J. Kane (1970). "Kronecker Matrices, Computer Implementation,
   and Generalized Spectra," *J. Assoc. Comput. Mach. 17*, 260–268.
C. de Boor (1979). "Efficient Computer Manipulation of Tensor Products," *ACM Trans.
   Math. Soft. 5*, 173–182.
A. Graham (1981). *Kronecker Products and Matrix Calculus with Applications*, Ellis
   Horwood Ltd., Chichester, England.
H.V. Henderson and S.R. Searle (1981). "The Vec-Permutation Matrix, The Vec Opera-
   tor, and Kronecker Products: A Review," *Linear and Multilinear Algebra 9*, 271–288.
P.A. Regalia and S. Mitra (1989). "Kronecker Products, Unitary Matrices, and Signal
   Processing Applications," *SIAM Review 31*, 586–613.

4.6   Vandermonde Systems and the FFT

Suppose $x(0{:}n) \in \mathbb{R}^{n+1}$. A matrix $V \in \mathbb{R}^{(n+1)\times(n+1)}$ of the form

$$
V = V(x_0,\ldots,x_n) = \begin{bmatrix} 1 & 1 & \cdots & 1 \\ x_0 & x_1 & \cdots & x_n \\ \vdots & \vdots & & \vdots \\ x_0^n & x_1^n & \cdots & x_n^n \end{bmatrix}
$$

is said to be a *Vandermonde matrix*. In this section, we show how the
systems $V^T a = f = f(0{:}n)$ and $Vz = b = b(0{:}n)$ can be solved in $O(n^2)$
flops. The discrete Fourier transform is briefly introduced. This special and
extremely important Vandermonde system has a a recursive block structure
and can be solved in $O(n \log n)$ flops. *In this section, vectors and matrices
are subscripted from 0.*

4.6.1   Polynomial Interpolation: $V^T a = f$

Vandermonde systems arise in many approximation and interpolation problems. Indeed, the key to obtaining a fast Vandermonde solver is to recognize
that solving $V^T a = f$ is equivalent to polynomial interpolation. This follows because if $V^T a = f$ and

$$p(x) = \sum_{j=0}^{n} a_j x^j \qquad (4.6.1)$$

then $p(x_i) = f_i$ for $i = 0{:}n$.

Recall that if the x_i are distinct then there is a unique polynomial of
degree n that interpolates $(x_0, f_0), \dots, (x_n, f_n)$. Consequently, V is nonsingular as long as the x_i are distinct. We assume this throughout the
section.

The first step in computing the a_j of (4.6.1) is to calculate the Newton
representation of the interpolating polynomial p:

$$p(x) = \sum_{k=0}^{n} c_k \left(\prod_{i=0}^{k-1} (x - x_i) \right). \qquad (4.6.2)$$

The constants c_k are divided differences and may be determined as follows:

$$
\begin{aligned}
& c(0{:}n) = f(0{:}n) \\
& \textbf{for } k = 0{:}n - 1 \\
& \qquad \textbf{for } i = n{:} -1{:}k+1 \\
& \qquad\qquad c_i = (c_i - c_{i-1})/(x_i - x_{i-k-1}) \\
& \qquad \textbf{end} \\
& \textbf{end}
\end{aligned}
\qquad (4.6.3)
$$

See Conte and de Boor (1980, chapter 2).

The next task is to generate $a(0{:}n)$ from $c(0{:}n)$. Define the polynomials
$p_n(x), \dots, p_0(x)$ by the iteration

$$
\begin{aligned}
& p_n(x) = c_n \\
& \textbf{for } k = n - 1{:} -1{:}0 \\
& \qquad p_k(x) = c_k + (x - x_k)p_{k+1}(x) \\
& \textbf{end}
\end{aligned}
$$

and observe that $p_0(x) = p(x)$. Writing

$$p_k(x) = a_k^{(k)} + a_{k+1}^{(k)}x + \cdots + a_n^{(k)}x^{n-k}$$

and equating like powers of x in the equation $p_k = c_k + (x - x_k)p_{k+1}$ gives the following recursion for the coefficients $a_i^{(k)}$:

$$a_n^{(n)} = c_n$$
for $k = n - 1: -1{:}0$
$\qquad a_k^{(k)} = c_k - x_k a_{k+1}^{(k+1)}$
$\qquad$ for $i = k + 1{:}n - 1$
$\qquad\qquad a_i^{(k)} = a_i^{(k+1)} - x_k a_{i+1}^{(k+1)}$
$\qquad$ end
$\qquad a_n^{(k)} = a_n^{(k+1)}$
end

Consequently, the coefficients $a_i = a_i^{(0)}$ can be calculated as follows:

$a(0{:}n) = c(0{:}n)$
for $k = n - 1: -1{:}0$
$\qquad$ for $i = k{:}n - 1$ (4.6.4)
$\qquad\qquad a_i = a_i - x_k a_{i+1}$
$\qquad$ end
end

Combining this iteration with (4.6.3) renders the following algorithm:

Algorithm 4.6.1   Given $x(0{:}n) \in \mathbb{R}^{n+1}$ with distinct entries and $f = f(0{:}n) \in \mathbb{R}^{n+1}$, the following algorithm overwrites f with the solution $a = a(0{:}n)$ to the Vandermonde system $V(x_0, \ldots, x_n)^T a = f$.

for $k = 0{:}n - 1$
$\qquad$ for $i = n: -1{:}k + 1$
$\qquad\qquad f(i) = (f(i) - f(i - 1))/(x(i) - x(i - k - 1))$
$\qquad$ end
end
for $k = n - 1: -1{:}0$
$\qquad$ for $i = k{:}n - 1$
$\qquad\qquad f(i) = f(i) - f(i + 1)x(k)$
$\qquad$ end
end

This algorithm requires $5n^2/2$ flops.

Example 4.6.1   Suppose Algorithm 4.6.1 is used to solve

$$\begin{bmatrix} 1 & 1 & 1 & 1 \\ 1 & 2 & 4 & 8 \\ 1 & 3 & 9 & 27 \\ 1 & 4 & 16 & 64 \end{bmatrix}^T \begin{bmatrix} a_0 \\ a_1 \\ a_2 \\ a_3 \end{bmatrix} = \begin{bmatrix} 10 \\ 26 \\ 58 \\ 112 \end{bmatrix}.$$

The first k-loop computes the Newton representation of $p(x)$:

$$p(x) = 10 + 16(x - 1) + 8(x - 1)(x - 2) + (x - 1)(x - 2)(x - 3).$$

The second k-loop computes $a = [4\ 3\ 2\ 1]^T$ from $[10\ 16\ 8\ 1]^T$.

4.6.2   The System $Vz = b$

Now consider the system $Vz = b$. To derive an efficient algorithm for this problem, we describe what Algorithm 4.6.1 does in matrix-vector language. Define the lower bidiagonal matrix $L_k(\alpha) \in \mathbb{R}^{(n+1)\times(n+1)}$ by

$$L_k(\alpha) = \begin{bmatrix} I_k & & & & & 0 \\ \hline & 1 & \cdots & & & 0 \\ & -\alpha & 1 & & & \\ 0 & & \ddots & \ddots & & \vdots \\ & \vdots & & \ddots & \ddots & \\ & & & & \ddots & 1 \\ & 0 & \cdots & & -\alpha & 1 \end{bmatrix}$$

and the diagonal matrix D_k by

$$D_k = \text{diag}(\underbrace{1, \ldots, 1}_{k+1}, x_{k+1} - x_0, \ldots, x_n - x_{n-k-1}).$$

With these definitions it is easy to verify from (4.6.3) that if $f = f(0{:}n)$ and $c = c(0{:}n)$ is the vector of divided differences then $c = U^T f$ where U is the upper triangular matrix defined by

$$U^T = D_{n-1}^{-1} L_{n-1}(1) \cdots D_0^{-1} L_0(1).$$

Similarly, from (4.6.4) we have

$$a = L^T c,$$

where L is the unit lower triangular matrix defined by:

$$L^T = L_0(x_0)^T \cdots L_{n-1}(x_{n-1})^T.$$

Thus, $a = L^T U^T f$ where $V^{-T} = L^T U^T$. In other words, Algorithm 4.6.1 solves $V^T a = f$ by tacitly computing the "UL" factorization of V^{-1}.

Consequently, the solution to the system $Vz = b$ is given by

$$z = V^{-1}b = U(Lb)$$
$$= \left(L_0(1)^T D_0^{-1} \cdots L_{n-1}(1)^T D_{n-1}^{-1}\right)\left(L_{n-1}(x_{n-1}) \cdots L_0(x_0)b\right)$$

This observation gives rise to the following algorithm:

Algorithm 4.6.2   Given $x(0{:}n) \in \mathbb{R}^{n+1}$ with distinct entries and $b = b(0{:}n) \in \mathbb{R}^{n+1}$, the following algorithm overwrites b with the solution $z = z(0{:}n)$ to the Vandermonde system $V(x_0, \ldots, x_n)z = b$.

> for $k = 0{:}n-1$
>   for $i = n{:}-1{:}k+1$
> $b(i) = b(i) - x(k)b(i-1)$
>   end
> end
> for $k = n-1{:}-1{:}0$
>   for $i = k+1{:}n$
> $b(i) = b(i)/(x(i) - x(i-k-1))$
>   end
>   for $i = k{:}n-1$
> $b(i) = b(i) - b(i+1)$
>   end
> end

This algorithm requires $5n^2/2$ flops.

Example 4.6.2 Suppose Algorithm 4.6.2 is used to solve

$$\begin{bmatrix} 1 & 1 & 1 & 1 \\ 1 & 2 & 3 & 4 \\ 1 & 4 & 9 & 16 \\ 1 & 8 & 27 & 64 \end{bmatrix} \begin{bmatrix} z_0 \\ z_1 \\ z_2 \\ z_3 \end{bmatrix} = \begin{bmatrix} 0 \\ -1 \\ 3 \\ 35 \end{bmatrix} .$$

The first k-loop computes the vector

$$L_3(3)L_2(2)L_1(1) \begin{bmatrix} 0 \\ -1 \\ 3 \\ 35 \end{bmatrix} = \begin{bmatrix} 0 \\ -1 \\ 6 \\ 6 \end{bmatrix} .$$

The second k-loop then calculates

$$L_0(1)^T D_0^{-1} L_1(1)^T D_1^{-1} L_2(1)^T D_2^{-1} \begin{bmatrix} 0 \\ -1 \\ 3 \\ 35 \end{bmatrix} = \begin{bmatrix} 3 \\ -4 \\ 0 \\ 1 \end{bmatrix} .$$

4.6.3   Stability

Algorithms 4.6.1 and 4.6.2 are discussed and analyzed in Björck and Pereyra (1970). Their experience is that these algorithms frequently produce surprisingly accurate solutions, even when V is ill-conditioned. They also show how to update the solution when a new coordinate pair (x_{n+1}, f_{n+1})

is added to the set of points to be interpolated, and how to solve *confluent Vandermonde systems*, i.e., systems involving matrices like

$$
V = V(x_0, x_1, x_1, x_3) = \begin{bmatrix} 1 & 1 & 0 & 1 \\ x_0 & x_1 & 1 & x_3 \\ x_0^2 & x_1^2 & 2x_1^2 & x_3^2 \\ x_0^3 & x_1^3 & 3x_1^2 & x_3^3 \end{bmatrix}.
$$

4.6.4   The Fast Fourier Transform

The discrete Fourier transform (DFT) matrix of order n is defined by

$$
F_n = (f_{jk}) \qquad f_{jk} = w_n^{jk}
$$

where

$$
w_n = \exp(-2\pi i/n) = \cos(2\pi/n) - i \cdot \sin(2\pi/n).
$$

The parameter w_n is an nth root of unity because $w_n^n = 1$. In the $n = 4$ case, $w_4 = -i$ and

$$
F_4 = \begin{bmatrix} 1 & 1 & 1 & 1 \\ 1 & w_4 & w_4^2 & w_4^3 \\ 1 & w_4^2 & w_4^4 & w_4^6 \\ 1 & w_4^3 & w_4^6 & w_4^9 \end{bmatrix} = \begin{bmatrix} 1 & 1 & 1 & 1 \\ 1 & -i & -1 & i \\ 1 & -1 & 1 & -1 \\ 1 & i & -1 & -i \end{bmatrix}.
$$

If $x \in \mathbb{C}^n$, then its DFT is the vector $F_n x$. The DFT has an extremely important role to play throughout applied mathematics and engineering.

If n is highly composite, then it is possible to carry out the DFT in many fewer than the $O(n^2)$ flops required by conventional matrix-vector multiplication. To illustrate this we set $n = 2^t$ and proceed to develop the *radix-2 fast Fourier transform (FFT)*. The starting point is to look at an even-order DFT matrix when we permute its columns so that the even-indexed columns come first. Consider the case $n = 8$. Noting that $w_8^{kj} = w_n^{kj \bmod 8}$ we have

$$
F_8 = \begin{bmatrix} 1 & 1 & 1 & 1 & 1 & 1 & 1 & 1 \\ 1 & w & w^2 & w^3 & w^4 & w^5 & w^6 & w^7 \\ 1 & w^2 & w^4 & w^6 & 1 & w^2 & w^4 & w^6 \\ 1 & w^3 & w^6 & w & w^4 & w^7 & w^2 & w^5 \\ 1 & w^4 & 1 & w^4 & 1 & w^4 & 1 & w^4 \\ 1 & w^5 & w^2 & w^7 & w^4 & w & w^6 & w^3 \\ 1 & w^6 & w^4 & w^2 & 1 & w^6 & w^4 & w^2 \\ 1 & w^7 & w^6 & w^5 & w^4 & w^3 & w^2 & w \end{bmatrix}, \qquad w = w_8.
$$

If we define the index vector $c = [0\ 2\ 4\ 6\ 1\ 3\ 5\ 7]$, then

$$F_8(:,c) = \left[\begin{array}{cccc|cccc}
1 & 1 & 1 & 1 & 1 & 1 & 1 & 1 \\
1 & \omega^2 & \omega^4 & \omega^6 & \omega & \omega^3 & \omega^5 & \omega^7 \\
1 & \omega^4 & 1 & \omega^4 & \omega^2 & \omega^6 & \omega^2 & \omega^6 \\
1 & \omega^6 & \omega^4 & \omega^2 & \omega^3 & \omega & \omega^7 & \omega^5 \\
\hline
1 & 1 & 1 & 1 & -1 & -1 & -1 & -1 \\
1 & \omega^2 & \omega^4 & \omega^6 & -\omega & -\omega^3 & -\omega^5 & -\omega^7 \\
1 & \omega^4 & 1 & \omega^4 & -\omega^2 & -\omega^6 & -\omega^2 & -\omega^6 \\
1 & \omega^6 & \omega^4 & \omega^2 & -\omega^3 & -\omega & -\omega^7 & -\omega^5
\end{array}\right].$$

The lines through the matrix are there to help us think of $F_n(:,c)$ as a 2-by-2 matrix with 4-by-4 blocks. Noting that $\omega^2 = \omega_8^2 = \omega_4$ we see that

$$F_8(:,c) = \left[\begin{array}{c|c} F_4 & \Omega_4 F_4 \\ \hline F_4 & -\Omega_4 F_4 \end{array}\right]$$

where

$$\Omega_4 = \left[\begin{array}{cccc}
1 & 0 & 0 & 0 \\
0 & \omega_8 & 0 & 0 \\
0 & 0 & \omega_8^2 & 0 \\
0 & 0 & 0 & \omega_8^3
\end{array}\right].$$

It follows that if x in an 8-vector, then

$$F_8 x = F(:,c)x(c) = \left[\begin{array}{c|c} F_4 & \Omega_4 F_4 \\ \hline F_4 & -\Omega_4 F_4 \end{array}\right]\left[\begin{array}{c} x(0{:}2{:}7) \\ x(1{:}2{:}7) \end{array}\right]$$

$$= \left[\begin{array}{c|c} I & \Omega_4 \\ \hline I & -\Omega_4 \end{array}\right]\left[\begin{array}{c} F_4 x(0{:}2{:}7) \\ F_4 x(1{:}2{:}7) \end{array}\right].$$

Thus, by simple scalings we can obtain the 8-point DFT $y = F_8 x$ from the 4-point DFTs $y_T = F_4 x(0{:}2{:}7)$ and $y_B = F_4 x(1{:}2{:}7)$:

$$y(0{:}4) = y_T + d.*y_B$$
$$y(4{:}7) = y_T - d.*y_B.$$

Here,

$$d = \left[\begin{array}{c} 1 \\ \omega \\ \omega^2 \\ \omega^3 \end{array}\right]$$

and ".*" indicates vector multiplication. In general, if $n = 2m$, then $y = F_n x$ is given by

$$y(0{:}m-1) = y_T + d.*y_B$$
$$y(m{:}n-1) = y_B - d.*y_B$$

where

$$
\begin{aligned}
d &= \left[\, 1,\, \omega, \cdots,\, \omega^{m-1}\,\right]^T \\
y_T &= F_m x(0{:}2{:}n-1), \\
y_B &= F_m x(1{:}2{:}n-1).
\end{aligned}
$$

For $n = 2^t$ we can recur on this process until $n = 1$ for which $F_1 x = x$:

> **function** $y = FFT(x, n)$
>   **if** $n = 1$
> $y = x$
>   **else**
> $m = n/2;\ \omega = e^{-2\pi i/n}$
> $y_T = FFT(x(0{:}2{:}n), m);\ y_B = FFT(x(1{:}2{:}n), m)$
> $d = \left[\, 1,\, \omega, \cdots,\, \omega^{m-1}\,\right]^T;\ z = d.*y_B$
> $y = \left[\begin{array}{c} y_T + z \\ y_T - z \end{array}\right]$
>   **end**

This is a member of the fast Fourier transform family of algorithms. It has a nonrecursive implementation that is best presented in terms of a factorization of F_n. Indeed, it can be shown that $F_n = A_t \cdots A_1 P_n$ where

$$
A_q = I_r \otimes B_L \qquad L = 2^q,\ r = n/L
$$

with

$$
B_L = \left[\begin{array}{cc} I_{L/2} & \Omega_{L/2} \\ I_{L/2} & -\Omega_{L/2} \end{array}\right] \quad \text{and} \quad \Omega_{L/2} = \mathrm{diag}(1, \omega_L, \ldots, \omega_L^{L/2-1}).
$$

The matrix P_n is called the *bit reversal permutation*, the description of which we omit. (Recall the definition of the Kronecker product "$\otimes$" from §4.5.5.) Note that with this factorization, $y = F_n x$ can be computed as follows:

$$
\begin{aligned}
&x = P_n x \\
&\textbf{for } q = 1{:}t \\
&\qquad L = 2^q,\ r = n/L \\
&\qquad x = (I_r \otimes B_L)x \\
&\textbf{end}
\end{aligned}
\qquad (4.6.5)
$$

The matrices $A_q = (I_r \otimes B_L)$ have 2 nonzeros per row and it is this sparsity that makes it possible to implement the DFT in $O(n \log n)$ flops. In fact, a careful implementation involves $5n \log_2 n$ flops.

The DFT matrix has the property that

$$
F_n^{-1} = \frac{1}{n} F_n^H = \frac{1}{n} \overline{F}_n. \qquad (4.6.6)
$$

That is, the inverse of F_n is obtained by conjugating its entries and scaling by n. A fast inverse DFT can be obtained from a (forward) FFT merely by replacing all root-of-unity references with their complex conjugate and scaling by n at the end.

The value of the DFT is that many "hard problems" are made simple by transforming into Fourier space (via F_n). The sought-after solution is then obtained by transforming the Fourier space solution into original coordinates (via F_n^{-1}).

Problems

P4.6.1 Show that if $V = V(x_0, \ldots, x_n)$, then

$$\det(V) = \prod_{n \geq i > j \geq 0} (x_i - x_j).$$

P4.6.2 (Gautschi 1975a) Verify the following inequality for the $n = 1$ case above:

$$\| V^{-1} \|_\infty \leq \max_{0 \leq k \leq n} \prod_{\substack{i=0 \\ i \neq k}}^{n} \frac{1 + |x_i|}{|x_k - x_i|} .$$

Equality results if the x_i are all on the same ray in the complex plane.

P4.6.3 Suppose $w = \left[1, \omega_n, \omega_n^2, \ldots, \omega_n^{n/2-1} \right]$ where $n = 2^t$. Using colon notation, express

$$\left[1, \omega_r, \omega_r^2, \ldots, \omega_r^{r/2-1} \right]$$

as a subvector of w where $r = 2^q$, $q = 1{:}t$.

P4.6.4 Prove (4.6.6).

P4.6.5 Expand the operation $x = (I \otimes B_L)x$ in (4.6.5) into a double loop and count the number of flops required by your implementation. (Ignore the details of $x = P_n x$.

P4.6.6 Suppose $n = 3m$ and examine

$$G = [\, F_n(:, 0{:}3{:}n-1) \quad F_n(1{:}3{:}n-1) \quad F_n(:, 2{:}3{:}n-1) \,]$$

as a 3-by-3 block matrix, looking for scaled copies of F_m. Based on what you find, develop a recursive radix-3 FFT analogous to the radix-2 implementation in the text.

Notes and References for Sec. 4.6

Our discussion of Vandermonde linear systems is drawn from the papers

A. Björck and V. Pereyra (1970). "Solution of Vandermonde Systems of Equations," *Math. Comp. 24*, 893–903.

A. Björck and T. Elfving (1973). "Algorithms for Confluent Vandermonde Systems," *Numer. Math. 21*, 130–37.

The divided difference computations we discussed are detailed in chapter 2 of

S.D. Conte and C. de Boor (1980). *Elementary Numerical Analysis: An Algorithmic Approach*, 3rd ed., McGraw-Hill, New York.

The latter reference includes an Algol procedure. Error analyses of Vandermonde system solvers include

N.J. Higham (1987b). "Error Analysis of the Björck-Pereyra Algorithms for Solving Vandermonde Systems," *Numer. Math. 50*, 613–632.

N.J. Higham (1988a). "Fast Solution of Vandermonde-like Systems Involving Orthogonal Polynomials," *IMA J. Num. Anal. 8*, 473-486.

N.J. Higham (1990). "Stability Analysis of Algorithms for Solving Confluent Vandermonde-like Systems," *SIAM J. Matrix Anal. Appl. 11*, 23–41.

S.G. Bartels and D.J. Higham (1992). "The Structured Sensitivity of Vandermonde-Like Systems," *Numer. Math. 62*, 17–34.

J.M. Varah (1993). "Errors and Perturbations in Vandermonde Systems," *IMA J. Num. Anal. 13*, 1–12.

Interesting theoretical results concerning the condition of Vandermonde systems may be found in

W. Gautschi (1975a). "Norm Estimates for Inverses of Vandermonde Matrices," *Numer. Math. 23*, 337–47.

W. Gautschi (1975b). "Optimally Conditioned Vandermonde Matrices," *Numer. Math. 24*, 1–12.

The basic algorithms presented can be extended to cover confluent Vandermonde systems, block Vandermonde systems, and Vandermonde systems that are based on other polynomial bases:

G. Galimberti and V. Pereyra (1970). "Numerical Differentiation and the Solution of Multidimensional Vandermonde Systems," *Math. Comp. 24*, 357–64.

G. Galimberti and V. Pereyra (1971). "Solving Confluent Vandermonde Systems of Hermitian Type," *Numer. Math. 18*, 44–60.

H. Van de Vel (1977). "Numerical Treatment of a Generalized Vandermonde systems of Equations," *Lin. Alg. and Its Applic. 17*, 149–74.

G.H. Golub and W.P Tang (1981). "The Block Decomposition of a Vandermonde Matrix and Its Applications," *BIT 21*, 505–17.

D. Calvetti and L. Reichel (1992). "A Chebychev-Vandermonde Solver," *Lin. Alg. and Its Applic. 172*, 219–229.

D. Calvetti and L. Reichel (1993). "Fast Inversion of Vandermonde-Like Matrices Involving Orthogonal Polynomials," *BIT 33*, 473–484.

H. Lu (1994). "Fast Solution of Confluent Vandermonde Linear Systems," *SIAM J. Matrix Anal. Appl. 15*, 1277–1289.

H. Lu (1996). "Solution of Vandermonde-like Systems and Confluent Vandermonde-like Systems," *SIAM J. Matrix Anal. Appl. 17*, 127–138.

The FFT literature is very extensive and scattered. For an overview of the area couched in Kronecker product notation, see

C.F. Van Loan (1992). *Computational Frameworks for the Fast Fourier Transform*, SIAM Publications, Philadelphia, PA.

The point of view in this text is that different FFTs correspond to different factorizations of the DFT matrix. These are *sparse* factorizations in that the factors have very few nonzeros per row.

4.7 Toeplitz and Related Systems

Matrices whose entries are constant along each diagonal arise in many applications and are called *Toeplitz matrices*. Formally, $T \in \mathbb{R}^{n \times n}$ is Toeplitz if there exist scalars $r_{-n+1}, \ldots, r_0, \ldots, r_{n-1}$ such that $a_{ij} = r_{j-i}$ for all i and j. Thus,

$$
T = \begin{bmatrix}
r_0 & r_1 & r_2 & r_3 \\
r_{-1} & r_0 & r_1 & r_2 \\
r_{-2} & r_{-1} & r_0 & r_1 \\
r_{-3} & r_{-2} & r_{-1} & r_0
\end{bmatrix}
=
\begin{bmatrix}
3 & 1 & 7 & 6 \\
4 & 3 & 1 & 7 \\
0 & 4 & 3 & 1 \\
9 & 0 & 4 & 3
\end{bmatrix}
$$

is Toeplitz.

Toeplitz matrices belong to the larger class of *persymmetric matrices*. We say that $B \in \mathbb{R}^{n \times n}$ is persymmetric if it symmetric about its northeast-southwest diagonal, i.e., $b_{ij} = b_{n-j+1,n-i+1}$ for all i and j. This is equivalent to requiring $B = E B^T E$ where $E = [\, e_n, \ldots, e_1 \,] = I_n(:, n: -1:1)$ is the *n-by-n exchange matrix*, i.e.,

$$
E = \begin{bmatrix}
0 & 0 & 0 & 1 \\
0 & 0 & 1 & 0 \\
0 & 1 & 0 & 0 \\
1 & 0 & 0 & 0
\end{bmatrix}.
$$

It is easy to verify that (a) Toeplitz matrices are persymmetric and (b) the inverse of a nonsingular Toeplitz matrix is persymmetric. In this section we show how the careful exploitation of (b) enables us to solve Toeplitz systems with $O(n^2)$ flops. The discussion focuses on the important case when T is also symmetric and positive definite. Unsymmetric Toeplitz systems and connections with circulant matrices and the discrete Fourier transform are briefly discussed.

4.7.1 Three Problems

Assume that we have scalars $r_1, \ldots, r_n$ such that for $k = 1:n$ the matrices

$$
T_k = \begin{bmatrix}
1 & r_1 & \cdots & r_{k-2} & r_{k-1} \\
r_1 & 1 & \ddots & & r_{k-2} \\
\vdots & \ddots & \ddots & \ddots & \vdots \\
r_{k-2} & & \ddots & \ddots & r_1 \\
r_{k-1} & r_{k-2} & \cdots & r_1 & 1
\end{bmatrix}
$$

are positive definite. (There is no loss of generality in normalizing the diagonal.) Three algorithms are described in this section:

- Durbin's algorithm for the *Yule-Walker problem* $T_n y = -[r_1, \ldots, r_n]^T$.

- Levinson's algorithm for the general righthand side problem $T_n x = b$.

- Trench's algorithm for computing $B = T_n^{-1}$.

In deriving these methods, we denote the k-by-k exchange matrix by E_k, i.e., $E_k = I_k(:, k:-1:1)$.

4.7.2   Solving the Yule-Walker Equations

We begin by presenting Durbin's algorithm for the Yule-Walker equations which arise in conjunction with certain linear prediction problems. Suppose for some k that satisfies $1 \le k \le n-1$ we have solved the k-th order Yule-Walker system $T_k y = -r = -(r_1, \ldots, r_k)^T$. We now show how the $(k+1)$-st order Yule-Walker system

$$\left[\begin{array}{cc} T_k & E_k r \\ r^T E_k & 1 \end{array} \right] \left[\begin{array}{c} z \\ \alpha \end{array} \right] = - \left[\begin{array}{c} r \\ r_{k+1} \end{array} \right]$$

can be solved in $O(k)$ flops. First observe that

$$z = T_k^{-1}(-r - \alpha E_k r) = y - \alpha T_k^{-1} E_k r$$

and

$$\alpha = -r_{k+1} - r^T E_k z.$$

Since T_k^{-1} is persymmetric, $T_k^{-1} E_k = E_k T_k^{-1}$ and thus,

$$z = y - \alpha E_k T_k^{-1} r = y + \alpha E_k y.$$

By substituting this into the above expression for α we find

$$\alpha = -r_{k+1} - r^T E_k (y + \alpha E_k y) = -(r_{k+1} + r^T E_k y)/(1 + r^T y).$$

The denominator is positive because T_{k+1} is positive definite and because

$$\left[\begin{array}{cc} I & E_k y \\ 0 & 1 \end{array} \right]^T \left[\begin{array}{cc} T_k & E_k r \\ r^T E_k & 1 \end{array} \right] \left[\begin{array}{cc} I & E_k y \\ 0 & 1 \end{array} \right] = \left[\begin{array}{cc} T_k & 0 \\ 0 & 1 + r^T y \end{array} \right].$$

We have illustrated the kth step of an algorithm proposed by Durbin (1960). It proceeds by solving the Yule-Walker systems

$$T_k y^{(k)} = -r^{(k)} = -[r_1, \ldots, r_k]^T$$

for $k = 1:n$ as follows:

$$y^{(1)} = -r_1$$

for $k = 1{:}n - 1$

$$\beta_k = 1 + [r^{(k)}]^T y^{(k)}$$
$$\alpha_k = -(r_{k+1} + r^{(k)T} E_k y^{(k)})/\beta_k \qquad (4.7.1)$$
$$z^{(k)} = y^{(k)} + \alpha_k E_k y^{(k)}$$
$$y^{(k+1)} = \begin{bmatrix} z^{(k)} \\ \alpha_k \end{bmatrix}$$

end

As it stands, this algorithm would require $3n^2$ flops to generate $y = y^{(n)}$. It is possible, however, to reduce the amount of work even further by exploiting some of the above expressions:

$$
\begin{aligned}
\beta_k &= 1 + [r^{(k)}]^T y^{(k)} \\
&= 1 + \begin{bmatrix} r^{(k-1)T} & r_k \end{bmatrix} \begin{bmatrix} y^{(k-1)} + \alpha_{k-1} E_{k-1} y^{(k-1)} \\ \alpha_{k-1} \end{bmatrix} \\
&= (1 + [r^{(k-1)}]^T y^{(k-1)}) + \alpha_{k-1} \left([r^{(k-1)}]^T E_{k-1} y^{(k-1)} + r_k \right) \\
&= \beta_{k-1} + \alpha_{k-1}(-\beta_{k-1}\alpha_{k-1}) \\
&= (1 - \alpha_{k-1}^2)\beta_{k-1}.
\end{aligned}
$$

Using this recursion we obtain the following algorithm:

Algorithm 4.7.1. (Durbin) Given real numbers $1 = r_0, r_1, \ldots, r_n$ such that $T = (r_{|i-j|}) \in \mathbb{R}^{n \times n}$ is positive definite, the following algorithm computes $y \in \mathbb{R}^n$ such that $Ty = -(r_1, \ldots, r_n)^T$.

$$y(1) = -r(1); \quad \beta = 1; \quad \alpha = -r(1)$$

for $k = 1{:}n - 1$

$$\beta = (1 - \alpha^2)\beta$$
$$\alpha = -\left(r(k+1) + r(k{:}-1{:}1)^T y(1{:}k)\right)/\beta$$
$$z(1{:}k) = y(1{:}k) + \alpha y(k{:}-1{:}1)$$
$$y(1{:}k+1) = \begin{bmatrix} z(1{:}k) \\ \alpha \end{bmatrix}$$

end

This algorithm requires $2n^2$ flops. We have included an auxiliary vector z for clarity, but it can be avoided.

Example 4.7.1 Suppose we wish to solve the Yule-Walker system

$$
\begin{bmatrix} 1 & .5 & .2 \\ .5 & 1 & .5 \\ .2 & .5 & 1 \end{bmatrix} \begin{bmatrix} y_1 \\ y_2 \\ y_3 \end{bmatrix} = - \begin{bmatrix} .5 \\ .2 \\ .1 \end{bmatrix}
$$

using Algorithm 4.7.1. After one pass through the loop we obtain

$$\alpha = 1/15, \qquad \beta = 3/4, \qquad y = \begin{bmatrix} -8/15 \\ 1/15 \end{bmatrix}.$$

We then compute

$$
\begin{aligned}
\beta &= (1 - \alpha^2)\beta = 56/75 \\
\alpha &= -(r_3 + r_2 y_1 + r_1 y_2)/\beta = -1/28 \\
z_1 &= y_1 + \alpha y_2 = -225/420 \\
z_2 &= y_2 + \alpha y_1 = -36/420,
\end{aligned}
$$

giving the final solution $y = [-75,\ 12,\ -5]^T/140$.

4.7.3   The General Right Hand Side Problem

With a little extra work, it is possible to solve a symmetric positive definite Toeplitz system that has an arbitrary right-hand side. Suppose that we have solved the system

$$
T_k x = b = (b_1, \ldots, b_k)^T \tag{4.7.2}
$$

for some k satisfying $1 \le k < n$ and that we now wish to solve

$$
\begin{bmatrix} T_k & E_k r \\ r^T E_k & 1 \end{bmatrix} \begin{bmatrix} v \\ \mu \end{bmatrix} = \begin{bmatrix} b \\ b_{k+1} \end{bmatrix}. \tag{4.7.3}
$$

Here, $r = (r_1, \ldots, r_k)^T$ as above. Assume also that the solution to the kth order Yule-Walker system $T_k y = -r$ is also available. From $T_k v + \mu E_k r = b$ it follows that

$$
v = T_k^{-1}(b - \mu E_k r) = x - \mu T_k^{-1} E_k = x + \mu E_k y
$$

and so

$$
\begin{aligned}
\mu &= b_{k+1} - r^T E_k v \\
&= b_{k+1} - r^T E_k x - \mu r^T y \\
&= \left(b_{k+1} - r^T E_k x \right) / \left(1 + r^T y \right).
\end{aligned}
$$

Consequently, we can effect the transition from (4.7.2) to (4.7.3) in $O(k)$ flops.

Overall, we can efficiently solve the system $T_n x = b$ by solving the systems $T_k x^{(k)} = b^{(k)} = (b_1, \ldots, b_k)^T$ and $T_k y^{(k)} = -r^{(k)} = (r_1, \ldots, r_k)^T$ "in parallel" for $k = 1{:}n$. This is the gist of the following algorithm:

Algorithm   4.7.2 (Levinson)   Given $b \in \mathbb{R}^n$ and real numbers $1 = r_0, r_1, \ldots, r_n$ such that $T = (r_{|i-j|}) \in \mathbb{R}^{n \times n}$ is positive definite, the following algorithm computes $x \in \mathbb{R}^n$ such that $Tx = b$.

$$
y(1) = -r(1); \quad x(1) = b(1); \quad \beta = 1; \quad \alpha = -r(1)
$$

for $k = 1{:}n - 1$
$\quad \beta = (1 - \alpha^2)\beta; \ \mu = \big(b(k+1) - r(1{:}k)^T x(k{:} - 1{:}1)\big)/\beta$
$\quad v(1{:}k) = x(1{:}k) + \mu y(k{:} - 1{:}1)$
$$x(1{:}k+1) = \left[\begin{array}{c} v(1{:}k) \\ \mu \end{array}\right]$$
$\quad$ if $k < n - 1$
$\qquad \alpha = \big(-r(k+1) + r(1{:}k)^T y(k{:} - 1{:}1)\big)/\beta$
$\qquad z(1{:}k) = y(1{:}k) + \alpha y(k{:} - 1{:}1)$
$$y(1{:}k+1) = \left[\begin{array}{c} z(1{:}k) \\ \alpha \end{array}\right]$$
$\quad$ end
end

This algorithm requires $4n^2$ flops. The vectors z and v are for clarity and can be avoided in a detailed implementation.

Example 4.7.2 Suppose we wish to solve the symmetric positive definite Toeplitz system
$$\left[\begin{array}{ccc} 1 & .5 & .2 \\ .5 & 1 & .5 \\ .2 & .5 & 1 \end{array}\right]\left[\begin{array}{c} x_1 \\ x_2 \\ x_3 \end{array}\right] = -\left[\begin{array}{c} 4 \\ -1 \\ 3 \end{array}\right]$$
using the above algorithm. After one pass through the loop we obtain
$$\alpha = 1/15, \qquad \beta = 3/4, \qquad y = \left[\begin{array}{c} -8/15 \\ 1/15 \end{array}\right] \qquad x = \left[\begin{array}{c} 6 \\ -4 \end{array}\right].$$
We then compute
$$\begin{array}{llll} \beta & = & (1-\alpha^2)\beta = 56/75 & \mu = (b_3 - r_1 x_2 - r_2 x_1)/\beta = 285/56 \\ v_1 & = & x_1 + \mu y_2 = 355/56 & v_2 = x_2 + \mu y_1 = -376/56 \end{array}$$
giving the final solution $x = [355, -376, 285]^T/56$.

4.7.4   Computing the Inverse

One of the most surprising properties of a symmetric positive definite Toeplitz matrix T_n is that its complete inverse can be calculated in $O(n^2)$ flops. To derive the algorithm for doing this, partition T_n^{-1} as follows

$$T_n^{-1} = \left[\begin{array}{cc} A & Er \\ r^T E & 1 \end{array}\right]^{-1} = \left[\begin{array}{cc} B & v \\ v^T & \gamma \end{array}\right] \tag{4.7.4}$$

where $A = T_{n-1}$, $E = E_{n-1}$, and $r = (r_1, \ldots, r_{n-1})^T$. From the equation

$$\left[\begin{array}{cc} A & Er \\ r^T E & 1 \end{array}\right]\left[\begin{array}{c} v \\ \gamma \end{array}\right] = \left[\begin{array}{c} 0 \\ 1 \end{array}\right]$$

it follows that $Av = -\gamma Er = -\gamma E(r_1, \ldots, r_{n-1})^T$ and $\gamma = 1 - r^T Ev$. If y solves the $(n-1)$-st order Yule-Walker system $Ay = -r$, then these expressions imply that

$$\begin{aligned} \gamma &= 1/(1 + r^T y) \\ v &= \gamma E y \,. \end{aligned}$$

Thus, the last row and column of T_n^{-1} are readily obtained.

It remains for us to develop working formulae for the entries of the submatrix B in (4.7.4). Since $AB + Erv^T = I_{n-1}$, it follows that

$$B = A^{-1} - (A^{-1}Er)v^T = A^{-1} + \frac{vv^T}{\gamma} \,.$$

Now since $A = T_{n-1}$ is nonsingular and Toeplitz, its inverse is persymmetric. Thus,

$$\begin{aligned} b_{ij} &= (A^{-1})_{ij} + \frac{v_i v_j}{\gamma} \\ &= (A^{-1})_{n-j,n-i} + \frac{v_i v_j}{\gamma} \qquad\qquad (4.7.5)\\ &= b_{n-j,n-i} - \frac{v_{n-j}v_{n-i}}{\gamma} + \frac{v_i v_j}{\gamma} \\ &= b_{n-j,n-i} + \frac{1}{\gamma}\left(v_i v_j - v_{n-j}v_{n-i}\right) \,. \end{aligned}$$

This indicates that although B is not persymmetric, we can readily compute an element b_{ij} from its reflection across the northeast-southwest axis. Coupling this with the fact that A^{-1} is persymmetric enables us to determine B from its "edges" to its "interior."

Because the order of operations is rather cumbersome to describe, we preview the formal specification of the algorithm pictorially. To this end, assume that we know the last column and row of T_n^{-1}:

$$T_n^{-1} = \begin{bmatrix} u & u & u & u & u & k \\ u & u & u & u & u & k \\ u & u & u & u & u & k \\ u & u & u & u & u & k \\ u & u & u & u & u & k \\ k & k & k & k & k & k \end{bmatrix}.$$

Here u and k denote the unknown and the known entries respectively, and $n = 6$. Alternately exploiting the persymmetry of T_n^{-1} and the recursion (4.7.5), we can compute B, the leading $(n-1)$-by-$(n-1)$ block of T_{n-1}^{-1}, as follows:

$$\xrightarrow{persym.} \begin{bmatrix} k & k & k & k & k & k \\ k & u & u & u & u & k \\ k & u & u & u & u & k \\ k & u & u & u & u & k \\ k & u & u & u & u & k \\ k & k & k & k & k & k \end{bmatrix} \xrightarrow{(4.7.5)} \begin{bmatrix} k & k & k & k & k & k \\ k & u & u & u & k & k \\ k & u & u & u & k & k \\ k & u & u & u & k & k \\ k & k & k & k & k & k \\ k & k & k & k & k & k \end{bmatrix}$$

$$
\text{persym.} \longrightarrow
\begin{bmatrix}
k & k & k & k & k & k \\
k & k & k & k & k & k \\
k & k & u & u & k & k \\
k & k & u & u & k & k \\
k & k & k & k & k & k \\
k & k & k & k & k & k
\end{bmatrix}
\quad (4.7.5) \longrightarrow \quad
\begin{bmatrix}
k & k & k & k & k & k \\
k & k & k & k & k & k \\
k & k & u & k & k & k \\
k & k & k & k & k & k \\
k & k & k & k & k & k \\
k & k & k & k & k & k
\end{bmatrix}
$$

$$
\text{persym.} \longrightarrow
\begin{bmatrix}
k & k & k & k & k & k \\
k & k & k & k & k & k \\
k & k & k & k & k & k \\
k & k & k & k & k & k \\
k & k & k & k & k & k \\
k & k & k & k & k & k
\end{bmatrix}.
$$

Of course, when computing a matrix that is both symmetric and persymmetric, such as T_n^{-1}, it is only necessary to compute the "upper wedge" of the matrix—e.g.,

$$
\begin{array}{cccccc}
\times & \times & \times & \times & \times & \times \\
\times & \times & \times & \times & & \\
\times & \times & & & &
\end{array}
\qquad (n = 6)
$$

With this last observation, we are ready to present the overall algorithm.

Algorithm 4.7.3 (Trench) Given real numbers $1 = r_0, r_1, \ldots, r_n$ such that $T = (r_{|i-j|}) \in \mathbb{R}^{n \times n}$ is positive definite, the following algorithm computes $B = T_n^{-1}$. Only those b_{ij} for which $i \leq j$ and $i + j \leq n + 1$ are computed.

> Use Algorithm 4.7.1 to solve $T_{n-1}y = -(r_1, \ldots, r_{n-1})^T$.
> $\gamma = 1/(1 + r(1{:}n - 1)^T y(n - 1{:} - 1{:}1))$
> $v(1{:}n - 1) = \gamma y(n - 1{:} - 1{:}1)$
> $B(1,1) = \gamma$
> $B(1, 2{:}n) = v(n - 1{:} - 1{:}1)^T$
> for $i = 2{:}\text{floor}((n - 1)/2) + 1$
>   for $j = i{:}n - i + 1$
> $B(i, j) = B(i - 1, j - 1) +$
> $(v(n + 1 - j)v(n + 1 - i) - v(i - 1)v(j - 1))/\gamma$
>   end
> end

This algorithm requires $13n^2/4$ flops.

Example 4.7.3 If the above algorithm is applied to compute the inverse B of the positive definite Toeplitz matrix

$$
\begin{bmatrix}
1 & .5 & .2 \\
.5 & 1 & .5 \\
.2 & .5 & 1
\end{bmatrix},
$$

then we obtain $\gamma = 75/56$, $b_{11} = 75/56$, $b_{12} = -5/7$, $b_{13} = 5/56$, and $b_{22} = 12/7$.

4.7.5   Stability Issues

Error analyses for the above algorithms have been performed by Cybenko (1978), and we briefly report on some of his findings.

The key quantities turn out to be the α_k in (4.7.1). In exact arithmetic these scalars satisfy

$$|\alpha_k| < 1$$

and can be used to bound $\| T^{-1} \|_1$:

$$\max \left\{ \frac{1}{\prod_{j=1}^{n-1}(1 - \alpha_j^2)} , \frac{1}{\prod_{j=1}^{n-1}(1 - \alpha_j)} \right\} \leq \| T_n^{-1} \| \leq \prod_{j=1}^{n-1} \frac{1 + |\alpha_j|}{1 - |\alpha_j|} \quad (4.7.6)$$

Moreover, the solution to the Yule-Walker system $T_n y = -r(1{:}n)$ satisfies

$$\| y \|_1 = \left(\prod_{k=1}^{n} (1 + \alpha_k) \right) - 1 \quad (4.7.7)$$

provided all the α_k are non-negative.

Now if $\hat{x}$ is the computed Durbin solution to the Yule-Walker equations then $r_D = T_n \hat{x} + r$ can be bounded as follows

$$\| r_D \| \approx \mathbf{u} \prod_{k=1}^{n} (1 + |\hat{\alpha}_k|)$$

where $\hat{\alpha}_k$ is the computed version of α_k. By way of comparison, since each $|r_i|$ is bounded by unity, it follows that $\| r_C \| \approx \mathbf{u} \| y \|_1$ where r_C is the residual associated with the computed solution obtained via Cholesky. Note that the two residuals are of comparable magnitude provided (4.7.7) holds. Experimental evidence suggests that this is the case even if some of the α_k are negative. Similar comments apply to the numerical behavior of the Levinson algorithm.

For the Trench method, the computed inverse $\hat{B}$ of T_n^{-1} can be shown to satisfy

$$\frac{\| T_n^{-1} - \hat{B} \|_1}{\| T_n^{-1} \|_1} \approx \mathbf{u} \prod_{k=1}^{n} \frac{1 + |\hat{\alpha}_k|}{1 - |\hat{\alpha}_k|} .$$

In light of (4.7.7) we see that the right-hand side is an approximate upper bound for $\mathbf{u} \| T_n^{-1} \|$ which is approximately the size of the relative error when T_n^{-1} is calculated using the Cholesky factorization.

4.7.6   The Unsymmetric Case

Similar recursions can be developed for the unsymmetric case. Suppose we are given scalars $r_1, \ldots, r_{n-1}, p_1, \ldots, p_{n-1}$, and $b_1, \ldots, b_n$ and that we want to solve a linear system $Tx = b$ of the form

$$
\begin{bmatrix}
1 & r_1 & r_2 & r_3 & r_4 \\
p_1 & 1 & r_1 & r_2 & r_3 \\
p_2 & p_1 & 1 & r_1 & r_2 \\
p_3 & p_2 & p_1 & 1 & r_1 \\
p_4 & p_3 & p_2 & p_1 & 1
\end{bmatrix}
\begin{bmatrix}
x_1 \\ x_2 \\ x_3 \\ x_4 \\ x_5
\end{bmatrix}
=
\begin{bmatrix}
b_1 \\ b_2 \\ b_3 \\ b_4 \\ b_5
\end{bmatrix}
\qquad (n = 5).
$$

In the process that follows we require the leading principle submatrices $T_k = T(1{:}k, 1{:}k)$, $k = 1{:}n$ to be nonsingular. Using the same notation as above, it can shown that if we have the solutions to the k-by-k systems

$$
\begin{aligned}
T_k^T y &= -r = -[r_1\ r_2\ \cdots\ r_k]^T \\
T_k w &= -p = -[p_1\ p_2\ \cdots\ p_k]^T \\
T_k x &= b = [b_1\ b_2\ \cdots\ b_k]^T,
\end{aligned}
\qquad (4.7.8)
$$

then we can obtain solutions to

$$
\begin{bmatrix} T_k & E_k r \\ p^T E_k & 1 \end{bmatrix}^T
\begin{bmatrix} z \\ \alpha \end{bmatrix}
= - \begin{bmatrix} r \\ r_{k+1} \end{bmatrix}
$$

$$
\begin{bmatrix} T_k & E_k r \\ p^T E_k & 1 \end{bmatrix}
\begin{bmatrix} u \\ \nu \end{bmatrix}
= - \begin{bmatrix} p \\ p_{k+1} \end{bmatrix}
\qquad (4.7.9)
$$

$$
\begin{bmatrix} T_k & E_k r \\ p^T E_k & 1 \end{bmatrix}
\begin{bmatrix} v \\ \mu \end{bmatrix}
= \begin{bmatrix} b \\ b_{k+1} \end{bmatrix}
$$

in $O(k)$ flops. This means that in principle it is possible to solve an unsymmetric Toeplitz system in $O(n^2)$ flops. However, the stability of the process cannot be assured unless the matrices $T_k = T(1{:}k, 1{:}k)$ are sufficiently well conditioned.

4.7.7   Circulant Systems

A very important class of Toeplitz matrices are the *circulant* matrices. Here is an example:

$$
C(v) =
\begin{bmatrix}
v_0 & v_4 & v_3 & v_2 & v_1 \\
v_1 & v_0 & v_4 & v_3 & v_2 \\
v_2 & v_1 & v_0 & v_4 & v_3 \\
v_3 & v_2 & v_1 & v_0 & v_4 \\
v_4 & v_3 & v_2 & v_1 & v_0
\end{bmatrix}.
$$

Notice that each column of a circulant is a "downshifted" version of its predecessor. In particular, if we define the downshift permutation S_n by

$$S_n = \begin{bmatrix} 0 & 0 & 0 & 0 & 1 \\ 1 & 0 & 0 & 0 & 0 \\ 0 & 1 & 0 & 0 & 0 \\ 0 & 0 & 1 & 0 & 0 \\ 0 & 0 & 0 & 1 & 0 \end{bmatrix} \qquad (n = 5)$$

and $v = [v_0 \ v_1 \ \cdots \ v_{n-1}]^T$, then $C(v) = [\, v, \ S_n v, \ S_n^2 v, \ldots, \ S_n^{n-1} v \,]$.

There are important connections between circulant matrices, Toeplitz matrices, and the DFT. First of all, it can be shown that

$$C(v) = F_n^{-1} \mathrm{diag}(F_n v) F_n. \qquad (4.7.10)$$

This means that a product of the form $y = C(v)x$ can be solved at "FFT speed":

$$\tilde{x} = F_n x$$
$$\tilde{v} = F_n v$$
$$z = \tilde{v} . * \tilde{x}$$
$$y = F_n^{-1} z$$

In other words, three DFTs and a vector multiply suffice to carry out the product of a circulant matrix and a vector. Products of this form are called *convolutions* and they are ubiquitous in signal processing and other areas.

Toeplitz-vector products can also be computed fast. The key idea is that any Toeplitz matrix can be "embedded" in a circulant. For example,

$$T = \begin{bmatrix} 5 & 2 & 7 \\ 4 & 5 & 2 \\ 9 & 4 & 5 \end{bmatrix}$$

is the leading 3-by-3 submatrix of

$$C = \left[\begin{array}{ccc|cc} 5 & 2 & 7 & 9 & 4 \\ 4 & 5 & 2 & 7 & 9 \\ 9 & 4 & 5 & 2 & 7 \\ \hline 7 & 9 & 4 & 5 & 2 \\ 2 & 7 & 9 & 4 & 5 \end{array} \right].$$

In general, if $T = (t_{ij})$ is an n-by-n Toeplitz matrix, then $T = C(1{:}n, 1{:}n)$ where $C \in \mathbb{R}^{(2n-1) \times (2n-1)}$ is a circulant with

$$C(:, 1) = \begin{bmatrix} T(1{:}n, 1) \\ T(1, n{:} - 1{:}2)^T \end{bmatrix}.$$

Note that if $y = Cx$ and $x(n+1{:}2n-1) = 0$, then $y(1{:}n) = Tx(1{:}n)$ showing that Toeplitz vector products can also be computed at "FFT speed."

Problems

P4.7.1 For any $v \in \mathbf{R}^n$ define the vectors $v_+ = (v + E_n v)/2$ and $v_- = (v - E_n v)/2$. Suppose $A \in \mathbf{R}^{n \times n}$ is symmetric and persymmetric. Show that if $Ax = b$ then $Ax_+ = b_+$ and $Ax_- = b_-$.

P4.7.2 Let $U \in \mathbf{R}^{n \times n}$ be the unit upper triangular matrix with the property that: $U(1{:}k - 1, k) = E_{k-1} y^{(k-1)}$ where $y^{(k)}$ is defined by (4.7.1). Show that

$$U^T T_n U = \mathrm{diag}(1, \beta_1, \ldots, \beta_{n-1}).$$

P4.7.3 Suppose $z \in \mathbf{R}^n$ and that $S \in \mathbf{R}^{n \times n}$ is orthogonal. Show that if

$$X = \begin{bmatrix} z, & Sz, & \ldots, & S^{n-1}z \end{bmatrix}$$

then $X^T X$ is Toeplitz.

P4.7.4 Consider the LDL^T factorization of an n-by-n symmetric, tridiagonal, positive definite Toeplitz matrix. Show that d_n and $\ell_{n,n-1}$ converge as $n \to \infty$.

P4.7.5 Show that the product of two lower triangular Toeplitz matrices is Toeplitz.

P4.7.6 Give an algorithm for determining $\mu \in \mathbf{R}$ such that

$$T_n + \mu \left(e_n e_1^T + e_1 e_n^T \right)$$

is singular. Assume $T_n = (r_{|i-j|})$ is positive definite, with $r_0 = 1$.

P4.7.7 Rewrite Algorithm 4.7.2 so that it does not require the vectors z and v.

P4.7.8 Give an algorithm for computing $\kappa_\infty(T_k)$ for $k = 1{:}n$.

P4.7.9 Suppose A_1, A_2, A_3 and A_4 are m-by-m matrices and that

$$A = \begin{bmatrix} A_0 & A_1 & A_2 & A_3 \\ A_3 & A_0 & A_1 & A_2 \\ A_2 & A_3 & A_0 & A_1 \\ A_1 & A_2 & A_3 & A_0 \end{bmatrix}.$$

Show that there is a permutation matrix Π such that $\Pi^T A \Pi = C = (C_{ij})$ where each C_{ij} is a 4-by-4 circulant matrix.

P4.7.10 A p-by-p block matrix $A = (A_{ij})$ with m-by-m blocks is *block Toeplitz* if there exist $A_{-p+1}, \ldots, A_{-1}, A_0, A_1, \ldots, A_{p-1} \in \mathbf{R}^{m \times m}$ so that $A_{ij} = A_{i-j}$, e.g.,

$$A = \begin{bmatrix} A_0 & A_1 & A_2 & A_3 \\ A_{-1} & A_0 & A_1 & A_2 \\ A_{-2} & A_{-1} & A_0 & A_1 \\ A_{-3} & A_{-2} & A_{-1} & A_0 \end{bmatrix}.$$

(a) Show that there is a permutation Π such that

$$\Pi^T A \Pi =: \begin{bmatrix} T_{11} & T_{12} & \cdots & T_{1m} \\ T_{21} & T_{22} & & \vdots \\ \vdots & & \ddots & \\ T_{m1} & \cdots & & T_{mm} \end{bmatrix}$$

where each T_{ij} is p-by-p and Toeplitz. Each T_{ij} should be "made up" of (i,j) entries selected from the A_k matrices. (b) What can you say about the T_{ij} if $A_k = A_{-k}$, $k = 1{:}p - 1$?

P4.7.11 Show how to compute the solutions to the systems in (4.7.9) given that the solutions to the systems in (4.7.8) are available. Assume that all the matrices involved are nonsingular. Proceed to develop a fast unsymmetric Toeplitz solver for $Tx = b$ assuming that T's leading principle submatrices are all nonsingular.

P4.7.12 A matrix $H \in \mathbb{R}^{n \times n}$ is *Hankel* if $H(n:-1{:}1,:)$ is Toeplitz. Show that if $A \in \mathbb{R}^{n \times n}$ is defined by

$$a_{ij} = \int_a^b \cos(k\theta) \cos(j\theta) d\theta$$

then A is the sum of a Hankel matrix and Toeplitz matrix. Hint. Make use of the identity $\cos(u + v) = \cos(u) \cos(v) - \sin(u) \sin(v)$.

P4.7.13 Verify that $F_n C(v) = \operatorname{diag}(F_n v) F_n$.

P4.7.14 Show that it is possible to embed a symmetric Toeplitz matrix into a symmetric circulant matrix.

P4.7.15 Consider the kth order Yule-Walker system $T_k y^{(k)} = -r^{(k)}$ that arises in (4.7.1):

$$T_k \begin{bmatrix} y_{k1} \\ \vdots \\ y_{kk} \end{bmatrix} = - \begin{bmatrix} r_1 \\ \vdots \\ r_k \end{bmatrix}.$$

Show that if

$$L = \begin{bmatrix}
1 & 0 & 0 & 0 & \cdots & 0 \\
y_{11} & 1 & 0 & 0 & \cdots & 0 \\
y_{22} & y_{21} & 1 & 0 & \cdots & 0 \\
y_{33} & y_{32} & y_{31} & 1 & \cdots & 0 \\
\vdots & \vdots & \vdots & \vdots & \ddots & \vdots \\
y_{n-1,n-1} & y_{n-1,n-2} & y_{n-1,n-3} & \cdots & y_{n-1,1} & 1
\end{bmatrix},$$

then $LT_n L^T = \operatorname{diag}(1, \beta_1, \ldots, \beta_{n-1})$ where $\beta_k = 1 + r^{(k)T} y^{(k)}$. Thus, the Durbin algorithm can be thought of as a fast method for computing the LDL^T factorization of T_n^{-1}.

Notes and References for Sec. 4.7

Anyone who ventures into the vast Toeplitz method literature should first read

J.R. Bunch (1985). "Stability of Methods for Solving Toeplitz Systems of Equations," *SIAM J. Sci. Stat. Comp. 6*, 349–364

for a clarification of stability issues. As is true with the "fast algorithms" area in general, unstable Toeplitz techniques abound and caution must be exercised. See also

G. Cybenko (1978). "Error Analysis of Some Signal Processing Algorithms," Ph.D. thesis, Princeton University.
G. Cybenko (1980). "The Numerical Stability of the Levinson-Durbin Algorithm for Toeplitz Systems of Equations," *SIAM J. Sci. and Stat. Comp. 1*, 303-19.
E. Linzer (1992). "On the Stability of Solution Methods for Band Toeplitz Systems," *Lin.Alg. and Its Application 170*, 1–32.
J.M. Varah (1994). "Backward Error Estimates for Toeplitz Systems," *SIAM J. Matrix Anal. Appl. 15*, 408–417.
A.W. Bojanczyk, R.P. Brent, F.R. de Hoog, and D.R. Sweet (1995). "On the Stability of the Bareiss and Related Toeplitz Factorization Algorithms," *SIAM J. Matrix Anal. Appl. 16*, 40–57.

M. Stewart and P. Van Dooren (1996). "Stability Issues in the Factorization of Structured Matrices," *SIAM J. Matrix Anal. Appl. 18*, to appear.

The original references for the three algorithms described in this section are:

J. Durbin (1960). "The Fitting of Time Series Models," *Rev. Inst. Int. Stat. 28* 233–43.
N. Levinson (1947). "The Weiner RMS Error Criterion in Filter Design and Prediction,"
J. Math. Phys. 25, 261–78.
W.F. Trench (1964). "An Algorithm for the Inversion of Finite Toeplitz Matrices," *J.
SIAM 12*, 515–22.

A more detailed description of the nonsymmetric Trench algorithm is given in

S. Zohar (1969). "Toeplitz Matrix Inversion: The Algorithm of W.F. Trench," *J. ACM
16*, 592–601.

Fast Toeplitz system solving has attracted an enormous amount of attention and a sampling of interesting algorithmic ideas may be found in

G. Ammar and W.B. Gragg (1988). "Superfast Solution of Real Positive Definite
Toeplitz Systems," *SIAM J. Matrix Anal. Appl. 9*, 61–76.
T.F. Chan and P. Hansen (1992). "A Look-Ahead Levinson Algorithm for Indefinite
Toeplitz Systems," *SIAM J. Matrix Anal. Appl. 13*, 490–506.
D.R. Sweet (1993). "The Use of Pivoting to Improve the Numerical Performance of
Algorithms for Toeplitz Matrices," *SIAM J. Matrix Anal. Appl. 14*, 468–493.
T. Kailath and J. Chun (1994). "Generalized Displacement Structure for Block-Toeplitz,
Toeplitz-Block, and Toeplitz-Derived Matrices," *SIAM J. Matrix Anal. Appl. 15*,
114–128.
T. Kailath and A.H. Sayed (1995). "Displacement Structure: Theory and Applications,"
SIAM Review 37, 297–386.

Important Toeplitz matrix applications are discussed in

J. Makhoul (1975). "Linear Prediction: A Tutorial Review," *Proc. IEEE 63*(4), 561–80.
J. Markel and A. Gray (1976). *Linear Prediction of Speech*, Springer-Verlag, Berlin and
New York.
A.V. Oppenheim (1978). *Applications of Digital Signal Processing* , Prentice-Hall, En-
glewood Cliffs.

Hankel matrices are constant along their antidiagonals and arise in several important
areas.

G. Heinig and P. Jankowski (1990). "Parallel and Superfast Algorithms for Hankel
Systems of Equations," *Numer. Math. 58*, 109–127.
R.W. Freund and H. Zha (1993). "A Look-Ahead Algorithm for the Solution of General
Hankel Systems," *Numer. Math. 64*, 295–322.

The DFT/Toeplitz/circulant connection is discussed in

C.F. Van Loan (1992). *Computational Frameworks for the Fast Fourier Transform*,
SIAM Publications, Philadelphia, PA.

Chapter 5

Orthogonalization and Least Squares

This chapter is primarily concerned with the least squares solution of overdetermined systems of equations, i.e., the minimization of $\| Ax - b \|_2$ where $A \in \mathbb{R}^{m \times n}$ with $m \geq n$ and $b \in \mathbb{R}^m$. The most reliable solution procedures for this problem involve the reduction of A to various canonical forms via orthogonal transformations. Householder reflections and Givens rotations are central to this process and we begin the chapter with a discussion of these important transformations. In §5.2 we discuss the computation of the factorization $A = QR$ where Q is orthogonal and R is upper triangular. This amounts to finding an orthonormal basis for ran(A). The QR factorization can be used to solve the full rank least squares problem as we show in §5.3. The technique is compared with the method of normal equations after a perturbation theory is developed. In §5.4 and §5.5 we consider methods for handling the difficult situation when A is rank deficient (or nearly so). QR with column pivoting and the SVD are featured. In §5.6 we discuss several steps that can be taken to improve the quality of a computed least squares solution. Some remarks about square and underdetermined systems are offered in §5.7.

Before You Begin

Chapters 1, 2, and 3 and §§4.1-4.3 are assumed. Within this chapter there are the following dependencies:

$$\S 5.1 \;\rightarrow\; \S 5.2 \;\rightarrow\; \S 5.3 \;\rightarrow\; \S 5.4 \;\rightarrow\; \begin{array}{c} \S 5.6 \\ \uparrow \\ \S 5.5 \\ \downarrow \\ \S 5.7 \end{array}$$

Complementary references include Lawson and Hanson (1974), Farebrother (1987), and Björck (1996). See also Stewart (1973), , Hager (1988), Stewart and Sun (1990), Watkins (1991), Gill, Murray, and Wright (1991), Higham (1996), Trefethen and Bau (1996), and Demmel (1996). Some MATLAB functions important to this chapter are qr, svd, pinv, orth, rank, and the "backslash" operator "\." LAPACK connections include

LAPACK: Householder/Givens Tools	
_LARFG	Generates a Householder matrix
_LARF	Householder times matrix
_LARFX	Small n Householder times matrix
_LARFB	Block Householder times matrix
_LARFT	Computes $I - VTV^H$ block reflector representation
_LARTG	Generates a plane rotation
_LARGV	Generates a vector of plane rotations
_LARTV	Applies a vector of plane rotations to a vector pair
_LASR	Applies rotation sequence to a matrix
CSROT	Real rotation times complex vector pair
CROT	Complex rotation (c real) times complex vector pair
CLACGV	Complex rotation (s real) times complex vector pair

LAPACK: Orthogonal Factorizations	
_GEQRF	$A = QR$
_GEQPF	$A\Pi = QR$
_ORMQR	Q (factored form) times matrix (real case)
_UNMQR	Q (factored form) times matrix (complex case)
_ORGQR	Generates Q (real case)
_UNGQR	Generates Q (complex case)
_GERQF	$A = RQ =$ (upper triangular)(orthogonal)
_GELQF	$A = QL =$ (orthogonal)(lower triangular)
_GEQLF	$A = LQ =$ (lower triangular)(orthogonal)
_TZRQF	$A = RQ$ where A is upper trapezoidal
_GESVD	$A = U\Sigma V^T$
_BDSQR	SVD of real bidiagonal matrix
_GEBRD	Bidiagonalization of general matrix
_ORGBR	Generates the orthogonal transformations
_GBBRD	Bidiagonalization of band matrix

LAPACK: Least Squares	
_GELS	Full rank min $\| AX - B \|_{2F}$ or min $\| A^H X - B \|_F$
_GELSS	SVD solution to min $\| AX - B \|_F$
_GELSX	Complete orthogonal decomposition solution to min $\| AX - B \|_F$.
_GEEQU	Equilibrates general matrix to reduce condition

5.1   Householder and Givens Matrices

Recall that $Q \in \mathbb{R}^{n \times n}$ is *orthogonal* if $Q^T Q = Q Q^T = I_n$. Orthogonal matrices have an important role to play in least squares and eigenvalue computations. In this section we introduce the key players in this game: Householder reflections and Givens rotations.

5.1.1   A 2-by-2 Preview

It is instructive to examine the geometry associated with rotations and reflections at the $n = 2$ level. A 2-by-2 orthogonal matrix Q is a *rotation* if it has the form

$$Q = \left[\begin{array}{cc} \cos(\theta) & \sin(\theta) \\ -\sin(\theta) & \cos(\theta) \end{array} \right].$$

If $y = Q^T x$, then y is obtained by rotating x counterclockwise through an angle θ.

A 2-by-2 orthogonal matrix Q is a *reflection* if it has the form

$$Q = \left[\begin{array}{cc} \cos(\theta) & \sin(\theta) \\ \sin(\theta) & -\cos(\theta) \end{array} \right].$$

If $y = Q^T x = Qx$, then y is obtained by reflecting the vector x across the line defined by

$$S = \text{span}\left\{ \left[\begin{array}{c} \cos(\theta/2) \\ \sin(\theta/2) \end{array} \right] \right\}.$$

Reflections and rotations are computationally attractive because they are easily constructed and because they can be used to introduce zeros in a vector by properly choosing the rotation angle or the reflection plane.

Example 5.1.1 Suppose $x = [\, 1,\ \sqrt{3}\,]^T$. If we set

$$Q = \left[\begin{array}{cc} \cos(-60^o) & \sin(-60^o) \\ -\sin(-60^o) & \cos(-60^o) \end{array} \right] = \left[\begin{array}{cc} 1/2 & -\sqrt{3}/2 \\ \sqrt{3}/2 & 1/2 \end{array} \right]$$

then $Q^T x = [\, 2,\ 0\,]^T$. Thus, a rotation of -60° zeros the second component of x. If

$$Q = \left[\begin{array}{cc} \cos(30^o) & \sin(30^o) \\ \sin(30^o) & -\cos(30^o) \end{array} \right] = \left[\begin{array}{cc} \sqrt{3}/2 & 1/2 \\ 1/2 & -\sqrt{3}/2 \end{array} \right]$$

then $Q^T x = [\, 2,\ 0\,]^T$. Thus, by reflecting x across the 30° line we can zero its second component.

5.1.2 Householder Reflections

Let $v \in \mathbb{R}^n$ be nonzero. An n-by-n matrix P of the form

$$P = I - \frac{2}{v^T v} v v^T \qquad (5.1.1)$$

is called a *Householder reflection*. (Synonyms: Householder matrix, Householder transformation.) The vector v is called a *Householder vector*. If a vector x is multiplied by P, then it is reflected in the hyperplane span$\{v\}^\perp$. It is easy to verify that Householder matrices are symmetric and orthogonal.

Householder reflections are similar in two ways to Gauss transformations, which we introduced in §3.2.1. They are rank-1 modifications of the identity and they can be used to zero selected components of a vector. In particular, suppose we are given $0 \neq x \in \mathbb{R}^n$ and want Px to be a multiple of $e_1 = I_n(:, 1)$. Note that

$$Px = \left(I - \frac{2 v v^T}{v^T v} \right) x = x - \frac{2 v^T x}{v^T v} v$$

and $Px \in \text{span}\{e_1\}$ imply $v \in \text{span}\{x, e_1\}$. Setting $v = x + \alpha e_1$ gives

$$v^T x = x^T x + \alpha x_1$$

and

$$v^T v = x^T x + 2 \alpha x_1 + \alpha^2,$$

and therefore

$$Px = \left(1 - 2 \frac{x^T x + \alpha x_1}{x^T x + 2 \alpha x_1 + \alpha^2} \right) x - 2 \alpha \frac{v^T x}{v^T v} e_1 .$$

In order for the coefficient of x to be zero, we set $\alpha = \pm \| x \|_2$ for then

$$v = x \pm \| x \|_2 e_1 \;\Rightarrow\; Px = \left(I - 2 \frac{v v^T}{v^T v} \right) x = \mp \| x \|_2 e_1. \qquad (5.1.2)$$

It is this simple determination of v that makes the Householder reflection so useful.

Example 5.1.2 If $x = [3, 1, 5, 1]^T$ and $v = [9, 1, 5, 1]^T$, then

$$P = I - 2 \frac{v v^T}{v^T v} = \frac{1}{54} \begin{bmatrix} -27 & -9 & -45 & -9 \\ -9 & 53 & -5 & -1 \\ -45 & -5 & 29 & -5 \\ -9 & -1 & -5 & 53 \end{bmatrix}$$

has the property that $Px = [-6, 0, 0, 0,]^T$.

5.1.3   Computing the Householder Vector

There are a number of important practical details associated with the determination of a Householder matrix, i.e., the determination of a Householder
vector. One concerns the choice of sign in the definition of v in (5.1.2).
Setting

$$v_1 = x_1 - \| \, x \, \|_2$$

has the nice property that Px is a positive multiple of e_1. But this recipe is
dangerous if x is close to a positive multiple of e_1 because severe cancellation
would occur. However, the formula

$$v_1 \; = \; x_1 - \| \, x \, \|_2 \; = \; \frac{x_1^2 - \| \, x \, \|_2^2}{x_1 + \| \, x \, \|_2} \; = \; \frac{-(x_2^2 + \cdots + x_n^2)}{x_1 + \| \, x \, \|_2}$$

suggested by Parlett (1971) does not suffer from this defect in the $x_1 > 0$
case.

In practice, it is handy to normalize the Householder vector so that
$v(1) = 1$. This permits the storage of $v(2{:}n)$ where the zeros have been
introduced in x, i.e., $x(2{:}n)$. We refer to $v(2{:}n)$ as the *essential part* of
the Householder vector. Recalling that $\beta = 2/v^T v$ and letting **length**(x)
specify vector dimension, we obtain the following encapsulation:

Algorithm 5.1.1 (Householder Vector)   Given $x \in \mathbb{R}^n$, this function
computes $v \in \mathbb{R}^n$ with $v(1) = 1$ and $\beta \in \mathbb{R}$ such that $P = I_n - \beta vv^T$ is
orthogonal and $Px = \| \, x \, \|_2 e_1$.

> **function:** $[v, \beta] = $ **house**(x)
> $n = $ **length**(x)
> $\sigma = x(2{:}n)^T x(2{:}n)$
> $v = \begin{bmatrix} 1 \\ x(2{:}n) \end{bmatrix}$
> **if** $\sigma = 0$
> > $\beta = 0$
>
> **else**
> > $\mu = \sqrt{x(1)^2 + \sigma}$
> > **if** $x(1) <= 0$
> > > $v(1) = x(1) - \mu$
> >
> > **else**
> > > $v(1) = -\sigma/(x(1) + \mu)$
> >
> > **end**
> > $\beta = 2v(1)^2/(\sigma + v(1)^2)$
> > $v = v/v(1)$
>
> **end**

This algorithm involves about $3n$ flops and renders a computed Householder
matrix that is orthogonal to machine precision, a concept discussed below.

A production version of Algorithm 5.1.1 may involve a preliminary scaling of the x vector $(x \leftarrow x/\| x \|)$ to avoid overflow.

5.1.4 Applying Householder Matrices

It is critical to exploit structure when applying a Householder reflection to a matrix. If $A \in \mathbb{R}^{m \times n}$ and $P = I - \beta vv^T \in \mathbb{R}^{m \times m}$, then

$$PA = \left(I - \beta vv^T\right) A = A - vw^T$$

where $w = \beta A^T v$. Likewise, if $P = I - \beta vv^T \in \mathbb{R}^{n \times n}$, then

$$AP = A\left(I - \beta vv^T\right) = A - wv^T$$

where $w = \beta Av$. Thus, an m-by-n Householder update involves a matrix-vector multiplication and an outer product update. It requires $4mn$ flops. Failure to recognize this and to treat P as a general matrix increases work by an order of magnitude. *Householder updates never entail the explicit formation of the Householder matrix.*

Both of the above Householder updates can be implemented in a way that exploits the fact that $v(1) = 1$. This feature can be important in the computation of PA when m is small and in the computation of AP when n is small.

As an example of a Householder matrix update, suppose we want to overwrite $A \in \mathbb{R}^{m \times n}$ $(m \geq n)$ with $B = Q^T A$ where Q is an orthogonal matrix chosen so that $B(j+1{:}m, j) = 0$ for some j that satisfies $1 \leq j \leq n$. In addition, suppose $A(j{:}m, 1{:}j - 1) = 0$ and that we want to store the essential part of the Householder vector in $A(j + 1{:}m, j)$. The following instructions accomplish this task:

$$[v, \beta] = \textbf{house}(A(j{:}m, j))$$
$$A(j{:}m, j{:}n) = (I_{m-j+1} - \beta vv^T)A(j{:}m, j{:}n)$$
$$A(j + 1{:}m, j) = v(2{:}m - j + 1)$$

From the computational point of view, we have applied an order $m - j + 1$ Householder matrix to the bottom $m - j + 1$ rows of A. However, mathematically we have also applied the m-by-m Householder matrix

$$\tilde{P} = \begin{bmatrix} I_{j-1} & 0 \\ 0 & P \end{bmatrix} = I_m - \beta \tilde{v}\tilde{v}^T \qquad \tilde{v} = \begin{bmatrix} 0 \\ v \end{bmatrix}$$

to A in its entirety. Regardless, the "essential" part of the Householder vector can be recorded in the zeroed portion of A.

5.1.5   Roundoff Properties

The roundoff properties associated with Householder matrices are very favorable. Wilkinson (1965, pp. 152-62) shows that **house** produces a Householder vector $\hat{v}$ very near the exact v. If $\hat{P} = I - 2\hat{v}\hat{v}^T/\hat{v}^T\hat{v}$ then

$$\| \hat{P} - P \|_2 = O(\mathbf{u})$$

meaning that $\hat{P}$ is *orthogonal to machine precision*. Moreover, the computed updates with $\hat{P}$ are close to the exact updates with P :

$$fl(\hat{P}A) \quad = \quad P(A+E) \qquad \| E \|_2 = O(\mathbf{u}\| A \|_2)$$

$$fl(A\hat{P}) \quad = \quad (A+E)P \qquad \| E \|_2 = O(\mathbf{u}\| A \|_2)$$

5.1.6   Factored Form Representation

Many Householder based factorization algorithms that are presented in the following sections compute products of Householder matrices

$$Q \;=\; Q_1 Q_2 \cdots Q_r \qquad Q_j = I - \beta_j v^{(j)} v^{(j)T} \tag{5.1.3}$$

where $r \leq n$ and each $v^{(j)}$ has the form

$$v^{(j)} \;=\; (\underbrace{0,\, 0, \cdots 0,}_{j-1}\, 1\, v_{j+1}^{(j)},\, \cdots ,v_n^{(j)})^T .$$

It is usually not necessary to compute Q explicitly even if it is involved in subsequent calculations. For example, if $C \in \mathbb{R}^{n \times q}$ and we wish to compute $Q^T C$, then we merely execute the loop

> **for** $j = 1{:}r$
> $\qquad C = Q_j C$
> **end**

The storage of the Householder vectors $v^{(1)} \cdots v^{(r)}$ and the corresponding β_j (if convenient) amounts to a *factored form* representation of Q. To illustrate the economies of the factored form representation, suppose that we have an array A and that $A(j+1{:}n, j)$ houses $v^{(j)}(j+1{:}n)$, the essential part of the jth Householder vector. The overwriting of $C \in \mathbb{R}^{n \times q}$ with $Q^T C$ can then be implemented as follows:

> **for** $j = 1{:}r$
> $$v(j{:}n) = \left[\begin{array}{c} 1 \\ A(j + 1{:}n, j) \end{array} \right] \tag{5.1.4}$$
> $\qquad C(j{:}n, :) = (I - \beta_j v(j{:}n)v(j{:}n)^T)C(j{:}n, :)$
> **end**

This involves about $2qr(2n - r)$ flops. If Q is explicitly represented as an n-by-n matrix, $Q^T C$ would involve $2n^2 q$ flops.

Of course, in some applications, it is necessary to explicitly form Q (or parts of it). Two possible algorithms for computing the Householder product matrix Q in (5.1.3) are *forward accumulation,*

$$Q = I_n$$
$$\textbf{for } j = 1{:}r$$
$$\qquad Q = QQ_j$$
$$\textbf{end}$$

and *backward accumulation,*

$$Q = I_n$$
$$\textbf{for } j = r{:} -1{:}1$$
$$\qquad Q = Q_j Q$$
$$\textbf{end}$$

Recall that the leading $(j-1)$-by-$(j-1)$ portion of Q_j is the identity. Thus, at the beginning of backward accumulation, Q is "mostly the identity" and it gradually becomes full as the iteration progresses. This pattern can be exploited to reduce the number of required flops. In contrast, Q is full in forward accumulation after the first step. For this reason, backward accumulation is cheaper and the strategy of choice:

$$Q = I_n$$
$$\textbf{for } j = r{:} -1{:}1$$
$$\qquad v(j{:}n) = \begin{bmatrix} 1 \\ A(j+1{:}n, j) \end{bmatrix} \qquad\qquad (5.1.5)$$
$$\qquad Q(j{:}n, j{:}n) = (I - \beta_j v(j{:}n)v(j{:}n)^T)Q(j{:}n, j{:}n)$$
$$\textbf{end}$$

This involves about $4(n^2 r - nr^2 + r^3/3)$ flops.

5.1.7   A Block Representation

Suppose $Q = Q_1 \cdots Q_r$ is a product of n-by-n Householder matrices as in (5.1.3). Since each Q_j is a rank-one modification of the identity, it follows from the structure of the Householder vectors that Q is a rank-r modification of the identity and can be written in the form

$$Q = I + WY^T \qquad\qquad (5.1.6)$$

where W and Y are n-by-r matrices. The key to computing the *block representation* (5.1.6) is the following lemma.

Lemma 5.1.1 *Suppose $Q = I + WY^T$ is an n-by-n orthogonal matrix with $W, Y \in \mathbb{R}^{n \times j}$. If $P = I - \beta v v^T$ with $v \in \mathbb{R}^n$ and $z = -\beta Q v$, then*

$$Q_+ = QP = I + W_+ Y_+^T$$

where $W_+ = [\, W \ z \,]$ and $Y_+ = [\, Y \ v \,]$ are each n-by-$(j+1)$.

Proof.

$$
\begin{aligned}
QP &= (I + WY^T)(I - \beta v v^T) = I + WY^T - \beta Q v v^T \\
&= I + WY^T + z v^T = I + [\, W \ z \,][\, Y \ v \,]^T \quad \square
\end{aligned}
$$

By repeatedly applying the lemma, we can generate the block representation of Q in (5.1.3) from the factored form representation as follows:

Algorithm 5.1.2 Suppose $Q = Q_1 \cdots Q_r$ is a product of n-by-n Householder matrices as described in (5.1.3). This algorithm computes matrices $W, Y \in \mathbb{R}^{n \times r}$ such that $Q = I + WY^T$.

$$Y = v^{(1)}$$
$$W = -\beta_1 v^{(1)}$$
$$\text{for } j = 2{:}r$$
$$\qquad z = -\beta_j (I + WY^T) v^{(j)}$$
$$\qquad W = [W \ z]$$
$$\qquad Y = [\, Y \ v^{(j)} \,]$$
$$\text{end}$$

This algorithm involves about $2r^2 n - 2r^3/3$ flops if the zeros in the $v^{(j)}$ are exploited. Note that Y is merely the matrix of Householder vectors and is therefore unit lower triangular. Clearly, the central task in the generation of the WY representation (5.1.6) is the computation of the W matrix.

The block representation for products of Householder matrices is attractive in situations where Q must be applied to a matrix. Suppose $C \in \mathbb{R}^{n \times q}$. It follows that the operation

$$C \leftarrow Q^T C = (I + WY^T)^T C = C + Y(W^T C)$$

is rich in level-3 operations. On the other hand, if Q is in factored form, $Q^T C$ is just rich in the level-2 operations of matrix-vector multiplication and outer product updates. Of course, in this context the distinction between level-2 and level-3 diminishes as C gets narrower.

We mention that the "WY" representation is not a generalized Householder transformation from the geometric point of view. True block reflectors have the form $Q = I - 2VV^T$ where $V \in \mathbb{R}^{n \times r}$ satisfies $V^T V = I_r$. See Schreiber and Parlett (1987) and also Schreiber and Van Loan (1989).

Example 5.1.3 If $n = 4$, $r = 2$, and $[\, 1, \ .6, \ 0, \ .8 \,]^T$ and $[\, 0, \ 1, \ .8, \ .6 \,]^T$ are the

Householder vectors associated with Q_1 and Q_2 respectively, then

$$Q_1 Q_2 = I_4 + WY^T \equiv I_4 + \begin{bmatrix} -1 & 1.080 \\ -.6 & -.352 \\ 0 & -.800 \\ -.8 & .264 \end{bmatrix} \begin{bmatrix} 1 & .6 & 0 & .8 \\ 0 & 1 & .8 & .6 \end{bmatrix}.$$

5.1.8   Givens Rotations

Householder reflections are exceedingly useful for introducing zeros on a grand scale, e.g., the annihilation of all but the first component of a vector. However, in calculations where it is necessary to zero elements more selectively, *Givens rotations* are the transformation of choice. These are rank-two corrections to the identity of the form

$$G(i,k,\theta) = \begin{bmatrix} 1 & \cdots & 0 & \cdots & 0 & \cdots & 0 \\ \vdots & \ddots & \vdots & & \vdots & & \vdots \\ 0 & \cdots & c & \cdots & s & \cdots & 0 \\ \vdots & & \vdots & \ddots & \vdots & & \vdots \\ 0 & \cdots & -s & \cdots & c & \cdots & 0 \\ \vdots & & \vdots & & \vdots & \ddots & \vdots \\ 0 & \cdots & 0 & \cdots & 0 & \cdots & 1 \end{bmatrix} \begin{matrix} \\ \\ i \\ \\ k \\ \\ \\ \end{matrix} \qquad (5.1.7)$$

$$\quad\quad\quad\quad\quad\quad\quad i \quad\quad\quad k$$

where $c = \cos(\theta)$ and $s = \sin(\theta)$ for some θ. Givens rotations are clearly orthogonal.

Premultiplication by $G(i,k,\theta)^T$ amounts to a counterclockwise rotation of θ radians in the (i,k) coordinate plane. Indeed, if $x \in \mathbb{R}^n$ and $y = G(i,k,\theta)^T x$, then

$$y_j = \begin{cases} cx_i - sx_k & j = i \\ sx_i + cx_k & j = k \\ x_j & j \neq i,k \end{cases}.$$

From these formulae it is clear that we can force y_k to be zero by setting

$$c = \frac{x_i}{\sqrt{x_i^2 + x_k^2}} \qquad s = \frac{-x_k}{\sqrt{x_i^2 + x_k^2}} \qquad (5.1.8)$$

Thus, it is a simple matter to zero a specified entry in a vector by using a Givens rotation. In practice, there are better ways to compute c and s than (5.1.8). The following algorithm, for example, guards against overflow.

Algorithm 5.1.3 Given scalars a and b, this function computes $c = \cos(\theta)$ and $s = \sin(\theta)$ so

$$\begin{bmatrix} c & s \\ -s & c \end{bmatrix}^T \begin{bmatrix} a \\ b \end{bmatrix} = \begin{bmatrix} r \\ 0 \end{bmatrix}.$$

> **function:** $[c, s] = \text{givens}(a, b)$
>   **if** $b = 0$
> $c = 1; \ s = 0$
>   **else**
>   **if** $|b| > |a|$
> $\tau = -a/b; \ s = 1/\sqrt{1 + \tau^2}; \ c = s\tau$
>   **else**
> $\tau = -b/a; \ c = 1/\sqrt{1 + \tau^2}; \ s = c\tau$
>   **end**
>   **end**

This algorithm requires 5 flops and a single square root. Note that it does not compute θ and so it does not involve inverse trigonometric functions.

Example 5.1.4 If $x = [1, \ 2, \ 3, \ 4]^T$, $\cos(\theta) = 1/\sqrt{5}$, and $\sin(\theta) = -2/\sqrt{5}$, then $G(2, 4, \theta)x = [1, \ \sqrt{20}, \ 3, \ 0]^T$.

5.1.9   Applying Givens Rotations

It is critical that the simple structure of a Givens rotation matrix be exploited when it is involved in a matrix multiplication. Suppose $A \in \mathbb{R}^{m \times n}$, $c = \cos(\theta)$, and $s = \sin(\theta)$. If $G(i, k, \theta) \in \mathbb{R}^{m \times m}$, then the update $A \leftarrow G(i, k, \theta)^T A$ effects just two rows of A,

$$A([i, k], :) = \begin{bmatrix} c & s \\ -s & c \end{bmatrix}^T A([i, k], :)$$

and requires just $6n$ flops:

> **for** $j = 1{:}n$
> $\tau_1 = A(i, j)$
> $\tau_2 = A(k, j)$
> $A(1, j) = c\tau_1 - s\tau_2$
> $A(2, j) = s\tau_1 + c\tau_2$
> **end**

Likewise, if $G(i, k, \theta) \in \mathbb{R}^{n \times n}$, then the update $A \leftarrow AG(i, k, \theta)$ effects just two columns of A,

$$A(:, [i, k]) = A(:, [i, k]) \begin{bmatrix} c & s \\ -s & c \end{bmatrix}$$

and requires just $6m$ flops:

> **for** $j = 1{:}m$
> $\quad \tau_1 = A(j, i)$
> $\quad \tau_2 = A(j, k)$
> $\quad A(j, i) = c\tau_1 - s\tau_2$
> $\quad A(j, k) = s\tau_1 + c\tau_2$
> **end**

5.1.10   Roundoff Properties

The numerical properties of Givens rotations are as favorable as those for Householder reflections. In particular, it can be shown that the computed $\hat{c}$ and $\hat{s}$ in **givens** satisfy

$$\begin{aligned}\hat{c} &= c(1 + \epsilon_c) & \epsilon_c &= O(\mathbf{u}) \\ \hat{s} &= s(1 + \epsilon_s) & \epsilon_s &= O(\mathbf{u}).\end{aligned}$$

If $\hat{c}$ and $\hat{s}$ are subsequently used in a Givens update, then the computed update is the exact update of a nearby matrix:

$$fl[\hat{G}(i, k, \theta)^T A] = G(i, k, \theta)^T (A + E) \qquad \| E \|_2 \approx \mathbf{u}\| A \|_2$$

$$fl[A\hat{G}(i, k, \theta)] = (A + E)G(i, k, \theta) \qquad \| E \|_2 \approx \mathbf{u}\| A \|_2.$$

A detailed error analysis of Givens rotations may be found in Wilkinson (1965, pp. 131-39).

5.1.11   Representing Products of Givens Rotations

Suppose $Q = G_1 \cdots G_t$ is a product of Givens rotations. As we have seen in connection with Householder reflections, it is more economical to keep the orthogonal matrix Q in factored form than to compute explicitly the product of the rotations. Using a technique demonstrated by Stewart (1976), it is possible to do this in a very compact way. The idea is to associate a single floating point number ρ with each rotation. Specifically, if

$$Z = \begin{bmatrix} c & s \\ -s & c \end{bmatrix} \qquad c^2 + s^2 = 1$$

then we define the scalar ρ by

> **if** $c = 0$
> $\quad \rho = 1$
> **elseif** $|s| < |c|$
> $\quad \rho = \text{sign}(c)s/2$ (5.1.9)
> **else**
> $\quad \rho = 2\text{sign}(s)/c$
> **end**

Essentially, this amounts to storing $s/2$ if the sine is smaller and $2/c$ if the cosine is smaller. With this encoding, it is possible to reconstruct $\pm Z$ as follows:

$$
\begin{aligned}
&\textbf{if } \rho = 1 \\
&\qquad c = 0; \ s = 1 \\
&\textbf{elseif } |\rho| < 1 \\
&\qquad s = 2\rho; \ c = \sqrt{1 - s^2} \\
&\textbf{else} \\
&\qquad c = 2/\rho; \ s = \sqrt{1 - c^2} \\
&\textbf{end}
\end{aligned}
\qquad (5.1.10)
$$

That $-Z$ may be generated is usually of no consequence for if Z zeros a particular matrix entry, so does $-Z$. The reason for essentially storing the smaller of c and s is that the formula $\sqrt{1 - x^2}$ renders poor results if x is near unity. More details may be found in Stewart (1976). Of course, to "reconstruct" $G(i, k, \theta)$ we need i and k in addition to the associated ρ. This usually poses no difficulty as we discuss in §5.2.3.

5.1.12   Error Propagation

We offer some remarks about the propagation of roundoff error in algorithms that involve sequences of Householder/Givens updates. To be precise, suppose $A = A_0 \in \mathbb{R}^{m \times n}$ is given and that matrices $A_1, \ldots, A_p = B$ are generated via the formula

$$
A_k = fl(\hat{Q}_k A_{k-1} \hat{Z}_k) \qquad k = 1{:}p .
$$

Assume that the above Householder and Givens algorithms are used for both the generation and application of the $\hat{Q}_k$ and $\hat{Z}_k$. Let Q_k and Z_k be the orthogonal matrices that would be produced in the absence of roundoff. It can be shown that

$$
B = (Q_p \cdots Q_1)(A + E)(Z_1 \cdots Z_p), \qquad (5.1.11)
$$

where $\| E \|_2 \le cu\| A \|_2$ and c is a constant that depends mildly on n, m, and p. In plain English, B is an exact orthogonal update of a matrix near to A.

5.1.13   Fast Givens Transformations

The ability to introduce zeros in a selective fashion makes Givens rotations an important zeroing tool in certain structured problems. This has led to the development of "fast Givens" procedures. The fast Givens idea amounts to a clever representation of Q when Q is the product of Givens rotations.

In particular, Q is represented by a matrix pair (M, D) where $M^T M = D = \operatorname{diag}(d_i)$ and each d_i is positive. The matrices Q, M, and D are connected through the formula

$$Q = MD^{-1/2} = M \operatorname{diag}(1/\sqrt{d_i}).$$

Note that $(MD^{-1/2})^T (MD^{-1/2}) = D^{-1/2}DD^{-1/2} = I$ and so the matrix $MD^{-1/2}$ is orthogonal. Moreover, if F is an n-by-n matrix with $F^T DF = D_{new}$ diagonal, then $M_{new}^T M_{new} = D_{new}$ where $M_{new} = MF$. Thus, it is possible to update the fast Givens representation (M, D) to obtain (M_{new}, D_{new}). For this idea to be of practical interest, we must show how to give F zeroing capabilities subject to the constraint that it "keeps" D diagonal.

The details are best explained at the 2-by-2 level. Let $x = [x_1 \ x_2]^T$ and $D = \operatorname{diag}(d_1, d_2)$ be given and assume that d_1 and d_2 are positive. Define

$$M_1 = \begin{bmatrix} \beta_1 & 1 \\ 1 & \alpha_1 \end{bmatrix} \tag{5.1.12}$$

and observe that

$$M_1^T x = \begin{bmatrix} \beta_1 x_1 + x_2 \\ x_1 + \alpha_1 x_2 \end{bmatrix}$$

and

$$M_1^T DM_1 = \begin{bmatrix} d_2 + \beta_1^2 d_1 & d_1 \beta_1 + d_2 \alpha_1 \\ d_1 \beta_1 + d_2 \alpha_1 & d_1 + \alpha_1^2 d_2 \end{bmatrix} \equiv D_1 .$$

If $x_2 \neq 0$, $\alpha_1 = -x_1/x_2$, and $\beta_1 = -\alpha_1 d_2/d_1$, then

$$M_1^T x = \begin{bmatrix} x_2(1 + \gamma_1) \\ 0 \end{bmatrix}$$

$$M_1^T DM_1 = \begin{bmatrix} d_2(1 + \gamma_1) & 0 \\ 0 & d_1(1 + \gamma_1) \end{bmatrix}$$

where $\gamma_1 = -\alpha_1 \beta_1 = (d_2/d_1)(x_1/x_2)^2$.

Analogously, if we assume $x_1 \neq 0$ and define M_2 by

$$M_2 = \begin{bmatrix} 1 & \alpha_2 \\ \beta_2 & 1 \end{bmatrix} \tag{5.1.13}$$

where $\alpha_2 = -x_2/x_1$ and $\beta_2 = -(d_1/d_2)\alpha_2$, then

$$M_2^T x = \begin{bmatrix} x_1(1 + \gamma_2) \\ 0 \end{bmatrix}$$

and

$$M_2^T DM_2 = \begin{bmatrix} d_1(1 + \gamma_2) & 0 \\ 0 & d_2(1 + \gamma_2) \end{bmatrix} \equiv D_2,$$

where $\gamma_2 = -\alpha_2\beta_2 = (d_1/d_2)(x_2/x_1)^2$.

It is easy to show that for either $i = 1$ or 2, the matrix $J = D^{1/2}M_iD_i^{-1/2}$ is orthogonal and that it is designed so that the second component of $J^T(D^{-1/2}x)$ is zero. (J may actually be a reflection and thus it is half-correct to use the popular term "fast Givens.")

Notice that the γ_i satisfy $\gamma_1\gamma_2 = 1$. Thus, we can always select M_i in the above so that the "growth factor" $(1 + \gamma_i)$ is bounded by 2. Matrices of the form

$$M_1 = \begin{bmatrix} \beta_1 & 1 \\ 1 & \alpha_1 \end{bmatrix} \qquad M_2 = \begin{bmatrix} 1 & \alpha_2 \\ \beta_2 & 1 \end{bmatrix}$$

that satisfy $-1 \le \alpha_i\beta_i \le 0$ are 2-by-2 *fast Givens transformations*. Notice that premultiplication by a fast Givens transformation involves half the number of multiplies as premultiplication by an "ordinary" Givens transformation. Also, the zeroing is carried out without an explicit square root.

In the n-by-n case, everything "scales up" as with ordinary Givens rotations. The "type 1" transformations have the form

$$F(i,k,\alpha,\beta) = \begin{bmatrix} 1 & \cdots & 0 & \cdots & 0 & \cdots & 0 \\ \vdots & \ddots & \vdots & & \vdots & & \vdots \\ 0 & \cdots & \beta & \cdots & 1 & \cdots & 0 \\ \vdots & & \vdots & \ddots & \vdots & & \vdots \\ 0 & \cdots & 1 & \cdots & \alpha & \cdots & 0 \\ \vdots & & \vdots & & \vdots & \ddots & \vdots \\ 0 & \cdots & 0 & \cdots & 0 & \cdots & 1 \end{bmatrix} \begin{matrix} \\ \\ i \\ \\ k \\ \\ \\ \end{matrix} \qquad (5.1.14)$$

while the "type 2" transformations are structured as follows:

$$F(i,k,\alpha,\beta) = \begin{bmatrix} 1 & \cdots & 0 & \cdots & 0 & \cdots & 0 \\ \vdots & \ddots & \vdots & & \vdots & & \vdots \\ 0 & \cdots & 1 & \cdots & \alpha & \cdots & 0 \\ \vdots & & \vdots & \ddots & \vdots & & \vdots \\ 0 & \cdots & \beta & \cdots & 1 & \cdots & 0 \\ \vdots & & \vdots & & \vdots & \ddots & \vdots \\ 0 & \cdots & 0 & \cdots & 0 & \cdots & 1 \end{bmatrix} \begin{matrix} \\ \\ i \\ \\ k \\ \\ \\ \end{matrix} \qquad (5.1.15)$$

Encapsulating all this we obtain

Algorithm 5.1.4   Given $x \in \mathbb{R}^2$ and positive $d \in \mathbb{R}^2$, the following algorithm computes a 2-by-2 fast Givens transformation M such that the second component of $M^T x$ is zero and $M^T D M = D_1$ is diagonal where D $= \text{diag}(d_1, d_2)$. If $type = 1$ then M has the form (5.1.12) while if $type = 2$ then M has the form (5.1.13). The diagonal elements of D_1 overwrite d.

> **function:** $[\alpha, \beta, type] = \textsf{fast.givens}(x, d)$
>   if $x(2) \neq 0$
> $\alpha = -x(1)/x(2); \quad \beta = -\alpha d(2)/d(1); \quad \gamma = -\alpha\beta$
>   if $\gamma \leq 1$
> $type = 1$
> $\tau = d(1); \quad d(1) = (1 + \gamma)d(2); \quad d(2) = (1 + \gamma)\tau$
>   else
> $type = 2$
> $\alpha = 1/\alpha; \quad \beta = 1/\beta; \quad \gamma = 1/\gamma$
> $d(1) = (1 + \gamma)d(1); \quad d(2) = (1 + \gamma)d(2)$
>   end
>   else
> $type = 2$
> $\alpha = 0; \quad \beta = 0$
>   end

The application of fast Givens transformations is analogous to that for ordinary Givens transformations. Even with the appropriate type of transformation used, the growth factor $1 + \gamma$ may still be as large as two. Thus, 2^s growth can occur in the entries of D and M after s updates. This means that the diagonal D must be monitored during a fast Givens procedure to avoid overflow. See Anda and Park (1994) for how to do this efficiently.

Nevertheless, element growth in M and D is controlled because at all times we have $M D^{-1/2}$ orthogonal. The roundoff properties of a fast givens procedure are what we would expect of a Givens matrix technique. For example, if we computed $\hat{Q} = fl(\hat{M}\hat{D}^{-1/2})$ where $\hat{M}$ and $\hat{D}$ are the computed M and D, then $\hat{Q}$ is orthogonal to working precision: $\| \hat{Q}^T \hat{Q} - I \|_2 \approx \mathbf{u}$.

Problems

P5.1.1 Execute **house** with $x = [\, 1, \ 7, \ 2, \ 3, \ -1\,]^T$.

P5.1.2 Let x and y be nonzero vectors in $\mathbb{R}^n$. Give an algorithm for determining a Householder matrix P such that Px is a multiple of y.

P5.1.3 Suppose $x \in \mathbb{C}^m$ and that $x_1 = |x_1|e^{i\theta}$ with $\theta \in \mathbb{R}$. Assume $x \neq 0$ and define $u = x + e^{i\theta}\| x \|_2 e_1$ Show that $P = I - 2uu^H/u^H u$ is unitary and that $Px = -e^{i\theta}\| x \|_2 e_1$.

P5.1.4 Use Householder matrices to show that $\det(I + xy^T) = 1 + x^T y$ where x and y are given n-vectors.

P5.1.5 Suppose $x \in \mathbb{C}^2$. Give an algorithm for determining a unitary matrix of the form

$$Q = \begin{bmatrix} c & \bar{s} \\ -s & c \end{bmatrix} \qquad c \in \mathbb{R}, \ c^2 + |s|^2 = 1$$

such that the second component of $Q^H x$ is zero.

P5.1.6   Suppose x and y are unit vectors in $\mathbf{R}^n$. Give an algorithm using Givens transformations which computes an orthogonal Q such that $Q^T x = y$.

P5.1.7   Determine $c = \cos(\theta)$ and $s = \sin(\theta)$ such that

$$
\begin{bmatrix} c & s \\ -s & c \end{bmatrix}^T \begin{bmatrix} 5 \\ 12 \end{bmatrix} = \begin{bmatrix} 13 \\ 0 \end{bmatrix} .
$$

P5.1.8   Suppose that $Q = I + YTY^T$ is orthogonal where $Y \in \mathbf{R}^{n \times j}$ and $T \in \mathbf{R}^{j \times j}$ is upper triangular. Show that if $Q_+ = QP$ where $P = I - 2vv^T / v^T v$ is a Householder matrix, then Q_+ can be expressed in the form $Q_+ = I + Y_+ T_+ Y_+^T$ where $Y_+ \in \mathbf{R}^{n \times (j+1)}$ and $T_+ \in \mathbf{R}^{(j+1) \times (j+1)}$ is upper triangular.

P5.1.9   Give a detailed implementation of Algorithm 5.1.2 with the assumption that $v^{(j)}(j+1{:}n)$, the essential part of the the jth Householder vector, is stored in $A(j+1{:}n, j)$. Since Y is effectively represented in A, your procedure need only set up the W matrix.

P5.1.10   Show that if S is skew-symmetric ($S^T = -S$), then $Q = (I + S)(I - S)^{-1}$ is orthogonal. (Q is called the *Cayley transform* of S.) Construct a rank-2 S so that if x is a vector then Qx is zero except in the first component.

P5.1.11   Suppose $P \in \mathbf{R}^{n \times n}$ satisfies $\| P^T P - I_n \|_2 = \epsilon < 1$. Show that all the singular values of P are in the interval $[1 - \epsilon, 1 + \epsilon]$ and that $\| P - UV^T \|_2 \leq \epsilon$ where $P = U \Sigma V^T$ is the SVD of P.

P5.1.12   Suppose $A \in \mathbf{R}^{2 \times 2}$. Under what conditions is the closest rotation to A closer than the closest reflection to A?

Notes and References for Sec. 5.1

Householder matrices are named after A.S. Householder, who popularized their use in numerical analysis. However, the properties of these matrices have been known for quite some time. See

H.W. Turnbull and A.C. Aitken (1961). *An Introduction to the Theory of Canonical Matrices*, Dover Publications, New York, pp. 102–5.

Other references concerned with Householder transformations include

A.R. Gourlay (1970). "Generalization of Elementary Hermitian Matrices," *Comp. J. 13*, 411–12.

B.N. Parlett (1971). "Analysis of Algorithms for Reflections in Bisectors," *SIAM Review 13*, 197–208.

N.K. Tsao (1975). "A Note on Implementing the Householder Transformations." *SIAM J. Num. Anal. 12*, 53–58.

B. Danloy (1976). "On the Choice of Signs for Householder Matrices," *J. Comp. Appl. Math. 2*, 67–69.

J.J.M. Cuppen (1984). "On Updating Triangular Products of Householder Matrices," *Numer. Math. 45*, 403–410.

L. Kaufman (1987). "The Generalized Householder Transformation and Sparse Matrices," *Lin. Alg. and Its Applic. 90*, 221–234.

A detailed error analysis of Householder transformations is given in Lawson and Hanson (1974, 83–89).

   The basic references for block Householder representations and the associated computations include

C.H. Bischof and C. Van Loan (1987). "The WY Representation for Products of Householder Matrices," *SIAM J. Sci. and Stat. Comp. 8*, s2–s13.

R. Schreiber and B.N. Parlett (1987). "Block Reflectors: Theory and Computation," *SIAM J. Numer. Anal. 25*, 189-205.

B.N. Parlett and R. Schreiber (1988). "Block Reflectors: Theory and Computation," *SIAM J. Num. Anal. 25*, 189–205.

R.S. Schreiber and C. Van Loan (1989). "A Storage-Efficient WY Representation for Products of Householder Transformations," *SIAM J. Sci. and Stat. Comp. 10*, 52–57.

C. Puglisi (1992). "Modification of the Householder Method Based on the Compact WY Representation," *SIAM J. Sci. and Stat. Comp. 13*, 723–726.

X. Sun and C.H. Bischof (1995). "A Basis-Kernel Representation of Orthogonal Matrices," *SIAM J. Matrix Anal. Appl. 16*, 1184–1196.

Givens rotations, named after W. Givens, are also referred to as Jacobi rotations. Jacobi devised a symmetric eigenvalue algorithm based on these transformations in 1846. See §8.4. The Givens rotation storage scheme discussed in the text is detailed in

G.W. Stewart (1976). "The Economical Storage of Plane Rotations," *Numer. Math. 25*, 137–38.

Fast Givens transformations are also referred to as "square-root-free" Givens transformations. (Recall that a square root must ordinarily be computed during the formation of Givens transformation.) There are several ways fast Givens calculations can be arranged. See

M. Gentleman (1973). "Least Squares Computations by Givens Transformations without Square Roots," *J. Inst. Math. Appl. 12*, 329–36.

C.F. Van Loan (1973). "Generalized Singular Values With Algorithms and Applications," Ph.D. thesis, University of Michigan, Ann Arbor.

S. Hammarling (1974). "A Note on Modifications to the Givens Plane Rotation," *J. Inst. Math. Appl. 13*, 215–18.

J.H. Wilkinson (1977). "Some Recent Advances in Numerical Linear Algebra," in *The State of the Art in Numerical Analysis*, ed. D.A.H. Jacobs, Academic Press, New York, pp. 1–53.

A.A. Anda and H. Park (1994). "Fast Plane Rotations with Dynamic Scaling," *SIAM J. Matrix Anal. Appl. 15*, 162–174.

5.2 The QR Factorization

We now show how Householder and Givens transformations can be used to compute various factorizations, beginning with the QR factorization. The QR factorization of an m-by-n matrix A is given by

$$A = QR$$

where $Q \in \mathbb{R}^{m \times m}$ is orthogonal and $R \in \mathbb{R}^{m \times n}$ is upper triangular. In this section we assume $m \geq n$. We will see that if A has full column rank, then the first n columns of Q form an orthonormal basis for ran(A). Thus, calculation of the QR factorization is one way to compute an orthonormal basis for a set of vectors. This computation can be arranged in several ways. We give methods based on Householder, block Householder, Givens, and fast Givens transformations. The Gram-Schmidt orthogonalization process and a numerically more stable variant called modified Gram-Schmidt are also discussed.

5.2.1   Householder QR

We begin with a QR factorization method that utilizes Householder transformations. The essence of the algorithm can be conveyed by a small example. Suppose $m = 6$, $n = 5$, and assume that Householder matrices H_1
and H_2 have been computed so that

$$
H_2 H_1 A \;=\;
\begin{bmatrix}
\times & \times & \times & \times & \times \\
0 & \times & \times & \times & \times \\
0 & 0 & \boxtimes & \times & \times \\
0 & 0 & \boxtimes & \times & \times \\
0 & 0 & \boxtimes & \times & \times \\
0 & 0 & \boxtimes & \times & \times
\end{bmatrix}.
$$

Concentrating on the highlighted entries, we determine a Householder matrix $\tilde{H}_3 \in \mathbb{R}^{4\times4}$ such that

$$
\tilde{H}_3
\begin{bmatrix}
\boxtimes \\
\boxtimes \\
\boxtimes \\
\boxtimes
\end{bmatrix}
\;=\;
\begin{bmatrix}
\times \\
0 \\
0 \\
0
\end{bmatrix}.
$$

If $H_3 = \mathrm{diag}(I_2, \tilde{H}_3)$, then

$$
H_3 H_2 H_1 A \;=\;
\begin{bmatrix}
\times & \times & \times & \times & \times \\
0 & \times & \times & \times & \times \\
0 & 0 & \times & \times & \times \\
0 & 0 & 0 & \times & \times \\
0 & 0 & 0 & \times & \times \\
0 & 0 & 0 & \times & \times
\end{bmatrix}.
$$

After n such steps we obtain an upper triangular $H_n H_{n-1} \cdots H_1 A = R$ and
so by setting $Q = H_1 \cdots H_n$ we obtain $A = QR$.

Algorithm 5.2.1 (Householder QR)   Given $A \in \mathbb{R}^{m\times n}$ with $m \geq n$,
the following algorithm finds Householder matrices $H_1, \ldots, H_n$ such that if
$Q = H_1 \cdots H_n$, then $Q^T A = R$ is upper triangular. The upper triangular
part of A is overwritten by the upper triangular part of R and components
$j + 1{:}m$ of the jth Householder vector are stored in $A(j+1{:}m, j), j < m$.

$$
\begin{aligned}
&\textbf{for } j = 1{:}n \\
&\qquad [v, \beta] = \textbf{house}(A(j{:}m, j)) \\
&\qquad A(j{:}m, j{:}n) = (I_{m-j+1} - \beta v v^T)A(j{:}m, j{:}n) \\
&\qquad \textbf{if } j < m \\
&\qquad\qquad A(j+1{:}m, j) = v(2{:}m - j + 1) \\
&\qquad \textbf{end} \\
&\textbf{end}
\end{aligned}
$$

This algorithm requires $2n^2(m - n/3)$ flops.

To clarify how A is overwritten, if

$$v^{(j)} = [\underbrace{0, \ldots, 0}_{j-1}, 1, v_{j+1}^{(j)}, \ldots, v_m^{(j)}]^T$$

is the jth Householder vector, then upon completion

$$A = \begin{bmatrix} r_{11} & r_{12} & r_{13} & r_{14} & r_{15} \\ v_2^{(1)} & r_{22} & r_{23} & r_{24} & r_{25} \\ v_3^{(1)} & v_3^{(2)} & r_{33} & r_{34} & r_{35} \\ v_4^{(1)} & v_4^{(2)} & v_4^{(3)} & r_{44} & r_{45} \\ v_5^{(1)} & v_5^{(2)} & v_5^{(3)} & v_5^{(4)} & r_{55} \\ v_6^{(1)} & v_6^{(2)} & v_6^{(3)} & v_6^{(4)} & v_6^{(5)} \end{bmatrix}.$$

If the matrix $Q = H_1 \cdots H_n$ is required, then it can be accumulated using (5.1.5). This accumulation requires $4(m^2 n - mn^2 + n^3/3)$ flops.

The computed upper triangular matrix $\hat{R}$ is the exact R for a nearby A in the sense that $Z^T(A + E) = \hat{R}$ where Z is some exact orthogonal matrix and $\| E \|_2 \approx \mathbf{u} \| A \|_2$.

5.2.2 Block Householder QR Factorization

Algorithm 5.2.1 is rich in the level-2 operations of matrix-vector multiplication and outer product updates. By reorganizing the computation and using the block Householder representation discussed in §5.1.7 we can obtain a level-3 procedure. The idea is to apply clusters of Householder transformations that are represented in the WY form of §5.1.7.

A small example illustrates the main idea. Suppose $n = 12$ and that the "blocking parameter" r has the value $r = 3$. The first step is to generate Householders H_1, H_2, and H_3 as in Algorithm 5.2.1. However, unlike Algorithm 5.2.1 where the H_i are applied to all of A, we only apply H_1, H_2, and H_3 to $A(:, 1{:}3)$. After this is accomplished we generate the block representation $H_1 H_2 H_3 = I + W_1 Y_1^T$ and then perform the level-3 update

$$A(:, 4{:}12) = (I + WY^T) A(:, 4{:}12).$$

Next, we generate H_4, H_5, and H_6 as in Algorithm 5.2.1. However, these transformations are not applied to $A(:, 7{:}12)$ until their block representation $H_4 H_5 H_6 = I + W_2 Y_2^T$ is found. This illustrates the general pattern.

$\lambda = 1; \; k = 0$
while $\lambda \leq n$
$\quad \tau = \min(\lambda + r - 1, n); \; k = k + 1$
$\quad$ Using Algorithm 5.2.1, upper triangularize $A(\lambda{:}m, \lambda{:}n)$
$\quad\quad$ generating Householder matrices $H_\lambda, \dots, H_\tau$. $\hspace{2em}$ (5.2.1)
$\quad$ Use Algorithm 5.1.2 to get the block representation
$\quad\quad I + W_k Y_k = H_\lambda, \dots, H_\tau.$
$\quad A(\lambda{:}m, \tau + 1{:}n) = (I + W_k Y_k^T)^T A(\lambda{:}m, \tau + 1{:}n)$
$\quad \lambda = \tau + 1$
end

The zero-nonzero structure of the Householder vectors that define the matrices $H_\lambda, \dots, H_\tau$ implies that the first $\lambda - 1$ rows of W_k and Y_k are zero. This fact would be exploited in a practical implementation.

The proper way to regard (5.2.1) is through the partitioning

$$A = [\, A_1, \dots, A_N \,] \hspace{2em} N = \text{ceil}(n/r)$$

where block column A_k is processed during the kth step. In the kth step of (5.2.1), a block Householder is formed that zeros the subdiagonal portion of A_k. The remaining block columns are then updated.

The roundoff properties of (5.2.1) are essentially the same as those for Algorithm 5.2.1. There is a slight increase in the number of flops required because of the W-matrix computations. However, as a result of the blocking, all but a small fraction of the flops occur in the context of matrix multiplication. In particular, the level-3 fraction of (5.2.1) is approximately $1 - 2/N$. See Bischof and Van Loan (1987) for further details.

5.2.3   Givens QR Methods

Givens rotations can also be used to compute the QR factorization. The 4-by-3 case illustrates the general idea:

Here we have highlighted the 2-vectors that define the underlying Givens rotations. Clearly, if G_j denotes the jth Givens rotation in the reduction, then $Q^T A = R$ is upper triangular where $Q = G_1 \cdots G_t$ and t is the total number of rotations. For general m and n we have:

Algorithm 5.2.2 (Givens QR) Given $A \in \mathbb{R}^{m \times n}$ with $m \geq n$, the following algorithm overwrites A with $Q^T A = R$, where R is upper triangular and Q is orthogonal.

> **for** $j = 1{:}n$
> > **for** $i = m{:} -1{:}j+1$
> > > $[c, s] = \mathbf{givens}(A(i-1, j), A(i, j))$
> > >
> > > $A(i-1{:}i, j{:}n) = \begin{bmatrix} c & s \\ -s & c \end{bmatrix}^T A(i-1{:}i, j{:}n)$
> >
> > **end**
>
> **end**

This algorithm requires $3n^2(m - n/3)$ flops. Note that we could use (5.1.9) to encode (c, s) in a single number ρ which could then be stored in the zeroed entry $A(i, j)$. An operation such as $x \leftarrow Q^T x$ could then be implemented by using (5.1.10), taking care to reconstruct the rotations in the proper order.

Other sequences of rotations can be used to upper triangularize A. For example, if we replace the **for** statements in Algorithm 5.2.2 with

> **for** $i = m{:} -1{:}2$
> > **for** $j = 1{:}\min\{i-1, n\}$

then the zeros in A are introduced row-by-row.

Another parameter in a Givens QR procedure concerns the planes of rotation that are involved in the zeroing of each a_{ij}. For example, instead of rotating rows $i - 1$ and i to zero a_{ij} as in Algorithm 5.2.2, we could use rows j and i:

> **for** $j = 1{:}n$
> > **for** $i = m{:} -1{:}j+1$
> > > $[c, s] = \mathbf{givens}(A(j, j), A(i, j))$
> > >
> > > $A([\,j\ i\,], j{:}n) = \begin{bmatrix} c & s \\ -s & c \end{bmatrix}^T A([\,j\ i\,], j{:}n)$
> >
> > **end**
>
> **end**

5.2.4 Hessenberg QR via Givens

As an example of how Givens rotations can be used in structured problems, we show how they can be employed to compute the QR factorization of an upper Hessenberg matrix. A small example illustrates the general idea.

Suppose $n = 6$ and that after two steps we have computed

$$G(2,3,\theta_2)^T G(1,2,\theta_1)^T A = \begin{bmatrix} \times & \times & \times & \times & \times & \times \\ 0 & \times & \times & \times & \times & \times \\ 0 & 0 & \times & \times & \times & \times \\ 0 & 0 & \times & \times & \times & \times \\ 0 & 0 & 0 & \times & \times & \times \\ 0 & 0 & 0 & 0 & \times & \times \end{bmatrix}.$$

We then compute $G(3, 4, \theta_3)$ to zero the current $(4,3)$ entry thereby obtaining

$$G(3,4,\theta_3)^T G(2,3,\theta_2)^T G(1,2,\theta_1)^T A = \begin{bmatrix} \times & \times & \times & \times & \times & \times \\ 0 & \times & \times & \times & \times & \times \\ 0 & 0 & \times & \times & \times & \times \\ 0 & 0 & 0 & \times & \times & \times \\ 0 & 0 & 0 & \times & \times & \times \\ 0 & 0 & 0 & 0 & \times & \times \end{bmatrix}.$$

Overall we have

Algorithm 5.2.3 (Hessenberg QR) If $A \in \mathbb{R}^{n \times n}$ is upper Hessenberg, then the following algorithm overwrites A with $Q^T A = R$ where Q is orthogonal and R is upper triangular. $Q = G_1 \cdots G_{n-1}$ is a product of Givens rotations where G_j has the form $G_j = G(j, j+1, \theta_j)$.

> **for** $j = 1{:}n - 1$
> $\qquad [\, c \; s \,] = \mathbf{givens}(A(j,j), A(j+1,j))$
> $\qquad A(j{:}j+1, j{:}n) = \begin{bmatrix} c & s \\ -s & c \end{bmatrix}^T A(j{:}j+1, j{:}n)$
> **end**

This algorithm requires about $3n^2$ flops.

5.2.5   Fast Givens QR

We can use the fast Givens transformations described in §5.1.13 to compute an (M, D) representation of Q. In particular, if M is nonsingular and D is diagonal such that $M^T A = T$ is upper triangular and $M^T M = D$ is diagonal, then $Q = MD^{-1/2}$ is orthogonal and $Q^T A = D^{-1/2}T \equiv R$ is upper triangular. Analogous to the Givens QR procedure we have:

Algorithm 5.2.4 (Fast Givens QR) Given $A \in \mathbb{R}^{m \times n}$ with $m \geq n$, the following algorithm computes nonsingular $M \in \mathbb{R}^{m \times m}$ and positive $d(1{:}m)$ such that $M^T A = T$ is upper triangular, and $M^T M = \text{diag}(d_1, \ldots, d_m)$. A is overwritten by T. Note: $A = (MD^{-1/2})(D^{1/2}T)$ is a QR factorization of A.

for $i = 1{:}m$
    $d(i) = 1$
end
for $j = 1{:}n$
    for $i = m{:} - 1{:}j + 1$
        $[\,\alpha,\ \beta,\ type\,] = $ **fast.givens**$(A(i-1{:}i, j), d(i-1{:}i))$
        if $type = 1$
$$A(i-1{:}i, j{:}n) = \begin{bmatrix} \beta & 1 \\ 1 & \alpha \end{bmatrix} A(i-1{:}i, j{:}n)$$
        else
$$A(i-1{:}i, j{:}n) = \begin{bmatrix} 1 & \alpha \\ \beta & 1 \end{bmatrix} A(i-1{:}i, j{:}n)$$
    end
end

This algorithm requires $2n^2(m - n/3)$ flops. As we mentioned in the previous section, it is necessary to guard against overflow in fast Givens algorithms such as the above. This means that M, D, and A must be periodically scaled if their entries become large.

If the QR factorization of a narrow band matrix is required, then the fast Givens approach is attractive because it involves no square roots. (We found LDL^T preferable to Cholesky in the narrow band case for the same reason; see §4.3.6.) In particular, if $A \in \mathbb{R}^{m \times n}$ has upper bandwidth q and lower bandwidth p, then $Q^T A = R$ has upper bandwidth $p + q$. In this case Givens QR requires about $O(np(p+q))$ flops and $O(np)$ square roots. Thus, the square roots are a significant portion of the overall computation if $p, q \ll n$.

5.2.6 Properties of the QR Factorization

The above algorithms "prove" that the QR factorization exists. Now we relate the columns of Q to $\text{ran}(A)$ and $\text{ran}(A)^{\perp}$ and examine the uniqueness question.

Theorem 5.2.1 *If $A = QR$ is a QR factorization of a full column rank $A \in \mathbb{R}^{m \times n}$ and $A = [\, a_1, \ldots, a_n \,]$ and $Q = [\, q_1, \ldots, q_m \,]$ are column partitionings, then*

$$\text{span}\{a_1, \ldots, a_k\} = \text{span}\{q_1, \ldots, q_k\} \qquad k = 1{:}n\,.$$

In particular, if $Q_1 = Q(1{:}m, 1{:}n)$ and $Q_2 = Q(1{:}m, n+1{:}m)$ then

$$\begin{aligned} \text{ran}(A) &= \text{ran}(Q_1) \\ \text{ran}(A)^{\perp} &= \text{ran}(Q_2) \end{aligned}$$

and $A = Q_1 R_1$ with $R_1 = R(1{:}n, 1{:}n)$.

Proof. Comparing kth columns in $A = QR$ we conclude that

$$a_k = \sum_{i=1}^{k} r_{ik} q_i \in \text{span}\{q_1, \ldots, q_k\} .   \qquad (5.2.2)$$

Thus, $\text{span}\{a_1, \ldots, a_k\} \subseteq \text{span}\{q_1, \ldots, q_k\}$. However, since $\text{rank}(A) = n$ it follows that $\text{span}\{a_1, \ldots, a_k\}$ has dimension k and so must equal $\text{span}\{q_1, \ldots, q_k\}$ The rest of the theorem follows trivially. $\square$

The matrices $Q_1 = Q(1{:}m, 1{:}n)$ and $Q_2 = Q(1{:}m, n+1{:}m)$ can be easily computed from a factored form representation of Q.

If $A = QR$ is a QR factorization of $A \in \mathbb{R}^{m \times n}$ and $m \geq n$, then we refer to $A = Q(:, 1{:}n)R(1{:}n, 1{:}n)$ as the *thin QR factorization*. The next result addresses the uniqueness issue for the *thin* QR factorization

Theorem 5.2.2 *Suppose $A \in \mathbb{R}^{m \times n}$ has full column rank. The thin QR factorization*

$$A = Q_1 R_1$$

is unique where $Q_1 \in \mathbb{R}^{m \times n}$ has orthonormal columns and R_1 is upper triangular with positive diagonal entries. Moreover, $R_1 = G^T$ where G is the lower triangular Cholesky factor of $A^T A$.

Proof. Since $A^T A = (Q_1 R_1)^T (Q_1 R_1) = R_1^T R_1$ we see that $G = R_1^T$ is the Cholesky factor of $A^T A$. This factor is unique by Theorem 4.2.5. Since $Q_1 = A R_1^{-1}$ it follows that Q_1 is also unique. $\square$

How are Q_1 and R_1 affected by perturbations in A? To answer this question we need to extend the notion of condition to rectangular matrices. Recall from §2.7.3 that the 2-norm condition of a square nonsingular matrix is the ratio of the largest and smallest singular values. For rectangular matrices with full column rank we continue with this definition:

$$A \in \mathbb{R}^{m \times n}, \text{rank}(A) = n \implies \kappa_2(A) = \frac{\sigma_{max}(A)}{\sigma_{min}(A)} .$$

If the columns of A are nearly dependent, then $\kappa_2(A)$ is large. Stewart (1993) has shown that $O(\epsilon)$ relative error in A induces $O(\epsilon \kappa_2(A))$ relative error in R and Q_1.

5.2.7   Classical Gram-Schmidt

We now discuss two alternative methods that can be used to compute the thin QR factorization $A = Q_1 R_1$ directly. If $\text{rank}(A) = n$, then equation (5.2.2) can be solved for q_k:

$$q_k = \left(a_k - \sum_{i=1}^{k-1} r_{ik} q_i \right) \bigg/ r_{kk} .$$

Thus, we can think of q_k as a unit 2-norm vector in the direction of

$$z_k = a_k - \sum_{i=1}^{k-1} r_{ik} q_i$$

where to ensure $z_k \in \text{span}\{q_1, \ldots, q_{k-1}\}^\perp$ we choose

$$r_{ik} = q_i^T a_k \qquad i = 1{:}k-1 .$$

This leads to the *classical Gram-Schmidt* (CGS) algorithm for computing $A = Q_1 R_1$.

$$
\begin{aligned}
&R(1,1) = \| A(:,1) \|_2 \\
&Q(:,1) = A(:,1)/R(1,1) \\
&\textbf{for } k = 2{:}n \\
&\qquad R(1{:}k-1,k) \;=\; Q(1{:}m,1{:}k-1)^T A(1{:}m,k) \\
&\qquad z \;=\; A(1{:}m,k) - Q(1{:}m,1{:}k-1)R(1{:}k-1,k) \qquad (5.2.3) \\
&\qquad R(k,k) \;=\; \| z \|_2 \\
&\qquad Q(1{:}m,k) \;=\; z/R(k,k) \\
&\textbf{end}
\end{aligned}
$$

In the kth step of CGS, the kth columns of both Q and R are generated.

5.2.8 Modified Gram-Schmidt

Unfortunately, the CGS method has very poor numerical properties in that there is typically a severe loss of orthogonality among the computed q_i. Interestingly, a rearrangement of the calculation, known as *modified Gram-Schmidt* (MGS), yields a much sounder computational procedure. In the kth step of MGS, the kth column of Q (denoted by q_k) and the kth row of R (denoted by r_k^T) are determined. To derive the MGS method, define the matrix $A^{(k)} \in \mathbb{R}^{m \times (n-k+1)}$ by

$$A - \sum_{i=1}^{k-1} q_i r_i^T = \sum_{i=k}^{n} q_i r_i^T = [\, 0 \; A^{(k)} \,] . \qquad (5.2.4)$$

It follows that if

$$A^{(k)} = \begin{bmatrix} z & B \\ 1 & n-k \end{bmatrix}$$

then $r_{kk} = \| z \|_2$, $q_k = z/r_{kk}$ and $(r_{k,k+1} \cdots r_{kn}) = q_k^T B$. We then compute the outer product $A^{(k+1)} = B - q_k (r_{k,k+1} \cdots r_{kn})$ and proceed to the next step. This completely describes the kth step of MGS.

Algorithm 5.2.5 (Modified Gram-Schmidt) Given $A \in \mathbb{R}^{m \times n}$ with $\text{rank}(A) = n$, the following algorithm computes the factorization $A = Q_1 R_1$ where $Q_1 \in \mathbb{R}^{m \times n}$ has orthonormal columns and $R_1 \in \mathbb{R}^{n \times n}$ is upper triangular.

for $k = 1{:}n$
$\quad R(k,k) = \| A(1{:}m, k) \|_2$
$\quad Q(1{:}m, k) = A(1{:}m, k)/R(k,k)$
$\quad$ **for** $j = k+1{:}n$
$\quad\quad R(k,j) = Q(1{:}m, k)^T A(1{:}m, j)$
$\quad\quad A(1{:}m, j) = A(1{:}m, j) - Q(1{:}m, k)R(k,j)$
$\quad$ **end**
end

This algorithm requires $2mn^2$ flops. It is not possible to overwrite A with both Q_1 and R_1. Typically, the MGS computation is arranged so that A is overwritten by Q_1 and the matrix R_1 is stored in a separate array.

5.2.9   Work and Accuracy

If one is interested in computing an orthonormal basis for $\text{ran}(A)$, then the Householder approach requires $2mn^2 - 2n^3/3$ flops to get Q in factored form and another $2mn^2 - 2n^3/3$ flops to get the first n columns of Q. (This requires "paying attention" to just the first n columns of Q in (5.1.5).) Therefore, for the problem of finding an orthonormal basis for $\text{ran}(A)$, MGS is about twice as efficient as Householder orthogonalization. However, Björck (1967) has shown that MGS produces a computed $\hat{Q}_1 = [\hat{q}_1, \ldots, \hat{q}_n]$ that satisfies

$$\hat{Q}_1^T \hat{Q}_1 = I + E_{MGS} \qquad \| E_{MGS} \|_2 \approx \mathbf{u}\kappa_2(A)$$

whereas the corresponding result for the Householder approach is of the form

$$\hat{Q}_1^T \hat{Q}_1 = I + E_H \qquad \| E_H \|_2 \approx \mathbf{u} .$$

Thus, if orthonormality is critical, then MGS should be used to compute orthonormal bases only when the vectors to be orthogonalized are fairly independent.

We also mention that the computed triangular factor $\hat{R}$ produced by MGS satisfies $\| A - \hat{Q}\hat{R} \| \approx \mathbf{u}\| A \|$ and that there exists a Q with perfectly orthonormal columns such that $\| A - Q\hat{R} \| \approx \mathbf{u}\| A \|$. See Higham (1996, p.379).

Example 5.2.1 If modified Gram-Schmidt is applied to

$$A = \begin{bmatrix} 1 & 1 \\ 10^{-3} & 0 \\ 0 & 10^{-3} \end{bmatrix} \qquad \kappa_2(A) \approx 1.4 \cdot 10^3$$

with 6-digit decimal arithmetic, then

$$[\hat{q}_1 \; \hat{q}_2] = \begin{bmatrix} 1.00000 & 0 \\ .001 & -.707107 \\ 0 & .707100 \end{bmatrix} .$$

5.2.10 A Note on Complex QR

Most of the algorithms that we present in this book have complex versions that are fairly straight forward to derive from their real counterparts. (This is *not* to say that everything is easy and obvious at the implementation level.) As an illustration we outline what a complex Householder QR factorization algorithm looks like.

Starting at the level of an individual Householder transformation, suppose $0 \neq x \in \mathbb{C}^n$ and that $x_1 = re^{i\theta}$ where $r, \theta \in \mathbb{R}$. If $v = x \pm e^{i\theta} \| x \|_2 e_1$ and $P = I_n - \beta v v^H$, $\beta = 2/v^H v$, then $Px = \mp e^{i\theta} \| x \|_2 e_1$. (See P5.1.3.) The sign can be determined to maximize $\| v \|_2$ for the sake of stability.

The upper triangularization of $A \in \mathbb{R}^{m \times n}$, $m \geq n$, proceeds as in Algorithm 5.2.1. In step j we zero the subdiagonal portion of $A(j{:}m, j)$:

> **for** $j = 1{:}n$
> $\quad x = A(j{:}m, j)$
> $\quad v = x \pm e^{i\theta} \| x \|_2 e_1$ where $x_1 = re^{i\theta}$.
> $\quad \beta = 2/v^H/v$
> $\quad A(j{:}m, j{:}n) = (I_{m-j+1} - \beta v v^H) A(j{:}m, j{:}n)$
> **end**

The reduction involves $8n^2(m - n/3)$ real flops, four times the number required to execute Algorithm 5.2.1. If $Q = P_1 \cdots P_n$ is the product of the Householder transformations, then Q is unitary and $Q^T A = R \in \mathbb{R}^{m \times n}$ is complex and upper triangular.

Problems

P5.2.1 Adapt the Householder QR algorithm so that it can efficiently handle the case when $A \in \mathbb{R}^{m \times n}$ has lower bandwidth p and upper bandwidth q.

P5.2.2 Adapt the Householder QR algorithm so that it computes the factorization $A = QL$ where L is lower triangular and Q is orthogonal. Assume that A is square. This involves rewriting the Householder vector function $v = \text{house}(x)$ so that $(I - 2vv^T/v^T v)x$ is zero everywhere but its bottom component.

P5.2.3 Adapt the Givens QR factorization algorithm so that the zeros are introduced by diagonal. That is, the entries are zeroed in the order $(m, 1)$, $(m-1, 1)$, $(m, 2)$, $(m-2, 1)$, $(m-1, 2)$, $(m, 3)$, etc.

P5.2.4 Adapt the fast Givens QR factorization algorithm so that it efficiently handles the case when A is n-by-n and tridiagonal. Assume that the subdiagonal, diagonal, and superdiagonal of A are stored in $e(1{:}n-1)$, $a(1{:}n)$, $f(1{:}n-1)$ respectively. Design your algorithm so that these vectors are overwritten by the nonzero portion of T.

P5.2.5 Suppose $L \in \mathbb{R}^{m \times n}$ with $m \geq n$ is lower triangular. Show how Householder matrices $H_1 \ldots H_n$ can be used to determine a lower triangular $L_1 \in \mathbb{R}^{n \times n}$ so that

$$H_n \cdots H_1 L = \begin{bmatrix} L_1 \\ 0 \end{bmatrix}$$

Hint: The second step in the 6-by-3 case involves finding H_2 so that

$$H_2 \begin{bmatrix} \times & 0 & 0 \\ \times & \times & 0 \\ \times & \times & \times \\ \times & \times & 0 \\ \times & \times & 0 \\ \times & \times & 0 \end{bmatrix} = \begin{bmatrix} \times & 0 & 0 \\ \times & \times & 0 \\ \times & \times & \times \\ \times & 0 & 0 \\ \times & 0 & 0 \\ \times & 0 & 0 \end{bmatrix}$$

with the property that rows 1 and 3 are left alone.

P5.2.6 Show that if

$$A = \begin{bmatrix} R & w \\ 0 & v \end{bmatrix} \begin{matrix} k \\ m-k \end{matrix} \qquad b = \begin{bmatrix} c \\ d \end{bmatrix} \begin{matrix} k \\ m-k \end{matrix}$$
$$\begin{matrix} k & n-k \end{matrix}$$

and A has full column rank, then min $\| Ax - b \|_2^2 = \| d \|_2^2 - \left(v^T d / \| v \|_2 \right)^2$.

P5.2.7 Suppose $A \in \mathbf{R}^{n \times n}$ and $D = \text{diag}(d_1, \ldots, d_n) \in \mathbf{R}^{n \times n}$. Show how to construct an orthogonal Q such that $Q^T A - DQ^T = R$ is upper triangular. Do not worry about efficiency—this is just an exercise in QR manipulation.

P5.2.8 Show how to compute the QR factorization of the product $A = A_p \cdots A_2 A_1$ without explicitly multiplying the matrices $A_1, \ldots, A_p$ together. Hint: In the $p = 3$ case, write $Q_3^T A = Q_3^T A_3 Q_2 Q_2^T A_2 Q_1 Q_1^T A_1$ and determine orthogonal Q_i so that $Q_i^T (A_i Q_{i-1})$ is upper triangular. $(Q_0 = I)$.

P5.2.9 Suppose $A \in \mathbf{R}^{n \times n}$ and let E be the permutation obtained by reversing the order of the rows in I_n. (This is just the exchange matrix of §4.7.) (a) Show that if $R \in \mathbf{R}^{n \times n}$ is upper triangular, then $L = ERE$ is lower triangular. (b) Show how to compute an orthogonal $Q \in \mathbf{R}^{n \times n}$ and a lower triangular $L \in \mathbf{R}^{n \times n}$ so that $A = QL$ assuming the availability of a procedure for computing the QR factorization.

P5.2.10 MGS applied to $A \in \mathbf{R}^{m \times n}$ is numerically equivalent to the first step in Householder QR applied to

$$\tilde{A} = \begin{bmatrix} O_n \\ A \end{bmatrix}$$

where O_n is the n-by-n zero matrix. Verify that this statement is true after the first step of each method is completed.

P5.2.11 Reverse the loop orders in Algorithm 5.2.5 (MGS QR) so that R is computed column-by-column.

P5.2.12 Develop a complex version of the Givens QR factorization. Refer to P5.1.5. where complex Givens rotations are the theme. Is it possible to organize the calculations so that the diagonal elements of R are nonnegative?

Notes and References for Sec. 5.2

The idea of using Householder transformations to solve the LS problem was proposed in

A.S. Householder (1958). "Unitary Triangularization of a Nonsymmetric Matrix," *J. ACM. 5*, 339–42.

The practical details were worked out in

P. Businger and G.H. Golub (1965). "Linear Least Squares Solutions by Householder Transformations," *Numer. Math. 7*, 269–76. See also Wilkinson and Reinsch (1971,111–18).

G.H. Golub (1965). "Numerical Methods for Solving Linear Least Squares Problems," *Numer. Math. 7*, 206–16.

The basic references on QR via Givens rotations include

W. Givens (1958). "Computation of Plane Unitary Rotations Transforming a General Matrix to Triangular Form," *SIAM J. App. Math. 6*, 26–50.

M. Gentleman (1973). "Error Analysis of QR Decompositions by Givens Transformations," *Lin. Alg. and Its Appl. 10*, 189–97.

For a discussion of how the QR factorization can be used to solve numerous problems in statistical computation, see

G.H. Golub (1969). "Matrix Decompositions and Statistical Computation," in *Statistical Computation* , ed. R.C. Milton and J.A. Nelder, Academic Press, New York, pp. 365–97.

The behavior of the Q and R factors when A is perturbed is discussed in

G.W. Stewart (1977). "Perturbation Bounds for the QR Factorization of a Matrix," *SIAM J. Num. Anal. 14*, 509–18.

H. Zha (1993). "A Componentwise Perturbation Analysis of the QR Decomposition," *SIAM J. Matrix Anal. Appl. 4*, 1124–1131.

G.W. Stewart (1993). "On the Perturbation of LU Cholesky, and QR Factorizations," *SIAM J. Matrix Anal. Appl. 14*, 1141–1145.

A. Barrlund (1994). "Perturbation Bounds for the Generalized QR Factorization," *Lin. Alg. and Its Applic. 207*, 251–271.

J.-G. Sun (1995). "On Perturbation Bounds for the QR Factorization," *Lin. Alg. and Its Applic. 215*, 95–112.

The main result is that the changes in Q and R are bounded by the condition of A times the relative change in A. Organizing the computation so that the entries in Q depend continuously on the entries in A is discussed in

T.F. Coleman and D.C. Sorensen (1984). "A Note on the Computation of an Orthonormal Basis for the Null Space of a Matrix," *Mathematical Programming 29*, 234–242.

References for the Gram-Schmidt process include include

J.R. Rice (1966). "Experiments on Gram-Schmidt Orthogonalization," *Math. Comp. 20*, 325–28.

A. Björck (1967). "Solving Linear Least Squares Problems by Gram-Schmidt Orthogonalization," *BIT 7*, 1–21.

N.N. Abdelmalek (1971). "Roundoff Error Analysis for Gram-Schmidt Method and Solution of Linear Least Squares Problems," *BIT 11*, 345–68.

J. Daniel, W.B. Gragg, L.Kaufman, and G.W. Stewart (1976). "Reorthogonalization and Stable Algorithms for Updating the Gram-Schmidt QR Factorization," *Math. Comp. 30*, 772–795.

A. Ruhe (1983). "Numerical Aspects of Gram-Schmidt Orthogonalization of Vectors," *Lin. Alg. and Its Applic. 52/53*, 591–601.

W. Jalby and B. Philippe (1991). "Stability Analysis and Improvement of the Block Gram-Schmidt Algorithm," *SIAM J. Sci. Stat. Comp. 12*, 1058–1073.

Å. Björck and C.C. Paige (1992). "Loss and Recapture of Orthogonality in the Modified Gram-Schmidt Algorithm," *SIAM J. Matrix Anal. Appl. 13*, 176–190.

A. Björck (1994). "Numerics of Gram-Schmidt Orthogonalization," *Lin. Alg. and Its Applic. 197/198*, 297–316.

The QR factorization of a structured matrix is usually structured itself. See

A.W. Bojanczyk, R.P. Brent, and F.R. de Hoog (1986). "QR Factorization of Toeplitz Matrices," *Numer. Math. 49*, 81–94.

S. Qiao(1986). "Hybrid Algorithm for Fast Toeplitz Orthogonalization," *Numer. Math.*
53, 351–366.

C.J. Demeure (1989). "Fast QR Factorization of Vandermonde Matrices," *Lin. Alg.
and Its Applic. 122/123/124*, 165–194.

L. Reichel (1991). "Fast QR Decomposition of Vandermonde-Like Matrices and Polynomial Least Squares Approximation," *SIAM J. Matrix Anal. Appl. 12*, 552–564.

D.R. Sweet (1991). "Fast Block Toeplitz Orthogonalization," *Numer. Math. 58*, 613–629.

Various high-performance issues pertaining to the QR factorization are discussed in

B. Mattingly, C. Meyer, and J. Ortega (1989). "Orthogonal Reduction on Vector Computers," *SIAM J. Sci. and Stat. Comp. 10*, 372–381.

P.A. Knight (1995). "Fast Rectangular Matrix Multiplication and the QR Decomposition," *Lin. Alg. and Its Applic. 221*, 69–81.

5.3   The Full Rank LS Problem

Consider the problem of finding a vector $x \in \mathbb{R}^n$ such that $Ax = b$ where the *data matrix* $A \in \mathbb{R}^{m \times n}$ and the *observation vector* $b \in \mathbb{R}^m$ are given and $m \geq n$. When there are more equations than unknowns, we say that the system $Ax = b$ is *overdetermined*. Usually an overdetermined system has no exact solution since b must be an element of $\text{ran}(A)$, a proper subspace of $\mathbb{R}^m$.

This suggests that we strive to minimize $\| Ax - b \|_p$ for some suitable choice of p. Different norms render different optimum solutions. For example, if $A = [\, 1, 1, 1\,]^T$ and $b = [\, b_1, b_2, b_3\,]^T$ with $b_1 \geq b_2 \geq b_3 \geq 0$, then it can be verified that

$$
\begin{aligned}
p &= 1 &\Rightarrow& \quad x_{opt} &= b_2 \\
p &= 2 &\Rightarrow& \quad x_{opt} &= (b_1 + b_2 + b_3)/3 \\
p &= \infty &\Rightarrow& \quad x_{opt} &= (b_1 + b_3)/2.
\end{aligned}
$$

Minimization in the 1-norm and ∞-norm is complicated by the fact that the function $f(x) = \| Ax - b \|_p$ is not differentiable for these values of p. However, much progress has been made in this area, and there are several good techniques available for 1-norm and ∞-norm minimization. See Coleman and Li (1992), Li (1993), and Zhang (1993).

In contrast to general p-norm minimization, the *least squares* (LS) problem

$$
\min_{x \in \mathbb{R}^n} \| Ax - b \|_2 \tag{5.3.1}
$$

is more tractable for two reasons:

- $\phi(x) = \frac{1}{2}\| Ax - b \|_2^2$ is a differentiable function of x and so the minimizers of ϕ satisfy the gradient equation $\nabla\phi(x) = 0$. This turns out to be an easily constructed symmetric linear system which is positive definite if A has full column rank.

- The 2-norm is preserved under orthogonal transformation. This means that we can seek an orthogonal Q such that the equivalent problem of minimizing $\| (Q^T A)x - (Q^T b) \|_2$ is "easy" to solve.

In this section we pursue these two solution approaches for the case when A has full column rank. Methods based on normal equations and the QR factorization are detailed and compared.

5.3.1 Implications of Full Rank

Suppose $x \in \mathbb{R}^n$, $z \in \mathbb{R}^n$, and $\alpha \in \mathbb{R}$ and consider the equality

$$\| A(x + \alpha z) - b \|_2^2 = \| Ax - b \|_2^2 + 2\alpha z^T A^T (Ax - b) + \alpha^2 \| Az \|_2^2$$

where $A \in \mathbb{R}^{m \times n}$ and $b \in \mathbb{R}^m$. If x solves the LS problem (5.3.1) then we must have $A^T(Ax - b) = 0$. Otherwise, if $z = -A^T(Ax - b)$ and we make α small enough, then we obtain the contradictory inequality $\| A(x + \alpha z) - b \|_2 < \| Ax - b \|_2$. We may also conclude that if x *and* $x + \alpha z$ are LS minimizers, then $z \in \text{null}(A)$.

Thus, if A has full column rank, then there is a unique LS solution x_{LS} and it solves the symmetric positive definite linear system

$$A^T A x_{LS} = A^T b.$$

These are called the *normal equations*. Since $\nabla \phi(x) = A^T(Ax - b)$ where $\phi(x) = \frac{1}{2} \| Ax - b \|_2^2$, we see that solving the normal equations is tantamount to solving the gradient equation $\nabla \phi = 0$. We call

$$r_{LS} = b - Ax_{LS}$$

the *minimum residual* and we use the notation

$$\rho_{LS} = \| Ax_{LS} - b \|_2$$

to denote its size. Note that if ρ_{LS} is small, then we can "predict" b with the columns of A.

So far we have been assuming that $A \in \mathbb{R}^{m \times n}$ has full column rank. This assumption is dropped in §5.5. However, even if $\text{rank}(A) = n$, then we can expect trouble in the above procedures if A is nearly rank deficient.

When assessing the quality of a computed LS solution $\hat{x}_{LS}$, there are two important issues to bear in mind:

- How close is $\hat{x}_{LS}$ to x_{LS}?

- How small is $\hat{r}_{LS} = b - A\hat{x}_{LS}$ compared to $r_{LS} = b - Ax_{LS}$?

The relative importance of these two criteria varies from application to application. In any case it is important to understand how x_{LS} and r_{LS} are affected by perturbations in A and b. Our intuition tells us that if the columns of A are nearly dependent, then these quantities may be quite sensitive.

Example 5.3.1 Suppose

$$
A = \begin{bmatrix} 1 & 0 \\ 0 & 10^{-6} \\ 0 & 0 \end{bmatrix}, \ \delta A = \begin{bmatrix} 0 & 0 \\ 0 & 0 \\ 0 & 10^{-8} \end{bmatrix}, \ b = \begin{bmatrix} 1 \\ 0 \\ 1 \end{bmatrix}, \ \delta b = \begin{bmatrix} 0 \\ 0 \\ 0 \end{bmatrix},
$$

and that x_{LS} and $\hat{x}_{LS}$ minimize $\| Ax - b \|_2$ and $\| (A + \delta A)x - (b + \delta b) \|_2$ respectively. Let r_{LS} and $\hat{r}_{LS}$ be the corresponding minimum residuals. Then

$$
x_{LS} = \begin{bmatrix} 1 \\ 0 \end{bmatrix}, \ \hat{x}_{LS} = \begin{bmatrix} 1 \\ .9999 \cdot 10^4 \end{bmatrix}, \ r_{LS} = \begin{bmatrix} 0 \\ 0 \\ 1 \end{bmatrix}, \ \hat{r}_{LS} = \begin{bmatrix} 0 \\ -.9999 \cdot 10^{-2} \\ .9999 \cdot 10^0 \end{bmatrix}.
$$

Since $\kappa_2(A) = 10^6$ we have

$$
\frac{\| \hat{x}_{LS} - x_{LS} \|_2}{\| x_{LS} \|_2} \approx .9999 \cdot 10^4 \leq \kappa_2(A)^2 \frac{\| \delta A \|_2}{\| A \|_2} = 10^{12} \cdot 10^{-8}
$$

and

$$
\frac{\| \hat{r}_{LS} - r_{LS} \|_2}{\| b \|_2} \approx .7070 \cdot 10^{-2} \leq \kappa_2(A) \frac{\| \delta A \|_2}{\| A \|_2} = 10^6 \cdot 10^{-8} .
$$

The example suggests that the sensitivity of x_{LS} depends upon $\kappa_2(A)^2$. At the end of this section we develop a perturbation theory for the LS problem and the $\kappa_2(A)^2$ factor will return.

5.3.2   The Method of Normal Equations

The most widely used method for solving the full rank LS problem is the method of normal equations.

Algorithm 5.3.1 (Normal Equations) Given $A \in \mathbb{R}^{m \times n}$ with the property that rank$(A) = n$ and $b \in \mathbb{R}^m$, this algorithm computes the solution x_{LS} to the LS problem min $\| Ax - b \|_2$ where $b \in \mathbb{R}^m$.

> Compute the lower triangular portion of $C = A^T A$.
> $d = A^T b$
> Compute the Cholesky factorization $C = GG^T$.
> Solve $Gy = d$ and $G^T x_{LS} = y$.

This algorithm requires $(m + n/3)n^2$ flops. The normal equation approach is convenient because it relies on standard algorithms: Cholesky factorization, matrix-matrix multiplication, and matrix-vector multiplication. The compression of the m-by-n data matrix A into the (typically) much smaller n-by-n cross-product matrix C is attractive.

Let us consider the accuracy of the computed normal equations solution $\hat{x}_{LS}$. For clarity, assume that no roundoff errors occur during the formation of $C = A^T A$ and $d = A^T b$. (On many computers inner products are accumulated in double precision and so this is not a terribly unfair assumption.) It follows from what we know about the roundoff properties of the Cholesky factorization (cf. §4.2.7) that

$$(A^T A + E)\hat{x}_{LS} = A^T b,$$

where $\| E \|_2 \approx \mathbf{u} \| A^T \|_2 \| A \|_2 \approx \mathbf{u} \| A^T A \|_2$ and thus we can expect

$$\frac{\| \hat{x}_{LS} - x_{LS} \|_2}{\| x_{LS} \|_2} \approx \mathbf{u}\kappa_2(A^T A) = \mathbf{u}\kappa_2(A)^2 . \tag{5.3.2}$$

In other words, the accuracy of the computed normal equations solution depends on the square of the condition. This seems to be consistent with Example 5.3.1 but more refined comments follow in §5.3.9.

Example 5.3.2 It should be noted that the formation of $A^T A$ can result in a severe loss of information.

$$A = \begin{bmatrix} 1 & 1 \\ 10^{-3} & 0 \\ 0 & 10^{-3} \end{bmatrix} \text{ and } b = \begin{bmatrix} 2 \\ 10^{-3} \\ 10^{-3} \end{bmatrix}$$

then $\kappa_2(A) \approx 1.4 \cdot 10^3$, $x_{LS} = [1\ 1]^T$, and $\rho_{LS} = 0$. If the normal equations method is executed with base 10, $t = 6$ arithmetic, then a divide-by-zero occurs during the solution process, since

$$fl(A^T A) = \begin{bmatrix} 1 & 1 \\ 1 & 1 \end{bmatrix}$$

is exactly singular.   On the other hand, if 7-digit,arithmetic is used, then $\hat{x}_{LS} = [\,2.000001\,,\,0\,]^T$ and $\| \hat{x}_{LS} - x_{LS} \|_2 / \| x_{LS} \|_2 \approx \mathbf{u}\kappa_2(A)^2$.

5.3.3   LS Solution Via QR Factorization

Let $A \in \mathbb{R}^{m \times n}$ with $m \geq n$ and $b \in \mathbb{R}^m$ be given and suppose that an orthogonal matrix $Q \in \mathbb{R}^{m \times m}$ has been computed such that

$$Q^T A = R = \begin{bmatrix} R_1 \\ 0 \end{bmatrix} \begin{matrix} n \\ m - n \end{matrix} \tag{5.3.3}$$

is upper triangular. If

$$Q^T b = \begin{bmatrix} c \\ d \end{bmatrix} \begin{matrix} n \\ m - n \end{matrix}$$

then

$$\| Ax - b \|_2^2 = \| Q^T Ax - Q^T b \|_2^2 = \| R_1 x - c \|_2^2 + \| d \|_2^2$$

for any $x \in \mathbb{R}^n$. Clearly, if $\text{rank}(A) = \text{rank}(R_1) = n$, then x_{LS} is defined by the upper triangular system $R_1 x_{LS} = c$. Note that

$$\rho_{LS} = \| d \|_2.$$

We conclude that the full rank LS problem can be readily solved once we have computed the QR factorization of A. Details depend on the exact QR procedure. If Householder matrices are used and Q^T is applied in factored form to b, then we obtain

Algorithm 5.3.2 (Householder LS Solution) If $A \in \mathbb{R}^{m \times n}$ has full column rank and $b \in \mathbb{R}^m$, then the following algorithm computes a vector $x_{LS} \in \mathbb{R}^n$ such that $\| Ax_{LS} - b \|_2$ is minimum.

> Use Algorithm 5.2.1 to overwrite A with its QR factorization.
> for $j = 1{:}n$
> $\quad v(j) = 1; \ v(j+1{:}m) = A(j+1{:}m, j)$
> $\quad b(j{:}m) = (I_{m-j+1} - \beta_j vv^T)b(j{:}m)$
> end
> Solve $R(1{:}n, 1{:}n)x_{LS} = b(1{:}n)$ using back substitution.

This method for solving the full rank LS problem requires $2n^2(m - n/3)$ flops. The $O(mn)$ flops associated with the updating of b and the $O(n^2)$ flops associated with the back substitution are not significant compared to the work required to factor A.

It can be shown that the computed $\hat{x}_{LS}$ solves

$$\min\| (A + \delta A)x - (b + \delta b) \|_2 \tag{5.3.4}$$

where

$$\| \delta A \|_F \leq (6m - 3n + 41)n\mathbf{u}\| A \|_F + O(\mathbf{u}^2) \tag{5.3.5}$$

and

$$\| \delta b \|_2 \leq (6m - 3n + 40)n\mathbf{u}\| b \|_2 + O(\mathbf{u}^2). \tag{5.3.6}$$

These inequalities are established in Lawson and Hanson (1974, p.90ff) and show that $\hat{x}_{LS}$ satisfies a "nearby" LS problem. (We cannot address the relative error in $\hat{x}_{LS}$ without an LS perturbation theory, to be discussed shortly.) We mention that similar results hold if Givens QR is used.

5.3.4 Breakdown in Near-Rank Deficient Case

Like the method of normal equations, the Householder method for solving the LS problem breaks down in the back substitution phase if $\text{rank}(A) < n$. Numerically, trouble can be expected whenever $\kappa_2(A) = \kappa_2(R) \approx 1/\mathbf{u}$. This is in contrast to the normal equations approach, where completion of the Cholesky factorization becomes problematical once $\kappa_2(A)$ is in the neighborhood of $1/\sqrt{u}$. (See Example 5.3.2.) Hence the claim in Lawson and Hanson (1974, 126–127) that for a fixed machine precision, a wider class of LS problems can be solved using Householder orthogonalization.

5.3.5   A Note on the MGS Approach

In principle, MGS computes the thin QR factorization $A = Q_1 R_1$. This is enough to solve the full rank LS problem because it transforms the normal equations $(A^T A)x = A^T b$ to the upper triangular system $R_1 x = Q_1^T b$. But an analysis of this approach when $Q_1^T b$ is explicitly formed introduces a $\kappa_2(A)^2$ term. This is because the computed factor $\hat{Q}_1$ satisfies $\| \hat{Q}_1^T \hat{Q}_1 - I_n \|_2 \approx u\kappa_2(A)$ as we mentioned in §5.2.9.

However, if MGS is applied to the augmented matrix

$$A_+ = [\, A\ b\,] = [\, Q_1\ q_{n+1}\,] \begin{bmatrix} R_1 & z \\ 0 & \rho \end{bmatrix},$$

then $z = Q_1^T b$. Computing $Q_1^T b$ in this fashion and solving $R_1 x_{LS} = z$ produces an LS solution $\hat{x}_{LS}$ that is "just as good" as the Householder QR method. That is to say, a result of the form (5.3.4)-(5.3.6) applies. See Björck and Paige (1992).

It should be noted that the MGS method is slightly more expensive than Householder QR because it always manipulates m-vectors whereas the latter procedure deals with ever shorter vectors.

5.3.6   Fast Givens LS Solver

The LS problem can also be solved using fast Givens transformations. Suppose $M^T M = D$ is diagonal and

$$M^T A = \begin{bmatrix} S_1 \\ 0 \end{bmatrix} \begin{matrix} n \\ m-n \end{matrix}$$

is upper triangular. If

$$M^T b = \begin{bmatrix} c \\ d \end{bmatrix} \begin{matrix} n \\ m-n \end{matrix}$$

then

$$\| Ax - b \|_2^2 = \| D^{-1/2} M^T (Ax - b) \|_2^2 = \left\| D^{-1/2} \left(\begin{bmatrix} S_1 \\ 0 \end{bmatrix} x - \begin{bmatrix} c \\ d \end{bmatrix} \right) \right\|_2^2$$

for any $x \in \mathbb{R}^n$. Clearly, x_{LS} is obtained by solving the nonsingular upper triangular system $S_1 x = c$.

The computed solution $\hat{x}_{LS}$ obtained in this fashion can be shown to solve a nearby LS problem in the sense of (5.3.4)-(5.3.6). This may seem surprising since large numbers can arise during the calculation. An entry in the scaling matrix D can double in magnitude after a single fast Givens update. However, largeness in D must be exactly compensated for by largeness in M, since $D^{-1/2}M$ is orthogonal at all stages of the computation. It is this phenomenon that enables one to push through a favorable error analysis.

5.3.7   The Sensitivity of the LS Problem

We now develop a perturbation theory that assists in the comparison of the normal equations and QR approaches to the LS problem. The theorem below examines how the LS solution and its residual are affected by changes in A and b. In so doing, the condition of the LS problem is identified.

Two easily established facts are required in the analysis:

$$\| A \|_2 \, \| (A^T A)^{-1} A^T \|_2 \;=\; \kappa_2(A)$$

$$\| A \|_2^2 \, \| (A^T A)^{-1} \|_2 \;=\; \kappa_2(A)^2$$

(5.3.7)

These equations can be verified using the SVD.

Theorem 5.3.1 *Suppose x, r, $\hat{x}$, and $\hat{r}$ satisfy*

$$\| Ax - b \|_2 \;=\; \min \qquad r = b - Ax$$

$$\| (A + \delta A)\hat{x} - (b + \delta b) \|_2 \;=\; \min \qquad \hat{r} = (b + \delta b) - (A + \delta A)\hat{x}$$

where A and δA are in $\mathbb{R}^{m \times n}$ with $m \geq n$ and $0 \neq b$ and δb are in $\mathbb{R}^m$. If

$$\epsilon \;=\; \max \left\{ \frac{\| \delta A \|_2}{\| A \|_2}, \, \frac{\| \delta b \|_2}{\| b \|_2} \right\} \;<\; \frac{\sigma_n(A)}{\sigma_1(A)}$$

and

$$\sin(\theta) \;=\; \frac{\rho_{LS}}{\| b \|_2} \;\neq\; 1$$

where $\rho_{LS} = \| Ax_{LS} - b \|_2$, then

$$\frac{\| \hat{x} - x \|_2}{\| x \|_2} \;\leq\; \epsilon \left\{ \frac{2\kappa_2(A)}{\cos(\theta)} + \tan(\theta)\kappa_2(A)^2 \right\} + O(\epsilon^2) \qquad (5.3.8)$$

$$\frac{\| \hat{r} - r \|_2}{\| b \|_2} \;\leq\; \epsilon \, (1 + 2\kappa_2(A)) \min(1, m - n) + O(\epsilon^2). \qquad (5.3.9)$$

Proof. Let E and f be defined by $E = \delta A/\epsilon$ and $f = \delta b/\epsilon$. By hypothesis $\| \delta A \|_2 < \sigma_n(A)$ and so by Theorem 2.5.2 we have $\mathrm{rank}(A + tE) = n$ for all $t \in [0, \epsilon]$. It follows that the solution $x(t)$ to

$$(A + tE)^T (A + tE)x(t) \; = \; (A + tE)^T (b + tf) \qquad (5.3.10)$$

is continuously differentiable for all $t \in [0, \epsilon]$. Since $x = x(0)$ and $\hat{x} = x(\epsilon)$, we have

$$\hat{x} = x + \epsilon \dot{x}(0) + O(\epsilon^2).$$

The assumptions $b \neq 0$ and $\sin(\theta) \neq 1$ ensure that x is nonzero and so

$$\frac{\| \hat{x} - x \|_2}{\| x \|_2} \; = \; \epsilon \frac{\| \dot{x}(0) \|_2}{\| x \|_2} + O(\epsilon^2). \qquad (5.3.11)$$

In order to bound $\| \dot{x}(0) \|_2$, we differentiate (5.3.10) and set $t = 0$ in the result. This gives

$$E^T Ax + A^T Ex + A^T A\dot{x}(0) \; = \; A^T f + E^T b$$

i.e.,

$$\dot{x}(0) \; = \; (A^T A)^{-1} A^T (f - Ex) + (A^T A)^{-1} E^T r. \qquad (5.3.12)$$

By substituting this result into (5.3.11), taking norms, and using the easily verified inequalities $\| f \|_2 \leq \| b \|_2$ and $\| E \|_2 \leq \| A \|_2$ we obtain

$$\frac{\| \hat{x} - x \|_2}{\| x \|_2} \; \leq \; \epsilon \left\{ \| A \|_2 \| (A^T A)^{-1} A^T \|_2 \left(\frac{\| b \|_2}{\| A \|_2 \| x \|_2} + 1 \right) \right.$$

$$\left. + \frac{\rho_{LS}}{\| A \|_2 \| x \|_2} \| A \|_2^2 \| (A^T A)^{-1} \|_2 \right\} + O(\epsilon^2).$$

Since $A^T (Ax - b) = 0$, Ax is orthogonal to $Ax - b$ and so

$$\| b - Ax \|_2^2 + \| Ax \|_2^2 \; = \; \| b \|_2^2 .$$

Thus,

$$\| A \|_2^2 \| x \|_2^2 \geq \| b \|_2^2 - \rho_{LS}^2$$

and so by using (5.3.7)

$$\frac{\| \hat{x} - x \|_2}{\| x \|_2} \leq \epsilon \left\{ \kappa_2(A) \left(\frac{1}{\cos(\theta)} + 1 \right) + \kappa_2(A)^2 \frac{\sin(\theta)}{\cos(\theta)} \right\} + O(\epsilon^2)$$

thereby establishing (5.3.8).

To prove (5.3.9), we define the differentiable vector function $r(t)$ by

$$r(t) = (b + tf) - (A + tE)x(t)$$

and observe that $r = r(0)$ and $\hat{r} = r(\epsilon)$. Using (5.3.12) it can be shown that

$$\dot{r}(0) = \left(I - A(A^T A)^{-1} A^T\right) (f - Ex) - A(A^T A)^{-1} E^T r .$$

Since $\| \hat{r} - r \|_2 = \epsilon \| \dot{r}(0) \|_2 + O(\epsilon^2)$ we have

$$
\begin{aligned}
\frac{\| \hat{r} - r \|_2}{\| b \|_2} &= \epsilon \frac{\| \dot{r}(0) \|_2}{\| b \|_2} + O(\epsilon^2) \\
&\leq \epsilon \left\{ \| I - A(A^T A)^{-1} A^T \|_2 \left(1 + \frac{\| A \|_2 \| x \|_2}{\| b \|_2} \right) \right. \\
&\quad \left. + \| A(A^T A)^{-1} \|_2 \| A \|_2 \frac{\rho_{LS}}{\| b \|_2} \right\} + O(\epsilon^2) .
\end{aligned}
$$

Inequality (5.3.9) now follows because

$$\| A \|_2 \| x \|_2 = \| A \|_2 \| A^+ b \|_2 \leq \kappa_2(A) \| b \|_2 ,$$

$$\rho_{LS} = \| (I - A(A^T A)^{-1} A^T) b \|_2 \leq \| I - A(A^T A)^{-1} A^T \|_2 \| b \|_2 ,$$

and

$$\| (I - A(A^T A)^{-1} A^T \|_2 = \min(m - n, 1). \quad \square$$

An interesting feature of the upper bound in (5.3.8) is the factor

$$\tan(\theta) \kappa_2(A)^2 = \frac{\rho_{LS}}{\sqrt{\| b \|_2^2 - \rho_{LS}^2}} \kappa_2(A)^2 .$$

Thus, in nonzero residual problems it is the square of the condition that measures the sensitivity of x_{LS}. In contrast, residual sensitivity depends just linearly on $\kappa_2(A)$. These dependencies are confirmed by Example 5.3.1.

5.3.8   Normal Equations Versus QR

It is instructive to compare the normal equation and QR approaches to the LS problem. Recall the following main points from our discussion:

- The sensitivity of the LS solution is roughly proportional to the quantity $\kappa_2(A) + \rho_{LS} \kappa_2(A)^2$.

- The method of normal equations produces an $\hat{x}_{LS}$ whose relative error depends on the square of the condition.

- The QR approach (Householder, Givens, careful MGS) solves a nearby LS problem and therefore produces a solution that has a relative error approximately given by $\mathbf{u}(\kappa_2(A) + \rho_{LS} \kappa_2(A)^2)$.

Thus, we may conclude that if ρ_{LS} is small and $\kappa_2(A)$ is large, then the method of normal equations does not solve a nearby problem and will usually render an LS solution that is less accurate than a stable QR approach. Conversely, the two methods produce comparably inaccurate results when applied to large residual, ill-conditioned problems.

Finally, we mention two other factors that figure in the debate about QR versus normal equations:

- The normal equations approach involves about half of the arithmetic when $m \gg n$ and does not require as much storage.

- QR approaches are applicable to a wider class of matrices because the Cholesky process applied to $A^T A$ breaks down "before" the back substitution process on $Q^T A = R$.

At the very minimum, this discussion should convince you how difficult it can be to choose the "right" algorithm!

Problems

P5.3.1 Assume $A^T A x = A^T b$, $(A^T A + F)\hat{x} = A^T b$, and $2\| F \|_2 \le \sigma_n(A)^2$. Show that if $r = b - Ax$ and $\hat{r} = b - A\hat{x}$, then $\hat{r} - r = A(A^T A + F)^{-1}Fx$ and

$$\| \hat{r} - r \|_2 \le 2\kappa_2(A)\frac{\| F \|_2}{\| A \|_2}\| x \|_2.$$

P5.3.2 Assume that $A^T A x = A^T b$ and that $A^T A\hat{x} = A^T b + f$ where $\| f \|_2 \le cu\| A^T \|_2\| b \|_2$ and A has full column rank. Show that

$$\frac{\| x - \hat{x} \|_2}{\| x \|_2} \le cu\kappa_2(A)^2\frac{\| A^T \|_2\| b \|_2}{\| A^T b \|}.$$

P5.3.3 Let $A \in \mathbf{R}^{m \times n}$ with $m > n$ and $y \in \mathbf{R}^m$ and define $\bar{A} = [A\ y] \in \mathbf{R}^{m \times (n+1)}$. Show that $\sigma_1(\bar{A}) \ge \sigma_1(A)$ and $\sigma_{n+1}(\bar{A}) \le \sigma_n(A)$. Thus, the condition grows if a column is added to a matrix.

P5.3.4 Let $A \in \mathbf{R}^{m \times n}$ $(m \ge n)$, $w \in \mathbf{R}^n$, and define

$$B = \left[\begin{array}{c} A \\ w^T \end{array} \right].$$

Show that $\sigma_n(B) \ge \sigma_n(A)$ and $\sigma_1(B) \le \sqrt{\| A \|_2^2 + \| w \|_2^2}$. Thus, the condition of a matrix may increase or decrease if a row is added.

P5.3.5 (Cline 1973) Suppose that $A \in \mathbf{R}^{m \times n}$ has rank n and that Gaussian elimination with partial pivoting is used to compute the factorization $PA = LU$, where $L \in \mathbf{R}^{m \times n}$ is unit lower triangular, $U \in \mathbf{R}^{n \times n}$ is upper triangular, and $P \in \mathbf{R}^{m \times m}$ is a permutation. Explain how the decomposition in P5.2.5 can be used to find a vector $z \in \mathbf{R}^n$ such that $\| Lz - Pb \|_2$ is minimized. Show that if $Ux = z$, then $\| Ax - b \|_2$ is minimum. Show that this method of solving the LS problem is more efficient than Householder QR from the flop point of view whenever $m \le 5n/3$.

P5.3.6 The matrix $C = (A^T A)^{-1}$, where $\text{rank}(A) = n$, arises in many statistical applications and is known as the *variance-covariance matrix*. Assume that the factorization

$A = QR$ is available. (a) Show $C = (R^T R)^{-1}$. (b) Give an algorithm for computing the diagonal of C that requires $n^3/3$ flops. (c) Show that

$$R = \begin{bmatrix} \alpha & v^T \\ 0 & S \end{bmatrix} \quad \Rightarrow \quad C = (R^T R)^{-1} = \begin{bmatrix} (1 + v^T C_1 v)/\alpha^2 & -v^T C_1/\alpha \\ -C_1 v/\alpha & C_1 \end{bmatrix}$$

where $C_1 = (S^T S)^{-1}$. (d) Using (c), give an algorithm that overwrites the upper triangular portion of R with the upper triangular portion of C. Your algorithm should require $2n^3/3$ flops.

P5.3.7 Suppose $A \in \mathbf{R}^{n \times n}$ is symmetric and that $r = b - Ax$ where $r, b, x \in \mathbf{R}^n$ and x is nonzero. Show how to compute a symmetric $E \in \mathbf{R}^{n \times n}$ with minimal Frobenius norm so that $(A + E)x = b$. Hint. Use the QR factorization of $[\, x \, , \, r \,]$ and note that $Ex = r \Rightarrow (Q^T E Q)(Q^T x) = Q^T r$.

P5.3.8 Show how to compute the nearest circulant matrix to a given Toeplitz matrix. Measure distance with the Frobenius norm.

Notes and References for Sec. 5.3

Our restriction to least squares approximation is not a vote against minimization in other norms. There are occasions when it is advisable to minimize $\| Ax - b \|_p$ for $p = 1$ and ∞. Some algorithms for doing this are described in

A.K. Cline (1976a). "A Descent Method for the Uniform Solution to Overdetermined Systems of Equations," *SIAM J. Num. Anal. 13*, 293–309.

R.H. Bartels, A.R. Conn, and C. Charalambous (1978). "On Cline's Direct Method for Solving Overdetermined Linear Systems in the L_∞ Sense," *SIAM J. Num. Anal. 15*, 255–70.

T.F. Coleman and Y. Li (1992). "A Globally and Quadratically Convergent Affine Scaling Method for Linear L_1 Problems," *Mathematical Programming, 56, Series A*, 189–222.

Y. Li (1993). "A Globally Convergent Method for L_p Problems," *SIAM J. Optimization 3*, 609–629.

Y. Zhang (1993). "A Primal-Dual Interior Point Approach for Computing the L_1 and L_∞ Solutions of Overdetermined Linear Systems," *J. Optimization Theory and Applications 77*, 323–341.

The use of Gauss transformations to solve the LS problem has attracted some attention because they are cheaper to use than Householder or Givens matrices. See

G. Peters and J.H. Wilkinson (1970). "The Least Squares Problem and Pseudo-Inverses," *Comp. J. 13*, 309–16.

A.K. Cline (1973). "An Elimination Method for the Solution of Linear Least Squares Problems," *SIAM J. Num. Anal. 10*, 283–89.

R.J. Plemmons (1974). "Linear Least Squares by Elimination and MGS," *J. Assoc. Comp. Mach. 21*, 581–85.

Important analyses of the LS problem and various solution approaches include

G.H. Golub and J.H. Wilkinson (1966). "Note on the Iterative Refinement of Least Squares Solution," *Numer. Math. 9*, 139–48.

A. van der Sluis (1975). "Stability of the Solutions of Linear Least Squares Problem," *Numer. Math. 23*, 241–54.

Y. Saad (1986). "On the Condition Number of Some Gram Matrices Arising from Least Squares Approximation in the Complex Plane," *Numer. Math. 48*, 337–348.

Å. Björck (1987). "Stability Analysis of the Method of Seminormal Equations," *Lin. Alg. and Its Applic. 88/89*, 31–48.

J. Gluchowska and A. Smoktunowicz (1990). "Solving the Linear Least Squares Problem with Very High Relative Accuracy," *Computing 45*, 345–354.

Å. Björck (1991). "Component-wise Perturbation Analysis and Error Bounds for Linear Least Squares Solutions," *BIT 31*, 238–244.

Å. Björck and C.C. Paige (1992). "Loss and Recapture of Orthogonality in the Modified Gram-Schmidt Algorithm," *SIAM J. Matrix Anal. Appl. 13*, 176–190.

B. Waldén, R. Karlson, J. Sun (1995). "Optimal Backward Perturbation Bounds for the Linear Least Squares Problem," *Numerical Lin. Alg. with Applic. 2*, 271–286.

The "seminormal" equations are given by $R^T R x = A^T b$ where $A = QR$. In the above paper it is shown that by solving the seminormal equations an acceptable LS solution is obtained if one step of fixed precision iterative improvement is performed.

An Algol implementation of the MGS method for solving the LS problem appears in

F.L. Bauer (1965). "Elimination with Weighted Row Combinations for Solving Linear Equations and Least Squares Problems," *Numer. Math. 7*, 338–52. See also Wilkinson and Reinsch (1971, 119–33).

Least squares problems often have special structure which, of course, should be exploited.

M.G. Cox (1981). "The Least Squares Solution of Overdetermined Linear Equations having Band or Augmented Band Structure," *IMA J. Num. Anal. 1*, 3–22.

G. Cybenko (1984). "The Numerical Stability of the Lattice Algorithm for Least Squares Linear Prediction Problems," *BIT 24*, 441–455.

P.C. Hansen and H. Gesmar (1993). "Fast Orthogonal Decomposition of Rank-Deficient Toeplitz Matrices," *Numerical Algorithms 4*, 151–166.

The use of Householder matrices to solve sparse LS problems requires careful attention to avoid excessive fill-in.

J.K. Reid (1967). "A Note on the Least Squares Solution of a Band System of Linear Equations by Householder Reductions," *Comp. J. 10*, 188–89.

I.S. Duff and J.K. Reid (1976). "A Comparison of Some Methods for the Solution of Sparse Over-Determined Systems of Linear Equations," *J. Inst. Math. Applic. 17*, 267–80.

P.E. Gill and W. Murray (1976). "The Orthogonal Factorization of a Large Sparse Matrix," in *Sparse Matrix Computations*, ed. J.R. Bunch and D.J. Rose, Academic Press, New York, pp. 177-200.

L. Kaufman (1979). "Application of Dense Householder Transformations to a Sparse Matrix," *ACM Trans. Math. Soft. 5*, 442–51.

Although the computation of the QR factorization is more efficient with Householder reflections, there are some settings where the Givens approach is advantageous. For example, if A is sparse, then the careful application of Givens rotations can minimize fill-in.

I.S. Duff (1974). "Pivot Selection and Row Ordering in Givens Reduction on Sparse Matrices," *Computing 13*, 239–48.

J.A. George and M.T. Heath (1980). "Solution of Sparse Linear Least Squares Problems Using Givens Rotations," *Lin. Alg. and Its Applic. 34*, 69–83.

5.4   Other Orthogonal Factorizations

If A is rank deficient, then the QR factorization need not give a basis for $\text{ran}(A)$. This problem can be corrected by computing the QR factorization of a column-permuted version of A, i.e., $A\Pi = QR$ where Π is a permutation.

The "data" in A can be compressed further if we permit right multiplication by a general orthogonal matrix Z:

$$Q^T A Z = T.$$

There are interesting choices for Q and Z and these, together with the column pivoted QR factorization, are discussed in this section.

5.4.1   Rank Deficiency: QR with Column Pivoting

If $A \in \mathbb{R}^{m \times n}$ and $\text{rank}(A) < n$, then the QR factorization does not necessarily produce an orthonormal basis for $\text{ran}(A)$. For example, if A has three columns and

$$A = [\, a_1, \, a_2, \, a_3 \,] = [\, q_1, \, q_2, \, q_3 \,] \begin{bmatrix} 1 & 1 & 1 \\ 0 & 0 & 1 \\ 0 & 0 & 1 \end{bmatrix}$$

is its QR factorization, then $\text{rank}(A) = 2$ but $\text{ran}(A)$ does not equal any of the subspaces $\text{span}\{q_1, q_2\}$, $\text{span}\{q_1, q_3\}$, or $\text{span}\{q_2, q_3\}$.

Fortunately, the Householder QR factorization procedure (Algorithm 5.2.1) can be modified in a simple way to produce an orthonormal basis for $\text{ran}(A)$. The modified algorithm computes the factorization

$$Q^T A \Pi = \begin{bmatrix} R_{11} & R_{12} \\ 0 & 0 \end{bmatrix} \begin{matrix} r \\ m-r \end{matrix} \qquad (5.4.1)$$
$$ \begin{matrix} r & n-r \end{matrix}$$

where $r = \text{rank}(A)$, Q is orthogonal, R_{11} is upper triangular and nonsingular, and Π is a permutation. If we have the column partitionings $A\Pi = [\, a_{c_1}, \ldots, a_{c_n} \,]$ and $Q = [\, q_1, \ldots, q_m \,]$, then for $k = 1{:}n$ we have

$$a_{c_k} = \sum_{i=1}^{\min\{r,k\}} r_{ik} q_i \in \text{span}\{q_1, \ldots, q_r\}$$

implying

$$\text{ran}(A) = \text{span}\{q_1, \ldots, q_r\}.$$

The matrices Q and Π are products of Householder matrices and interchange matrices respectively. Assume for some k that we have computed

Householder matrices $H_1, \ldots, H_{k-1}$ and permutations $\Pi_1, \ldots, \Pi_{k-1}$ such that

$$(H_{k-1} \cdots H_1) A (\Pi_1 \cdots \Pi_{k-1}) = \tag{5.4.2}$$

$$R^{(k-1)} = \left[\begin{array}{cc} R_{11}^{(k-1)} & R_{12}^{(k-1)} \\ 0 & R_{22}^{(k-1)} \end{array} \right] \begin{array}{c} k-1 \\ m-k+1 \end{array}$$
$$\qquad\qquad\quad k-1 \quad n-k+1$$

where $R_{11}^{(k-1)}$ is a nonsingular and upper triangular matrix. Now suppose that

$$R_{22}^{(k-1)} = \left[z_k^{(k-1)}, \ldots, z_n^{(k-1)} \right]$$

is a column partitioning and let $p \geq k$ be the smallest index such that

$$\| z_p^{(k-1)} \|_2 = \max \left\{ \| z_k^{(k-1)} \|_2, \cdots, \| z_n^{(k-1)} \|_2 \right\} . \tag{5.4.3}$$

Note that if $k-1 = \text{rank}(A)$, then this maximum is zero and we are finished. Otherwise, let Π_k be the n-by-n identity with columns p and k interchanged and determine a Householder matrix H_k such that if $R^{(k)} = H_k R^{(k-1)} \Pi_k$, then $R^{(k)}(k+1{:}m, k) = 0$. In other words, Π_k moves the largest column in $R_{22}^{(k-1)}$ to the lead position and $\tilde{H}_k$ zeroes all of its subdiagonal components.

The column norms do not have to be recomputed at each stage if we exploit the property

$$Q^T z = \left[\begin{array}{c} \alpha \\ w \end{array} \right] \begin{array}{c} 1 \\ s-1 \end{array} \qquad \Longrightarrow \qquad \| w \|_2^2 = \| z \|_2^2 - \alpha^2,$$

which holds for any orthogonal matrix $Q \in \mathbb{R}^{s \times s}$. This reduces the overhead associated with column pivoting from $O(mn^2)$ flops to $O(mn)$ flops because we can get the new column norms by updating the old column norms, e.g.,

$$\| z^{(j)} \|_2^2 = \| z^{(j-1)} \|_2^2 - r_{kj}^2.$$

Combining all of the above we obtain the following algorithm established by Businger and Golub (1965):

Algorithm 5.4.1 (Householder QR With Column Pivoting) Given $A \in \mathbb{R}^{m \times n}$ with $m \geq n$, the following algorithm computes $r = \text{rank}(A)$ and the factorization (5.4.1) with $Q = H_1 \cdots H_r$ and $\Pi = \Pi_1 \cdots \Pi_r$. The upper triangular part of A is overwritten by the upper triangular part of R and components $j + 1{:}m$ of the jth Householder vector are stored in $A(j+1{:}m, j)$. The permutation Π is encoded in an integer vector piv. In particular, Π_j is the identity with rows j and $piv(j)$ interchanged.

for $j = 1{:}n$
$\qquad c(j) = A(1{:}m, j)^T A(1{:}m, j)$
end
$r = 0; \ \tau = \max\{c(1), \ldots, c(n)\}$
Find smallest k with $1 \le k \le n$ so $c(k) = \tau$
while $\tau > 0$
$\qquad r = r + 1$
$\qquad piv(r) = k; \ A(1{:}m, r) \leftrightarrow A(1{:}m, k); \ c(r) \leftrightarrow c(k)$
$\qquad [v, \beta] = \textbf{house}(A(r{:}m, r))$
$\qquad A(r{:}m, r{:}n) = (I_{m-r+1} - \beta v v^T) A(r{:}m, r{:}n)$
$\qquad A(r + 1{:}m, r) = v(2{:}m - r + 1)$
$\qquad$ **for** $i = r + 1{:}n$
$\qquad\qquad c(i) = c(i) - A(r, i)^2$
$\qquad$ **end**
$\qquad$ **if** $r < n$
$\qquad\qquad \tau = \max\{c(r + 1), \ldots, c(n)\}$
$\qquad\qquad$ Find smallest k with $r + 1 \le k \le n$ so $c(k) = \tau$.
$\qquad$ **else**
$\qquad\qquad \tau = 0$
$\qquad$ **end**
end

This algorithm requires $4mnr - 2r^2(m+n) + 4r^3/3$ flops where $r = \text{rank}(A)$. As with the nonpivoting procedure, Algorithm 5.2.1, the orthogonal matrix Q is stored in factored form in the subdiagonal portion of A.

Example 5.4.1 If Algorithm 5.4.1 is applied to

$$A = \begin{bmatrix} 1 & 2 & 3 \\ 1 & 5 & 6 \\ 1 & 8 & 9 \\ 1 & 11 & 12 \end{bmatrix},$$

then $\Pi = [e_3 \ e_2 \ e_1]$ and to three significant digits we obtain

$$A\Pi = QR = \begin{bmatrix} -.182 & -.816 & .514 & .191 \\ -.365 & .408 & -.827 & .129 \\ .548 & .000 & .113 & -.829 \\ -.730 & .408 & .200 & .510 \end{bmatrix} \begin{bmatrix} -16.4 & -14.600 & -1.820 \\ 0.0 & .816 & -.816 \\ 0.0 & .000 & 0.000 \end{bmatrix}.$$

5.4.2   Complete Orthogonal Decompositions

The matrix R produced by Algorithm 5.4.1 can be further reduced if it is post-multiplied by an appropriate sequence of Householder matrices. In particular, we can use Algorithm 5.2.1 to compute

$$Z_r \cdots Z_1 \begin{bmatrix} R_{11}^T \\ R_{12}^T \end{bmatrix} = \begin{bmatrix} T_{11}^T \\ 0 \end{bmatrix} \begin{matrix} r \\ n - r \end{matrix} \qquad (5.4.4)$$

where the Z_i are Householder transformations and T_{11}^T is upper triangular. It then follows that

$$
Q^T A Z = T = \begin{bmatrix} T_{11} & 0 \\ 0 & 0 \end{bmatrix} \begin{matrix} r \\ m-r \end{matrix} \quad .
\qquad (5.4.5)
$$
$$
\begin{matrix} r & n-r \end{matrix}
$$

where $Z = \Pi Z_1 \cdots Z_r$. We refer to any decomposition of this form as a *complete orthogonal decomposition*. Note that $\text{null}(A) = \text{ran}(Z(1{:}n, r+1{:}n))$. See P5.2.5 for details about the exploitation of structure in (5.4.4).

5.4.3 Bidiagonalization

Suppose $A \in \mathbb{R}^{m \times n}$ and $m \geq n$. We next show how to compute orthogonal U_B (m-by-m) and V_B (n-by-n) such that

$$
U_B^T A V_B = \begin{bmatrix} d_1 & f_1 & 0 & \cdots & 0 \\ 0 & d_2 & f_2 & & 0 \\ \vdots & \ddots & \ddots & \ddots & \vdots \\ 0 & \cdots & & d_{n-1} & f_{n-1} \\ 0 & \cdots & & 0 & d_n \\ \hline & & \mathbf{0} & & \end{bmatrix} .
\qquad (5.4.6)
$$

$U_B = U_1 \cdots U_n$ and $V_B = V_1 \cdots V_{n-2}$ can each be determined as a product of Householder matrices:

$$
\begin{bmatrix} \times & \times & \times & \times \\ \times & \times & \times & \times \\ \times & \times & \times & \times \\ \times & \times & \times & \times \\ \times & \times & \times & \times \end{bmatrix} \xrightarrow{U_1} \begin{bmatrix} \times & \times & \times & \times \\ 0 & \times & \times & \times \\ 0 & \times & \times & \times \\ 0 & \times & \times & \times \\ 0 & \times & \times & \times \end{bmatrix} \xrightarrow{V_1}
$$

$$
\begin{bmatrix} \times & \times & 0 & 0 \\ 0 & \times & \times & \times \\ 0 & \times & \times & \times \\ 0 & \times & \times & \times \\ 0 & \times & \times & \times \end{bmatrix} \xrightarrow{U_2} \begin{bmatrix} \times & \times & 0 & 0 \\ 0 & \times & \times & \times \\ 0 & 0 & \times & \times \\ 0 & 0 & \times & \times \\ 0 & 0 & \times & \times \end{bmatrix} \xrightarrow{V_2}
$$

$$
\begin{bmatrix} \times & \times & 0 & 0 \\ 0 & \times & \times & 0 \\ 0 & 0 & \times & \times \\ 0 & 0 & \times & \times \\ 0 & 0 & \times & \times \end{bmatrix} \xrightarrow{U_3} \begin{bmatrix} \times & \times & 0 & 0 \\ 0 & \times & \times & 0 \\ 0 & 0 & \times & \times \\ 0 & 0 & 0 & \times \\ 0 & 0 & 0 & \times \end{bmatrix} \xrightarrow{U_4} \begin{bmatrix} \times & \times & 0 & 0 \\ 0 & \times & \times & 0 \\ 0 & 0 & \times & \times \\ 0 & 0 & 0 & \times \\ 0 & 0 & 0 & 0 \end{bmatrix} .
$$

In general, U_k introduces zeros into the kth column, while V_k zeros the appropriate entries in row k. Overall we have:

Algorithm 5.4.2 (Householder Bidiagonalization) Given $A \in \mathbb{R}^{m \times n}$ with $m \geq n$, the following algorithm overwrites A with $U_B^T A V_B = B$ where B is upper bidiagonal and $U_B = U_1 \cdots U_n$ and $V_B = V_1 \cdots V_{n-2}$. The essential part of U_j's Householder vector is stored in $A(j+1{:}m, j)$ and the essential part of V_j's Householder vector is stored in $A(j, j+2{:}n)$.

$\quad$ **for** $j = 1{:}n$
$\qquad [v, \beta] = \textbf{house}(A(j{:}m, j))$
$\qquad A(j{:}m, j{:}n) = (I_{m-j+1} - \beta v v^T)A(j{:}m, j{:}n)$
$\qquad A(j+1{:}m, j) = v(2{:}m - j + 1)$
$\qquad$ **if** $j \leq n - 2$
$\qquad\quad [v, \beta] = \textbf{house}(A(j, j+1{:}n)^T)$
$\qquad\quad A(j{:}m, j+1{:}n) = A(j{:}m, j+1{:}n)(I_{n-j} - \beta v v^T)$
$\qquad\quad A(j, j+2{:}n) = v(2{:}n - j)^T$
$\qquad$ **end**
$\quad$ **end**

This algorithm requires $4mn^2 - 4n^3/3$ flops. Such a technique is used in Golub and Kahan (1965), where bidiagonalization is first described. If the matrices U_B and V_B are explicitly desired, then they can be accumulated in $4m^2 n - 4n^3/3$ and $4n^3/3$ flops, respectively. The bidiagonalization of A is related to the tridiagonalization of $A^T A$. See §8.2.1.

Example 5.4.2 If Algorithm 5.4.2 is applied to

$$
A = \begin{bmatrix} 1 & 2 & 3 \\ 4 & 5 & 6 \\ 7 & 8 & 9 \\ 10 & 11 & 12 \end{bmatrix},
$$

then to three significant digits we obtain

$$
\hat{B} = \begin{bmatrix} 12.8 & 21.8 & 0 \\ 0 & 2.24 & -.613 \\ 0 & 0 & 0 \\ 0 & 0 & 0 \end{bmatrix} \qquad \hat{V}_B = \begin{bmatrix} 1.00 & 0.00 & 0.00 \\ 0.00 & -.667 & -.745 \\ 0.00 & -.745 & .667 \end{bmatrix}
$$

$$
\hat{U}_B = \begin{bmatrix} -.0776 & -.833 & .392 & -.383 \\ -.3110 & -.451 & -.238 & .802 \\ -.5430 & -.069 & .701 & -.457 \\ -.7760 & .312 & .547 & .037 \end{bmatrix}.
$$

5.4.4   R-Bidiagonalization

A faster method of bidiagonalizing when $m \gg n$ results if we upper triangularize A first before applying Algorithm 5.4.2. In particular, suppose we compute an orthogonal $Q \in \mathbb{R}^{m \times m}$ such that

$$Q^T A = \begin{bmatrix} R_1 \\ 0 \end{bmatrix}$$

is upper triangular. We then bidiagonalize the square matrix R_1,

$$U_R^T R_1 V_B = B_1 .$$

Here U_R and V_B are n-by-n orthogonal and B_1 is n-by-n upper bidiagonal. If $U_B = Q \operatorname{diag}(U_R, I_{m-n})$ then

$$U^T A V = \begin{bmatrix} B_1 \\ 0 \end{bmatrix} \equiv B$$

is a bidiagonalization of A.

The idea of computing the bidiagonalization in this manner is mentioned in Lawson and Hanson (1974, p.119) and more fully analyzed in Chan (1982a). We refer to this method as R-bidiagonalization. By comparing its flop count $(2mn^2 + 2n^3)$ with that for Algorithm 5.4.2 $(4mn^2 - 4n^3/3)$ we see that it involves fewer computations (approximately) whenever $m \geq 5n/3$.

5.4.5   The SVD and its Computation

Once the bidiagonalization of A has been achieved, the next step in the Golub-Reinsch SVD algorithm is to zero the superdiagonal elements in B. This is an iterative process and is accomplished by an algorithm due to Golub and Kahan (1965). Unfortunately, we must defer our discussion of this iteration until §8.6 as it requires an understanding of the symmetric eigenvalue problem. Suffice it to say here that it computes orthogonal matrices U_Σ and V_Σ such that

$$U_\Sigma^T B V_\Sigma = \Sigma = \operatorname{diag}(\sigma_1, \ldots, \sigma_n) \in \mathbb{R}^{m \times n} .$$

By defining $U = U_B U_\Sigma$ and $V = V_B V_\Sigma$ we see that $U^T A V = \Sigma$ is the SVD of A. The flop counts associated with this portion of the algorithm depend upon "how much" of the SVD is required. For example, when solving the LS problem, U^T need never be explicitly formed but merely applied to b as it is developed. In other applications, only the matrix $U_1 = U(:, 1:n)$ is required. Altogether there are six possibilities and the total amount of work required by the SVD algorithm in each case is summarized in the table below. Because of the two possible bidiagonalization schemes, there are two columns of flop counts. If the bidiagonalization is achieved via Algorithm 5.4.2, the Golub-Reinsch (1970) SVD algorithm results, while if R-bidiagonalization is invoked we obtain the R-SVD algorithm detailed in Chan (1982a). By comparing the entries in this table (which are meant only as approximate estimates of work), we conclude that the R-SVD approach is more efficient unless $m \approx n$.

Required	Golub-Reinsch SVD	R-SVD
Σ	$4mn^2 - 4n^3/3$	$2mn^2 + 2n^3$
Σ, V	$4mn^2 + 8n^3$	$2mn^2 + 11n^3$
Σ, U	$4m^2n - 8mn^2$	$4m^2n + 13n^3$
Σ, U_1	$14mn^2 - 2n^3$	$6mn^2 + 11n^3$
Σ, U, V	$4m^2n + 8mn^2 + 9n^3$	$4m^2n + 22n^3$
Σ, U_1, V	$14mn^2 + 8n^3$	$6mn^2 + 20n^3$

Problems

P5.4.1 Suppose $A \in \mathbf{R}^{m \times n}$ with $m < n$. Give an algorithm for computing the factorization

$$U^T A V = [\, B \; O \,]$$

where B is an m-by-m upper bidiagonal matrix. (Hint: Obtain the form

$$\begin{bmatrix} \times & \times & 0 & 0 & 0 & 0 \\ 0 & \times & \times & 0 & 0 & 0 \\ 0 & 0 & \times & \times & 0 & 0 \\ 0 & 0 & 0 & \times & \times & 0 \end{bmatrix}.$$

using Householder matrices and then "chase" the $(m, m+1)$ entry up the $(m+1)$st column by applying Givens rotations from the right.)

P5.4.2 Show how to efficiently bidiagonalize an n-by-n upper triangular matrix using Givens rotations.

P5.4.3 Show how to upper bidiagonalize a tridiagonal matrix $T \in \mathbf{R}^{n \times n}$ using Givens rotations.

P5.4.4 Let $A \in \mathbf{R}^{m \times n}$ and assume that $0 \neq v$ satisfies $\| Av \|_2 = \sigma_n(A) \| v \|_2$ Let Π be a permutation such that if $\Pi^T v = w$, then $|w_n| = \| w \|_\infty$. Show that if $A\Pi = QR$ is the QR factorization of $A\Pi$, then $|r_{nn}| \le \sqrt{n}\sigma_n(A)$. Thus, there always exists a permutation Π such that the QR factorization of $A\Pi$ "displays" near rank deficiency.

P5.4.5 Let $x, y \in \mathbf{R}^m$ and $Q \in \mathbf{R}^{m \times m}$ be given with Q orthogonal. Show that if

$$Q^T x \;=\; \begin{bmatrix} \alpha \\ u \end{bmatrix} \begin{matrix} 1 \\ m-1 \end{matrix} \qquad Q^T y \;=\; \begin{bmatrix} \beta \\ v \end{bmatrix} \begin{matrix} 1 \\ m-1 \end{matrix}$$

then $u^T v = x^T y - \alpha\beta$.

P5.4.6 Let $A = [\, a_1, \ldots, a_n \,] \in \mathbf{R}^{m \times n}$ and $b \in \mathbf{R}^m$ be given. For any subset of A's columns $\{a_{c_1}, \ldots, a_{c_k}\}$ define

$$\mathrm{res}\,[\, a_{c_1}, \ldots, a_{c_k} \,] \;=\; \min_{x \,\in\, \mathbf{R}^k} \; \| \, [\, a_{c_1}, \ldots, a_{c_k} \,] x - b \, \|_2$$

Describe an alternative pivot selection procedure for Algorithm 5.4.1 such that if $QR = A\Pi = [\, a_{c_1}, \ldots, a_{c_n} \,]$ in the final factorization, then for $k = 1{:}n$:

$$\mathrm{res}\,[\, a_{c_1}, \ldots, a_{c_k} \,] \;=\; \min_{i \ge k} \; \mathrm{res}[a_{c_1}, \ldots, a_{c_{k-1}}, a_{c_i}]$$

Notes and References for Sec. 5.4

Aspects of the complete orthogonal decomposition are discussed in

R.J. Hanson and C.L. Lawson (1969). "Extensions and Applications of the Householder Algorithm for Solving Linear Least Square Problems," *Math. Comp. 23*, 787–812.

P.A. Wedin (1973). "On the Almost Rank-Deficient Case of the Least Squares Problem," *BIT 13*, 344–54.

G.H. Golub and V. Pereyra (1976). "Differentiation of Pseudo-Inverses, Separable Nonlinear Least Squares Problems and Other Tales," in *Generalized Inverses and Applications* , ed. M.Z. Nashed, Academic Press, New York, pp. 303–24.

The computation of the SVD is detailed in §8.6. But here are some of the standard references concerned with its calculation:

G.H. Golub and W. Kahan (1965). "Calculating the Singular Values and Pseudo-Inverse of a Matrix," *SIAM J. Num. Anal. 2*, 205–24.

P.A. Businger and G.H. Golub (1969). "Algorithm 358: Singular Value Decomposition of the Complex Matrix," *Comm. ACM 12*, 564–65.

G.H. Golub and C. Reinsch (1970). "Singular Value Decomposition and Least Squares Solutions," *Numer. Math. 14*, 403–20. See also Wilkinson and Reinsch(1971, pp. 1334–51).

T.F. Chan (1982). "An Improved Algorithm for Computing the Singular Value Decomposition," *ACM Trans. Math. Soft. 8*, 72–83.

QR with column pivoting was first discussed in

P.A. Businger and G.H. Golub (1965). "Linear Least Squares Solutions by Householder Transformations," *Numer. Math. 7*, 269–76. See also Wilkinson and Reinsch (1971, pp. 11–18).

Knowing when to stop in the algorithm is difficult. In questions of rank deficiency, it is helpful to obtain information about the smallest singular value of the upper triangular matrix R. This can be done using the techniques of §3.5.4 or those that are discussed in

I. Karasalo (1974). "A Criterion for Truncation of the QR Decomposition Algorithm for the Singular Linear Least Squares Problem," *BIT 14*, 156–66.

N. Anderson and I. Karasalo (1975). "On Computing Bounds for the Least Singular Value of a Triangular Matrix," *BIT 15*, 1–4.

Other aspects of rank estimation with QR are discussed in

L.V. Foster (1986). "Rank and Null Space Calculations Using Matrix Decomposition without Column Interchanges," *Lin. Alg. and Its Applic. 74*, 47–71.

T.F. Chan (1987). "Rank Revealing QR Factorizations," *Lin. Alg. and Its Applic. 88/89*, 67–82.

T.F. Chan and P. Hansen (1992). "Some Applications of the Rank Revealing QR Factorization," *SIAM J. Sci. and Stat. Comp. 13*, 727–741.

J.L. Barlow and U.B. Vemulapati (1992). "Rank Detection Methods for Sparse Matrices," *SIAM J. Matrix. Anal. Appl. 13*, 1279–1297.

T-M. Hwang, W-W. Lin, and E.K. Yang (1992). "Rank-Revealing LU Factorizations," *Lin. Alg. and Its Applic. 175*, 115–141.

C.H. Bischof and P.C. Hansen (1992). "A Block Algorithm for Computing Rank-Revealing QR Factorizations," *Numerical Algorithms 2*, 371-392.

S. Chandrasekaren and I.C.F. Ipsen (1994). "On Rank-Revealing Factorizations," *SIAM J. Matrix Anal. Appl. 15*, 592–622.

R.D. Fierro and P.C. Hansen (1995). "Accuracy of TSVD Solutions Computed from Rank-Revealing Decompositions," *Numer. Math. 70*, 453–472.

5.5   The Rank Deficient LS Problem

If A is rank deficient, then there are an infinite number of solutions to the LS problem and we must resort to special techniques. These techniques must address the difficult problem of numerical rank determination.

After some SVD preliminaries, we show how QR with column pivoting can be used to determine a minimizer x_B with the property that Ax_B is a linear combination of $r = \text{rank}(A)$ columns. We then discuss the minimum 2-norm solution that can be obtained from the SVD.

5.5.1   The Minimum Norm Solution

Suppose $A \in \mathbb{R}^{m \times n}$ and $\text{rank}(A) = r < n$. The rank deficient LS problem has an infinite number of solutions, for if x is a minimizer and $z \in \text{null}(A)$ then $x + z$ is also a minimizer. The set of all minimizers

$$\mathcal{X} = \{x \in \mathbb{R}^n : \| Ax - b \|_2 = \min \}$$

is convex, for if $x_1, x_2 \in \mathcal{X}$ and $\lambda \in [0, 1]$, then

$$\begin{aligned} \| A(\lambda x_1 + (1 - \lambda)x_2) - b \|_2 &\leq \lambda \| Ax_1 - b \|_2 + (1 - \lambda) \| Ax_2 - b \|_2 \\ &= \min \| Ax - b \|_2. \end{aligned}$$

Thus, $\lambda x_1 + (1 - \lambda)x_2 \in \mathcal{X}$. It follows that $\mathcal{X}$ has a unique element having minimum 2-norm and we denote this solution by x_{LS}. (Note that in the full rank case, there is only one LS solution and so it must have minimal 2-norm. Thus, we are consistent with the notation in §5.3.)

5.5.2   Complete Orthogonal Factorization and x_{LS}

Any complete orthogonal factorization can be used to compute x_{LS}. In particular, if Q and Z are orthogonal matrices such that

$$Q^T A Z = T = \begin{bmatrix} T_{11} & 0 \\ 0 & 0 \end{bmatrix} \begin{matrix} r \\ m - r \end{matrix} \qquad r = \text{rank}(A)$$
$$ \begin{matrix} r & n - r \end{matrix}$$

then

$$\| Ax - b \|_2^2 = \| (Q^T A Z)Z^T x - Q^T b \|_2^2 = \| T_{11}w - c \|_2^2 + \| d \|_2^2$$

where

$$Z^T x = \begin{bmatrix} w \\ y \end{bmatrix} \begin{matrix} r \\ n - r \end{matrix} \qquad Q^T b = \begin{bmatrix} c \\ d \end{bmatrix} \begin{matrix} r \\ m - r \end{matrix} .$$

Clearly, if x is to minimize the sum of squares, then we must have $w = T_{11}^{-1}c$. For x to have minimal 2-norm, y must be zero, and thus,

$$x_{LS} = Z \begin{bmatrix} T_{11}^{-1}c \\ 0 \end{bmatrix} .$$

5.5.3   The SVD and the LS Problem

Of course, the SVD is a particularly revealing complete orthogonal decomposition. It provides a neat expression for x_{LS} and the norm of the minimum residual $\rho_{LS} = \| Ax_{LS} - b \|_2$.

Theorem 5.5.1 *Suppose $U^T A V = \Sigma$ is the SVD of $A \in \mathbb{R}^{m \times n}$ with $r = \text{rank}(A)$. If $U = [\, u_1, \ldots, u_m \,]$ and $V = [\, v_1, \ldots, v_n \,]$ are column partitionings and $b \in \mathbb{R}^m$, then*

$$x_{LS} = \sum_{i=1}^{r} \frac{u_i^T b}{\sigma_i} v_i \qquad (5.5.1)$$

minimizes $\| Ax - b \|_2$ and has the smallest 2-norm of all minimizers. Moreover

$$\rho_{LS}^2 = \| Ax_{LS} - b \|_2^2 = \sum_{i=r+1}^{m} (u_i^T b)^2. \qquad (5.5.2)$$

Proof. For any $x \in \mathbb{R}^n$ we have:

$$
\begin{aligned}
\| Ax - b \|_2^2 &= \| (U^T A V)(V^T x) - U^T b \|_2^2 = \| \Sigma \alpha - U^T b \|_2^2 \\
&= \sum_{i=1}^{r} (\sigma_i \alpha_i - u_i^T b)^2 + \sum_{i=r+1}^{m} (u_i^T b)^2
\end{aligned}
$$

where $\alpha = V^T x$. Clearly, if x solves the LS problem, then $\alpha_i = (u_i^T b / \sigma_i)$ for $i = 1{:}r$. If we set $\alpha(r + 1{:}n) = 0$, then the resulting x clearly has minimal 2-norm. $\square$

5.5.4   The Pseudo-Inverse

Note that if we define the matrix $A^+ \in \mathbb{R}^{n \times m}$ by $A^+ = V \Sigma^+ U^T$ where

$$\Sigma^+ = \text{diag}\left(\frac{1}{\sigma_1}, \ldots, \frac{1}{\sigma_r}, 0, \ldots, 0 \right) \in \mathbb{R}^{n \times m} \qquad r = \text{rank}(A)$$

then $x_{LS} = A^+ b$ and $\rho_{LS} = \| (I - AA^+)b \|_2$. A^+ is referred to as the *pseudo-inverse* of A. It is the unique minimal Frobenius norm solution to the problem

$$\min_{X \,\in\, \mathbb{R}^{n \times m}} \| AX - I_m \|_F . \qquad (5.5.3)$$

If $\text{rank}(A) = n$, then $A^+ = (A^T A)^{-1} A^T$, while if $m = n = \text{rank}(A)$, then $A^+ = A^{-1}$. Typically, A^+ is defined to be the unique matrix $X \in \mathbb{R}^{n \times m}$ that satisfies the four *Moore-Penrose conditions:*

$$
\begin{array}{llll}
\text{(i)} & AXA = A & \text{(iii)} & (AX)^T = AX \\
\text{(ii)} & XAX = X & \text{(iv)} & (XA)^T = XA.
\end{array}
$$

These conditions amount to the requirement that AA^+ and A^+A be orthogonal projections onto $\text{ran}(A)$ and $\text{ran}(A^T)$, respectively. Indeed, $AA^+ = U_1U_1^T$ where $U_1 = U(1{:}m, 1{:}r)$ and $A^+A = V_1V_1^T$ where $V_1 = V(1{:}n, 1{:}r)$.

5.5.5   Some Sensitivity Issues

In §5.3 we examined the sensitivity of the full rank LS problem. The behavior of x_{LS} in this situation is summarized in Theorem 5.3.1. If we drop the full rank assumptions then x_{LS} is not even a continuous function of the data and small changes in A and b can induce arbitrarily large changes in $x_{LS} = A^+b$. The easiest way to see this is to consider the behavior of the pseudo inverse. If A and δA are in $\mathbb{R}^{m \times n}$, then Wedin (1973) and Stewart (1975) show that

$$\| (A + \delta A)^+ - A^+ \|_F \leq 2\| \delta A \|_F \max \left\{ \| A^+ \|_2^2 \, , \, \| (A + \delta A)^+ \|_2^2 \right\}.$$

This inequality is a generalization of Theorem 2.3.4 in which perturbations in the matrix inverse are bounded. However, unlike the square nonsingular case, the upper bound does not necessarily tend to zero as δA tends to zero. If

$$A = \begin{bmatrix} 1 & 0 \\ 0 & 0 \\ 0 & 0 \end{bmatrix} \quad \text{and} \quad \delta A = \begin{bmatrix} 0 & 0 \\ 0 & \epsilon \\ 0 & 0 \end{bmatrix}$$

then

$$A^+ = \begin{bmatrix} 1 & 0 & 0 \\ 0 & 0 & 0 \end{bmatrix} \quad \text{and} \quad (A + \delta A)^+ = \begin{bmatrix} 1 & 0 & 0 \\ 1 & 1/\epsilon & 0 \end{bmatrix}$$

and $\| A^+ - (A + \delta A)^+ \|_2 = 1/\epsilon$. The numerical determination of an LS minimizer in the presence of such discontinuities is a major challenge.

5.5.6   QR with Column Pivoting and Basic Solutions

Suppose $A \in \mathbb{R}^{m \times n}$ has rank r. QR with column pivoting (Algorithm 5.4.1) produces the factorization $A\Pi = QR$ where

$$R = \begin{bmatrix} R_{11} & R_{12} \\ 0 & 0 \end{bmatrix} \begin{matrix} r \\ m-r \end{matrix} \quad .$$
$$ \begin{matrix} r & n-r \end{matrix}$$

Given this reduction, the LS problem can be readily solved. Indeed, for any $x \in \mathbb{R}^n$ we have

$$
\begin{aligned}
\| Ax - b \|_2^2 &= \| (Q^T A\Pi)(\Pi^T x) - (Q^T b) \|_2^2 \\
&= \| R_{11}y - (c - R_{12}z) \|_2^2 + \| d \|_2^2 ,
\end{aligned}
$$

where

$$\Pi^T x = \begin{bmatrix} y \\ z \end{bmatrix} \begin{matrix} r \\ n-r \end{matrix} \quad \text{and} \quad Q^T b = \begin{bmatrix} c \\ d \end{bmatrix} \begin{matrix} r \\ m-r \end{matrix} .$$

Thus, if x is an LS minimizer, then we must have

$$x = \Pi \begin{bmatrix} R_{11}^{-1}(c - R_{12}z) \\ z \end{bmatrix} .$$

If z is set to zero in this expression, then we obtain the *basic solution*

$$x_B = \Pi \begin{bmatrix} R_{11}^{-1}c \\ 0 \end{bmatrix} .$$

Notice that x_B has at most r nonzero components and so Ax_B involves a subset of A's columns.

The basic solution is not the minimal 2-norm solution unless the submatrix R_{12} is zero since

$$\| x_{LS} \|_2 = \min_{z \in \mathbb{R}^{n-r}} \left\| x_B - \Pi \begin{bmatrix} R_{11}^{-1}R_{12} \\ -I_{n-r} \end{bmatrix} z \right\|_2 . \tag{5.5.4}$$

Indeed, this characterization of $\| x_{LS} \|_2$ can be used to show

$$1 \le \frac{\| x_B \|_2}{\| x_{LS} \|_2} \le \sqrt{1 + \| R_{11}^{-1}R_{12} \|_2^2} . \tag{5.5.5}$$

See Golub and Pereyra (1976) for details.

5.5.7 Numerical Rank Determination with $A\Pi = QR$

If Algorithm 5.4.1 is used to compute x_B, then care must be exercised in the determination of rank(A). In order to appreciate the difficulty of this, suppose

$$fl(H_k \cdots H_1 A \Pi_1 \cdots \Pi_k) = \hat{R}^{(k)} = \begin{bmatrix} \hat{R}_{11}^{(k)} & \hat{R}_{12}^{(k)} \\ 0 & \hat{R}_{22}^{(k)} \end{bmatrix} \begin{matrix} k \\ m-k \end{matrix}$$
$$\quad\quad\quad\quad\quad\quad\quad\quad\quad\quad\quad k \quad\quad n-k$$

is the matrix computed after k steps of the algorithm have been executed in floating point. Suppose rank(A) = k. Because of roundoff error, $\hat{R}_{22}^{(k)}$ will not be exactly zero. However, if $\hat{R}_{22}^{(k)}$ is suitably small in norm then it is reasonable to terminate the reduction and declare A to have rank k. A typical termination criteria might be

$$\| \hat{R}_{22}^{(k)} \|_2 \le \epsilon_1 \| A \|_2 \tag{5.5.6}$$

for some small machine-dependent parameter ϵ_1. In view of the roundoff properties associated with Householder matrix computation (cf. §5.1.12), we know that $\hat{R}^{(k)}$ is the exact R factor of a matrix $A + E_k$, where

$$\| E_k \|_2 \leq \epsilon_2 \| A \|_2 \qquad \epsilon_2 = O(\mathbf{u}) \,.$$

Using Theorem 2.5.2 we have

$$\sigma_{k+1}(A + E_k) = \sigma_{k+1}(\hat{R}^{(k)}) \leq \| \hat{R}_{22}^{(k)} \|_2 \,.$$

Since $\sigma_{k+1}(A) \leq \sigma_{k+1}(A + E_k) + \| E_k \|_2$, it follows that

$$\sigma_{k+1}(A) \leq (\epsilon_1 + \epsilon_2) \| A \|_2.$$

In other words, a relative perturbation of $O(\epsilon_1 + \epsilon_2)$ in A can yield a rank-k matrix. With this termination criterion, we conclude that QR with column pivoting "discovers" rank degeneracy if in the course of the reduction $\hat{R}_{22}^{(k)}$ is small for some $k < n$.

Unfortunately, this is not always the case. A matrix can be nearly rank deficient without a single $\hat{R}_{22}^{(k)}$ being particularly small. Thus, QR with column pivoting *by itself* is not entirely reliable as a method for detecting near rank deficiency. However, if a good condition estimator is applied to R it is practically impossible for near rank deficiency to go unnoticed.

Example 5.5.1 Let $T_n(c)$ be the matrix

$$T_n(c) \;=\; \text{diag}(1, s, \ldots, s^{n-1}) \begin{bmatrix} 1 & -c & -c & \cdots & -c \\ 0 & 1 & -c & \cdots & -c \\ & & \ddots & & \vdots & \vdots \\ \vdots & & & 1 & -c \\ 0 & & \cdots & & 1 \end{bmatrix}$$

with $c^2 + s^2 = 1$ with $c, s > 0$ (See Lawson and Hanson (1974, p.31).) These matrices are unaltered by Algorithm 5.4.1 and thus $\| R_{22}^{(k)} \|_2 \geq s^{n-1}$ for $k = 1{:}n-1$. This inequality implies (for example) that the matrix $T_{100}(.2)$ has no particularly small trailing principal submatrix since $s^{99} \approx .13$. However, it can be shown that $\sigma_n = O(10^{-8})$.

5.5.8   Numerical Rank and the SVD

We now focus our attention on the ability of the SVD to handle rank-deficiency in the presence of roundoff. Recall that if $A = U\Sigma V^T$ is the SVD of A, then

$$x_{LS} = \sum_{i=1}^{r} \frac{u_i^T b}{\sigma_i} v_i \tag{5.5.7}$$

where $r = \text{rank}(A)$. Denote the computed versions of U, V, and $\Sigma = \text{diag}(\sigma_i)$ by $\hat{U}$, $\hat{V}$, and $\hat{\Sigma} = \text{diag}(\hat{\sigma}_i)$. Assume that both sequences of singular values range from largest to smallest. For a reasonably implemented SVD algorithm it can be shown that

$$\hat{U} = W + \Delta U \qquad W^T W = I_m \qquad \| \Delta U \|_2 \le \epsilon \qquad (5.5.8)$$

$$\hat{V} = Z + \Delta V \qquad Z^T Z = I_n \qquad \| \Delta V \|_2 \le \epsilon \qquad (5.5.9)$$

$$\hat{\Sigma} = W^T (A + \Delta A) Z \qquad \| \Delta A \|_2 \le \epsilon \| A \|_2 \qquad (5.5.10)$$

where ϵ is a small multiple of u, the machine precision. In plain English, the SVD algorithm computes the singular values of a "nearby" matrix $A + \Delta A$.

Note that $\hat{U}$ and $\hat{V}$ are not necessarily close to their exact counterparts. However, we can show that $\hat{\sigma}_k$ is close to σ_k. Using (5.5.10) and Theorem 2.5.2 we have

$$\sigma_k \quad = \quad \min_{\text{rank}(B)=k-1} \| A - B \|_2$$

$$= \quad \min_{\text{rank}(B)=k-1} \| (\hat{\Sigma} - B) - W^T (\Delta A) Z \|_2 .$$

Since $\| W^T (\Delta A) Z \|_2 \le \epsilon \| A \|_2 = \epsilon \sigma_1$ and

$$\min_{\text{rank}(B)=k-1} \| \hat{\Sigma}_k - B \|_2 = \hat{\sigma}_k$$

it follows that $|\sigma_k - \hat{\sigma}_k| \le \epsilon \sigma_1$ for $k = 1{:}n$. Thus, if A has rank r then we can expect $n - r$ of the computed singular values to be small. Near rank deficiency in A cannot escape detection when the SVD of A is computed.

Example 5.5.2 For the matrix $T_{100}(.2)$ in Example 5.5.1, $\sigma_n \approx .367 \cdot 10^{-8}$.

One approach to estimating $r = \text{rank}(A)$ from the computed singular values is to have a tolerance $\delta > 0$ and a convention that A has "numerical rank" $\hat{r}$ if the $\hat{\sigma}_i$ satisfy

$$\hat{\sigma}_1 \ge \cdots \ge \hat{\sigma}_r > \delta \ge \hat{\sigma}_{r+1} \ge \cdots \ge \hat{\sigma}_n .$$

The tolerance δ should be consistent with the machine precision, e.g. $\delta = u \| A \|_\infty$. However, if the general level of relative error in the data is larger than u, then δ should be correspondingly bigger, e.g., $\delta = 10^{-2} \| A \|_\infty$ if the entries in A are correct to two digits.

If $\hat{r}$ is accepted as the numerical rank then we can regard

$$x_{\hat{r}} = \sum_{i=1}^{\hat{r}} \frac{\hat{u}_i^T b}{\hat{\sigma}_i} \hat{v}_i$$

as an approximation to x_{LS}. Since $\| x_{\hat{r}} \|_2 \approx 1/\sigma_{\hat{r}} \leq 1/\delta$ then δ may also be chosen with the intention of producing an approximate LS solution with suitably small norm. In §12.1, we discuss more sophisticated methods for doing this.

If $\hat{\sigma}_{\hat{r}} \gg \delta$, then we have reason to be comfortable with $x_{\hat{r}}$ because A can then be unambiguously regarded as a rank($A_{\hat{r}}$) matrix (modulo δ).

On the other hand, $\{\hat{\sigma}_1, \ldots, \hat{\sigma}_n\}$ might not clearly split into subsets of small and large singular values, making the determination of $\hat{r}$ by this means somewhat arbitrary. This leads to more complicated methods for estimating rank which we now discuss in the context of the LS problem.

For example, suppose $r = n$, and assume for the moment that $\Delta A = 0$ in (5.5.10). Thus $\sigma_i = \hat{\sigma}_i$ for $i = 1{:}n$. Denote the ith columns of the matrices $\hat{U}$, $\hat{W}$, $\hat{V}$, and Z by u_i, w_i, v_i, and z_i, respectively. Subtracting $x_{\hat{r}}$ from x_{LS} and taking norms we obtain

$$\| x_{\hat{r}} - x_{LS} \|_2 \leq \sum_{i=1}^{\hat{r}} \frac{\| (w_i^T b)z_i - (u_i^T b)v_i \|_2}{\sigma_i} + \sqrt{\sum_{i=\hat{r}+1}^{n} \left(\frac{w_i^T b}{\sigma_i} \right)^2}.$$

From (5.5.8) and (5.5.9) it is easy to verify that

$$\| (w_i^T b)z_i - (u_i^T b)v_i \|_2 \leq 2(1 + \epsilon)\epsilon \| b \|_2 \qquad (5.5.11)$$

and therefore

$$\| x_{\hat{r}} - x_{LS} \|_2 \leq \frac{\hat{r}}{\sigma_{\hat{r}}} 2(1 + \epsilon)\epsilon \| b \|_2 + \sqrt{\sum_{i=\hat{r}+1}^{n} \left(\frac{w_i^T b}{\sigma_i} \right)^2}.$$

The parameter $\hat{r}$ can be determined as that integer which minimizes the upper bound. Notice that the first term in the bound increases with $\hat{r}$, while the second decreases.

On occasions when minimizing the residual is more important than accuracy in the solution, we can determine $\hat{r}$ on the basis of how close we surmise $\| b - Ax_{\hat{r}} \|_2$ is to the true minimum. Paralleling the above analysis, it can be shown that

$$\| b - Ax_{\hat{r}} \|_2 - \| b - Ax_{LS} \|_2 \leq (n - \hat{r})\| b \|_2 + \epsilon \| b \|_2 \left(\hat{r} + \frac{\hat{\sigma}_1}{\hat{\sigma}_{\hat{r}}}(1 + \epsilon) \right).$$

Again $\hat{r}$ could be chosen to minimize the upper bound. See Varah (1973) for practical details and also the LAPACK manual.

5.5.9   Some Comparisons

As we mentioned, when solving the LS problem via the SVD, only Σ and V have to be computed. The following table compares the efficiency of this approach with the other algorithms that we have presented.

LS Algorithm	Flop Count
Normal Equations	$mn^2 + n^3/3$
Householder Orthogonalization	$2mn^2 - 2n^3/3$
Modified Gram Schmidt	$2mn^2$
Givens Orthogonalization	$3mn^2 - n^3$
Householder Bidiagonalization	$4mn^2 - 4n^3/2$
R-Bidiagonalization	$2mn^2 + 2n^3$
Golub-Reinsch SVD	$4mn^2 + 8n^3$
R-SVD	$2mn^2 + 11n^3$

Problems

P5.5.1 Show that if
$$A = \begin{bmatrix} T & S \\ 0 & 0 \end{bmatrix} \begin{matrix} r \\ m-r \end{matrix}$$
$$\begin{matrix} r & n-r \end{matrix}$$
where $r = \text{rank}(A)$ and T is nonsingular, then
$$X = \begin{bmatrix} T^{-1} & 0 \\ 0 & 0 \end{bmatrix} \begin{matrix} r \\ n-r \end{matrix}$$
$$\begin{matrix} r & m-r \end{matrix}$$
satisfies $AXA = A$ and $(AX)^T = (AX)$. In this case, we say that X is a (1,3) *pseudo-inverse* of A. Show that for general A, $x_B = Xb$ where X is a (1,3) pseudo-inverse of A.

P5.5.2 Define $B(\lambda) \in \mathbf{R}^{n \times m}$ by $B(\lambda) = (A^T A + \lambda I)^{-1} A^T$, where $\lambda > 0$. Show
$$\| B(\lambda) - A^+ \|_2 = \frac{\lambda}{\sigma_r(A)[\sigma_r(A)^2 + \lambda]} \qquad r = \text{rank}(A)$$
and therefore that $B(\lambda) \to A^+$ as $\lambda \to 0$.

P5.5.3 Consider the rank deficient LS problem
$$\min_{\substack{y \in \mathbf{R}^r \\ z \in \mathbf{R}^{n-r}}} \left\| \begin{bmatrix} R & S \\ 0 & 0 \end{bmatrix} \begin{bmatrix} y \\ z \end{bmatrix} - \begin{bmatrix} c \\ d \end{bmatrix} \right\|_2$$
where $R \in \mathbf{R}^{r \times r}$, $S \in \mathbf{R}^{r \times n-r}$, $y \in \mathbf{R}^r$, and $z \in \mathbf{R}^{n-r}$. Assume that R is upper triangular and nonsingular. Show how to obtain the minimum norm solution to this problem by computing an appropriate QR factorization without pivoting and then solving for the appropriate y and z.

P5.5.4 Show that if $A_k \to A$ and $A_k^+ \to A^+$, then there exists an integer k_0 such that $\text{rank}(A_k)$ is constant for all $k \geq k_0$.

P5.5.5 Show that if $A \in \mathbf{R}^{m \times n}$ has rank n, then so does $A + E$ if we have the inequality $\| E \|_2 \| A^+ \|_2 < 1$.

Notes and References for Sec. 5.5

The pseudo-inverse literature is vast, as evidenced by the 1,775 references in

M.Z. Nashed (1976). *Generalized Inverses and Applications*, Academic Press, New York.

The differentiation of the pseudo-inverse is further discussed in

C.L. Lawson and R.J. Hanson (1969). "Extensions and Applications of the Householder
Algorithm for Solving Linear Least Squares Problems," *Math. Comp. 23*, 787–812.
G.H. Golub and V. Pereyra (1973). "The Differentiation of Pseudo-Inverses and Nonlinear Least Squares Problems Whose Variables Separate," *SIAM J. Num. Anal. 10*,
413–32.

Survey treatments of LS perturbation theory may be found in Lawson and Hanson
(1974), Stewart and Sun (1991), Björck (1996), and

P.A. Wedin (1973). "Perturbation Theory for Pseudo-Inverses," *BIT 13*, 217–32.
G.W. Stewart (1977). "On the Perturbation of Pseudo-Inverses, Projections, and Linear
Least Squares," *SIAM Review 19*, 634–62.

Even for full rank problems, column pivoting seems to produce more accurate solutions.
The error analysis in the following paper attempts to explain why.

L.S. Jennings and M.R. Osborne (1974). "A Direct Error Analysis for Least Squares,"
Numer. Math. 22, 322–32.

Various other aspects rank deficiency are discussed in

J.M. Varah (1973). "On the Numerical Solution of Ill-Conditioned Linear Systems with
Applications to Ill-Posed Problems," *SIAM J. Num. Anal. 10*, 257–67.
G.W. Stewart (1984). "Rank Degeneracy," *SIAM J. Sci. and Stat. Comp. 5*, 403–413.
P.C. Hansen (1987). "The Truncated SVD as a Method for Regularization," *BIT 27*,
534–553.
G.W. Stewart (1987). "Collinearity and Least Squares Regression," *Statistical Science
2*, 68–100.

We have more to say on the subject in §12.1 and §12.2.

5.6   Weighting and Iterative Improvement

The concepts of scaling and iterative improvement were introduced in the
Chapter 3 context of square linear systems. Generalizations of these ideas
that are applicable to the least squares problem are now offered.

5.6.1   Column Weighting

Suppose $G \in \mathbb{R}^{n \times n}$ is nonsingular. A solution to the LS problem

$$\min \| Ax - b \|_2 \qquad A \in \mathbb{R}^{m \times n}, \ b \in \mathbb{R}^m \qquad (5.6.1)$$

can be obtained by finding the minimum 2-norm solution y_{LS} to

$$\min \| (AG)y - b \|_2 \qquad\qquad (5.6.2)$$

and then setting $x_G = Gy_{LS}$. If rank$(A) = n$, then $x_G = x_{LS}$. Otherwise,
x_G is the minimum G-norm solution to (5.6.1), where the G-norm is defined
by $\| z \|_G = \| G^{-1}z \|_2$.

The choice of G is important. Sometimes its selection can be based on à priori knowledge of the uncertainties in A. On other occasions, it may be desirable to normalize the columns of A by setting

$$G = G_0 \equiv \text{diag}(1/\| A(:,1) \|_2, \ldots, 1/\| A(:,n) \|_2).$$

Van der Sluis (1969) has shown that with this choice, $\kappa_2(AG)$ is approximately minimized. Since the computed accuracy of y_{LS} depends on $\kappa_2(AG)$, a case can be made for setting $G = G_0$.

We remark that column weighting affects singular values. Consequently, a scheme for determining numerical rank may not return the same estimates when applied to A and AG. See Stewart (1984b).

5.6.2   Row Weighting

Let $D = \text{diag}(d_1, \ldots, d_m)$ be nonsingular and consider the *weighted least squares problem*

$$\text{minimize } \| D(Ax - b) \|_2 \qquad A \in \mathbb{R}^{m \times n}, \ b \in \mathbb{R}^m. \tag{5.6.3}$$

Assume $\text{rank}(A) = n$ and that x_D solves (5.6.3). It follows that the solution x_{LS} to (5.6.1) satisfies

$$x_D - x_{LS} = (A^T D^2 A)^{-1} A^T (D^2 - I)(b - Ax_{LS}). \tag{5.6.4}$$

This shows that row weighting in the LS problem affects the solution. (An important exception occurs when $b \in \text{ran}(A)$ for then $x_D = x_{LS}$.)

One way of determining D is to let d_k be some measure of the uncertainty in b_k, e.g., the reciprocal of the standard deviation in b_k. The tendency is for $r_k = e_k^T(b - Ax_D)$ to be small whenever d_k is large. The precise effect of d_k on r_k can be clarified as follows. Define

$$D(\delta) = \text{diag}(d_1, \ldots, d_{k-1}, d_k \sqrt{1 + \delta}, d_{k+1}, \ldots, d_m)$$

where $\delta > -1$. If $x(\delta)$ minimizes $\| D(\delta)(Ax - b) \|_2$ and $r_k(\delta)$ is the k-th component of $b - Ax(\delta)$, then it can be shown that

$$r_k(\delta) = \frac{r_k}{1 + \delta d_k^2 e_k^T A (A^T D^2 A)^{-1} A^T e_k}. \tag{5.6.5}$$

This explicit expression shows that $r_k(\delta)$ is a monotone decreasing function of δ. Of course, how r_k changes when all the weights are varied is much more complicated.

Example 5.6.1 Suppose

$$A = \begin{bmatrix} 1 & 2 \\ 3 & 4 \\ 5 & 6 \\ 7 & 8 \end{bmatrix} \qquad b = \begin{bmatrix} 1 \\ 0 \\ 0 \\ 0 \end{bmatrix}.$$

If $D = I_4$ then $x_D = [-1, .85]^T$ and $r = b - Ax_D = [.3, -.4, -.1, .2]^T$. On
the other hand, if $D = \text{diag}(1000, 1, 1, 1)$ then we have $x_D \approx [-1.43, 1.21]^T$ and
$r = b - Ax_D = [.000428 \ -.571428 \ -.142853 \ .285714]^T$.

5.6.3   Generalized Least Squares

In many estimation problems, the vector of observations b is related to x
through the equation

$$b = Ax + w \qquad (5.6.6)$$

where the *noise vector* w has zero mean and a symmetric positive definite *variance-covariance* matrix $\sigma^2 W$. Assume that W is known and that
$W = BB^T$ for some $B \in \mathbb{R}^{m \times m}$. The matrix B might be given or it might
be W's Cholesky triangle. In order that all the equations in (5.6.6) contribute equally to the determination of x, statisticians frequently solve the
LS problem

$$\min \| B^{-1}(Ax - b) \|_2 . \qquad (5.6.7)$$

An obvious computational approach to this problem is to form $\tilde{A} = B^{-1}A$
and $\tilde{b} = B^{-1}b$ and then apply any of our previous techniques to minimize
$\| \tilde{A}x - \tilde{b} \|_2$. Unfortunately, x will be poorly determined by such a procedure if B is ill-conditioned.

A much more stable way of solving (5.6.7) using orthogonal transformations has been suggested by Paige (1979a, 1979b). It is based on the idea
that (5.6.7) is equivalent to the *generalized least squares* problem,

$$\min_{b=Ax+Bv} v^T v . \qquad (5.6.8)$$

Notice that this problem is defined even if A and B are rank deficient.
Although Paige's technique can be applied when this is the case, we shall
describe it under the assumption that both these matrices have full rank.
The first step is to compute the QR factorization of A:

$$Q^T A = \begin{bmatrix} R_1 \\ 0 \end{bmatrix} \qquad Q = [\underset{n}{Q_1} \quad \underset{m-n}{Q_2}] .$$

An orthogonal matrix $Z \in \mathbb{R}^{m \times m}$ is then determined so that

$$Q_2^T BZ = [\underset{n}{0} \quad \underset{m-n}{S}] \qquad Z = [\underset{n}{Z_1} \quad \underset{m-n}{Z_2}]$$

where S is upper triangular. With the use of these orthogonal matrices the
constraint in (5.6.8) transforms to

$$\begin{bmatrix} Q_1^T b \\ Q_2^T b \end{bmatrix} = \begin{bmatrix} R_1 \\ 0 \end{bmatrix} x + \begin{bmatrix} Q_1^T BZ_1 & Q_1^T BZ_2 \\ 0 & S \end{bmatrix} \begin{bmatrix} Z_1^T v \\ Z_2^T v \end{bmatrix} .$$

Notice that the "bottom half" of this equation determines v,

$$Su = Q_2^T b \qquad v = Z_2 u, \tag{5.6.9}$$

while the "top half" prescribes x:

$$R_1 x = Q_1^T b - (Q_1^T B Z_1 Z_1^T + Q_1^T B Z_2 Z_2^T) v = Q_1^T b - Q_1^T B Z_2 u. \tag{5.6.10}$$

The attractiveness of this method is that all potential ill-conditioning is concentrated in triangular systems (5.6.9) and (5.6.10). Moreover, Paige (1979b) has shown that the above procedure is numerically stable, something that is not true of any method that explicitly forms $B^{-1}A$.

5.6.4   Iterative Improvement

A technique for refining an approximate LS solution has been analyzed by Björck (1967, 1968). It is based on the idea that if

$$\begin{bmatrix} I_m & A \\ A^T & 0 \end{bmatrix} \begin{bmatrix} r \\ x \end{bmatrix} = \begin{bmatrix} b \\ 0 \end{bmatrix} \qquad A \in \mathbb{R}^{m \times n}, \ b \in \mathbb{R}^m \tag{5.6.11}$$

then $\| b - Ax \|_2 = \min$. This follows because $r + Ax = b$ and $A^T r = 0$ imply $A^T A x = A^T b$. The above augmented system is nonsingular if $\mathrm{rank}(A) = n$, which we hereafter assume.

By casting the LS problem in the form of a square linear system, the iterative improvement scheme (3.5.5) can be applied:

$$r^{(0)} = 0; \ x^{(0)} = 0$$
$$\textbf{for } k = 0, 1,$$

$$\begin{bmatrix} f^{(k)} \\ g^{(k)} \end{bmatrix} = \begin{bmatrix} b \\ 0 \end{bmatrix} - \begin{bmatrix} I & A \\ A^T & 0 \end{bmatrix} \begin{bmatrix} r^{(k)} \\ x^{(k)} \end{bmatrix}$$

$$\begin{bmatrix} I & A \\ A^T & 0 \end{bmatrix} \begin{bmatrix} p^{(k)} \\ z^{(k)} \end{bmatrix} = \begin{bmatrix} f^{(k)} \\ g^{(k)} \end{bmatrix}$$

$$\begin{bmatrix} r^{(k+1)} \\ x^{(k+1)} \end{bmatrix} = \begin{bmatrix} r^{(k)} \\ x^{(k)} \end{bmatrix} + \begin{bmatrix} p^{(k)} \\ z^{(k)} \end{bmatrix}$$

end

The residuals $f^{(k)}$ and $g^{(k)}$ must be computed in higher precision and an original copy of A must be around for this purpose.

If the QR factorization of A is available, then the solution of the augmented system is readily obtained. In particular, if $A = QR$ and $R_1 = R(1{:}n, 1{:}n)$, then a system of the form

$$\begin{bmatrix} I & A \\ A^T & 0 \end{bmatrix} \begin{bmatrix} p \\ z \end{bmatrix} = \begin{bmatrix} f \\ g \end{bmatrix}$$

transforms to

$$\begin{bmatrix} I_n & 0 & R_1 \\ 0 & I_{m-n} & 0 \\ R_1^T & 0 & 0 \end{bmatrix} \begin{bmatrix} h \\ f_2 \\ z \end{bmatrix} = \begin{bmatrix} f_1 \\ f_2 \\ g \end{bmatrix}$$

where

$$Q^T f = \begin{bmatrix} f_1 \\ f_2 \end{bmatrix} \begin{matrix} n \\ m-n \end{matrix} \qquad Q^T p = \begin{bmatrix} h \\ f_2 \end{bmatrix} \begin{matrix} n \\ m-n \end{matrix} \ .$$

Thus, p and z can be determined by solving the triangular systems $R_1^T h = g$ and $R_1 z = f_1 - h$ and setting $p = Q \begin{bmatrix} h \\ f_2 \end{bmatrix}$. Assuming that Q is stored in factored form, each iteration requires $8mn - 2n^2$ flops.

The key to the iteration's success is that both the LS residual and solution are updated—not just the solution. Björck (1968) shows that if $\kappa_2(A) \approx \beta^q$ and t-digit, β-base arithmetic is used, then $x^{(k)}$ has approximately $k(t - q)$ correct base β digits, provided the residuals are computed in double precision. Notice that it is $\kappa_2(A)$, not $\kappa_2(A)^2$, that appears in this heuristic.

Problems

P5.6.1 Verify (5.6.4).

P5.6.2 Let $A \in \mathbb{R}^{m \times n}$ have full rank and define the diagonal matrix

$$\Delta = \text{diag}(\ \underbrace{1, \ldots, 1}_{k-1}, (1 + \delta), \underbrace{1, \ldots, 1}_{m-k}\ )$$

for $\delta > -1$. Denote the LS solution to min $\| \Delta(Ax - b) \|_2$ by $x(\delta)$ and its residual by $r(\delta) = b - Ax(\delta)$. (a) Show

$$r(\delta) = \left(I - \delta \frac{A(A^T A)^{-1} A^T e_k e_k^T}{1 + \delta e_k^T A(A^T A)^{-1} A^T e_k} \right) r(0) .$$

(b) Letting $r_k(\delta)$ stand for the kth component of $r(\delta)$, show

$$r_k(\delta) = \frac{r_k(0)}{1 + \delta e_k^T A(A^T A)^{-1} A^T e_k} .$$

(c) Use (b) to verify (5.6.5).

P5.6.3 Show how the SVD can be used to solve the generalized LS problem when the matrices A and B in (5.6.8) are rank deficient.

P5.6.4 Let $A \in \mathbf{R}^{m \times n}$ have rank n and for $\alpha \geq 0$ define

$$M(\alpha) = \begin{bmatrix} \alpha I_m & A \\ A^T & 0 \end{bmatrix}.$$

Show that

$$\sigma_{m+n}(M(\alpha)) = \min \left\{ \alpha , \, -\frac{\alpha}{2} + \sqrt{\sigma_n(A)^2 + \left(\frac{\alpha}{2}\right)^2} \right\}$$

and determine the value of α that minimizes $\kappa_2(M(\alpha))$.

P5.6.5 Another iterative improvement method for LS problems is the following:

$$x^{(0)} = 0$$
for $k = 0, 1, \ldots$
$\quad r^{(k)} = b - Ax^{(k)} \quad$ (double precision)
$\quad \| Az^{(k)} - r^{(k)} \|_2 = \min$
$\quad x^{(k+1)} = x^{(k)} + z^{(k)}$
end

(a) Assuming that the QR factorization of A is available, how many flops per iteration are required? (b) Show that the above iteration results by setting $g^{(k)} = 0$ in the iterative improvement scheme given in §5.6.4.

Notes and References for Sec. 5.6

Row and column weighting in the LS problem is discussed in Lawson and Hanson (SLS, pp. 180-88). The various effects of scaling are discussed in

A. van der Sluis (1969). "Condition Numbers and Equilibration of Matrices," *Numer. Math. 14*, 14–23.
G.W. Stewart (1984b). "On the Asymptotic Behavior of Scaled Singular Value and QR Decompositions," *Math. Comp. 43*, 483–490.

The theoretical and computational aspects of the generalized least squares problem appear in

S. Kourouklis and C.C. Paige (1981). "A Constrained Least Squares Approach to the General Gauss-Markov Linear Model," *J. Amer. Stat. Assoc. 76*, 620–25.
C.C. Paige (1979a). "Computer Solution and Perturbation Analysis of Generalized Least Squares Problems," *Math. Comp. 33*, 171–84.
C.C. Paige (1979b). "Fast Numerically Stable Computations for Generalized Linear Least Squares Problems," *SIAM J. Num. Anal. 16*, 165–71.
C.C. Paige (1985). "The General Limit Model and the Generalized Singular Value Decomposition," *Lin. Alg. and Its Applic. 70*, 269–284.

Iterative improvement in the least squares context is discussed in

G.H. Golub and J.H. Wilkinson (1966). "Note on Iterative Refinement of Least Squares Solutions," *Numer. Math. 9*, 139–48.
Å. Björck and G.H. Golub (1967). "Iterative Refinement of Linear Least Squares Solutions by Householder Transformation," *BIT 7*, 322–37.
Å. Björck (1967). "Iterative Refinement of Linear Least Squares Solutions I," *BIT 7*, 257–78.
Å. Björck (1968). "Iterative Refinement of Linear Least Squares Solutions II," *BIT 8*, 8–30.
Å. Björck (1987). "Stability Analysis of the Method of Seminormal Equations for Linear Least Squares Problems," *Linear Alg. and Its Applic. 88/89*, 31–48.

5.7   Square and Underdetermined Systems

The orthogonalization methods developed in this chapter can be applied to square systems and also to systems in which there are fewer equations than unknowns. In this brief section we discuss some of the various possibilities.

5.7.1   Using QR and SVD to Solve Square Systems

The least squares solvers based on the QR factorization and the SVD can be used to solve square linear systems: just set $m = n$. However, from the flop point of view, Gaussian elimination is the cheapest way to solve a square linear system as shown in the following table which assumes that the right hand side is available at the time of factorization:

Method	Flops
Gaussian Elimination	$2n^3/3$
Householder Orthogonalization	$4n^3/3$
Modified Gram-Schmidt	$2n^3$
Bidiagonalization	$8n^3/3$
Singular Value Decomposition	$12n^3$

Nevertheless, there are three reasons why orthogonalization methods might be considered:

- The flop counts tend to exaggerate the Gaussian elimination advantage. When memory traffic and vectorization overheads are considered, the QR approach is comparable in efficiency.

- The orthogonalization methods have guaranteed stability; there is no "growth factor" to worry about as in Gaussian elimination.

- In cases of ill-conditioning, the orthogonal methods give an added measure of reliability. QR with condition estimation is very dependable and, of course, SVD is unsurpassed when it comes to producing a meaningful solution to a nearly singular system.

We are not expressing a strong preference for orthogonalization methods but merely suggesting viable alternatives to Gaussian elimination.

We also mention that the SVD entry in Table 5.7.1 assumes the availability of b at the time of decomposition. Otherwise, $20n^3$ flops are required because it then becomes necessary to accumulate the U matrix.

If the QR factorization is used to solve $Ax = b$, then we ordinarily have to carry out a back substitution: $Rx = Q^Tb$. However, this can be avoided by "preprocessing" b. Suppose H is a Householder matrix such that $Hb = \beta e_n$ where e_n is the last column of I_n. If we compute the QR factorization of $(HA)^T$, then $A = H^T R^T Q^T$ and the system transforms to

$$R^T y = \beta e_n$$

where $y = Q^T x$. Since R^T is lower triangular, $y = (\beta / r_{nn}) e_n$ and so

$$x = \frac{\beta}{r_{nn}} Q(:, n).$$

5.7.2   Underdetermined Systems

We say that a linear system

$$Ax = b \qquad A \in \mathbb{R}^{m \times n}, \; b \in \mathbb{R}^m \tag{5.7.1}$$

is *underdetermined* whenever $m < n$. Notice that such a system either has no solution or has an infinity of solutions. In the second case, it is important to distinguish between algorithms that find the minimum 2-norm solution and those that do not necessarily do so. The first algorithm we present is in the latter category. Assume that A has full row rank and that we apply QR with column pivoting to obtain:

$$Q^T A \Pi = [\, R_1 \; R_2 \,]$$

where $R_1 \in \mathbb{R}^{m \times m}$ is upper triangular and $R_2 \in \mathbb{R}^{m \times (n-m)}$. Thus, $Ax = b$ transforms to

$$(Q^T A \Pi)(\Pi^T x) = [\, R_1 \; R_2 \,] \begin{bmatrix} z_1 \\ z_2 \end{bmatrix} = Q^T b$$

where

$$\Pi^T x = \begin{bmatrix} z_1 \\ z_2 \end{bmatrix}$$

with $z_1 \in \mathbb{R}^m$ and $z_2 \in \mathbb{R}^{(n-m)}$. By virtue of the column pivoting, R_1 is nonsingular because we are assuming that A has full row rank. One solution to the problem is therefore obtained by setting $z_1 = R_1^{-1} Q^T b$ and $z_2 = 0$.

Algorithm 5.7.1   Given $A \in \mathbb{R}^{m \times n}$ with $\text{rank}(A) = m$ and $b \in \mathbb{R}^m$, the following algorithm finds an $x \in \mathbb{R}^n$ such that $Ax = b$.

$Q^T A \Pi = R$ (QR with column pivoting.)
Solve $R(1{:}m, 1{:}m) z_1 = Q^T b$.
Set $x = \Pi \begin{bmatrix} z_1 \\ 0 \end{bmatrix}$.

This algorithm requires $2m^2n - m^3/3$ flops. The minimum norm solution is not guaranteed. (A different Π would render a smaller z_1.) However, if we compute the QR factorization

$$A^T = QR = Q \begin{bmatrix} R_1 \\ 0 \end{bmatrix}$$

with $R_1 \in \mathbb{R}^{m \times m}$, then $Ax = b$ becomes

$$(QR)^T x = \begin{bmatrix} R_1^T & 0 \end{bmatrix} \begin{bmatrix} z_1 \\ z_2 \end{bmatrix} = b$$

where

$$Q^T x = \begin{bmatrix} z_1 \\ z_2 \end{bmatrix} \qquad z_1 \in \mathbb{R}^m, \; z_2 \in \mathbb{R}^{n-m}.$$

Now the minimum norm solution *does* follow by setting $z_2 = 0$.

Algorithm 5.7.2 Given $A \in \mathbb{R}^{m \times n}$ with $\text{rank}(A) = m$ and $b \in \mathbb{R}^m$, the following algorithm finds the minimal 2-norm solution to $Ax = b$.

$A^T = QR$ (QR factorization)
Solve $R(1{:}m, 1{:}m)^T z = b$.
$x = Q(:, 1{:}m)z$

This algorithm requires at most $2m^2n - 2m^3/3$
     The SVD can also be used to compute the minimal norm solution of an underdetermined $Ax = b$ problem. If

$$A = \sum_{i=1}^{r} \sigma_i u_i v_i^T \qquad r = \text{rank}(A)$$

is A's singular value expansion, then

$$x = \sum_{i=1}^{r} \frac{u_i^T b}{\sigma_i} v_i .$$

As in the least squares problem, the SVD approach is desirable whenever A is nearly rank deficient.

5.7.3   Perturbed Underdetermined Systems

We conclude this section with a perturbation result for full-rank underdetermined systems.

Theorem 5.7.1 *Suppose* $\mathrm{rank}(A) = m \leq n$ *and that* $A \in \mathbb{R}^{m \times n}$, $\delta A \in \mathbb{R}^{m \times n}$, $0 \neq b \in \mathbb{R}^m$, *and* $\delta b \in \mathbb{R}^m$ *satisfy*

$$\epsilon = \max\{\epsilon_A, \epsilon_b\} < \sigma_m(A),$$

where $\epsilon_A = \| \delta A \|_2 / \| A \|_2$ *and* $\epsilon_b = \| \delta b \|_2 / \| b \|_2$. *If* x *and* $\hat{x}$ *are minimum norm solutions that satisfy*

$$Ax = b \qquad\qquad (A + \delta A)\hat{x} = b + \delta b$$

then

$$\frac{\| \hat{x} - x \|_2}{\| x \|_2} \leq \kappa_2(A)\,(\epsilon_A \min\{2, n - m + 1\} + \epsilon_b) + O(\epsilon^2).$$

Proof. Let E and f be defined by $\delta A/\epsilon$ and $\delta b/\epsilon$. Note that $\mathrm{rank}(A + tE) = m$ for all $0 < t < \epsilon$ and that

$$x(t) = (A + tE)^T \left((A + tE)(A + tE)^T \right)^{-1} (b + tf)$$

satisfies $(A + tE)x(t) = b + tf$. By differentiating this expression with respect to t and setting $t = 0$ in the result we obtain

$$\dot{x}(0) = \left(I - A^T (AA^T)^{-1}A \right) E^T (AA^T)^{-1}b + A^T (AA^T)^{-1}(f - Ex).$$

Since

$$\| x \|_2 = \| A^T (AA^T)^{-1}b \|_2 \geq \sigma_m(A)\| (AA^T)^{-1}b \|_2,$$
$$\| I - A^T (AA^T)^{-1}A \|_2 = \min(1, n - m),$$

and

$$\frac{\| f \|_2}{\| x \|_2} \leq \frac{\| f \|_2 \| A \|_2}{\| b \|_2},$$

we have

$$\frac{\| \hat{x} - x \|_2}{\| x \|_2} = \frac{x(\epsilon) - x(0)}{\| x(0) \|_2} = \epsilon \frac{\| \dot{x}(0) \|_2}{\| x \|_2} + O(\epsilon^2)$$

$$\leq \epsilon \min(1, n - m) \left\{ \frac{\| E \|_2}{\| A \|_2} + \frac{\| f \|_2}{\| b \|_2} + \frac{\| E \|_2}{\| A \|_2} \right\} \kappa_2(A) + O(\epsilon^2)$$

from which the theorem follows. $\square$

Note that there is no $\kappa_2(A)^2$ factor as in the case of overdetermined systems.

Problems

P5.7.1 Derive the above expression for $\dot{x}(0)$.

P5.7.2 Find the minimal norm solution to the system $Ax = b$ where $A = [\,1\ 2\ 3\,]$ and $b = 1$.

P5.7.3 Show how triangular system solving can be avoided when using the QR factorization to solve an underdetermined system.

P5.7.4 Suppose $b, x \in \mathbf{R}^n$ are given. Consider the following problems:

(a) Find an unsymmetric Toeplitz matrix T so $Tx = b$.

(b) Find a symmetric Toeplitz matrix T so $Tx = b$.

(c) Find a circulant matrix C so $Cx = b$.

Pose each problem in the form $Ap = b$ where A is a matrix made up of entries from x and p is the vector of sought-after parameters.

Notes and References for Sec. 5.7

Interesting aspects concerning singular systems are discussed in

T.F. Chan (1984). "Deflated Decomposition Solutions of Nearly Singular Systems," *SIAM J. Num. Anal. 21*, 738–754.

G.H. Golub and C.D. Meyer (1986). "Using the QR Factorization and Group Inversion to Compute, Differentiate, and estimate the Sensitivity of Stationary Probabilities for Markov Chains," *SIAM J. Alg. and Dis. Methods, 7*, 273–281.

Papers on underdetermined systems include

R.E. Cline and R.J. Plemmons (1976). "L_2-Solutions to Underdetermined Linear Systems," *SIAM Review 18*, 92-106.

M. Arioli and A. Laratta (1985). "Error Analysis of an Algorithm for Solving an Underdetermined System," *Numer. Math. 46*, 255–268.

J.W. Demmel and N.J. Higham (1993). "Improved Error Bounds for Underdetermined System Solvers," *SIAM J. Matrix Anal. Appl. 14*, 1–14.

The QR factorization can of course be used to solve linear systems. See

N.J. Higham (1991). "Iterative Refinement Enhances the Stability of QR Factorization Methods for Solving Linear Equations," *BIT 31*, 447–468.

Chapter 6

Parallel Matrix Computations

The parallel matrix computation area has been the focus of intense research. Although much of the work is machine/system dependent, a number of basic strategies have emerged. Our aim is to present these along with a picture of what it is like to "think parallel" during the design of a matrix computation.

The distributed and shared memory paradigms are considered. We use matrix-vector multiplication to introduce the notion of a node program in §6.1. Load balancing, speed-up, and synchronization are also discussed. In §6.2 matrix-matrix multiplication is used to show the effect of blocking on granularity and to convey the spirit of two-dimensional data flow. Two parallel implementations of the Cholesky factorization are given in §6.3.

Before You Begin

Chapter 1, §4.1, and §4.2 are assumed. Within this chapter there are the following dependencies:

$$\S6.1 \quad \rightarrow \quad \S6.2 \quad \rightarrow \quad \S6.3$$

Complementary references include the books by Schönauer (1987), Hockney and Jesshope (1988), Modi (1988), Ortega (1988), Dongarra, Duff,

Sorensen, and van der Vorst (1991), and Golub and Ortega (1993) and the excellent review papers by Heller (1978), Ortega and Voight (1985), Gallivan, Plemmons, and Sameh (1990), and Demmel, Heath, and van der Vorst (1993).

6.1   Basic Concepts

In this section we introduce the distributed and shared memory paradigms using the gaxpy operation

$$z = y + Ax, \qquad A \in \mathbb{R}^{n \times n}, x, y, z \in \mathbb{R}^n \tag{6.1.1}$$

as an example. In practice, there is a fuzzy line between these two styles of parallel computing and typically a blend of our comments apply to any particular machine.

6.1.1   Distributed Memory Systems

In a distributed memory multiprocessor each processor has a *local memory* and executes its own *node program*. The program can alter values in the executing processor's local memory and can send data in the form of *messages* to the other processors in the network. The interconnection of the processors defines the *network topology* and one simple example that is good enough for our introduction is the *ring*. See FIGURE 6.1.1. Other

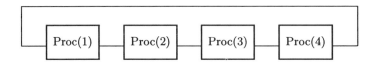

FIGURE 6.1.1 *A Four-Processor Ring*

important interconnection schemes include the mesh and torus (for their close correspondence with two-dimensional arrays), the hypercube (for its generality and optimality), and the tree (for its handling of divide and conquer procedures). See Ortega and Voigt (1985) for a discussion of the possibilities. Our immediate goal is to develop a ring algorithm for (6.1.1). Matrix multiplication on a torus is discussed in §6.2.

Each processor has an *identification number*. The μth processor is designated by Proc(μ). We say that Proc(λ) is a *neighbor* of Proc(μ) if there is a direct physical connection between them. Thus, in a p-processor ring, Proc($p - 1$) and Proc(1) are neighbors of Proc(p).

Important factors in the design of an effective distributed memory algorithm include (a) the number of processors and the capacity of the local memories, (b) how the processors are interconnected, (c) the speed of computation relative to the speed of interprocessor communication, and (d) whether or not a node is able to compute and communicate at the same time.

6.1.2 Communication

To describe the sending and receiving of messages we adopt a simple notation:

$$\textbf{send}(\ \{matrix\}\ ,\ \{\text{id of the receiving processor}\}\ )$$

$$\textbf{recv}(\ \{matrix\}\ ,\ \{\text{id of the sending processor}\}\ )$$

Scalars and vectors are matrices and therefore messages. In our model, if $\text{Proc}(\mu)$ executes the instruction $\textbf{send}(V_{loc}, \lambda)$, then a copy of the local matrix V_{loc} is sent to $\text{Proc}(\lambda)$ and the execution of $\text{Proc}(\mu)$'s node program resumes immediately. It is legal for a processor to send a message to itself. To emphasize that a matrix is stored in a local memory we use the subscript "*loc*."

If $\text{Proc}(\mu)$ executes the instruction $\textbf{recv}(U_{loc}, \lambda)$, then the execution of its node program is suspended until a message is received from $\text{Proc}(\lambda)$. Once received, the message is placed in a local matrix U_{loc} and $\text{Proc}(\mu)$ resumes execution of its node program.

Although the syntax and semantics of our send/receive notation is adequate for our purposes, it does suppress a number of important details:

- Message assembly overhead. In practice, there may be a penalty associated with the transmission of a matrix whose entries are not contiguous in the sender's local memory. We ignore this detail.

- Message tagging. Messages need not arrive in the order they are sent, and a system of message tagging is necessary so that the receiver is not "confused." We ignore this detail by assuming that messages *do* arrive in the order that they are sent.

- Message interpretation overhead. In practice a message is a bit string, and a header must be provided that indicates to the receiver the dimensions of the matrix and the format of the floating point words that are used to represent its entries. Going from message to stored matrix takes time, but it is an overhead that we do not try to quantify.

These simplifications enable us to focus on high-level algorithmic ideas. But it should be remembered that the success of a particular implementation may hinge upon the control of these hidden overheads.

6.1.3   Some Distributed Data Structures

Before we can specify our first distributed memory algorithm, we must
consider the matter of *data layout*. How are the participating matrices and
vectors distributed around the network?

Suppose $x \in \mathbb{R}^n$ is to be distributed among the local memories of a p-
processor network. Assume for the moment that $n = rp$. Two "canonical"
approaches to this problem are store-by-row and store-by-column.

In store-by-column we regard the vector x as an r-by-p matrix,

$$x_{r \times p} = \left[\; x(1{:}r) \quad x(r+1{:}2r) \quad \cdots \quad x(1+(p-1)r{:}n) \; \right] ,$$

and store each column in a processor, i.e., $x(1+(\mu-1)r{:}\mu r) \in \text{Proc}(\mu)$.
(In this context "$\in$" means "is stored in.") Note that each processor houses
a contiguous portion of x.

In the store-by-row scheme we regard x as a p-by-r matrix

$$x_{p \times r} = \left[\; x(1{:}p) \quad x(p+1{:}2p) \quad \cdots \quad x((r-1)p+1{:}n) \; \right] ,$$

and store each *row* in a processor, i.e., $x(\mu{:}p{:}n) \in \text{Proc}(\mu)$. Store-by-row is
sometimes referred to as the *wrap* method of distributing a vector because
the components of x can be thought of as cards in a deck that are "dealt"
to the processors in wrap-around fashion.

If n is not an exact multiple of p, then these ideas go through with minor
modification. Consider store-by-column with $n = 14$ and $p = 4$:

$$x^T = \left[\underbrace{x_1\, x_2\, x_3\, x_4}_{\text{Proc}(1)} \mid \underbrace{x_5\, x_6\, x_7\, x_8}_{\text{Proc}(2)} \mid \underbrace{x_9\, x_{10}\, x_{11}}_{\text{Proc}(3)} \mid \underbrace{x_{12}\, x_{13}\, x_{14}}_{\text{Proc}(4)} \right].$$

In general, if $n = pr + q$ with $0 \le q < p$, then $\text{Proc}(1),\ldots,\text{Proc}(q)$ can
each house $r+1$ components and $\text{Proc}(q+1),\ldots,\text{Proc}(p)$ can house r
components. In store-by-row we simply let $\text{Proc}(\mu)$ house $x(\mu{:}p{:}n)$.

Similar options apply to the layout of a matrix. There are four obvious
possibilities if $A \in \mathbb{R}^{n \times n}$ and (for simplicity) $n = rp$:

Orientation	Style	What is in Proc(μ)
Column	Contiguous	$A(:,1+(\mu-1)r{:}\mu r)$
Column	Wrap	$A(:,\mu{:}p{:}n)$
Row	Contiguous	$A(1+(\mu-1)r{:}\mu r,:)$
Row	Wrap	$A(\mu{:}p{:}n,:)$

These strategies have block analogs. For example, if $A = [A_1,\ldots,A_N]$ is
a block column partitioning, then we could arrange to have $\text{Proc}(\mu)$ store
A_i for $i = \mu{:}p{:}N$.

6.1.4 Gaxpy on a Ring

We are now set to develop a ring algorithm for the gaxpy $z = y + Ax$ ($A \in \mathbb{R}^{n \times n}$, $x, y \in \mathbb{R}^n$). For clarity, assume that $n = rp$ where p is the size of the ring. Partition the gaxpy as

$$\begin{bmatrix} z_1 \\ \vdots \\ z_p \end{bmatrix} = \begin{bmatrix} y_1 \\ \vdots \\ y_p \end{bmatrix} + \begin{bmatrix} A_{11} & \cdots & A_{1p} \\ \vdots & & \vdots \\ A_{p1} & \cdots & A_{pp} \end{bmatrix} \begin{bmatrix} x_1 \\ \vdots \\ x_p \end{bmatrix}. \tag{6.1.2}$$

where $A_{ij} \in \mathbb{R}^{r \times r}$ and $x_i, y_i, z_i \in \mathbb{R}^r$. We assume that at the start of computation $\text{Proc}(\mu)$ houses x_μ, y_μ, and the μth block row of A. Upon completion we set as our goal the overwriting of y_μ by z_μ. From the $\text{Proc}(\mu)$ perspective, the computation of

$$z_\mu = y_\mu + \sum_{\tau=1}^{p} A_{\mu\tau} x_\tau$$

involves local data $(A_{\mu\tau}, y_\mu, x_\mu)$ and nonlocal data $(x_\tau, \tau \neq \mu)$. To make the nonlocal portions of x available, we circulate its subvectors around the ring. For example, in the $p = 3$ case we rotate the x_1, x_2, and x_3 as follows:

step	Proc(1)	Proc(2)	Proc(3)
1	x_3	x_1	x_2
2	x_2	x_3	x_1
3	x_1	x_2	x_3

When a subvector of x "visits", the host processor must incorporate the appropriate term into its running sum:

step	Proc(1)	Proc(2)	Proc(3)
1	$y_1 = y_1 + A_{13}x_3$	$y_2 = y_2 + A_{21}x_1$	$y_3 = y_3 + A_{32}x_2$
2	$y_1 = y_1 + A_{12}x_2$	$y_2 = y_2 + A_{23}x_3$	$y_3 = y_3 + A_{31}x_1$
3	$y_1 = y_1 + A_{11}x_1$	$y_2 = y_2 + A_{22}x_2$	$y_3 = y_3 + A_{33}x_3$

In general, the "merry-go-round" of x subvectors makes p "stops." For each received x-subvector, a processor performs an r-by-r gaxpy.

Algorithm 6.1.1 Suppose $A \in \mathbb{R}^{n \times n}$, $x \in \mathbb{R}^n$, and $y \in \mathbb{R}^n$ are given and that $z = y + Ax$. If each processor in a p-processor ring executes the following node program and $n = rp$, then upon completion $\text{Proc}(\mu)$ houses $z(1+(\mu-1)r{:}\mu r)$ in y_{loc}. Assume the following local memory initializations: p, μ (the node id), *left* and *right* (the neighbor id's), n, $row = 1+(\mu-1)r{:}\mu r$, $A_{loc} = A(row, :)$, $x_{loc} = x(row)$, $y_{loc} = y(row)$.

for $t = 1{:}p$
    send$(x_{loc}, right)$
    recv$(x_{loc}, left)$
    $\tau = \mu - t$
    if $\tau \le 0$
        $\tau = \tau + p$
    end
    $\{ \ x_{loc} = x(1 + (\tau - 1)r{:}\tau r) \ \}$
    $y_{loc} = y_{loc} + A_{loc}(:, 1 + (\tau - 1)r{:}\tau r)x_{loc}$
end

The index τ names the currently available x subvector. Once it is computed it is possible to carry out the update of the locally housed portion of y. The **send-recv** pair passes the currently housed x subvector to the right and waits to receive the next one from the left. Synchronization is achieved because the local y update cannot begin until the "new" x subvector arrives. It is impossible for one processor to "race ahead" of the others or for an x subvector to pass another in the merry-go-round. The algorithm is tailored to the ring topology in that only nearest neighbor communication is involved. The computation is also perfectly *load balanced* meaning that each processor has the same amount of computation and communication. Load imbalance is discussed further in §6.1.7.

The design of a parallel program involves subtleties that do not arise in the uniprocessor setting. For example, if we inadvertently reverse the order of the **send** and the **recv**, then each processor starts its node program by waiting for a message from its *left* neighbor. Since that neighbor in turn is waiting for a message from *its* left neighbor, a state of *deadlock* results.

6.1.5   The Cost of Communication

Communication overheads can be estimated if we model the cost of sending and receiving a message. To that end we assume that a **send** or **recv** involving m floating point numbers requires

$$\tau(m) = \alpha_d + \beta_d m \tag{6.1.3}$$

seconds to carry out. Here α_d is the time required to initiate the **send** or **recv** and β_d is the reciprocal of the rate that a message can be transferred. Note that this model does not take into consideration the "distance" between the sender and receiver. Clearly, it takes longer to pass a message halfway around a ring than to a neighbor. That is why it is always desirable to arrange (if possible) a distributed computation so that communication is just between neighbors.

During each step in Algorithm 6.1.1 an r-vector is sent and received and $2r^2$ flops are performed. If the computation proceeds at R flops per second and there is no idle waiting associated with the **recv**, then each y_{loc} update requires approximately $(2r^2/R) + 2(\alpha_d + \beta_d r)$ seconds.

Another instructive statistic is the *computation-to-communication ratio*. For Algorithm 6.1.1 this is prescribed by

$$\frac{\text{Time spent computing}}{\text{Time spent communicating}} \approx \frac{2r^2/R}{2(\alpha_d + \beta_d r)}.$$

This fraction quantifies the overhead of communication relative to the volume of computation. Clearly, as $r = n/p$ grows, the fraction of time spent computing increases.[1]

6.1.6   Efficiency and Speed-Up

The *efficiency* of a p-processor parallel algorithm is given by

$$E = \frac{T(1)}{pT(p)}$$

where $T(k)$ is the time required to execute the program on k processors. If computation proceeds at R flops/sec and communication is modeled by (6.1.3), then a reasonable estimate of $T(k)$ for Algorithm 6.1.1 is given by

$$T(k) = \sum_{t=1}^{k} 2(n/k)^2/R + 2(\alpha_d + \beta_d(n/k)) = \frac{2n^2}{Rk} + 2\alpha_d k + 2\beta_d n$$

for $k > 1$. This assumes no idle waiting. If $k = 1$, then no communication is required and $T(1) = 2n^2/R$. It follows that the efficiency

$$E = \frac{1}{1 + \frac{pR}{n}\left(\alpha_d \frac{p}{n} + \beta\right)}.$$

improves with increasing n and degradates with increasing p or R. In practice, benchmarking is the only dependable way to assess efficiency.

A concept related to efficiency is *speed-up*. We say that a parallel algorithm for a particular problem achieves speed-up S if

$$S = T_{seq}/T_{par}$$

where T_{par} is the time required for execution of the parallel program and T_{seq} is the time required by one processor when the best uniprocessor procedure is used. For some problems, the fastest sequential algorithm does not parallelize and so two distinct algorithms are involved in the speed-up assessment.

---

[1] We mention that these simple measures are not particularly illuminating in systems where the nodes are able to overlap computation and communication.

6.1.7   The Challenge of Load Balancing

If we apply Algorithm 6.1.1 to a matrix $A \in \mathbb{R}^{n \times n}$ that is lower triangular, then approximately half of the flops associated with the y_{loc} updates are unnecessary because half of the A_{ij} in (6.1.2) are zero. In particular, in the μth processor, $A_{loc}(:, 1 + (\tau - 1)r{:}\tau r)$ is zero if $\tau > \mu$. Thus, if we guard the y_{loc} update as follows,

> if $\tau \leq \mu$
> $\quad y_{loc} = y_{loc} + A_{loc}(:, 1 + (\tau - 1)r{:}\tau r)x_{loc}$
> end

then the overall number of flops is halved. This solves the superfluous flops problem but it creates a load imbalance problem. $\text{Proc}(\mu)$ oversees about $\mu r^2/2$ flops, an increasing function of the processor id μ. Consider the following $r = p = 3$ example:

$$
\begin{bmatrix} z_1 \\ z_2 \\ z_3 \\ z_4 \\ z_5 \\ z_6 \\ z_7 \\ z_8 \\ z_9 \end{bmatrix}
=
\left[\begin{array}{ccc|ccc|ccc}
\alpha & 0 & 0 & 0 & 0 & 0 & 0 & 0 & 0 \\
\alpha & \alpha & 0 & 0 & 0 & 0 & 0 & 0 & 0 \\
\alpha & \alpha & \alpha & 0 & 0 & 0 & 0 & 0 & 0 \\
\beta & \beta & \beta & \beta & 0 & 0 & 0 & 0 & 0 \\
\beta & \beta & \beta & \beta & \beta & 0 & 0 & 0 & 0 \\
\beta & \beta & \beta & \beta & \beta & \beta & 0 & 0 & 0 \\
\gamma & \gamma & \gamma & \gamma & \gamma & \gamma & \gamma & 0 & 0 \\
\gamma & \gamma & \gamma & \gamma & \gamma & \gamma & \gamma & \gamma & 0 \\
\gamma & \gamma & \gamma & \gamma & \gamma & \gamma & \gamma & \gamma & \gamma
\end{array} \right]
\begin{bmatrix} x_1 \\ x_2 \\ x_3 \\ x_4 \\ x_5 \\ x_6 \\ x_7 \\ x_8 \\ x_9 \end{bmatrix}
+
\begin{bmatrix} y_1 \\ y_2 \\ y_3 \\ y_4 \\ y_5 \\ y_6 \\ y_7 \\ y_8 \\ y_9 \end{bmatrix}
$$

Here, $\text{Proc}(1)$ handles the α part, $\text{Proc}(2)$ handles the β part, and $\text{Proc}(3)$ handles the γ part.

However, if processors 1, 2, and 3 compute (z_1, z_4, z_7), (z_2, z_5, z_8), and (z_3, z_6, z_9), respectively, then approximate load balancing results:

$$
\begin{bmatrix} z_1 \\ z_4 \\ z_7 \\ z_2 \\ z_5 \\ z_8 \\ z_3 \\ z_6 \\ z_9 \end{bmatrix}
=
\left[\begin{array}{ccc|ccc|ccc}
\alpha & 0 & 0 & 0 & 0 & 0 & 0 & 0 & 0 \\
\beta & \beta & \beta & \beta & 0 & 0 & 0 & 0 & 0 \\
\gamma & \gamma & \gamma & \gamma & \gamma & \gamma & \gamma & 0 & 0 \\
\alpha & \alpha & 0 & 0 & 0 & 0 & 0 & 0 & 0 \\
\beta & \beta & \beta & \beta & \beta & 0 & 0 & 0 & 0 \\
\gamma & \gamma & \gamma & \gamma & \gamma & \gamma & \gamma & \gamma & 0 \\
\alpha & \alpha & \alpha & 0 & 0 & 0 & 0 & 0 & 0 \\
\beta & \beta & \beta & \beta & \beta & \beta & 0 & 0 & 0 \\
\gamma & \gamma & \gamma & \gamma & \gamma & \gamma & \gamma & \gamma & \gamma
\end{array} \right]
\begin{bmatrix} x_1 \\ x_2 \\ x_3 \\ x_4 \\ x_5 \\ x_6 \\ x_7 \\ x_8 \\ x_9 \end{bmatrix}
+
\begin{bmatrix} y_1 \\ y_4 \\ y_7 \\ y_2 \\ y_5 \\ y_8 \\ y_3 \\ y_6 \\ y_9 \end{bmatrix}
$$

The amount of arithmetic still increases with μ, but the effect is not noticeable if $n \gg p$.

The development of the general algorithm requires some index manipulation. Assume that $\text{Proc}(\mu)$ is initialized with $A_{loc} = A(\mu{:}p{:}n, :)$ and

$y_{loc} = y(\mu{:}p{:}n)$, and assume that the contiguous x-subvectors circulate as before. If at some stage x_{loc} contains $x(1 + (\tau - 1)r{:}\tau r)$, then the update

$$y_{loc} = y_{loc} + A_{loc}(:, 1 + (\tau - 1)r{:}\tau r)x_{loc}$$

implements

$$y(\mu{:}p{:}n) = y(\mu{:}p{:}n) + A(\mu{:}p{:}n, 1 + (\tau - 1)r{:}\tau r)x(1 + (\tau - 1)r{:}\tau r).$$

To exploit the triangular structure of A in the y_{loc} computation, we express the gaxpy as a double loop:

```
for α = 1:r
    for β = 1:r
        y_loc(α) = y_loc(α) + A_loc(α, β + (τ − 1)r)x_loc(β)
    end
end
```

The A_{loc} reference refers to $A(\mu+(\alpha-1)p, \beta+(\tau-1)r)$ which is zero unless the column index is less than or equal to the row index. Abbreviating the inner loop range with this in mind we obtain

Algorithm 6.1.2 Suppose $A \in \mathbb{R}^{n \times n}$, $x \in \mathbb{R}^n$ and $y \in \mathbb{R}^n$ are given and that $z = y + Ax$. Assume that $n = rp$ and that A is lower triangular. If each processor in a p-processor ring executes the following node program, then upon completion $\text{Proc}(\mu)$ houses $z(\mu{:}p{:}n)$ in y_{loc}. Assume the following local memory initializations: p, μ (the node id), *left* and *right* (the neighbor id's), n, $A_{loc} = A(\mu{:}p{:}n, :)$, $y_{loc} = y(\mu{:}p{:}n)$, and $x_{loc} = x(1 + (\mu - 1)r{:}\mu r)$.

```
r = n/p
for t = 1:p
    send(x_loc, right)
    recv(x_loc, left)
    τ = μ − t
    if τ ≤ 0
        τ = τ + p
    end
    {x_loc = x(1 + (τ − 1)r:τr)}
    for α = 1:r
        for β = 1:μ + (α − 1)p − (τ − 1)r
            y_loc(α) = y_loc(α) + A_loc(α, β + (τ − 1)r)x_loc(β)
        end
    end
end
```

Having to map indices back and forth between "node space" and "global space" is one aspect of distributed matrix computations that requires care and (hopefully) compiler assistance.

6.1.8   Tradeoffs

As we did in §1.1, let us develop a column-oriented gaxpy and anticipate its performance. With the block column partitioning

$$A = [A_1, \ldots, A_p] \qquad A_i \in \mathbb{R}^{n \times r}, \ r = n/p$$

the gaxpy $z = y + Ax$ becomes

$$z = y + \sum_{\mu=1}^{p} A_\mu x_\mu$$

where $x_\mu = x(1 + (\mu - 1)r : \mu r)$. Assume that $\mathrm{Proc}(\mu)$ contains A_μ and x_μ. Its contribution to the gaxpy is the product $A_\mu x_\mu$ and involves local data. However, these products must be summed. We assign this task to $\mathrm{Proc}(1)$ which we assume contains y. The strategy is thus for each processor to compute $A_\mu x_\mu$ and to send the result to $\mathrm{Proc}(1)$.

Algorithm 6.1.3 Suppose $A \in \mathbb{R}^{n \times n}$, $x \in \mathbb{R}^n$ and $y \in \mathbb{R}^n$ are given and that $z = y + Ax$. If each processor in a p-processor network executes the following node program and $n = rp$, then upon completion $\mathrm{Proc}(1)$ houses z. Assume the following local memory initializations: p, μ (the node id), n, $x_{loc} = x(1 + (\mu - 1)r : \mu r)$, $A_{loc} = A(:, 1 + (\mu - 1)r : \mu r)$, and (in $\mathrm{Proc}(1)$ only) $y_{loc} = y$.

> **if** $\mu = 1$
>> $y_{loc} = y_{loc} + A_{loc} x_{loc}$
>> **for** $t = 2:p$
>>> **recv**(w_{loc}, t)
>>> $y_{loc} = y_{loc} + w_{loc}$
>> **end**
> **else**
>> $w_{loc} = A_{loc} x_{loc}$
>> **send**$(w_{loc}, 1)$
> **end**

At first glance this seems to be much less attractive than the row-oriented Algorithm 6.1.1. The additional responsibilities of $\mathrm{Proc}(1)$ mean that it has more arithmetic to perform by a factor of about

$$\frac{2n^2/p + np}{2n^2/p} = 1 + \frac{p^2}{2n}$$

and more messages to process by a factor of about p. This imbalance becomes less critical if $n \gg p$ and the communication parameters α_d and β_d factors are small enough. Another possible mitigating factor is that Algorithm 6.1.3 manipulates length n vectors whereas Algorithm 6.1.1 works with length n/p vectors. If the nodes are capable of vector arithmetic, then the longer vectors may raise the level of performance.

This brief comparison of Algorithms 6.1.1 and 6.1.3 reminds us once again that different implementations of the same computation can have very different performance characteristics.

6.1.9   Shared Memory Systems

We now discuss the gaxpy problem for a shared memory multiprocessor. In this environment each processor has access to a common, global memory as depicted in Figure 6.1.2. Communication between processors is achieved

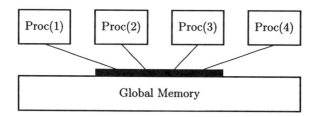

FIGURE 6.1.2 *A Four-Processor Shared Memory System*

by reading and writing to *global variables* that reside in the global memory. Each processor executes its own *local program* and has its own *local memory*. Data flows to and from the global memory during execution.

All the concerns that attend distributed memory computation are with us in modified form. The overall procedure should be *load balanced* and the computations should be arranged so that the individual processors have to wait as little as possible for something useful to compute. The traffic between the global and local memories must be managed carefully, because the extent of such data transfers is typically a significant overhead. (It corresponds to interprocessor communication in the distributed memory setting and to data motion up and down a memory hierarchy as discussed in §1.4.5.) The nature of the physical connection between the processors and the shared memory is very important and can effect algorithmic development. However, for simplicity we regard this aspect of the system as a black box as shown in Figure 6.1.2.

6.1.10   A Shared Memory Gaxpy

Consider the following partitioning of the n-by-n gaxpy problem $z = y + Ax$:

$$
\begin{bmatrix} z_1 \\ \vdots \\ z_p \end{bmatrix} = \begin{bmatrix} y_1 \\ \vdots \\ y_p \end{bmatrix} + \begin{bmatrix} A_1 \\ \vdots \\ A_p \end{bmatrix} x \,. \tag{6.1.4}
$$

Here we assume that $n = rp$ and that $A_\mu \in \mathbb{R}^{r \times n}$, $y_\mu \in \mathbb{R}^r$, and $z_\mu \in \mathbb{R}^r$. We use the following algorithm to introduce the basic ideas and notations.

Algorithm 6.1.4   Suppose $A \in \mathbb{R}^{n \times n}$, $x \in \mathbb{R}^n$, and $y \in \mathbb{R}^n$ reside in a global memory accessible to p processors. If $n = rp$ and each processor executes the following algorithm, then upon completion, y is overwritten by $z = y + Ax$. Assume the following initializations in each local memory: p, μ (the node id), and n.

$$
\begin{aligned}
&r = n/p \\
&row = 1 + (\mu - 1)r : \mu r \\
&x_{loc} = x \\
&y_{loc} = y(row) \\
&\textbf{for } j = 1{:}n \\
&\qquad a_{loc} = A(row, j) \\
&\qquad y_{loc} = y_{loc} + a_{loc} x_{loc}(j) \\
&\textbf{end} \\
&y(row) = y_{loc}
\end{aligned}
$$

We assume that a copy of this program resides in each processor. Floating point variables that are local to an individual processor have a *"loc"* subscript.

Data is transferred to and from the global memory during the execution of Algorithm 6.1.4. There are two global memory reads before the loop ($x_{loc} = x$ and $y_{loc} = y(row)$), one read each time through the loop ($a_{loc} = A(row, j)$), and one write after the loop ($y(row) = y_{loc}$).

Only one processor writes to a given global memory location in y, and so there is no need to synchronize the participating processors. Each has a completely independent part of the overall gaxpy operation and does not have to monitor the progress of the other processors. The computation is *statically scheduled* because the partitioning of work is determined before execution.

If A is lower triangular, then steps have to be taken to preserve the load balancing in Algorithm 6.1.4. As we discovered in §6.1.7, the wrap mapping is a vehicle for doing this. Assigning Proc(μ) the computation of $z(\mu{:}p{:}n) = y(\mu{:}p{:}n) + A(\mu{:}p{:}n, :)x$ effectively partitions the n^2 flops among the p processors.

6.1.11 Memory Traffic Overhead

It is important to recognize that overall performance depends strongly on the overheads associated with the reads and writes to the global memory. If such a data transfer involves m floating point numbers, then we model the transfer time by

$$\tau(m) = \alpha_s + \beta_s m. \tag{6.1.5}$$

The parameter α_s represents a start-up overhead and β_s is the reciprocal transfer rate. We modelled interprocessor communication in the distributed environment exactly the same way. (See (6.1.3).)

Accounting for all the shared memory reads and writes in Algorithm 6.1.4 we see that each processor spends time

$$T \approx (n+3)\alpha_s + \frac{n^2}{p}\beta_s$$

communicating with global memory.

We organized the computation so that one column of $A(row.:)$ is read at a time from shared memory. If the local memory is large enough, then the loop in Algorithm 6.1.4 can be replaced with

$A_{loc} = A(row, :)$
$y_{loc} = y_{loc} + A_{loc}x_{loc}$

This changes the communication overhead to

$$\tilde{T} \approx 3\alpha_s + \frac{n^2}{p}\beta_s,$$

a significant improvement if the start-up parameter α_s is large.

6.1.12 Barrier Synchronization

Let us consider the shared memory version of Algorithm 6.1.4 in which the gaxpy is column oriented. Assume $n = rp$ and $col = 1 + (\mu - 1)r{:}\mu r$. A reasonable idea is to use a global array $W(1{:}n, 1{:}p)$ to house the products $A(:, col)x(col)$ produced by each processor, and then have some chosen processor (say Proc(1)) add its columns:

$A_{loc} = A(:, col);\ x_{loc} = x(col);\ w_{loc} = A_{loc}x_{loc};\ W(:, \mu) = w_{loc}$
if $\mu = 1$
    $y_{loc} = y$
    for $j = 1{:}p$
        $w_{loc} = W(:, j)$
        $y_{loc} = y_{loc} + w_{loc}$
    end
    $y = y_{loc}$
end

However, this strategy is seriously flawed because there is no guarantee that $W(1{:}n, 1{:}p)$ is fully initialized when Proc(1) begins the summation process.

What we need is a synchronization construct that can delay the Proc(1) summation until all the processors have computed and stored their contributions in the W array. For this purpose many shared memory systems support some version of the **barrier** construct which we introduce in the following algorithm:

Algorithm 6.1.5 Suppose $A \in \mathbb{R}^{n \times n}$, $x \in \mathbb{R}^n$, and $y \in \mathbb{R}^n$ reside in a global memory accessible to p processors. If $n = rp$ and each processor executes the following algorithm, then upon completion y is overwritten by $y + Ax$. Assume the following initializations in each local memory: p, μ (the node id), and n.

$$r = n/p; \ col = 1 + (\mu - 1)r{:}\mu r; \ A_{loc} = A(:, col); \ x_{loc} = x(col)$$
$$w_{loc} = A_{loc} x_{loc}$$
$$W(:, \mu) = w_{loc}$$
barrier
if $\mu = 1$
$\qquad y_{loc} = y$
$\qquad$ **for** $j = 1{:}p$
$\qquad\qquad w_{loc} = W(:, j)$
$\qquad\qquad y_{loc} = y_{loc} + w_{loc}$
$\qquad$ **end**
$\qquad y = y_{loc}$
end

To understand the **barrier**, it is convenient to regard a processor as either blocked or free. A processor is blocked and suspends execution when it executes the **barrier**. After the pth processor is blocked, all the processors return to the "free state" and resume execution. Think of the **barrier** as treacherous stream to be traversed by all p processors. For safety, they all congregate on the bank before attempting to cross. When the last member of the party arrives, they ford the stream in unison and resume their individual treks.

In Algorithm 6.1.5, the processors are blocked after computing their portion of the matrix-vector product. We cannot predict the order in which these blockings occur, but once the last processor reaches the **barrier**, they are all released and Proc(1) can carry out the vector summation.

6.1.13 Dynamic Scheduling

Instead of having one processor in charge of the vector summation, it is tempting to have each processor add its contribution directly to the global variable y. For Proc(μ), this means executing the following:

$$r = n/p; \; col = 1 + (\mu - 1)r{:}\mu r; \; A_{loc} = A(:, col); \; x_{loc} = x(col)$$
$$w_{loc} = A_{loc} x_{loc}$$
$$y_{loc} = y; \; y_{loc} = y_{loc} + w_{loc}; \; y = y_{loc}$$

However, a problem concerns the read-update-write triplet

$$y_{loc} = y; \; y_{loc} = y_{loc} + w_{loc}; \; y = y_{loc}$$

Indeed, if more than one processor is executing this code fragment at the same time, then there may be a loss of information. Consider the following sequence:

Proc(1) reads y
Proc(2) reads y
Proc(1) writes y
Proc(2) writes y

The contribution of Proc(1) is lost because Proc(1) and Proc(2) obtain the same version of y. As a result, the effect of the Proc(1) write is erased by the Proc(2) write.

To prevent this kind of thing from happening most shared memory systems support the idea of a *critical section*. These are special, isolated portions of a node program that require a "key" to enter. Throughout the system, there is only one key and so the net effect is that only one processor can be executing in a critical section at any given time.

Algorithm 6.1.6 Suppose $A \in \mathbb{R}^{n \times n}$, $x \in \mathbb{R}^n$, and $y \in \mathbb{R}^n$ reside in a global memory accessible to p processors. If $n = pr$ and each processor executes the following algorithm, then upon completion, y is overwritten by $y + Ax$. Assume the following initializations in each local memory: p, μ (the node id), and n.

$$r = n/p; \; col = 1 + (\mu - 1)r{:}\mu r; \; A_{loc} = A(:, col); \; x_{loc} = x(col)$$
$$w_{loc} = A_{loc} x_{loc}$$
begin critical section
$\quad y_{loc} = y$
$\quad y_{loc} = y_{loc} + w_{loc}$
$\quad y = y_{loc}$
end critical section

This use of the critical section concept controls the update of y in a way that ensures correctness. The algorithm is *dynamically scheduled* because the order in which the summations occur is determined as the computation unfolds. Dynamic scheduling is very important in problems with irregular structure.

Problems

P6.1.1 Modify Algorithm 6.1.1 so that it can handle arbitrary n.

P6.1.2 Modify Algorithm 6.1.2 so that it efficiently handles the upper triangular case.

P6.1.3 (a) Modify Algorithms 6.1.3 and 6.1.4 so that they overwrite y with $z = y + A^m x$ for a given positive integer m that is available to each processor. (b) Modify Algorithms 6.1.3 and 6.1.4 so that y is overwritten by $z = y + A^T A x$.

P6.1.4 Modify Algorithm 6.1.3 so that upon completion, the local array A_{loc} in Proc(μ) houses the μth block column of $A + xy^T$.

P6.1.5 Modify Algorithm 6.1.4 so that (a) A is overwritten by the outer product update $A + xy^T$, (b) x is overwritten with $A^2 x$, (c) y is overwritten by a unit 2-norm vector in the direction of $y + A^k x$, and (d) it efficiently handles the case when A is lower triangular.

Notes and References for Sec. 6.1

General references on parallel computations that include several chapters on matrix computations include

G. C. Fox, M. A. Johnson, G. A. Lyzenga, S. W. Otto, J. K. Salmon and D. W. Walker (1988). *Solving Problems on Concurrent Processors, Volume 1.* Prentice Hall, Englewood Cliffs, NJ.

D. P. Bertsekas and J. N. Tsitsiklis (1989). *Parallel and Distributed Computation: Numerical Methods*, Prentice Hall, Englewood Cliffs, NJ.

S. Lakshmivarahan and S. K. Dhall (1990). *Analysis and Design of Parallel Algorithms: Arithmetic and Matrix Problems*, McGraw-Hill, New York.

T. L. Freeman and C. Phillips (1992). *Parallel Numerical Algorithms*, Prentice Hall, New York.

F.T. Leighton (1992). *Introduction to Parallel Algorithms and Architectures*, Morgan Kaufmann, San Mateo, CA.

G. C. Fox, R. D. Williams, and P. C. Messina (1994). *Parallel Computing Works!*, Morgan Kaufmann, San Francisco.

V. Kumar, A. Grama, A. Gupta and G. Karypis (1994). *Introduction to Parallel Computing: Design and Analysis of Algorithms*, Benjamin/Cummings, Reading, MA.

E.F. Van de Velde (1994). *Concurrent Scientific Computing*, Springer-Verlag, New York.

M. Cosnard and D. Trystram (1995). *Parallel Algorithms and Architectures*, International Thomson Computer Press, New York.

Here are some general references that are more specific to parallel matrix computations:

V. Fadeeva and D. Fadeev (1977). "Parallel Computations in Linear Algebra," *Kibernetica 6*, 28–40.

D. Heller (1978). "A Survey of Parallel Algorithms in Numerical Linear Algebra," *SIAM Review 20*, 740-777.

J.M. Ortega and R.G. Voigt (1985). "Solution of Partial Differential Equations on Vector and Parallel Computers," *SIAM Review 27*, 149-240.

J.J. Dongarra and D.C. Sorensen (1986). "Linear Algebra on High Performance Computers," *Appl. Math. and Comp. 20*, 57–88.

K.A. Gallivan, R.J. Plemmons, and A.H. Sameh (1990). "Parallel Algorithms for Dense Linear Algebra Computations," *SIAM Review 32*, 54–135.

J.W. Demmel, M.T. Heath, and H.A. van der Vorst (1993) "Parallel Numerical Linear Algebra," in *Acta Numerica 1993*, Cambridge University Press.

See also

B.N. Datta (1989). "Parallel and Large-Scale Matrix Computations in Control: Some Ideas," *Lin. Alg. and Its Applic. 121*, 243–264.

A. Edelman (1993). "Large Dense Numerical Linear Algebra in 1993: The Parallel Computing Influence," *Int'l J. Supercomputer Appl. 7*, 113–128.

Managing and modelling communication in a distributed memory environment is an important, difficult problem. See

L. Adams and T. Crockett (1984). "Modeling Algorithm Execution Time on Processor Arrays," *Computer 17*, 38–43.

D. Gannon and J. Van Rosendale (1984). "On the Impact of Communication Complexity on the Design of Parallel Numerical Algorithms," *IEEE Trans. Comp. C-33*, 1180–1194.

S.L. Johnsson (1987). "Communication Efficient Basic Linear Algebra Computations on Hypercube Multiprocessors," *J. Parallel and Distributed Computing*, No. 4, 133–172.

Y. Saad and M. Schultz (1989). "Data Communication in Hypercubes," *J. Dist. Parallel Comp. 6*, 115–135.

Y. Saad and M.H. Schultz (1989). "Data Communication in Parallel Architectures," *J. Dist. Parallel Comp. 11*, 131–150.

For snapshots of basic linear algebra computation on a distributed memory system, see

O. McBryan and E.F. van de Velde (1987). "Hypercube Algorithms and Implementations," *SIAM J. Sci. and Stat. Comp. 8*, s227–s287.

S.L. Johnsson and C.T. Ho (1988). "Matrix Transposition on Boolean n-cube Configured Ensemble Architectures," *SIAM J. Matrix Anal. Appl. 9*, 419–454.

T. Dehn, M. Eiermann, K. Giebermann, and V. Sperling (1995). "Structured Sparse Matrix Vector Multiplication on Massively Parallel SIMD Architectures," *Parallel Computing 21*, 1867–1894.

J. Choi, J.J. Dongarra, and D.W. Walker (1995). "Parallel Matrix Transpose Algorithms on Distributed Memory Concurrent Computers," *Parallel Computing 21*, 1387–1406.

L. Colombet, Ph. Michallon, and D. Trystram (1996). "Parallel Matrix-Vector Product on Rings with a Minimum of Communication," *Parallel Computing 22*, 289–310.

The implementation of a parallel algorithm is usually very challenging. It is important to have compilers and related tools that are able to handle the details. See

D.P. O'Leary and G.W. Stewart (1986). "Assignment and Scheduling in Parallel Matrix Factorization," *Lin. Alg. and Its Applic. 77*, 275–300.

J. Dongarra and D.C. Sorensen (1987). "A Portable Environment for Developing Parallel Programs," *Parallel Computing 5*, 175–186.

K. Connolly, J.J. Dongarra, D. Sorensen, and J. Patterson (1988). "Programming Methodology and Performance Issues for Advanced Computer Architectures," *Parallel Computing 5*, 41-58.

P. Jacobson, B. Kagstrom, and M. Rannar (1992). "Algorithm Development for Distributed Memory Multicomputers Using Conlab," *Scientific Programming, 1*, 185–203.

C. Ancourt, F. Coelho, F. Irigoin, and R. Keryell (1993). "A Linear Algebra Framework for Static HPF Code Distribution," *Proceedings of the 4th Workshop on Compilers for Parallel Computers*, Delft, The Netherlands.

D. Bau, I. Kodukula, V. Kotlyar, K. Pingali, and P. Stodghil (1993). "Solving Alignment Using Elementary Linear Algebra," in *Proceedings of the 7th International Workshop on Languages and Compilers for Parallel Computing*, Lecture Notes in Computer Science 892, Springer-Verlag, New York, 46–60.

M. Wolfe (1996). *High Performance Compilers for Parallel Computers*, Addison-Wesley, Reading MA.

6.2 Matrix Multiplication

In this section we develop two parallel algorithms for matrix-matrix multiplication. A shared memory implementation is used to illustrate the effect of blocking on granularity and load balancing. A torus implementation is designed to convey the spirit of two-dimensional data flow.

6.2.1 A Block Gaxpy Procedure

Suppose $A, B, C \in \mathbb{R}^{n \times n}$ with B upper triangular and consider the computation of the matrix multiply update

$$D = C + AB \tag{6.2.1}$$

on a shared memory computer with p processors. Assume that $n = rkp$ and partition the update

$$[\, D_1, \ldots, D_{k \cdot p} \,] \;=\; [\, C_1, \ldots, C_{k \cdot p} \,] + [\, A_1, \ldots, A_{k \cdot p} \,][\, B_1, \ldots, B_{k \cdot p} \,] \tag{6.2.2}$$

where each block column has width $r = n/(kp)$. If

$$B_j \;=\; \begin{bmatrix} B_{1j} \\ \vdots \\ B_{jj} \\ 0 \\ \vdots \\ 0 \end{bmatrix}, \qquad B_{ij} \in \mathbb{R}^{r \times r},$$

then

$$D_j \;=\; C_j + AB_j \;=\; C_j + \sum_{\tau=1}^{j} A_\tau B_{\tau j}. \tag{6.2.3}$$

The number of flops required to compute D_j is given by

$$f_j \;=\; 2nr^2 j \;=\; \left(\frac{2n^3}{k^2 p^2} \right) j.$$

This is an increasing function of j because B is upper triangular. As we discovered in the previous section, the wrap mapping is the way to solve load imbalance problems that result from triangular matrix structure. This suggests that we assign $\text{Proc}(\mu)$ the task of computing D_j for $j = \mu{:}p{:}kp$.

Algorithm 6.2.1 Suppose A, B, and C are n-by-n matrices that reside in a global memory accessible to p processors. Assume that B is upper triangular and $n = rkp$. If each processor executes the following algorithm, then upon completion C is overwritten by $D = C + AB$. Assume the following initializations in each local memory: n, r, k, p and μ (the node id).

> **for** $j = \mu{:}p{:}kp$
> > {Compute D_j.}
> > $B_{loc} = B(1{:}jr, 1 + (j-1)r{:}jr)$
> > $C_{loc} = C({:}, 1 + (j-1)r{:}jr)$
> > **for** $\tau = 1{:}j$
> > > $col = 1 + (\tau - 1)r{:}\tau r$
> > > $A_{loc} = A({:}, col)$
> > > $C_{loc} = C_{loc} + A_{loc}B_{loc}(col, {:})$
> >
> > **end**
> > $C({:}, 1 + (j-1)r{:}jr) = C_{loc}$
>
> **end**

Let us examine the degree of load balancing as a function of the parameter k. For $\mathrm{Proc}(\mu)$, the number of flops required is given by

$$F(\mu) = \sum_{i=1}^{k} f_{\mu+(i-1)p} \approx \left(k\mu + \frac{k^2 p}{2} \right) \frac{2n^3}{k^2 p^2}.$$

The quotient $F(p)/F(1)$ is a measure of load balancing from the flop point of view. Since

$$\frac{F(p)}{F(1)} = \frac{kp + k^2 p/2}{k + k^2 p/2} = 1 + \frac{2(p-1)}{2 + kp}$$

we see that arithmetic balance improves with increasing k. A similar analysis shows that the communication overheads are well balanced as k increases.

On the other hand, the total number of global memory reads and writes associated with Algorithm 6.2.1 increases with the square of k. If the start-up parameter α_s in (6.1.5) is large, then performance can degrade with increased k.

The optimum choice for k given these two opposing forces is system dependent. If communication is fast, then smaller tasks can be supported without penalty and this makes it easier to achieve load balancing. A multiprocessor with this attribute supports *fine-grained parallelism*. However, if granularity is too fine in a system with high-performance nodes, then it may be impossible for the node programs to perform at level-2 or level-3 speeds simply because there just is not enough local linear algebra. Again, benchmarking is the only way to clarify these issues.

6.2.2 Torus

A torus is a two-dimensional processor array in which each row and column is a ring. See FIGURE 6.2.1. A Processor *id* in this context is an ordered pair and each processor has four neighbors. In the displayed exam-

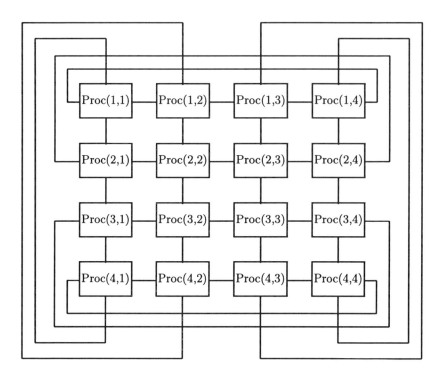

FIGURE 6.2.1 *A Four-by-Four Torus*

ple, Proc(1,3) has *west* neighbor Proc(1,2), *east* neighbor Proc(1,4), *south* neighbor Proc(2,3), and *north* neighbor Proc(4,3).

To show what it is like to organize a toroidal matrix computation, we develop an algorithm for the matrix multiplication $D = C + AB$ where $A, B, C \in \mathbb{R}^{n \times n}$. Assume that the torus is p_1-by-p_1 and that $n = rp_1$. Regard $A = (A_{ij})$, $B = (B_{ij})$, and $C = (C_{ij})$ as p_1-by-p_1 block matrices with r-by-r blocks. Assume that Proc(i, j) contains A_{ij}, B_{ij}, and C_{ij} and that its mission is to overwrite C_{ij} with

$$D_{ij} = C_{ij} + \sum_{k=1}^{p_1} A_{ik} B_{kj}.$$

We develop the general algorithm from the $p_1 = 3$ case, displaying the torus in cellular form as follows:

Proc(1,1)	Proc(1,2)	Proc(1,3)
Proc(2,1)	Proc(2,2)	Proc(2,3)
Proc(3,1)	Proc(3,2)	Proc(3,3)

Let us focus attention on Proc(1,1) and the calculation of

$$D_{11} = C_{11} + A_{11}B_{11} + A_{12}B_{21} + A_{13}B_{31}.$$

Suppose the six inputs that define this block dot product are positioned within the torus as follows:

A_{11}	B_{11}	A_{12}	·	A_{13}	·
·	B_{21}	·	·	·	∘
·	B_{31}	·	·	·	·

(Pay no attention to the "dots." They are later replaced by various A_{ij} and B_{ij}).

Our plan is to "ratchet" the first block row of A and the first block column of B through Proc(1,1) in a coordinated fashion. The pairs A_{11} and B_{11}, A_{12} and B_{21}, and A_{13} and B_{31} meet, are multiplied, and added into a running sum array C_{loc}:

A_{12}	B_{21}	A_{13}	·	A_{11}	·
·	B_{31}	·	·	·	·
·	B_{11}	·	·	·	·

$$C_{loc} = C_{loc} + A_{12}B_{21}$$

A_{13}	B_{31}	A_{11}	·	A_{12}	·
·	B_{11}	·	·	·	·
·	B_{21}	·	·	·	·

$$C_{loc} = C_{loc} + A_{13}B_{31}$$

A_{11}	B_{11}	A_{12}	·	A_{13}	·
·	B_{21}	·	·	·	·
·	B_{31}	·	·	·	·

$$C_{loc} = C_{loc} + A_{11}B_{11}$$

Thus, after three steps, the local array C_{loc} in Proc(1,1) houses D_{11}.

We have organized the flow of data so that the A_{1j} migrate westwards and the B_{i1} migrate northwards through the torus. It is thus apparent that Proc(1,1) must execute a node program of the form:

> **for** $t = 1:3$
>> **send**($A_{loc}, west$)
>> **send**($B_{loc}, north$)
>> **recv**($A_{loc}, east$)
>> **recv**($B_{loc}, south$)
>> $C_{loc} = C_{loc} + A_{loc}B_{loc}$
>
> **end**

The **send-recv-send-recv** sequence

> **for** $t = 1:3$
>> **send**($A_{loc}, west$)
>> **recv**($A_{loc}, east$)
>> **send**($B_{loc}, north$)
>> **recv**($B_{loc}, south$)
>> $C_{loc} = C_{loc} + A_{loc}B_{loc}$
>
> **end**

also works. However, this induces unnecessary delays into the process because the B submatrix is not sent until the new A submatrix arrives.

We next consider the activity in Proc(1,2), Proc(1,3), Proc(2,1), and Proc(3,1). At this point in the development, these processors merely help circulate blocks A_{11}, A_{12}, and A_{13} and B_{11}, B_{21}, and B_{31}, respectively. If B_{32}, B_{12}, and B_{22} flowed through Proc(1,2) during these steps, then

$$D_{12} = C_{12} + A_{13}B_{32} + A_{11}B_{12} + A_{12}B_{22}$$

could be formed. Likewise, Proc(1,3) could compute

$$D_{13} = C_{13} + A_{11}B_{13} + A_{12}B_{23} + A_{13}B_{33}$$

if B_{13}, B_{23}, and B_{33} are available during $t = 1:3$. To this end we initialize the torus as follows

A_{11}	B_{11}	A_{12}	B_{22}	A_{13}	B_{33}
$\cdot$	B_{21}	$\cdot$	B_{32}	$\cdot$	B_{13}
$\cdot$	B_{31}	$\cdot$	B_{12}	$\cdot$	B_{23}

With northward flow of the B_{ij} we get

A_{12} B_{21}'	A_{13} B_{32}	A_{11} B_{13}	
· B_{31}	· B_{12}	· B_{23}	$t = 1$
· B_{11}	· B_{22}	· B_{33}	

A_{13} B_{31}	A_{11} B_{12}	a_{12} B_{23}	
· B_{11}	· B_{22}	· B_{33}	$t = 2$
· B_{21}	· B_{32}	· B_{13}	

A_{11} B_{11}	A_{12} B_{22}	a_{13} B_{33}	
· B_{21}	· B_{32}	· B_{13}	$t = 3$
· B_{31}	· B_{12}	· B_{23}	

Thus, if B is mapped onto the torus in a "staggered start" fashion, we can arrange for the first row of processors to compute the first row of C.

If we stagger the second and third rows of A in a similar fashion, then we can arrange for all nine processors to perform a multiply-add at each step. In particular, if we set

A_{11} B_{11}	A_{12} B_{22}	A_{13} B_{33}
A_{22} B_{21}	A_{23} B_{32}	A_{21} B_{13}
A_{33} B_{31}	A_{31} B_{12}	A_{32} B_{23}

then with westward flow of the A_{ij} and northward flow of the B_{ij} we obtain

A_{12} B_{21}	A_{13} B_{32}	A_{11} B_{13}	
A_{23} B_{31}	A_{21} B_{12}	A_{22} B_{23}	$t = 1$
A_{31} B_{11}	A_{32} B_{22}	A_{33} B_{33}	

A_{13} B_{31}	A_{11} B_{12}	A_{12} B_{23}
A_{21} B_{11}	A_{22} B_{22}	A_{23} B_{33}
A_{32} B_{21}	A_{33} B_{32}	A_{31} B_{13}

$t = 2$

A_{11} B_{11}	A_{12} B_{22}	A_{13} B_{33}
A_{22} B_{21}	A_{23} B_{32}	A_{21} B_{13}
A_{33} B_{31}	A_{31} B_{12}	A_{32} B_{23}

$t = 3$

From this example we are ready to specify the general algorithm. We assume that at the start, $\text{Proc}(i, j)$ houses A_{ij}, B_{ij}, and C_{ij}. To obtain the necessary staggering of the A data, we note that in processor row i the A_{ij} should be circulated westward $i - 1$ positions. Likewise, in the jth column of processors, the B_{ij} should be circulated northward $j - 1$ positions. This gives the following algorithm:

Algorithm 6.2.2 Suppose $A \in \mathbb{R}^{n \times n}$, $B \in \mathbb{R}^{n \times n}$, and $C \in \mathbb{R}^{n \times n}$ are given and that $D = C + AB$. If each processor in a p_1-by-p_1 torus executes the following algorithm and $n = p_1 r$, then upon completion $\text{Proc}(\mu, \lambda)$ houses $D_{\mu\lambda}$ in local variable C_{loc}. Assume the following local memory initializations: p_1, (μ, λ) (the node id), *north*, *east*, *south*, and *west*, (the four neighbor *id*'s), $row = 1 + (\mu - 1)r{:}\mu r$, $col = 1 + (\lambda - 1)r{:}\lambda r$, $A_{loc} = A(row, col)$, $B_{loc} = B(row, col)$, and $C_{loc} = C(row, col)$.

> {Stagger the $A_{\mu j}$ and $B_{i\lambda}$. }
> for $k = 1{:}\mu - 1$
>   **send**$(A_{loc}, west)$; **recv**$(A_{loc}, east)$
> end
> for $k = 1{:}\lambda - 1$
>   **send**$(B_{loc}, north)$; **recv**$(B_{loc}, south)$
> end
> for $k = 1{:}p_1$
> $C_{loc} = C_{loc} + A_{loc}B_{loc}$
>   **send**$(A_{loc}, west)$
>   **send**$(B_{loc}, north)$
>   **recv**$(A_{loc}, east)$
>   **recv**$(B_{loc}, south)$
> end

{Unstagger the $A_{\mu j}$ and $B_{i\lambda}$.}
for $k = 1{:}\mu - 1$
    send$(A_{loc}, east)$; **recv**$(A_{loc}, west)$
end
for $k = 1{:}\lambda - 1$
    send$(B_{loc}, south)$; **recv**$(B_{loc}, north)$
end

It is not hard to show that the computation-to-communication ratio for this algorithm goes to zero as n/p_1 increases.

Problems

P6.2.1 Develop a ring implementation for Algorithm 6.2.1.

P6.2.2 An upper triangular matrix can be overwritten with its square without any additional workspace. Write a dynamically scheduled, shared-memory procedure for doing this.

Notes and References for Sec. 6.2

Matrix computations on 2-dimensional arrays are discussed in

H.T. Kung (1982). "Why Systolic Architectures?," *Computer 15*, 37-46.

D.P. O'Leary and G.W. Stewart (1985). "Data Flow Algorithms for Parallel Matrix Computations," *Comm. ACM 28*, 841-853.

B. Hendrickson and D. Womble (1994). "The Torus-Wrap Mapping for Dense Matrix Calculations on Massively Parallel Computers," *SIAM J. Sci. Comput. 15*, 1201–1226.

Various aspects of parallel matrix multiplication are discussed in

L.E. Cannon (1969). *A Cellular Computer to Implement the Kalman Filter Algorithm*, Ph.D. Thesis, Montana State University.

K.H. Cheng and S. Sahni (1987). "VLSI Systems for Band Matrix Multiplication," *Parallel Computing 4*, 239–258.

G. Fox, S.W. Otto, and A.J. Hey (1987). "Matrix Algorithms on a Hypercube I: Matrix Multiplication," *Parallel Computing 4*, 17–31.

J. Berntsen (1989). "Communication Efficient Matrix Multiplication on Hypercubes," *Parallel Computing 12*, 335–342.

H.J. Jagadish and T. Kailath (1989). "A Family of New Efficient Arrays for Matrix Multiplication," *IEEE Trans. Comput. 38*, 149–155.

P. Bjørstad, F. Manne, T.Sørevik, and M. Vajteršic (1992). "Efficient Matrix Multiplication on SIMD Computers," *SIAM J. Matrix Anal. Appl. 13*, 386–401.

K. Mathur and S.L. Johnsson (1994). "Multiplication of Matrices of Arbitrary Shape on a Data Parallel Computer," *Parallel Computing 20*, 919–952.

R. Mathias (1995). "The Instability of Parallel Prefix Matrix Multiplication," *SIAM J. Sci. Comp. 16*, 956–973.

6.3   Factorizations

In this section we present a pair of parallel Cholesky factorizations. To illustrate what a distributed memory factorization looks like, we implement the gaxpy Cholesky algorithm on a ring. A shared memory implementation of outer product Cholesky is also detailed.

6.3.1   A Ring Cholesky

Let us see how the Cholesky factorization procedure can be distributed on a ring of p processors. The starting point is the equation

$$G(\mu,\mu)G(\mu{:}n,\mu) \;=\; A(\mu{:}n,\mu) - \sum_{j=1}^{\mu-1} G(\mu,j)G(\mu{:}n,j) \;\equiv\; v(\mu{:}n) \,.$$

This equation is obtained by equating the μth column in the n-by-n equation $A = GG^T$. Once the vector $v(\mu{:}n)$ is found then $G(\mu{:}n,\mu)$ is a simple scaling:

$$G(\mu{:}n,\mu) \;=\; v(\mu{:}n)/\sqrt{v(\mu)}.$$

For clarity, we first assume that $n = p$ and that $\text{Proc}(\mu)$ initially houses $A(\mu{:}n,\mu)$. Upon completion, each processor overwrites its A-column with the corresponding G-column. For $\text{Proc}(\mu)$ this process involves $\mu - 1$ saxpy updates of the form

$$A(\mu{:}n,\mu) \longleftarrow A(\mu{:}n,\mu) - G(\mu,j)G(\mu{:}n,j)$$

followed by a square root and a scaling. The general structure of $\text{Proc}(\mu)$'s node program is therefore as follows:

> **for** $j = 1{:}\mu - 1$
> > Receive a G-column from the left neighbor.
> > If necessary, send a copy of the received G-column to
> > > the right neighbor.
> > Update $A(\mu{:}n,\mu)$.
> **end**
> Generate $G(\mu{:}n,\mu)$ and, if necessary, send it to the
> > right neighbor.

Thus $\text{Proc}(1)$ immediately computes $G(1{:}n,1) = A(1{:}n,1)/\sqrt{A(1,1)}$ and sends it to $\text{Proc}(2)$. As soon as $\text{Proc}(2)$ receives this column it can generate $G(2{:}n,2)$ and pass it to $\text{Proc}(3)$ etc.. With this pipelining arrangement we can assert that once a processor computes its G-column, it can quit. It also follows that each processor receives G-columns in ascending order, i.e., $G(1{:}n,1)$, $G(2{:}n,2)$, etc. Based on these observations we have

$j = 1$
while $j < \mu$
      recv$(g_{loc}(j{:}n), left)$
      if $\mu < n$
          send$(g_{loc}(j{:}n), right)$
      end
      $A_{loc}(\mu{:}n) = A_{loc}(\mu{:}n) - g_{loc}(\mu)g(\mu{:}n)$
      $j = j + 1$
end
$A_{loc}(\mu{:}n) = A_{loc}(\mu{:}n)/\sqrt{A_{loc}(\mu)}$
if $\mu < n$
      send$(A_{loc}(\mu{:}n), right)$
end

Note that the number of received G-columns is given by $j - 1$. If $j = \mu$, then it is time for Proc(μ) to generate and send $G(\mu{:}n, \mu)$.

We now extend this strategy to the general n case. There are two obvious ways to distribute the computation. We could require each processor to compute a contiguous set of G-columns. For example, if $n = 11$, $p = 3$, and $A = [\, a_1, \ldots, a_{11} \,]$, then we could distribute A as follows

$$\left[\underbrace{a_1\ a_2\ a_3\ a_4}_{\text{Proc}(1)} \,\middle|\, \underbrace{a_5\ a_6\ a_7\ a_8}_{\text{Proc}(2)} \,\middle|\, \underbrace{a_9\ a_{10}\ a_{11}}_{\text{Proc}(3)} \right] .$$

Each processor could then proceed to find the corresponding G columns. The trouble with this approach is that (for example) Proc(1) is idle after the fourth column of G is found even though much work remains.

Greater load balancing results if we distribute the computational tasks using the wrap mapping, i.e.,

$$\left[\underbrace{a_1\ a_4\ a_7\ a_{10}}_{\text{Proc}(1)} \,\middle|\, \underbrace{a_2\ a_5\ a_8\ a_{11}}_{\text{Proc}(2)} \,\middle|\, \underbrace{a_3\ a_6\ a_9}_{\text{Proc}(3)} \right] .$$

In this scheme Proc(μ) carries out the construction of $G(:, \mu{:}p{:}n)$. When a given processor finishes computing its G-columns, each of the other processors has at most one more G column to find. Thus if $n/p \gg 1$, then all of the processors are busy most of the time.

Let us examine the details of a wrap-distributed Cholesky procedure. Each processor maintains a pair of counters. The counter j is the index of the next G-column to be received by Proc(μ). A processor also needs to know the index of the next G-column that it is to produce. Note that if $col = \mu{:}p{:}n$, then Proc(μ) is responsible for $G(:, col)$ and that $L = \textbf{length}(col)$ is the number of the G-columns that it must compute. We use q to indicate the status of G-column production. At any instant,

$col(q)$ is the index of the next G-column to be produced.

Algorithm 6.3.1 Suppose $A \in \mathbb{R}^{n \times n}$ is symmetric and positive definite and that $A = GG^T$ is its Cholesky factorization. If each node in a p-processor ring executes the following program, then upon completion $\text{Proc}(\mu)$ houses $G(k:n, k)$ for $k = \mu:p:n$ in a local array $A_{loc}(1:n, L)$ where $L = \textbf{length}(col)$ and $col = \mu:p:n$. In particular, $G(col(q):n, col(q))$ is housed in $A_{loc}(col(q):n, q)$ for $q = 1:L$. Assume the following local memory initializations: p, μ (the node id), *left* and *right* (the neighbor id's), n, and $A_{loc} = A(\mu:p:n, :)$.

$$j = 1; \ q = 1; \ col = \mu:p:n; \ L = \textbf{length}(col)$$
$$\textbf{while } q \leq L$$
$$\quad \textbf{if } j = col(q)$$
$$\quad\quad \{ \text{ Form } G(j:n, j) \}$$
$$\quad\quad A_{loc}(j:n, q) = A_{loc}(j:n, q)/\sqrt{A_{loc}(j, q)}$$
$$\quad\quad \textbf{if } j < n$$
$$\quad\quad\quad \textbf{send}(A_{loc}(j:n, q), right)$$
$$\quad\quad \textbf{end}$$
$$\quad\quad j = j + 1$$
$$\quad\quad \{ \text{ Update local columns. } \}$$
$$\quad\quad \textbf{for } k = q + 1:L$$
$$\quad\quad\quad r = col(k)$$
$$\quad\quad\quad A_{loc}(r:n, k) = A_{loc}(r:n, k) - A_{loc}(r, q)A_{loc}(r:n, q)$$
$$\quad\quad \textbf{end}$$
$$\quad\quad q = q + 1$$
$$\quad \textbf{else}$$
$$\quad\quad \textbf{recv}(g_{loc}(j:n), left)$$
$$\quad\quad \text{Compute } \alpha, \text{ the id of the processor that generated the}$$
$$\quad\quad\quad \text{received } G\text{-column.}$$
$$\quad\quad \text{Compute } \beta, \text{ the index of } \text{Proc}(right)\text{'s final column.}$$
$$\quad\quad \textbf{if } right \neq \alpha \ \wedge \ j < \beta$$
$$\quad\quad\quad \textbf{send}(g_{loc}(j:n), right)$$
$$\quad\quad \textbf{end}$$
$$\quad\quad \{ \text{ Update local columns. } \}$$
$$\quad\quad \textbf{for } k = q:L$$
$$\quad\quad\quad r = col(k)$$
$$\quad\quad\quad A.loc(r:n, k) = A_{loc}(r:n, k) - g_{loc}(r)g_{loc}(r:n)$$
$$\quad\quad \textbf{end}$$
$$\quad\quad j = j + 1$$
$$\quad \textbf{end}$$
$$\textbf{end}$$

To illustrate the logic of the pointer system we consider a sample 3-processor situation with $n = 10$. Assume that the three local values of q are 3,2, and

2 and that the corresponding values of $col(q)$ are 7, 5, and 6 :

$$\left[\underbrace{a_1\ a_4\ \overset{\downarrow}{a_7}\ a_{10}}_{\text{Proc}(1)}\ \Big|\ \underbrace{a_2\ \overset{\downarrow}{a_5}\ a_8\ a_{11}}_{\text{Proc}(2)}\ \Big|\ \underbrace{a_3\ \overset{\downarrow}{a_6}\ a_9}_{\text{Proc}(3)}\right]$$

Proc(2) now generates the fifth G-column and increment its q to 3.

The decision to pass a received G-column to the right neighbor needs to be explained. Two conditions must be fulfilled:

- The right neighbor must not be the processor which generated the G column. This way the circulation of the received G-column is properly terminated.

- The right neighbor must still have more G-columns to generate. Otherwise, a G-column will be sent to an inactive processor.

This kind of reasoning is quite typical in distributed memory matrix computations.

Let us examine the behavior of Algorithm 6.3.1 under the assumption that $n \gg p$. It is not hard to show that $\text{Proc}(\mu)$ performs

$$F(\mu) = \sum_{k=1}^{L} 2(n - (\mu + (k-1)p))(\mu + (k-1)p) \approx \frac{n^3}{3p}$$

flops. Each processor receives and sends just about every G-column. Using our communication overhead model (6.1.3), we see that the time each processor spends communicating is given by

$$m_\mu = \sum_{j=1}^{n} 2(\alpha_d + \beta_d(n-j)) \approx 2\alpha_d n + \beta_d n^2 \,.$$

If we assume that computation proceeds at R flops per second, then the computation/communication ratio for Algorithm 6.3.1 is approximately given by $(n/p)(1/3R\beta_d)$. Thus, communication overheads diminish in importance as n/p grows.

6.3.2 A Shared Memory Cholesky

Next we consider a shared memory implementation of the outer product Cholesky algorithm:

```
for k = 1:n
    A(k:n, k) = A(k:n, k)/√A(k, k)
    for j = k + 1:n
        A(j:n, j) = A(j:n, j) − A(j:n, k)A(j, k)
    end
end
```

The j-loop oversees an outer product update. The $n - k$ saxpy operations that make up its body are independent and easily parallelized. The scaling $A(k{:}n, k)$ can be carried out by a single processor with no threat to load balancing.

Algorithm   6.3.2   Suppose $A \in \mathbb{R}^{n \times n}$ is a symmetric positive definite matrix stored in a shared memory accessible to p processors. If each processor executes the following algorithm, then upon completion the lower triangular part of A is overwritten with its Cholesky factor. Assume the following initializations in each local memory: n, p and μ (the node id).

> for $k = 1{:}n$
>> if $\mu = 1$
>>> $v_{loc}(k{:}n) = A(k{:}n)$
>>> $v_{loc}(k{:}n) = v_{loc}(k{:}n)/\sqrt{v_{loc}(k)}$
>>> $A(k{:}n, k) = v_{loc}(k{:}n)$
>> end
>> barrier
>> $v_{loc}(k + 1{:}n) = A(k + 1{:}n, k)$
>> for $j = (k + \mu){:}p{:}n$
>>> $w_{loc}(j{:}n) = A(j{:}n, j)$
>>> $w_{loc}(j{:}n) = w_{loc}(j{:}n) - v_{loc}(j)v_{loc}(j{:}n)$
>>> $A(j{:}n, j) = w_{loc}(j{:}n)$
>> end
>> barrier
> end

The scaling before the j-loop represents very little work compared to the outer product update and so it is reasonable to assign that portion of the computation to a single processor. Notice that two **barrier** statements are required. The first ensures that a processor does not begin working on the kth outer product update until the kth column of G is made available by Proc(1). The second **barrier** prevents the processing of the $k+1$st step to begin until the kth step is completely finished.

Problems

P6.3.1 It is possible to formulate a block version of Algorithm 6.3.1. Suppose $n = rN$. For $k = 1{:}N$ we (a) have Proc(1) generate $G(:, 1+(k-1)r{:}kr)$ and (b) have all processors participate in the rank r update of the trailing submatrix $A(kr+1{:}n, kr+1{:}n)$. See §4.2.6. The coarser granularity may improve performance if the individual processors like level-3 operations.

P6.3.2 Develop a shared memory QR factorization patterned after Algorithm 6.3.2. Proc(1) should generate the Householder vectors and all processors should share in the ensuing Householder update.

Notes and References for Sec. 6.3

General comments on distributed memory factorization procedures may be found in

G.A. Geist and M.T. Heath (1986). "Matrix Factorization on a Hypercube," in M.T. Heath (ed) (1986). *Proceedings of First SIAM Conference on Hypercube Multiprocessors*, SIAM Publications, Philadelphia, Pa.

I.C.F. Ipsen, Y. Saad, and M. Schultz (1986). "Dense Linear Systems on a Ring of Processors," *Lin. Alg. and Its Applic. 77*, 205–239.

D.P. O'Leary and G.W. Stewart (1986). "Assignment and Scheduling in Parallel Matrix Factorization," *Lin. Alg. and Its Applic. 77*, 275–300.

R.S. Schreiber (1988). "Block Algorithms for Parallel Machines," in *Numerical Algorithms for Modern Parallel Computer Architectures*, M.H. Schultz (ed), IMA Volumes in Mathematics and Its Applications, Number 13, Springer-Verlag, Berlin, 197–207.

S.L. Johnsson and W. Lichtenstein (1993). "Block Cyclic Dense Linear Algebra," *SIAM J. Sci.Comp. 14*, 1257–1286.

Papers specifically concerned with LU, Cholesky and QR include

R.N. Kapur and J.C. Browne (1984). "Techniques for Solving Block Tridiagonal Systems on Reconfigurable Array Computers," *SIAM J. Sci. and Stat. Comp. 5*, 701-719.

G.J. Davis (1986). "Column LU Pivoting on a Hypercube Multiprocessor," *SIAM J. Alg. and Disc. Methods 7*, 538–550.

J.M. Delosme and I.C.F. Ipsen (1986). "Parallel Solution of Symmetric Positive Definite Systems with Hyperbolic Rotations," *Lin. Alg. and Its Applic. 77*, 75–112.

A. Pothen, S. Jha, and U. Vemapulati (1987). "Orthogonal Factorization on a Distributed Memory Multiprocessor," in *Hypercube Multiprocessors*, ed. M.T. Heath, SIAM Press, 1987.

C.H. Bischof (1988). "QR Factorization Algorithms for Coarse Grain Distributed Systems," PhD Thesis, Dept. of Computer Science, Cornell University, Ithaca, NY.

G.A. Geist and C.H. Romine (1988). "LU Factorization Algorithms on Distributed Memory Multiprocessor Architectures," *SIAM J. Sci. and Stat. Comp. 9*, 639–649.

J.M. Ortega and C.H. Romine (1988). "The *ijk* Forms of Factorization Methods II: Parallel Systems," *Parallel Computing 7*, 149–162.

M. Marrakchi and Y. Robert (1989). "Optimal Algorithms for Gaussian Elimination on an MIMD Computer," *Parallel Computing 12*, 183–194.

Parallel triangular system solving is discussed in

R. Montoye and D. Laurie (1982). "A Practical Algorithm for the Solution of Triangular Systems on a Parallel Processing System," *IEEE Trans. Comp. C-31*, 1076–1082.

D.J. Evans and R. Dunbar (1983). "The Parallel Solution of Triangular Systems of Equations," *IEEE Trans. Comp. C-32*, 201–204.

C.H. Romine and J.M. Ortega (1988). "Parallel Solution of Triangular Systems of Equations," *Parallel Computing 6*, 109–114.

M.T. Heath and C.H. Romine (1988). "Parallel Solution of Triangular Systems on Distributed Memory Multiprocessors," *SIAM J. Sci. and Stat. Comp. 9*, 558–588.

G. Li and T. Coleman (1988). "A Parallel Triangular Solver for a Distributed-Memory Multiprocessor," *SIAM J. Sci. and Stat. Comp. 9*, 485–502.

S.C. Eisenstat, M.T. Heath, C.S. Henkel, and C.H. Romine (1988). "Modified Cyclic Algorithms for Solving Triangular Systems on Distributed Memory Multiprocessors," *SIAM J. Sci. and Stat. Comp. 9*, 589–600.

N.J. Higham (1995). "Stability of Parallel Triangular System Solvers," *SIAM J. Sci. Comp. 16*, 400–413.

Papers on the parallel computation of the LU and Cholesky factorization include

R.P. Brent and F.T. Luk (1982) "Computing the Cholesky Factorization Using a Systolic Architecture," *Proc. 6th Australian Computer Science Conf.* 295–302.

D.P. O'Leary and G.W. Stewart (1985). "Data Flow Algorithms for Parallel Matrix Computations," *Comm. of the ACM 28*, 841–853.

J.M. Delosme and I.C.F. Ipsen (1986). "Parallel Solution of Symmetric Positive Definite Systems with Hyperbolic Rotations," *Lin. Alg. and Its Applic. 77*, 75–112.

R.E. Funderlic and A. Geist (1986). "Torus Data Flow for Parallel Computation of Missized Matrix Problems," *Lin. Alg. and Its Applic. 77*, 149–164.

M. Costnard, M. Marrakchi, and Y. Robert (1988). "Parallel Gaussian Elimination on an MIMD Computer," *Parallel Computing 6*, 275–296.

Parallel methods for banded and sparse systems include

S.L. Johnsson (1985). "Solving Narrow Banded Systems on Ensemble Architectures," *ACM Trans. Math. Soft. 11*, 271–288.

S.L. Johnsson (1986). "Band Matrix System Solvers on Ensemble Architectures," in *Supercomputers: Algorithms, Architectures, and Scientific Computation*, eds. F.A. Matsen and T. Tajima, University of Texas Press, Austin TX., 196–216.

S.L. Johnsson (1987). "Solving Tridiagonal Systems on Ensemble Architectures," *SIAM J. Sci. and Stat. Comp. 8*, 354–392.

U. Meier (1985). "A Parallel Partition Method for Solving Banded Systems of Linear Equations," *Parallel Computers 2*, 33–43.

H. van der Vorst (1987). "Large Tridiagonal and Block Tridiagonal Linear Systems on Vector and Parallel Computers," *Parallel Comput. 5*, 45–54.

R. Bevilacqua, B. Codenotti, and F. Romani (1988). "Parallel Solution of Block Tridiagonal Linear Systems," *Lin.Alg. and Its Applic. 104*, 39–57.

E. Gallopoulos and Y. Saad (1989). "A Parallel Block Cyclic Reduction Algorithm for the Fast Solution of Elliptic Equations," *Parallel Computing 10*, 143–160.

J.M. Conroy (1989). "A Note on the Parallel Cholesky Factorization of Wide Banded Matrices," *Parallel Computing 10*, 239–246.

M. Hegland (1991). "On the Parallel Solution of Tridiagonal Systems by Wrap-Around Partitioning and Incomplete LU Factorization," *Numer. Math. 59*, 453–472.

M.T. Heath, E. Ng, and B.W. Peyton (1991). "Parallel Algorithms for Sparse Linear Systems," *SIAM Review 33*, 420–460.

V. Mehrmann (1993). "Divide and Conquer Methods for Block Tridiagonal Systems," *Parallel Computing 19*, 257–280.

P. Raghavan (1995). "Distributed Sparse Gaussian Elimination and Orthogonal Factorization," *SIAM J. Sci. Comp. 16*, 1462–1477.

Parallel QR factorization procedures are of interest in real-time signal processing. Details may be found in

W.M. Gentleman and H.T. Kung (1981). "Matrix Triangularization by Systolic Arrays," SPIE Proceedings, Vol. 298, 19–26.

D.E. Heller and I.C.F. Ipsen (1983). "Systolic Networks for Orthogonal Decompositions," *SIAM J. Sci. and Stat. Comp. 4*, 261–269.

M. Costnard, J.M. Muller, and Y. Robert (1986). "Parallel QR Decomposition of a Rectangular Matrix," *Numer. Math. 48*, 239–250.

L. Eldin and R. Schreiber (1986). "An Application of Systolic Arrays to Linear Discrete Ill-Posed Problems," *SIAM J. Sci. and Stat. Comp. 7*, 892–903.

F.T. Luk (1986). "A Rotation Method for Computing the QR Factorization," *SIAM J. Sci. and Stat. Comp. 7*, 452–459.

J.J. Modi and M.R.B. Clarke (1986). "An Alternative Givens Ordering," *Numer. Math. 43*, 83–90.

P. Amodio and L. Brugnano (1995). "The Parallel QR Factorization Algorithm for Tridiagonal Linear Systems," *Parallel Computing 21*, 1097–1110.

Parallel factorization methods for shared memory machines are discussed in

S. Chen, D. Kuck, and A. Sameh (1978). "Practical Parallel Band Triangular Systems Solvers," *ACM Trans. Math. Soft. 4*, 270–277.

A. Sameh and D. Kuck (1978). "On Stable Parallel Linear System Solvers," *J. Assoc. Comp. Mach. 25*, 81-91.

P. Swarztrauber (1979). "A Parallel Algorithm for Solving General Tridiagonal Equations," *Math. Comp. 33*, 185-199.

S. Chen, J. Dongarra, and C. Hsuing (1984). "Multiprocessing Linear Algebra Algorithms on the Cray X-MP-2: Experiences with Small Granularity," *J. Parallel and Distributed Computing 1*, 22–31.

J.J. Dongarra and A.H. Sameh (1984). "On Some Parallel Banded System Solvers," *Parallel Computing 1*, 223–235.

J.J. Dongarra and R.E. Hiromoto (1984). "A Collection of Parallel Linear Equation Routines for the Denelcor HEP,' *Parallel Computing 1*, 133–142.

J.J. Dongarra and T. Hewitt (1986). "Implementing Dense Linear Algebra Algorithms Using Multitasking on the Cray X-MP-4 (or Approaching the Gigaflop)," *SIAM J. Sci. and Stat. Comp. 7*, 347–350.

J.J. Dongarra, A. Sameh, and D. Sorensen (1986). "Implementation of Some Concurrent Algorithms for Matrix Factorization," *Parallel Computing 3*, 25–34.

A. George, M.T. Heath, and J. Liu (1986). "Parallel Cholesky Factorization on a Shared Memory Multiprocessor," *Lin. Alg. and Its Applic. 77*, 165–187.

J.J. Dongarra and D.C. Sorensen (1987). "Linear Algebra on High Performance Computers," *Appl. Math. and Comp. 20*, 57–88.

K. Dackland, E. Elmroth, and B. Kagstrom (1992). "Parallel Block Factorizations on the Shared Memory Multiprocessor IBM 3090 VF/600J," *International J. Supercomputer Applications, 6*, 69–97.

Chapter 7

The Unsymmetric Eigenvalue Problem

Having discussed linear equations and least squares, we now direct our attention to the third major problem area in matrix computations, the algebraic eigenvalue problem. The unsymmetric problem is considered in this chapter and the more agreeable symmetric case in the next.

Our first task is to present the decompositions of Schur and Jordan along with the basic properties of eigenvalues and invariant subspaces. The contrasting behavior of these two decompositions sets the stage for §7.2 in which we investigate how the eigenvalues and invariant subspaces of a matrix are affected by perturbation. Condition numbers are developed that permit estimation of the errors that can be expected to arise because of roundoff.

The key algorithm of the chapter is the justly famous QR algorithm. This procedure is the most complex algorithm presented in this book and its development is spread over three sections. We derive the basic QR iteration in §7.3 as a natural generalization of the simple power method. The next two sections are devoted to making this basic iteration computationally feasible. This involves the introduction of the Hessenberg decomposition in §7.4 and the notion of origin shifts in §7.5.

The QR algorithm computes the real Schur form of a matrix, a canonical form that displays eigenvalues but not eigenvectors. Consequently, additional computations usually must be performed if information regarding invariant subspaces is desired. In §7.6, which could be subtitled, "What to Do after the Real Schur Form is Calculated," we discuss various invariant subspace calculations that can follow the QR algorithm.

Finally, in the last section we consider the generalized eigenvalue problem $Ax = \lambda Bx$ and a variant of the QR algorithm that has been devised to solve it. This algorithm, called the QZ algorithm, underscores the importance of orthogonal matrices in the eigenproblem, a central theme of the chapter.

It is appropriate at this time to make a remark about complex versus real arithmetic. In this book, we focus on the development of real arithmetic algorithms for real matrix problems. This chapter is no exception even though a real unsymmetric matrix can have complex eigenvalues. However, in the derivation of the practical, real arithmetic QR algorithm and in the mathematical analysis of the eigenproblem itself, it is convenient to work in the complex field. Thus, the reader will find that we have switched to complex notation in §7.1, §7.2, and §7.3. In these sections, we use complex versions of the QR factorization, the singular value decomposition, and the CS decomposition.

Before You Begin

Chapters 1-3 and §§5.1-5.2 are assumed. Within this chapter there are the following dependencies:

$$§7.1 \;\;\rightarrow\;\; §7.2 \;\;\rightarrow\;\; §7.3 \;\;\rightarrow\;\; §7.4 \;\;\rightarrow\;\; §7.5 \;\;\rightarrow\;\; §7.6 \;\;\rightarrow\;\; §7.7$$

Complementary references include Fox (1964), Wilkinson (1965), Gourlay and Watson (1973), Stewart (1973), Hager (1988), Ciarlet (1989), Stewart and Sun (1990), Watkins (1991), Saad (1992), Jennings and Mc Keowen (1992), Datta (1995), Trefethen and Bau (1997), and Demmel (1996). Some Matlab functions important to this chapter are **eig**, **poly**, **polyeig**, **hess**, **qz**, **rsf2csf**, **cdf2rdf**, **schur**, and **balance**. LAPACK connections include

LAPACK: Unsymmetric Eigenproblem	
_GEBAL	Balance transform
_GEBAK	Undo balance transform
_GEHRD	Hessenberg reduction $U^H AV = H$
_ORMHR	U (factored form) times matrix (real case)
_ORGHR	Generates U (real case)
_UNMHR	U (factored form) times matrix (complex case)
_UNGHR	Generates U (complex case)
_HSEQR	Schur decomposition of Hessenberg matrix
_HSEIN	Eigenvectors of Hessenberg matrix by inverse iteration
_GEES	Schur decomp of general matrix with e.value ordering
_GEESX	Same but with condition estimates
_GEEV	Eigenvalues and left and right eigenvectors of general matrix
_GEEVX	Same but with condition estimates
_TREVC	Selected eigenvectors of upper quasitriangular matrix
_TRSNA	Cond. estimates of selected eigenvalues of upper quasitriangular matrix
_TREXC	Unitary reordering of Schur decomposition
_TRSEN	Same but with condition estimates
_TRSYL	Solves $AX + XB = C$ for upper quasitriangular A and B

LAPACK: Unsymmetric Generalized Eigenproblem	
_GGBAL	Balance transform
_GGHRD	Reduction to Hessenberg-Triangular form
_HGEQZ	Generalized Schur decomposition
_TGEVC	Eigenvectors
_GGBAK	Undo balance transform

7.1   Properties and Decompositions

In this section we survey the mathematical background necessary to develop and analyze the eigenvalue algorithms that follow.

7.1.1   Eigenvalues and Invariant Subspaces

The *eigenvalues* of a matrix $A \in \mathbb{C}^{n \times n}$ are the n roots of its *characteristic polynomial* $p(z) = \det(zI - A)$. The set of these roots is called the *spectrum* and is denoted by $\lambda(A)$. If $\lambda(A) = \{\lambda_1, \ldots, \lambda_n\}$, then it follows that

$$\det(A) = \lambda_1 \lambda_2 \cdots \lambda_n .$$

Moreover, if we define the *trace* of A by

$$\text{tr}(A) = \sum_{i=1}^{n} a_{ii},$$

then $\text{tr}(A) = \lambda_1 + \cdots + \lambda_n$. This follows by looking at the coefficient of z^{n-1} in the characteristic polynomial.

If $\lambda \in \lambda(A)$, then the nonzero vectors $x \in \mathbb{C}^n$ that satisfy

$$Ax = \lambda x$$

are referred to as *eigenvectors*. More precisely, x is a *right eigenvector* for λ if $Ax = \lambda x$ and a *left eigenvector* if $x^H A = \lambda x^H$. Unless otherwise stated, "eigenvector" means "right eigenvector."

An eigenvector defines a one-dimensional subspace that is invariant with respect to premultiplication by A. More generally, a subspace $S \subseteq \mathbb{C}^n$ with the property that

$$x \in S \Longrightarrow Ax \in S$$

is said to be *invariant* (for A). Note that if

$$AX = XB, \qquad B \in \mathbb{C}^{k \times k}, \; X \in \mathbb{C}^{n \times k},$$

then $\text{ran}(X)$ is invariant and $By = \lambda y \Rightarrow A(Xy) = \lambda(Xy)$. Thus, if X has full column rank, then $AX = XB$ implies that $\lambda(B) \subseteq \lambda(A)$. If X is square and nonsingular, then $\lambda(A) = \lambda(B)$ and we say that A and $B = X^{-1}AX$ are *similar*. In this context, X is called a *similarity transformation*.

7.1.2 Decoupling

Many eigenvalue computations involve breaking the given problem down into a collection of smaller eigenproblems. The following result is the basis for these reductions.

Lemma 7.1.1 *If $T \in \mathbb{C}^{n \times n}$ is partitioned as follows,*

$$T = \begin{bmatrix} T_{11} & T_{12} \\ 0 & T_{22} \end{bmatrix} \begin{matrix} p \\ q \end{matrix}$$
$$\phantom{T = \begin{bmatrix}}p \qquad q$$

then $\lambda(T) = \lambda(T_{11}) \cup \lambda(T_{22})$.

Proof. Suppose

$$Tx = \begin{bmatrix} T_{11} & T_{12} \\ 0 & T_{22} \end{bmatrix} \begin{bmatrix} x_1 \\ x_2 \end{bmatrix} = \lambda \begin{bmatrix} x_1 \\ x_2 \end{bmatrix}$$

where $x_1 \in \mathbb{C}^p$ and $x_2 \in \mathbb{C}^q$. If $x_2 \neq 0$, then $T_{22}x_2 = \lambda x_2$ and so $\lambda \in \lambda(T_{22})$. If $x_2 = 0$, then $T_{11}x_1 = \lambda x_1$ and so $\lambda \in \lambda(T_{11})$. It follows that $\lambda(T) \subset \lambda(T_{11}) \cup \lambda(T_{22})$. But since both $\lambda(T)$ and $\lambda(T_{11}) \cup \lambda(T_{22})$ have the same cardinality, the two sets are equal. $\square$

7.1.3 The Basic Unitary Decompositions

By using similarity transformations, it is possible to reduce a given matrix to any one of several canonical forms. The canonical forms differ in how they display the eigenvalues and in the kind of invariant subspace information that they provide. Because of their numerical stability we begin by discussing the reductions that can be achieved with unitary similarity.

Lemma 7.1.2 *If $A \in \mathbb{C}^{n \times n}$, $B \in \mathbb{C}^{p \times p}$, and $X \in \mathbb{C}^{n \times p}$ satisfy*

$$AX = XB, \qquad \mathrm{rank}(X) = p, \tag{7.1.1}$$

then there exists a unitary $Q \in \mathbb{C}^{n \times n}$ such that

$$Q^H A Q = T = \begin{bmatrix} T_{11} & T_{12} \\ 0 & T_{22} \end{bmatrix} \begin{matrix} p \\ n-p \end{matrix} \tag{7.1.2}$$
$$\qquad\qquad\quad\; p \quad n-p$$

where $\lambda(T_{11}) = \lambda(A) \cap \lambda(B)$.

Proof. Let

$$X = Q \begin{bmatrix} R_1 \\ 0 \end{bmatrix} \qquad Q \in \mathbb{C}^{n \times n}, \; R_1 \in \mathbb{C}^{p \times p}$$

be a QR factorization of X. By substituting this into (7.1.1) and rearranging we have

$$\begin{bmatrix} T_{11} & T_{12} \\ T_{21} & T_{22} \end{bmatrix} \begin{bmatrix} R_1 \\ 0 \end{bmatrix} = \begin{bmatrix} R_1 \\ 0 \end{bmatrix} B$$

where

$$Q^H A Q = \begin{bmatrix} T_{11} & T_{12} \\ T_{21} & T_{22} \end{bmatrix} \begin{matrix} p \\ n-p \end{matrix} \quad .$$
$$\qquad\qquad\quad\; p \quad n-p$$

By using the nonsingularity of R_1 and the equations $T_{21} R_1 = 0$ and $T_{11} R_1 = R_1 B$, we can conclude that $T_{21} = 0$ and $\lambda(T_{11}) = \lambda(B)$. The conclusion now follows because from Lemma 7.1.1 $\lambda(A) = \lambda(T) = \lambda(T_{11}) \cup \lambda(T_{22})$. $\square$

Example 7.1.1 If

$$A = \begin{bmatrix} 67.00 & 177.60 & -63.20 \\ -20.40 & 95.88 & -87.16 \\ 22.80 & 67.84 & 12.12 \end{bmatrix},$$

$X = [20, -9, -12]^T$ and $B = [25]$, then $AX = XB$. Moreover, if the orthogonal matrix Q is defined by

$$Q = \begin{bmatrix} -.800 & .360 & .480 \\ .360 & .928 & -.096 \\ .480 & -.096 & .872 \end{bmatrix},$$

then $Q^T X = [-25, 0, 0]^T$ and

$$Q^T A Q = T = \begin{bmatrix} 25 & -90 & 5 \\ 0 & 147 & -104 \\ 0 & 146 & 3 \end{bmatrix}.$$

A calculation shows that $\lambda(A) = \{25, 75 + 100i, 75 - 100i\}$.

Lemma 7.1.2 says that a matrix can be reduced to block triangular form using unitary similarity transformations if we know one of its invariant subspaces. By induction we can readily establish the decomposition of Schur (1909).

Theorem 7.1.3 (Schur Decomposition) *If $A \in \mathbb{C}^{n \times n}$, then there exists a unitary $Q \in \mathbb{C}^{n \times n}$ such that*

$$Q^H A Q = T = D + N \tag{7.1.3}$$

where $D = \mathrm{diag}(\lambda_1, \ldots, \lambda_n)$ and $N \in \mathbb{C}^{n \times n}$ is strictly upper triangular. Furthermore, Q can be chosen so that the eigenvalues λ_i appear in any order along the diagonal.

Proof. The theorem obviously holds when $n = 1$. Suppose it holds for all matrices of order $n - 1$ or less. If $Ax = \lambda x$, where $x \neq 0$, then by Lemma 7.1.2 (with $B = (\lambda)$) there exists a unitary U such that:

$$U^H A U = \begin{bmatrix} \lambda & w^H \\ 0 & C \end{bmatrix} \begin{matrix} 1 \\ n-1 \end{matrix} .$$
$$ \begin{matrix} 1 & n-1 \end{matrix}$$

By induction there is a unitary $\tilde{U}$ such that $\tilde{U}^H C \tilde{U}$ is upper triangular. Thus, if $Q = U\mathrm{diag}(1, \tilde{U})$, then $Q^H A Q$ is upper triangular. $\square$

Example 7.1.2 If

$$A = \begin{bmatrix} 3 & 8 \\ -2 & 3 \end{bmatrix} \quad \text{and} \quad Q = \begin{bmatrix} .8944i & .4472 \\ -.4472 & -.8944i \end{bmatrix},$$

then Q is unitary and

$$Q^H A Q = \begin{bmatrix} 3+4i & -6 \\ 0 & 3-4i \end{bmatrix} .$$

If $Q = [\, q_1, \ldots, q_n \,]$ is a column partitioning of the unitary matrix Q in (7.1.3), then the q_i are referred to as *Schur vectors*. By equating columns in the equations $AQ = QT$ we see that the Schur vectors satisfy

$$Aq_k = \lambda_k q_k + \sum_{i=1}^{k-1} n_{ik} q_i \qquad k = 1{:}n . \tag{7.1.4}$$

From this we conclude that the subspaces

$$S_k = \mathrm{span}\{q_1, \ldots, q_k\} \qquad k = 1{:}n$$

are invariant. Moreover, it is not hard to show that if $Q_k = [\, q_1, \ldots, q_k \,]$, then $\lambda(Q_k^H A Q_k) = \{\lambda_1, \ldots, \lambda_k\}$. Since the eigenvalues in (7.1.3) can be arbitrarily ordered, it follows that there is at least one k-dimensional invariant subspace associated with each subset of k eigenvalues.

Another conclusion to be drawn from (7.1.4) is that the Schur vector q_k is an eigenvector if and only if the k-th column of N is zero. This turns out to be the case for $k = 1{:}n$ whenever $A^H A = AA^H$. Matrices that satisfy this property are called *normal*.

Corollary 7.1.4 $A \in \mathbb{C}^{n \times n}$ *is normal if and only if there exists a unitary* $Q \in \mathbb{C}^{n \times n}$ *such that* $Q^H A Q = \mathrm{diag}(\lambda_1, \ldots, \lambda_n)$.

Proof. It is easy to show that if A is unitarily similar to a diagonal matrix, then A is normal. On the other hand, if A is normal and $Q^H A Q = T$ is its Schur decomposition, then T is also normal. The corollary follows by showing that a normal, upper triangular matrix is diagonal. □

Note that if $Q^H A Q = T = \mathrm{diag}(\lambda_i) + N$ is a Schur decomposition of a general n-by-n matrix A, then $\| N \|_F$ is independent of the choice of Q:

$$\| N \|_F^2 = \| A \|_F^2 - \sum_{i=1}^{n} |\lambda_i|^2 \equiv \Delta^2(A).$$

This quantity is referred to as A's *departure from normality*. Thus, to make T "more diagonal," it is necessary to rely on nonunitary similarity transformations.

7.1.4   Nonunitary Reductions

To see what is involved in nonunitary similarity reduction, we examine the block diagonalization of a 2-by-2 block triangular matrix.

Lemma 7.1.5 *Let* $T \in \mathbb{C}^{n \times n}$ *be partitioned as follows:*

$$T = \begin{bmatrix} T_{11} & T_{12} \\ 0 & T_{22} \end{bmatrix} \begin{matrix} p \\ q \end{matrix} \quad .$$
$$\quad\;\; p \quad\;\; q$$

Define the linear transformation $\phi : \mathbb{C}^{p \times q} \to \mathbb{C}^{p \times q}$ *by*

$$\phi(X) = T_{11} X - X T_{22}$$

where $X \in \mathbb{C}^{p \times q}$. *Then* ϕ *is nonsingular if and only if* $\lambda(T_{11}) \cap \lambda(T_{22}) = \emptyset$. *If* ϕ *is nonsingular and Y is defined by*

$$Y = \begin{bmatrix} I_p & Z \\ 0 & I_q \end{bmatrix} \qquad \phi(Z) = -T_{12}$$

then $Y^{-1} T Y = \mathrm{diag}(T_{11}, T_{22})$.

Proof. Suppose $\phi(X) = 0$ for $X \neq 0$ and that

$$U^H X V = \begin{bmatrix} \Sigma_r & 0 \\ 0 & 0 \end{bmatrix} \begin{matrix} r \\ p - r \end{matrix}$$
$$\quad\;\; r \quad\; q - r$$

is the SVD of X with $\Sigma_r = \text{diag}(\sigma_i)$, $r = \text{rank}(X)$. Substituting this into the equation $T_{11}X = XT_{22}$ gives

$$
\begin{bmatrix} A_{11} & A_{12} \\ A_{21} & A_{22} \end{bmatrix} \begin{bmatrix} \Sigma_r & 0 \\ 0 & 0 \end{bmatrix} = \begin{bmatrix} \Sigma_r & 0 \\ 0 & 0 \end{bmatrix} \begin{bmatrix} B_{11} & B_{12} \\ B_{21} & B_{22} \end{bmatrix}
$$

where $U^H T_{11} U = (A_{ij})$ and $V^H T_{22} V = (B_{ij})$. By comparing blocks we see that $A_{21} = 0$, $B_{12} = 0$, and $\lambda(A_{11}) = \lambda(B_{11})$. Consequently,

$$
\emptyset \neq \lambda(A_{11}) = \lambda(B_{11}) \subseteq \lambda(T_{11}) \cap \lambda(T_{22}).
$$

On the other hand, if $\lambda \in \lambda(T_{11}) \cap \lambda(T_{22})$ then we have nonzero vectors x and y so $T_{11}x = \lambda x$ and $y^H T_{22} = \lambda y^H$. A calculation shows that $\phi(xy^H) = 0$. Finally, if ϕ is nonsingular then the matrix Z above exists and

$$
\begin{aligned}
Y^{-1}TY &= \begin{bmatrix} I & -Z \\ 0 & I \end{bmatrix} \begin{bmatrix} T_{11} & T_{12} \\ 0 & T_{22} \end{bmatrix} \begin{bmatrix} I & Z \\ 0 & I \end{bmatrix} \\
&= \begin{bmatrix} T_{11} & T_{11}Z - ZT_{22} + T_{12} \\ 0 & T_{22} \end{bmatrix} = \begin{bmatrix} T_{11} & 0 \\ 0 & T_{22} \end{bmatrix}. \quad \square
\end{aligned}
$$

Example 7.1.3 If

$$
T = \begin{bmatrix} 1 & 2 & 3 \\ 0 & 3 & 8 \\ 0 & -2 & 3 \end{bmatrix} \quad \text{and} \quad Y = \begin{bmatrix} 1.0 & 0.5 & -0.5 \\ 0.0 & 1.0 & 0.0 \\ 0.0 & 0.0 & 1.0 \end{bmatrix}
$$

then

$$
Y^{-1}TY = \begin{bmatrix} 1 & 0 & 0 \\ 0 & 3 & 8 \\ 0 & -2 & 3 \end{bmatrix}.
$$

By repeatedly applying Lemma 7.1.5, we can establish the following more general result:

Theorem 7.1.6 (Block Diagonal Decomposition) *Suppose*

$$
Q^H AQ = T = \begin{bmatrix} T_{11} & T_{12} & \cdots & T_{1q} \\ 0 & T_{22} & \cdots & T_{2q} \\ \vdots & \vdots & \ddots & \vdots \\ 0 & 0 & \cdots & T_{qq} \end{bmatrix} \tag{7.1.5}
$$

is a Schur decomposition of $A \in \mathbb{C}^{n \times n}$ and assume that the T_{ii} are square. If $\lambda(T_{ii}) \cap \lambda(T_{jj}) = \emptyset$ whenever $i \neq j$, then there exists a nonsingular matrix $Y \in \mathbb{C}^{n \times n}$ such that

$$
(QY)^{-1} A(QY) = \text{diag}(T_{11}, \ldots, T_{qq}). \tag{7.1.6}
$$

Proof. A proof can be obtained by using Lemma 7.1.5 and induction. □

If each diagonal block T_{ii} is associated with a distinct eigenvalue, then we obtain

Corollary 7.1.7 *If $A \in \mathbb{C}^{n \times n}$ then there exists a nonsingular X such that*

$$X^{-1}AX = \operatorname{diag}(\lambda_1 I + N_1, \ldots, \lambda_q I + N_q) \qquad N_i \in \mathbb{C}^{n_i \times n_i} \qquad (7.1.7)$$

where $\lambda_1, \ldots, \lambda_q$ are distinct, the integers $n_1, \ldots, n_q$ satisfy $n_1 + \cdots + n_q = n$, and each N_i is strictly upper triangular.

A number of important terms are connected with decomposition (7.1.7). The integer n_i is referred to as the *algebraic multiplicity* of λ_i. If $n_i = 1$, then λ_i is said to be *simple* . The *geometric multiplicity* of λ_i equals the dimensions of null(N_i), i.e., the number of linearly independent eigenvectors associated with λ_i. If the algebraic multiplicity of λ_i exceeds its geometric multiplicity, then λ_i is said to be a *defective eigenvalue*. A matrix with a defective eigenvalue is referred to as a *defective matrix*. Nondefective matrices are also said to be *diagonalizable* in light of the following result:

Corollary 7.1.8 (Diagonal Form) $A \in \mathbb{C}^{n \times n}$ *is nondefective if and only if there exists a nonsingular $X \in \mathbb{C}^{n \times n}$ such that*

$$X^{-1}AX = \operatorname{diag}(\lambda_1, \ldots, \lambda_n). \qquad (7.1.8)$$

Proof. A is nondefective if and only if there exist independent vectors $x_1 \ldots x_n \in \mathbb{C}^n$ and scalars $\lambda_1, \ldots, \lambda_n$ such that $Ax_i = \lambda_i x_i$ for $i = 1{:}n$. This is equivalent to the existence of a nonsingular $X = [\, x_1, \ldots, x_n \,] \in \mathbb{C}^{n \times n}$ such that $AX = XD$ where $D = \operatorname{diag}(\lambda_1, \ldots, \lambda_n)$. □

Note that if y_i^H is the ith row of X^{-1}, then $y_i^H A = \lambda_i y_i^H$. Thus, the columns of X^{-T} are left eigenvectors and the columns of X are right eigenvectors.

Example 7.1.4 If

$$A = \begin{bmatrix} 5 & -1 \\ -2 & 6 \end{bmatrix} \qquad \text{and} \qquad X = \begin{bmatrix} 1 & 1 \\ 1 & -2 \end{bmatrix}$$

then $X^{-1}AX = \operatorname{diag}(4, 7)$.

If we partition the matrix X in (7.1.7),

$$X = \begin{bmatrix} X_1 & , \ldots, & X_q \end{bmatrix}$$
$$\qquad n_1 \qquad\qquad n_q$$

then $\mathbb{C}^n = \operatorname{ran}(X_1) \oplus \ldots \oplus \operatorname{ran}(X_q)$, a direct sum of invariant subspaces. If the bases for these subspaces are chosen in a special way, then it is possible to introduce even more zeroes into the upper triangular portion of $X^{-1}AX$.

Theorem 7.1.9 (Jordan Decomposition) *If $A \in \mathbb{C}^{n \times n}$, then there exists a nonsingular $X \in \mathbb{C}^{n \times n}$ such that $X^{-1}AX = \text{diag}(J_1, \ldots, J_t)$ where*

$$
J_i = \begin{bmatrix}
\lambda_i & 1 & & \cdots & 0 \\
0 & \lambda_i & \ddots & & \vdots \\
& \ddots & \ddots & \ddots & \\
\vdots & & \ddots & \ddots & 1 \\
0 & \cdots & & 0 & \lambda_i
\end{bmatrix}
$$

is m_i-by-m_i and $m_1 + \cdots + m_t = n$.

Proof. See Halmos (1958, pp. 112 ff.) □

The J_i are referred to as *Jordan blocks* . The number and dimensions of the Jordan blocks associated with each distinct eigenvalue is unique, although their ordering along the diagonal is not.

7.1.5 Some Comments on Nonunitary Similarity

The Jordan block structure of a defective matrix is difficult to determine numerically. The set of n-by-n diagonalizable matrices is dense in $\mathbb{C}^{n \times n}$, and thus, small changes in a defective matrix can radically alter its Jordan form. We have more to say about this in §7.6.5.

A related difficulty that arises in the eigenvalue problem is that a nearly defective matrix can have a poorly conditioned matrix of eigenvectors. For example, any matrix X that diagonalizes

$$
A = \begin{bmatrix} 1+\epsilon & 1 \\ 0 & 1-\epsilon \end{bmatrix} \qquad 0 < \epsilon \ll 1 \tag{7.1.9}
$$

has a 2-norm condition of order $1/\epsilon$.

These observations serve to highlight the difficulties associated with ill-conditioned similarity transformations. Since

$$
fl(X^{-1}AX) = X^{-1}AX + E, \tag{7.1.10}
$$

where

$$
\| E \|_2 \approx \mathbf{u}\kappa_2(X)\| A \|_2 \tag{7.1.11}
$$

is it clear that large errors can be introduced into an eigenvalue calculation when we depart from unitary similarity.

7.1.6   Singular Values and Eigenvalues

Since the singular values of A and its Schur decomposition $Q^H A Q = \text{diag}(\lambda_i) + N$ are the same, it follows that

$$\sigma_{min}(A) \leq \min_i |\lambda_i| \leq \max_i |\lambda_i| \leq \sigma_{max}(A).$$

From what we know about the condition of triangular matrices, it may be the case that

$$\max_{i,j} \frac{|\lambda_i|}{|\lambda_j|} \ll \kappa_2(A).$$

This is a reminder that for nonnormal matrices, eigenvalues do not have the "predictive power" of singular values when it comes to $Ax = b$ sensitivity matters. Eigenvalues of nonnormal matrices have other shortcomings. See §11.3.4.

Problems

P7.1.1 Show that if $T \in \mathbb{C}^{n \times n}$ is upper triangular and normal, then T is diagonal.

P7.1.2 Verify that if X diagonalizes the 2-by-2 matrix in (7.1.9) and $\epsilon \leq 1/2$ then $\kappa_1(X) \geq 1/\epsilon$.

P7.1.3 Suppose $A \in \mathbb{C}^{n \times n}$ has distinct eigenvalues. Show that if $Q^H A Q = T$ is its Schur decomposition and $AB = BA$, then $Q^H B Q$ is upper triangular.

P7.1.4 Show that if A and B^H are in $\mathbb{C}^{m \times n}$ with $m \geq n$, then:

$$\lambda(AB) = \lambda(BA) \cup \{ \underbrace{0, \ldots, 0}_{m-n} \}.$$

P7.1.5 Given $A \in \mathbb{C}^{n \times n}$, use the Schur decomposition to show that for every $\epsilon > 0$, there exists a diagonalizable matrix B such that $\| A - B \|_2 \leq \epsilon$. This shows that the set of diagonalizable matrices is dense in $\mathbb{C}^{n \times n}$ and that the Jordan canonical form is not a continuous matrix decomposition.

P7.1.6 Suppose $A_k \to A$ and that $Q_k^H A_k Q_k = T_k$ is a Schur decomposition of A_k. Show that $\{Q_k\}$ has a converging subsequence $\{Q_{k_i}\}$ with the property that

$$\lim_{i \to \infty} Q_{k_i} = Q$$

where $Q^H A Q = T$ is upper triangular. This shows that the eigenvalues of a matrix are continuous functions of its entries.

P7.1.7 Justify (7.1.10) and (7.1.11).

P7.1.8 . Show how to compute the eigenvalues of

$$M = \begin{bmatrix} A & C \\ B & D \end{bmatrix} \begin{matrix} k \\ j \end{matrix}$$
$$\phantom{M = \begin{bmatrix} A & C \\ B & D \end{bmatrix}} \begin{matrix} k & j \end{matrix}$$

where A, B, C, and D are given real diagonal matrices.

P7.1.9 Use the JCF to show that if all the eigenvalues of a matrix A are strictly less than unity, then $\lim_{k \to \infty} A^k = 0$.

P7.1.10 The initial value problem

$$
\begin{array}{rclcl}
\dot{x}(t) & = & y(t) & \quad & x(0) = 1 \\
\dot{y}(t) & = & -x(t) & \quad & y(0) = 0
\end{array}
$$

has solution $x(t) = \cos(t)$ and $y(t) = \sin(t)$. Let $h > 0$. Here are three reasonable iterations that can be used to compute approximations $x_k \approx x(kh)$ and $y_k \approx y(kh)$ assuming that $x_0 = 1$ and $y_k = 0$:

$$
\text{Method 1:} \quad
\begin{array}{rcl}
x_{k+1} & = & 1 + h y_k \\
y_{k+1} & = & 1 - h x_k
\end{array}
$$

$$
\text{Method 2:} \quad
\begin{array}{rcl}
x_{k+1} & = & 1 + h y_k \\
y_{k+1} & = & 1 - h x_{k+1}
\end{array}
$$

$$
\text{Method 3:} \quad
\begin{array}{rcl}
x_{k+1} & = & 1 + h y_{k+1} \\
y_{k+1} & = & 1 - h x_{k+1}
\end{array}
$$

Express each method in the form

$$
\begin{bmatrix} x_{k+1} \\ y_{k+1} \end{bmatrix} = A_h \begin{bmatrix} x_k \\ y_k \end{bmatrix}
$$

where A_h is a 2-by-2 matrix. For each case, compute $\lambda(A_h)$ and use the previous problem to discuss $\lim x_k$ and $\lim y_k$ as $k \to \infty$.

P7.1.11 If $J \in \mathbf{R}^{d \times d}$ is a Jordan block, what is $\kappa_\infty(J)$?

P7.1.12 Show that if

$$
R = \begin{bmatrix} R_{11} & R_{12} \\ 0 & R_{22} \end{bmatrix} \begin{matrix} p \\ q \end{matrix}
$$
$$
\phantom{R = \begin{bmatrix}} p \quad\ \ q
$$

is normal and $\lambda(R_{11}) \cap \lambda(R_{22}) = \emptyset$, then $R_{12} = 0$.

Notes and References for Sec. 7.1

The mathematical properties of the algebraic eigenvalue problem are elegantly covered in Wilkinson (1965, chapter 1) and Stewart (1973, chapter 6). For those who need further review we also recommend

R. Bellman (1970). *Introduction to Matrix Analysis*, 2nd ed., McGraw-Hill, New York.
I.C. Gohberg, P. Lancaster, and L. Rodman (1986). *Invariant Subspaces of Matrices With Applications*, John Wiley and Sons, New York.
M. Marcus and H. Minc (1964). *A Survey of Matrix Theory and Matrix Inequalities*, Allyn and Bacon, Boston.
L. Mirsky (1963). *An Introduction to Linear Algebra* , Oxford University Press, Oxford.

The Schur decomposition originally appeared in

I. Schur (1909). "On the Characteristic Roots of a Linear Substitution with an Application to the Theory of Integral Equations." *Math. Ann.* **66**, 488-510 (German).

A proof very similar to ours is given on page 105 of

H.W. Turnbull and A.C. Aitken (1961). *An Introduction to the Theory of Canonical Forms*, Dover, New York.

Connections between singular values, eigenvalues, and pseudoeigenvalues (see §11.3.4) are discussed in

K-C. Toh and L.N. Trefethen (1994). "Pseudozeros of Polynomials and Pseudospectra of Companion Matrices," *Numer. Math. 68*, 403–425.

F. Kittaneh (1995). "Singular Values of Companion Matrices and Bounds on Zeros of Polynomials," *SIAM J. Matrix Anal. Appl. 16*, 333–340.

7.2   Perturbation Theory

The act of computing eigenvalues is the act of computing zeros of the characteristic polynomial. Galois theory tells us that such a process has to be iterative if $n > 4$ and so errors will arise because of finite termination. In order to develop intelligent stopping criteria we need an informative perturbation theory that tells us how to think about approximate eigenvalues and invariant subspaces.

7.2.1   Eigenvalue Sensitivity

Several eigenvalue routines produce a sequence of similarity transformations X_k with the property that the matrices $X_k^{-1} A X_k$ are progressively "more diagonal." The question naturally arises, how well do the diagonal elements of a matrix approximate its eigenvalues?

Theorem 7.2.1 (Gershgorin Circle Theorem) *If* $X^{-1} A X = D + F$ *where* $D = \text{diag}(d_1, \dots, d_n)$ *and* F *has zero diagonal entries, then*

$$\lambda(A) \subseteq \bigcup_{i=1}^{n} D_i$$

where $D_i = \{ z \in \mathbb{C} : |z - d_i| \leq \sum_{j=1}^{n} |f_{ij}| \}$.

Proof. Suppose $\lambda \in \lambda(A)$ and assume without loss of generality that $\lambda \neq d_i$ for $i = 1{:}n$. Since $(D - \lambda I) + F$ is singular, it follows from Lemma 2.3.3 that

$$1 \leq \| (D - \lambda I)^{-1} F \|_\infty = \sum_{j=1}^{n} \frac{|f_{kj}|}{|d_k - \lambda|}$$

for some k, $1 \leq k \leq n$. But this implies that $\lambda \in D_k$. $\square$

It can also be shown that if the Gershgorin disk D_i is isolated from the other disks, then it contains precisely one of A's eigenvalues. See Wilkinson (1965, pp.71ff.).

Example 7.2.1 If

$$A = \begin{bmatrix} 10 & 2 & 3 \\ -1 & 0 & 2 \\ 1 & -2 & 1 \end{bmatrix}$$

then $\lambda(A) \approx \{10.226, .3870 + 2.2216i, .3870 - 2.2216i\}$ and the Gershgorin disks are $D_1 = \{\,|z| : |z - 10| \le 5\,\}$, $D_2 = \{\,|z| : |z| \le 3\,\}$, and $D_3 = \{\,|z| : |z - 1| \le 3\,\}$.

For some very important eigenvalue routines it is possible to show that the computed eigenvalues are the exact eigenvalues of a matrix $A + E$ where E is small in norm. Consequently, we must understand how the eigenvalues of a matrix can be affected by small perturbations. A sample result that sheds light on this issue is the following theorem.

Theorem 7.2.2 (Bauer-Fike) *If μ is an eigenvalue of $A + E \in \mathbb{C}^{n \times n}$ and $X^{-1}AX = D = \operatorname{diag}(\lambda_1, \ldots, \lambda_n)$, then*

$$\min_{\lambda \in \lambda(A)} |\lambda - \mu| \le \kappa_p(X) \| E \|_p$$

where $\| \cdot \|_p$ denotes any of the p-norms.

Proof. We need only consider the case when μ is not in $\lambda(A)$. If the matrix $X^{-1}(A + E - \mu I)X$ is singular, then so is $I + (D - \mu I)^{-1}(X^{-1}EX)$. Thus, from Lemma 2.3.3 we obtain

$$1 \le \| (D - \mu I)^{-1}(X^{-1}EX) \|_p \le \| (D - \mu I)^{-1} \|_p \| X \|_p \| E \|_p \| X^{-1} \|_p .$$

Since $(D - \mu I)^{-1}$ is diagonal and the p-norm of a diagonal matrix is the absolute value of the largest diagonal entry, it follows that

$$\| (D - \mu I)^{-1} \|_p = \min_{\lambda \in \lambda(A)} \frac{1}{|\lambda - \mu|}$$

from which the theorem follows. $\square$

An analogous result can be obtained via the Schur decomposition:

Theorem 7.2.3 *Let $Q^H AQ = D + N$ be a Schur decomposition of $A \in \mathbb{C}^{n \times n}$ as in (7.1.3). If $\mu \in \lambda(A + E)$ and p is the smallest positive integer such that $|N|^p = 0$, then*

$$\min_{\lambda \in \lambda(A)} |\lambda - \mu| \le \max(\theta, \theta^{1/p})$$

where

$$\theta = \| E \|_2 \sum_{k=0}^{p-1} \| N \|_2^k .$$

Proof. Define

$$\delta = \min_{\lambda \in \lambda(A)} |\lambda - \mu| = \frac{1}{\| (\mu I - D)^{-1} \|_2}.$$

The theorem is clearly true if $\delta = 0$. If $\delta > 0$ then $I - (\mu I - A)^{-1}E$ is singular and by Lemma 2.3.3 we have

$$\begin{aligned} 1 &\leq \| (\mu I - A)^{-1}E \|_2 \leq \| (\mu I - A)^{-1} \|_2 \| E \|_2 \qquad (7.2.1) \\ &= \| ((\mu I - D) - N)^{-1} \|_2 \| E \|_2 \, . \end{aligned}$$

Since $(\mu I - D)^{-1}$ is diagonal and $|N|^p = 0$ it is not hard to show that $((\mu I - D)^{-1}N)^p = 0$. Thus,

$$((\mu I - D) - N)^{-1} = \sum_{k=0}^{p-1} ((\mu I - D)^{-1}N)^k (\mu I - D)^{-1}$$

and so

$$\| ((\mu I - D) - N)^{-1} \|_2 \leq \frac{1}{\delta} \sum_{k=0}^{p-1} \left(\frac{\| N \|_2}{\delta} \right)^k .$$

If $\delta > 1$ then

$$\| (\mu I - D) - N)^{-1} \|_2 \leq \frac{1}{\delta} \sum_{k=0}^{p-1} \| N \|_2^k$$

and so from (7.2.1), $\delta \leq \theta$. If $\delta \leq 1$ then

$$\| (\mu I - D) - N)^{-1} \|_2 \leq \frac{1}{\delta^p} \sum_{k=0}^{p-1} \| N \|_2^k$$

and so from (7.2.1), $\delta^p \leq \theta$. Thus, $\delta \leq \max(\theta, \theta^{1/p})$. $\square$

Example 7.2.2 If

$$A = \begin{bmatrix} 1 & 2 & 3 \\ 0 & 4 & 5 \\ 0 & 0 & 4.001 \end{bmatrix} \qquad \text{and} \qquad E = \begin{bmatrix} 0 & 0 & 0 \\ 0 & 0 & 0 \\ .001 & 0 & 0 \end{bmatrix},$$

then $\lambda(A + E) \approx \{1.0001, \ 4.0582, \ 3.9427\}$ and A's matrix of eigenvectors satisfies $\kappa_2(X) \approx 10^7$. The Bauer-Fike bound in Theorem 7.2.2 has order 10^4, while the Schur bound in Theorem 7.2.3 has order 10^0.

Theorems 7.2.2 and 7.2.3 each indicate potential eigenvalue sensitivity if A is nonnormal. Specifically, if $\kappa_2(X)$ or $\| N \|_2^{p-1}$ is large, then small changes in A can induce large changes in the eigenvalues.

Example 7.2.3 If

$$A = \begin{bmatrix} 0 & I_9 \\ 0 & 0 \end{bmatrix} \qquad \text{and} \qquad E = \begin{bmatrix} 0 & 0 \\ 10^{-10} & 0 \end{bmatrix},$$

then for all $\lambda \in \lambda(A)$ and $\mu \in \lambda(A + E)$, $|\lambda - \mu| = 10^{-1}$. In this example a change of order 10^{-10} in A results in a change of order 10^{-1} in its eigenvalues.

7.2.2 The Condition of a Simple Eigenvalue

Extreme eigenvalue sensitivity for a matrix A cannot occur if A is normal. On the other hand, nonnormality does not necessarily imply eigenvalue sensitivity. Indeed, a nonnormal matrix can have a mixture of well-conditioned and ill-conditioned eigenvalues. For this reason, it is beneficial to refine our perturbation theory so that it is applicable to individual eigenvalues and not the spectrum as a whole.

To this end, suppose that λ is a simple eigenvalue of $A \in \mathbb{C}^{n \times n}$ and that x and y satisfy $Ax = \lambda x$ and $y^H A = \lambda y^H$ with $\| x \|_2 = \| y \|_2 = 1$. If $Y^H AX = J$ is the Jordan decomposition with $Y^H = X^{-1}$, then y and x are nonzero multiples of $X(:, i)$ and $Y(:, i)$ for some i. It follows from $1 = Y(:, i)^H X(:, i)$ that $y^H x \neq 0$, a fact that we shall use shortly.

Using classical results from function theory, it can be shown that in a neighborhood of the origin there exist differentiable $x(\epsilon)$ and $\lambda(\epsilon)$ such that

$$(A + \epsilon F)x(\epsilon) = \lambda(\epsilon)x(\epsilon) \qquad \| F \|_2 = 1$$

where $\lambda(0) = \lambda$ and $x(0) = x$. By differentiating this equation with respect to ϵ and setting $\epsilon = 0$ in the result, we obtain

$$A\dot{x}(0) + Fx = \dot{\lambda}(0)x + \lambda\dot{x}(0) .$$

Applying y^H to both sides of this equation, dividing by $y^H x$, and taking absolute values gives

$$|\dot{\lambda}(0)| = \left| \frac{y^H Fx}{y^H x} \right| \leq \frac{1}{|y^H x|} .$$

The upper bound is attained if $F = yx^H$. For this reason we refer to the reciprocal of

$$s(\lambda) = |y^H x|$$

as the *condition of the eigenvalue* λ.

Roughly speaking, the above analysis shows that if order ϵ perturbations are made in A, then an eigenvalue λ may be perturbed by an amount $\epsilon/s(\lambda)$. Thus, if $s(\lambda)$ is small, then λ is appropriately regarded as ill-conditioned. Note that $s(\lambda)$ is the cosine of the angle between the left and right eigenvectors associated with λ and is unique only if λ is simple.

A small $s(\lambda)$ implies that A is near a matrix having a multiple eigenvalue. In particular, if λ is distinct and $s(\lambda) < 1$, then there exists an E such that λ is a repeated eigenvalue of $A + E$ and

$$\frac{\| E \|_2}{\| A \|_2} \leq \frac{s(\lambda)}{\sqrt{1 - s(\lambda)^2}} .$$

This result is proved in Wilkinson (1972).

Example 7.2.4 If

$$
A = \begin{bmatrix} 1 & 2 & 3 \\ 0 & 4 & 5 \\ 0 & 0 & 4.001 \end{bmatrix} \quad \text{and} \quad E = \begin{bmatrix} 0 & 0 & 0 \\ 0 & 0 & 0 \\ .001 & 0 & 0 \end{bmatrix},
$$

then $\lambda(A + E) \approx \{1.0001,\ 4.0582,\ 3.9427\}$ and $s(1) \approx .8 \times 10^0$, $s(4) \approx .2 \times 10^{-3}$, and $s(4.001) \approx .2 \times 10^{-3}$. Observe that $\| E \|_2/s(\lambda)$ is a good estimate of the perturbation that each eigenvalue undergoes.

7.2.3   Sensitivity of Repeated Eigenvalues

If λ is a repeated eigenvalue, then the eigenvalue sensitivity question is more complicated. For example, if

$$
A = \begin{bmatrix} 1 & a \\ 0 & 1 \end{bmatrix} \quad \text{and} \quad F = \begin{bmatrix} 0 & 0 \\ 1 & 0 \end{bmatrix},
$$

then $\lambda(A + \epsilon F) = \{1 \pm \sqrt{\epsilon a}\}$. Note that if $a \neq 0$, then it follows that the eigenvalues of $A + \epsilon F$ are not differentiable at zero; their rate of change at the origin is infinite. In general, if λ is a defective eigenvalue of A, then $O(\epsilon)$ perturbations in A can result in $O(\epsilon^{1/p})$ perturbations in λ if λ is associated with a p-dimensional Jordan block. See Wilkinson (1965, pp. 77ff.) for a more detailed discussion.

7.2.4   Invariant Subspace Sensitivity

A collection of sensitive eigenvectors can define an insensitive invariant subspace provided the corresponding cluster of eigenvalues is isolated. To be precise, suppose

$$
Q^H A Q = \begin{bmatrix} T_{11} & T_{12} \\ 0 & T_{22} \end{bmatrix} \begin{matrix} r \\ n - r \end{matrix} \qquad (7.2.2)
$$
$$
\phantom{Q^H A Q = \begin{bmatrix} T_{11} \end{bmatrix}} r \quad n - r
$$

is a Schur decomposition of A with

$$
Q = [\ Q_1 \quad Q_2\ ] \atop r \quad n - r \qquad (7.2.3)
$$

It is clear from our discussion of eigenvector perturbation that the sensitivity of the invariant subspace $\mathrm{ran}(Q_1)$ depends on the distance between $\lambda(T_{11})$ and $\lambda(T_{22})$. The proper measure of this distance turns out to be the smallest singular value of the linear transformation $X \rightarrow T_{11}X - XT_{22}$.

(Recall that this transformation figures in Lemma 7.1.5.) In particular, if we define the *separation* between the matrices T_{11} and T_{22} by

$$\text{sep}(T_{11}, T_{22}) = \min_{X \neq 0} \frac{\| T_{11}X - XT_{22} \|_F}{\| X \|_F}, \qquad (7.2.4)$$

then we have the following general result:

Theorem 7.2.4 *Suppose that (7.2.2) and (7.2.3) hold and that for any matrix $E \in \mathbb{C}^{n \times n}$ we partition $Q^H EQ$ as follows:*

$$Q^H EQ = \begin{bmatrix} E_{11} & E_{12} \\ E_{21} & E_{22} \end{bmatrix} \begin{matrix} r \\ n-r \end{matrix} \ .$$
$$\qquad\qquad\ \ r \quad\ n-r$$

If $\text{sep}(T_{11}, T_{22}) > 0$ *and*

$$\| E \|_2 \left(1 + \frac{5\| T_{12} \|_2}{\text{sep}(T_{11}, T_{22})} \right) \leq \frac{\text{sep}(T_{11}, T_{22})}{5},$$

then there exists a $P \in \mathbb{C}^{(n-r) \times r}$ *with*

$$\| P \|_2 \leq 4\frac{\| E_{21} \|_2}{\text{sep}(T_{11}, T_{22})}$$

such that the columns of $\hat{Q}_1 = (Q_1 + Q_2 P)(I + P^H P)^{-1/2}$ *are an orthonormal basis for a subspace invariant for* $A + E$.

Proof. This result is a slight recasting of Theorem 4.11 in Stewart (1973) which should be consulted for proof details. See also Stewart and Sun (1990, p.230). The matrix $(I + P^H P)^{-1/2}$ is the inverse of the square root of the symmetric positive definite matrix $I + P^H P$. See §4.2.10. □

Corollary 7.2.5 *If the assumptions in Theorem 7.2.4 hold, then*

$$\text{dist}(\text{ran}(Q_1), \text{ran}(\hat{Q}_1)) \leq 4\frac{\| E_{21} \|_2}{\text{sep}(T_{11}, T_{22})}.$$

Proof. Using the SVD of P, it can be shown that

$$\| P(I + P^H P)^{-1/2} \|_2 \leq \| P \|_2. \qquad (7.2.5)$$

The corollary follows because the required distance is the norm of $Q_2^H \hat{Q}_1 = P(I + P^H P)^{-1/2}$. □

Thus, the reciprocal of $\text{sep}(T_{11}, T_{22})$ can be thought of as a condition number that measures the sensitivity of $\text{ran}(Q_1)$ as an invariant subspace.

Example 7.2.5 Suppose

$$T_{11} = \begin{bmatrix} 3 & 10 \\ 0 & 1 \end{bmatrix}, \quad T_{22} = \begin{bmatrix} 0 & -20 \\ 0 & 3.01 \end{bmatrix}, \quad \text{and} \quad T_{12} = \begin{bmatrix} 1 & -1 \\ -1 & 1 \end{bmatrix}$$

and that
$$A = T = \begin{bmatrix} T_{11} & T_{12} \\ 0 & T_{22} \end{bmatrix}.$$

Observe that $AQ_1 = Q_1 T_{11}$ where $Q_1 = [e_1, e_2] \in \mathbf{R}^{4 \times 2}$.   A calculation shows that $\text{sep}(T_{11}, T_{22}) \approx .0003$.   If
$$E_{21} = 10^{-6} \begin{bmatrix} 1 & 1 \\ 1 & 1 \end{bmatrix}$$

and we examine the Schur decomposition of
$$A + E = \begin{bmatrix} T_{11} & T_{12} \\ E_{21} & T_{22} \end{bmatrix},$$

then we find that Q_1 gets perturbed to
$$\hat{Q}_1 = \begin{bmatrix} -.9999 & -.0003 \\ .0003 & -.9999 \\ -.0005 & -.0026 \\ .0000 & .0003 \end{bmatrix}.$$

Thus, we have $\text{dist}(\text{ran}(\hat{Q}_1), \text{ran}(Q_1)) \approx .0027 \approx 10^{-6}/\text{sep}(T_{11}, T_{22})$.

7.2.5   Eigenvector Sensitivity

If we set $r = 1$ in the preceding subsection, then the analysis addresses the issue of eigenvector sensitivity.

Corollary 7.2.6 *Suppose* $A, E \in \mathbf{C}^{n \times n}$ *and that* $Q = [\, q_1 \; Q_2 \,] \in \mathbf{C}^{n \times n}$ *is unitary with* $q_1 \in \mathbf{C}^n$. *Assume*
$$Q^H A Q = \begin{bmatrix} \lambda & v^H \\ 0 & T_{22} \end{bmatrix} \begin{matrix} 1 \\ n-1 \end{matrix} \qquad Q^H E Q = \begin{bmatrix} \epsilon & \gamma^H \\ \delta & E_{22} \end{bmatrix} \begin{matrix} 1 \\ n-1 \end{matrix} .$$
$$\begin{matrix} 1 & n-1 \end{matrix} \qquad\qquad\qquad \begin{matrix} 1 & n-1 \end{matrix}$$

(Thus, q_1 *is an eigenvector.) If* $\sigma = \sigma_{min}(T_{22} - \lambda I) > 0$ *and*
$$\| E \|_2 \left(1 + \frac{5\| v \|_2}{\sigma} \right) \leq \frac{\sigma}{5},$$

then there exists $p \in \mathbf{C}^{n-1}$ *with*
$$\| p \|_2 \leq 4\frac{\| \delta \|_2}{\sigma}$$

such that $\hat{q}_1 = (q_1 + Q_2 p)/\sqrt{1 + p^H p}$ *is a unit 2-norm eigenvector for* $A + E$. *Moreover,*
$$\text{dist}(\text{span}\{q_1\}, \text{span}\{\hat{q}_1\}) \leq 4\frac{\| \delta \|_2}{\sigma}.$$

Proof.   The result follows from Theorem 7.2.4, Corollary 7.2.5 and the observation that if $T_{11} = \lambda$, then $\text{sep}(T_{11}, T_{22}) = \sigma_{min}(T_{22} - \lambda I)$. $\square$

Note that $\sigma_{min}(T_{22} - \lambda I)$ roughly measures the separation of λ from the eigenvalues of T_{22}. We have to say "roughly" because

$$\text{sep}(\lambda, T_{22}) = \sigma_{min}(T_{22} - \lambda I) \leq \min_{\mu \in \lambda(T_{22})} |\mu - \lambda|$$

and the upper bound can be a gross overestimate.

That the separation of the eigenvalues should have a bearing upon eigenvector sensitivity should come as no surprise. Indeed, if λ is a nondefective, repeated eigenvalue, then there are an infinite number of possible eigenvector bases for the associated invariant subspace. The preceding analysis merely indicates that this indeterminancy begins to be felt as the eigenvalues coalesce. In other words, the eigenvectors associated with nearby eigenvalues are "wobbly."

Example 7.2.6 If

$$A = \begin{bmatrix} 1.01 & 0.01 \\ 0.00 & 0.99 \end{bmatrix}$$

then the eigenvalue $\lambda = .99$ has condition $1/s(.99) \approx 1.118$ and associated eigenvector $x = [.4472, \ -.8944]^T$. On the other hand, the eigenvalue $\hat{\lambda} = 1.00$ of the "nearby" matrix

$$A + E = \begin{bmatrix} 1.01 & 0.01 \\ 0.00 & 1.00 \end{bmatrix}$$

has an eigenvector $\hat{x} = [.7071, \ -.7071]^T$.

Problems

P7.2.1 Suppose $Q^H A Q = \text{diag}(\lambda_1) + N$ is a Schur decomposition of $A \in \mathbb{C}^{n \times n}$ and define $\nu(A) = \| A^H A - A A^H \|_F$. The upper and lower bounds in

$$\frac{\nu(A)^2}{6 \| A \|_F^2} \leq \| N \|_F^2 \leq \sqrt{\frac{n^3 - n}{12}} \nu(A)$$

are established by Henrici (1962) and Eberlein (1965), respectively. Verify these results for the case $n = 2$.

P7.2.2 Suppose $A \in \mathbb{C}^{n \times n}$ and $X^{-1} A X = \text{diag}(\lambda_1, \ldots, \lambda_n)$ with distinct λ_i. Show that if the columns of X have unit 2-norm, then $\kappa_F(X)^2 = n \sum_{i=1}^{n} (1/s(\lambda_i))^2$

P7.2.3 Suppose $Q^H A Q = \text{diag}(\lambda_i) + N$ is a Schur decomposition of A and that $X^{-1} A X = \text{diag}(\lambda_i)$. Show $\kappa_2(X)^2 \geq 1 + (\| N \|_F / \| A \|_F)^2$. See Loizou (1969).

P7.2.4 If $X^{-1} A X = \text{diag}(\lambda_i)$ and $|\lambda_1| \geq \cdots \geq |\lambda_n|$, then

$$\frac{\sigma_i(A)}{\kappa_2(X)} \leq |\lambda_i| \leq \kappa_2(X) \sigma_i(A).$$

Prove this result for the $n = 2$ case. See Ruhe (1975).

P7.2.5 Show that if $A = \begin{bmatrix} a & c \\ 0 & b \end{bmatrix}$ and $a \neq b$, then $s(a) = s(b) = (1 + |c/(a-b)|^2)^{-1/2}$.

P7.2.6 Suppose

$$A = \begin{bmatrix} \lambda & v^T \\ 0 & T_{22} \end{bmatrix}$$

and that $\lambda \notin \lambda(T_{22})$. Show that if $\sigma = \text{sep}(\lambda, T_{22})$, then

$$s(\lambda) = \frac{1}{\sqrt{1 + \| (T_{22} - \lambda I)^{-1} v \|_2^2}} \leq \frac{\sigma}{\sqrt{\sigma^2 + \| v \|_2^2}}.$$

P7.2.7   Show that the condition of a simple eigenvalue is preserved under unitary similarity transformations.

P7.2.8   With the same hypothesis as in the Bauer-Fike theorem (Theorem 7.2.2), show

that $\min_{\lambda \in \lambda(A)} |\lambda - \mu| \leq \| |X^{-1}| |E| |X| \|_p.$

P7.2.9   Verify (7.2.5).

P7.2.10   Show that if $B \in \mathbb{C}^{m \times m}$ and $C \in \mathbb{C}^{n \times n}$, then $\text{sep}(B, C)$ is less than or equal to $|\lambda - \mu|$ for all $\lambda \in \lambda(B)$ and $\mu \in \lambda(C)$.

Notes and References for Sec. 7.2

Many of the results presented in this section may be found in Wilkinson (1965, chapter 2), Stewart and Sun (1990) as well as in

F.L. Bauer and C.T. Fike (1960). "Norms and Exclusion Theorems," *Numer. Math. 2*, 123–44.
A.S. Householder (1964). *The Theory of Matrices in Numerical Analysis*. Blaisdell, New York.

The following papers are concerned with the effect of perturbations on the eigenvalues of a general matrix:

A. Ruhe (1970). "Perturbation Bounds for Means of Eigenvalues and Invariant Subspaces," *BIT 10*, 343–54.
A. Ruhe (1970). "Properties of a Matrix with a Very Ill-Conditioned Eigenproblem," *Numer. Math. 15*, 57–60.
J.H. Wilkinson (1972). "Note on Matrices with a Very Ill-Conditioned Eigenproblem," *Numer. Math. 19*, 176–78.
W. Kahan, B.N. Parlett, and E. Jiang (1982). "Residual Bounds on Approximate Eigensystems of Nonnormal Matrices," *SIAM J. Numer. Anal. 19*, 470–484.
J.H. Wilkinson (1984). "On Neighboring Matrices with Quadratic Elementary Divisors," *Numer. Math. 44*, 1-21.
J.V. Burke and M.L. Overton (1992). "Stable Perturbations of Nonsymmetric Matrices," *Lin.Alg. and Its Application 171*, 249–273.

Wilkinson's work on nearest defective matrices is typical of a growing body of literature that is concerned with "nearness" problems. See

N.J. Higham (1985). "Nearness Problems in Numerical Linear Algebra," PhD Thesis, University of Manchester, England.
C. Van Loan (1985). "How Near is a Stable Matrix to an Unstable Matrix?," *Contemporary Mathematics, Vol. 47*, 465–477.
J.W. Demmel (1987). "On the Distance to the Nearest Ill-Posed Problem," *Numer. Math. 51*, 251–289.

J.W. Demmel (1987). "A Counterexample for two Conjectures About Stability," *IEEE Trans. Auto. Cont. AC-32*, 340–342.

A. Ruhe (1987). "Closest Normal Matrix Found!," *BIT 27*, 585–598.

R. Byers (1988). "A Bisection Method for Measuring the Distance of a Stable Matrix to the Unstable Matrices," *SIAM J. Sci. and Stat. Comp. 9*, 875–881.

J.W. Demmel (1988). "The Probability that a Numerical Analysis Problem is Difficult," *Math. Comp. 50*, 449–480.

N.J. Higham (1989). "Matrix Nearness Problems and Applications," in *Applications of Matrix Theory*, M.J.C. Gover and S. Barnett (eds), Oxford University Press, Oxford UK, 1–27.

Aspects of eigenvalue condition are discussed in

C. Van Loan (1987). "On Estimating the Condition of Eigenvalues and Eigenvectors," *Lin. Alg. and Its Applic. 88/89*, 715–732.

C.D. Meyer and G.W. Stewart (1988). "Derivatives and Perturbations of Eigenvectors," *SIAM J. Num. Anal. 25*, 679–691.

G.W. Stewart and G. Zheng (1991). "Eigenvalues of Graded Matrices and the Condition Numbers of Multiple Eigenvalues," *Numer. Math. 58*, 703–712.

J.-G. Sun (1992). "On Condition Numbers of a Nondefective Multiple Eigenvalue," *Numer. Math. 61*, 265–276.

The relationship between the eigenvalue condition number, the departure from normality, and the condition of the eigenvector matrix is discussed in

P. Henrici (1962). "Bounds for Iterates, Inverses, Spectral Variation and Fields of Values of Non-normal Matrices," *Numer. Math. 4*, 24–40.

P. Eberlein (1965). "On Measures of Non-Normality for Matrices," *Amer. Math. Soc. Monthly 72*, 995–96.

R.A. Smith (1967). "The Condition Numbers of the Matrix Eigenvalue Problem," *Numer. Math. 10* 232–40.

G. Loizou (1969). "Nonnormality and Jordan Condition Numbers of Matrices," *J. ACM 16*, 580–40.

A. van der Sluis (1975). "Perturbations of Eigenvalues of Non-normal Matrices," *Comm. ACM 18*, 30–36.

The paper by Henrici also contains a result similar to Theorem 7.2.3. Penetrating treatments of invariant subspace perturbation include

T. Kato (1966). *Perturbation Theory for Linear Operators*, Springer-Verlag, New York.

C. Davis and W.M. Kahan (1970). "The Rotation of Eigenvectors by a Perturbation, III," *SIAM J. Num. Anal. 7*, 1–46.

G.W. Stewart (1971). "Error Bounds for Approximate Invariant Subspaces of Closed Linear Operators," *SIAM. J. Num. Anal. 8*, 796–808.

G.W. Stewart (1973). "Error and Perturbation Bounds for Subspaces Associated with Certain Eigenvalue Problems," *SIAM Review 15*, 727–64.

Detailed analyses of the function sep(.,.) and the map $X \rightarrow AX + XA^T$ are given in

J. Varah (1979). "On the Separation of Two Matrices," *SIAM J. Num. Anal. 16*, 216–22.

R. Byers and S.G. Nash (1987). "On the Singular Vectors of the Lyapunov Operator," *SIAM J. Alg. and Disc. Methods 8*, 59–66.

Gershgorin's Theorem can be used to derive a comprehensive perturbation theory. See Wilkinson (1965, chapter 2). The theorem itself can be generalized and extended in various ways; see

R.S. Varga (1970). "Minimal Gershgorin Sets for Partitioned Matrices," *SIAM J. Num. Anal. 7*, 493–507.

R.J. Johnston (1971). "Gershgorin Theorems for Partitioned Matrices," *Lin. Alg. and Its Applic. 4*, 205–20.

7.3   Power Iterations

Suppose that we are given $A \in \mathbb{C}^{n \times n}$ and a unitary $U_0 \in \mathbb{C}^{n \times n}$. Assume that Householder orthogonalization (Algorithm 5.2.1) can be extended to complex matrices (it can) and consider the following iteration:

$$
\begin{aligned}
&T_0 = U_0^H A U_0 \\
&\textbf{for } k = 1, 2, \ldots \\
&\qquad T_{k-1} = U_k R_k \quad \text{(QR factorization)} \\
&\qquad T_k = R_k U_k \\
&\textbf{end}
\end{aligned}
\qquad (7.3.1)
$$

Since $T_k = R_k U_k = U_k^H (U_k R_k) U_k = U_k^H T_{k-1} U_k$ it follows by induction that

$$
T_k = (U_0 U_1 \cdots U_k)^H A (U_0 U_1 \cdots U_k). \qquad (7.3.2)
$$

Thus, each T_k is unitarily similar to A. Not so obvious, and what is the central theme of this section, is that the T_k almost always converge to upper triangular form. That is, (7.3.2) almost always "converges" to a Schur decomposition of A.

Iteration (7.3.1) is called the *QR iteration*, and it forms the backbone of the most effective algorithm for computing the Schur decomposition. In order to motivate the method and to derive its convergence properties, two other eigenvalue iterations that are important in their own right are presented first: the power method and the method of orthogonal iteration.

7.3.1   The Power Method

Suppose $A \in \mathbb{C}^{n \times n}$ is diagonalizable, that $X^{-1} A X = \text{diag}(\lambda_1, \ldots, \lambda_n)$ with $X = [x_1, \ldots, x_n]$, and $|\lambda_1| > |\lambda_2| \geq \cdots \geq |\lambda_n|$. Given a unit 2-norm $q^{(0)} \in \mathbb{C}^n$, the *power method* produces a sequence of vectors $q^{(k)}$ as follows:

$$
\begin{aligned}
&\textbf{for } k = 1, 2, \ldots \\
&\qquad z^{(k)} = A q^{(k-1)} \\
&\qquad q^{(k)} = z^{(k)} / \| z^{(k)} \|_2 \\
&\qquad \lambda^{(k)} = [q^{(k)}]^H A q^{(k)} \\
&\textbf{end}
\end{aligned}
\qquad (7.3.3)
$$

There is nothing special about doing a 2-norm normalization except that it imparts a greater unity on the overall discussion in this section.

Let us examine the convergence properties of the power iteration. If

$$q^{(0)} = a_1 x_1 + a_2 x_2 + \cdots + a_n x_n$$

and $a_1 \neq 0$, then it follows that

$$A^k q^{(0)} = a_1 \lambda_1^k \left(x_1 + \sum_{j=2}^{n} \frac{a_j}{a_1} \left(\frac{\lambda_j}{\lambda_1} \right)^k x_j \right).$$

Since $q^{(k)} \in \text{span}\{A^k q^{(0)}\}$ we conclude that

$$\text{dist}\left(\text{span}\{q^{(k)}\}, \text{span}\{x_1\}\right) = O\left(\left| \frac{\lambda_2}{\lambda_1} \right|^k \right)$$

and moreover,

$$| \lambda_1 - \lambda^{(k)} | = O\left(\left| \frac{\lambda_2}{\lambda_1} \right|^k \right).$$

If $|\lambda_1| > |\lambda_2| \geq \cdots \geq |\lambda_n|$ then we say that λ_1 is a *dominant eigenvalue*. Thus, the power method converges if λ_1 is dominant and if $q^{(0)}$ has a component in the direction of the corresponding *dominant eigenvector* x_1.

The behavior of the iteration without these assumptions is discussed in Wilkinson (1965, p.570) and Parlett and Poole (1973).

Example 7.3.1 If

$$A = \begin{bmatrix} -261 & 209 & -49 \\ -530 & 422 & -98 \\ -800 & 631 & -144 \end{bmatrix}$$

then $\lambda(A) = \{10, 4, 3\}$. Applying (7.3.3) with $q^{(0)} = [1, 0, 0]^T$ we find

k	$\lambda^{(k)}$
1	13.0606
2	10.7191
3	10.2073
4	10.0633
5	10.0198
6	10.0063
7	10.0020
8	10.0007
9	10.0002

In practice, the usefulness of the power method depends upon the ratio $|\lambda_2|/|\lambda_1|$, since it dictates the rate of convergence. The danger that $q^{(0)}$ is deficient in x_1 is a less worrisome matter because rounding errors sustained during the iteration typically ensure that the subsequent $q^{(k)}$ have a component in this direction. Moreover, it is typically the case in applications where the dominant eigenvalue and eigenvector are desired that an à priori estimate of x_1 is known. Normally, by setting $q^{(0)}$ to be this estimate, the dangers of a small a_1 are minimized.

Note that the only thing required to implement the power method is a subroutine capable of computing matrix-vector products of the form Aq. It is not necessary to store A in an n-by-n array. For this reason, the algorithm can be of interest when A is large and sparse and when there is a sufficient gap between $|\lambda_1|$ and $|\lambda_2|$.

Estimates for the error $|\lambda^{(k)} - \lambda_1|$ can be obtained by applying the perturbation theory developed in the previous section. Define the vector $r^{(k)} = Aq^{(k)} - \lambda^{(k)}q^{(k)}$ and observe that $(A + E^{(k)})q^{(k)} = \lambda^{(k)}q^{(k)}$ where $E^{(k)} = -r^{(k)}[q^{(k)}]^H$. Thus $\lambda^{(k)}$ is an eigenvalue of $A + E^{(k)}$ and

$$| \lambda^{(k)} - \lambda_1 | \approx \frac{\| E^{(k)} \|_2}{s(\lambda_1)} = \frac{\| r^{(k)} \|_2}{s(\lambda_1)} .$$

If we use the power method to generate approximate right *and* left dominant eigenvectors, then it is possible to obtain an estimate of $s(\lambda_1)$. In particular, if $w^{(k)}$ is a unit 2-norm vector in the direction of $(A^H)^k w^{(0)}$, then we can use the approximation $s(\lambda_1) \approx | w^{(k)^H} q^{(k)} |$.

7.3.2   Orthogonal Iteration

A straightforward generalization of the power method can be used to compute higher-dimensional invariant subspaces. Let r be a chosen integer satisfying $1 \leq r \leq n$. Given an n-by-r matrix Q_0 with orthonormal columns, the method of *orthogonal iteration* generates a sequence of matrices $\{Q_k\} \subseteq \mathbb{C}^{n \times r}$ as follows:

$$
\begin{aligned}
&\textbf{for } k = 1, 2, \ldots \\
&\qquad Z_k = A Q_{k-1} \\
&\qquad Q_k R_k = Z_k \qquad \text{(QR factorization)} \\
&\textbf{end}
\end{aligned}
\tag{7.3.4}
$$

Note that if $r = 1$, then this is just the power method. Moreover, the sequence $\{Q_k e_1\}$ is precisely the sequence of vectors produced by the power iteration with starting vector $q^{(0)} = Q_0 e_1$.

In order to analyze the behavior of this iteration, suppose that

$$Q^H A Q = T = \text{diag}(\lambda_i) + N \qquad |\lambda_1| \geq |\lambda_2| \geq \cdots \geq |\lambda_n| \tag{7.3.5}$$

is a Schur decomposition of $A \in \mathbb{C}^{n \times n}$. Assume that $1 \leq r < n$ and partition Q, T, and N as follows:

$$
Q = [\, Q_\alpha \quad Q_\beta \,] \qquad\qquad T = \begin{bmatrix} T_{11} & T_{12} \\ 0 & T_{22} \end{bmatrix} \begin{matrix} r \\ n-r \end{matrix}
$$
$$
r \quad\;\; n-r \qquad\qquad\qquad\quad r \quad\; n-r
$$

$$\tag{7.3.6}$$

$$N = \begin{bmatrix} N_{11} & N_{12} \\ 0 & N_{22} \end{bmatrix} \begin{matrix} r \\ n-r \end{matrix} \quad .$$
$$\quad r \quad n-r$$

If $|\lambda_r| > |\lambda_{r+1}|$, then the subspace $D_r(A) = \mathrm{ran}(Q_\alpha)$ is said to be a *dominant* invariant subspace. It is the unique invariant subspace associated with the eigenvalues $\lambda_1, \ldots, \lambda_r$. The following theorem shows that with reasonable assumptions, the subspaces $\mathrm{ran}(Q_k)$ generated by (7.3.4) converge to $D_r(A)$ at a rate proportional to $|\lambda_{r+1}/\lambda_r|^k$.

Theorem 7.3.1 *Let the Schur decomposition of $A \in \mathbb{C}^{n \times n}$ be given by (7.3.5) and (7.3.6) with $n \geq 2$. Assume that $|\lambda_r| > |\lambda_{r+1}|$ and that $\theta \geq 0$ satisfies*

$$(1 + \theta)|\lambda_r| > \| N \|_F .$$

If $Q_0 \in \mathbb{C}^{n \times r}$ has orthonormal columns and

$$d = \mathrm{dist}(D_r(A^H), \mathrm{ran}(Q_0)) < 1,$$

then the matrices Q_k generated by (7.3.4) satisfy

$$\mathrm{dist}(D_r(A), \mathrm{ran}(Q_k)) \leq$$
$$\frac{(1 + \theta)^{n-2}}{\sqrt{1 - d^2}} \left(1 + \frac{\| T_{12} \|_F}{\mathrm{sep}(T_{11}, T_{22})} \right) \left(\frac{|\lambda_{r+1}| + \| N \|_F/(1 + \theta)}{|\lambda_r| - \| N \|_F/(1 + \theta)} \right)^k .$$

Proof. The proof is given in an appendix at the end of this section. □

The condition $d < 1$ in Theorem 7.3.1 ensures that the initial Q matrix is not deficient in certain eigendirections:

$$d < 1 \quad \leftrightarrow \quad D_r(A^H)^\perp \cap \mathrm{ran}(Q_0) = \{0\}.$$

The theorem essentially says that if this condition holds and if θ is chosen large enough, then

$$\mathrm{dist}(D_r(A), \mathrm{ran}(Q_k)) \leq c \left| \frac{\lambda_{r+1}}{\lambda_r} \right|^k$$

where c depends on $\mathrm{sep}(T_{11}, T_{22})$ and A's departure from normality. Needless to say, convergence can be very slow if the gap between $|\lambda_r|$ and $|\lambda_{r+1}|$ is not sufficiently wide.

Example 7.3.2 If (7.3.4) is applied to the matrix A in Example 7.3.1, with $Q_0 = [e_1, e_2]$, we find:

k	$\mathrm{dist}(D_2(A), \mathrm{ran}(Q_k))$
1	.0052
2	.0047
3	.0039
4	.0030
5	.0023
6	.0017
7	.0013

The error is tending to zero with rate $(\lambda_3/\lambda_2)^k = (3/4)^k$.

It is possible to accelerate the convergence in orthogonal iteration using a technique described in Stewart (1976). In the accelerated scheme, the approximate eigenvalue $\lambda_i^{(k)}$ satisfies

$$|\lambda_i^{(k)} - \lambda_i| \approx \left|\frac{\lambda_{r+1}}{\lambda_i}\right|^k \qquad i = 1{:}r\,.$$

(Without the acceleration, the right-hand side is $|\lambda_{i+1}/\lambda_i|^k$.) Stewart's algorithm involves computing the Schur decomposition of the matrices $Q_k^T A Q_k$ every so often. The method can be very useful in situations where A is large and sparse and a few of its largest eigenvalues are required.

7.3.3   The QR Iteration

We now "derive" the QR iteration (7.3.1) and examine its convergence. Suppose $r = n$ in (7.3.4) and the eigenvalues of A satisfy

$$|\lambda_1| > |\lambda_2| > \cdots > |\lambda_n|.$$

Partition the matrix Q in (7.3.5) and Q_k in (7.3.4) as follows:

$$Q = [\,q_1, \ldots, q_n\,] \qquad Q_k = \left[\,q_1^{(k)}, \ldots, q_n^{(k)}\,\right]$$

If

$$\mathrm{dist}(D_i(A^H), \mathrm{span}\{q_1^{(0)}, \ldots, q_i^{(0)}\}) < 1 \qquad i = 1{:}n \tag{7.3.7}$$

then it follows from Theorem 7.3.1 that

$$\mathrm{dist}(\mathrm{span}\{q_1^{(k)}, \ldots, q_i^{(k)}\}, \mathrm{span}\{q_1, \ldots, q_i\}) \to 0$$

for $i = 1{:}n$. This implies that the matrices T_k defined by

$$T_k = Q_k^H A Q_k$$

are converging to upper triangular form. Thus, it can be said that the method of orthogonal iteration computes a Schur decomposition provided the original iterate $Q_0 \in \mathbb{C}^{n \times n}$ is not deficient in the sense of (7.3.7).

The QR iteration arises naturally by considering how to compute the matrix T_k directly from its predecessor T_{k-1}. On the one hand, we have from (7.3.4) and the definition of T_{k-1} that

$$T_{k-1} = Q_{k-1}^H A Q_{k-1} = Q_{k-1}^H (A Q_{k-1}) = (Q_{k-1}^H Q_k) R_k.$$

On the other hand

$$T_k = Q_k^H A Q_k = (Q_k^H A Q_{k-1})(Q_{k-1}^H Q_k) = R_k (Q_{k-1}^H Q_k).$$

Thus, T_k is determined by computing the QR factorization of T_{k-1} and then multiplying the factors together in reverse order. This is precisely what is done in (7.3.1).

Example 7.3.3 If the iteration:

> **for** $k = 1, 2, \ldots$
> $\quad A = QR$
> $\quad A = RQ$
> **end**

is applied to the matrix of Example 7.3.1, then the strictly lower triangular elements diminish as follows:

k	$O(|a_{21}|)$	$O(|a_{31}|)$	$O(|a_{32}|)$
1	10^{-1}	10^{-1}	10^{-2}
2	10^{-2}	10^{-2}	10^{-3}
3	10^{-2}	10^{-3}	10^{-3}
4	10^{-3}	10^{-3}	10^{-3}
5	10^{-3}	10^{-4}	10^{-3}
6	10^{-4}	10^{-5}	10^{-3}
7	10^{-4}	10^{-5}	10^{-3}
8	10^{-5}	10^{-6}	10^{-4}
9	10^{-5}	10^{-7}	10^{-4}
10	10^{-6}	10^{-8}	10^{-4}

Note that a single QR iteration is an $O(n^3)$ calculation. Moreover, since convergence is only linear (when it exists), it is clear that the method is a prohibitively expensive way to compute Schur decompositions. Fortunately these practical difficulties can be overcome, as we show in §7.4 and §7.5.

7.3.4 LR Iterations

We conclude with some remarks about power iterations that rely on the LU factorization rather than the QR factorizaton. Let $G_0 \in \mathbb{C}^{n \times r}$ have rank r. Corresponding to (7.3.4) we have the following iteration:

> **for** $k = 1, 2, \ldots$
> $\quad Z_k = AG_{k-1}$ $\qquad\qquad\qquad\qquad\qquad$ (7.3.8)
> $\quad Z_k = G_k R_k \qquad$ (LU factorization)
> **end**

Suppose $r = n$ and that we define the matrices T_k by

$$T_k = G_k^{-1} A G_k. \qquad\qquad (7.3.9)$$

It can be shown that if we set $L_0 = G_0$, then the T_k can be generated as follows:

$$T_0 = L_0^{-1}AL_0$$
$$\text{for } k = 1, 2, \ldots$$
$$\qquad T_{k-1} = L_k R_k \qquad \text{(LU factorization)}$$
$$\qquad T_k = R_k L_k$$
$$\text{end}$$

(7.3.10)

Iterations (7.3.8) and (7.3.10) are known as *treppeniteration* and the *LR iteration*, respectively. Under reasonable assumptions, the T_k converge to upper triangular form. To successfully implement either method, it is necessary to pivot. See Wilkinson (1965, p.602).

Appendix

In order to establish Theorem 7.3.1 we need the following lemma which is concerned with bounding the powers of a matrix and its inverse.

Lemma 7.3.2 *Let $Q^H A Q = T = D + N$ be a Schur decomposition of $A \in \mathbb{C}^{n \times n}$ where D is diagonal and N strictly upper triangular. Let λ and μ denote the largest and smallest eigenvalues of A in absolute value. If $\theta \geq 0$ then for all $k \geq 0$ we have*

$$\| A^k \|_2 \leq (1+\theta)^{n-1} \left(|\lambda| + \frac{\| N \|_F}{1+\theta} \right)^k . \tag{7.3.11}$$

If A is nonsingular and $\theta \geq 0$ satisfies $(1+\theta)|\mu| > \| N \|_F$, then for all $k \geq 0$ we also have

$$\| A^{-k} \|_2 \leq (1+\theta)^{n-1} \left(\frac{1}{|\mu| - \| N \|_F/(1+\theta)} \right)^k . \tag{7.3.12}$$

Proof. For $\theta \geq 0$, define the diagonal matrix Δ by

$$\Delta = \text{diag} \left(1, (1+\theta), (1+\theta)^2, \ldots, (1+\theta)^{n-1} \right)$$

and note that $\kappa_2(\Delta) = (1+\theta)^{n-1}$. Since N is strictly upper triangular, it is easy to verify that $\| \Delta N \Delta^{-1} \|_F \leq \| N \|_F/(1+\theta)$. Thus,

$$\begin{aligned}
\| A^k \|_2 &= \| T^k \|_2 = \| \Delta^{-1}(D + \Delta N \Delta^{-1})^k \Delta \|_2 \\
&\leq \kappa_2(\Delta) \left(\| D \|_2 + \| \Delta N \Delta^{-1} \|_2 \right)^k \\
&\leq (1+\theta)^{n-1} \left(|\lambda| + \frac{\| N \|_F}{1+\theta} \right)^k .
\end{aligned}$$

On the other hand, if A is nonsingular and $(1+\theta)|\mu| > \| N \|_F$, then $\| \Delta D^{-1} N \Delta^{-1} \|_2 < 1$ and using Lemma 2.3.3 we obtain

$$\| A^{-k} \|_2 = \| T^{-k} \|_2 = \left\| \Delta^{-1}[(I + \Delta D^{-1} N \Delta^{-1})^{-1} D^{-1}]^k \Delta \right\|_2$$

$$\leq \kappa_2(\Delta) \left(\frac{\| D^{-1} \|_2}{1 - \| \Delta D^{-1} N \Delta^{-1} \|_2} \right)^k$$

$$\leq (1 + \theta)^{n-1} \left(\frac{1}{|\mu| - \| N \|_F / (1 + \theta)} \right)^k . \square$$

Proof of Theorem 7.3.1

It is easy to show by induction that $A^k Q_0 = Q_k(R_k \cdots R_1)$. By substituting (7.3.5) and 7.3.6) into this equality we obtain

$$T^k \begin{bmatrix} V_0 \\ W_0 \end{bmatrix} = \begin{bmatrix} V_k \\ W_k \end{bmatrix} (R_k \cdots R_1)$$

where $V_k = Q_\alpha^H Q_k$ and $W_k = Q_\beta^H Q_k$. Using Lemma 7.1.5 we know that a matrix $X \in \mathbb{C}^{r \times (n-r)}$ exists such that

$$\begin{bmatrix} I_r & X \\ 0 & I_{n-r} \end{bmatrix}^{-1} \begin{bmatrix} T_{11} & T_{12} \\ 0 & T_{22} \end{bmatrix} \begin{bmatrix} I_r & X \\ 0 & I_{n-r} \end{bmatrix} = \begin{bmatrix} T_{11} & 0 \\ 0 & T_{22} \end{bmatrix}$$

and so

$$\begin{bmatrix} T_{11}^k & 0 \\ 0 & T_{22}^k \end{bmatrix} \begin{bmatrix} V_0 - X W_0 \\ W_0 \end{bmatrix} = \begin{bmatrix} V_k - X W_k \\ W_k \end{bmatrix} (R_k \cdots R_1) .$$

Below we establish that the matrix $V_0 - \dot{X} W_0$ is nonsingular and this enables us to obtain the following expression:

$$W_k = T_{22}^k W_0 (V_0 - X W_0)^{-1} T_{11}^{-k} [I_r, -X] \begin{bmatrix} V_k \\ W_k \end{bmatrix} .$$

Recalling the definition of distance between subspaces from §2.6.3,

$$\text{dist}(D_r(A), \text{ran}(Q_k)) = \| Q_\beta^H Q_k \|_2 = \| W_k \|_2.$$

Since

$$\| [I_r, -X] \|_2 \leq 1 + \| X \|_F$$

we have

$$\text{dist}(D_r(A), \text{ran}(Q_k)) \leq \qquad \qquad (7.3.13)$$
$$\| T_{22}^k \|_2 \| (V_0 - X W_0)^{-1} \|_2 \| T_{11}^{-k} \|_2 (1 + \| X \|_F) .$$

To prove the theorem we must look at each of the four factors in the upper bound.

Since $\text{sep}(T_{11}, T_{22})$ is the smallest singular value of the linear transformation $\phi(X) = T_{11} X - X T_{22}$ it readily follows from $\phi(X) = -T_{12}$ that

$$\| X \|_F \leq \frac{\| T_{12} \|_F}{\text{sep}(T_{11}, T_{22})} . \qquad \qquad (7.3.14)$$

Using Lemma 7.3.2, it can be shown that

$$\| T_{22}^k \|_2 \le (1 + \theta)^{n-r-1} \left(|\lambda_{r+1}| + \frac{\| N \|_F}{1 + \theta} \right)^k \qquad (7.3.15)$$

and

$$\| T_{11}^{-k} \|_2 \le (1 + \theta)^{r-1} / \left(|\lambda_r| - \frac{\| N \|_F}{1 + \theta} \right)^k . \qquad (7.3.16)$$

Finally, we turn our attention to the $\| (V_0 - XW_0)^{-1} \|$ factor. Note that

$$
\begin{aligned}
V_0 - XW_0 &= Q_\alpha^H Q_0 - X Q_\beta^H Q_0 \\
&= [\, I_r , \, -X \,] \begin{bmatrix} Q_\alpha^H \\ Q_\beta^H \end{bmatrix} Q_0 \\
&= \left[[\, Q_\alpha \, Q_\beta \,] \begin{bmatrix} I_r \\ -X^H \end{bmatrix} \right]^H Q_0 \\
&= (I_r + XX^H)^{1/2} (Z^H Q_0)
\end{aligned}
$$

where

$$
\begin{aligned}
Z &= [\, Q_\alpha \, Q_\beta \,] \begin{bmatrix} I_r \\ -X^H \end{bmatrix} (I_r + XX^H)^{-1/2} \\
&= (Q_\alpha - Q_\beta X^H)(I_r + XX^H)^{-1/2}.
\end{aligned}
$$

The columns of this matrix are orthonormal. They are also a basis for $D_r(A^H)$ because

$$A^H (Q_\alpha - Q_\beta X^H) = (Q_\alpha - Q_\beta X^H) T_{11}^H.$$

This last fact follows from the equation $A^H Q = Q T^H$.

From Theorem 2.6.1

$$d = \operatorname{dist}(D_r(A^H), \operatorname{range}(Q_0)) = \sqrt{1 - \sigma_r(Z^H Q_0)^2}$$

and since $d < 1$ by hypothesis,

$$\sigma_r(Z^H Q_0) > 0.$$

This shows that

$$(V_0 - XW_0) = (I_r + XX^H)^{1/2}(Z^H Q_0)$$

is nonsingular and thus,

$$
\begin{aligned}
\| (V_0 - XW_0)^{-1} \|_2 &\le \| (I_r + XX^H)^{-1/2} \|_2 \| (Z^H Q_0)^{-1} \|_2 \\
&\le 1/\sqrt{1 - d^2}. \qquad (7.3.17)
\end{aligned}
$$

The theorem follows by substituting (7.3.14)-(7.3.17) into (7.3.13). □

Problems

P7.3.1 (a) Show that if $X \in \mathbb{C}^{n \times n}$ is nonsingular, then $\| A \|_X = \| X^{-1} A X \|_2$ defines a matrix norm with the property that $\| AB \|_X \leq \| A \|_X \| B \|_X$. (b) Let $A \in \mathbb{C}^{n \times n}$ and set $\rho = \max |\lambda_i|$. Show that for any $\epsilon > 0$ there exists a nonsingular $X \in \mathbb{C}^{n \times n}$ such that $\| A \|_X = \| X^{-1} A X \|_2 \leq \rho + \epsilon$. Conclude that there is a constant M such that $\| A^k \|_2 \leq M(\rho + \epsilon)^k$ for all non-negative integers k. (Hint: Set $X = Q \, \mathrm{diag}(1, a, \ldots, a^{n-1})$ where $Q^H A Q = D + N$ is A's Schur decomposition.)

P7.3.2 Verify that (7.3.10) calculates the matrices T_k defined by (7.3.9).

P7.3.3 Suppose $A \in \mathbb{C}^{n \times n}$ is nonsingular and that $Q_0 \in \mathbb{C}^{n \times p}$ has orthonormal columns. The following iteration is referred to as *inverse orthogonal iteration*.

> for $k = 1, 2, \ldots$
>> Solve $A Z_k = Q_{k-1}$ for $Z_k \in \mathbb{C}^{n \times p}$
>> $Z_k = Q_k R_k$ (QR factorization)
> end

Explain why this iteration can usually be used to compute the p smallest eigenvalues of A in absolute value. Note that to implement this iteration it is necessary to be able to solve linear systems that involve A. When $p = 1$, the method is referred to as the *inverse power method*.

P7.3.4 Assume $A \in \mathbb{R}^{n \times n}$ has eigenvalues $\lambda_1, \ldots, \lambda_n$ that satisfy

$$\lambda = \lambda_1 = \lambda_2 = \lambda_3 = \lambda_4 > |\lambda_5| \geq \cdots \geq |\lambda_n|$$

where λ is positive. Assume that A has two Jordan blocks of the form.

$$\begin{bmatrix} \lambda & 1 \\ 0 & \lambda \end{bmatrix}.$$

Discuss the convergence properties of the power method when applied to this matrix. Discuss how the convergence might be accelerated.

Notes and References for Sec. 7.3

A detailed, practical discussion of the power method is given in Wilkinson (1965, chapter 10). Methods are discussed for accelerating the basic iteration, for calculating nondominant eigenvalues, and for handling complex conjugate eigenvalue pairs. The connections among the various power iterations are discussed in

B.N. Parlett and W.G. Poole (1973). "A Geometric Theory for the QR, LU, and Power Iterations," *SIAM J. Num. Anal. 10*, 389–412.

The QR iteration was concurrently developed in

J.G.F. Francis (1961). "The QR Transformation: A Unitary Analogue to the LR Transformation," *Comp. J. 4*, 265–71, 332–34.

V.N. Kublanovskaya (1961). "On Some Algorithms for the Solution of the Complete Eigenvalue Problem," *USSR Comp. Math. Phys. 3*, 637–57.

As can be deduced from the title of the first paper, the LR iteration predates the QR iteration. The former very fundamental algorithm was proposed by

H. Rutishauser (1958). "Solution of Eigenvalue Problems with the LR Transformation," *Nat. Bur. Stand. App. Math. Ser. 49*, 47–81.

B.N. Parlett (1995). "The New qd Algorithms," *ACTA Numerica 5*, 459–491.

Numerous papers on the convergence of the QR iteration have appeared. Several of these are

J.H. Wilkinson (1965). "Convergence of the LR, QR, and Related Algorithms," *Comp. J. 8*, 77–84.

B.N. Parlett (1965). "Convergence of the Q-R Algorithm," *Numer. Math. 7*, 187–93. (Correction in *Numer. Math. 10*, 163–64.)

B.N. Parlett (1966). "Singular and Invariant Matrices Under the QR Algorithm," *Math. Comp. 20*, 611–15.

B.N. Parlett (1968). "Global Convergence of the Basic QR Algorithm on Hessenberg Matrices," *Math. Comp. 22*, 803–17.

Wilkinson (AEP, chapter 9) also discusses the convergence theory for this important algorithm.

Deeper insight into the convergence of the QR algorithm and its connection to other important algorithms can be attained by reading

D.S. Watkins (1982). "Understanding the QR Algorithm," *SIAM Review 24*, 427–440.

T. Nanda (1985). "Differential Equations and the QR Algorithm," *SIAM J. Numer. Anal. 22*, 310–321.

D.S. Watkins (1993). "Some Perspectives on the Eigenvalue Problem," *SIAM Review 35*, 430–471.

The following papers are concerned with various practical and theoretical aspects of simultaneous iteration:

H. Rutishauser (1970). "Simultaneous Iteration Method for Symmetric Matrices," *Numer. Math. 16*, 205–23. See also (Wilkinson and Reinsch(1971, pp. 284–302.

M. Clint and A. Jennings (1971). "A Simultaneous Iteration Method for the Unsymmetric Eigenvalue Problem," *J. Inst. Math. Applic. 8*, 111–21.

A. Jennings and D.R.L. Orr (1971). "Application of the Simultaneous Iteration Method to Undamped Vibration Problems," *Inst. J. Numer. Math. Eng. 3*, 13–24.

A. Jennings and W.J. Stewart (1975). "Simultaneous Iteration for the Partial Eigensolution of Real Matrices," *J. Inst. Math. Applic. 15*, 351–62.

G.W. Stewart (1975). "Methods of Simultaneous Iteration for Calculating Eigenvectors of Matrices," in *Topics in Numerical Analysis II* , ed. John J.H. Miller, Academic Press, New York, pp. 185–96.

G.W. Stewart (1976). "Simultaneous Iteration for Computing Invariant Subspaces of Non-Hermitian Matrices," *Numer. Math. 25*, 123–36.

See also chapter 10 of

A. Jennings (1977). *Matrix Computation for Engineers and Scientists*, John Wiley and Sons, New York.

Simultaneous iteration and the Lanczos algorithm (cf. Chapter 9) are the principal methods for finding a few eigenvalues of a general sparse matrix.

7.4 The Hessenberg and Real Schur Forms

In this and the next section we show how to make the QR iteration (7.3.1) a fast, effective method for computing Schur decompositions. Because the majority of eigenvalue/invariant subspace problems involve real data, we concentrate on developing the real analog of (7.3.1) which we write as follows:

$$
\begin{aligned}
&H_0 = U_0^T A U_0 \\
&\text{for } k = 1, 2, \ldots \\
&\qquad H_{k-1} = U_k R_k \qquad \text{(QR factorization)} \\
&\qquad H_k = R_k U_k \\
&\text{end}
\end{aligned}
\qquad (7.4.1)
$$

Here, $A \in \mathbb{R}^{n \times n}$, each $U_k \in \mathbb{R}^{n \times n}$ is orthogonal, and each $R_k \in \mathbb{R}^{n \times n}$ is upper triangular. A difficulty associated with this real iteration is that the H_k can never converge to strict, "eigenvalue revealing," triangular form in the event that A has complex eigenvalues. For this reason, we must lower our expectations and be content with the calculation of an alternative decomposition known as the *real Schur decomposition*.

In order to compute the real Schur decomposition efficiently we must carefully choose the initial orthogonal similarity transformation U_0 in (7.4.1). In particular, if we choose U_0 so that H_0 is upper Hessenberg, then the amount of work per iteration is reduced from $O(n^3)$ to $O(n^2)$. The initial reduction to Hessenberg form (the U_0 computation) is a very important computation in its own right and can be realized by a sequence of Householder matrix operations.

7.4.1 The Real Schur Decomposition

A block upper triangular matrix with either 1-by-1 or 2-by-2 diagonal blocks is upper *quasi-triangular*. The real Schur decomposition amounts to a real reduction to upper quasi-triangular form.

Theorem 7.4.1 (Real Schur Decomposition) *If $A \in \mathbb{R}^{n \times n}$, then there exists an orthogonal $Q \in \mathbb{R}^{n \times n}$ such that*

$$
Q^T A Q =
\begin{bmatrix}
R_{11} & R_{12} & \cdots & R_{1m} \\
0 & R_{22} & \cdots & R_{2m} \\
\vdots & \vdots & \ddots & \vdots \\
0 & 0 & \cdots & R_{mm}
\end{bmatrix}
\qquad (7.4.2)
$$

where each R_{ii} is either a 1-by-1 matrix or a 2-by-2 matrix having complex conjugate eigenvalues.

Proof. The complex eigenvalues of A must come in conjugate pairs, since the characteristic polynomial $\det(zI - A)$ has real coefficients. Let k be the number of complex conjugate pairs in $\lambda(A)$. We prove the theorem by induction on k. Observe first that Lemma 7.1.2 and Theorem 7.1.3 have obvious real analogs. Thus, the theorem holds if $k = 0$. Now suppose that $k \geq 1$. If $\lambda = \gamma + i\mu \in \lambda(A)$ and $\mu \neq 0$, then there exist vectors y and z in $\mathbb{R}^n (z \neq 0)$ such that $A(y + iz) = (\gamma + i\mu)(y + iz)$, i.e.,

$$A \begin{bmatrix} y & z \end{bmatrix} = \begin{bmatrix} y & z \end{bmatrix} \begin{bmatrix} \gamma & \mu \\ -\mu & \gamma \end{bmatrix}.$$

The assumption that $\mu \neq 0$ implies that y and z span a two-dimensional, real invariant subspace for A. It then follows from Lemma 7.1.2 that an orthogonal $U \in \mathbb{R}^{n \times n}$ exists such that

$$U^T A U = \begin{bmatrix} T_{11} & T_{12} \\ 0 & T_{22} \end{bmatrix} \begin{matrix} 2 \\ n-2 \end{matrix}$$
$$\phantom{U^T A U = \begin{bmatrix}}2 \quad n-2$$

where $\lambda(T_{11}) = \{\lambda, \bar{\lambda}\}$. By induction, there exists an orthogonal $\tilde{U}$ so $\tilde{U}^T T_{22} \tilde{U}$ has the required structure. The theorem follows by setting $Q = U$ diag$(I_2, \tilde{U})$. □

The theorem shows that any real matrix is orthogonally similar to an upper quasi-triangular matrix. It is clear that the real and imaginary part of the complex eigenvalues can be easily obtained from the 2-by-2 diagonal blocks.

7.4.2   A Hessenberg QR Step

We now turn our attention to the speedy calculation of a single QR step in (7.4.1). In this regard, the most glaring shortcoming associated with (7.4.1) is that each step requires a full QR factorization costing $O(n^3)$ flops. Fortunately, the amount of work per iteration can be reduced by an order of magnitude if the orthogonal matrix U_0 is judiciously chosen. In particular, if $U_0^T A U_0 = H_0 = (h_{ij})$ is upper Hessenberg ($h_{ij} = 0$, $i > j+1$), then each subsequent H_k requires only $O(n^2)$ flops to calculate. To see this we look at the computations $H = QR$ and $H_+ = RQ$ when H is upper Hessenberg. As described in §5.2.4, we can upper triangularize H with a sequence of $n-1$ Givens rotations: $Q^T H \equiv G_{n-1}^T \cdots G_1^T H = R$. Here, $G_i = G(i, i+1, \theta_i)$. For the $n = 4$ case there are three Givens premultiplications:

$$\begin{bmatrix} \times & \times & \times & \times \\ \times & \times & \times & \times \\ 0 & \times & \times & \times \\ 0 & 0 & \times & \times \end{bmatrix} \longrightarrow \begin{bmatrix} \times & \times & \times & \times \\ 0 & \times & \times & \times \\ 0 & \times & \times & \times \\ 0 & 0 & \times & \times \end{bmatrix} \longrightarrow \begin{bmatrix} \times & \times & \times & \times \\ 0 & \times & \times & \times \\ 0 & 0 & \times & \times \\ 0 & 0 & \times & \times \end{bmatrix}$$

$$\longrightarrow \begin{bmatrix} \times & \times & \times & \times \\ 0 & \times & \times & \times \\ 0 & 0 & \times & \times \\ 0 & 0 & 0 & \times \end{bmatrix}.$$

See Algorithm 5.2.3.

The computation $RQ = R(G_1 \cdots G_{n-1})$ is equally easy to implement. In the $n = 4$ case there are three Givens post-multiplications:

$$
\begin{bmatrix} \times & \times & \times & \times \\ 0 & \times & \times & \times \\ 0 & 0 & \times & \times \\ 0 & 0 & 0 & \times \end{bmatrix}
\longrightarrow
\begin{bmatrix} \times & \times & \times & \times \\ \times & \times & \times & \times \\ 0 & 0 & \times & \times \\ 0 & 0 & 0 & \times \end{bmatrix}
\longrightarrow
\begin{bmatrix} \times & \times & \times & \times \\ \times & \times & \times & \times \\ 0 & \times & \times & \times \\ 0 & 0 & 0 & \times \end{bmatrix}
$$

$$
\longrightarrow
\begin{bmatrix} \times & \times & \times & \times \\ \times & \times & \times & \times \\ 0 & \times & \times & \times \\ 0 & 0 & \times & \times \end{bmatrix}
$$

Overall we obtain the following algorithm:

Algorithm 7.4.1 If H is an n-by-n upper Hessenberg matrix, then this algorithm overwrites H with $H_+ = RQ$ where $H = QR$ is the QR factorization of H.

> **for** $k = 1{:}n - 1$
> $\quad [\, c(k),\ s(k) \,] = \textbf{givens}(H(k,k), H(k+1,k))$
> $\quad H(k{:}k+1, k{:}n) = \begin{bmatrix} c(k) & s(k) \\ -s(k) & c(k) \end{bmatrix}^T H(k{:}k+1, k{:}n)$
> **end**
> **for** $k = 1{:}n - 1$
> $\quad H(1{:}k+1, k{:}k+1) = H(1{:}k+1, k{:}k+1) \begin{bmatrix} c(k) & s(k) \\ -s(k) & c(k) \end{bmatrix}$
> **end**

Let $G_k = G(k, k+1, \theta_k)$ be the kth Givens rotation. It is easy to confirm that the matrix $Q = G_1 \cdots G_{n-1}$ is upper Hessenberg. Thus, $RQ = H_+$ is also upper Hessenberg. The algorithm requires about $6n^2$ flops and thus is an order-of-magnitude quicker than a full matrix QR step (7.3.1).

Example 7.4.1 If Algorithm 7.4.1 is applied to:

$$
H = \begin{bmatrix} 3 & 1 & 2 \\ 4 & 2 & 3 \\ 0 & .01 & 1 \end{bmatrix},
$$

then

$$
G_1 = \begin{bmatrix} .6 & -.8 & 0 \\ .8 & .6 & 0 \\ 0 & 0 & 1 \end{bmatrix}, \qquad
G_2 = \begin{bmatrix} 1 & 0 & 0 \\ 0 & .9996 & -.0249 \\ 0 & .0249 & .9996 \end{bmatrix},
$$

and

$$
H_+ = \begin{bmatrix} 4.7600 & -2.5442 & 5.4653 \\ .3200 & .1856 & -2.1796 \\ .0000 & .0263 & 1.0540 \end{bmatrix}.
$$

7.4.3 The Hessenberg Reduction

It remains for us to show how the *Hessenberg decomposition*

$$U_0^T A U_0 = H \qquad U_0^T U_0 = I \qquad (7.4.3)$$

can be computed. The transformation U_0 can be computed as a product of Householder matrices $P_1, \ldots, P_{n-2}$. The role of P_k is to zero the kth column below the subdiagonal. In the $n = 6$ case, we have

$$
\begin{bmatrix}
\times & \times & \times & \times & \times & \times \\
\times & \times & \times & \times & \times & \times \\
\times & \times & \times & \times & \times & \times \\
\times & \times & \times & \times & \times & \times \\
\times & \times & \times & \times & \times & \times \\
\times & \times & \times & \times & \times & \times
\end{bmatrix}
\xrightarrow{P_1}
\begin{bmatrix}
\times & \times & \times & \times & \times & \times \\
\times & \times & \times & \times & \times & \times \\
0 & \times & \times & \times & \times & \times \\
0 & \times & \times & \times & \times & \times \\
0 & \times & \times & \times & \times & \times \\
0 & \times & \times & \times & \times & \times
\end{bmatrix}
\xrightarrow{P_2}
$$

$$
\begin{bmatrix}
\times & \times & \times & \times & \times & \times \\
\times & \times & \times & \times & \times & \times \\
0 & \times & \times & \times & \times & \times \\
0 & 0 & \times & \times & \times & \times \\
0 & 0 & \times & \times & \times & \times \\
0 & 0 & \times & \times & \times & \times
\end{bmatrix}
\xrightarrow{P_3}
\begin{bmatrix}
\times & \times & \times & \times & \times & \times \\
\times & \times & \times & \times & \times & \times \\
0 & \times & \times & \times & \times & \times \\
0 & 0 & \times & \times & \times & \times \\
0 & 0 & 0 & \times & \times & \times \\
0 & 0 & 0 & \times & \times & \times
\end{bmatrix}
\xrightarrow{P_4}
$$

$$
\begin{bmatrix}
\times & \times & \times & \times & \times & \times \\
\times & \times & \times & \times & \times & \times \\
0 & \times & \times & \times & \times & \times \\
0 & 0 & \times & \times & \times & \times \\
0 & 0 & 0 & \times & \times & \times \\
0 & 0 & 0 & 0 & \times & \times
\end{bmatrix}.
$$

In general, after $k - 1$ steps we have computed $k - 1$ Householder matrices $P_1, \ldots, P_{k-1}$ such that

$$
(P_1 \cdots P_{k-1})^T A (P_1 \cdots P_{k-1}) =
\begin{array}{c}
\begin{bmatrix}
B_{11} & B_{12} & B_{13} \\
B_{21} & B_{22} & B_{23} \\
0 & B_{32} & B_{33}
\end{bmatrix}
\begin{array}{l}
k - 1 \\
1 \\
n - k
\end{array} \\
\begin{array}{ccc}
k - 1 & \;1\; & n - k
\end{array}
\end{array}
$$

is upper Hessenberg through its first $k-1$ columns. Suppose $\bar{P}_k$ is an order $n - k$ Householder matrix such that $\bar{P}_k B_{32}$ is a multiple of $e_1^{(n-k)}$. If $P_k = \mathrm{diag}(I_k, \bar{P}_k)$, then

$$
(P_1 \cdots P_k)^T A (P_1 \cdots P_k) =
\begin{bmatrix}
B_{11} & B_{12} & B_{13}\bar{P}_k \\
B_{21} & B_{22} & B_{23}\bar{P}_k \\
0 & \bar{P}_k B_{32} & \bar{P}_k B_{33}\bar{P}_k
\end{bmatrix}
$$

is upper Hessenberg through its first k columns. Repeating this for $k = 1{:}n - 2$ we obtain

Algorithm 7.4.2 (Householder Reduction to Hessenberg Form)
Given $A \in \mathbb{R}^{n \times n}$, the following algorithm overwrites A with $H = U_0^T A U_0$ where H is upper Hessenberg and U_0 is product of Householder matrices.

$$
\begin{aligned}
&\textbf{for } k = 1{:}n - 2 \\
&\qquad [v, \beta] = \textbf{house}(A(k+1{:}n, k)) \\
&\qquad A(k+1{:}n, k{:}n) = (I - \beta v v^T) A(k+1{:}n, k{:}n) \\
&\qquad A(1{:}n, k+1{:}n) = A(1{:}n, k+1{:}n)(I - \beta v v^T) \\
&\textbf{end}
\end{aligned}
$$

This algorithm requires $10n^3/3$ flops. If U_0 is explicitly formed, an additional $4n^3/3$ flops are required. The kth Householder matrix can be represented in $A(k+2{:}n, k)$. See Martin and Wilkinson (1968d) for a detailed description.

The roundoff properties of this method for reducing A to Hessenberg form are very desirable. Wilkinson (1965, p.351) states that the computed Hessenberg matrix $\hat{H}$ satisfies $\hat{H} = Q^T(A + E)Q$, where Q is orthogonal and $\| E \|_F \leq cn^2 \mathbf{u} \| A \|_F$ with c a small constant.

Example 7.4.2 If

$$
A = \begin{bmatrix} 1 & 5 & 7 \\ 3 & 0 & 6 \\ 4 & 3 & 1 \end{bmatrix} \quad \text{and} \quad U_0 = \begin{bmatrix} 1 & 0 & 0 \\ 0 & .6 & .8 \\ 0 & .8 & -.6 \end{bmatrix}
$$

then

$$
U_0^T A U_0 = H = \begin{bmatrix} 1.00 & 8.60 & -.20 \\ 5.00 & 4.96 & -.72 \\ 0.00 & 2.28 & -3.96 \end{bmatrix} .
$$

7.4.4 Level-3 Aspects

The Hessenberg reduction (Algorithm 7.4.2) is rich in level-2 operations: half gaxpys and half outer product updates. We briefly discuss two methods for introducing level-3 computations into the process.

The first approach involves a block reduction to block Hessenberg form and is quite straightforward. Suppose (for clarity) that $n = rN$ and write

$$
A = \begin{bmatrix} A_{11} & A_{12} \\ A_{21} & A_{22} \end{bmatrix} \begin{matrix} r \\ n-r \end{matrix} \quad .
$$
$$
 \;\; r \quad\;\; n-r
$$

Suppose that we have computed the QR factorization $A_{21} = \bar{Q}_1 R_1$ and that $\bar{Q}_1$ is in WY form. That is, we have $W_1, Y_1 \in \mathbb{R}^{(n-r) \times r}$ such that

$\bar{Q}_1 = I - W_1 Y_1^T$. (See §5.2.2 for details.) If $Q_1 = \text{diag}(I_r, \bar{Q}_1)$ then

$$Q_1^T A Q_1 = \begin{bmatrix} A_{11} & A_{12} \bar{Q}_1 \\ R_1 & \bar{Q}_1^T A_{22} \bar{Q}_1 \end{bmatrix}.$$

Notice that the updates of the (1,2) and (2,2) blocks are rich in level-3 operations given that $\bar{Q}_1$ is in WY form. This fully illustrates the overall process as $Q_1^T A Q_1$ is block upper Hessenberg through its first block column. We next repeat the computations on the first r columns of $\bar{Q}_1^T A_{22} \bar{Q}_1$. After $N - 2$ such steps we obtain

$$H = U_0^T A U_0 = \begin{bmatrix} H_{11} & H_{12} & \cdots & \cdots & H_{1N} \\ H_{21} & H_{22} & \cdots & \cdots & H_{2N} \\ 0 & \ddots & \ddots & \cdots & \vdots \\ \vdots & \vdots & \ddots & \ddots & \vdots \\ 0 & 0 & \cdots & H_{N,N-1} & H_{NN} \end{bmatrix}$$

where each H_{ij} is r-by-r and $U_0 = Q_1 \cdots Q_{N-2}$ with with each Q_i in WY form. The overall algorithm has a level-3 fraction of the form $1 - O(1/N)$.

Note that the subdiagonal blocks in H are upper triangular and so the matrix has lower bandwidth p. It is possible to reduce H to actual Hessenberg form by using Givens rotations to zero all but the first subdiagonal.

Dongarra, Hammarling and Sorensen (1987) have shown how to proceed directly to Hessenberg form using a mixture of gaxpy's and level-3 updates. Their idea involves minimal updating after each Householder transformation is generated. For example, suppose the first Householder P_1 has been computed. To generate P_2 we need just the second column of $P_1 A P_1$, not the full outer product update. To generate P_3 we need just the 3rd column of $P_2 P_1 A P_1 P_2$, etc. In this way, the Householder matrices can be determined using only gaxpy operations. No outer product updates are involved. Once a suitable number of Householder matrices are known they can be aggregated and applied in a level-3 fashion.

7.4.5   Important Hessenberg Matrix Properties

The Hessenberg decomposition is not unique. If Z is any n-by-n orthogonal matrix and we apply Algorithm 7.4.2 to $Z^T A Z$, then $Q^T A Q = H$ is upper Hessenberg where $Q = Z U_0$. However, $Q e_1 = Z(U_0 e_1) = Z e_1$ suggesting that H is unique once the first column of Q is specified. This is essentially the case provided H has no zero subdiagonal entries. Hessenberg matrices with this property are said to be *unreduced*. Here is a very important theorem that clarifies the uniqueness of the Hessenberg reduction.

Theorem 7.4.2 (Implicit Q Theorem) *Suppose* $Q = [q_1, \ldots, q_n]$ *and* $V = [v_1, \ldots, v_n]$ *are orthogonal matrices with the property that both* $Q^T A Q$

$= H$ and $V^T A V = G$ are upper Hessenberg where $A \in \mathbb{R}^{n \times n}$. Let k denote the smallest positive integer for which $h_{k+1,k} = 0$, with the convention that $k = n$ if H is unreduced. If $q_1 = v_1$, then $q_i = \pm v_i$ and $|h_{i,i-1}| = |g_{i,i-1}|$ for $i = 2{:}k$. Moreover, if $k < n$, then $g_{k+1,k} = 0$.

Proof. Define the orthogonal matrix $W = [w_1, \ldots, w_n] = V^T Q$ and observe that $GW = WH$. By comparing column $i - 1$ in this equation for $i = 2{:}k$ we see that

$$h_{i,i-1} w_i = G w_{i-1} - \sum_{j=1}^{i-1} h_{j,i-1} w_j .$$

Since $w_1 = e_1$, it follows that $[w_1, \ldots, w_k]$ is upper triangular and thus $w_i = \pm I_n(:,i) = \pm e_i$ for $i = 2{:}k$. Since $w_i = V^T q_i$ and $h_{i,i-1} = w_i^T G w_{i-1}$ it follows that $v_i = \pm q_i$ and

$$|h_{i,i-1}| = |q_i^T A q_{i-1}| = |v_i^T A v_{i-1}| = |g_{i,i-1}|$$

for $i = 2{:}k$. If $k < n$, then

$$g_{k+1,k} = e_{k+1}^T G e_k = e_{k+1}^T G W e_k = e_{k+1}^T W H e_k$$
$$= e_{k+1}^T \sum_{i=1}^{k} h_{ik} W e_i = \sum_{i=1}^{k} h_{ik} e_{k+1}^T e_i = 0. \square$$

The gist of the implicit Q theorem is that if $Q^T A Q = H$ and $Z^T A Z = G$ are each unreduced upper Hessenberg matrices and Q and Z have the same first column, then G and H are "essentially equal" in the sense that $G = D^{-1} H D$ where $D = \mathrm{diag}(\pm 1, \ldots, \pm 1)$.

Our next theorem involves a new type of matrix called a *Krylov* matrix. If $A \in \mathbb{R}^{n \times n}$ and $v \in \mathbb{R}^n$, then the Krylov matrix $K(A, v, j) \in \mathbb{R}^{n \times j}$ is defined by

$$K(A, v, j) = [v, Av, \cdots, A^{j-1}v].$$

It turns out that there is a connection between the Hessenberg reduction $Q^T A Q = H$ and the QR factorization of the Krylov matrix $K(A, Q(:,1), n)$.

Theorem 7.4.3 *Suppose $Q \in \mathbb{R}^{n \times n}$ is an orthogonal matrix and $A \in \mathbb{R}^{n \times n}$. Then $Q^T A Q = H$ is an unreduced upper Hessenberg matrix if and only if $Q^T K(A, Q(:,1), n) = R$ is nonsingular and upper triangular.*

Proof. Suppose $Q \in \mathbb{R}^{n \times n}$ is orthogonal and set $H = Q^T A Q$. Consider the identity

$$Q^T K(A, Q(:,1), n) = [e_1, He_1, \ldots, H^{n-1}e_1] \equiv R.$$

If H is an unreduced upper Hessenberg matrix, then it is clear that R is upper triangular with $r_{ii} = h_{21} h_{32} \cdots h_{i,i-1}$ for $i = 2{:}n$. Since $r_{11} = 1$ it follows that R is nonsingular.

To prove the converse, suppose R is upper triangular and nonsingular. Since $R(:, k+1) = HR(:, k)$ it follows that $H(:, k) \in \text{span}\{ e_1, \ldots, e_{k+1} \}$. This implies that H is upper Hessenberg. Since $r_{nn} = h_{21}h_{32} \cdots h_{n,n-1} \neq 0$ it follows that H is also unreduced. $\square$

Thus, there is more or less a correspondence between nonsingular Krylov matrices and orthogonal similarity reductions to unreduced Hessenberg form.

Our last result concerns eigenvalues of an unreduced upper Hessenberg matrix.

Theorem 7.4.4 *If λ is an eigenvalue of an unreduced upper Hessenberg matrix $H \in \mathbb{R}^{n \times n}$, then its geometric multiplicity is one.*

Proof. For any $\lambda \in \mathbb{C}$ we have $\text{rank}(A - \lambda I) \geq n - 1$ because the first $n - 1$ columns of $H - \lambda I$ are independent. $\square$

7.4.6   Companion Matrix Form

Just as the Schur decomposition has a nonunitary analog in the Jordan decomposition, so does the Hessenberg decomposition have a nonunitary analog in the *companion matrix decomposition*. Let $x \in \mathbb{R}^n$ and suppose that the Krylov matrix $K = K(A, x, n)$ is nonsingular. If $c = c(0:n-1)$ solves the linear system $Kc = -A^n x$, then it follows that $AK = KC$ where C has the form:

$$ C = \begin{bmatrix} 0 & 0 & \cdots & 0 & -c_0 \\ 1 & 0 & \cdots & 0 & -c_1 \\ 0 & 1 & \cdots & 0 & -c_2 \\ \vdots & \vdots & \vdots & \vdots & \vdots \\ 0 & 0 & \cdots & 1 & -c_{n-1} \end{bmatrix}. \tag{7.4.4} $$

The matrix C is said to be a *companion matrix*. Since

$$ \det(zI - C) = c_0 + c_1 z + \cdots + c_{n-1} z^{n-1} + z^n $$

it follows that if K is nonsingular, then the decomposition $K^{-1}AK = C$ displays A's characteristic polynomial. This, coupled with the sparseness of C, has led to "companion matrix methods" in various application areas. These techniques typically involve:

- Computing the Hessenberg decomposition $U_0^T A U_0 = H$.

- Hoping H is unreduced and setting $Y = \begin{bmatrix} e_1, & He_1, & \ldots, & H^{n-1}e_1 \end{bmatrix}$.

- Solving $YC = HY$ for C.

Unfortunately, this calculation can be highly unstable. A is similar to an unreduced Hessenberg matrix only if each eigenvalue has unit geometric multiplicity. Matrices that have this property are called *nonderogatory*. It follows that the matrix Y above can be very poorly conditioned if A is close to a derogatory matrix.

A full discussion of the dangers associated with companion matrix computation can be found in Wilkinson (1965, pp. 405 ff.).

7.4.7 Hessenberg Reduction Via Gauss Transforms

While we are on the subject of nonorthogonal reduction to Hessenberg form, we should mention that Gauss transformations can be used in lieu of Householder matrices in Algorithm 7.4.2. In particular, suppose permutations $\Pi_1, \ldots, \Pi_{k-1}$ and Gauss transformations $M_1, \ldots, M_{k-1}$ have been determined such that

$$(M_{k-1}\Pi_{k-1} \cdots M_1\Pi_1)A(M_{k-1}\Pi_{k-1} \cdots M_1\Pi_1)^{-1} = B$$

where

$$B = \begin{bmatrix} B_{11} & B_{12} & B_{13} \\ B_{21} & B_{22} & B_{23} \\ 0 & B_{32} & B_{33} \end{bmatrix} \begin{matrix} k-1 \\ 1 \\ n-k \end{matrix}$$
$$\quad\quad\; k-1 \quad\; 1 \quad\; n-k$$

is upper Hessenberg through its first $k-1$ columns. A permutation $\bar{\Pi}_k$ of order $n-k$ is then determined such that the first element of $\bar{\Pi}_k B_{32}$ is maximal in absolute value. This makes it possible to determine a stable Gauss transformation $\bar{M}_k = I - z_k e_1^T$ also of order $n-k$, such that all but the first component of $\bar{M}_k(\bar{\Pi}_k B_{32})$ is zero. Defining $\Pi_k = \text{diag}(I_k, \bar{\Pi}_k)$ and $M_k = \text{diag}(I_k, \bar{M}_k)$, we see that

$$(M_k\Pi_k \cdots M_1\Pi_1)A(M_k\Pi_k \cdots M_1\Pi_1)^{-1} =$$
$$\begin{bmatrix} B_{11} & B_{12} & B_{13}\bar{\Pi}_k^T\bar{M}_k^{-1} \\ B_{21} & B_{22} & B_{23}\bar{\Pi}_k^T\bar{M}_k^{-1} \\ 0 & \bar{M}_k\bar{\Pi}_k B_{32} & \bar{M}_k\bar{\Pi}_k B_{33}\bar{\Pi}_k^T\bar{M}_k^{-1} \end{bmatrix}$$

is upper Hessenberg through its first k columns. Note that $\bar{M}_k^{-1} = I + z_k e_1^T$ and so some very simple rank-one updates are involved in the reduction.

A careful operation count reveals that the Gauss reduction to Hessenberg form requires only half the number of flops of the Householder method. However, as in the case of Gaussian elimination with partial pivoting, there is a (fairly remote) chance of 2^n growth. See Businger (1969). Another difficulty associated with the Gauss approach is that the eigenvalue condition numbers — the $s(\lambda)^{-1}$ — are not preserved with nonorthogonal similarity transformations and this complicates the error estimation process.

Problems

P7.4.1 Suppose $A \in \mathbf{R}^{n \times n}$ and $z \in \mathbf{R}^n$. Give a detailed algorithm for computing an orthogonal Q such that $Q^T A Q$ is upper Hessenberg and $Q^T z$ is a multiple of e_1. (Hint: Reduce z first and then apply Algorithm 7.4.2.)

P7.4.2 Specify a complete reduction to Hessenberg form using Gauss transformations and verify that it only requires $5n^3/3$ flops.

P7.4.3 In some situations, it is necessary to solve the linear system $(A + zI)x = b$ for many different values of $z \in \mathbf{R}$ and $b \in \mathbf{R}^n$. Show how this problem can be efficiently and stably solved using the Hessenberg decomposition.

P7.4.4 Give a detailed algorithm for explicitly computing the matrix U_0 in Algorithm 7.4.2. Design your algorithm so that H is overwritten by U_0.

P7.4.5 Suppose $H \in \mathbf{R}^{n \times n}$ is an unreduced upper Hessenberg matrix. Show that there exists a diagonal matrix D such that each subdiagonal element of $D^{-1}HD$ is equal to one. What is $\kappa_2(D)$?

P7.4.6 Suppose $W, Y \in \mathbf{R}^{n \times n}$ and define the matrices C and B by

$$C = W + iY, \qquad B = \begin{bmatrix} W & -Y \\ Y & W \end{bmatrix}$$

Show that if $\lambda \in \lambda(C)$ is real, then $\lambda \in \lambda(B)$. Relate the corresponding eigenvectors.

P7.4.7 Suppose $A = \begin{bmatrix} w & x \\ y & z \end{bmatrix}$ is a real matrix having eigenvalues $\lambda \pm i\mu$, where μ is nonzero. Give an algorithm that stably determines $c = \cos(\theta)$ and $s = \sin(\theta)$ such that

$$\begin{bmatrix} c & s \\ -s & c \end{bmatrix}^T \begin{bmatrix} w & x \\ y & z \end{bmatrix} \begin{bmatrix} c & s \\ -s & c \end{bmatrix} = \begin{bmatrix} \lambda & \beta \\ \alpha & \lambda \end{bmatrix}$$

where $\alpha\beta = -\mu^2$.

P7.4.8 Suppose (λ, x) is a known eigenvalue-eigenvector pair for the upper Hessenberg matrix $H \in \mathbf{R}^{n \times n}$. Give an algorithm for computing an orthogonal matrix P such that

$$P^T H P = \begin{bmatrix} \lambda & w^T \\ 0 & H_1 \end{bmatrix}$$

where $H_1 \in \mathbf{R}^{(n-1) \times (n-1)}$ is upper Hessenberg. Compute P as a product of Givens rotations.

P7.4.9 Suppose $H \in \mathbf{R}^{n \times n}$ has lower bandwidth p. Show how to compute $Q \in \mathbf{R}^{n \times n}$, a product of Givens rotations, such that $Q^T H Q$ is upper Hessenberg. How many flops are required?

P7.4.10 Show that if C is a companion matrix with distinct eigenvalues $\lambda_1, \ldots, \lambda_n$, then $VCV^{-1} = \mathrm{diag}(\lambda_1, \ldots, \lambda_n)$ where

$$V = \begin{bmatrix} 1 & \lambda_1 & \cdots & \lambda_1^{n-1} \\ 1 & \lambda_2 & \cdots & \lambda_2^{n-1} \\ \vdots & \vdots & \ddots & \vdots \\ 1 & \lambda_n & \cdots & \lambda_n^{n-1} \end{bmatrix}.$$

Notes and References for Sec. 7.4

The real Schur decomposition was originally presented in

F.D. Murnaghan and A. Wintner (1931). "A Canonical Form for Real Matrices Under Orthogonal Transformations," *Proc. Nat. Acad. Sci. 17*, 417–20.

A thorough treatment of the reduction to Hessenberg form is given in Wilkinson (1965, chapter 6), and Algol procedures for both the Householder and Gauss methods appear in

R.S. Martin and J.H. Wilkinson (1968). "Similarity Reduction of a General Matrix to Hessenberg Form," *Numer. Math. 12*, 349–68. See also Wilkinson and Reinsch (1971,pp.339–58).

Fortran versions of the Algol procedures in the last reference are in Eispack.
    Givens rotations can also be used to compute the Hessenberg decomposition. See

W. Rath (1982). "Fast Givens Rotations for Orthogonal Similarity," *Numer. Math. 40*, 47–56.

The high performance computation of the Hessenberg reduction is discussed in

J.J. Dongarra, L. Kaufman, and S. Hammarling (1986). "Squeezing the Most Out of Eigenvalue Solvers on High Performance Computers," *Lin. Alg. and Its Applic. 77*, 113–136.

J.J. Dongarra, S. Hammarling, and D.C. Sorensen (1989). "Block Reduction of Matrices to Condensed Forms for Eigenvalue Computations," *JACM 27*, 215–227.

M.W. Berry, J.J. Dongarra, and Y. Kim (1995). "A Parallel Algorithm for the Reduction of a Nonsymmetric Matrix to Block Upper Hessenberg Form," *Parallel Computing 21*, 1189–1211.

The possibility of exponential growth in the Gauss transformation approach was first pointed out in

P. Businger (1969). "Reducing a Matrix to Hessenberg Form," *Math. Comp. 23*, 819–21.

However, the algorithm should be regarded in the same light as Gaussian elimination with partial pivoting—stable for all practical purposes. See Eispack, pp. 56–58.

Aspects of the Hessenberg decomposition for sparse matrices are discussed in

I.S. Duff and J.K. Reid (1975). "On the Reduction of Sparse Matrices to Condensed Forms by Similarity Transformations," *J. Inst. Math. Applic. 15*, 217–24.

Once an eigenvalue of an unreduced upper Hessenberg matrix is known, it is possible to zero the last subdiagonal entry using Givens similarity transformations. See

P.A. Businger (1971). "Numerically Stable Deflation of Hessenberg and Symmetric Tridiagonal Matrices,*BIT 11*, 262–70.

Some interesting mathematical properties of the Hessenberg form may be found in

B.N. Parlett (1967). "Canonical Decomposition of Hessenberg Matrices," *Math. Comp. 21*, 223–27.

Y. Ikebe (1979). "On Inverses of Hessenberg Matrices," *Lin. Alg. and Its Applic. 24*, 93–97.

Although the Hessenberg decomposition is largely appreciated as a "front end" decomposition for the QR iteration, it is increasingly popular as a cheap alternative to the more expensive Schur decomposition in certain problems. For a sampling of applications where it has proven to be very useful, consult

W. Enright (1979). "On the Efficient and Reliable Numerical Solution of Large Linear Systems of O.D.E.'s," *IEEE Trans. Auto. Cont. AC-24*, 905–8.

G.H. Golub, S. Nash and C. Van Loan (1979). "A Hessenberg-Schur Method for the Problem $AX + XB = C$," *IEEE Trans. Auto. Cont. AC-24,* 909–13.

A. Laub (1981). "Efficient Multivariable Frequency Response Computations," *IEEE Trans. Auto. Cont. AC-26,* 407–8.

C.C. Paige (1981). "Properties of Numerical Algorithms Related to Computing Controllability," *IEEE Trans. Auto. Cont. AC-26,* 130–38.

G. Miminis and C.C. Paige (1982). "An Algorithm for Pole Assignment of Time Invariant Linear Systems," *International J. of Control 35,* 341–354.

C. Van Loan (1982). "Using the Hessenberg Decomposition in Control Theory," in *Algorithms and Theory in Filtering and Control* , D.C. Sorensen and R.J. Wets (eds), Mathematical Programming Study No. 18, North Holland, Amsterdam, pp. 102–11.

The advisability of posing polynomial root problems as companion matrix eigenvalue problem is discussed in

K.-C. Toh and L.N. Trefethen (1994). "Pseudozeros of Polynomials and Pseudospectra of Companion Matrices," *Numer. Math. 68,* 403–425.

A. Edelman and H. Murakami (1995). "Polynomial Roots from Companion Matrix Eigenvalues," *Math. Comp. 64,* 763–776.

7.5   The Practical QR Algorithm

We return to the Hessenberg QR iteration which we write as follows:

$$
\begin{aligned}
&H = U_0^T A U_0 \qquad \text{(Hessenberg Reduction)} \\
&\textbf{for } k = 1, 2, \ldots \\
&\qquad H = UR \qquad \text{(QR factorization)} \\
&\qquad H = RU \\
&\textbf{end}
\end{aligned}
\qquad (7.5.1)
$$

Our aim in this section is to describe how the H's converge to upper quasi-triangular form and to show how the convergence rate can be accelerated by incorporating *shifts*.

7.5.1   Deflation

Without loss of generality we may assume that each Hessenberg matrix H in (7.5.1) is unreduced. If not, then at some stage we have

$$
H = \begin{bmatrix} H_{11} & H_{12} \\ 0 & H_{22} \end{bmatrix} \begin{matrix} p \\ n-p \end{matrix}
$$
$$
\qquad\quad p \quad\; n-p
$$

where $1 \le p < n$ and the problem *decouples* into two smaller problems involving H_{11} and H_{22}. The term *deflation* is also used in this context, usually when $p = n - 1$ or $n - 2$.

In practice, decoupling occurs whenever a subdiagonal entry in H is suitably small. For example, in Eispack if

$$
|h_{p+1,p}| \le \mathbf{cu}(|h_{pp}| + |h_{p+1,p+1}|) \qquad (7.5.2)
$$

for a small constant c, then $h_{p+1,p}$ is "declared" to be zero. This is justified since rounding errors of order $\mathbf{u}\| H \|$ are already present throughout the matrix.

7.5.2   The Shifted QR Iteration

Let $\mu \in \mathbb{R}$ and consider the iteration:

$$H = U_0^T A U_0 \qquad \text{(Hessenberg Reduction)}$$
$$\textbf{for } k = 1,2,\ldots$$
$$\qquad \text{Determine a scalar } \mu.$$
$$\qquad H - \mu I = U R \qquad \text{(QR factorization)} \qquad (7.5.3)$$
$$\qquad H = R U + \mu I$$
$$\textbf{end}$$

The scalar μ is referred to as a *shift* . Each matrix H generated in (7.5.3) is similar to A, since $RU + \mu I = U^T(UR + \mu I)U = U^T HU$. If we order the eigenvalues λ_i of A so that

$$|\lambda_1 - \mu| \geq \cdots \geq |\lambda_n - \mu|,$$

and μ is fixed from iteration to iteration, then the theory of §7.3 says that the pth subdiagonal entry in H converges to zero with rate

$$\left| \frac{\lambda_{p+1} - \mu}{\lambda_p - \mu} \right|^k .$$

Of course, if $\lambda_p = \lambda_{p+1}$, then there is no convergence at all. But if, for example, μ is much closer to λ_n than to the other eigenvalues, then the zeroing of the $(n, n-1)$ entry is rapid. In the extreme case we have the following:

Theorem 7.5.1 *Let μ be an eigenvalue of an n-by-n unreduced Hessenberg matrix H. If $\bar{H} = RU + \mu I$, where $H - \mu I = UR$ is the QR factorization of $H - \mu I$, then $\bar{h}_{n,n-1} = 0$ and $\bar{h}_{nn} = \mu$.*

Proof. Since H is an unreduced Hessenberg matrix the first $n-1$ columns of $H - \mu I$ are independent, regardless of μ. Thus, if $UR = (H - \mu I)$ is the QR factorization then $r_{ii} \neq 0$ for $i = 1{:}n-1$. But if $H - \mu I$ is singular then $r_{11} \cdots r_{nn} = 0$. Thus, $r_{nn} = 0$ and $\bar{H}(n,:) = [\,0, \ldots, 0, \mu\,]$. $\square$

The theorem says that if we shift by an exact eigenvalue, then in exact arithmetic deflation occurs in one step.

Example 7.5.1 If

$$H = \begin{bmatrix} 9 & -1 & -2 \\ 2 & 6 & -2 \\ 0 & 1 & 5 \end{bmatrix},$$

then $6 \in \lambda(H)$. If $UR = H - 6I$ is the QR factorization, then $\bar{H} = RU + 6I$ is given by

$$\bar{H} = \begin{bmatrix} 8.5384 & -3.7313 & -1.0090 \\ 0.6343 & 5.4615 & 1.3867 \\ 0.0000 & 0.0000 & 6.0000 \end{bmatrix}.$$

7.5.3   The Single Shift Strategy

Now let us consider varying μ from iteration to iteration incorporating new information about $\lambda(A)$ as the subdiagonal entries converge to zero. A good heuristic is to regard h_{nn} as the best approximate eigenvalue along the diagonal. If we shift by this quantity during each iteration, we obtain the *single-shift QR iteration*:

> **for** $k = 1, 2, \ldots$
> $\qquad \mu = H(n, n)$
> $\qquad H - \mu I = UR \qquad$ (QR Factorization) $\qquad\qquad$ (7.5.4)
> $\qquad H = RU + \mu I$
> **end**

If the $(n, n - 1)$ entry converges to zero, it is likely to do so at a quadratic rate. To see this, we borrow an example from Stewart (1973, p. 366). Suppose H is an unreduced upper Hessenberg matrix of the form

$$H = \begin{bmatrix} \times & \times & \times & \times & \times \\ \times & \times & \times & \times & \times \\ 0 & \times & \times & \times & \times \\ 0 & 0 & \times & \times & \times \\ 0 & 0 & 0 & \epsilon & h_{nn} \end{bmatrix}$$

and that we perform one step of the single-shift QR algorithm: $UR = H - h_{nn}I$, $\bar{H} = RU + h_{nn}I$. After $n - 2$ steps in the reduction of $H - h_{nn}I$ to upper triangular form we obtain a matrix with the following structure:

$$H = \begin{bmatrix} \times & \times & \times & \times & \times \\ 0 & \times & \times & \times & \times \\ 0 & 0 & \times & \times & \times \\ 0 & 0 & 0 & a & b \\ 0 & 0 & 0 & \epsilon & 0 \end{bmatrix}$$

It is not hard to show that the $(n, n - 1)$ entry in $\bar{H} = RU + h_{nn}I$ is given by $-\epsilon^2 b / (\epsilon^2 + a^2)$. If we assume that $\epsilon \ll a$, then it is clear that the new $(n, n - 1)$ entry has order ϵ^2, precisely what we would expect of a quadratically converging algorithm.

Example 7.5.2 If

$$H = \begin{bmatrix} 1 & 2 & 3 \\ 4 & 5 & 6 \\ 0 & .001 & 7 \end{bmatrix}$$

and $UR = H - 7I$ is the QR factorization, then $\bar{H} = RU + 7I$ is given by

$$\bar{H} \approx \begin{bmatrix} -0.5384 & 1.6908 & 0.8351 \\ 0.3076 & 6.5264 & -6.6555 \\ 0.0000 & 2 \cdot 10^{-5} & 7.0119 \end{bmatrix}.$$

Near-perfect shifts as above almost always ensure a small $\bar{h}_{n,n-1}$. However, this is just a heuristic. There are examples in which $\bar{h}_{n,n-1}$ is a relatively large matrix entry even though $\sigma_{min}(H - \mu I) \approx \mathbf{u}$.

7.5.4   The Double Shift Strategy

Unfortunately, difficulties with (7.5.4) can be expected if at some stage the eigenvalues a_1 and a_2 of

$$G = \begin{bmatrix} h_{mm} & h_{mn} \\ h_{nm} & h_{nn} \end{bmatrix} \qquad m = n - 1 \qquad (7.5.5)$$

are complex for then h_{nn} would tend to be a poor approximate eigenvalue.

A way around this difficulty is to perform two single-shift QR steps in succession using a_1 and a_2 as shifts:

$$\begin{aligned} H - a_1 I &= U_1 R_1 \\ H_1 &= R_1 U_1 + a_1 I \\ H_1 - a_2 I &= U_2 R_2 \\ H_2 &= R_2 U_2 + a_2 I \end{aligned} \qquad (7.5.6)$$

These equations can be manipulated to show that

$$(U_1 U_2)(R_2 R_1) = M \qquad (7.5.7)$$

where M is defined by

$$M = (H - a_1 I)(H - a_2 I). \qquad (7.5.8)$$

Note that M is a real matrix even if G's eigenvalues are complex since

$$M = H^2 - sH + tI$$

where

$$s = a_1 + a_2 = h_{mm} + h_{nn} = \text{trace}(G) \in \mathbb{R}$$

and
$$t = a_1 a_2 = h_{mm} h_{nn} - h_{mn} h_{nm} = \det(G) \in \mathbb{R}.$$

Thus, (7.5.7) is the QR factorization of a real matrix and we may choose U_1 and U_2 so that $Z = U_1 U_2$ is real orthogonal. It then follows that

$$H_2 = U_2^H H_1 U_2 = U_2^H (U_1^H H U_1) U_2 = (U_1 U_2)^H H (U_1 U_2) = Z^T H Z.$$

is real.

Unfortunately, roundoff error almost always prevents an exact return to the real field. A real H_2 could be guaranteed if we

- explicitly form the real matrix $M = H^2 - sH + tI$,

- compute the real QR factorization $M = ZR$, and

- set $H_2 = Z^T H Z$.

But since the first of these steps requires $O(n^3)$ flops, this is not a practical course of action.

7.5.5   The Double Implicit Shift Strategy

Fortunately, it turns out that we can implement the double shift step with $O(n^2)$ flops by appealing to the Implicit Q Theorem of §7.4.5. In particular we can effect the transition from H to H_2 in $O(n^2)$ flops if we

- compute Me_1, the first column of M;

- determine a Householder matrix P_0 such that $P_0(Me_1)$ is a multiple of e_1;

- compute Householder matrices $P_1, \ldots, P_{n-2}$ such that if Z_1 is the product $Z_1 = P_0 P_1 \cdots P_{n-2}$, then $Z_1^T H Z_1$ is upper Hessenberg and the first columns of Z and Z_1 are the same.

Under these circumstances, the Implicit Q theorem permits us to conclude that if $Z^T H Z$ and $Z_1^T H Z_1$ are both unreduced upper Hessenberg matrices, then they are essentially equal. Note that if these Hessenberg matrices are not unreduced, then we can effect a decoupling and proceed with smaller unreduced subproblems.

Let us work out the details. Observe first that P_0 can be determined in $O(1)$ flops since $Me_1 = [x, \ y, \ z, \ 0, \ldots, 0]^T$ where

$$x = h_{11}^2 + h_{12} h_{21} - s h_{11} + t$$
$$y = h_{21}(h_{11} + h_{22} - s)$$
$$z = h_{21} h_{32}.$$

Since a similarity transformation with P_0 only changes rows and columns 1, 2, and 3, we see that

$$
P_0 H P_0 = \begin{bmatrix}
\times & \times & \times & \times & \times & \times \\
\times & \times & \times & \times & \times & \times \\
\times & \times & \times & \times & \times & \times \\
\times & \times & \times & \times & \times & \times \\
0 & 0 & 0 & \times & \times & \times \\
0 & 0 & 0 & 0 & \times & \times
\end{bmatrix}
$$

Now the mission of the Householder matrices $P_1, \ldots, P_{n-2}$ is to restore this matrix to upper Hessenberg form. The calculation proceeds as follows:

$$
\begin{bmatrix}
\times & \times & \times & \times & \times & \times \\
\times & \times & \times & \times & \times & \times \\
\times & \times & \times & \times & \times & \times \\
\times & \times & \times & \times & \times & \times \\
0 & 0 & 0 & \times & \times & \times \\
0 & 0 & 0 & 0 & \times & \times
\end{bmatrix}
\xrightarrow{P_1}
\begin{bmatrix}
\times & \times & \times & \times & \times & \times \\
\times & \times & \times & \times & \times & \times \\
0 & \times & \times & \times & \times & \times \\
0 & \times & \times & \times & \times & \times \\
0 & \times & \times & \times & \times & \times \\
0 & 0 & 0 & 0 & \times & \times
\end{bmatrix}
\xrightarrow{P_2}
$$

$$
\begin{bmatrix}
\times & \times & \times & \times & \times & \times \\
\times & \times & \times & \times & \times & \times \\
0 & \times & \times & \times & \times & \times \\
0 & 0 & \times & \times & \times & \times \\
0 & 0 & \times & \times & \times & \times \\
0 & 0 & \times & \times & \times & \times
\end{bmatrix}
\xrightarrow{P_3}
\begin{bmatrix}
\times & \times & \times & \times & \times & \times \\
\times & \times & \times & \times & \times & \times \\
0 & \times & \times & \times & \times & \times \\
0 & 0 & \times & \times & \times & \times \\
0 & 0 & 0 & \times & \times & \times \\
0 & 0 & 0 & \times & \times & \times
\end{bmatrix}
\xrightarrow{P_4}
$$

$$
\begin{bmatrix}
\times & \times & \times & \times & \times & \times \\
\times & \times & \times & \times & \times & \times \\
0 & \times & \times & \times & \times & \times \\
0 & 0 & \times & \times & \times & \times \\
0 & 0 & 0 & \times & \times & \times \\
0 & 0 & 0 & 0 & \times & \times
\end{bmatrix}
$$

Clearly, the general P_k has the form $P_k = \mathrm{diag}(I_k, \bar{P}_k, I_{n-k-3})$ where $\bar{P}_k$ is a 3-by-3 Householder matrix. For example,

$$
P_2 = \begin{bmatrix}
1 & 0 & 0 & 0 & 0 & 0 \\
0 & 1 & 0 & 0 & 0 & 0 \\
0 & 0 & \times & \times & \times & 0 \\
0 & 0 & \times & \times & \times & 0 \\
0 & 0 & \times & \times & \times & 0 \\
0 & 0 & 0 & 0 & 0 & 1
\end{bmatrix}.
$$

Note that P_{n-2} is an exception to this since $P_{n-2}= \mathrm{diag}(I_{n-2}, \bar{P}_{n-2})$.

The applicability of Theorem 7.4.3 (the Implicit Q theorem) follows from the observation that $P_k e_1 = e_1$ for $k = 1{:}n - 2$ and that P_0 and Z

have the same first column. Hence, $Z_1 e_1 = Z e_1$, and we can assert that Z_1 essentially equals Z provided that the upper Hessenberg matrices $Z^T H Z$ and $Z_1^T H Z_1$ are each unreduced.

The implicit determination of H_2 from H outlined above was first described by Francis (1961) and we refer to it as a *Francis QR step*. The complete Francis step is summarized as follows:

Algorithm 7.5.1 (Francis QR Step) Given the unreduced upper Hessenberg matrix $H \in \mathbb{R}^{n \times n}$ whose trailing 2-by-2 principal submatrix has eigenvalues a_1 and a_2, this algorithm overwrites H with $Z^T H Z$, where $Z = P_1 \cdots P_{n-2}$ is a product of Householder matrices and $Z^T (H - a_1 I)(H - a_2 I)$ is upper triangular.

> $m = n - 1$
> {Compute first column of $(H - a_1 I)(H - a_2 I)$.}
> $s = H(m, m) + H(n, n)$
> $t = H(m, m)H(n, n) - H(m, n)H(n, m)$
> $x = H(1, 1)H(1, 1) + H(1, 2)H(2, 1) - sH(1, 1) + t$
> $y = H(2, 1)(H(1, 1) + H(2, 2) - s)$
> $z = H(2, 1)H(3, 2)$
> **for** $k = 0{:}n - 3$
> > $[v, \beta] = \textbf{house}([x \ y \ z]^T)$
> > $q = \max\{1, k\}.$
> > $H(k + 1{:}k + 3, q{:}n) = (I - \beta v v^T)H(k + 1{:}k + 3, q{:}n)$
> > $r = \min\{k + 4, n\}$
> > $H(1{:}r, k + 1{:}k + 3) = H(1{:}r, k + 1{:}k + 3)(I - \beta v v^T)$
> > $x = H(k + 2, k + 1)$
> > $y = H(k + 3, k + 1)$
> > **if** $k < n - 3$
> > > $z = H(k + 4, k + 1)$
> >
> > **end**
>
> **end**
> $[v, \beta] = \textbf{house}([x \ y]^T)$
> $H(n - 1{:}n, n - 2{:}n) = (I - \beta v v^T)H(n - 1{:}n, n - 2{:}n)$
> $H(1{:}n, n - 1, n) = H(1{:}n, n - 1, n)(I - \beta v v^T)$

This algorithm requires $10n^2$ flops. If Z is accumulated into a given orthogonal matrix, an additional $10n^2$ flops are necessary.

7.5.6 The Overall Process

Reducing A to Hessenberg form using Algorithm 7.4.2 and then iterating with Algorithm 7.5.1 to produce the real Schur form is the standard means by which the dense unsymmetric eigenproblem is solved. During the iteration it is necessary to monitor the subdiagonal elements in H in order to spot any possible decoupling. How this is done is illustrated in the following algorithm:

Algorithm 7.5.2 (QR Algorithm)   Given $A \in \mathbb{R}^{n \times n}$ and a tolerance *tol* greater than the unit roundoff, this algorithm computes the real Schur canonical form $Q^T A Q = T$. A is overwritten with the Hessenberg decomposition. If Q and T are desired, then T is stored in H. If only the eigenvalues are desired, then diagonal blocks in T are stored in the corresponding positions in H.

Use Algorithm 7.4.2 to compute the Hessenberg reduction
  $H = U_0^T A U_0$ where $U_0 = P_1 \cdots P_{n-2}$.
If Q is desired form $Q = P_1 \cdots P_{n-2}$. See§5.1.6.
until $q = n$
   Set to zero all subdiagonal elements that satisfy:
   $|h_{i,i-1}| \leq tol(|h_{ii}| + |h_{i-1,i-1}|)$.
   Find the largest non-negative q and the smallest
   non-negative p such that

$$
H = \begin{bmatrix} H_{11} & H_{12} & H_{13} \\ 0 & H_{22} & H_{23} \\ 0 & 0 & H_{33} \end{bmatrix} \begin{matrix} p \\ n-p-q \\ q \end{matrix}
$$
$$
 \begin{matrix} p & n-p-q & q \end{matrix}
$$

   where H_{33} is upper quasi-triangular and H_{22} is
   unreduced. (Note: either p or q may be zero.)
   if $q < n$
      Perform a Francis QR step on H_{22}: $H_{22} = Z^T H_{22} Z$
      if Q is desired
         $Q = Q\mathrm{diag}(I_p, Z, I_q)$
         $H_{12} = H_{12} Z$
         $H_{23} = Z^T H_{23}$
      end
   end
end
Upper triangularize all 2-by-2 diagonal blocks in H that have
   real eigenvalues and accumulate the transformations
   if necessary.

This algorithm requires $25n^3$ flops if Q and T are computed. If only the eigenvalues are desired, then $10n^3$ flops are necessary. These flops counts are very approximate and are based on the empirical observation that on average only two Francis iterations are required before the lower 1-by-1 or

2-by-2 decouples.

Example 7.5.3 If Algorithm 7.5.2 is applied to

$$A = \begin{bmatrix} 2 & 3 & 4 & 5 & 6 \\ 4 & 4 & 5 & 6 & 7 \\ 0 & 3 & 6 & 7 & 8 \\ 0 & 0 & 2 & 8 & 9 \\ 0 & 0 & 0 & 1 & 10 \end{bmatrix},$$

then the subdiagonal entries converge as follows

Iteration	$O(|h_{21}|)$	$O(|h_{32}|)$	$O(|h_{43}|)$	$O(|h_{54}|)$
1	10^0	10^0	10^0	10^0
2	10^0	10^0	10^0	10^0
3	10^0	10^0	10^{-1}	10^0
4	10^0	10^0	10^{-3}	10^{-3}
5	10^0	10^0	10^{-6}	10^{-5}
6	10^{-1}	10^0	10^{-13}	10^{-13}
7	10^{-1}	10^0	10^{-28}	10^{-13}
8	10^{-4}	10^0	converg.	converg.
9	10^{-8}	10^0		
10	10^{-8}	10^0		
11	10^{-16}	10^0		
12	10^{-32}	10^0		
13	converg.	converg.		

The roundoff properties of the QR algorithm are what one would expect of any orthogonal matrix technique. The computed real Schur form $\hat{T}$ is orthogonally similar to a matrix near to A, i.e.,

$$Q^T(A + E)Q = \hat{T}$$

where $Q^TQ = I$ and $\| E \|_2 \approx \mathbf{u}\| A \|_2$. The computed $\hat{Q}$ is almost orthogonal in the sense that $\hat{Q}^T\hat{Q} = I + F$ where $\| F \|_2 \approx \mathbf{u}$.

The order of the eigenvalues along $\hat{T}$ is somewhat arbitrary. But as we discuss in §7.6, any ordering can be achieved by using a simple procedure for swapping two adjacent diagonal entries.

7.5.7   Balancing

Finally, we mention that if the elements of A have widely varying magnitudes, then A should be *balanced* before applying the QR algorithm. This is an $O(n^2)$ calculation in which a diagonal matrix D is computed so that if

$$D^{-1}AD = [\, c_1, \ldots, c_n \,] = \begin{bmatrix} r_1^T \\ \vdots \\ r_n^T \end{bmatrix}$$

then $\| r_i \|_\infty \approx \| c_i \|_\infty$ for $i = 1{:}n$. The diagonal matrix D is chosen to have the form $D = \text{diag}(\beta^{i_1}, \ldots, \beta^{i_n})$ where β is the floating point base. Note that $D^{-1}AD$ can be calculated without roundoff. When A is balanced, the computed eigenvalues are often more accurate. See Parlett and Reinsch (1969).

Problems

P7.5.1 Show that if $\bar{H} = Q^T H Q$ is obtained by performing a single-shift QR step with $H = \begin{bmatrix} w & x \\ y & z \end{bmatrix}$, then $|\bar{h}_{21}| \le |y^2 x| / [(w-z)^2 + y^2]$.

P7.5.2 Give a formula for the 2-by-2 diagonal matrix D that minimizes $\| D^{-1} A D \|_F$ where $A = \begin{bmatrix} w & x \\ y & z \end{bmatrix}$.

P7.5.3 Explain how the single-shift QR step $H - \mu I = UR$, $\bar{H} = RU + \mu I$ can be carried out implicitly. That is, show how the transition from $\bar{H}$ to H can be carried out without subtracting the shift μ from the diagonal of H.

P7.5.4 Suppose H is upper Hessenberg and that we compute the factorization $PH = LU$ via Gaussian elimination with partial pivoting. (See Algorithm 4.3.4.) Show that $H_1 = U(P^T L)$ is upper Hessenberg and similar to H. (This is the basis of the *modified LR algorithm*.)

P7.5.5 Show that if $H = H_0$ is given and we generate the matrices H_k via $H_k - \mu_k I = U_k R_k$, $H_{k+1} = R_k U_k + \mu_k I$, then

$$(U_1 \cdots U_j)(R_j \cdots R_1) = (H - \mu_1 I) \cdots (H - \mu_j I).$$

Notes and References for Sec. 7.5

The development of the practical QR algorithm began with the important paper

H. Rutishauser (1958). "Solution of Eigenvalue Problems with the LR Transformation," *Nat. Bur. Stand. App. Math. Ser. 49*, 47–81.

The algorithm described here was then "orthogonalized" in

J.G.F. Francis (1961). "The QR Transformation: A Unitary Analogue to the LR Transformation, Parts I and II" *Comp. J. 4*, 265–72, 332–45.

Descriptions of the practical QR algorithm may be found in Wilkinson (1965) and Stewart (1973), and Watkins (1991). See also

D. Watkins and L. Elsner (1991). "Chasing Algorithms for the Eigenvalue Problem," *SIAM J. Matrix Anal. Appl. 12*, 374–384.

D.S. Watkins and L. Elsner (1991). "Convergence of Algorithms of Decomposition Type for the Eigenvalue Problem," *Lin.Alg. and Its Application 143*, 19–47.

J. Erxiong (1992). "A Note on the Double-Shift QL Algorithm," *Lin.Alg. and Its Application 171*, 121–132.

Algol procedures for LR and QR methods are given in

R.S. Martin and J.H. Wilkinson (1968). "The Modified LR Algorithm for Complex Hessenberg Matrices," *Numer. Math. 12*, 369–76. See also Wilkinson and Reinsch(1971, pp. 396–403).

R.S. Martin, G. Peters, and J.H. Wilkinson (1970). "The QR Algorithm for Real Hessenberg Matrices," *Numer. Math. 14*, 219–31. See also Wilkinson and Reinsch(1971, pp. 359–71).

Aspects of the balancing problem are discussed in

E.E. Osborne (1960). "On Preconditioning of Matrices," *JACM 7*, 338–45.
B.N. Parlett and C. Reinsch (1969). "Balancing a Matrix for Calculation of Eigenvalues and Eigenvectors," *Numer. Math. 13*, 292–304. See also Wilkinson and Reinsch(1971, pp. 315-26).

High performance eigenvalue solver papers include

Z. Bai and J.W. Demmel (1989). "On a Block Implementation of Hessenberg Multishift QR Iteration," *Int'l J. of High Speed Comput. 1*, 97–112.
G. Shroff (1991). "A Parallel Algorithm for the Eigenvalues and Eigenvectors of a General Complex Matrix," *Numer. Math. 58*, 779–806.
R.A. Van De Geijn (1993). "Deferred Shifting Schemes for Parallel QR Methods," *SIAM J. Matrix Anal. Appl. 14*, 180–194.
A.A. Dubrulle and G.H. Golub (1994). "A Multishift QR Iteration Without Computation of the Shifts," *Numerical Algorithms 7*, 173–181.

7.6   Invariant Subspace Computations

Several important invariant subspace problems can be solved once the real Schur decomposition $Q^T A Q = T$ has been computed. In this section we discuss how to

- compute the eigenvectors associated with some subset of $\lambda(A)$,

- compute an orthonormal basis for a given invariant subspace,

- block-diagonalize A using well-conditioned similarity transformations,

- compute a basis of eigenvectors regardless of their condition, and

- compute an approximate Jordan canonical form of A.

Eigenvector/invariant subspace computation for sparse matrices is discussed elsewhere. See §7.3 as well as portions of Chapters 8 and 9.

7.6.1   Selected Eigenvectors via Inverse Iteration

Let $q^{(0)} \in \mathbb{C}^n$ be a given unit 2-norm vector and assume that $A - \mu I \in \mathbb{R}^{n \times n}$ is nonsingular. The following is referred to as *inverse iteration*:

$$
\begin{aligned}
&\textbf{for } k = 1, 2, \ldots \\
&\qquad \text{Solve } (A - \mu I) z^{(k)} = q^{(k-1)} \\
&\qquad q^{(k)} = z^{(k)} / \| z^{(k)} \|_2 \\
&\qquad \lambda^{(k)} = {q^{(k)}}^T A q^{(k)} \\
&\textbf{end}
\end{aligned}
\qquad (7.6.1)
$$

Inverse iteration is just the power method applied to $(A - \mu I)^{-1}$.

To analyze the behavior of (7.6.1), assume that A has a basis of eigenvectors $\{x_1, \ldots, x_n\}$ and that $Ax_i = \lambda_i x_i$ for $i = 1{:}n$. If

$$q^{(0)} = \sum_{i=1}^{n} \beta_i x_i$$

then $q^{(k)}$ is a unit vector in the direction of

$$(A - \mu I)^{-k} q^{(0)} = \sum_{i=1}^{n} \frac{\beta_i}{(\lambda_i - \mu)^k} x_i \; .$$

Clearly, if μ is much closer to an eigenvalue λ_j than to the other eigenvalues, then $q^{(k)}$ is rich in the direction of x_j provided $\beta_j \neq 0$.

A sample stopping criterion for (7.6.1) might be to quit as soon as the residual

$$r^{(k)} = (A - \mu I) q^{(k)}$$

satisfies

$$\| r^{(k)} \|_\infty \leq c \mathbf{u} \| A \|_\infty \tag{7.6.2}$$

where c is a constant of order unity. Since

$$(A + E_k) q^{(k)} = \mu q^{(k)}$$

with $E_k = -r^{(k)} q^{(k)T}$, it follows that (7.6.2) forces μ and $q^{(k)}$ to be an exact eigenpair for a nearby matrix.

Inverse iteration can be used in conjunction with the QR algorithm as follows:

- Compute the Hessenberg decomposition $U_0^T A U_0 = H$.

- Apply the double implicit shift Francis iteration to H *without* accumulating transformations.

- For each computed eigenvalue λ whose corresponding eigenvector x is sought, apply (7.6.1) with $A = H$ and $\mu = \lambda$ to produce a vector z such that $Hz \approx \mu z$.

- Set $x = U_0 z$.

Inverse iteration with H is very economical because (1) we do not have to accumulate transformations during the double Francis iteration; (2) we can factor matrices of the form $H - \lambda I$ in $O(n^2)$ flops, and (3) only one iteration is typically required to produce an adequate approximate eigenvector.

This last point is perhaps the most interesting aspect of inverse iteration and requires some justification since λ can be comparatively inaccurate if it is ill-conditioned. Assume for simplicity that λ is real and let

$$H - \lambda I = \sum_{i=1}^{n} \sigma_i u_i v_i^T = U\Sigma V^T$$

be the SVD of $H - \lambda I$. From what we said about the roundoff properties of the QR algorithm in §7.5.6, there exists a matrix $E \in \mathbb{R}^{n \times n}$ such that $H + E - \lambda I$ is singular and $\| E \|_2 \approx \mathbf{u}\| H \|_2$. It follows that $\sigma_n \approx \mathbf{u}\sigma_1$ and $\| (H - \hat{\lambda}I)v_n \|_2 \approx \mathbf{u}\sigma_1$, i.e., v_n is a good approximate eigenvector. Clearly if the starting vector $q^{(0)}$ has the expansion

$$q^{(0)} = \sum_{i=1}^{n} \gamma_i u_i$$

then

$$z^{(1)} = \sum_{i=1}^{n} \frac{\gamma_i}{\sigma_i} v_i$$

is "rich" in the direction v_n. Note that if $s(\lambda) \approx |u_n^T v_n|$ is small, then $z^{(1)}$ is rather deficient in the direction u_n. This explains (heuristically) why another step of inverse iteration is not likely to produce an improved eigenvector approximate, especially if λ is ill-conditioned. For more details, see Peters and Wilkinson (1979).

Example 7.6.1 The matrix

$$A = \begin{bmatrix} 1 & 1 \\ 10^{-10} & 1 \end{bmatrix}$$

has eigenvalues $\lambda_1 = .99999$ and $\lambda_2 = 1.00001$ and corresponding eigenvectors $x_1 = [1, -10^{-5}]^T$ and $x_2 = [1, 10^{-5}]^T$. The condition of both eigenvalues is of order 10^5. The approximate eigenvalue $\mu = 1$ is an exact eigenvalue of $A + E$ where

$$E = \begin{bmatrix} 0 & 0 \\ -10^{-10} & 0 \end{bmatrix}.$$

Thus, the quality of μ is typical of the quality of an eigenvalue produced by the QR algorithm when executed in 10-digit floating point.

If (7.6.1) is applied with starting vector $q^{(0)} = [0, 1]^T$, then $q^{(1)} = [1, 0]^T$ and $\| Aq^{(1)} - \mu q^{(1)} \|_2 = 10^{-10}$. However, one more step produces $q^{(2)} = [0, 1]^T$ for which $\| Aq^{(2)} - \mu q^{(2)} \|_2 = 1$. This example is discussed in Peters and Wilkinson (1979).

7.6.2 Ordering Eigenvalues in the Real Schur Form

Recall that the real Schur decomposition provides information about invariant subspaces. If

$$Q^T A Q = T = \begin{bmatrix} T_{11} & T_{12} \\ 0 & T_{22} \end{bmatrix} \begin{matrix} p \\ q \end{matrix}$$
$$\begin{matrix} p & q \end{matrix}$$

and $\lambda(T_{11}) \cap \lambda(T_{22}) = \emptyset$, then the first p columns of Q span the unique invariant subspace associated with $\lambda(T_{11})$. (See §7.1.4.) Unfortunately, the Francis iteration supplies us with a real Schur decomposition $Q_F^T A Q_F = T_F$ in which the eigenvalues appear somewhat randomly along the diagonal of T_F. This poses a problem if we want an orthonormal basis for an invariant subspace whose associated eigenvalues are not at the top of T_F's diagonal. Clearly, we need a method for computing an orthogonal matrix Q_D such that $Q_D^T T_F Q_D$ is upper quasi-triangular with appropriate eigenvalue ordering.

A look at the 2-by-2 case suggests how this can be accomplished. Suppose

$$Q_F^T A Q_F = T_F = \begin{bmatrix} \lambda_1 & t_{12} \\ 0 & \lambda_2 \end{bmatrix} \qquad \lambda_1 \neq \lambda_2$$

and that we wish to reverse the order of the eigenvalues. Note that $T_F x = \lambda_2 x$ where

$$x = \begin{bmatrix} t_{12} \\ \lambda_2 - \lambda_1 \end{bmatrix}.$$

Let Q_D be a Givens rotation such that the second component of $Q_D^T x$ is zero. If $Q = Q_F Q_D$ then

$$(Q^T A Q) e_1 = Q_D^T T_F (Q_D e_1) = \lambda_2 Q_D^T (Q_D e_1) = \lambda_2 e_1$$

and so $Q^T A Q$ must have the form

$$Q^T A Q = \begin{bmatrix} \lambda_2 & \pm t_{12} \\ 0 & \lambda_1 \end{bmatrix}.$$

By systematically interchanging adjacent pairs of eigenvalues using this technique, we can move any subset of $\lambda(A)$ to the top of T's diagonal assuming that no 2-by-2 bumps are encountered along the way.

Algorithm 7.6.1 Given an orthogonal matrix $Q \in \mathbb{R}^{n \times n}$, an upper triangular matrix $T = Q^T A Q$, and a subset $\Delta = \{\lambda_1, \ldots, \lambda_p\}$ of $\lambda(A)$, the following algorithm computes an orthogonal matrix Q_D such that $Q_D^T T Q_D = S$ is upper triangular and $\{s_{11}, \ldots, s_{pp}\} = \Delta$. The matrices Q and T are overwritten by $Q Q_D$ and S respectively.

while $\{t_{11}, \ldots, t_{pp}\} \neq \Delta$
    for $k = 1{:}n - 1$
        if $t_{kk} \notin \Delta$ and $t_{k+1,k+1} \in \Delta$

$$[\, c, \ s\,] = \mathbf{givens}(T(k, k + 1), T(k + 1, k + 1) - T(k, k))$$

$$T(k{:}k + 1, k{:}n) = \begin{bmatrix} c & s \\ -s & c \end{bmatrix}^{T} T(k{:}k + 1, k{:}n)$$

$$T(1{:}k + 1, k{:}k + 1) = T(1{:}k + 1, k{:}k + 1) \begin{bmatrix} c & s \\ -s & c \end{bmatrix}$$

$$Q(1{:}n, k{:}k + 1) = Q(1{:}n, k{:}k + 1) \begin{bmatrix} c & s \\ -s & c \end{bmatrix}$$

        end
    end
end

This algorithm requires $k(12n)$ flops, where k is the total number of required swaps. The integer k is never greater than $(n - p)p$.

The swapping gets a little more complicated when T has 2-by-2 blocks along its diagonal. See Ruhe (1970) and Stewart (1976) for details. Of course, these interchanging techniques can be used to sort the eigenvalues, say from maximum to minimum modulus.

Computing invariant subspaces by manipulating the real Schur decomposition is extremely stable. If $\hat{Q} = [\, \hat{q}_1, \ldots, \hat{q}_n \,]$ denotes the computed orthogonal matrix Q, then $\| \hat{Q}^T \hat{Q} - I \|_2 \approx \mathbf{u}$ and there exists a matrix E satisfying $\| E \|_2 \approx \mathbf{u} \| A \|_2$ such that $(A + E)\hat{q}_i \in \text{span}\{\hat{q}_1, \ldots, \hat{q}_p\}$ for $i = 1{:}p$.

7.6.3   Block Diagonalization

Let

$$T = \begin{bmatrix} T_{11} & T_{12} & \cdots & T_{1q} \\ 0 & T_{22} & \cdots & T_{2q} \\ \vdots & \vdots & \ddots & \vdots \\ 0 & 0 & \cdots & T_{qq} \end{bmatrix} \begin{matrix} n_1 \\ n_2 \\ \\ n_q \end{matrix} \qquad (7.6.3)$$
$$\begin{matrix} n_1 & n_2 & & n_q \end{matrix}$$

be a partitioning of some real Schur canonical form $Q^T A Q = T \in \mathbb{R}^{n \times n}$ such that $\lambda(T_{11}), \ldots, \lambda(T_{qq})$ are disjoint. By Theorem 7.1.6 there exists a matrix Y such that $Y^{-1}TY = \text{diag}(T_{11}, \ldots, T_{qq})$. A practical procedure for determining Y is now given together with an analysis of Y's sensitivity as a function of the above partitioning.

Partition $I_n = [\, E_1, \ldots, E_q \,]$ conformally with T and define the matrix

$Y_{ij} \in \mathbb{R}^{n \times n}$ as follows:

$$Y_{ij} = I_n + E_i Z_{ij} E_j^T. \qquad i < j, \; Z_{ij} \in \mathbb{R}^{n_i \times n_j}$$

In other words, Y_{ij} looks just like the identity except that Z_{ij} occupies the (i, j) block position. It follows that if $Y_{ij}^{-1} T Y_{ij} = \bar{T} = (\bar{T}_{ij})$ then T and $\bar{T}$ are identical except that

$$
\begin{aligned}
\bar{T}_{ij} &= T_{ii} Z_{ij} - Z_{ij} T_{jj} + T_{ij} \\
\bar{T}_{ik} &= T_{ik} - Z_{ij} T_{jk} & (k = j + 1{:}q) \\
\bar{T}_{kj} &= T_{ki} Z_{ij} + T_{kj} & (k = 1{:}i - 1)
\end{aligned}
$$

Thus, T_{ij} can be zeroed provided we have an algorithm for solving the *Sylvester equation*

$$FZ - ZG = C \qquad (7.6.4)$$

where $F \in \mathbb{R}^{p \times p}$ and $G \in \mathbb{R}^{r \times r}$ are given upper quasi-triangular matrices and $C \in \mathbb{R}^{p \times r}$.

Bartels and Stewart (1972) have devised a method for doing this. Let $C = [\, c_1, \ldots, c_r \,]$ and $Z = [\, z_1, \ldots, z_r \,]$ be column partitionings. If $g_{k+1,k} = 0$, then by comparing columns in (7.6.4) we find

$$Fz_k - \sum_{i=1}^{k} g_{ik} z_i = c_k.$$

Thus, once we know $z_1, \ldots, z_{k-1}$ then we can solve the quasi-triangular system

$$(F - g_{kk} I) z_k = c_k + \sum_{i=1}^{k-1} g_{ik} z_i$$

for z_k. If $g_{k+1,k} \neq 0$, then z_k and z_{k+1} can be simultaneously found by solving the $2p$-by-$2p$ system

$$
\begin{bmatrix} F - g_{kk}I & -g_{mk}I \\ -g_{km}I & F - g_{mm}I \end{bmatrix}
\begin{bmatrix} z_k \\ z_m \end{bmatrix}
=
\begin{bmatrix} c_k \\ c_m \end{bmatrix}
+
\sum_{i=1}^{k-1}
\begin{bmatrix} g_{ik} z_i \\ g_{im} z_i \end{bmatrix}
\qquad (7.6.5)
$$

where $m = k+1$. By reordering the equations according to the permutation $(1, p+1, 2, p+2, \ldots, p, 2p)$, a banded system is obtained that can be solved in $O(p^2)$ flops. The details may be found in Bartels and Stewart (1972). Here is the overall process for the case when F and G are each triangular.

Algorithm 7.6.2 (Bartels-Stewart Algorithm) Given $C \in \mathbb{R}^{p \times r}$ and upper triangular matrices $F \in \mathbb{R}^{p \times p}$ and $G \in \mathbb{R}^{r \times r}$ that satisfy $\lambda(F) \cap \lambda(G) = \emptyset$, the following algorithm overwrites C with the solution to the equation $FZ - ZG = C$.

for $k = 1{:}r$
$\quad C(1{:}p, k) = C(1{:}p, k) + C(1{:}p, 1{:}k - 1)G(1{:}k - 1, k)$
$\quad$ Solve $(F - G(k, k)I)z = C(1{:}p, k)$ for z.
$\quad C(1{:}p, k) = z$
end

This algorithm requires $pr(p + r)$ flops.

By zeroing the super diagonal blocks in T in the appropriate order, the entire matrix can be reduced to block diagonal form.

Algorithm 7.6.3 Given an orthogonal matrix $Q \in \mathbb{R}^{n \times n}$, an upper quasi-triangular matrix $T = Q^T A Q$, and the partitioning (7.6.3), the following algorithm overwrites Q with QY where $Y^{-1}TY = \mathrm{diag}(T_{11}, \ldots, T_{qq})$.

for $j = 2{:}q$
$\quad$ **for** $i = 1{:}j - 1$
$\quad\quad$ Solve $T_{ii}Z - ZT_{jj} = -T_{ij}$ for Z using Algorithm 7.6.2.
$\quad\quad$ **for** $k = j + 1{:}q$
$\quad\quad\quad T_{ik} = T_{ik} - ZT_{jk}$
$\quad\quad$ **end**
$\quad\quad$ **for** $k = 1{:}q$
$\quad\quad\quad Q_{kj} = Q_{ki}Z + Q_{kj}$
$\quad\quad$ **end**
$\quad$ **end**
end

The number of flops required by this algorithm is a complicated function of the block sizes in (7.6.3).

The choice of the real Schur form T and its partitioning in (7.6.3) determines the sensitivity of the Sylvester equations that must be solved in Algorithm 7.6.3. This in turn affects the condition of the matrix Y and the overall usefulness of the block diagonalization. The reason for these dependencies is that the relative error of the computed solution $\hat{Z}$ to

$$T_{ii}Z - ZT_{jj} = -T_{ij} \tag{7.6.6}$$

satisfies

$$\frac{\|\hat{Z} - Z\|_F}{\|Z\|_F} \approx \mathbf{u}\frac{\|T\|_F}{\mathrm{sep}(T_{ii}, T_{jj})}.$$

For details, see Golub, Nash, and Van Loan (1979). Since

$$\mathrm{sep}(T_{ii}, T_{jj}) = \min_{X \neq 0} \frac{\|T_{ii}X - XT_{jj}\|_F}{\|X\|_F} \leq \min_{\substack{\lambda \in \lambda(T_{ii}) \\ \mu \in \lambda(T_{jj})}} |\lambda - \mu|.$$

there can be a substantial loss of accuracy whenever the subsets $\lambda(T_{ii})$ are insufficiently separated. Moreover, if Z satisfies (7.6.6) then

$$\| Z \|_F \leq \frac{\| T_{ij} \|_F}{\text{sep}(T_{ii}, T_{jj})}.$$

Thus, large-norm solutions can be expected if $\text{sep}(T_{ii}, T_{jj})$ is small. This tends to make the matrix Y in Algorithm 7.6.3 ill-conditioned since it is the product of the matrices

$$Y_{ij} = \begin{bmatrix} I & Z \\ 0 & I \end{bmatrix}.$$

Note: $\kappa_F(Y_{ij}) = 2n + \| Z \|_F^2$.

Confronted with these difficulties, Bavely and Stewart (1979) develop an algorithm for block diagonalizing that dynamically determines the eigenvalue ordering and partitioning in (7.6.3) so that all the Z matrices in Algorithm 7.6.3 are bounded in norm by some user-supplied tolerance. They find that the condition of Y can be controlled by controlling the condition of the Y_{ij}.

7.6.4 Eigenvector Bases

If the blocks in the partitioning (7.6.3) are all 1-by-1, then Algorithm 7.6.3 produces a basis of eigenvectors. As with the method of inverse iteration, the computed eigenvalue-eigenvector pairs are exact for some "nearby" matrix. A widely followed rule of thumb for deciding upon a suitable eigenvector method is to use inverse iteration whenever fewer than 25% of the eigenvectors are desired.

We point out, however, that the real Schur form can be used to determine selected eigenvectors. Suppose

$$Q^T A Q = \begin{bmatrix} T_{11} & u & T_{13} \\ 0 & \lambda & v^T \\ 0 & 0 & T_{33} \end{bmatrix} \begin{matrix} k-1 \\ 1 \\ n-k \end{matrix}$$
$$\begin{matrix} k-1 & 1 & n-k \end{matrix}$$

is upper quasi-triangular and that $\lambda \notin \lambda(T_{11}) \cup \lambda(T_{33})$. It follows that if we solve the linear systems $(T_{11} - \lambda I)w = -u$ and $(T_{33} - \lambda I)^T z = -v$ then

$$x = Q \begin{bmatrix} w \\ 1 \\ 0 \end{bmatrix} \quad \text{and} \quad y = Q \begin{bmatrix} 0 \\ 1 \\ z \end{bmatrix}$$

are the associated right and left eigenvectors, respectively. Note that the condition of λ is prescribed by $1/s(\lambda) = \sqrt{(1 + w^T w)(1 + z^T z)}$.

7.6.5   Ascertaining Jordan Block Structures

Suppose that we have computed the real Schur decomposition $A = QTQ^T$, identified clusters of "equal" eigenvalues, and calculated the corresponding block diagonalization $T = Y\text{diag}(T_{11}, \ldots, T_{qq})Y^{-1}$. As we have seen, this can be a formidable task. However, even greater numerical problems confront us if we attempt to ascertain the Jordan block structure of each T_{ii}. A brief examination of these difficulties will serve to highlight the limitations of the Jordan decomposition.

Assume for clarity that $\lambda(T_{ii})$ is real. The reduction of T_{ii} to Jordan form begins by replacing it with a matrix of the form $C = \lambda I + N$, where N is the strictly upper triangular portion of T_{ii} and where λ, say, is the mean of its eigenvalues.

Recall that the dimension of a Jordan block $J(\lambda)$ is the smallest nonnegative integer k for which $[J(\lambda) - \lambda I]^k = 0$. Thus, if $p_i = \text{dim}[\text{null}(N^i)]$, for $i = 0{:}n$, then $p_i - p_{i-1}$ equals the number of blocks in C's Jordan form that have dimension i or greater. A concrete example helps to make this assertion clear and to illustrate the role of the SVD in Jordan form computations.

Assume that C is 7-by-7. Suppose we compute the SVD $U_1^T N V_1 = \Sigma_1$ and "discover" that N has rank 3. If we order the singular values from small to large then it follows that the matrix $N_1 = V_1^T N V_1$ has the form

$$
N_1 = \begin{array}{c} \begin{bmatrix} 0 & K \\ 0 & L \end{bmatrix} & \begin{array}{c} 4 \\ 3 \end{array} \\ \begin{array}{cc} 4 & 3 \end{array} \end{array}
$$

At this point, we know that the geometric multiplicity of λ is 4—i.e, C's Jordan form has 4 blocks ($p_1 - p_0 = 4 - 0 = 4$).

Now suppose $\tilde{U}_2^T L \tilde{V}_2 = \Sigma_2$ is the SVD of L and that we find that L has unit rank. If we again order the singular values from small to large, then $L_2 = \tilde{V}_2^T L \tilde{V}_2$ clearly has the following structure:

$$
L_2 = \begin{bmatrix} 0 & 0 & a \\ 0 & 0 & b \\ 0 & 0 & c \end{bmatrix}.
$$

However $\lambda(L_2) = \lambda(L) = \{0, 0, 0\}$ and so $c = 0$. Thus, if

$$
V_2 = \text{diag}(I_4, \tilde{V}_2)
$$

then $N_2 = V_2^T N_1 V_2$ has the following form:

$$N_2 = \begin{bmatrix} 0 & 0 & 0 & 0 & \times & \times & \times \\ 0 & 0 & 0 & 0 & \times & \times & \times \\ 0 & 0 & 0 & 0 & \times & \times & \times \\ 0 & 0 & 0 & 0 & \times & \times & \times \\ 0 & 0 & 0 & 0 & 0 & 0 & a \\ 0 & 0 & 0 & 0 & 0 & 0 & b \\ 0 & 0 & 0 & 0 & 0 & 0 & 0 \end{bmatrix}.$$

Besides allowing us to introduce more zeroes into the upper triangle, the SVD of L also enables us to deduce the dimension of the null space of N^2. Since

$$N_1^2 = \begin{bmatrix} 0 & KL \\ 0 & L^2 \end{bmatrix} = \begin{bmatrix} 0 & K \\ 0 & L \end{bmatrix} \begin{bmatrix} 0 & K \\ 0 & L \end{bmatrix}$$

and $\begin{bmatrix} K \\ L \end{bmatrix}$ has full column rank,

$$p_2 = \dim(\text{null}(N^2)) = \dim(\text{null}(N_1^2)) = 4 + \dim(\text{null}(L)) = p_1 + 2.$$

Hence, we can conclude at this stage that the Jordan form of C has at least two blocks of dimension 2 or greater.

Finally, it is easy to see that $N_1^3 = 0$, from which we conclude that there is $p_3 - p_2 = 7 - 6 = 1$ block of dimension 3 or larger. If we define $V = V_1 V_2$ then it follows that the decomposition

$$V^T C V = \begin{bmatrix} \lambda & 0 & 0 & 0 & \times & \times & \times \\ 0 & \lambda & 0 & 0 & \times & \times & \times \\ 0 & 0 & \lambda & 0 & \times & \times & \times \\ 0 & 0 & 0 & \lambda & \times & \times & \times \\ 0 & 0 & 0 & 0 & \lambda & \times & a \\ 0 & 0 & 0 & 0 & 0 & \lambda & 0 \\ 0 & 0 & 0 & 0 & 0 & 0 & \lambda \end{bmatrix} \begin{matrix} \left.\vphantom{\begin{matrix}1\\1\\1\\1\end{matrix}}\right\} \text{4 blocks of order 1 or larger} \\ \\ \left.\vphantom{\begin{matrix}1\\1\end{matrix}}\right\} \text{2 blocks of order 2 or larger} \\ \left.\vphantom{1}\right\} \text{1 block of order 3 or larger} \end{matrix}$$

"displays" C's Jordan block structure: 2 blocks of order 1, 1 block of order 2, and 1 block of order 3.

To compute the Jordan decomposition it is necessary to resort to non-orthogonal transformations. We refer the reader to either Golub and Wilkinson (1976) or Kågström and Ruhe (1980a, 1980b) for how to proceed with this phase of the reduction.

The above calculations with the SVD amply illustrate that difficult rank decisions must be made at each stage and that the final computed block structure depends critically on those decisions. Fortunately, the stable Schur decomposition can almost always be used in lieu of the Jordan decomposition in practical applications.

Problems

P7.6.1 Give a complete algorithm for solving a real, n-by-n, upper quasi-triangular system $Tx = b$.

P7.6.2 Suppose $U^{-1}AU = \text{diag}(\alpha_1, \ldots, \alpha_m)$ and $V^{-1}BV = \text{diag}(\beta_1, \ldots, \beta_n)$. Show that if $\phi(X) = AX + XB$, then $\lambda(\phi) = \{\, \alpha_i + \beta_j : i = 1{:}m, \ j = 1{:}n \,\}$. What are the corresponding eigenvectors? How can these decompositions be used to solve $AX + XB = C$?

P7.6.3 Show that if $Y = \begin{bmatrix} I & Z \\ 0 & I \end{bmatrix}$ then $\kappa_2(Y) = [2 + \sigma^2 + \sqrt{4\sigma^2 + \sigma^4}\,]/2$ where $\sigma = \| Z \|_2$.

P7.6.4 Derive the system (7.6.5).

P7.6.5 Assume that $T \in \mathbf{R}^{n \times n}$ is block upper triangular and partitioned as follows:

$$ T = \begin{bmatrix} T_{11} & T_{12} & T_{13} \\ 0 & T_{22} & T_{23} \\ 0 & 0 & T_{33} \end{bmatrix} \qquad T \in \mathbf{R}^{n \times n} $$

Suppose that the diagonal block T_{22} is 2-by-2 with complex eigenvalues that are disjoint from $\lambda(T_{11})$ and $\lambda(T_{33})$. Give an algorithm for computing the 2-dimensional real invariant subspace associated with T_{22}'s eigenvalues.

P7.6.6 Suppose $H \in \mathbf{R}^{n \times n}$ is upper Hessenberg with a complex eigenvalue $\lambda + i \cdot \mu$. How could inverse iteration be used to compute $x, y \in \mathbf{R}^n$ so that $H(x + iy) = (\lambda + i\mu)(x + iy)$? Hint: compare real and imaginary parts in this equation and obtain a $2n$-by-$2n$ real system.

P7.6.6 (a) Prove that if $\mu_0 \in \mathbf{C}$ has nonzero real part, then the iteration

$$ \mu_{k+1} = \frac{1}{2}\left(\mu_k + \frac{1}{\mu_k}\right) $$

converges to 1 if $\text{Re}(\mu_0) > 0$ and to -1 if $\text{Re}(\mu_0) < 0$. (b) Suppose $A \in \mathbf{C}^{n \times n}$ is diagonalizable and that

$$ A = X \begin{bmatrix} D_+ & 0 \\ 0 & D_- \end{bmatrix} X^{-1} $$

where $D_+ \in \mathbf{C}^{p \times p}$ and $D_- \in \mathbf{C}^{(n-p) \times (n-p)}$ are diagonal with eigenvalues in the open right half plane and open left half plane, respectively. Show that the iteration

$$ A_{k+1} = \frac{1}{2}\left(A_k + A_k^{-1}\right) \qquad A_0 = A $$

converges to

$$ \text{sign}(A) \equiv X \begin{bmatrix} I_p & 0 \\ 0 & -I_{n-p} \end{bmatrix} X^{-1}. $$

(c) Suppose

$$ M = \begin{bmatrix} M_{11} & M_{12} \\ 0 & M_{22} \end{bmatrix} \begin{matrix} p \\ n-p \end{matrix} $$
$$ \begin{matrix} p & \ n-p \end{matrix} $$

with the property that $\lambda(M_{11})$ is in the open right half plane and $\lambda(M_{22})$ is in the open left half plane. Show that

$$ \text{sign}(M) = \begin{bmatrix} I_p & Z \\ 0 & -I_{n-p} \end{bmatrix} $$

and that $-Z/2$ solves $M_{11}X - XM_{22} = -M_{12}$. Thus,

$$ U = \begin{bmatrix} I_p & -Z/2 \\ 0 & I_{n-p} \end{bmatrix} \quad \Rightarrow U^{-1}MU = \begin{bmatrix} M_{11} & 0 \\ 0 & M_{22} \end{bmatrix}. $$

Notes and References for Sec. 7.6

Much of the material discussed in this section may be found in the survey paper

G.H. Golub and J.H. Wilkinson (1976). "Ill-Conditioned Eigensystems and the Computation of the Jordan Canonical Form," *SIAM Review 18*, 578–619.

Papers that specifically analyze the method of inverse iteration for computing eigenvectors include

J. Varah (1968). "The Calculation of the Eigenvectors of a General Complex Matrix by Inverse Iteration," *Math. Comp. 22*, 785–91.
J. Varah (1968). "Rigorous Machine Bounds for the Eigensystem of a General Complex Matrix," *Math. Comp. 22*, 793–801.
J. Varah (1970). "Computing Invariant Subspaces of a General Matrix When the Eigensystem is Poorly Determined," *Math. Comp. 24*, 137–49.
G. Peters and J.H. Wilkinson (1979). "Inverse Iteration, Ill-Conditioned Equations, and Newton's Method," *SIAM Review 21*, 339–60.

The Algol version of the Eispack inverse iteration subroutine is given in

G. Peters and J.H. Wilkinson (1971). "The Calculation of Specified Eigenvectors by Inverse Iteration," in Wilkinson and Reinsch (1971,pp.418–39).

The problem of ordering the eigenvalues in the real Schur form is the subject of

A. Ruhe (1970). "An Algorithm for Numerical Determination of the Structure of a General Matrix," *BIT 10*, 196–216.
G.W. Stewart (1976). "Algorithm 406: HQR3 and EXCHNG: Fortran Subroutines for Calculating and Ordering the Eigenvalues of a Real Upper Hessenberg Matrix," *ACM Trans. Math. Soft. 2*, 275–80.
J.J. Dongarra, S. Hammarling, and J.H. Wilkinson (1992). "Numerical Considerations in Computing Invariant Subspaces," *SIAM J. Matrix Anal. Appl. 13*, 145–161.
Z. Bai and J.W. Demmel (1993). "On Swapping Diagonal Blocks in Real Schur Form," *Lin. Alg. and Its Applic. 186*, 73–95

Fortran programs for computing block diagonalizations and Jordan forms are described in

C. Bavely and G.W. Stewart (1979). "An Algorithm for Computing Reducing Subspaces by Block Diagonalization," *SIAM J. Num. Anal. 16*, 359–67.
B. Kågström and A. Ruhe (1980a). "An Algorithm for Numerical Computation of the Jordan Normal Form of a Complex Matrix," *ACM Trans. Math. Soft. 6*, 398–419.
B. Kågström and A. Ruhe (1980b). "Algorithm 560 JNF: An Algorithm for Numerical Computation of the Jordan Normal Form of a Complex Matrix," *ACM Trans. Math. Soft. 6*, 437–43.
J.W. Demmel (1983). "A Numerical Analyst's Jordan Canonical Form," Ph.D. Thesis, Berkeley.

Papers that are concerned with estimating the error in a computed eigenvalue and/or eigenvector include

S.P. Chan and B.N. Parlett (1977). "Algorithm 517: A Program for Computing the Condition Numbers of Matrix Eigenvalues Without Computing Eigenvectors," *ACM Trans. Math. Soft. 3*, 186–203.
H.J. Symm and J.H. Wilkinson (1980). "Realistic Error Bounds for a Simple Eigenvalue and Its Associated Eigenvector," *Numer. Math. 35*, 113–26.

C. Van Loan (1987). "On Estimating the Condition of Eigenvalues and Eigenvectors,"
  Lin. Alg. and Its Applic. 88/89, 715–732.
Z. Bai, J. Demmel, and A. McKenney (1993). "On Computing Condition Numbers for
  the Nonsymmetric Eigenproblem," *ACM Trans. Math. Soft.* 19, 202–223.

As we have seen, the sep(.,.) function is of great importance in the assessment of a computed invariant subspace. Aspects of this quantity and the associated Sylvester equation are discussed in

J. Varah (1979). "On the Separation of Two Matrices," *SIAM J. Num. Anal.* 16,
  212–22.
R. Byers (1984). "A Linpack-Style Condition Estimator for the Equation $AX - XB^T =
  C$," *IEEE Trans. Auto. Cont.* AC-29, 926–928.
K. Datta (1988). "The Matrix Equation $XA - BX = R$ and Its Applications," *Lin. Alg.
  and Its Appl.* 109, 91–105.
N.J. Higham (1993). "Perturbation Theory and Backward Error for AX - XB = C,"
  BIT 33, 124–136.
J. Gardiner, M.R. Wette, A.J. Laub, J.J. Amato, and C.B. Moler (1992). "Algorithm
  705: A FORTRAN-77 Software Package for Solving the Sylvester Matrix Equation
  $AXB^T + CXD^T = E$," *ACM Trans. Math. Soft.* 18, 232–238.

Numerous algorithms have been proposed for the Sylvester equation, but those described in

R.H. Bartels and G.W. Stewart (1972). "Solution of the Equation $AX + XB = C$,"
  Comm. ACM 15, 820–26.
G.H. Golub, S. Nash, and C. Van Loan (1979). "A Hessenberg-Schur Method for the
  Matrix Problem $AX + XB = C$," *IEEE Trans. Auto. Cont.* AC-24, 909–13.

are among the more reliable in that they rely on orthogonal transformations. A constrained Sylvester equation problem is considerd in

J.B. Barlow, M.M. Monahemi, and D.P. O'Leary (1992). "Constrained Matrix Sylvester
  Equations," *SIAM J. Matrix Anal. Appl.* 13, 1–9.
The Lyapunov problem $FX + XF^T = -C$ where C is non-negative definite has a very important role to play in control theory. See

S. Barnett and C. Storey (1968). "Some Applications of the Lyapunov Matrix Equation,"
  J. Inst. Math. Applic. 4, 33–42.
G. Hewer and C. Kenney (1988). "The Sensitivity of the Stable Lyapunov Equation,"
  SIAM J. Control Optim 26, 321–344.
A.R. Ghavimi and A.J. Laub (1995). "Residual Bounds for Discrete-Time Lyapunov
  Equations," *IEEE Trans. Auto. Cont. 40*, 1244–1249.

Several authors have considered generalizations of the Sylvester equation, i.e., $\Sigma F_i X G_i = C$. These include

P. Lancaster (1970). "Explicit Solution of Linear Matrix Equations," *SIAM Review 12*,
  544–66.
H. Wimmer and A.D. Ziebur (1972). "Solving the Matrix Equations $\Sigma f_p(A)g_p(A) = C$,"
  SIAM Review 14, 318–23.
W.J. Vetter (1975). "Vector Structures and Solutions of Linear Matrix Equations," *Lin.
  Alg. and Its Applic.* 10, 181–88.

Some ideas about improving computed eigenvalues, eigenvectors, and invariant subspaces may be found in

J.J. Dongarra, C.B. Moler, and J.H. Wilkinson (1983). "Improving the Accuracy of Computed Eigenvalues and Eigenvectors," *SIAM J. Numer. Anal. 20*, 23–46.

J.W. Demmel (1987). "Three Methods for Refining Estimates of Invariant Subspaces," *Computing 38*, 43–57.

Hessenberg/QR iteration techniques are fast, but not very amenable to parallel computation. Because of this there is a hunger for radically new approaches to the eigenproblem. Here are some papers that focus on the matrix sign function and related ideas that have high performance potential:

C.S. Kenney and A.J. Laub (1991). "Rational Iterative Methods for the Matrix Sign Function," *SIAM J. Matrix Anal. Appl. 12*, 273–291.

C.S. Kenney, A.J. Laub, and P.M. Papadopouos (1992). "Matrix Sign Algorithms for Riccati Equations," *IMA J. of Math. Control Inform. 9*, 331–344.

C.S. Kenney and A.J. Laub (1992). "On Scaling Newton's Method for Polar Decomposition and the Matrix Sign Function," *SIAM J. Matrix Anal. Appl. 13*, 688–706.

N.J. Higham (1994). "The Matrix Sign Decomposition and Its Relation to the Polar Decomposition," *Lin. Alg. and Its Applic 212/213*, 3–20.

L. Adams and P. Arbenz (1994). "Towards a Divide and Conquer Algorithm for the Real Nonsymmetric Eigenvalue Problem," *SIAM J. Matrix Anal. Appl. 15*, 1333–1353.

7.7 The QZ Method for $Ax = \lambda Bx$

Let A and B be two n-by-n matrices. The set of all matrices of the form $A - \lambda B$ with $\lambda \in \mathbb{C}$ is said to be a *pencil*. The eigenvalues of the pencil are elements of the set $\lambda(A, B)$ defined by

$$\lambda(A, B) = \{ z \in \mathbb{C} : \det(A - zB) = 0 \}.$$

If $\lambda \in \lambda(A, B)$ and

$$Ax = \lambda Bx \qquad x \neq 0 \tag{7.7.1}$$

then x is referred to as an eigenvector of $A - \lambda B$.

In this section we briefly survey some of the mathematical properties of the generalized eigenproblem (7.7.1) and present a stable method for its solution. The important case when A and B are symmetric with the latter positive definite is discussed in §8.7.2.

7.7.1 Background

The first thing to observe about the generalized eigenvalue problem is that there are n eigenvalues if and only if $\text{rank}(B) = n$. If B is rank deficient then $\lambda(A, B)$ may be finite, empty, or infinite:

$$A = \begin{bmatrix} 1 & 2 \\ 0 & 3 \end{bmatrix} \quad B = \begin{bmatrix} 1 & 0 \\ 0 & 0 \end{bmatrix} \quad \Rightarrow \quad \lambda(A, B) = \{1\}$$

$$A = \begin{bmatrix} 1 & 2 \\ 0 & 3 \end{bmatrix} \quad B = \begin{bmatrix} 0 & 1 \\ 0 & 0 \end{bmatrix} \quad \Rightarrow \quad \lambda(A, B) = \emptyset$$

$$A = \begin{bmatrix} 1 & 2 \\ 0 & 0 \end{bmatrix} \quad B = \begin{bmatrix} 1 & 0 \\ 0 & 0 \end{bmatrix} \quad \Rightarrow \quad \lambda(A, B) = \mathbb{C}$$

Note that if $0 \neq \lambda \in \lambda(A, B)$ then $(1/\lambda) \in \lambda(B, A)$. Moreover, if B is nonsingular then $\lambda(A, B) = \lambda(B^{-1}A, I) = \lambda(B^{-1}A)$.

This last observation suggests one method for solving the $A - \lambda B$ problem when B is nonsingular:

- Solve $BC = A$ for C using (say) Gaussian elimination with pivoting.

- Use the QR algorithm to compute the eigenvalues of C.

Note that C will be affected by roundoff errors of order $\mathbf{u}\| A \|_2 \| B^{-1} \|_2$. If B is ill-conditioned, then this can rule out the possibility of computing any generalized eigenvalue accurately—even those eigenvalues that may be regarded as well-conditioned.

Example 7.7.1 If

$$A = \begin{bmatrix} 1.746 & .940 \\ 1.246 & 1.898 \end{bmatrix} \quad \text{and} \quad B = \begin{bmatrix} .780 & .563 \\ .913 & .659 \end{bmatrix}$$

then $\lambda(A, B) = \{2, 1.07 \times 10^6\}$. With 7-digit floating point arithmetic, we find $\lambda(fl(AB^{-1}))$
$= \{1.562539, 1.01 \times 10^6\}$. The poor quality of the small eigenvalue is because $\kappa_2(B) \approx 2 \times 10^6$. On the other hand, we find that

$$\lambda(I, fl(A^{-1}B)) \approx \{2.000001, 1.06 \times 10^6\}.$$

The accuracy of the small eigenvalue is improved because $\kappa_2(A) \approx 4$.

Example 7.7.1 suggests that we seek an alternative approach to the $A - \lambda B$ problem. One idea is to compute well-conditioned Q and Z such that the matrices

$$A_1 = Q^{-1}AZ \qquad B_1 = Q^{-1}BZ \tag{7.7.2}$$

are each in canonical form. Note that $\lambda(A, B) = \lambda(A_1, B_1)$ since

$$Ax = \lambda Bx \quad \Leftrightarrow \quad A_1 y = \lambda B_1 y \qquad x = Zy$$

We say that the pencils $A - \lambda B$ and $A_1 - \lambda B_1$ are *equivalent* if (7.7.2) holds with nonsingular Q and Z.

7.7.2 The Generalized Schur Decomposition

As in the standard eigenproblem $A - \lambda I$ there is a choice between canonical forms. Analogous to the Jordan form is a decomposition of Kronecker in which both A_1 and B_1 are block diagonal. The blocks are similar to Jordan blocks. The Kronecker canonical form poses the same numerical difficulties as the Jordan form. However, this decomposition does provide insight into the mathematical properties of the pencil $A - \lambda B$. See Wilkinson (1978) and Demmel and Kågström (1987) for details.

More attractive from the numerical point of view is the following decomposition described in Moler and Stewart (1973).

Theorem 7.7.1 (Generalized Schur Decomposition) *If A and B are in $\mathbb{C}^{n \times n}$, then there exist unitary Q and Z such that $Q^H A Z = T$ and $Q^H B Z = S$ are upper triangular. If for some k, t_{kk} and s_{kk} are both zero, then $\lambda(A, B) = \mathbb{C}$. Otherwise*

$$\lambda(A, B) = \{t_{ii}/s_{ii} : s_{ii} \neq 0\}.$$

Proof. Let $\{B_k\}$ be a sequence of nonsingular matrices that converge to B. For each k, let $Q_k^H (AB_k^{-1}) Q_k = R_k$ be a Schur decomposition of AB_k^{-1}. Let Z_k be unitary such that $Z_k^H (B_k^{-1} Q_k) \equiv S_k^{-1}$ is upper triangular. It follows that both $Q_k^H A Z_k = R_k S_k$ and $Q_k^H B_k Z_k = S_k$ are also upper triangular.

Using the Bolzano-Weierstrass theorem, we know that the bounded sequence $\{(Q_k, Z_k)\}$ has a converging subsequence, $\lim(Q_{k_i}, Z_{k_i}) = (Q, Z)$. It is easy to show that Q and Z are unitary and that $Q^H A Z$ and $Q^H B Z$ are upper triangular. The assertions about $\lambda(A, B)$ follow from the identity

$$\det(A - \lambda B) = \det(QZ^H) \prod_{i=1}^{n} (t_{ii} - \lambda s_{ii}). \quad \square$$

If A and B are real then the following decomposition, which corresponds to the real schur decomposition (Theorem 7.4.1), is of interest.

Theorem 7.7.2 (Generalized Real Schur Decomposition) *If A and B are in $\mathbb{R}^{n \times n}$ then there exist orthogonal matrices Q and Z such that $Q^T A Z$ is upper quasi-triangular and $Q^T B Z$ is upper triangular.*

Proof. See Stewart (1972). $\square$

In the remainder of this section we are concerned with the computation of this decomposition and the mathematical insight that it provides.

7.7.3 Sensitivity Issues

The generalized Schur decomposition sheds light on the issue of eigenvalue sensitivity for the $A - \lambda B$ problem. Clearly, small changes in A and B can induce large changes in the eigenvalue $\lambda_i = t_{ii}/s_{ii}$ if s_{ii} is small. However, as Stewart (1978) argues, it may not be appropriate to regard such an eigenvalue as "ill-conditioned." The reason is that the reciprocal $\mu_i = s_{ii}/t_{ii}$ might be a very well behaved eigenvalue for the pencil $\mu A - B$. In the Stewart analysis, A and B are treated symmetrically and the eigenvalues are regarded more as ordered pairs (t_{ii}, s_{ii}) than as quotients. With this point of view it becomes appropriate to measure eigenvalue perturbations in the *chordal metric* chord (a, b) defined by

$$\text{chord}(a,b) = \frac{|a - b|}{\sqrt{1 + a^2}\sqrt{1 + b^2}}.$$

Stewart shows that if λ is a distinct eigenvalue of $A - \lambda B$ and λ_ϵ is the corresponding eigenvalue of the perturbed pencil $\tilde{A} - \lambda \tilde{B}$ with $\| A - \tilde{A} \|_2 \approx \| B - \tilde{B} \|_2 \approx \epsilon$, then

$$\text{chord}(\lambda, \lambda_\epsilon) \leq \frac{\epsilon}{(y^H A x)^2 + (y^H B x)^2} + O(\epsilon^2)$$

where x and y have unit 2-norm and satisfy $Ax = \lambda Bx$ and $y^H = \lambda y^H B$. Note that the denominator in the upper bound is symmetric in A and B. The "truly" ill-conditioned eigenvalues are those for which this denominator is small.

The extreme case when $t_{kk} = s_{kk} = 0$ for some k has been studied by Wilkinson (1979). He makes the interesting observation that when this occurs, the remaining quotients t_{ii}/s_{ii} can assume arbitrary values.

7.7.4   Hessenberg-Triangular Form

The first step in computing the generalized Schur decomposition of the pair (A, B) is to reduce A to upper Hessenberg form and B to upper triangular form via orthogonal transformations. We first determine an orthogonal U such that $U^T B$ is upper triangular. Of course, to preserve eigenvalues, we must also update A in exactly the same way. Let's trace what happens in the $n = 5$ case.

$$A = U^T A = \begin{bmatrix} \times & \times & \times & \times & \times \\ \times & \times & \times & \times & \times \\ \times & \times & \times & \times & \times \\ \times & \times & \times & \times & \times \\ \times & \times & \times & \times & \times \end{bmatrix}, B = U^T B = \begin{bmatrix} \times & \times & \times & \times & \times \\ 0 & \times & \times & \times & \times \\ 0 & 0 & \times & \times & \times \\ 0 & 0 & 0 & \times & \times \\ 0 & 0 & 0 & 0 & \times \end{bmatrix}$$

Next, we reduce A to upper Hessenberg form while preserving B's upper triangular form. First, a Givens rotation Q_{45} is determined to zero a_{51}:

$$A = Q_{45}^T A = \begin{bmatrix} \times & \times & \times & \times & \times \\ \times & \times & \times & \times & \times \\ \times & \times & \times & \times & \times \\ \times & \times & \times & \times & \times \\ 0 & \times & \times & \times & \times \end{bmatrix}, B = Q_{45}^T B = \begin{bmatrix} \times & \times & \times & \times & \times \\ 0 & \times & \times & \times & \times \\ 0 & 0 & \times & \times & \times \\ 0 & 0 & 0 & \times & \times \\ 0 & 0 & 0 & \times & \times \end{bmatrix}$$

The nonzero entry arising in the (5,4) position in B can be zeroed by postmultiplying with an appropriate Givens rotation Z_{45} :

$$A = AZ_{45} = \begin{bmatrix} \times & \times & \times & \times & \times \\ \times & \times & \times & \times & \times \\ \times & \times & \times & \times & \times \\ \times & \times & \times & \times & \times \\ 0 & \times & \times & \times & \times \end{bmatrix}, B = BZ_{45} = \begin{bmatrix} \times & \times & \times & \times & \times \\ 0 & \times & \times & \times & \times \\ 0 & 0 & \times & \times & \times \\ 0 & 0 & 0 & \times & \times \\ 0 & 0 & 0 & 0 & \times \end{bmatrix}$$

Zeros are similarly introduced into the (4, 1) and (3, 1) positions in A:

$$A = Q_{34}^T A = \begin{bmatrix} \times & \times & \times & \times & \times \\ \times & \times & \times & \times & \times \\ \times & \times & \times & \times & \times \\ 0 & \times & \times & \times & \times \\ 0 & \times & \times & \times & \times \end{bmatrix}, B = Q_{34}^T B = \begin{bmatrix} \times & \times & \times & \times & \times \\ 0 & \times & \times & \times & \times \\ 0 & 0 & \times & \times & \times \\ 0 & 0 & \times & \times & \times \\ 0 & 0 & 0 & 0 & \times \end{bmatrix}$$

$$A = AZ_{34} = \begin{bmatrix} \times & \times & \times & \times & \times \\ \times & \times & \times & \times & \times \\ \times & \times & \times & \times & \times \\ 0 & \times & \times & \times & \times \\ 0 & \times & \times & \times & \times \end{bmatrix}, B = BZ_{34} = \begin{bmatrix} \times & \times & \times & \times & \times \\ 0 & \times & \times & \times & \times \\ 0 & 0 & \times & \times & \times \\ 0 & 0 & 0 & \times & \times \\ 0 & 0 & 0 & 0 & \times \end{bmatrix}$$

$$A = Q_{23}^T A = \begin{bmatrix} \times & \times & \times & \times & \times \\ \times & \times & \times & \times & \times \\ 0 & \times & \times & \times & \times \\ 0 & \times & \times & \times & \times \\ 0 & \times & \times & \times & \times \end{bmatrix}, B = Q_{23}^T B = \begin{bmatrix} \times & \times & \times & \times & \times \\ 0 & \times & \times & \times & \times \\ 0 & \times & \times & \times & \times \\ 0 & 0 & 0 & \times & \times \\ 0 & 0 & 0 & 0 & \times \end{bmatrix}$$

$$A = AZ_{23} = \begin{bmatrix} \times & \times & \times & \times & \times \\ \times & \times & \times & \times & \times \\ 0 & \times & \times & \times & \times \\ 0 & \times & \times & \times & \times \\ 0 & \times & \times & \times & \times \end{bmatrix}, B = BZ_{23} = \begin{bmatrix} \times & \times & \times & \times & \times \\ 0 & \times & \times & \times & \times \\ 0 & 0 & \times & \times & \times \\ 0 & 0 & 0 & \times & \times \\ 0 & 0 & 0 & 0 & \times \end{bmatrix}$$

A is now upper Hessenberg through its first column. The reduction is completed by zeroing a_{52}, a_{42}, and a_{53}. As is evident above, two orthogonal transformations are required for each a_{ij} that is zeroed—one to do the zeroing and the other to restore B's triangularity. Either Givens rotations or 2-by-2 modified Householder transformations can be used. Overall we have:

Algorithm 7.7.1 (Hessenberg-Triangular Reduction) Given A and B in $\mathbb{R}^{n \times n}$, the following algorithm overwrites A with an upper Hessenberg matrix $Q^T A Z$ and B with an upper triangular matrix $Q^T B Z$ where both Q and Z are orthogonal.

Using Algorithm 5.2.1, overwrite B with $Q^T B = R$ where
$\quad$ Q is orthogonal and R is upper triangular.
$A = Q^T A$
for $j = 1{:}n - 2$
$\quad$ for $i = n{:} - 1{:}j + 2$
$\quad\quad$ $[c, s] = \mathbf{givens}(A(i - 1, j), A(i, j))$
$$A(i - 1{:}i, j{:}n) = \begin{bmatrix} c & s \\ -s & c \end{bmatrix}^T A(i - 1{:}i, j{:}n)$$
$$B(i - 1{:}i, i - 1{:}n) = \begin{bmatrix} c & s \\ -s & c \end{bmatrix}^T B(i - 1{:}i, i - 1{:}n)$$
$\quad\quad$ $[c, s] = \mathbf{givens}(-B(i, i), B(i, i - 1))$
$$B(1{:}i, i - 1{:}i) = B(1{:}i, i - 1{:}i) \begin{bmatrix} c & s \\ -s & c \end{bmatrix}$$
$$A(1{:}n, i - 1{:}i) = A(1{:}n, i - 1{:}i) \begin{bmatrix} c & s \\ -s & c \end{bmatrix}$$
$\quad$ end
end

This algorithm requires about $8n^3$ flops. The accumulation of Q and Z requires about $4n^3$ and $3n^3$ flops, respectively.

The reduction of $A - \lambda B$ to Hessenberg-triangular form serves as a "front end" decomposition for a generalized QR iteration known as the QZ iteration which we describe next.

Example 7.7.3 If
$$A = \begin{bmatrix} 10 & 1 & 2 \\ 1 & 2 & -1 \\ 1 & 1 & 2 \end{bmatrix} \quad \text{and} \quad B = \begin{bmatrix} 1 & 2 & 3 \\ 4 & 5 & 6 \\ 7 & 8 & 9 \end{bmatrix}$$
and orthogonal matrices Q and Z are defined by
$$Q = \begin{bmatrix} -.1231 & -.9917 & .0378 \\ -.4924 & .0279 & -.8699 \\ -.8616 & .1257 & .4917 \end{bmatrix} \quad \text{and} \quad Z = \begin{bmatrix} 1.0000 & 0.0000 & 0.0000 \\ 0.0000 & -.8944 & -.4472 \\ 0.0000 & .4472 & -.8944 \end{bmatrix}$$
then $A_1 = Q^T A Z$ and $B_1 = Q^T B Z$ are given by
$$A_1 = \begin{bmatrix} -2.5849 & 1.5413 & 2.4221 \\ -9.7631 & .0874 & 1.9239 \\ 0.0000 & 2.7233 & -.7612 \end{bmatrix} \quad \text{and} \quad B_1 = \begin{bmatrix} -8.1240 & 3.6332 & 14.2024 \\ 0.0000 & 0.0000 & 1.8739 \\ 0.0000 & 0.0000 & .7612 \end{bmatrix}$$

7.7.5 Deflation

In describing the QZ iteration we may assume without loss of generality that A is an unreduced upper Hessenberg matrix and that B is a nonsingular upper triangular matrix. The first of these assertions is obvious, for if $a_{k+1,k} = 0$ then

$$A - \lambda B \;=\; \begin{bmatrix} A_{11} - \lambda B_{11} & A_{12} - \lambda B_{12} \\ 0 & A_{22} - \lambda B_{22} \\ k & n-k \end{bmatrix} \begin{matrix} k \\ n-k \end{matrix}$$

and we may proceed to solve the two smaller problems $A_{11} - \lambda B_{11}$ and $A_{22} - \lambda B_{22}$. On the other hand, if $b_{kk} = 0$ for some k, then it is possible to introduce a zero in A's $(n, n-1)$ position and thereby deflate. Illustrating by example, suppose $n = 5$ and $k = 3$:

$$A = \begin{bmatrix} \times & \times & \times & \times & \times \\ \times & \times & \times & \times & \times \\ 0 & \times & \times & \times & \times \\ 0 & 0 & \times & \times & \times \\ 0 & 0 & 0 & \times & \times \end{bmatrix}, \qquad B = \begin{bmatrix} \times & \times & \times & \times & \times \\ 0 & \times & \times & \times & \times \\ 0 & 0 & 0 & \times & \times \\ 0 & 0 & 0 & \times & \times \\ 0 & 0 & 0 & 0 & \times \end{bmatrix}$$

The zero on B's diagonal can be "pushed down" to the (5,5) position as follows using Givens rotations:

$$A = Q_{34}^T A = \begin{bmatrix} \times & \times & \times & \times & \times \\ \times & \times & \times & \times & \times \\ 0 & \times & \times & \times & \times \\ 0 & \times & \times & \times & \times \\ 0 & 0 & 0 & \times & \times \end{bmatrix}, B = Q_{34}^T B = \begin{bmatrix} \times & \times & \times & \times & \times \\ 0 & \times & \times & \times & \times \\ 0 & 0 & 0 & \times & \times \\ 0 & 0 & 0 & 0 & \times \\ 0 & 0 & 0 & 0 & \times \end{bmatrix}$$

$$A = A Z_{23} = \begin{bmatrix} \times & \times & \times & \times & \times \\ \times & \times & \times & \times & \times \\ 0 & \times & \times & \times & \times \\ 0 & 0 & \times & \times & \times \\ 0 & 0 & 0 & \times & \times \end{bmatrix}, B = B Z_{23} = \begin{bmatrix} \times & \times & \times & \times & \times \\ 0 & \times & \times & \times & \times \\ 0 & 0 & 0 & \times & \times \\ 0 & 0 & 0 & 0 & \times \\ 0 & 0 & 0 & 0 & \times \end{bmatrix}$$

$$A = Q_{45}^T A = \begin{bmatrix} \times & \times & \times & \times & \times \\ \times & \times & \times & \times & \times \\ 0 & \times & \times & \times & \times \\ 0 & 0 & \times & \times & \times \\ 0 & 0 & \times & \times & \times \end{bmatrix}, B = Q_{45}^T B = \begin{bmatrix} \times & \times & \times & \times & \times \\ 0 & \times & \times & \times & \times \\ 0 & 0 & 0 & \times & \times \\ 0 & 0 & 0 & 0 & \times \\ 0 & 0 & 0 & 0 & 0 \end{bmatrix}$$

$$A = A Z_{34} = \begin{bmatrix} \times & \times & \times & \times & \times \\ \times & \times & \times & \times & \times \\ 0 & \times & \times & \times & \times \\ 0 & 0 & \times & \times & \times \\ 0 & 0 & 0 & \times & \times \end{bmatrix}, B = B Z_{34}^T = \begin{bmatrix} \times & \times & \times & \times & \times \\ 0 & \times & \times & \times & \times \\ 0 & 0 & \times & \times & \times \\ 0 & 0 & 0 & 0 & \times \\ 0 & 0 & 0 & 0 & 0 \end{bmatrix}$$

$$A = AZ_{45} = \begin{bmatrix} \times & \times & \times & \times & \times \\ \times & \times & \times & \times & \times \\ 0 & \times & \times & \times & \times \\ 0 & 0 & \times & \times & \times \\ 0 & 0 & 0 & 0 & \times \end{bmatrix}, B = BZ_{45} = \begin{bmatrix} \times & \times & \times & \times & \times \\ 0 & \times & \times & \times & \times \\ 0 & 0 & \times & \times & \times \\ 0 & 0 & 0 & \times & \times \\ 0 & 0 & 0 & 0 & 0 \end{bmatrix}$$

This zero-chasing technique is perfectly general and can be used to zero $a_{n,n-1}$ regardless of where the zero appears along B's diagonal.

7.7.6   The QZ Step

We are now in a position to describe a QZ step. The basic idea is to update A and B as follows

$$(\bar{A} - \lambda \bar{B}) \; = \; \bar{Q}^T (A - \lambda B) \bar{Z},$$

where $\bar{A}$ is upper Hessenberg, $\bar{B}$ is upper triangular, $\bar{Q}$ and $\bar{Z}$ are each orthogonal, and $\bar{A}\bar{B}^{-1}$ is essentially the same matrix that would result if a Francis QR step (Algorithm 7.5.2) were explicitly applied to AB^{-1}. This can be done with some clever zero-chasing and an appeal to the implicit Q theorem.

Let $M = AB^{-1}$ (upper Hessenberg) and let v be the first column of the matrix $(M - aI)(M - bI)$, where a and b are the eigenvalues of M's lower 2-by-2 submatrix. Note that v can be calculated in $O(1)$ flops. If P_0 is a Householder matrix such that $P_0 v$ is a multiple of e_1, then

$$A \; = \; \dot{P}_0 A \; = \; \begin{bmatrix} \times & \times & \times & \times & \times & \times \\ \times & \times & \times & \times & \times & \times \\ \times & \times & \times & \times & \times & \times \\ 0 & 0 & \times & \times & \times & \times \\ 0 & 0 & 0 & \times & \times & \times \\ 0 & 0 & 0 & 0 & \times & \times \end{bmatrix}$$

$$B \; = \; P_0 B \; = \; \begin{bmatrix} \times & \times & \times & \times & \times & \times \\ \times & \times & \times & \times & \times & \times \\ \times & \times & \times & \times & \times & \times \\ 0 & 0 & 0 & \times & \times & \times \\ 0 & 0 & 0 & 0 & \times & \times \\ 0 & 0 & 0 & 0 & 0 & \times \end{bmatrix}.$$

The idea now is to restore these matrices to Hessenberg-triangular form by chasing the unwanted nonzero elements down the diagonal.

To this end, we first determine a pair of Householder matrices Z_1 and

Z_2 to zero b_{31}, b_{32}, and b_{21}:

$$A = AZ_1Z_2 = \begin{bmatrix} \times & \times & \times & \times & \times & \times \\ \times & \times & \times & \times & \times & \times \\ \times & \times & \times & \times & \times & \times \\ \times & \times & \times & \times & \times & \times \\ 0 & 0 & 0 & \times & \times & \times \\ 0 & 0 & 0 & 0 & \times & \times \end{bmatrix}$$

$$B = BZ_1Z_2 = \begin{bmatrix} \times & \times & \times & \times & \times & \times \\ 0 & \times & \times & \times & \times & \times \\ 0 & 0 & \times & \times & \times & \times \\ 0 & 0 & 0 & \times & \times & \times \\ 0 & 0 & 0 & 0 & \times & \times \\ 0 & 0 & 0 & 0 & 0 & \times \end{bmatrix}$$

Then a Householder matrix P_1 is used to zero a_{31} and a_{41}:

$$A = P_1A = \begin{bmatrix} \times & \times & \times & \times & \times & \times \\ \times & \times & \times & \times & \times & \times \\ 0 & \times & \times & \times & \times & \times \\ 0 & \times & \times & \times & \times & \times \\ 0 & 0 & 0 & \times & \times & \times \\ 0 & 0 & 0 & 0 & \times & \times \end{bmatrix}$$

$$B = P_1B = \begin{bmatrix} \times & \times & \times & \times & \times & \times \\ 0 & \times & \times & \times & \times & \times \\ 0 & \times & \times & \times & \times & \times \\ 0 & \times & \times & \times & \times & \times \\ 0 & 0 & 0 & 0 & \times & \times \\ 0 & 0 & 0 & 0 & 0 & \times \end{bmatrix}$$

Notice that with this step the unwanted nonzero elements have been shifted down and to the right from their original position. This illustrates a typical step in the QZ iteration. Notice that $Q = Q_0Q_1 \cdots Q_{n-2}$ has the same first column as Q_0. By the way the initial Householder matrix was determined, we can apply the implicit Q theorem and assert that $AB^{-1} = Q^T(AB^{-1})Q$ is indeed essentially the same matrix that we would obtain by applying the Francis iteration to $M = AB^{-1}$ directly. Overall we have:

Algorithm 7.7.2 (The QZ Step) Given an unreduced upper Hessenberg matrix $A \in \mathbb{R}^{n \times n}$ and a nonsingular upper triangular matrix $B \in \mathbb{R}^{n \times n}$, the following algorithm overwrites A with the upper Hessenberg matrix Q^TAZ and B with the upper triangular matrix Q^TBZ where Q and Z are orthogonal and Q has the same first column as the orthogonal similarity transformation in Algorithm 7.5.1 when it is applied to AB^{-1}.

Let $M = AB^{-1}$ and compute $(M - aI)(M - bI)e_1 = (x, y, z, 0, \ldots, 0)^T$
where a and b are the eigenvalues of M's lower 2-by-2.
for $k = 1{:}n - 2$
    Find Householder Q_k so $Q_k[\, x \; y \; z \,]^T = [\, * \; 0 \; 0 \,]^T$.
    $A = \text{diag}(I_{k-1}, Q_k, I_{n-k-2})A$
    $B = \text{diag}(I_{k-1}, Q_k, I_{n-k-2})B$
    Find Householder Z_{k1} so
$$\begin{bmatrix} b_{k+2,k} & b_{k+2,k+1} & b_{k+2,k+2} \end{bmatrix} Z_{k1} = \begin{bmatrix} 0 & 0 & * \end{bmatrix}.$$
    $A = A\text{diag}(I_{k-1}, Z_{k1}, I_{n-k-2})$
    $B = B\text{diag}(I_{k-1}, Z_{k1}, I_{n-k-2})$
    Find Householder Z_{k2} so
$$\begin{bmatrix} b_{k+1,k} & b_{k+1,k+1} \end{bmatrix} Z_{k2} = \begin{bmatrix} 0 & * \end{bmatrix}.$$
    $A = A\text{diag}(I_{k-1}, Z_{k2}, I_{n-k-1})$
    $B = B\text{diag}(I_{k-1}, Z_{k2}, I_{n-k-1})$
    $x = a_{k+1,k}; \; y = a_{k+1,k}$
    if $k < n - 2$
        $z = a_{k+3,k}$
    end
end
Find Householder Q_{n-1} so $Q_{n-1}\begin{bmatrix} x \\ y \end{bmatrix} = \begin{bmatrix} * \\ 0 \end{bmatrix}$
$A = \text{diag}(I_{n-2}, Q_{n-1})A$
$B = \text{diag}(I_{n-2}, Q_{n-1})B$
Find Householder Z_{n-1} so
$$\begin{bmatrix} b_{n,n-1} & b_{nn} \end{bmatrix} Z_{n-1} = \begin{bmatrix} 0 & * \end{bmatrix}$$
$A = A\text{diag}(I_{n-2}, Z_{n-1})$
$B = B\text{diag}(I_{n-2}, Z_{n-1})$

This algorithm requires $22n^2$ flops. Q and Z can be accumulated for an additional $8n^2$ flops and $13n^2$ flops, respectively.

7.7.7   The Overall QZ Process

By applying a sequence of QZ steps to the Hessenberg-triangular pencil $A - \lambda B$, it is possible to reduce A to quasi-triangular form. In doing this it is necessary to monitor A's subdiagonal and B's diagonal in order to bring about decoupling whenever possible. The complete process, due to Moler and Stewart (1973), is as follows:

Algorithm 7.7.3 Given $A \in \mathbb{R}^{n \times n}$ and $B \in \mathbb{R}^{n \times n}$, the following algorithm computes orthogonal Q and Z such that $Q^T AZ = T$ is upper quasi-triangular and $Q^T BZ = S$ is upper triangular. A is overwritten by T and B by S.

Using Algorithm 7.7.1, overwrite A with $Q^T A Z$ (upper Hessenberg) and B with $Q^T B Z$ (upper triangular).

until $q = n$

Set all subdiagonal elements in A to zero that satisfy
$$|a_{i,i-1}| \le \epsilon(|a_{i-1,i-1}| + |a_{ii}|)$$
Find the largest nonnegative q and the smallest nonnegative p such that if

$$A = \begin{bmatrix} A_{11} & A_{12} & A_{13} \\ 0 & A_{22} & A_{23} \\ 0 & 0 & A_{33} \end{bmatrix} \begin{matrix} p \\ n-p-q \\ q \end{matrix}$$
$$\begin{matrix} p & n-p-q & q \end{matrix}$$

then A_{33} is upper quasi-triangular and A_{22} is unreduced upper Hessenberg.

Partition B conformably:

$$B = \begin{bmatrix} B_{11} & B_{12} & B_{13} \\ 0 & B_{22} & B_{23} \\ 0 & 0 & B_{33} \end{bmatrix} \begin{matrix} p \\ n-p-q \\ q \end{matrix}$$
$$\begin{matrix} p & n-p-q & q \end{matrix}$$

if $q < n$

  if B_{22} is singular

    Zero $a_{n-q,n-q-1}$

  else

    Apply Algorithm 7.7.2 to A_{22} and B_{22}
$$A = \text{diag}(I_p, Q, I_q)^T A \text{diag}(I_p, Q, I_q)$$
$$B = \text{diag}(I_p, Q, I_q)^T B \text{diag}(I_p, Q, I_q)$$

  end

end

end

This algorithm requires $30n^3$ flops. If Q is desired, an additional $16n^3$ are necessary. If Z is required, an additional $20n^3$ are needed. These estimates of work are based on the experience that about two QZ iterations per eigenvalue are necessary. Thus, the convergence properties of QZ are the same as for QR. The speed of the QZ algorithm is not affected by rank deficiency in B.

The computed S and T can be shown to satisfy

$$Q_0^T(A + E)Z_0 = T \qquad Q_0^T(B + F)Z_0 = S$$

where Q_0 and Z_0 are exactly orthogonal and $\| E \|_2 \approx u\| A \|_2$ and $\| F \|_2 \approx u\| B \|_2$.

Example 7.7.5 If the QZ algorithm is applied to

$$
A = \begin{bmatrix} 2 & 3 & 4 & 5 & 6 \\ 4 & 4 & 5 & 6 & 7 \\ 0 & 3 & 6 & 7 & 8 \\ 0 & 0 & 2 & 8 & 9 \\ 0 & 0 & 0 & 1 & 10 \end{bmatrix} \quad \text{and} \quad B = \begin{bmatrix} 1 & -1 & -1 & -1 & -1 \\ 0 & 1 & -1 & -1 & -1 \\ 0 & 0 & 1 & -1 & -1 \\ 0 & 0 & 0 & 1 & -1 \\ 0 & 0 & 0 & 0 & 1 \end{bmatrix}
$$

then the subdiagonal elements of A converge as follows

Iteration	$O(|h_{21}|)$	$O(|h_{32}|)$	$O(|h_{43}|)$	$O(|h_{54}|)$
1	10_0	10^1	10^0	10^{-1}
2	10^0	10^0	10^0	10^{-1}
3	10^0	10^1	10^{-1}	10^{-3}
4	10^0	10^0	10^{-1}	10^{-8}
5	10^0	10^1	10^{-1}	10^{-16}
6	10^0	10^0	10^{-2}	converg.
7	10^0	10^{-1}	10^{-4}	
8	10^1	10^{-1}	10^{-8}	
9	10^0	10^{-1}	10^{-19}	
10	10^0	10^{-2}	converg.	
11	10^{-1}	10^{-4}		
12	10^{-2}	10^{-11}		
13	10^{-3}	10^{-27}		
14	converg.	converg.		

7.7.8   Generalized Invariant Subspace Computations

Many of the invariant subspace computations discussed in §7.6 carry over to the generalized eigenvalue problem. For example, approximate eigenvectors can be found via inverse iteration:

$q^{(0)} \in \mathbb{C}^{n \times n}$ given.
for $k = 1, 2, \ldots$
    Solve $(A - \mu B)z^{(k)} = Bq^{(k-1)}$
    Normalize: $q^{(k)} = z^{(k)}/\| z^{(k)} \|_2$
    $\lambda^{(k)} = [q^{(k)}]^H A q^{(k)} / [q^{(k)}]^H A q^{(k)}$
end

When B is nonsingular, this is equivalent to applying (7.6.1) with the matrix $B^{-1}A$. Typically, only a single iteration is required if μ is an approximate eigenvalue computed by the QZ algorithm. By inverse iterating with the Hessenberg-triangular pencil, costly accumulation of the Z-transformations during the QZ iteration can be avoided.

Corresponding to the notion of an invariant subspace for a single matrix, we have the notion of a *deflating* subspace for the pencil $A - \lambda B$. In particular, we say that a k-dimensional subspace $S \subseteq \mathbb{R}^n$ is "deflating" for the pencil $A - \lambda B$ if the subspace $\{ Ax + By : x, y \in S \}$ has dimension k or less. Note that the columns of the matrix Z in the generalized Schur decomposition define a family of deflating subspaces, for if $Q = [\, q_1, \ldots, q_n \,]$ and $Z = [\, z_1, \ldots, z_n \,]$ then we have $\mathrm{span}\{Az_1, \ldots, Az_k\} \subseteq \mathrm{span}\{q_1, \ldots, q_k\}$ and $\mathrm{span}\{Bz_1, \ldots, Bz_k\} \subseteq \mathrm{span}\{q_1, \ldots, q_k\}$. Properties of deflating subspaces and their behavior under perturbation are described in Stewart (1972).

Problems

P7.7.1 Suppose A and B are in $\mathbb{R}^{n \times n}$ and that

$$U^T B V = \begin{bmatrix} D & 0 \\ 0 & 0 \end{bmatrix} \begin{matrix} r \\ n-r \end{matrix} \qquad U = [\, U_1 \quad U_2 \,] \qquad V = [\, V_1 \quad V_2 \,]$$
$$\begin{matrix} r & n-r \end{matrix} \qquad\qquad \begin{matrix} r & n-r \end{matrix} \qquad\quad \begin{matrix} r & n-r \end{matrix}$$

is the SVD of B, where D is r-by-r and $r = \mathrm{rank}(B)$. Show that if $\lambda(A, B) = \mathbb{C}$ then $U_2^T A V_2$ is singular.

P7.7.2 Define F : $\mathbb{R}^n \to \mathbb{R}$ by

$$F(x) = \frac{1}{2} \left\| Ax - \frac{x^T B^T Ax}{x^T B^T Bx} Bx \right\|_2^2$$

where A and B are in $\mathbb{R}^{m \times n}$. Show that if $\nabla F(x) = 0$, then Ax is a multiple of Bx.

P7.7.3 Suppose A and B are in $\mathbb{R}^{n \times n}$. Give an algorithm for computing orthogonal Q and Z such that $Q^T A Z$ is upper Hessenberg and $Z^T B Q$ is upper triangular.

P7.7.4 Suppose

$$A = \begin{bmatrix} A_{11} & A_{12} \\ 0 & A_{22} \end{bmatrix} \quad \text{and} \quad B = \begin{bmatrix} B_{11} & B_{12} \\ 0 & B_{22} \end{bmatrix}$$

with $A_{11}, B_{11} \in \mathbb{R}^{k \times k}$ and $A_{22}, B_{22} \in \mathbb{R}^{j \times j}$. Under what circumstances do there exist

$$X = \begin{bmatrix} I_k & X_{12} \\ 0 & I_j \end{bmatrix} \quad \text{and} \quad Y = \begin{bmatrix} I_k & Y_{12} \\ 0 & I_j \end{bmatrix}$$

so that $Y^{-1} A X$ and $Y^{-1} B X$ are both block diagonal? This is the *generalized Sylvester equation problem*. Specify an algorithm for the case when A_{11}, A_{22}, B_{11}, and B_{22} are upper triangular. See Kågström (1994).

P7.7.5 Suppose $\mu \notin \lambda(A, B)$. Relate the eigenvalues and eigenvectors of $A_1 = (A - \mu B)^{-1} A$ and $B_1 = (A - \mu B)^{-1} B$ to the generalized eigenvalues and eigenvectors of $A - \lambda B$.

P7.7.6 Suppose $A, B, C, D \in \mathbb{R}^{n \times n}$. Show how to compute orthogonal matrices Q, Z, U, and V such that $Q^T A U$ is upper Hessenberg and $V^T C Z$, $Q^T B V$, and $V^T D Z$ are all upper triangular. Note that this converts the pencil $AC - \lambda BD$ to Hessenberg-triangular form. Your algorithm should not form the products AC or BD explicitly and not should not compute any matrix inverse. See Van Loan (1975).

Notes and References for Sec. 7.7

Mathematical aspects of the generalized eigenvalue problem are covered in

F. Gantmacher (1959). *The Theory of Matrices* , vol. 2, Chelsea, New York.

H.W. Turnbill and A.C. Aitken (1961). *An Introduction to the Theory of Canonical Matrices*, Dover, New York.

I. Erdelyi (1967). "On the Matrix Equation $Ax = \lambda Bx$," *J. Math. Anal. and Applic.* *17*, 119–32.

A good general volume that covers many aspects of the $A - \lambda B$ problem is

B. Kågström and A. Ruhe (1983). *Matrix Pencils*, Proc. Pite Havsbad, 1982, Lecture Notes in Mathematics 973, Springer-Verlag, New York and Berlin.

The perturbation theory for the generalized eigenvalue problem is treated in

G.W. Stewart (1972). "On the Sensitivity of the Eigenvalue Problem $Ax = \lambda Bx$," *SIAM J. Num. Anal. 9*, 669–86.

G.W. Stewart (1973). "Error and Perturbation Bounds for Subspaces Associated with Certain Eigenvalue Problems," *SIAM Review 15*, 727–64.

G.W. Stewart (1975). "Gershgorin Theory for the Generalized Eigenvalue Problem $Ax = \lambda Bx$," *Math. Comp. 29*, 600–606.

G.W. Stewart (1978). "Perturbation Theory for the Generalized Eigenvalue Problem'", in *Recent Advances in Numerical Analysis* , ed. C. de Boor and G.H. Golub, Academic Press, New York.

A. Pokrzywa (1986). "On Perturbations and the Equivalence Orbit of a Matrix Pencil," *Lin. Alg. and Applic. 82*, 99–121.

The QZ-related papers include

C.B. Moler and G.W. Stewart (1973). "An Algorithm for Generalized Matrix Eigenvalue Problems," *SIAM J. Num. Anal. 10*, 241–56.

L. Kaufman (1974). "The LZ Algorithm to Solve the Generalized Eigenvalue Problem," *SIAM J. Num. Anal. 11*, 997–1024.

R.C. Ward (1975). "The Combination Shift QZ Algorithm," *SIAM J. Num. Anal. 12*, 835–853.

C.F. Van Loan (1975). "A General Matrix Eigenvalue Algorithm," *SIAM J. Num. Anal. 12*, 819–834.

L. Kaufman (1977). "Some Thoughts on the QZ Algorithm for Solving the Generalized Eigenvalue Problem," *ACM Trans. Math. Soft. 3*, 65–75.

R.C. Ward (1981). "Balancing the Generalized Eigenvalue Problem," *SIAM J. Sci and Stat. Comp. 2*, 141–152.

P. Van Dooren (1982). "Algorithm 590: DSUBSP and EXCHQZ: Fortran Routines for Computing Deflating Subspaces with Specified Spectrum," *ACM Trans. Math. Software 8*, 376–382.

D. Watkins and L. Elsner (1994). "Theory of Decomposition and Bulge-Chasing Algorithms for the Generalized Eigenvalue Problem," *SIAM J. Matrix Anal. Appl. 15*, 943–967.

Just as the Hessenberg decomposition is important in its own right, so is the Hessenberg-triangular decomposition that serves as a QZ front end. See

W. Enright and S. Serbin (1978). "A Note on the Efficient Solution of Matrix Pencil Systems," *BIT 18*, 276–81.

Other solution frameworks are proposed in

V.N. Kublanovskaja and V.N. Fadeeva (1964). "Computational Methods for the Solution of a Generalized Eigenvalue Problem," *Amer. Math. Soc. Transl. 2*, 271–90.

G. Peters and J.H. Wilkinson (1970a). "$Ax = \lambda Bx$ and the Generalized Eigenproblem," *SIAM J. Num. Anal. 7*, 479–92.

G. Rodrigue (1973). "A Gradient Method for the Matrix Eigenvalue Problem $Ax = \lambda Bx$," *Numer. Math. 22*, 1–16.

H.R. Schwartz (1974). "The Method of Coordinate Relaxation for $(A - \lambda B)x = 0$," *Num. Math. 23*, 135–52.

A. Jennings and M.R. Osborne (1977). "Generalized Eigenvalue Problems for Certain Unsymmetric Band Matrices," *Lin. Alg. and Its Applic. 29*, 139–50.

V.N. Kublanovskaya (1984). "AB Algorithm and Its Modifications for the Spectral Problem of Linear Pencils of Matrices," *Numer. Math. 43*, 329–342.

C. Oara (1994). "Proper Deflating Subspaces: Properties, Algorithms, and Applications," *Numerical Algorithms 7*, 355–373.

The general $Ax = \lambda Bx$ problem is central to some important control theory applications. See

P. Van Dooren (1981). "A Generalized Eigenvalue Approach for Solving Riccati Equations," *SIAM J. Sci. and Stat. Comp. 2*, 121–135.

P. Van Dooren (1981). "The Generalized Eigenstructure Problem in Linear System Theory," *IEEE Trans. Auto. Cont. AC-26*, 111–128.

W.F. Arnold and A.J. Laub (1984). "Generalized Eigenproblem Algorithms and Software for Algebraic Riccati Equations," *Proc. IEEE 72*, 1746–1754.

J.W. Demmel and B. Kågström (1988). "Accurate Solutions of Ill-Posed Problems in Control Theory," *SIAM J. Matrix Anal. Appl.* 126–145.

U. Flaschka, W-W. Li, and J-L. Wu (1992). "A KQZ Algorithm for Solving Linear-Response Eigenvalue Equations," *Lin. Alg. and Its Applic. 165*, 93–123.

Rectangular generalized eigenvalue problems arise in certain applications. See

G.L. Thompson and R.L. Weil (1970). "Reducing the Rank of $A - \lambda B$," *Proc. Amer. Math. Sec. 26*, 548–54.

G.L. Thompson and R.L. Weil (1972). "Roots of Matrix Pencils $Ay = \lambda By$: Existence, Calculations, and Relations to Game Theory," *Lin. Alg. and Its Applic. 5*, 207–26.

G.W. Stewart (1994). "Perturbation Theory for Rectangular Matrix Pencils," *Lin. Alg. and Applic. 208/209*, 297–301.

The *Kronecker Structure* of the pencil $A - \lambda B$ is analogous to Jordan structure of $A - \lambda I$: it provides very useful information about the underlying application.

J.H. Wilkinson (1978). "Linear Differential Equations and Kronecker's Canonical Form," in *Recent Advances in Numerical Analysis*, ed. C. de Boor and G.H. Golub, Academic Press, New York, pp. 231–65.

Interest in the Kronecker structure has led to a host of new algorithms and analyses.

J.H. Wilkinson (1979). "Kronecker's Canonical Form and the QZ Algorithm," *Lin. Alg. and Its Applic. 28*, 285–303.

P. Van Dooren (1979). "The Computation of Kronecker's Canonical Form of a Singular Pencil," *Lin. Alg. and Its Applic. 27*, 103–40.

J.W. Demmel (1983). "The Condition Number of Equivalence Transformations that Block Diagonalize Matrix Pencils," *SIAM J. Numer. Anal. 20*, 599–610.

J.W. Demmel and B. Kågström (1987). "Computing Stable Eigendecompositions of Matrix Pencils," *Linear Alg. and Its Applic 88/89*, 139–186.

B. Kågström (1985). "The Generalized Singular Value Decomposition and the General $A - \lambda B$ Problem," *BIT 24*, 568–583.

B. Kågström (1986). "RGSVD: An Algorithm for Computing the Kronecker Structure and Reducing Subspaces of Singular $A - \lambda B$ Pencils," *SIAM J. Sci. and Stat. Comp. 7*, 185–211.

J. Demmel and B. Kågström (1986). "Stably Computing the Kronecker Structure and Reducing Subspaces of Singular Pencils $A - \lambda B$ for Uncertain Data," in *Large Scale*

Eigenvalue Problems, J. Cullum and R.A. Willoughby (eds), North-Holland, Amsterdam.

T. Beelen and P. Van Dooren (1988). "An Improved Algorithm for the Computation of Kronecker's Canonical Form of a Singular Pencil," *Lin. Alg. and Its Applic. 105*, 9–65.

B. Kågström and L. Westin (1989). "Generalized Schur Methods with Condition Estimators for Solving the Generalized Sylvester Equation," *IEEE Trans. Auto. Cont. AC-34*, 745–751.

B. Kågström and P. Poromaa (1992). "Distributed and Shared Memory Block Algorithms for the Triangular Sylvester Equation with sep^{-1} Estimators," *SIAM J. Matrix Anal. Appl. 13*, 90–101.

B. Kågström (1994). "A Perturbation Analysis of the Generalized Sylvester Equation $(AR - LB, DR - LE) = (C, F)$," *SIAM J. Matrix Anal. Appl. 15*, 1045–1060.

E. Elmroth and B. Kågström (1996). "The Set of 2-by-3 Matrix Pencils–Kronecker Structure and their Transitions under Perturbations," *SIAM J. Matrix Anal.*, to appear.

A. Edelman, E. Elmroth, and B. Kågström (1996). "A Geometric Approach to Perturbation Theory of Matrices and Matrix Pencils," *SIAM J. Matrix Anal.*, to appear.

Chapter 8

The Symmetric Eigenvalue Problem

The symmetric eigenvalue problem with its rich mathematical structure is one of the most aesthetically pleasing problems in numerical linear algebra. We begin our presentation with a brief discussion of the mathematical properties that underlie this computation. In §8.2 and §8.3 we develop various power iterations eventually focusing on the symmetric QR algorithm.

In §8.4 we discuss Jacobi's method, one of the earliest matrix algorithms to appear in the literature. This technique is of current interest because it is amenable to parallel computation and because under certain circumstances it has superior accuracy.

Various methods for the tridiagonal case are presented in §8.5. These include the method of bisection and a divide and conquer technique.

The computation of the singular value decomposition is detailed in §8.6. The central algorithm is a variant of the symmetric QR iteration that works on bidiagonal matrices.

In the final section we discuss the generalized eigenvalue problem $Ax = \lambda Bx$ for the important case when A is symmetric and B is symmetric positive definite. No suitable analog of the orthogonally-based QZ algorithm (see §7.7) exists for this specially structured, generalized eigenproblem. However, there are several successful methods that can be applied and these are presented along with a discussion of the generalized singular value decomposition.

Before You Begin

Chapter 1, §§2.1-2.5, and §2.7, Chapter 3, §§4.1-4.3, §§5.1-5.5 and §7.1.1 are assumed. Within this chapter there are the following dependencies:

$$\begin{array}{ccccccc}
§8.4 & & & & & & \\
\uparrow & & & & & & \\
§8.1 & \rightarrow & §8.2 & \rightarrow & §8.3 & \rightarrow & §8.6 & \rightarrow & §8.7 \\
\downarrow & & & & & & \\
§8.5 & & & & & &
\end{array}$$

Many of the algorithms and theorems in this chapter have unsymmetric counterparts in Chapter 7. However, except for a few concepts and definitions, our treatment of the symmetric eigenproblem can be studied before reading Chapter 7.

Complementary references include Wilkinson (1965), Stewart (1973), Gourlay and Watson (1973), Hager (1988), Chatelin (1993), Parlett (1980), Stewart and Sun (1990), Watkins (1991), Jennings and McKeowen (1992), and Datta (1995). Some Matlab functions important to this chapter are schur and svd. LAPACK connections include

LAPACK: Symmetric Eigenproblem	
_SYEV	All eigenvalues and vectors
_SYEVD	Same but uses divide and conquer for eigenvectors
_SYEVX	Selected eigenvalues and vectors
_SYTRD	Householder tridiagonalization
_SBTRD	Householder tridiagonalization (A banded)
_SPTRD	Householder tridiagonalization (A in packed storage)
_STEQR	All eigenvalues and vectors of tridiagonal by implicit QR
_STEDC	All eigenvalues and vectors of tridiagonal by divide and conquer
_STERF	All eigenvalues of tridiagonal by root-free QR
_PTEQR	All eigenvalues and eigenvectors of positive definite tridiagonal
_STEBZ	Selected eigenvalues of tridiagonal by bisection
_STEIN	Selected eigenvectors of tridiagonal by inverse iteration

LAPACK: Symmetric-Definite Eigenproblems	
_SYGST	Converts $A - \lambda B$ to $C - \lambda I$ form
_PBSTF	Split Cholesky factorization
_SBGST	Converts banded $A - \lambda B$ to $C - \lambda I$ form via split Cholesky

LAPACK: SVD	
_GESVD	$A = U\Sigma V^T$
_BDSQR	SVD of real bidiagonal matrix
_GEBRD	bidiagonalization of general matrix
_ORGBR	generates the orthogonal transformations
_GBBRD	bidiagonalization of band matrix

LAPACK: The Generalized Singular Value Problem	
_GGSVP	Converts $A^T A - \mu^2 B^T B$ to triangular $A_1^T A_1 - \mu^2 B_1^T B_1$
_TGSJA	Computes GSVD of a pair of triangular matrices.

8.1 Properties and Decompositions

In this section we set down the mathematics that is required to develop and analyze algorithms for the symmetric eigenvalue problem.

8.1.1 Eigenvalues and Eigenvectors

Symmetry guarantees that all of A's eigenvalues are real and that there is an orthonormal basis of eigenvectors.

Theorem 8.1.1 (Symmetric Schur Decomposition) *If $A \in \mathbb{R}^{n \times n}$ is symmetric, then there exists a real orthogonal Q such that*

$$Q^T A Q = \Lambda = \text{diag}(\lambda_1, \ldots, \lambda_n).$$

Moreover, for $k = 1{:}n$, $AQ(:,k) = \lambda_k Q(:,k)$. See Theorem 7.1.3.

Proof. Suppose $\lambda_1 \in \lambda(A)$ and that $x \in \mathbb{C}^n$ is a unit 2-norm eigenvector with $Ax = \lambda_1 x$. Since $\lambda_1 = x^H A x = x^H A^H x = \overline{x^H A x} = \overline{\lambda_1}$ it follows that $\lambda_1 \in \mathbb{R}$. Thus, we may assume that $x \in \mathbb{R}^n$. Let $P_1 \in \mathbb{R}^{n \times n}$ be a Householder matrix such that $P_1^T x = e_1 = I_n(:,1)$. It follows from $Ax = \lambda_1 x$ that $(P_1^T A P_1)e_1 = \lambda e_1$. This says that the first column of $P_1^T A P_1$ is a multiple of e_1. But since $P_1^T A P_1$ is symmetric it must have the form

$$P_1^T A P_1 = \begin{bmatrix} \lambda_1 & 0 \\ 0 & A_1 \end{bmatrix}$$

where $A_1 \in \mathbb{R}^{(n-1) \times (n-1)}$ is symmetric. By induction we may assume that there is an orthogonal $Q_1 \in \mathbb{R}^{(n-1) \times (n-1)}$ such that $Q_1^T A_1 Q_1 = \Lambda_1$ is diagonal. The theorem follows by setting

$$Q = P_1 \begin{bmatrix} 1 & 0 \\ 0 & Q_1 \end{bmatrix} \quad \text{and} \quad \Lambda = \begin{bmatrix} \lambda_1 & 0 \\ 0 & \Lambda_1 \end{bmatrix}$$

and comparing columns in the matrix equation $AQ = Q\Lambda$. $\square$

Example 8.1.1 If

$$A = \begin{bmatrix} 6.8 & 2.4 \\ 2.4 & 8.2 \end{bmatrix} \quad \text{and} \quad Q = \begin{bmatrix} .6 & -.8 \\ .8 & .6 \end{bmatrix},$$

then Q is orthogonal and $Q^T A Q = \text{diag}(10,5)$.

For a symmetric matrix A we shall use the notation $\lambda_k(A)$ to designate the kth largest eigenvalue. Thus,

$$\lambda_n(A) \leq \cdots \leq \lambda_2(A) \leq \lambda_1(A).$$

It follows from the orthogonal invariance of the 2-norm that A has singular values $\{|\lambda_1(A)|, \ldots, |\lambda_n(A)|\}$ and so

$$\| A \|_2 \; = \; \max\{\, |\lambda_1(A)| \,,\; |\lambda_n(A)| \,\}.$$

The eigenvalues of a symmetric matrix have a "minimax" characterization based on the values that can be assumed by the quadratic form ratio $x^T A x / x^T x$.

Theorem 8.1.2 (Courant-Fischer Minimax Theorem) *If* $A \in \mathbb{R}^{n \times n}$ *is symmetric, then*

$$\lambda_k(A) \; = \; \max_{\dim(S)=k} \; \min_{0 \neq y \in S} \; \frac{y^T A y}{y^T y}$$

for $k = 1{:}n$.

Proof. Let $Q^T A Q = \text{diag}(\lambda_i)$ be the Schur decomposition with $\lambda_k = \lambda_k(A)$ and $Q = [\, q_1, \, q_2, \ldots, \, q_n \,]$. Define

$$S_k = \text{span}\{q_1, \ldots, q_k\},$$

the invariant subspace associated with $\lambda_1, \ldots, \lambda_k$. It is easy to show that

$$\max_{\dim(S)=k} \; \min_{0 \neq y \in S} \; \frac{y^T A y}{y^T y} \; \geq \; \min_{0 \neq y \in S_k} \; \frac{y^T A y}{y^T y} \; = \; q_k^T A q_k \; = \; \lambda_k(A).$$

To establish the reverse inequality, let S be any k-dimensional subspace and note that it must intersect $\text{span}\{q_k, \ldots, q_n\}$, a subspace that has dimension $n - k + 1$. If $y_* = \alpha_k q_k + \cdots + \alpha_n q_n$ is in this intersection, then

$$\min_{0 \neq y \in S} \; \frac{y^T A y}{y^T y} \; \leq \; \frac{y_*^T A y_*}{y_*^T y_*} \; \leq \; \lambda_k(A).$$

Since this inequality holds for all k-dimensional subspaces,

$$\max_{\dim(S)=k} \; \min_{0 \neq y \in S} \; \frac{y^T A y}{y^T y} \; \leq \; \lambda_k(A)$$

thereby completing the proof of the theorem. $\square$

If $A \in \mathbb{R}^{n \times n}$ is symmetric positive definite, then $\lambda_n(A) > 0$.

8.1.2 Eigenvalue Sensitivity

An important solution framework for the symmetric eigenproblem involves the production of a sequence of orthogonal transformations $\{Q_k\}$ with the property that the matrices $Q_k^T A Q_k$ are progressively "more diagonal." The question naturally arises, how well do the diagonal elements of a matrix approximate its eigenvalues?

Theorem 8.1.3 (Gershgorin) *Suppose $A \in \mathbb{R}^{n \times n}$ is symmetric and that $Q \in \mathbb{R}^{n \times n}$ is orthogonal. If $Q^T A Q = D + F$ where $D = \text{diag}(d_1, \ldots, d_n)$ and F has zero diagonal entries, then*

$$\lambda(A) \subseteq \bigcup_{i=1}^{n} [d_i - r_i, d_i + r_i]$$

where $r_i = \sum_{j=1}^{n} |f_{ij}|$ for $i = 1{:}n$. See Theorem 7.2.1.

Proof. Suppose $\lambda \in \lambda(A)$ and assume without loss of generality that $\lambda \neq d_i$ for $i = 1{:}n$. Since $(D - \lambda I) + F$ is singular, it follows from Lemma 2.3.3 that

$$1 \leq \| (D - \lambda I)^{-1} F \|_\infty = \sum_{j=1}^{n} \frac{|f_{kj}|}{|d_k - \lambda|} = \frac{r_k}{|d_k - \lambda|}$$

for some k, $1 \leq k \leq n$. But this implies that $\lambda \in [d_k - r_k, d_k + r_k]$. $\square$

Example 8.1.2 The matrix

$$A = \begin{bmatrix} 2.0000 & 0.1000 & 0.2000 \\ 0.2000 & 5.0000 & 0.3000 \\ 0.1000 & 0.3000 & -1.0000 \end{bmatrix}$$

has Gerschgorin intervals $[1.7, 2.3]$, $[4.5, 5.5]$, and $[-1.4, -.6]$ and eigenvalues 1.9984, 5.0224, and -1.0208.

The next results show that if A is perturbed by a symmetric matrix E, then its eigenvalues do not move by more than $\| E \|$.

Theorem 8.1.4 (Wielandt-Hoffman) *If A and $A + E$ are n-by-n symmetric matrices, then*

$$\sum_{i=1}^{n} (\lambda_i(A + E) - \lambda_i(A))^2 \leq \| E \|_F^2 .$$

Proof. A proof can be found in Wilkinson (1965, pp.104–8) or Stewart and Sun (1991, pp.189–191). See also P8.1.5. $\square$

Example 8.1.3 If

$$A = \begin{bmatrix} 6.8 & 2.4 \\ 2.4 & 8.2 \end{bmatrix} \quad \text{and} \quad E = \begin{bmatrix} .002 & .003 \\ .003 & .001 \end{bmatrix},$$

then $\lambda(A) = \{5, 10\}$ and $\lambda(A + E) = \{4.9988, 10.004\}$ confirming that

$$1.95 \times 10^{-5} = |4.9988 - 5|^2 + |10.004 - 10|^2 \leq \| E \|_F^2 = 2.3 \times 10^{-5}.$$

Theorem 8.1.5 *If A and $A + E$ are n-by-n symmetric matrices, then*

$$\lambda_k(A) + \lambda_n(E) \leq \lambda_k(A + E) \leq \lambda_k(A) + \lambda_1(E) \qquad k = 1{:}n.$$

Proof. This follows from the minimax characterization. See Wilkinson (1965, pp.101–2) or Stewart and Sun (1990, p.203). $\square$

Example 8.1.4 If

$$A = \begin{bmatrix} 6.8 & 2.4 \\ 2.4 & 8.2 \end{bmatrix} \quad \text{and} \quad E = \begin{bmatrix} .002 & .003 \\ .003 & .001 \end{bmatrix},$$

then $\lambda(A) = \{5, 10\}$, $\lambda(E) = \{-.0015, .0045\}$, and $\lambda(A + E) = \{4.9988, 10.0042\}$. confirming that

$$\begin{aligned} 5 - .0015 & \leq 4.9988 \leq & 5 + .0045 \\ 10 - .0015 & \leq 10.0042 \leq & 10 + .0045. \end{aligned}$$

Corollary 8.1.6 *If A and $A + E$ are n-by-n symmetric matrices, then*

$$|\lambda_k(A + E) - \lambda_k(A)| \leq \| E \|_2$$

for $k = 1{:}n$.

Proof.

$$|\lambda_k(A + E) - \lambda_k(A)| \leq \max\{|\lambda_n(E)|, |\lambda_1(E)\|\} = \| E \|_2. \square$$

Several more useful perturbation results follow from the minimax property.

Theorem 8.1.7 (Interlacing Property) *If $A \in \mathbb{R}^{n \times n}$ is symmetric and $A_r = A(1{:}r, 1{:}r)$, then*

$$\lambda_{r+1}(A_{r+1}) \leq \lambda_r(A_r) \leq \lambda_r(A_{r+1}) \leq \cdots \leq \lambda_2(A_{r+1}) \leq \lambda_1(A_r) \leq \lambda_1(A_{r+1})$$

for $r = 1{:}n - 1$.

Proof. Wilkinson (1965, pp.103–4). □

Example 8.1.5 If

$$
A = \begin{bmatrix} 1 & 1 & 1 & 1 \\ 1 & 2 & 3 & 4 \\ 1 & 3 & 6 & 10 \\ 1 & 4 & 10 & 20 \end{bmatrix}
$$

then $\lambda(A_1) = \{1\}$, $\lambda(A_2) = \{.3820,\ 2.6180\}$, $\lambda(A_3) = \{.1270,\ 1.0000,\ 7.873\}$, and $\lambda(A_4) = \{.0380,\ .4538,\ 2.2034, 26.3047\}$.

Theorem 8.1.8 *Suppose* $B = A + \tau cc^T$ *where* $A \in \mathbb{R}^{n \times n}$ *is symmetric,* $c \in \mathbb{R}^n$ *has unit 2-norm and* $\tau \in \mathbb{R}$. *If* $\tau \geq 0$, *then*

$$
\lambda_i(B) \in [\lambda_i(A),\ \lambda_{i-1}(A)] \qquad i = 2{:}n
$$

while if $\tau \leq 0$ *then*

$$
\lambda_i(B) \in [\lambda_{i+1}(A), \lambda_i(A)], \qquad i = 1{:}n - 1\,.
$$

In either case, there exist nonnegative $m_1, \ldots, m_n$ *such that*

$$
\lambda_i(B) = \lambda_i(A) + m_i\tau, \qquad i = 1{:}n
$$

with $m_1 + \cdots + m_n = 1$.

Proof. Wilkinson (1965, pp.94–97). See also P8.1.8. □

8.1.3 Invariant Subspaces

Many eigenvalue computations proceed by breaking the original problem into a collection of smaller subproblems. The following result is the basis for this solution framework.

Theorem 8.1.9 *Suppose* $A \in \mathbb{R}^{n \times n}$ *is symmetric and that*

$$
Q = \begin{array}{c} [\ Q_1 \quad Q_2\ ] \\ r \quad\ \ n-r \end{array}
$$

is orthogonal. If $\mathrm{ran}(Q_1)$ *is an invariant subspace, then*

$$
Q^T AQ = D = \begin{bmatrix} D_1 & 0 \\ 0 & D_2 \end{bmatrix} \begin{array}{c} r \\ n-r \end{array} \tag{8.1.1}
$$
$$
\quad\ \ r \quad\ n-r
$$

and $\lambda(A) = \lambda(D_1) \cup \lambda(D_2)$. *See also Lemma 7.1.2.*

Proof. If

$$Q^T A Q = \begin{bmatrix} D_1 & E_{21}^T \\ E_{21} & D_2 \end{bmatrix},$$

then from $AQ = QD$ we have $AQ_1 - Q_1 D_1 = Q_2 E_{21}$. Since $\operatorname{ran}(Q_1)$ is invariant, the columns of $Q_2 E_{21}$ are also in $\operatorname{ran}(Q_1)$ and therefore perpendicular to the columns of Q_2. Thus,

$$0 = Q_2^T (AQ_1 - Q_1 D_1) = Q_2^T Q_2 E_{21} = E_{21}.$$

and so (8.1.1) holds. It is easy to show

$$\det(A - \lambda I_n) = \det(Q^T A Q - \lambda I_n) = \det(D_1 - \lambda I_r)\det(D_2 - \lambda I_{n-r})$$

confirming that $\lambda(A) = \lambda(D_1) \cup \lambda(D_2)$. $\square$

The sensitivity to perturbation of an invariant subspace depends upon the separation of the associated eigenvalues from the rest of the spectrum. The appropriate measure of separation between the eigenvalues of two symmetric matrices B and C is given by

$$\operatorname{sep}(B, C) = \min_{\substack{\lambda \in \lambda(B) \\ \mu \in \lambda(C)}} |\lambda - \mu|. \tag{8.1.2}$$

With this definition we have

Theorem 8.1.10 *Suppose A and $A + E$ are n-by-n symmetric matrices and that*

$$Q = [\; Q_1 \quad Q_2 \;]$$
$$\quad\;\; r \quad\; n - r$$

is an orthogonal matrix such that $\operatorname{ran}(Q_1)$ is an invariant subspace for A. Partition the matrices $Q^T A Q$ and $Q^T E Q$ as follows:

$$Q^T A Q = \begin{bmatrix} D_1 & 0 \\ 0 & D_2 \end{bmatrix} \begin{matrix} r \\ n-r \end{matrix} \qquad Q^T E Q = \begin{bmatrix} E_{11} & E_{21}^T \\ E_{21} & E_{22} \end{bmatrix} \begin{matrix} r \\ n-r \end{matrix} \;.$$
$$\quad\; r \quad n-r \qquad\qquad\qquad\qquad r \quad n-r$$

If $\operatorname{sep}(D_1, D_2) > 0$ and

$$\| E \|_2 \le \frac{\operatorname{sep}(D_1, D_2)}{5},$$

then there exists a matrix $P \in \mathbb{R}^{(n-r) \times r}$ with

$$\| P \|_2 \le \frac{4}{\operatorname{sep}(D_1, D_2)} \| E_{21} \|_2$$

such that the columns of $\hat{Q}_1 = (Q_1 + Q_2 P)(I + P^T P)^{-1/2}$ define an orthonormal basis for a subspace that is invariant for $A + E$. See also Theorem 7.2.4.

Proof. This result is a slight adaptation of of Theorem 4.11 in Stewart (1973). The matrix $(I + P^T P)^{-1/2}$ is the inverse of the square root of $I + P^T P$. See §4.2.10. □

Corollary 8.1.11 *If the conditions of the theorem hold, then*

$$\text{dist}(\text{ran}(Q_1), \text{ran}(\hat{Q}_1)) \leq \frac{4}{\text{sep}(D_1, D_2)} \| E_{21} \|_2.$$

See also Corollary 7.2.5.

Proof. It can be shown using the SVD that

$$\| P(I + P^T P)^{-1/2} \|_2 \leq \| P \|_2. \tag{8.1.3}$$

Since $Q_2^T \hat{Q}_1 = P(I + P^H P)^{-1/2}$ it follows that

$$\text{dist}(\text{ran}(Q_1), \text{ran}(\hat{Q}_1)) = \| Q_2^T \hat{Q}_1 \|_2 = \| P(I + P^H P)^{-1/2} \|_2$$

$$\leq \| P \|_2 \leq \| E_{21} \|_2 / \text{sep}(D_1, D_2). \ \square$$

Thus, the reciprocal of $\text{sep}(D_1, D_2)$ can be thought of as a condition number that measures the sensitivity of $\text{ran}(Q_1)$ as an invariant subspace.

The effect of perturbations on a single eigenvector is sufficiently important that we specialize the above results to this important case.

Theorem 8.1.12 *Suppose A and $A + E$ are n-by-n symmetric matrices and that*

$$Q = \begin{array}{c} [\ q_1 \quad Q_2\ ] \\ 1 \quad n-1 \end{array}$$

is an orthogonal matrix such that q_1 is an eigenvector for A. Partition the matrices $Q^T AQ$ and $Q^T EQ$ as follows:

$$Q^T AQ = \begin{array}{c} \begin{bmatrix} \lambda & 0 \\ 0 & D_2 \end{bmatrix} \begin{array}{c} 1 \\ n-1 \end{array} \\ 1 \quad n-1 \end{array} \qquad Q^T EQ = \begin{array}{c} \begin{bmatrix} \epsilon & e^T \\ e & E_{22} \end{bmatrix} \begin{array}{c} 1 \\ n-1 \end{array} \\ 1 \quad n-1 \end{array}.$$

If $d = \min\limits_{\mu \in \lambda(D_2)} |\lambda - \mu| > 0$ and

$$\| E \|_2 \leq \frac{d}{4},$$

then there exists $p \in \mathbb{R}^{n-1}$ satisfying

$$\| p \|_2 \leq \frac{4}{d} \| e \|_2$$

such that $\hat{q}_1 = (q_1 + Q_2 p)/\sqrt{1 + p^T p}$ *is a unit 2-norm eigenvector for* $A + E$. *Moreover,*

$$\text{dist}(\text{span}\{q_1\}, \text{span}\{\hat{q}_1\}) = \sqrt{1 - (q_1^T \hat{q}_1)^2} \leq \frac{4}{d} \parallel e \parallel_2.$$

See also Corollary 7.2.6.

Proof. Apply Theorem 8.1.10 and Corollary 8.1.11 with $r = 1$ and observe that if $D_1 = (\lambda)$, then $d = \text{sep}(D_1, D_2)$. $\square$

Example 8.1.6 If $A = \text{diag}(.999, \ 1.001, \ 2.)$, and

$$E = \begin{bmatrix} 0.00 & 0.01 & 0.01 \\ 0.01 & 0.00 & 0.01 \\ 0.01 & 0.01 & 0.00 \end{bmatrix},$$

then $\hat{Q}^T (A + E)\hat{Q} = \text{diag}(.9899, \ 1.0098, \ 2.0002)$ where

$$\hat{Q} = \begin{bmatrix} -.7418 & .6706 & .0101 \\ .6708 & .7417 & .0101 \\ .0007 & -.0143 & .9999 \end{bmatrix}$$

is orthogonal. Let $\hat{q}_i = \hat{Q} e_i$, $i = 1, 2, 3$. Thus, $\hat{q}_i$ is the perturbation of A's eigenvector $q_i = e_i$. A calculation shows that

$$\text{dist}\{\text{span}\{q_1\}, \text{span}\{\hat{q}_1\}\} = \text{dist}\{\text{span}\{q_2\}, \text{span}\{\hat{q}_2\}\} = .67$$

Thus, because they are associated with nearby eigenvalues, the eigenvectors q_1 and q_2 cannot be computed accurately. On the other hand, since λ_1 and λ_2 are well separated from λ_3, they define a two-dimensional subspace that is not particularly sensitive as $\text{dist}\{\text{span}\{q_1, q_2\}, \text{span}\{\hat{q}_1, \hat{q}_2\}\} = .01$.

8.1.4   Approximate Invariant Subspaces

If the columns of $Q_1 \in \mathbb{R}^{n \times r}$ are independent and the *residual matrix* $R = AQ_1 - Q_1 S$ is small for some $S \in \mathbb{R}^{r \times r}$, then the columns of Q_1 define an approximate invariant subspace. Let us discover what we can say about the eigensystem of A when in the possession of such a matrix.

Theorem 8.1.13 *Suppose* $A \in \mathbb{R}^{n \times n}$ *and* $S \in \mathbb{R}^{r \times r}$ *are symmetric and that*

$$AQ_1 - Q_1 S = E_1$$

where $Q_1 \in \mathbb{R}^{n \times r}$ *satisfies* $Q_1^T Q_1 = I_r$. *Then there exist* $\mu_1, \dots, \mu_r \in \lambda(A)$ *such that*

$$|\mu_k - \lambda_k(S)| \leq \sqrt{2} \parallel E_1 \parallel_2$$

for $k = 1{:}r$.

Proof. Let $Q_2 \in \mathbb{R}^{n \times (n-r)}$ be any matrix such that $Q = [\, Q_1, \ Q_2 \,]$ is orthogonal. It follows that

$$Q^T A Q = \begin{bmatrix} S & 0 \\ 0 & Q_2^T A Q_2 \end{bmatrix} + \begin{bmatrix} Q_1^T E_1 & E_1^T Q_2 \\ Q_2^T E_1 & 0 \end{bmatrix} \equiv B + E$$

and so by using Corollary 8.1.6 we have $|\lambda_k(A) - \lambda_k(B)| \leq \| E \|_2$ for $k = 1{:}n$. Since $\lambda(S) \subseteq \lambda(B)$, there exist $\mu_1, \dots, \mu_r \in \lambda(A)$ such that

$$|\mu_k - \lambda_k(S)| \leq \| E \|_2$$

for $k = 1{:}r$. The theorem follows by noting that for any $x \in \mathbb{R}^r$ and $y \in \mathbb{R}^{n-r}$ we have

$$\left\| E \begin{bmatrix} x \\ y \end{bmatrix} \right\|_2 \leq \| E_1 x \|_2 + \| E_1^T Q_2 y \|_2 \leq \| E_1 \|_2 \| x \|_2 + \| E_1 \|_2 \| y \|_2$$

from which we readily conclude that $\| E \|_2 \leq \sqrt{2} \| E_1 \|_2$. $\square$

Example 8.1.7 If

$$A = \begin{bmatrix} 6.8 & 2.4 \\ 2.4 & 8.2 \end{bmatrix}, \qquad Q_1 = \begin{bmatrix} .7994 \\ .6007 \end{bmatrix}, \text{ and } S = (5.1) \in \mathbb{R}$$

then

$$A Q_1 - Q_1 S = \begin{bmatrix} -.0828 \\ -.0562 \end{bmatrix} = E_1.$$

The theorem predicts that A has an eigenvalue within $\sqrt{2} \| E_1 \|_2 \approx .1415$ of 5.1. This is true since $\lambda(A) = \{5, 10\}$.

The eigenvalue bounds in Theorem 8.1.13 depend on $\| A Q_1 - Q_1 S \|_2$. Given A and Q_1, the following theorem indicates how to choose S so that this quantity is minimized in the Frobenius norm.

Theorem 8.1.14 *If $A \in \mathbb{R}^{n \times n}$ is symmetric and $Q_1 \in \mathbb{R}^{n \times r}$ has orthonormal columns, then*

$$\min_{S \in \mathbb{R}^{r \times r}} \| A Q_1 - Q_1 S \|_F = \| (I - Q_1 Q_1^T) A Q_1 \|_F$$

and $S = Q_1^T A Q_1$ is the minimizer.

Proof. Let $Q_2 \in \mathbb{R}^{n \times (n-r)}$ be such that $Q = [\, Q_1, \ Q_2 \,]$ is orthogonal. For any $S \in \mathbb{R}^{r \times r}$ we have

$$\begin{aligned} \| A Q_1 - Q_1 S \|_F^2 &= \| Q^T A Q_1 - Q^T Q_1 S \|_F^2 \\ &= \| Q_1^T A Q_1 - S \|_F^2 + \| Q_2^T A Q_1 \|_F^2. \end{aligned}$$

Clearly, the minimizing S is given by $S = Q_1^T A Q_1$. $\square$

This result enables us to associate any r-dimensional subspace $\mathrm{ran}(Q_1)$, with a set of r "optimal" eigenvalue-eigenvector approximates.

Theorem 8.1.15 *Suppose $A \in \mathbb{R}^{n \times n}$ is symmetric and that $Q_1 \in \mathbb{R}^{n \times r}$ satisfies $Q_1^T Q_1 = I_r$. If*

$$Z^T (Q_1^T A Q_1) Z \;=\; \mathrm{diag}(\theta_1, \dots, \theta_r) \;=\; D$$

is the Schur decomposition of $Q_1^T A Q_1$ and $Q_1 Z = [\, y_1, \dots, y_r \,]$, then

$$\| \, A y_k - \theta_k y_k \, \|_2 \;=\; \| \, (I - Q_1 Q_1^T) A Q_1 Z e_k \, \|_2 \;\leq\; \| \, (I - Q_1 Q_1^T) A Q_1 \, \|_2$$

for $k = 1{:}r$.

Proof.

$$A y_k - \theta_k y_k \;=\; A Q_1 Z e_k = Q_1 Z D e_k \;=\; (A Q_1 - Q_1 (Q_1^T A Q_1)) Z e_k.$$

The theorem follows by taking norms. $\square$

In Theorem 8.1.15, the θ_k are called *Ritz values*, the y_k are called *Ritz vectors*, and the (θ_k, y_k) are called *Ritz pairs*.

The usefulness of Theorem 8.1.13 is enhanced if we weaken the assumption that the columns of Q_1 are orthonormal. As can be expected, the bounds deteriorate with the loss of orthogonality.

Theorem 8.1.16 *Suppose $A \in \mathbb{R}^{n \times n}$ is symmetric and that*

$$A X_1 - X_1 S \;=\; F_1,$$

where $X_1 \in \mathbb{R}^{n \times r}$ and $S = X_1^T A X_1$. If

$$\| \, X_1^T X_1 - I_r \, \|_2 = \tau < 1, \tag{8.1.4}$$

then there exist $\mu_1, \dots, \mu_r \in \lambda(A)$ such that

$$|\mu_k - \lambda_k(S)| \;\leq\; \sqrt{2} \, (\| \, F_1 \, \|_2 \, + \, \tau(2 + \tau) \| \, A \, \|_2)$$

for $k = 1{:}r$.

Proof. Let $X_1 = ZP$ be the polar decomposition of X_1. Recall from §4.2.10 that this means $Z \in \mathbb{R}^{n \times r}$ has orthonormal columns and $P \in \mathbb{R}^{k \times k}$ is a symmetric positive semidefinite matrix that satisfies $P^2 = X_1^T X_1$. Taking norms in the equation

$$
\begin{aligned}
E_1 \;\equiv\; A Z - Z S \;&=\; (A X_1 - X_1 S) + A(Z - X_1) - (Z - X_1) S \\
&=\; F_1 + A Z(I - P) - Z(I - P) X_1^T A X_1
\end{aligned}
$$

gives

$$\| E_1 \|_2 \le \| F_1 \|_2 + \| A \|_2 \| I - P \|_2 \left(1 + \| X_1 \|_2^2 \right). \quad (8.1.5)$$

Equation (8.1.4) implies that

$$\| X_1 \|_2^2 \le 1 + \tau. \quad (8.1.6)$$

Since P is positive semidefinite, $(I + P)$ is nonsingular and so

$$I - P = (I + P)^{-1}(I - P^2) = (I + P)^{-1}(I - X_1^T X_1)$$

which implies $\| I - P \|_2 \le \tau$. By substituting this inequality and (8.1.6) into (8.1.5) we have $\| E_1 \|_2 \le \| F_1 \|_2 + \tau(2 + \tau)\| A \|_2$. The proof is completed by noting that we can use Theorem 8.1.13 with $Q_1 = Z$ to relate the eigenvalues of A and S via the residual E_1. $\square$

8.1.5 The Law of Inertia

The *inertia* of a symmetric matrix A is a triplet of nonnegative integers (m, z, p) where m, z, and p are respectively the number of negative, zero, and positive elements of $\lambda(A)$.

Theorem 8.1.17 (Sylvester Law of Inertia) *If $A \in \mathbb{R}^{n \times n}$ is symmetric and $X \in \mathbb{R}^{n \times n}$ is nonsingular, then A and $X^T AX$ have the same inertia.*

Proof. Suppose for some r that $\lambda_r(A) > 0$ and define the subspace $S_0 \subseteq \mathbb{R}^n$ by

$$S_0 = \text{span}\{X^{-1}q_1, \ldots, X^{-1}q_r\}, \quad q_i \neq 0$$

where $Aq_i = \lambda_i(A)q_i$ and $i = 1{:}r$. From the minimax characterization of $\lambda_r(X^T AX)$ we have

$$\lambda_r(X^T AX) = \max_{\dim(S)=r} \min_{y \in S} \frac{y^T(X^T AX)y}{y^T y} \ge \min_{y \in S_0} \frac{y^T(X^T AX)y}{y^T y}.$$

Since

$$y \in \mathbb{R}^n \Rightarrow \frac{y^T(X^T X)y}{y^T y} \ge \sigma_n(X)^2$$

$$y \in S_0 \Rightarrow \frac{y^T(X^T AX)y}{y^T y} \ge \lambda_r(A)$$

it follows that

$$\lambda_r(X^T AX) \ge \min_{y \in S_0} \left\{ \frac{y^T(X^T AX)y}{y^T(X^T X)y} \frac{y^T(X^T X)y}{y^T y} \right\} \ge \lambda_r(A)\sigma_n(X)^2.$$

An analogous argument with the roles of A and $X^T A X$ reversed shows that

$$\lambda_r(A) \geq \lambda_r(X^T A X)\sigma_n(X^{-1})^2 = \frac{\lambda_r(X^T A X)}{\sigma_1(X)^2}.$$

Thus, $\lambda_r(A)$ and $\lambda_r(X^T A X)$ have the same sign and so we have shown that A and $X^T A X$ have the same number of positive eigenvalues. If we apply this result to $-A$, we conclude that A and $X^T A X$ have the same number of negative eigenvalues. Obviously, the number of zero eigenvalues possessed by each matrix is also the same. $\square$

Example 8.1.8 If $A = \text{diag}(3, 2, -1)$ and

$$X = \begin{bmatrix} 1 & 4 & 5 \\ 0 & 1 & 2 \\ 0 & 0 & 1 \end{bmatrix},$$

then

$$X^T A X = \begin{bmatrix} 3 & 12 & 15 \\ 12 & 50 & 64 \\ 15 & 64 & 82 \end{bmatrix}$$

and $\lambda(X^T A X) = \{134.769, .3555, -.1252\}$.

Problems

P8.1.1 Without using any of the results in this section, show that the eigenvalues of a 2-by-2 symmetric matrix must be real.

P8.1.2 Compute the Schur decomposition of $A = \begin{bmatrix} 1 & 2 \\ 2 & 3 \end{bmatrix}$.

P8.1.3 Show that the eigenvalues of a Hermitian matrix ($A^H = A$) are real. For each theorem and corollary in this section, state and prove the corresponding result for Hermitian matrices. Which results have analogs when A is skew-symmetric? (Hint: If $A^T = -A$, then iA is Hermitian.)

P8.1.4 Show that if $X \in \mathbf{R}^{n \times r}$, $r \leq n$, and $\| X^T X - I \| = \tau < 1$, then $\sigma_{min}(X) \geq 1-\tau$.

P8.1.5 Suppose $A, E \in \mathbf{R}^{n \times n}$ are symmetric and consider the Schur decomposition $A + tE = QDQ^T$ where we *assume* that $Q = Q(t)$ and $D = D(t)$ are continuously differentiable functions of $t \in \mathbf{R}$. Show that $\dot{D}(t) = \text{diag}(Q(t)^T EQ(t))$ where the matrix on the right is the diagonal part of $Q(t)^T EQ(t)$. Establish the Wielandt-Hoffman theorem by integrating both sides of this equation from 0 to 1 and taking Frobenius norms to show that

$$\| D(1) - D(0) \|_F \leq \int_0^1 \| \text{diag}(Q(t)^T EQ(t) \|_F dt \leq \| E \|_F.$$

P8.1.6 Prove Theorem 8.1.5.

P8.1.7 Prove Theorem 8.1.7.

P8.1.8 If $C \in \mathbf{R}^{n \times n}$ then the *trace function* $\text{tr}(C) = c_{11} + \cdots + c_{nn}$ equals the sum of C's eigenvalues. Use this to prove Theorem 8.1.8.

P8.1.9 Show that if $B \in \mathbf{R}^{m \times m}$ and $C \in \mathbf{R}^{n \times n}$ are symmetric, then $\text{sep}(B, C) = \min$

$\| BX - XC \|_F$ where the min is taken over all matrices in $\mathbf{R}^{m \times n}$.

P8.1.10 Prove the inequality (8.1.3).

P8.1.11 Suppose $A \in \mathbf{R}^{n \times n}$ is symmetric and $C \in \mathbf{R}^{n \times r}$ has full column rank and assume that $r \ll n$. By using Theorem 8.1.8 relate the eigenvalues of $A + CC^T$ to the eigenvalues of A.

Notes and References for Sec. 8.1

The perturbation theory for the symmetric eigenvalue problem is surveyed in Wilkinson (1965, chapter 2), Parlett (1980, chapters 10 and 11), and Stewart and Sun (1990, chapters 4 and 5). Some representative papers in this well-researched area include

G.W. Stewart (1973). "Error and Perturbation Bounds for Subspaces Associated with Certain Eigenvalue Problems," *SIAM Review 15*, 727–64.

C.C. Paige (1974). "Eigenvalues of Perturbed Hermitian Matrices," *Lin. Alg. and Its Applic . 8*, 1–10.

A. Ruhe (1975). "On the Closeness of Eigenvalues and Singular Values for Almost Normal Matrices," *Lin. Alg. and Its Applic. 11*, 87–94.

W. Kahan (1975). "Spectra of Nearly Hermitian Matrices," *Proc. Amer. Math. Soc. 48*, 11–17.

A. Schonhage (1979). "Arbitrary Perturbations of Hermitian Matrices," *Lin. Alg. and Its Applic. 24*, 143–49.

P. Deift, T. Nanda, and C. Tomei (1983). "Ordinary Differential Equations and the Symmetric Eigenvalue Problem," *SIAM J. Numer. Anal. 20*, 1–22.

D.S. Scott (1985). "On the Accuracy of the Gershgorin Circle Theorem for Bounding the Spread of a Real Symmetric Matrix," *Lin. Alg. and Its Applic. 65*, 147–155

J.-G. Sun (1995). "A Note on Backward Error Perturbations for the Hermitian Eigenvalue Problem," *BIT 35*, 385–393.

R.-C. Li (1996). "Relative Perturbation Theory (I) Eigenvalue and Singular Value Variations," Technical Report UCB//CSD-94-855, Department of EECS, University of California at Berkeley.

R.-C. Li (1996). "Relative Perturbation Theory (II) Eigenspace and Singular Subspace Variations," Technical Report UCB//CSD-94-856, Department of EECS, University of California at Berkeley.

8.2 Power Iterations

Assume that $A \in \mathbb{R}^{n \times n}$ is symmetric and that $U_0 \in \mathbb{R}^{n \times n}$ is orthogonal. Consider the following *QR iteration*:

$$
\begin{aligned}
&T_0 = U_0^T A U_0 \\
&\textbf{for } k = 1, 2, \ldots \\
&\qquad T_{k-1} = U_k R_k \quad \text{(QR factorization)} \\
&\qquad T_k = R_k U_k \\
&\textbf{end}
\end{aligned}
\tag{8.2.1}
$$

Since $T_k = R_k U_k = U_k^T (U_k R_k) U_k = U_k^T T_{k-1} U_k$ it follows by induction that

$$
T_k = (U_0 U_1 \cdots U_k)^T A (U_0 U_1 \cdots U_k).
\tag{8.2.2}
$$

Thus, each T_k is orthogonally similar to A. Moreover, the T_k almost always converge to diagonal form and so it can be said that (8.2.1) almost always "converges" to a Schur decomposition of A. In order to establish this remarkable result we first consider the power method and the method of orthogonal iteration.

8.2.1   The Power Method

Given a unit 2-norm $q^{(0)} \in \mathbb{R}^n$, the *power method* produces a sequence of vectors $q^{(k)}$ as follows:

$$
\begin{aligned}
&\textbf{for } k = 1, 2, \ldots \\
&\qquad z^{(k)} = A q^{(k-1)} \\
&\qquad q^{(k)} = z^{(k)} / \| z^{(k)} \|_2 \\
&\qquad \lambda^{(k)} = [q^{(k)}]^T A q^{(k)} \\
&\textbf{end}
\end{aligned}
\tag{8.2.3}
$$

If $q^{(0)}$ is not "deficient" and A's eigenvalue of maximum modulus is unique, then the $q^{(k)}$ converge to an eigenvector.

Theorem 8.2.1 *Suppose $A \in \mathbb{R}^{n \times n}$ is symmetric and that*

$$
Q^T A Q = \mathrm{diag}(\lambda_1, \ldots, \lambda_n)
$$

where $Q = [\, q_1, \ldots, q_n \,]$ is orthogonal and $|\lambda_1| > |\lambda_2| \geq \cdots \geq |\lambda_n|$. Let the vectors q_k be specified by (8.2.3) and define $\theta_k \in [0, \pi/2]$ by

$$
\cos(\theta_k) = \left| q_1^T q^{(k)} \right|.
$$

If $\cos(\theta_0) \neq 0$, then

$$
|\sin(\theta_k)| \;\leq\; \tan(\theta_0) \left| \frac{\lambda_2}{\lambda_1} \right|^k
\tag{8.2.4}
$$

$$
|\lambda^{(k)} - \lambda| \;\leq\; |\lambda_1 - \lambda_n| \tan(\theta_0)^2 \left| \frac{\lambda_2}{\lambda_1} \right|^{2k}.
\tag{8.2.5}
$$

Proof. From the definition of the iteration, it follows that $q^{(k)}$ is a multiple of $A^k q^{(0)}$ and so

$$
|\sin(\theta_k)|^2 \;=\; 1 - \left(q_1^T q^{(k)} \right)^2 \;=\; 1 - \left(\frac{q_1^T A^k q^{(0)}}{\| A^k q^{(0)} \|_2} \right)^2.
$$

If $q^{(0)}$ has the eigenvector expansion $q^{(0)} = a_1 q_1 + \cdots + a_n q_n$, then

$$
|a_1| = |q_1^T q^{(0)}| = \cos(\theta_0) \neq 0,
$$

$$a_1^2 + \cdots + a_n^2 = 1,$$

and

$$A^k q^{(0)} = a_1 \lambda_1^k q_1 + a_2 \lambda_2^k q_2 + \cdots + a_n \lambda_n^k q_n .$$

Thus,

$$|\sin(\theta_k)|^2 = 1 - \frac{a_1^2 \lambda_1^{2k}}{\displaystyle\sum_{i=1}^{n} a_i^2 \lambda_i^{2k}} = \frac{\displaystyle\sum_{i=2}^{n} a_i^2 \lambda_i^{2k}}{\displaystyle\sum_{i=1}^{n} a_i^2 \lambda_i^{2k}}$$

$$\leq \frac{\displaystyle\sum_{i=2}^{n} a_i^2 \lambda_i^{2k}}{a_1^2 \lambda_1^{2k}} = \frac{1}{a_1^2} \sum_{i=2}^{n} a_i^2 \left(\frac{\lambda_i}{\lambda_1}\right)^{2k}$$

$$\leq \frac{1}{a_1^2} \left(\sum_{i=2}^{n} a_i^2\right) \left(\frac{\lambda_2}{\lambda_1}\right)^{2k} = \frac{1 - a_1^2}{a_1^2} \left(\frac{\lambda_2}{\lambda_1}\right)^{2k}$$

$$= \tan(\theta_0)^2 \left(\frac{\lambda_2}{\lambda_1}\right)^{2k}.$$

This proves (8.2.4). Likewise,

$$\lambda^{(k)} = \left[q^{(k)}\right]^T A q^{(k)} = \frac{\left[q^{(0)}\right]^T A^{2k+1} q^{(0)}}{\left[q^{(0)}\right]^T A^{2k} q^{(0)}} = \frac{\displaystyle\sum_{i=1}^{n} a_i^2 \lambda_i^{2k+1}}{\displaystyle\sum_{i=1}^{n} a_i^2 \lambda_i^{2k}}$$

and so

$$\left|\lambda^{(k)} - \lambda_1\right| = \left|\frac{\displaystyle\sum_{i=2}^{n} a_i^2 \lambda_i^{2k} (\lambda_i - \lambda_1)}{\displaystyle\sum_{i=1}^{n} a_i^2 \lambda_i^{2k}}\right| \leq |\lambda_1 - \lambda_n| \frac{1}{a_1^2} \sum_{i=2}^{n} a_i^2 \left(\frac{\lambda_i}{\lambda_1}\right)^{2k}$$

$$\leq |\lambda_1 - \lambda_n| \tan(\theta_0)^2 \left(\frac{\lambda_2}{\lambda_1}\right)^{2k}. \ \square$$

Example 8.2.1 The eigenvalues of

$$A = \begin{bmatrix} -1.6407 & 1.0814 & 1.2014 & 1.1539 \\ 1.0814 & 4.1573 & 7.4035 & -1.0463 \\ 1.2014 & 7.4035 & 2.7890 & -1.5737 \\ 1.1539 & -1.0463 & -1.5737 & 8.6944 \end{bmatrix}$$

are given by $\lambda(A) = \{12, 8, -4, -2\}$. If (8.2.3) is applied to this matrix with $q^{(0)} = [1\ 0\ 0\ 0]^T$, then

k	$\lambda^{(k)}$
1	2.3156
2	8.6802
3	10.3163
4	11.0663
5	11.5259
6	11.7747
7	11.8967
8	11.9534
9	11.9792
10	11.9907

Observe the convergence to $\lambda_1 = 12$ with rate $|\lambda_2/\lambda_1|^{2k} = (8/12)^{2k} = (4/9)^k$.

Computable error bounds for the power method can be obtained by using Theorem 8.1.13. If

$$\| Aq^{(k)} - \lambda^{(k)}q^{(k)} \|_2 = \delta,$$

then there exists $\lambda \in \lambda(A)$ such that $|\lambda^{(k)} - \lambda| \leq \sqrt{2}\delta$.

8.2.2   Inverse Iteration

Suppose the power method is applied with A replaced by $(A - \lambda I)^{-1}$. If λ is very close to a distinct eigenvalue of A, then the next iterate vector will be very rich in the corresponding eigendirection:

$$\left.\begin{array}{l} x = \displaystyle\sum_{i=1}^{n} a_i q_i \\[2em] Aq_i = \lambda_i q_i,\ i = 1{:}n \end{array}\right\} \Rightarrow (A - \lambda I)^{-1}x = \sum_{i=1}^{n} \frac{a_i}{\lambda_i - \lambda} q_i.$$

Thus, if $\lambda \approx \lambda_j$ and a_j is not too small, then this vector has a strong component in the direction of q_j. This process is called *inverse iteration* and it requires the solution of a linear system with matrix of coefficients $A - \lambda I$.

8.2.3   Rayleigh Quotient Iteration

Suppose $A \in \mathbb{R}^{n \times n}$ is symmetric and that x is a given nonzero n-vector. A simple differentiation reveals that

$$\lambda = r(x) \equiv \frac{x^T A x}{x^T x}$$

minimizes $\| (A - \lambda I)x \|_2$. (See also Theorem 8.1.14.) The scalar $r(x)$ is called the *Rayleigh quotient* of x. Clearly, if x is an approximate eigenvector, then $r(x)$ is a reasonable choice for the corresponding eigenvalue.

Combining this idea with inverse iteration gives rise to the *Rayleigh quotient iteration:*

$$
\begin{aligned}
&x_0 \text{ given, } \| x_0 \|_2 = 1 \\
&\text{for } k = 0, 1, \ldots \\
&\qquad \mu_k = r(x_k) \\
&\qquad \text{Solve } (A - \mu_k I)z_{k+1} = x_k \text{ for } z_{k+1} \\
&\qquad x_{k+1} = z_{k+1}/\| z_{k+1} \|_2 \\
&\text{end}
\end{aligned}
\tag{8.2.6}
$$

Example 8.2.2 If (8.2.6) is applied to

$$
A = \begin{bmatrix}
1 & 1 & 1 & 1 & 1 & 1 \\
1 & 2 & 3 & 4 & 5 & 6 \\
1 & 3 & 6 & 10 & 15 & 21 \\
1 & 4 & 10 & 20 & 35 & 56 \\
1 & 5 & 15 & 35 & 70 & 126 \\
1 & 6 & 21 & 56 & 126 & 252
\end{bmatrix}
$$

with $x_0 = [1,\ 1,\ 1,\ 1,\ 1,\ 1]^T/6$, then

k	μ_k
0	153.8333
1	120.0571
2	49.5011
3	13.8687
4	15.4959
5	15.5534

The iteration is converging to the eigenvalue $\lambda = 15.5534732737$.

The Rayleigh quotient iteration almost always converges and when it does, the rate of convergence is cubic. We demonstrate this for the case $n = 2$. Without loss of generality, we may assume that $A = \text{diag}(\lambda_1, \lambda_2)$, with $\lambda_1 > \lambda_2$. Denoting x_k by

$$
x_k = \begin{bmatrix} c_k \\ s_k \end{bmatrix} \qquad c_k^2 + s_k^2 = 1
$$

it follows that $\mu_k = \lambda_1 c_k^2 + \lambda_1 s_k^2$ in (8.2.6) and

$$
z_{k+1} = \frac{1}{\lambda_1 - \lambda_2} \begin{bmatrix} c_k/s_k^2 \\ -s_k/c_k^2 \end{bmatrix}.
$$

A calculation shows that

$$
c_{k+1} = \frac{c_k^3}{\sqrt{c_k^6 + s_k^6}} \qquad s_{k+1} = \frac{-s_k^3}{\sqrt{c_k^6 + s_k^6}}.
\tag{8.2.7}
$$

From these equations it is clear that the x_k converge cubically to either span$\{e_1\}$ or span$\{e_2\}$ provided $|c_k| \neq |s_k|$.

Details associated with the practical implementation of the Rayleigh quotient iteration may be found in Parlett (1974).

8.2.4   Orthogonal Iteration

A straightforward generalization of the power method can be used to compute higher-dimensional invariant subspaces. Let r be a chosen integer satisfying $1 \le r \le n$. Given an n-by-r matrix Q_0 with orthonormal columns, the method of *orthogonal iteration* generates a sequence of matrices $\{Q_k\} \subseteq \mathbb{R}^{n \times r}$ as follows:

$$
\begin{aligned}
&\textbf{for } k = 1, 2, \ldots \\
&\qquad Z_k = A Q_{k-1} \\
&\qquad Q_k R_k = Z_k \qquad \text{(QR factorization)} \\
&\textbf{end}
\end{aligned}
\qquad (8.2.8)
$$

Note that if $r = 1$, then this is just the power method. Moreover, the sequence $\{Q_k e_1\}$ is precisely the sequence of vectors produced by the power iteration with starting vector $q^{(0)} = Q_0 e_1$.

In order to analyze the behavior of (8.2.8), assume that

$$
Q^T A Q = D = \text{diag}(\lambda_i) \qquad |\lambda_1| \ge |\lambda_2| \ge \cdots \ge |\lambda_n| \qquad (8.2.9)
$$

is a Schur decomposition of $A \in \mathbb{R}^{n \times n}$. Partition Q and D as follows:

$$
Q = [\, Q_\alpha \quad Q_\beta \,] \qquad\qquad D = \begin{bmatrix} D_1 & 0 \\ 0 & D_2 \end{bmatrix} \begin{matrix} r \\ n-r \end{matrix} \qquad (8.2.10)
$$
$$
 r \quad\;\; n-r \phantom{\,] \qquad\qquad D = \begin{bmatrix} D_1 & 0 \\ 0 & D_2 \end{bmatrix}} r \quad n-r
$$

If $|\lambda_r| > |\lambda_{r+1}|$, then

$$
D_r(A) = \text{ran}(Q_\alpha)
$$

is the *dominant* invariant subspace of dimension r. It is the unique invariant subspace associated with the eigenvalues $\lambda_1, \ldots, \lambda_r$.

The following theorem shows that with reasonable assumptions, the subspaces $\text{ran}(Q_k)$ generated by (8.2.8) converge to $D_r(A)$ at a rate proportional to $|\lambda_{r+1}/\lambda_r|^k$.

Theorem 8.2.2 *Let the Schur decomposition of $A \in \mathbb{R}^{n \times n}$ be given by (8.2.9) and (8.2.10) with $n \ge 2$. Assume that $|\lambda_r| > |\lambda_{r+1}|$ and that the n-by-r matrices $\{Q_k\}$ are defined by (8.2.8). If $\theta \in [0, \pi/2]$ is specified by*

$$
\cos(\theta) = \min_{\substack{u \in D_r(A) \\ v \in \text{ran}(Q_0)}} \frac{|u^T v|}{\|\, u \,\|_2 \|\, v \,\|_2} > 0 ,
$$

then

$$
\text{dist}(D_r(A), \text{ran}(Q_k)) \le \tan(\theta) \left| \frac{\lambda_{r+1}}{\lambda_r} \right|^k .
$$

See also Theorem 7.3.1.

Proof. By induction it can be shown that

$$A^k Q_0 = Q_k (R_k \cdots R_1)$$

and so with the partitionings (8.2.10) we have

$$\begin{bmatrix} D_1^k & 0 \\ 0 & D_2^k \end{bmatrix} \begin{bmatrix} Q_\alpha^T Q_0 \\ Q_\beta^T Q_0 \end{bmatrix} = \begin{bmatrix} Q_\alpha^T Q_k \\ Q_\beta^T Q_k \end{bmatrix} (R_k \cdots R_1).$$

If

$$Q^T Q_k = [Q_\alpha, Q_\beta]^T Q_k = \begin{bmatrix} Q_\alpha^T Q_k \\ Q_\beta^T Q_k \end{bmatrix} \equiv \begin{bmatrix} V_k \\ W_k \end{bmatrix},$$

then

$$\begin{aligned}
\cos(\theta_{min}) &= \sigma_r(V_0) = \sqrt{1 - \| W_0 \|_2^2} \\
\mathrm{dist}(D_r(A), \mathrm{ran}(Q_k)) &= \| W_k \|_2 \\
D_1^k V_0 &= V_k (R_k \cdots R_1) \\
D_2^k W_0 &= W_k (R_k \cdots R_1)
\end{aligned}$$

It follows that V_0 is nonsingular which in turn implies that V_k and $(R_k \cdots R_1)$ are also nonsingular. Thus,

$$\begin{aligned}
W_k &= D_2^k W_0 (R_k \cdots R_1)^{-1} = D_2^k W_0 (V_k^{-1} D_1^k V_0)^{-1} \\
&= D_2^k W_0 V_0^{-1} D_1^{-k} V_k
\end{aligned}$$

and so

$$\begin{aligned}
\| W_k \|_2 &\leq \| D_2^k \|_2 \| W_0 \|_2 \| V_0^{-1} \|_2 \| D_1^{-k} \|_2 \| V_k \|_2 \\
&\leq |\lambda_{r+1}|^k \sin(\theta) \frac{1}{\cos(\theta)} \frac{1}{|\lambda_r|^k} = \tan(\theta) \left| \frac{\lambda_{r+1}}{\lambda_r} \right|^k. \quad \square
\end{aligned}$$

Example 8.2.3 If (8.2.8) is applied to the matrix of Example 8.2.1 with $r = 2$ and $Q_0 = I_4(:, 1{:}2)$, then

k	$\mathrm{dist}(D_2(A), \mathrm{ran}(Q_k))$
1	0.8806
2	0.4091
3	0.1121
4	0.0313
5	0.0106
6	0.0044
7	0.0020
8	0.0010
9	0.0005
10	0.0002

8.2.5   The QR Iteration

Consider what happens when we apply the method of orthogonal iteration
(8.2.8) with $r = n$. Let $Q^T AQ = \text{diag}(\lambda_1, \ldots, \lambda_n)$ be the Schur decomposition and assume

$$|\lambda_1| > |\lambda_2| > \cdots > |\lambda_n|.$$

If $Q = [\,q_1, \ldots, q_n\,]$ and $Q_k = \left[\, q_1^{(k)}, \ldots, q_n^{(k)} \,\right]$ and

$$\text{dist}(D_i(A), \text{span}\{q_1^{(0)}, \ldots, q_i^{(0)}\}) < 1 \qquad (8.2.11)$$

for $i = 1{:}n - 1$, then it follows from Theorem 8.2.2 that

$$\text{dist}(\text{span}\{q_1^{(k)}, \ldots, q_i^{(k)}\}, \text{span}\{q_1, \ldots, q_i\}) = 0\left(\left|\frac{\lambda_{i+1}}{\lambda_i}\right|^k\right).$$

for $i = 1{:}n - 1$. This implies that the matrices T_k defined by

$$T_k = Q_k^T AQ_k$$

are converging to diagonal form. Thus, it can be said that the method
of orthogonal iteration computes a Schur decomposition if $r = n$ and the
original iterate $Q_0 \in \mathbb{R}^{n \times n}$ is not deficient in the sense of (8.2.11).

The QR iteration arises by considering how to compute the matrix T_k
directly from its predecessor T_{k-1}. On the one hand, we have from (8.2.1)
and the definition of T_{k-1} that

$$T_{k-1} = Q_{k-1}^T AQ_{k-1} = Q_{k-1}^T (AQ_{k-1}) = (Q_{k-1}^T Q_k)R_k.$$

On the other hand,

$$T_k = Q_k^T AQ_k = (Q_k^T AQ_{k-1})(Q_{k-1}^T Q_k) = R_k(Q_{k-1}^T Q_k).$$

Thus, T_k is determined by computing the QR factorization of T_{k-1} and
then multiplying the factors together in reverse order. This is precisely
what is done in (8.2.1).

Example 8.2.4 If the QR iteration (8.2.1) is applied to the matrix in Example 8.2.1,
then after 10 iterations

$$T_{10} = \begin{bmatrix} 11.9907 & -0.1926 & -0.0004 & 0.0000 \\ -0.1926 & 8.0093 & -0.0029 & 0.0001 \\ -0.0004 & -0.0029 & -4.0000 & 0.0007 \\ 0.0000 & 0.0001 & 0.0007 & -2.0000 \end{bmatrix}.$$

The off-diagonal entries of the T_k matrices go to zero as follows:

k	$\|T_k(2,1)\|$	$\|T_k(3,1)\|$	$\|T_k(4,1)\|$	$\|T_k(3,2)\|$	$\|T_k(4,2)\|$	$\|T_k(4,3)\|$
1	3.9254	1.8122	3.3892	4.2492	2.8367	1.1679
2	2.6491	1.2841	2.1908	1.1587	3.1473	0.2294
3	2.0147	0.6154	0.5082	0.0997	0.9859	0.0748
4	1.6930	0.2408	0.0970	0.0723	0.2596	0.0440
5	1.2928	0.0866	0.0173	0.0665	0.0667	0.0233
6	0.9222	0.0299	0.0030	0.0405	0.0169	0.0118
7	0.6346	0.0101	0.0005	0.0219	0.0043	0.0059
8	0.4292	0.0034	0.0001	0.0113	0.0011	0.0030
9	0.2880	0.0011	0.0000	0.0057	0.0003	0.0015
10	0.1926	0.0004	0.0000	0.0029	0.0001	0.0007

Note that a single QR iteration involves $O(n^3)$ flops. Moreover, since convergence is only linear (when it exists), it is clear that the method is a prohibitively expensive way to compute Schur decompositions. Fortunately, these practical difficulties can be overcome as we show in the next section.

Problems

P8.2.1 Suppose $A_0 \in \mathbf{R}^{n \times n}$ is symmetric and positive definite and consider the following iteration:

$$\text{for } k = 1, 2, \ldots$$
$$A_{k-1} = G_k G_k^T \quad \text{(Cholesky)}$$
$$A_k = G_k^T G_k$$
$$\text{end}$$

(a) Show that this iteration is defined. (b) Show that if $A_0 = \begin{bmatrix} a & b \\ b & c \end{bmatrix}$ with $a \geq c$ has eigenvalues $\lambda_1 \geq \lambda_2 > 0$, then the A_k converge to $\text{diag}(\lambda_1, \lambda_2)$.

P8.2.2 Prove (8.2.7).

P8.2.3 Suppose $A \in \mathbf{R}^{n \times n}$ is symmetric and define the function $f : \mathbf{R}^{n+1} \to \mathbf{R}^{n+1}$ by

$$f\left(\begin{array}{c} x \\ \lambda \end{array} \right) = \left[\begin{array}{c} Ax - \lambda x \\ (x^T x - 1)/2 \end{array} \right]$$

where $x \in \mathbf{R}^n$ and $\lambda \in \mathbf{R}$. Suppose x_+ and λ_+ are produced by applying Newton's method to f at the "current point" defined by x_c and λ_c. Give expressions for x_+ and λ_+ assuming that $\| x_c \|_2 = 1$ and $\lambda_c = x_c^T A x_c$.

Notes and References for Sec. 8.2

The following references are concerned with the method of orthogonal iteration (a.k.a. the method of simultaneous iteration):

G.W. Stewart (1969). "Accelerating The Orthogonal Iteration for the Eigenvalues of a Hermitian Matrix," *Numer. Math. 13*, 362–76.

M. Clint and A. Jennings (1970). "The Evaluation of Eigenvalues and Eigenvectors of Real Symmetric Matrices by Simultaneous Iteration," *Comp. J. 13*, 76–80.

H. Rutishauser (1970). "Simultaneous Iteration Method for Symmetric Matrices," *Numer. Math. 16*, 205–23. See also Wilkinson and Reinsch (1971,pp.284–302).

References for the Rayleigh quotient method include

J. Vandergraft (1971). "Generalized Rayleigh Methods with Applications to Finding Eigenvalues of Large Matrices," *Lin. Alg. and Its Applic. 4*, 353–68.

B.N. Parlett (1974). "The Rayleigh Quotient Iteration and Some Generalizations for Nonnormal Matrices," *Math. Comp. 28*, 679-93.

R.A. Tapia and D.L. Whitley (1988). "The Projected Newton Method Has Order $1+\sqrt{2}$ for the Symmetric Eigenvalue Problem," *SIAM J. Num. Anal. 25*, 1376–1382.

S. Batterson and J. Smillie (1989). "The Dynamics of Rayleigh Quotient Iteration," *SIAM J. Num. Anal. 26*, 624–636.

C. Beattie and D.W. Fox (1989). "Localization Criteria and Containment for Rayleigh Quotient Iteration," *SIAM J. Matrix Anal. Appl. 10*, 80–93.

P.T.P. Tang (1994). "Dynamic Condition Estimation and Rayleigh-Ritz Approximation," *SIAM J. Matrix Anal. Appl. 15*, 331–346.

8.3 The Symmetric QR Algorithm

The symmetric QR iteration (8.2.1) can be made very efficient in two ways. First, we show how to compute an orthogonal U_0 such that $U_0^T A U = T$ is tridiagonal. With this reduction, the iterates produced by (8.2.1) are all tridiagonal and this reduces the work per step to $O(n^2)$. Second, the idea of shifts are introduced and with this change the convergence to diagonal form proceeds at a cubic rate. This is far better than having the off-diagonal entries going to to zero like $|\lambda_{i+1}/\lambda_i|^k$ as discussed in §8.2.5.

8.3.1 Reduction to Tridiagonal Form

If A is symmetric, then it is possible to find an orthogonal Q such that

$$Q^T A Q = T \qquad (8.3.1)$$

is tridiagonal. We call this the *tridiagonal decomposition* and as a compression of data, it represents a very big step towards diagonalization.

We show how to compute (8.3.1) with Householder matrices. Suppose that Householder matrices $P_1, \ldots, P_{k-1}$ have been determined such that if $A_{k-1} = (P_1 \cdots P_{k-1})^T A (P_1 \cdots P_{k-1})$, then

$$A_{k-1} = \begin{bmatrix} B_{11} & B_{12} & 0 \\ B_{21} & B_{22} & B_{23} \\ 0 & B_{32} & B_{33} \end{bmatrix} \begin{matrix} k-1 \\ 1 \\ n-k \end{matrix}$$
$$\phantom{A_{k-1} = } \begin{matrix} k-1 & 1 & n-k \end{matrix}$$

is tridiagonal through its first $k-1$ columns. If $\bar{P}_k$ is an order $n-k$ Householder matrix such that $\bar{P}_k B_{32}$ is a multiple of $I_{n-k}(:,1)$ and if $P_k =$

$\operatorname{diag}(I_k, \bar{P}_k)$, then the leading k-by-k principal submatrix of

$$
A_k = P_k A_{k-1} P_k =
\begin{array}{c}
\left[
\begin{array}{ccc}
B_{11} & B_{12} & 0 \\
B_{21} & B_{22} & B_{23}\bar{P}_k \\
0 & \bar{P}_k B_{32} & \bar{P}_k B_{33}\bar{P}_k
\end{array}
\right]
&
\begin{array}{c}
k-1 \\
1 \\
n-k
\end{array}
\\
\begin{array}{ccc}
k-1 & 1 & n-k
\end{array}
\end{array}
$$

is tridiagonal. Clearly, if $U_0 = P_1 \cdots P_{n-2}$, then $U_0^T A U_0 = T$ is tridiagonal.

In the calculation of A_k it is important to exploit symmetry during the formation of the matrix $\bar{P}_k B_{33} \bar{P}_k$. To be specific, suppose that $\bar{P}_k$ has the form

$$
\bar{P}_k = I - \beta v v^T \qquad \beta = 2/v^T v, \quad 0 \neq v \in \mathbb{R}^{n-k}.
$$

Note that if $p = \beta B_{33} v$ and $w = p - (\beta p^T v/2)v$, then

$$
\bar{P}_k B_{33} \bar{P}_k = B_{33} - v w^T - w v^T.
$$

Since only the upper triangular portion of this matrix needs to be calculated, we see that the transition from A_{k-1} to A_k can be accomplished in only $4(n-k)^2$ flops.

Algorithm 8.3.1 (Householder Tridiagonalization) Given a symmetric $A \in \mathbb{R}^{n \times n}$, the following algorithm overwrites A with $T = Q^T A Q$, where T is tridiagonal and $Q = H_1 \cdots H_{n-2}$ is the product of Householder transformations.

for $k = 1:n-2$
$\quad [v, \beta] = \textbf{house}(A(k+1:n, k))$
$\quad p = \beta A(k+1:n, k+1:n)v$
$\quad w = p - (\beta p^T v/2)v$
$\quad A(k+1, k) = \| A(k+1:n, k) \|_2; \; A(k, k+1) = A(k+1.k)$
$\quad A(k+1:n, k+1:n) = A(k+1:n, k+1:n) - v w^T - w v^T$
end

This algorithm requires $4n^3/3$ flops when symmetry is exploited in calculating the rank-2 update. The matrix Q can be stored in factored form in the subdiagonal portion of A. If Q is explicitly required, then it can be formed with an additional $4n^3/3$ flops.

Example 8.3.1

$$
\begin{bmatrix}
1 & 0 & 0 \\
0 & .6 & .8 \\
0 & .8 & -.6
\end{bmatrix}^T
\begin{bmatrix}
1 & 3 & 4 \\
3 & 2 & 8 \\
4 & 8 & 3
\end{bmatrix}
\begin{bmatrix}
1 & 0 & 0 \\
0 & .6 & .8 \\
0 & .8 & -.6
\end{bmatrix}
=
\begin{bmatrix}
1 & 5 & 0 \\
5 & 10.32 & 1.76 \\
0 & 1.76 & -5.32
\end{bmatrix}.
$$

Note that if T has a zero subdiagonal, then the eigenproblem splits into a pair of smaller eigenproblems. In particular, if $t_{k+1,k} = 0$, then $\lambda(T) =$

$\lambda(T(1{:}k, 1{:}k)) \cup \lambda(T(k+1{:}n, k+1{:}n))$. If T has no zero subdiagonal entries, then it is said to be *unreduced*.

Let $\hat{T}$ denote the computed version of T obtained by Algorithm 8.3.1. It can be shown that $\hat{T} = \tilde{Q}^T(A + E)\tilde{Q}$ where $\tilde{Q}$ is exactly orthogonal and E is a symmetric matrix satisfying $\| E \|_F \leq cu\| A \|_F$ where c is a small constant. See Wilkinson (1965, p. 297).

8.3.2   Properties of the Tridiagonal Decomposition

We prove two theorems about the tridiagonal decomposition both of which have key roles to play in the sequel. The first connects (8.3.1) to the QR factorization of a certain *Krylov matrix*. These matrices have the form

$$K(A, v, k) = [v, \; Av, \cdots, \; A^{k-1}v] \qquad A \in \mathbb{R}^{n \times n}, \; v \in \mathbb{R}^n.$$

Theorem 8.3.1 *If $Q^T AQ = T$ is the tridiagonal decomposition of the symmetric matrix $A \in \mathbb{R}^{n \times n}$, then $Q^T K(A, Q(:, 1), n) = R$ is upper triangular. If R is nonsingular, then T is unreduced. If R is singular and k is the smallest index so $r_{kk} = 0$, then k is also the smallest index so $t_{k,k-1}$ is zero. See also Theorem 7.4.3.*

Proof. It is clear that if $q_1 = Q(:, 1)$, then

$$
\begin{aligned}
Q^T K(A, Q(:, 1), n) &= \left[Q^T q_1, \; (Q^T AQ)(Q^T q_1), \ldots, (Q^T AQ)^{n-1}(Q^T q_1) \right] \\
&= \left[e_1, \; Te_1, \ldots, T^{n-1}e_1 \right] = R
\end{aligned}
$$

is upper triangular with the property that $r_{11} = 1$ and $r_{ii} = t_{21} t_{32} \cdots t_{i,i-1}$ for $i = 2{:}n$. Clearly, if R is nonsingular, then T is unreduced. If R is singular and r_{kk} is its first zero diagonal entry, then $k \geq 2$ and $t_{k,k-1}$ is the first zero subdiagonal entry. $\square$

The next result shows that Q is essentially unique once $Q(:, 1)$ is specified.

Theorem 8.3.2 (Implicit Q Theorem) *Suppose $Q = [\, q_1, \ldots, q_n \,]$ and $V = [\, v_1, \ldots, v_n \,]$ are orthogonal matrices with the property that both $Q^T AQ = T$ and $V^T AV = S$ are tridiagonal where $A \in \mathbb{R}^{n \times n}$ is symmetric. Let k denote the smallest positive integer for which $t_{k+1,k} = 0$, with the convention that $k = n$ if T is unreduced. If $v_1 = q_1$, then $v_i = \pm q_i$ and $|t_{i,i-1}| = |s_{i,i-1}|$ for $i = 2{:}k$. Moreover, if $k < n$, then $s_{k+1,k} = 0$. See also Theorem 7.4.2.*

Proof. Define the orthogonal matrix $W = Q^T V$ and observe that $W(:, 1) = I_n(:, 1) = e_1$ and $W^T TW = S$. By Theorem 8.3.1, $W^T K(T, e_1, k)$ is upper triangular with full column rank. But $K(T, e_1, k)$ is upper triangular and so by the essential uniqueness of the thin QR factorization,

$$W(:, 1{:}k) = I_n(:, 1{:}k)\text{diag}(\pm 1, \ldots, \pm 1).$$

This says that $Q(:,i) = \pm V(:,i)$ for $i = 1{:}k$. The comments about the subdiagonal entries follows from this since $t_{i+1,i} = Q(:,i+1)^T A Q(:,i)$ and $s_{i+1,i} = V(:,i+1)^T A V(:,i)$ for $i = 1{:}n-1$. $\square$

8.3.3 The QR Iteration and Tridiagonal Matrices

We quickly state four facts that pertain to the QR iteration and tridiagonal matrices. Complete verifications are straight forward.

1. *Preservation of Form.* If $T = QR$ is the QR factorization of a symmetric tridiagonal matrix $T \in \mathbb{R}^{n \times n}$, then Q has lower bandwidth 1 and R has upper bandwidth 2 and it follows that

$$T_+ = RQ = Q^T(QR)Q = Q^T TQ$$

is also symmetric and tridiagonal.

2. *Shifts.* If $s \in \mathbb{R}$ and $T - sI = QR$ is the QR factorization, then

$$T_+ = RQ + sI = Q^T TQ$$

is also tridiagonal. This is called a *shifted* QR step.

3. *Perfect Shifts.* If T is unreduced, then the first $n-1$ columns of $T - sI$ are independent regardless of s. Thus, if $s \in \lambda(T)$ and

$$QR = T - sI$$

is a QR factorization, then $r_{nn} = 0$ and the last column of $T_+ = RQ + sI$ equals $sI_n(:,n) = se_n$.

4. *Cost.* If $T \in \mathbb{R}^{n \times n}$ is tridiagonal, then its QR factorization can be computed by applying a sequence of $n-1$ Givens rotations:

> **for** $k = 1{:}n-1$
> $[c,s] = \textbf{givens}(t_{kk}, t_{k+1,k})$
> $m = \min\{k+2, n\}$
> $T(k{:}k+1, k{:}m) = \begin{bmatrix} c & s \\ -s & c \end{bmatrix}^T T(k{:}k+1, k{:}m)$
> **end**

This requires $O(n)$ flops. If the rotations are accumulated, then $O(n^2)$ flops are needed.

8.3.4   Explicit Single Shift QR Iteration

If s is a good approximate eigenvalue, then we suspect that the $(n, n-1)$ will be small after a QR step with shift s. This is the philosophy behind the following iteration:

$$
\begin{aligned}
&T = U_0^T A U_0 \qquad \text{(tridiagonal)} \\
&\textbf{for } k = 0, 1, \ldots \\
&\qquad \text{Determine real shift } \mu. \\
&\qquad T - \mu I \; = \; UR \qquad \text{(QR factorization)} \\
&\qquad T = RU + \mu I \\
&\textbf{end}
\end{aligned}
\qquad (8.3.2)
$$

If

$$
T \; = \;
\begin{bmatrix}
a_1 & b_1 & & \cdots & & 0 \\
b_1 & a_2 & \ddots & & & \vdots \\
& \ddots & \ddots & \ddots & & \\
\vdots & & \ddots & \ddots & & b_{n-1} \\
0 & \cdots & & & b_{n-1} & a_n
\end{bmatrix} .
$$

then one reasonable choice for the shift is $\mu = a_n$. However, a more effective choice is to shift by the eigenvalue of

$$
T(n-1{:}n, n-1{:}n) \; = \;
\begin{bmatrix}
a_{n-1} & b_{n-1} \\
b_{n-1} & a_n
\end{bmatrix}
$$

that is closer to a_n. This is known as the *Wilkinson shift* and it is given by

$$
\mu \; = \; a_n + d - \text{sign}(d)\sqrt{d^2 + b_{n-1}^2} \qquad (8.3.3)
$$

where $d = (a_{n-1} - a_n)/2$. Wilkinson (1968b) has shown that (8.3.2) is cubically convergent with either shift strategy, but gives heuristic reasons why (8.3.3) is preferred.

8.3.5   Implicit Shift Version

It is possible to execute the transition from T to $T_+ = RU + \mu I = U^T T U$ without explicitly forming the matrix $T - \mu I$. This has advantages when the shift is much larger than some of the a_i. Let $c = \cos(\theta)$ and $s = \sin(\theta)$ be computed such that

$$
\begin{bmatrix} c & s \\ -s & c \end{bmatrix}^T
\begin{bmatrix} a_1 - \mu \\ b_1 \end{bmatrix}
=
\begin{bmatrix} \times \\ 0 \end{bmatrix} .
$$

If we set $G_1 = G(1, 2, \theta)$ then $G_1 e_1 = U e_1$ and

$$T \leftarrow G_1^T T G_1 = \begin{bmatrix} \times & \times & + & 0 & 0 & 0 \\ \times & \times & \times & 0 & 0 & 0 \\ + & \times & \times & \times & 0 & 0 \\ 0 & 0 & \times & \times & \times & 0 \\ 0 & 0 & 0 & \times & \times & \times \\ 0 & 0 & 0 & 0 & \times & \times \end{bmatrix}.$$

We are thus in a position to apply the Implicit Q theorem provided we can compute rotations $G_2, \ldots, G_{n-1}$ with the property that if $Z = G_1 G_2 \cdots G_{n-1}$ then $Z e_1 = G_1 e_1 = U e_1$ and $Z^T T Z$ is tridiagonal.

Note that the first column of Z and U are identical provided we take each G_i to be of the form $G_i = G(i, i+1, \theta_i)$, $i = 2{:}n - 1$. But G_i of this form can be used to chase the unwanted nonzero element "$+$" out of the matrix $G_1^T T G_1$ as follows:

$$\xrightarrow{G_2} \begin{bmatrix} \times & \times & 0 & 0 & 0 & 0 \\ \times & \times & \times & + & 0 & 0 \\ 0 & \times & \times & \times & 0 & 0 \\ 0 & + & \times & \times & \times & 0 \\ 0 & 0 & 0 & \times & \times & \times \\ 0 & 0 & 0 & 0 & \times & \times \end{bmatrix} \quad \xrightarrow{G_3} \begin{bmatrix} \times & \times & 0 & 0 & 0 & 0 \\ \times & \times & \times & 0 & 0 & 0 \\ 0 & \times & \times & \times & + & 0 \\ 0 & 0 & \times & \times & \times & 0 \\ 0 & 0 & + & \times & \times & \times \\ 0 & 0 & 0 & 0 & \times & \times \end{bmatrix}$$

$$\xrightarrow{G_4} \begin{bmatrix} \times & \times & 0 & 0 & 0 & 0 \\ \times & \times & \times & 0 & 0 & 0 \\ 0 & \times & \times & \times & 0 & 0 \\ 0 & 0 & \times & \times & \times & + \\ 0 & 0 & 0 & \times & \times & \times \\ 0 & 0 & 0 & + & \times & \times \end{bmatrix} \quad \xrightarrow{G_5} \begin{bmatrix} \times & \times & 0 & 0 & 0 & 0 \\ \times & \times & \times & 0 & 0 & 0 \\ 0 & \times & \times & \times & 0 & 0 \\ 0 & 0 & \times & \times & \times & 0 \\ 0 & 0 & 0 & \times & \times & \times \\ 0 & 0 & 0 & 0 & \times & \times \end{bmatrix}$$

Thus, it follows from the Implicit Q theorem that the tridiagonal matrix $Z^T T Z$ produced by this zero-chasing technique is essentially the same as the tridiagonal matrix T obtained by the explicit method. (We may assume that all tridiagonal matrices in question are unreduced for otherwise the problem decouples.)

Note that at any stage of the zero-chasing, there is only one nonzero entry outside the tridiagonal band. How this nonzero entry moves down the matrix during the update $T \leftarrow G_k^T T G_k$ is illustrated in the following:

$$\begin{bmatrix} 1 & 0 & 0 & 0 \\ 0 & c & s & 0 \\ 0 & -s & c & 0 \\ 0 & 0 & 0 & 1 \end{bmatrix}^T \begin{bmatrix} a_k & b_k & z_k & 0 \\ b_k & a_p & b_p & 0 \\ z_k & b_p & a_q & b_q \\ 0 & 0 & b_q & a_r \end{bmatrix} \begin{bmatrix} 1 & 0 & 0 & 0 \\ 0 & c & s & 0 \\ 0 & -s & c & 0 \\ 0 & 0 & 0 & 1 \end{bmatrix} = \begin{bmatrix} a_k & b_k & 0 & 0 \\ b_k & a_p & b_p & z_p \\ 0 & b_p & a_q & b_q \\ 0 & z_p & b_q & a_r \end{bmatrix}.$$

Here $(p, q, r) = (k+1, k+2, k+3)$. This update can be performed in about

26 flops once c and s have been determined from the equation $b_k s + z_k c = 0$. Overall we obtain

Algorithm 8.3.2 (Implicit Symmetric QR Step with Wilkinson Shift) Given an unreduced symmetric tridiagonal matrix $T \in \mathbb{R}^{n \times n}$, the following algorithm overwrites T with $Z^T T Z$, where $Z = G_1 \cdots G_{n-1}$ is a product of Givens rotations with the property that $Z^T(T - \mu I)$ is upper triangular and μ is that eigenvalue of T's trailing 2-by-2 principal submatrix closer to t_{nn}.

$$d = (t_{n-1,n-1} - t_{nn})/2$$
$$\mu = t_{nn} - t_{n,n-1}^2 \Big/ \left(d + \text{sign}(d)\sqrt{d^2 + t_{n,n-1}^2} \right)$$
$$x = t_{11} - \mu$$
$$z = t_{21}$$
$$\text{for } k = 1{:}n - 1$$
$$\quad [c, s] = \textbf{givens}(x, z)$$
$$\quad T = G_k^T T G_k, \quad \text{where } G_k = G(k, k+1, \theta)$$
$$\quad \text{if } k < n - 1$$
$$\quad\quad x = t_{k+1,k}$$
$$\quad\quad z = t_{k+2,k}$$
$$\quad \text{end}$$
$$\text{end}$$

This algorithm requires about $30n$ flops and n square roots. If a given orthogonal matrix Q is overwritten with $QG_1 \cdots G_{n-1}$, then an additional $6n^2$ flops are needed. Of course, in any practical implementation the tridiagonal matrix T would be stored in a pair of n-vectors and not in an n-by-n array.

Example 8.3.2 If Algorithm 8.3.2 is applied to

$$T = \begin{bmatrix} 1 & 1 & 0 & 0 \\ 1 & 2 & 1 & 0 \\ 0 & 1 & 3 & .01 \\ 0 & 0 & .01 & 4 \end{bmatrix},$$

then the new tridiagonal matrix T is given by

$$T = \begin{bmatrix} .5000 & .5916 & 0 & 0 \\ .5916 & 1.785 & .1808 & 0 \\ 0 & .1808 & 3.7140 & .0000044 \\ 0 & 0 & .0000044 & 4.002497 \end{bmatrix}.$$

Algorithm 8.3.2 is the basis of the symmetric QR algorithm—the standard means for computing the Schur decomposition of a dense symmetric matrix.

Algorithm 8.3.3 (Symmetric QR Algorithm) Given $A \in \mathbb{R}^{n \times n}$ (symmetric) and a tolerance tol greater than the unit roundoff, this algorithm computes an approximate symmetric Schur decomposition $Q^T A Q = D$. A is overwritten with the tridiagonal decomposition.

Use Algorithm 8.3.1, compute the tridiagonalization
$T = (P_1 \cdots P_{n-2})^T A (P_1 \cdots P_{n-2})$.
Set $D = T$ and if Q is desired, form $Q = P_1 \cdots P_{n-2}$. See §5.1.6.
until $q = n$
    For $i = 1{:}n - 1$, set $d_{i+1,i}$ and $d_{i,i+1}$ to zero if
    $|d_{i+1,i}| = |d_{i,i+1}| \leq tol(|d_{ii}| + |d_{i+1,i+1}|)$
    Find the largest q and the smallest p such that if

$$
D = \begin{bmatrix} D_{11} & 0 & 0 \\ 0 & D_{22} & 0 \\ 0 & 0 & D_{33} \end{bmatrix} \begin{matrix} p \\ n-p-q \\ q \end{matrix}
$$
$$
\begin{matrix} p & n-p-q & q \end{matrix}
$$

    then D_{33} is diagonal and D_{22} is unreduced.
    if $q < n$
        Apply Algorithm 8.3.2 to D_{22}:
        $D = \mathrm{diag}(I_p, \bar{Z}, I_q)^T D \, \mathrm{diag}(I_p, \bar{Z}, I_q)$
        If Q is desired, then $Q = Q \, \mathrm{diag}(I_p, \bar{Z}, I_q)$.
    end
end

This algorithm requires about $4n^3/3$ flops if Q is not accumulated and about $9n^3$ flops if Q is accumulated.

Example 8.3.3 Suppose Algorithm 8.3.3 is applied to the tridiagonal matrix

$$
A = \begin{bmatrix} 1 & 2 & 0 & 0 \\ 2 & 3 & 4 & 0 \\ 0 & 4 & 5 & 6 \\ 0 & 0 & 6 & 7 \end{bmatrix}
$$

The subdiagonal entries change as follows during the execution of Algorithm 8.3.3:

Iteration	a_{21}	a_{32}	a_{43}
1	1.6817	3.2344	.8649
2	1.6142	2.5755	.0006
3	1.6245	1.6965	10^{-13}
4	1.6245	1.6965	converg.
5	1.5117	.0150	
6	1.1195	10^{-9}	
7	.7071	converg.	
8	converg.		

Upon completion we find $\lambda(A) = \{-2.4848, .7046, 4.9366, 12.831\}$.

The computed eigenvalues $\hat{\lambda}_i$ obtained via Algorithm 8.3.3 are the exact eigenvalues of a matrix that is near to A, i.e., $Q_0^T(A + E)Q_0 = \text{diag}(\hat{\lambda}_i)$ where $Q_0^T Q_0 = I$ and $\| E \|_2 \approx \mathbf{u} \| A \|_2$. Using Corollary 8.1.6 we know that the absolute error in each $\hat{\lambda}_i$ is small in the sense that $|\hat{\lambda}_i - \lambda_i| \approx \mathbf{u} \| A \|_2$. If $\hat{Q} = [\hat{q}_1, \ldots, \hat{q}_n]$ is the computed matrix of orthonormal eigenvectors, then the accuracy of $\hat{q}_i$ depends on the separation of λ_i from the remainder of the spectrum. See Theorem 8.1.12.

If all of the eigenvalues and a few of the eigenvectors are desired, then it is cheaper not to accumulate Q in Algorithm 8.3.3. Instead, the desired eigenvectors can be found via inverse iteration with T. See §8.2.2. Usually just one step is sufficient to get a good eigenvector, even with a random initial vector.

If just a few eigenvalues and eigenvectors are required, then the special techniques in §8.5 are appropriate.

It is interesting to note the connection between Rayleigh quotient iteration and the symmetric QR algorithm. Suppose we apply the latter to the tridiagonal matrix $T \in \mathbb{R}^{n \times n}$ with shift $\sigma = e_n^T T e_n = t_{nn}$ where $e_n = I_n(:, n)$. If $T - \sigma I = QR$, then we obtain $T = RQ + \sigma I$. From the equation $(T - \sigma I)Q = R^T$ it follows that

$$(T - \sigma I)q_n = r_{nn}e_n,$$

where q_n is the last column of the orthogonal matrix Q. Thus, if we apply (8.2.6) with $x_0 = e_n$, then $x_1 = q_n$.

8.3.6 Orthogonal Iteration with Ritz Acceleration

Recall from §8.2.4 that an orthogonal iteration step involves a matrix-matrix product and a QR factorization:

$$Z_k = A\tilde{Q}_{k-1}$$
$$\tilde{Q}_k R_k = Z_k \quad \text{(QR factorization)}$$

Theorem 8.1.14 says that we can minimize $\| A\tilde{Q}_k - \tilde{Q}_k S \|_F$ by setting $S = S_k \equiv \tilde{Q}_k^T A\tilde{Q}_k$. If $U_k^T S_k U_k = D_k$ is the Schur decomposition of $S_k \in \mathbb{R}^{r \times r}$ and $Q_k = \tilde{Q}_k U_k$, then

$$\| AQ_k - Q_k D_k \|_F = \| A\tilde{Q}_k - \tilde{Q}_k S_k \|_F$$

showing that the columns of Q_k are the best possible basis to take after k steps from the standpoint of minimizing the residual. This defines the *Ritz acceleration* idea:

$Q_0 \in \mathbb{R}^{n \times p}$ given with $Q_0^T Q_0 = I_p$
for $k = 1, 2, \ldots$
  $\quad Z_k = A Q_{k-1}$
  $\quad \tilde{Q}_k R_k = Z_k \qquad$ (QR factorization)
  $\quad S_k = \tilde{Q}_k^T A \tilde{Q}_k$
  $\quad U_k^T S_k U_k = D_k \qquad$ (Schur decomposition)
  $\quad Q_k = \tilde{Q}_k U_k$
end

$\hfill (8.3.6)$

It can be shown that if

$$D_k = \mathrm{diag}(\theta_1^{(k)}, \ldots, \theta_r^{(k)}) \qquad |\theta_1^{(k)}| \geq \cdots \geq |\theta_r^{(k)}|$$

then

$$|\theta_i^{(k)} - \lambda_i(A)| = O\left(\left| \frac{\lambda_{r+1}}{\lambda_i} \right|^k \right) \qquad i = 1{:}r$$

Recall that Theorem 8.2.2 says the eigenvalues of $\tilde{Q}_k^T A \tilde{Q}_k$ converge with rate $|\lambda_{r+1}/\lambda_r|^k$. Thus, the Ritz values converge at a more favorable rate. For details, see Stewart (1969).

Example 8.3.4 If we apply (8.3.6) with

$$A = \begin{bmatrix} 100 & 1 & 1 & 1 \\ 1 & 99 & 1 & 1 \\ 1 & 1 & 2 & 1 \\ 1 & 1 & 1 & 1 \end{bmatrix} \quad \text{and} \quad Q_0 = \begin{bmatrix} 1 & 0 \\ 0 & 1 \\ 0 & 0 \\ 0 & 0 \end{bmatrix}$$

then

k	dist$\{D_2(A), Q_k\}$
0	$.2 \times 10^{-1}$
1	$.5 \times 10^{-3}$
2	$.1 \times 10^{-4}$
3	$.3 \times 10^{-6}$
4	$.8 \times 10^{-8}$

Clearly, convergence is taking place at the rate $(2/99)^k$.

Problems

P8.3.1 Suppose λ is an eigenvalue of a symmetric tridiagonal matrix T. Show that if λ has algebraic multiplicity k, then at least $k - 1$ of T's subdiagonal elements are zero.

P8.3.2 Suppose A is symmetric and has bandwidth p. Show that if we perform the shifted QR step $A - \mu I = QR$, $A = RQ + \mu I$, then A has bandwidth p.

P8.3.3 Suppose $B \in \mathbb{R}^{n \times n}$ is upper bidiagonal with diagonal entries $d(1{:}n)$ and superdiagonal entries $f(1{:}n - 1)$. State and prove a singular value version of Theorem 8.3.1.

P8.3.4 Let $A = \begin{bmatrix} w & x \\ x & z \end{bmatrix}$ be real and suppose we perform the following shifted QR step: $A - zI = UR$, $\bar{A} = RU + zI$. Show that if $\bar{A} = \begin{bmatrix} \bar{w} & \bar{x} \\ \bar{x} & z \end{bmatrix}$ then

$$\bar{w} = w + x^2(w - z)/[(w - z)^2 + x^2]$$
$$\bar{z} = z - x^2(w - z)/[(w - z)^2 + x^2]$$
$$\bar{x} = -x^3/[(w - z)^2 + x^2].$$

P8.3.5 Suppose $A \in \mathbb{C}^{n \times n}$ is Hermitian. Show how to construct unitary Q such that $Q^H A Q = T$ is real, symmetric, and tridiagonal.

P8.3.6 Show that if $A = B + iC$ is Hermitian, then $M = \begin{bmatrix} B & -C \\ C & B \end{bmatrix}$ is symmetric. Relate the eigenvalues and eigenvectors of A and M.

P8.3.7 Rewrite Algorithm 8.2.2 for the case when A is stored in two n-vectors. Justify the given flop count.

P8.3.8 Suppose $A = S + \sigma u u^T$ where $S \in \mathbb{R}^{n \times n}$ is skew-symmetric ($A^T = -A$, $u \in \mathbb{R}^n$ has unit 2-norm, and $\sigma \in \mathbb{R}$. Show how to compute an orthogonal Q such that $Q^T A Q$ is tridiagonal and $Q^T u = I_n(:, 1) = e_1$.

Notes and References for Sec. 8.3

The tridiagonalization of a symmetric matrix is discussed in

R.S. Martin and J.H. Wilkinson (1968). "Householder's Tridiagonalization of a Symmetric Matrix," *Numer. Math. 11*, 181-95. See also Wilkinson and Reinsch (1971, pp.212–26).

H.R. Schwartz (1968). "Tridiagonalization of a Symmetric Band Matrix," *Numer. Math. 12*, 231–41. See also Wilkinson and Reinsch (1971, pp.273–83).

N.E. Gibbs and W.G. Poole, Jr. (1974). "Tridiagonalization by Permutations," *Comm. ACM 17*, 20–24.

The first two references contain Algol programs. Algol procedures for the explicit and implicit tridiagonal QR algorithm are given in

H. Bowdler, R.S. Martin, C. Reinsch, and J.H. Wilkinson (1968). "The QR and QL Algorithms for Symmetric Matrices," *Numer. Math. 11*, 293–306. See also Wilkinson and Reinsch (1971, pp.227–40).

A. Dubrulle, R.S. Martin, and J.H. Wilkinson (1968). "The Implicit QL Algorithm," *Numer. Math. 12*, 377-83. see also Wilkinson and Reinsch (1971, pp.241–48).

The "QL" algorithm is identical to the QR algorithm except that at each step the matrix $T - \lambda I$ is factored into a product of an orthogonal matrix and a lower triangular matrix. Other papers concerned with these methods include

G.W. Stewart (1970). "Incorporating Original Shifts into the QR Algorithm for Symmetric Tridiagonal Matrices," *Comm. ACM 13*, 365–67.

A. Dubrulle (1970). "A Short Note on the Implicit QL Algorithm for Symmetric Tridiagonal Matrices," *Numer. Math. 15*, 450.

Extensions to Hermitian and skew-symmetric matrices are described in

D. Mueller (1966). "Householder's Method for Complex Matrices and Hermitian Matrices," *Numer. Math. 8*, 72–92.

R.C. Ward and L.J. Gray (1978). "Eigensystem Computation for Skew-Symmetric and A Class of Symmetric Matrices," *ACM Trans. Math. Soft. 4*, 278–85.

The convergence properties of Algorithm 8.2.3 are detailed in Lawson and Hanson (1974, Appendix B), as well as in

J.H. Wilkinson (1968b). "Global Convergence of Tridiagonal QR Algorithm With Origin Shifts," *Lin. Alg. and Its Applic. 1*, 409–20.
T.J. Dekker and J.F. Traub (1971). "The Shifted QR Algorithm for Hermitian Matrices," *Lin. Alg. and Its Applic. 4*, 137–54.
W. Hoffman and B.N. Parlett (1978). "A New Proof of Global Convergence for the Tridiagonal QL Algorithm," *SIAM J. Num. Anal. 15*, 929–37.
S. Batterson (1994). "Convergence of the Francis Shifted QR Algorithm on Normal Matrices," *Lin. Alg. and Its Applic. 207*, 181–195.

For an analysis of the method when it is applied to normal matrices see

C.P. Huang (1981). "On the Convergence of the QR Algorithm with Origin Shifts for Normal Matrices," *IMA J. Num. Anal. 1*, 127–33.

Interesting papers concerned with shifting in the tridiagonal QR algorithm include

F.L. Bauer and C. Reinsch (1968). "Rational QR Transformations with Newton Shift for Symmetric Tridiagonal Matrices," *Numer. Math. 11*, 264–72. See also Wilkinson and Reinsch (1971, pp.257–65).
G.W. Stewart (1970). "Incorporating Origin Shifts into the QR Algorithm for Symmetric Tridiagonal Matrices," *Comm. Assoc. Comp. Mach. 13*, 365–67.

Some parallel computation possibilities for the algorithms in this section are discussed in

S. Lo, B. Philippe, and A. Sameh (1987). "A Multiprocessor Algorithm for the Symmetric Tridiagonal Eigenvalue Problem," *SIAM J. Sci. and Stat. Comp. 8*, s155–s165.
H.Y. Chang and M. Salama (1988). "A Parallel Householder Tridiagonalization Strategy Using Scattered Square Decomposition," *Parallel Computing 6*, 297–312.

Another way to compute a specified subset of eigenvalues is via the rational QR algorithm. In this method, the shift is determined using Newton's method. This makes it possible to "steer" the iteration towards desired eigenvalues. See

C. Reinsch and F.L. Bauer (1968). "Rational QR Transformation with Newton's Shift for Symmetric Tridiagonal Matrices," *Numer. Math. 11*, 264–72. See also Wilkinson and Reinsch (1971, pp.257–65).

Papers concerned with the symmetric QR algorithm for banded matrices include

R.S. Martin and J.H. Wilkinson (1967). "Solution of Symmetric and Unsymmetric Band Equations and the Calculation of Eigenvectors of Band Matrices," *Numer. Math. 9*, 279–301. See also See also Wilkinson and Reinsch (1971, pp.70–92).
R.S. Martin, C. Reinsch, and J.H. Wilkinson (1970). "The QR Algorithm for Band Symmetric Matrices," *Numer. Math. 16*, 85–92. See also Wilkinson and Reinsch (1971, pp.266–72).

8.4   Jacobi Methods

Jacobi methods for the symmetric eigenvalue problem attract current attention because they are inherently parallel. They work by performing a sequence of orthogonal similarity updates $A \leftarrow Q^T A Q$ with the property that each new A, although full, is "more diagonal" than its predecessor. Eventually, the off-diagonal entries are small enough to be declared zero.

After surveying the basic ideas behind the Jacobi approach we develop a parallel Jacobi procedure.

8.4.1   The Jacobi Idea

The idea behind Jacobi's method is to systematically reduce the quantity

$$\text{off}(A) \;=\; \sqrt{\sum_{i=1}^{n} \sum_{\substack{j=1 \\ j \neq i}}^{n} a_{ij}^2} \,,$$

i.e., the "norm" of the off-diagonal elements. The tools for doing this are rotations of the form

$$J(p,q,\theta) \;=\; \begin{bmatrix} 1 & \cdots & 0 & \cdots & 0 & \cdots & 0 \\ \vdots & \ddots & \vdots & & \vdots & & \vdots \\ 0 & \cdots & c & \cdots & s & \cdots & 0 \\ \vdots & & \vdots & \ddots & \vdots & & \vdots \\ 0 & \cdots & -s & \cdots & c & \cdots & 0 \\ \vdots & & \vdots & & \vdots & \ddots & \vdots \\ 0 & \cdots & 0 & \cdots & 0 & \cdots & 1 \end{bmatrix} \begin{matrix} \\ \\ p \\ \\ q \\ \\ \\ \end{matrix}$$

$$\begin{matrix} \quad\quad\quad\quad\quad p \quad\quad\quad\quad q \end{matrix}$$

which we call *Jacobi rotations*. Jacobi rotations are no different from Givens rotations, c.f. §5.1.8. We submit to the name change in this section to honor the inventor.

The basic step in a Jacobi eigenvalue procedure involves (1) choosing an index pair (p,q) that satisfies $1 \leq p < q \leq n$, (2) computing a cosine-sine pair (c,s) such that

$$\begin{bmatrix} b_{pp} & b_{pq} \\ b_{qp} & b_{qq} \end{bmatrix} = \begin{bmatrix} c & s \\ -s & c \end{bmatrix}^T \begin{bmatrix} a_{pp} & a_{pq} \\ a_{qp} & a_{qq} \end{bmatrix} \begin{bmatrix} c & s \\ -s & c \end{bmatrix} \tag{8.4.1}$$

is diagonal, and (3) overwriting A with $B = J^T A J$ where $J = J(p,q,\theta)$. Observe that the matrix B agrees with A except in rows and columns p

and q. Moreover, since the Frobenius norm is preserved by orthogonal transformations we find that

$$a_{pp}^2 + a_{qq}^2 + 2a_{pq}^2 = b_{pp}^2 + b_{qq}^2 + 2b_{pq}^2 = b_{pp}^2 + b_{qq}^2$$

and so

$$\text{off}(B)^2 = \| B \|_F^2 - \sum_{i=1}^{n} b_{ii}^2 \qquad (8.4.2)$$

$$= \| A \|_F^2 - \sum_{i=1}^{n} a_{ii}^2 + (a_{pp}^2 + a_{qq}^2 - b_{pp}^2 - b_{qq}^2)$$

$$= \text{off}(A)^2 - 2a_{pq}^2 .$$

It is in this sense that A moves closer to diagonal form with each Jacobi step.

Before we discuss how the index pair (p, q) can be chosen, let us look at the actual computations associated with the (p, q) subproblem.

8.4.2   The 2-by-2 Symmetric Schur Decomposition

To say that we diagonalize in (8.4.1) is to say that

$$0 = b_{pq} = a_{pq}(c^2 - s^2) + (a_{pp} - a_{qq})cs. \qquad (8.4.3)$$

If $a_{pq} = 0$, then we just set $(c, s) = (1, 0)$. Otherwise define

$$\tau = \frac{a_{qq} - a_{pp}}{2a_{pq}} \quad \text{and} \quad t = s/c$$

and conclude from (8.4.3) that $t = \tan(\theta)$ solves the quadratic

$$t^2 + 2\tau t - 1 = 0 .$$

It turns out to be important to select the smaller of the two roots,

$$t = -\tau \pm \sqrt{1 + \tau^2}$$

whereupon c and s can be resolved from the formulae

$$c = 1/\sqrt{1 + t^2} \qquad s = tc .$$

Choosing t to be the smaller of the two roots ensures that $|\theta| \le \pi/4$ and has the effect of minimizing the difference between B and A because

$$\| B - A \|_F^2 = 4(1 - c) \sum_{\substack{i=1 \\ i \ne p, q}}^{n} (a_{ip}^2 + a_{iq}^2) + 2a_{pq}^2/c^2$$

We summarize the 2-by-2 computations as follows:

Algorithm 8.4.1 Given an n-by-n symmetric A and integers p and q that satisfy $1 \le p < q \le n$, this algorithm computes a cosine-sine pair (c, s) such that if $B = J(p, q, \theta)^T AJ(p, q, \theta)$ then $b_{pq} = b_{qp} = 0$.

> **function:** $[c, s] = $ **sym.schur2**(A, p, q)
>   **if** $A(p, q) \ne 0$
> $\tau = (A(q, q) - A(p, p))/(2A(p, q))$
>   **if** $\tau \ge 0$
> $t = 1/(\tau + \sqrt{1 + \tau^2})$;
>   **else**
> $t = -1/(-\tau + \sqrt{1 + \tau^2})$;
>   **end**
> $c = 1/\sqrt{1 + t^2}$
> $s = tc$
>   **else**
> $c = 1$
> $s = 0$
>   **end**

8.4.3   The Classical Jacobi Algorithm

As we mentioned above, only rows and columns p and q are altered when the (p, q) subproblem is solved. Once **sym.schur2** determines the 2-by-2 rotation, then the update $A \leftarrow J(p, q, \theta)^T AJ(p, q, \theta)$ can be implemented in $6n$ flops if symmetry is exploited.

How do we choose the indices p and q? From the standpoint of maximizing the reduction of off(A) in (8.4.2), it makes sense to choose (p, q) so that a_{pq}^2 is maximal. This is the basis of the *classical* Jacobi algorithm.

Algorithm 8.4.2 (Classical Jacobi) Given a symmetric $A \in \mathbb{R}^{n \times n}$ and a tolerance $tol > 0$, this algorithm overwrites A with $V^T AV$ where V is orthogonal and off$(V^T AV) \le tol\| A \|_F$.

> $V = I_n$; $eps = tol\| A \|_F$
> **while** off$(A) > eps$
>   Choose (p, q) so $|a_{pq}| = \max_{i \ne j} |a_{ij}|$.
> $(c, s) = $ **sym.schur2**(A, p, q)
> $A = J(p, q, \theta)^T AJ(p, q, \theta)$
> $V = VJ(p, q, \theta)$
> **end**

Since $|a_{pq}|$ is the largest off-diagonal entry, off$(A)^2 \le N(a_{pq}^2 + a_{qp}^2)$ where

$N = n(n-1)/2$. From (8.4.2) it follows that

$$\text{off}(B)^2 \leq \left(1 - \frac{1}{N}\right)\text{off}(A)^2 .$$

By induction, if $A^{(k)}$ denotes the matrix A after k Jacobi updates, then

$$\text{off}(A^{(k)})^2 \leq \left(1 - \frac{1}{N}\right)^k \text{off}(A^{(0)})^2.$$

This implies that the classical Jacobi procedure converges at a linear rate.

However, the asymptotic convergence rate of the method is considerably better than linear. Schonhage (1964) and van Kempen (1966) show that for k large enough, there is a constant c such that

$$\text{off}(A^{(k+N)}) \leq c \cdot \text{off}(A^{(k)})^2$$

i.e., quadratic convergence. An earlier paper by Henrici (1958) established the same result for the special case when A has distinct eigenvalues. In the convergence theory for the Jacobi iteration, it is critical that $|\theta| \leq \pi/4$. Among other things this precludes the possibility of "interchanging" nearly converged diagonal entries. This follows from the formulae $b_{pp} = a_{pp} - ta_{pq}$ and $b_{qq} = a_{qq} + ta_{pq}$, which can be derived from equations (8.4.1) and the definition $t = \sin(\theta)/\cos(\theta)$.

It is customary to refer to N Jacobi updates as a *sweep*. Thus, after a sufficient number of iterations, quadratic convergence is observed when examining off(A) after every sweep.

Example 8.4.1 Applying the classical Jacobi iteration to

$$A = \begin{bmatrix} 1 & 1 & 1 & 1 \\ 1 & 2 & 3 & 4 \\ 1 & 3 & 6 & 10 \\ 1 & 4 & 10 & 20 \end{bmatrix}$$

we find

sweep	$O(\text{off}(A))$
0	10^2
1	10^1
2	10^{-2}
3	10^{-11}
4	10^{-17}

There is no rigorous theory that enables one to predict the number of sweeps that are required to achieve a specified reduction in off(A). However, Brent and Luk (1985) have argued heuristically that the number of sweeps is proportional to $\log(n)$ and this seems to be the case in practice.

8.4.4   The Cyclic-by-Row Algorithm

The trouble with the classical Jacobi method is that the updates involve $O(n)$ flops while the search for the optimal (p, q) is $O(n^2)$. One way to address this imbalance is to fix the sequence of subproblems to be solved in advance. A reasonable possibility is to step through all the subproblems in row-by-row fashion. For example, if $n = 4$ we cycle as follows:

$$(p, q) \; = \; (1, 2), (1, 3), (1, 4), (2, 3), (2, 4), (3, 4), (1, 2), \ldots$$

This ordering scheme is referred to as *cyclic-by-row* and it results in the following procedure:

Algorithm 8.4.3 (Cyclic Jacobi)   Given a symmetric $A \in \mathbb{R}^{n \times n}$ and a tolerance $tol > 0$, this algorithm overwrites A with $V^T A V$ where V is orthogonal and off$(V^T A V) \le tol \| A \|_F$.

$V = I_n$
$eps = tol \| A \|_F$
while off$(A) > eps$
  for $p = 1{:}n - 1$
    for $q = p + 1{:}n$
      $(c\,,\, s) = \textbf{sym.schur2}(A, p, q)$
      $A = J(p, q, \theta)^T A J(p, q, \theta)$
      $V = V J(p, q, \theta)$
    end
  end
end

Cyclic Jacobi converges also quadratically. (See Wilkinson (1962) and van Kempen (1966).) However, since it does not require off-diagonal search, it is considerably faster than Jacobi's original algorithm.

Example 8.4.2 If the cyclic Jacobi method is applied to the matrix in Example 8.4.1 we find

Sweep	$O(\text{off}(A))$
0	10^2
1	10^1
2	10^{-1}
3	10^{-6}
4	10^{-16}

8.4.5 Error Analysis

Using Wilkinson's error analysis it is possible to show that if r sweeps are needed in Algorithm 8.4.3 then the computed d_i satisfy

$$\sum_{i=1}^{n}(d_i - \lambda_i)^2 \leq (\delta + k_r)\| A \|_F \mathbf{u}$$

for some ordering of A's eigenvalues λ_i. The parameter k_r depends mildly on r.

Although the cyclic Jacobi method converges quadratically, it is not generally competitive with the symmetric QR algorithm. For example, if we just count flops, then 2 sweeps of Jacobi is roughly equivalent to a complete QR reduction to diagonal form with accumulation of transformations. However, for small n this liability is not very dramatic. Moreover, if an approximate eigenvector matrix V is known, then $V^T A V$ is almost diagonal, a situation that Jacobi can exploit but not QR.

Another interesting feature of the Jacobi method is that it can a compute the eigenvalues with small *relative* error if A is positive definite. To appreciate this point, note that the Wilkinson analysis cited above coupled the §8.1 perturbation theory ensures that the computed eigenvalues $\hat{\lambda}_1 \geq \cdots \geq \hat{\lambda}_n$ satisfy

$$\frac{|\hat{\lambda}_i - \lambda_i(A)|}{\lambda_i(A)} \approx \mathbf{u}\frac{\| A \|_2}{\lambda_i(A)} \leq \mathbf{u}\kappa_2(A).$$

However, a refined, componentwise error analysis by Demmel and Veselič (1992) shows that in the positive definite case,

$$\frac{|\hat{\lambda}_i - \lambda_i(A)|}{\lambda_i(A)} \approx \mathbf{u}\kappa_2(D^{-1}AD^{-1}). \qquad (8.4.4)$$

where $D = \text{diag}(\sqrt{a_{11}}, \ldots, \sqrt{a_{nn}})$ and this is generally a much smaller approximating bound. The key to establishing this result is some new perturbation theory and a demonstration that if A_+ is a computed Jacobi update obtained from the current matrix A_c, then the eigenvalues of A_+ are relatively close to the eigenvalues of A_c in the sense of (8.4.4). To make the whole thing work in practice, the termination criteria is not based upon the comparison of off(A) with $\mathbf{u}\| A \|_F$ but rather on the size of each $|a_{ij}|$ compared to $\mathbf{u}\sqrt{a_{ii}a_{jj}}$. This work is typical of a new genre of research concerned with high-accuracy algorithms based upon careful, componentwise error analysis. See Mathias (1995).

8.4.6 Parallel Jacobi

Perhaps the most interesting distinction between the QR and Jacobi approaches to the symmetric eigenvalue problem is the rich inherent parallelism of the latter algorithm. To illustrate this, suppose $n = 4$ and group the six subproblems into three *rotation sets* as follows:

$$\begin{aligned}
rot.set(1) &= \{(1,2),(3,4)\} \\
rot.set(2) &= \{(1,3),(2,4)\} \\
rot.set(3) &= \{(1,4),(2,3)\}
\end{aligned}$$

Note that all the rotations within each of the three rotation sets are "non-conflicting." That is, subproblems (1,2) and (3,4) can be carried out in parallel. Likewise the (1,3) and (2,4) subproblems can be executed in parallel as can subproblems (1,4) and (2,3). In general, we say that

$$(i_1,j_1),(i_2,j_2),\ldots,(i_N,j_N) \qquad N = (n-1)n/2$$

is a *parallel ordering* of the set $\{(i,j) \mid 1 \leq i < j \leq n\}$ if for $s = 1{:}n-1$ the rotation set $rot.set(s) = \{\,(i_r,j_r) : r = 1 + n(s-1)/2{:}ns/2\,\}$ consists of nonconflicting rotations. This requires n to be even, which we assume throughout this section. (The odd n case can be handled by bordering A with a row and column of zeros and being careful when solving the subproblems that involve these augmented zeros.)

A good way to generate a parallel ordering is to visualize a chess tournament with n players in which everybody must play everybody else exactly once. In the $n = 8$ case this entails 7 "rounds." During round one we have the following four games:

1	3	5	7
2	4	6	8

$rot.set(1) = \{\,(1,2),(3,4),(5,6),(7,8)\,\}$

i.e., 1 plays 2, 3 plays 4, etc. To set up rounds 2 through 7, player 1 stays put and players 2 through 8 embark on a merry-go-round:

1	2	3	5
4	6	8	7

$rot.set(2) = \{(1,4),(2,6),(3,8),(5,7)\}$

1	4	2	3
6	8	7	5

$rot.set(3) = \{(1,6),(4,8),(2,7),(3,5)\}$

1	6	4	2
8	7	5	3

$rot.set(4) = \{(1,8),(6,7),(4,5),(2,3)\}$

1	8	6	4
7	5	3	2

$rot.set(5) = \{(1,7),(5,8),(3,6),(2,4)\}$

1	7	8	6
5	3	2	4

$rot.set(6) = \{(1,5),(3,7),(2,8),(4,6)\}$

1	5	7	8
3	2	4	6

$$rot.set(7) = \{(1,3),(2,5),(4,7),(6,8)\}$$

We can encode these operations in a pair of integer vectors $top(1:n/2)$ and $bot(1:n/2)$. During a given round $top(k)$ plays $bot(k)$, $k = 1:n/2$. The pairings for the next round is obtained by updating top and bot as follows:

function: $[new.top, new.bot] = $ **music**(top, bot, n)
    $m = n/2$
    for $k = 1{:}m$
        if $k = 1$
            $new.top(1) = 1$
        else if $k = 2$
            $new.top(k) = bot(1)$
        elseif $k > 2$
            $new.top(k) = top(k-1)$
        end
        if $k = m$
            $new.bot(k) = top(k)$
        else
            $new.bot(k) = bot(k+1)$
        end
    end

Using **music** we obtain the following parallel order Jacobi procedure.

Algorithm 8.4.4 (Parallel Order Jacobi) Given a symmetric $A \in \mathbb{R}^{n \times n}$ and a tolerance $tol > 0$, this algorithm overwrites A with $V^T A V$ where V is orthogonal and off$(V^T A V) \le tol\| A \|_F$. It is assumed that n is even.

$V = I_n$
$eps = tol\| A \|_F$
$top = 1{:}2{:}n; bot = 2{:}2{:}n$
while off$(A) > eps$
    for $set = 1{:}n - 1$
        for $k = 1{:}n/2$
            $p = \min(top(k), bot(k))$
            $q = \max(top(k), bot(k))$
            $(c , s) = $ **sym.schur2**(A, p, q)
            $A = J(p, q, \theta)^T A J(p, q, \theta)$
            $V = V J(p, q, \theta)$
        end
        $[top, bot] = $ **music**(top, bot, n)
    end
end

Notice that the k-loop steps through $n/2$ independent, nonconflicting sub-problems.

8.4.7   A Ring Procedure

We now discuss how Algorithm 8.4.4 could be implemented on a ring of p processors. We assume that $p = n/2$ for clarity. At any instant, Proc(μ) houses two columns of A and the corresponding V columns. For example, if $n = 8$ then here is how the column distribution of A proceeds from step to step:

	Proc(1)	Proc(2)	Proc(3)	Proc(4)
Step 1:	[1 2]	[3 4]	[5 6]	[7 8]
Step 2:	[1 4]	[2 6]	[3 8]	[5 7]
Step 3:	[1 6]	[4 8]	[2 7]	[3 5]
		etc.		

The ordered pairs denote the indices of the housed columns. The first index names the *left* column and the second index names the *right* column. Thus, the *left* and *right* columns in Proc(3) during step 3 are 2 and 7 respectively.

Note that in between steps, the columns are shuffled according to the permutation implicit in **music** and that nearest neighbor communication prevails. At each step, each processor oversees a single subproblem. This involves (a) computing an orthogonal $V_{small} \in \mathbb{R}^{2 \times 2}$ that solves a local 2-by-2 Schur problem, (b) using the 2-by-2 V_{small} to update the two housed columns of A and V, (c) sending the 2-by-2 V_{small} to all the other processors, and (d) receiving the V_{small} matrices from the other processors and updating the local portions of A and V accordingly. Since A is stored by column, communication is necessary to carry out the V_{small} updates because they effect rows of A. For example, in the second step of the $n = 8$ problem, Proc(2) must receive the 2-by-2 rotations associated with sub-problems (1,4), (3,8), and (5,7). These come from Proc(1), Proc(3), and Proc(4) respectively. In general, the sharing of the rotation matrices can be conveniently implemented by circulating the 2-by-2 V_{small} matrices in "merry go round" fashion around the ring. Each processor copies a passing 2-by-2 V_{small} into its local memory and then appropriately updates the locally housed portions of A and V.

The termination criteria in Algorithm 8.4.4 poses something of a problem in a distributed memory environment in that the value of off($\cdot$) and $\| A \|_F$ require access to all of A. However, these global quantities can be computed during the V matrix merry-go-round phase. Before the circulation of the V's begins, each processor can compute its contribution to $\| A \|_F$ and off($\cdot$) . These quantities can then be summed by each processor if they are placed on the merry-go-round and read at each stop. By the end of one revolution each processor has its own copy of $\| A \|_F$ and off($\cdot$).

8.4.8 Block Jacobi Procedures

It is usually the case when solving the symmetric eigenvalue problem on a
p-processor machine that $n \gg p$. In this case a block version of the Jacobi
algorithm may be appropriate. Block versions of the above procedures are
straightforward. Suppose that $n = rN$ and that we partition the n-by-n
matrix A as follows:

$$
A = \begin{bmatrix} A_{11} & \cdots & A_{1N} \\ \vdots & & \vdots \\ A_{N1} & \cdots & A_{NN} \end{bmatrix}.
$$

Here, each A_{ij} is r-by-r. In block Jacobi the (p,q) subproblem involves
computing the $2r$-by-$2r$ Schur decomposition

$$
\begin{bmatrix} V_{pp} & V_{pq} \\ V_{qp} & V_{qq} \end{bmatrix}^T \begin{bmatrix} A_{pp} & A_{pq} \\ A_{qp} & A_{qq} \end{bmatrix} \begin{bmatrix} V_{pp} & V_{pq} \\ V_{qp} & V_{qq} \end{bmatrix} = \begin{bmatrix} D_{pp} & O \\ O & D_{qq} \end{bmatrix}
$$

and then applying to A the block Jacobi rotation made up of the V_{ij}. If
we call this block rotation V then it is easy to show that

$$
\text{off}(V^T AV)^2 = \text{off}(A)^2 - \left(2\| A_{pq} \|_F^2 + \text{off}(A_{pp})^2 + \text{off}(A_{qq})^2 \right).
$$

Block Jacobi procedures have many interesting computational aspects. For
example, there are many ways to solve the subproblems and the choice
appears to be critical. See Bischof (1987).

Problems

P8.4.1 Let the scalar γ be given along with the matrix

$$
A = \begin{bmatrix} w & x \\ x & z \end{bmatrix}.
$$

It is desired to compute an orthogonal matrix

$$
J = \begin{bmatrix} c & s \\ -s & c \end{bmatrix}
$$

such that the $(1,1)$ entry of $J^T AJ$ equals γ. Show that this requirement leads to the
equation
$$
(w - \gamma)\tau^2 - 2x\tau + (z - \gamma) = 0,
$$
where $\tau = c/s$. Verify that this quadratic has real roots if γ satisfies $\lambda_2 \le \gamma \le \lambda_1$, where
λ_1 and λ_2 are the eigenvalues of A.

P8.4.2 Let $A \in \mathbf{R}^{n \times n}$ be symmetric. Give an algorithm that computes the factorization
$$
Q^T AQ = \gamma I + F
$$
where Q is a product of Jacobi rotations, $\gamma = \text{trace}(A)/n$, and F has zero diagonal
entries. Discuss the uniqueness of Q.

P8.4.3 Formulate Jacobi procedures for (a) skew symmetric matrices and (b) complex

Hermitian matrices.

P8.4.4 Partition the n-by-n real symmetric matrix A as follows:

$$A = \begin{bmatrix} a & v^T \\ v & A_1 \end{bmatrix} \begin{matrix} 1 \\ n-1 \end{matrix}$$
$$ \begin{matrix} 1 & n-1 \end{matrix}$$

Let Q be a Householder matrix such that if $B = Q^T A Q$, then $B(3{:}n,1) = 0$. Let $J = J(1,2,\theta)$ be determined such that if $C = J^T B J$, then $c_{12} = 0$ and $c_{11} \geq c_{22}$. Show $c_{11} \geq a + \| v \|_2$. La Budde (1964) formulated an algorithm for the symmetric eigenvalue probem based upon repetition of this Householder-Jacobi computation.

P8.4.5 Organize function **music** so that it involves minimum workspace.

P8.4.6 When implementing cyclic Jacobi, it is sensible to skip the annihilation of a_{pq} if its modulus is less than some small, sweep-dependent parameter, because the net reduction in off(A) is not worth the cost. This leads to what is called the *threshold Jacobi method*. Details concerning this variant of Jacobi's algorithm may be found in Wilkinson (1965, p.277). Show that appropriate thresholding can guarantee convergence.

Notes and References for Sec. 8.4

Jacobi's original paper is one of the earliest references found in the numerical analysis literature

C.G.J. Jacobi (1846). "Uber ein Leichtes Verfahren Die in der Theorie der Sacularstroungen Vorkommendern Gleichungen Numerisch Aufzulosen," *Crelle's J. 30*, 51–94.

Prior to the QR algorithm, the Jacobi technique was the standard method for solving dense symmetric eigenvalue problems. Early attempts to improve upon it include

M. Lotkin (1956). "Characteristic Values of Arbitrary Matrices," *Quart. Appl. Math. 14*, 267–75.

D.A. Pope and C. Tompkins (1957). "Maximizing Functions of Rotations: Experiments Concerning Speed of Diagonalization of Symmetric Matrices Using Jacobi's Method," *J. ACM 4*, 459–66.

C.D. La Budde (1964). "Two Classes of Algorithms for Finding the Eigenvalues and Eigenvectors of Real Symmetric Matrices," *J. ACM 11*, 53–58.

The computational aspects of Jacobi method are described in Wilkinson (1965,p.265). See also

H. Rutishauser (1966). "The Jacobi Method for Real Symmetric Matrices," *Numer. Math. 9*, 1–10. See also Wilkinson and Reinsch (1971, pp. 202–11).

N. Mackey (1995). "Hamilton and Jacobi Meet Again: Quaternions and the Eigenvalue Problem," *SIAM J. Matrix Anal. Applic. 16*, 421–435.

The method is also useful when a nearly diagonal matrix must be diagonalized. See

J.H. Wilkinson (1968). "Almost Diagonal Matrices with Multiple or Close Eigenvalues," *Lin. Alg. and Its Applic. I*, 1–12.

Establishing the quadratic convergence of the classical and cyclic Jacobi iterations has attracted much attention:

P. Henrici (1958). "On the Speed of Convergence of Cyclic and Quasicyclic Jacobi Methods for Computing the Eigenvalues of Hermitian Matrices," *SIAM J. Appl. Math. 6*, 144–62.

E.R. Hansen (1962). "On Quasicyclic Jacobi Methods," *ACM J. 9*, 118–35.

J.H. Wilkinson (1962). "Note on the Quadratic Convergence of the Cyclic Jacobi Process," *Numer. Math. 6*, 296–300.
E.R. Hansen (1963). "On Cyclic Jacobi Methods," *SIAM J. Appl. Math. 11*, 448–59.
A. Schonhage (1964). "On the Quadratic Convergence of the Jacobi Process," *Numer. Math. 6*, 410–12.
H.P.M. van Kempen (1966). "On Quadratic Convergence of the Special Cyclic Jacobi Method," *Numer. Math. 9*, 19–22.
P. Henrici and K. Zimmermann (1968). "An Estimate for the Norms of Certain Cyclic Jacobi Operators," *Lin. Alg. and Its Applic. 1*, 489–501.
K.W. Brodlie and M.J.D. Powell (1975). "On the Convergence of Cyclic Jacobi Methods," *J. Inst. Math. Applic. 15*, 279–87.

Detailed error analyses that establish imprtant componentwise error bounds include

J. Barlow and J. Demmel (1990). "Computing Accurate Eigensystems of Scaled Diagonally Dominant Matrices," *SIAM J. Numer. Anal. 27*, 762–791.
J.W. Demmel and K. Veselič (1992). "Jacobi's Method is More Accurate than QR," *SIAM J. Matrix Anal. Appl. 13*, 1204–1245.
Z. Drmač (1994). *The Generalized Singular Value Problem*, Ph.D. Thesis, FernUniversitat, Hagen, Germany.
W.F. Mascarenhas (1994). "A Note on Jacobi Being More Accurate than QR," *SIAM J. Matrix Anal. Appl. 15*, 215–218.
R. Mathias (1995). "Accurate Eigensystem Computations by Jacobi Methods," *SIAM J. Matrix Anal. Appl. 16*, 977–1003.

Attempts have been made to extend the Jacobi iteration to other classes of matrices and to push through corresponding convergence results. The case of normal matrices is discussed in

H.H. Goldstine and L.P. Horowitz (1959). "A Procedure for the Diagonalization of Normal Matrices," *J. Assoc. Comp. Mach. 6*, 176–95.
G. Loizou (1972). "On the Quadratic Convergence of the Jacobi Method for Normal Matrices," *Comp. J. 15*, 274–76.
A. Ruhe (1972). "On the Quadratic Convergence of the Jacobi Method for Normal Matrices," *BIT 7*, 305–13.

See also

M.H.C. Paardekooper (1971). "An Eigenvalue Algorithm for Skew Symmetric Matrices," *Numer. Math. 17*, 189–202.
D. Hacon (1993). "Jacobi's Method for Skew-Symmetric Matrices," *SIAM J. Matrix Anal. Appl. 14*, 619–628.

Essentially, the analysis and algorithmic developments presented in the text carry over to the normal case with minor modification. For non-normal matrices, the situation is considerably more difficult. Consult

J. Greenstadt (1955). "A Method for Finding Roots of Arbitrary Matrices," *Math. Tables and Other Aids to Comp. 9*, 47–52.
C.E. Froberg (1965). "On Triangularization of Complex Matrices by Two Dimensional Unitary Tranformations," *BIT 5*, 230–34.
J. Boothroyd and P.J. Eberlein (1968). "Solution to the Eigenproblem by a Norm-Reducing Jacobi-Type Method (Handbook)," *Numer. Math. 11*, 1–12. See also Wilkinson and Reinsch (1971, pp.327–38).
A. Ruhe (1968). On the Quadratic Convergence of a Generalization of the Jacobi Method to Arbitrary Matrices," *BIT 8*, 210–31.
A. Ruhe (1969). "The Norm of a Matrix After a Similarity Transformation," *BIT 9*, 53–58.

P.J. Eberlein (1970). "Solution to the Complex Eigenproblem by a Norm-Reducing Jacobi-type Method," *Numer. Math. 14*, 232–45. See also Wilkinson and Reinsch (1971, pp.404–17).

C.P. Huang (1975). "A Jacobi-Type Method for Triangularizing an Arbitrary Matrix," *SIAM J. Num. Anal. 12*, 566–70.

V. Hari (1982). "On the Global Convergence of the Eberlein Method for Real Matrices," *Numer. Math. 39*, 361–370.

G.W. Stewart (1985). "A Jacobi-Like Algorithm for Computing the Schur Decomposition of a Nonhermitian Matrix," *SIAM J. Sci. and Stat. Comp. 6*, 853–862.

W-W. Lin and C.W. Chen (1991). "An Acceleration Method for Computing the Generalized Eigenvalue Problem on a Parallel Computer," *Lin.Alg. and Its Applic. 146*, 49–65.

Jacobi methods for complex symmetric matrices have also been developed. See

J.J. Seaton (1969). "Diagonalization of Complex Symmetric Matrices Using a Modified Jacobi Method," *Comp. J. 12*, 156–57.

P.J. Eberlein (1971). "On the Diagonalization of Complex Symmetric Matrices," *J. Inst. Math. Applic. 7*, 377–83.

P. Anderson and G. Loizou (1973). "On the Quadratic Convergence of an Algorithm Which Diagonalizes a Complex Symmetric Matrix," *J. Inst. Math. Applic. 12*, 261–71.

P. Anderson and G. Loizou (1976). "A Jacobi-Type Method for Complex Symmetric Matrices (Handbook)," *Numer. Math. 25*, 347–63.

Although the symmetric QR algorithm is generally much faster than the Jacobi method, there are special settings where the latter technique is of interest. As we illustrated, on a parallel-computer it is possible to perform several rotations concurrently, thereby accelerating the reduction of the off-diagonal elements. See

A. Sameh (1971). "On Jacobi and Jacobi-like Algorithms for a Parallel Computer," *Math. Comp. 25*, 579–90.

J.J. Modi and J.D. Pryce (1985). "Efficient Implementation of Jacobi's Diagonalization Method on the DAP," *Numer. Math. 46*, 443–454.

D.S. Scott, M.T. Heath, and R.C. Ward (1986). "Parallel Block Jacobi Eigenvalue Algorithms Using Systolic Arrays," *Lin. Alg. and Its Applic. 77*, 345–356.

P.J. Eberlein (1987). "On Using the Jacobi Method on a Hypercube," in *Hypercube Multiprocessors*, ed. M.T. Heath, SIAM Publications, Philadelphia.

G. Shroff and R. Schreiber (1989). "On the Convergence of the Cyclic Jacobi Method for Parallel Block Orderings," *SIAM J. Matrix Anal. Appl. 10*, 326–346.

M.H.C. Paardekooper (1991). "A Quadratically Convergent Parallel Jacobi Process for Diagonally Dominant Matrices with Nondistinct Eigenvalues," *Lin.Alg. and Its Applic. 145*, 71–88.

8.5 Tridiagonal Methods

In this section we develop special methods for the symmetric tridiagonal eigenproblem. The tridiagonal form

$$
T = \begin{bmatrix}
a_1 & b_1 & & \cdots & & 0 \\
b_1 & a_2 & \ddots & & & \vdots \\
& \ddots & \ddots & \ddots & & \\
\vdots & & \ddots & \ddots & b_{n-1} \\
0 & \cdots & & & b_{n-1} & a_n
\end{bmatrix}
\tag{8.5.1}
$$

can be obtained by Householder reduction (cf. §8.3.1). However, symmetric tridiagonal eigenproblems arise naturally in many settings.

We first discuss bisection methods that are of interest when selected portions of the eigensystem are required. This is followed by the presentation of a divide and conquer algorithm that can be used to acquire the full symmetric Schur decomposition in a way that is amenable to parallel processing.

8.5.1 Eigenvalues by Bisection

Let T_r denote the leading r-by-r principal submatrix of the matrix T in (8.5.1). Define the polynomials $p_r(x) = \det(T_r - xI)$, $r = 1{:}n$. A simple determinantal expansion shows that

$$
p_r(x) = (a_r - x)p_{r-1}(x) - b_{r-1}^2 p_{r-2}(x)
\tag{8.5.2}
$$

for $r = 2{:}n$ if we set $p_0(x) = 1$. Because $p_n(x)$ can be evaluated in $O(n)$ flops, it is feasible to find its roots using the method of bisection. For example, if $p_n(y)p_n(z) < 0$ and $y < z$, then the iteration

> **while** $|y - z| > \epsilon(|y| + |z|)$
> $\quad x = (y + z)/2$
> $\quad$ **if** $p_n(x)p_n(y) < 0$
> $\quad\quad z = x$
> $\quad$ **else**
> $\quad\quad y = x$
> $\quad$ **end**
> **end**

is guaranteed to terminate with $(y + z)/2$ an approximate zero of $p_n(x)$, i.e., an approximate eigenvalue of T. The iteration converges linearly in that the error is approximately halved at each step.

8.5.2   Sturm Sequence Methods

Sometimes it is necessary to compute the kth largest eigenvalue of T for some prescribed value of k. This can be done efficiently by using the bisection idea and the following classical result:

Theorem 8.5.1 (Sturm Sequence Property) *If the tridiagonal matrix in (8.5.1) has no zero subdiagonal entries, then the eigenvalues of T_{r-1} strictly separate the eigenvalues of T_r:*

$$\lambda_r(T_r) < \lambda_{r-1}(T_{r-1}) < \lambda_{r-1}(T_r) < \cdots < \lambda_2(T_r) < \lambda_1(T_{r-1}) < \lambda_1(T_r).$$

Moreover, if $a(\lambda)$ denotes the number of sign changes in the sequence

$$\{\, p_0(\lambda),\ p_1(\lambda), \ldots,\ p_n(\lambda) \,\}$$

then $a(\lambda)$ equals the number of T's eigenvalues that are less than λ. Here, the polynomials $p_r(x)$ are defined by (8.5.2) and we have the convention that $p_r(\lambda)$ has the opposite sign of $p_{r-1}(\lambda)$ if $p_r(\lambda) = 0$.

Proof. It follows from Theorem 8.1.7 that the eigenvalues of T_{r-1} weakly separate those of T_r. To prove that the separation must be strict, suppose that $p_r(\mu) = p_{r-1}(\mu) = 0$ for some r and μ. It then follows from (8.5.2) and the assumption that T is unreduced that $p_0(\mu) = p_1(\mu) = \cdots = p_r(\mu) = 0$, a contradiction. Thus, we must have strict separation.

The assertion about $a(\lambda)$ is established in Wilkinson (1965, 300-301). We mention that if $p_r(\lambda) = 0$, then its sign is assumed to be opposite the sign of $p_{r-1}(\lambda)$. $\square$

Example 8.5.1 If

$$T = \begin{bmatrix} 1 & -1 & 0 & 0 \\ -1 & 2 & -1 & 0 \\ 0 & -1 & 3 & -1 \\ 0 & 0 & -1 & 4 \end{bmatrix}$$

then $\lambda(T) \approx \{.254,\ 1.82,\ 3.18,\ 4.74\}$. The sequence

$$\{\, p_0(2),\ p_1(2),\ p_2(2),\ p_3(2),\ p_4(2) \,\}\ =\ \{\, 1,\ -1,\ -1,\ 0,\ 1 \,\}$$

confirms that there are two eigenvalues less than $\lambda = 2$.

Suppose we wish to compute $\lambda_k(T)$. From the Gershgorin theorem (Theorem 8.1.3) it follows that $\lambda_k(T) \in [y, z]$ where

$$y = \min_{1 \le i \le n} a_i - |b_i| - |b_{i-1}| \qquad z = \max_{1 \le i \le n} a_i + |b_i| + |b_{i-1}|$$

if we define $b_0 = b_n = 0$. With these starting values, it is clear from the Sturm sequence property that the iteration

```
while |z − y| > u(|y| + |z|)
    x = (y + z)/2
    if a(x) ≥ n − k                                    (8.5.3)
        z = x
    else
        y = x
    end
end
```

produces a sequence of subintervals that are repeatedly halved in length but which always contain $\lambda_k(T)$.

Example 8.5.2 If (8.5.3) is applied to the matrix of Example 8.5.1 with $k = 3$, then the values shown in the following table are generated:

y	z	x	$a(x)$
0.0000	5.0000	2.5000	2
0.0000	2.5000	1.2500	1
1.2500	2.5000	1.3750	1
1.3750	2.5000	1.9375	2
1.3750	1.9375	1.6563	1
1.6563	1.9375	1.7969	1

We conclude from the output that $\lambda_3(T) \in [1.7969, 1.9375]$. Note: $\lambda_3(T) \approx 1.82$.

During the execution of (8.5.3), information about the location of other eigenvalues is obtained. By systematically keeping track of this information it is possible to devise an efficient scheme for computing "contiguous" subsets of $\lambda(T)$, e.g., $\lambda_k(T), \lambda_{k+1}(T), \ldots, \lambda_{k+j}(T)$. See Barth, Martin, and Wilkinson (1967).

If selected eigenvalues of a general symmetric matrix A are desired, then it is necessary first to compute the tridiagonalization $T = U_0^T T U_0$ before the above bisection schemes can be applied. This can be done using Algorithm 8.3.1 or by the Lanczos algorithm discussed in the next chapter. In either case, the corresponding eigenvectors can be readily found via inverse iteration since tridiagonal systems can be solved in $O(n)$ flops. See §4.3.6 and §8.2.2.

In those applications where the original matrix A already has tridiagonal form, bisection computes eigenvalues with small relative error, regardless of their magnitude. This is in contrast to the tridiagonal QR iteration, where the computed eigenvalues $\tilde{\lambda}_i$ can be guaranteed only to have small absolute error: $|\tilde{\lambda}_i - \lambda_i(T)| \approx u\| T \|_2$

Finally, it is possible to compute specific eigenvalues of a symmetric matrix by using the LDL^T factorization (see §4.2) and exploiting the Sylvester inertia theorem (Theorem 8.1.17). If

$$A - \mu I = LDL^T \qquad A = A^T \in \mathbb{R}^{n \times n}$$

is the LDL^T factorization of $A - \mu I$ with $D = \text{diag}(d_1, \ldots, d_n)$, then the number of negative d_i equals the number of $\lambda_i(A)$ that are less than μ. See Parlett (1980, p.46) for details.

8.5.3   Eigensystems of Diagonal Plus Rank-1 Matrices

Our next method for the symmetric tridiagonal eigenproblem requires that we be able to compute efficiently the eigenvalues and eigenvectors of a matrix of the form $D + \rho z z^T$ where $D \in \mathbb{R}^{n \times n}$ is diagonal, $z \in \mathbb{R}^z$, and $\rho \in \mathbb{R}$. This problem is important in its own right and the key computations rest upon the following pair of results.

Lemma 8.5.2 *Suppose* $D = \text{diag}(d_1, \ldots, d_n) \in \mathbb{R}^{n \times n}$ *has the property that* $d_1 > \cdots > d_n$. *Assume that* $\rho \neq 0$ *and that* $z \in \mathbb{R}^n$ *has no zero components. If*

$$(D + \rho z z^T)v = \lambda v \qquad v \neq 0$$

then $z^T v \neq 0$ *and* $D - \lambda I$ *is nonsingular.*

Proof. If $\lambda \in \lambda(D)$, then $\lambda = d_i$ for some i and thus

$$0 = e_i^T[(D - \lambda I)v + \rho(z^T v)z] = \rho(z^T v)z_i.$$

Since ρ and z_i are nonzero we must have $0 = z^T v$ and so $Dv = \lambda v$. However, D has distinct eigenvalues and therefore, $v \in \text{span}\{e_i\}$. But then $0 = z^T v = z_i$, a contradiction. Thus, D and $D + \rho z z^T$ do not have any common eigenvalues and $z^T v \neq 0$. $\square$

Theorem 8.5.3 *Suppose* $D = \text{diag}(d_1, \ldots, d_n) \in \mathbb{R}^{n \times n}$ *and that the diagonal entries satisfy* $d_1 > \cdots > d_n$. *Assume that* $\rho \neq 0$ *and that* $z \in \mathbb{R}^n$ *has no zero components. If* $V \in \mathbb{R}^{n \times n}$ *is orthogonal such that*

$$V^T(D + \rho z z^T)V = \text{diag}(\lambda_1, \ldots, \lambda_n)$$

with $\lambda_1 \geq \cdots \geq \lambda_n$ *and* $V = [v_1, \ldots, v_n]$, *then*

(a) *The* λ_i *are the* n *zeros of* $f(\lambda) = 1 + \rho z^T (D - \lambda I)^{-1} z$.

(b) *If* $\rho > 0$, *then* $\lambda_1 > d_1 > \lambda_2 > \cdots > d_n$.
     If $\rho < 0$, *then* $d_1 > \lambda_1 > d_2 > \cdots > d_n > \lambda_n$.

(c) *The eigenvector* v_i *is a multiple of* $(D - \lambda_i I)^{-1} z$.

Proof. If $(D + \rho z z^T)v = \lambda v$, then

$$(D - \lambda I)v + \rho(z^T v)z = 0. \tag{8.5.4}$$

We know from Lemma 8.5.2 that $D - \lambda I$ is nonsingular. Thus,

$$v \in \text{span}\{(D - \lambda I)^{-1}z\}$$

thereby establishing (c). Moreover, if we apply $z^T(D - \lambda I)^{-1}$ to both sides of equation (8.5.4) we obtain

$$z^T v \left(1 + \rho z^T (D - \lambda I)^{-1} z\right) = 0.$$

By Lemma 8.5.2, $z^T v \neq 0$ and so this shows that if $\lambda \in \lambda(D + \rho z z^T)$, then $f(\lambda) = 0$. We must show that all the zeros of f are eigenvalues of $D + \rho z z^T$ and that the interlacing relations (b) hold.

To do this we look more carefully at the equations

$$f(\lambda) = 1 + \rho \left(\frac{z_1^2}{d_1 - \lambda} + \cdots + \frac{z_n^2}{d_n - \lambda}\right)$$

$$f'(\lambda) = \rho \left(\frac{z_1^2}{(d_1 - \lambda)^2} + \cdots + \frac{z_n^2}{(d_n - \lambda)^2}\right)$$

Note that f is monotone in between its poles. This allows us to conclude that if $\rho > 0$, then f has precisely n roots, one in each of the intervals

$$(d_n, d_{n-1}), \ldots, (d_2, d_1), (d_1, \infty).$$

If $\rho < 0$ then f has exactly n roots, one in each of the intervals

$$(-\infty, d_n), (d_n, d_{n-1}), \ldots, (d_2, d_1).$$

In either case, it follows that the zeros of f are precisely the eigenvalues of $D + \rho v v^T$. $\square$

The theorem suggests that to compute V we (a) find the roots $\lambda_1, \ldots, \lambda_n$ of f using a Newton-like procedure and then (b) compute the columns of V by normalizing the vectors $(D - \lambda_i I)^{-1}z$ for $i = 1{:}n$. The same plan of attack can be followed even if there are repeated d_i and zero z_i.

Theorem 8.5.4 *If $D = \text{diag}(d_1, ..., d_n)$ and $z \in \mathbb{R}^n$, then there exists an orthogonal matrix V_1 such that if $V_1^T D V_1 = \text{diag}(\mu_1, \ldots, \mu_n)$ and $w = V_1^T z$ then*

$$\mu_1 > \mu_2 > \cdots > \mu_r \geq \mu_{r+1} \geq \cdots \geq \mu_n,$$

$w_i \neq 0$ for $i = 1{:}r$, and $w_i = 0$ for $i = r + 1{:}n$.

Proof. We give a constructive proof based upon two elementary operations. (a) Suppose $d_i = d_j$ for some $i < j$. Let $J(i, j, \theta)$ be a Jacobi rotation in the (i, j) plane with the property that the jth component of $J(i, j, \theta)^T z$ is zero. It is not hard to show that $J(i, j, \theta)^T D J(i, j, \theta) = D$. Thus, we can zero a component of z if there is a repeated d_i. (b) If $z_i = 0$,

$z_j \neq 0$, and $i < j$, then let P be the identity with columns i and j interchanged. It follows that $P^T D P$ is diagonal, $(P^T z)_i \neq 0$, and $(P^T z)_j = 0$. Thus, we can permute all the zero z_i to the "bottom." Clearly, repetition of (a) and (b) eventually renders the desired canonical structure. V_1 is the product of the rotations. $\square$

See Barlow (1993) and the references therein for a discussion of the solution procedures that we have outlined above.

8.5.4   A Divide and Conquer Method

We now present a divide-and-conquer method for computing the Schur decomposition

$$Q^T T Q = \Lambda = \operatorname{diag}(\lambda_1, \ldots, \lambda_n) \qquad Q^T Q = I \qquad (8.5.5)$$

for tridiagonal T that involves (a) "tearing" T in half, (b) computing the the Schur decompositions of the two parts, and (c) combining the two half-sized Schur decompositions into the required full size Schur decomposition. The overall procedure, developed by Dongarra and Sorensen (1987), is suitable for parallel computation.

We first show how T can be "torn" in half with a rank-one modification. For simplicity, assume $n = 2m$. Define $v \in \mathbb{R}^n$ as follows

$$v = \begin{bmatrix} e_m^{(m)} \\ \theta e_1^{(m)} \end{bmatrix}. \qquad (8.5.6)$$

Note that for all $\rho \in \mathbb{R}$ the matrix $\tilde{T} = T - \rho v v^T$ is identical to T except in its "middle four" entries:

$$\tilde{T}(m{:}m+1, m{:}m+1) = \begin{bmatrix} a_m - \rho & b_m - \rho\theta \\ b_m - \rho\theta & a_{m+1} - \rho\theta^2 \end{bmatrix}.$$

If we set $\rho\theta = b_m$ then

$$T = \begin{bmatrix} T_1 & 0 \\ 0 & T_2 \end{bmatrix} + \rho v v^T$$

where

$$T_1 = \begin{bmatrix} a_1 & b_1 & & \cdots & & 0 \\ b_1 & a_2 & \ddots & & & \vdots \\ & \ddots & \ddots & \ddots & & \\ \vdots & & \ddots & \ddots & & b_{m-1} \\ 0 & \cdots & & & b_{m-1} & \tilde{a}_m \end{bmatrix},$$

$$T_2 = \begin{bmatrix} \tilde{a}_{m+1} & b_{m+1} & & \cdots & & 0 \\ b_{m+1} & a_{m+2} & \ddots & & & \vdots \\ & \ddots & \ddots & \ddots & & \\ \vdots & & \ddots & \ddots & & b_{n-1} \\ 0 & \cdots & & & b_{n-1} & a_n \end{bmatrix},$$

and $\tilde{a}_m = a_m - \rho$ and $\tilde{a}_{m+1} = a_{m+1} - \rho\theta^2$.

Now suppose that we have m-by-m orthogonal matrices Q_1 and Q_2 such that $Q_1^T T_1 Q_1 = D_1$ and $Q_2^T T_2 Q_2 = D_2$ are each diagonal. If we set

$$U = \begin{bmatrix} Q_1 & 0 \\ 0 & Q_2 \end{bmatrix},$$

then

$$U^T T U = U^T \left(\begin{bmatrix} T_1 & 0 \\ 0 & T_2 \end{bmatrix} + \rho v v^T \right) U = D + \rho z z^T$$

where

$$D = \begin{bmatrix} D_1 & 0 \\ 0 & D_2 \end{bmatrix}$$

is diagonal and

$$z = U^T v = \begin{bmatrix} Q_1^T e_m \\ \theta Q_2^T e_1 \end{bmatrix}.$$

Comparing these equations we see that the effective synthesis of the two half-sized Schur decompositions requires the quick and stable computation of an orthogonal V such that

$$V^T (D + \rho z z^T) V = \Lambda = \text{diag}(\lambda_1, \ldots, \lambda_n)$$

which we discussed in §8.5.3.

8.5.5 A Parallel Implementation

Having stepped through the tearing and synthesis operations, we are ready to illustrate the overall process and how it can be implemented on a multiprocessor. For clarity, assume that $n = 8N$ for some positive integer N and that three levels of tearing are performed. We can depict this with a binary tree as shown in FIG. 8.5.1. The indices are specified in binary. FIG. 8.5.2 depicts a single node and should be interpreted to mean that the eigensystem for the tridiagonal $T(b)$ is obtained from the eigensystems of the tridiagonals $T(b0)$ and $T(b1)$. For example, the eigensystems for the N-by-N matrices $T(110)$ and $T(111)$ are combined to produce the eigensystem for the $2N$-by-$2N$ tridiagonal matrix $T(11)$.

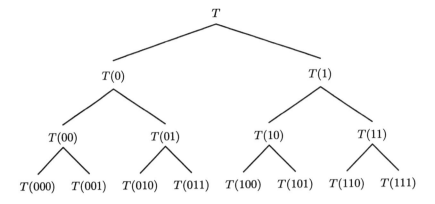

FIGURE 8.5.1 *Computation Tree*

FIGURE 8.5.2 *Synthesis at a Node*

With tree-structured algorithms there is always the danger that parallelism is lost as the tree is "climbed" towards the root, but this is not the case in our problem. To see this suppose we have 8 processors and that the first task of Proc(b) is to compute the Schur decomposition of $T(b)$ where $b = 000, 001, 010, 011, 100, 101, 110, 111$. This portion of the computation is perfectly load balanced and does not involve interprocessor communication. (We are ignoring the Theorem 8.5.4 deflations, which are unlikely to cause significant load imbalance.)

At the next level there are four gluing operations to perform: $T(00)$, $T(01)$, $T(10)$, $T(11)$. However, each of these computations neatly subdivides and we can assign two processors to each task. For example, once the secular equation that underlies the $T(00)$ synthesis is known to both Proc(000) and Proc(001), then they each can go about getting half of the eigenvalues and corresponding eigenvectors. Likewise, 4 processors can each be assigned to the $T(0)$ and $T(1)$ problem. All 8 processors can participate in computing the eigensystem of T. Thus, at every level full parallelism can be maintained because the eigenvalue/eigenvector computations are independent of one another.

Problems

P8.5.1 Suppose λ is an eigenvalue of a symmetric tridiagonal matrix T. Show that if λ has algebraic multiplicity k, then at least $k - 1$ of T's subdiagonal elements are zero.

P8.5.2 Give an algorithm for determining ρ and θ in (8.5.6) with the property that $\theta \in \{-1, 1\}$ and $\min\{|a_r - \rho|, |a_{r+1} - \rho|\}$ is maximized.

P8.5.3 Let $p_r(\lambda) = \det(T(1{:}r, 1{:}r) - \lambda I_r)$ where T is given by (8.5.1). Derive a recursion for evaluating $p'_n(\lambda)$ and use it to develop a Newton iteration that can compute eigenvalues of T.

P8.5.4 What communication is necessary between the processors assigned to a particular T_b? Is it possible to share the work associated with the processing of repeated d_i and zero z_i ?

P8.5.5 If T is positive definite, does it follow that the matrices T_1 and T_2 in §8.5.4 are positive definite?

P8.5.6 Suppose that

$$A = \begin{bmatrix} D & v \\ v^T & d_{nn} \end{bmatrix}$$

where $D = \mathrm{diag}(d_1, \ldots, d_{n-1})$ has distinct diagonal entries and $v \in \mathbf{R}^{n-1}$ has no zero entries. (a) Show that if $\lambda \in \lambda(A)$, then $D - \lambda I_{n-1}$ is nonsingular. (b) Show that if $\lambda \in \lambda(A)$, then λ is a zero of

$$f(\lambda) = \lambda + \sum_{k=1}^{n-1} \frac{v_k^2}{d_k - \lambda} - d_n.$$

P8.5.7 Suppose $A = S + \sigma u u^T$ where $S \in \mathbf{R}^{n \times n}$ is skew-symmetric, $u \in \mathbf{R}^n$, and $\sigma \in \mathbf{R}$. Show how to compute an orthogonal Q such that $Q^T A Q = T + \sigma e_1 e_1^T$ where T is tridiagonal and skew-symmetric and e_1 is the first column of I_n.

P8.5.8 It is known that $\lambda \in \lambda(T)$ where $T \in \mathbf{R}^{n \times n}$ is symmetric and tridiagonal with no zero subdiagonal entries. Show how to compute $x(1{:}n-1)$ from the equation $Tx = \lambda x$ given that $x_n = 1$.

Notes and References for Sec. 8.5

Bisection/ Strum sequence methods are discussed in

W. Barth, R.S. Martin, and J.H. Wilkinson (1967). "Calculation of the Eigenvalues of a Symmetric Tridiagonal Matrix by the Method of Bisection," *Numer. Math. 9*, 386–93. See also Wilkinson and Reinsch (1971, 249–256).

K.K. Gupta (1972). "Solution of Eigenvalue Problems by Sturm Sequence Method," *Int. J. Numer. Meth. Eng. 4*, 379–404.

Various aspects of the divide and conquer algorithm discussed in this section is detailed in

G.H. Golub (1973). "Some Modified Matrix Eigenvalue Problems," *SIAM Review 15*, 318–44.

J.R. Bunch, C.P. Nielsen, and D.C. Sorensen (1978). "Rank-One Modification of the Symmetric Eigenproblem," *Numer. Math. 31*, 31–48.

J.J.M. Cuppen (1981). "A Divide and Conquer Method for the Symmetric Eigenproblem," *Numer. Math. 36*, 177–95.

J.J. Dongarra and D.C. Sorensen (1987). "A Fully Parallel Algorithm for the Symmetric Eigenvalue Problem," *SIAM J. Sci. and Stat. Comp. 8*, S139–S154.
S. Crivelli and E.R. Jessup (1995). "The Cost of Eigenvalue Computation on Distributed Memory MIMD Computers," *Parallel Computing 21*, 401–422.

The very delicate computations required by the method are carefully analyzed in

J.L. Barlow (1993). "Error Analysis of Update Methods for the Symmetric Eigenvalue Problem," *SIAM J. Matrix Anal. Appl. 14*, 598–618.

Various generalizations to banded symmetric eigenproblems have been explored.

P. Arbenz, W. Gander, and G.H. Golub (1988). "Restricted Rank Modification of the Symmetric Eigenvalue Problem: Theoretical Considerations," *Lin. Alg. and Its Applic. 104*, 75–95.
P. Arbenz and G.H. Golub (1988). "On the Spectral Decomposition of Hermitian Matrices Subject to Indefinite Low Rank Perturbations with Applications," *SIAM J. Matrix Anal. Appl. 9*, 40–58.

A related divide and conquer method based on the "arrowhead" matrix (see P8.5.7) is given in

M. Gu and S.C. Eisenstat (1995). "A Divide-and-Conquer Algorithm for the Symmetric Tridiagonal Eigenproblem," *SIAM J. Matrix Anal. Appl. 16*, 172–191.

8.6   Computing the SVD

There are important relationships between the singular value decomposition of a matrix A and the Schur decompositions of the symmetric matrices $A^T A$, AA^T, and $\begin{bmatrix} 0 & A^T \\ A & 0 \end{bmatrix}$. Indeed, if

$$U^T AV = \text{diag}(\sigma_1, \ldots, \sigma_n)$$

is the SVD of $A \in \mathbb{R}^{m \times n}$ $(m \geq n)$, then

$$V^T(A^T A)V = \text{diag}(\sigma_1^2, \ldots, \sigma_n^2) \in \mathbb{R}^{n \times n} \tag{8.6.1}$$

and

$$U^T(AA^T)U = \text{diag}(\sigma_1^2, \ldots, \sigma_n^2, \underbrace{0, \ldots, 0}_{m-n}) \in \mathbb{R}^{m \times m} \tag{8.6.2}$$

Moreover, if

$$U = \begin{array}{cc} [\, U_1 & U_2 \,] \\ n & m-n \end{array}$$

and we define the orthogonal matrix $Q \in \mathbb{R}^{(m+n) \times (m+n)}$ by

$$Q = \frac{1}{\sqrt{2}} \begin{bmatrix} V & V & 0 \\ U_1 & -U_1 & \sqrt{2}\,U_2 \end{bmatrix}$$

then

$$Q^T \begin{bmatrix} 0 & A^T \\ A & 0 \end{bmatrix} Q = \mathrm{diag}(\sigma_1, \ldots, \sigma_n, -\sigma_1, \ldots, -\sigma_n, \underbrace{0, \ldots, 0}_{m-n}). \quad (8.6.3)$$

These connections to the symmetric eigenproblem allow us to adapt the mathematical and algorithmic developments of the previous sections to the singular value problem. Good references for this section include Lawson and Hanson (1974) and Stewart and Sun (1990).

8.6.1   Perturbation Theory and Properties

We first establish perturbation results for the SVD based on the theorems of §8.1. Recall that $\sigma_i(A)$ denotes the ith largest singular value of A.

Theorem 8.6.1 *If $A \in \mathbb{R}^{m \times n}$, then for $k = 1{:}\min\{m, n\}$*

$$\sigma_k(A) = \max_{\substack{\dim(S)=k \\ \dim(T)=k}} \min_{\substack{x \in S \\ y \in T}} \frac{y^T A x}{\| x \|_2 \| y \|_2} = \max_{\dim(S)=k} \min_{x \in S} \frac{\| A x \|_2}{\| x \|_2}.$$

Note that in this expression $S \subseteq \mathbb{R}^n$ and $T \subseteq \mathbb{R}^m$ are subspaces.

Proof. The right-most characterization follows by applying Theorem 8.1.2 to $A^T A$. The remainder of the proof we leave as an exercise. $\square$

Corollary 8.6.2 *If A and $A+E$ are in $\mathbb{R}^{m \times n}$ with $m \geq n$, then for $k = 1{:}n$*

$$|\sigma_k(A + E) - \sigma_k(A)| \leq \sigma_1(E) = \| E \|_2.$$

Proof. Apply Corollary 8.1.6 to

$$\begin{bmatrix} 0 & A^T \\ A & 0 \end{bmatrix} \quad \text{and} \quad \begin{bmatrix} 0 & (A+E)^T \\ A+E & 0 \end{bmatrix}. \ \square$$

Example 8.6.1 If

$$A = \begin{bmatrix} 1 & 4 \\ 2 & 5 \\ 3 & 6 \end{bmatrix} \quad \text{and} \quad A + E = \begin{bmatrix} 1 & 4 \\ 2 & 5 \\ 3 & 6.01 \end{bmatrix}$$

then $\sigma(A) = \{9.5080, .7729\}$ and $\sigma(A + E) = \{9.5145, .7706\}$. It is clear that for $i = 1{:}2$ we have $|\sigma_i(A + E) - \sigma_i(A)| \leq \| E \|_2 = .01$.

Corollary 8.6.3 *Let $A = [\, a_1, \ldots, a_n \,] \in \mathbb{R}^{m \times n}$ be a column partitioning with $m \geq n$. If $A_r = [\, a_1, \ldots, a_r \,]$, then for $r = 1{:}n - 1$*

$$\sigma_1(A_{r+1}) \geq \sigma_1(A_r) \geq \sigma_2(A_{r+1}) \geq \cdots \geq \sigma_r(A_{r+1}) \geq \sigma_r(A_r) \geq \sigma_{r+1}(A_{r+1}).$$

Proof. Apply Corollary 8.1.7 to $A^T A$. $\square$

This last result says that by adding a column to a matrix, the largest
singular value increases and the smallest singular value is diminished.

Example 8.3.2

$$
A = \begin{bmatrix} 1 & 6 & 11 \\ 2 & 7 & 12 \\ 3 & 8 & 13 \\ 4 & 9 & 14 \\ 5 & 10 & 15 \end{bmatrix} \quad \Rightarrow \quad \begin{cases} \sigma(A_1) = \{7.4162\} \\ \sigma(A_2) = \{19.5377, 1.8095\} \\ \sigma(A_3) = \{35.1272, 2.4654, 0.0000\} \end{cases}
$$

thereby confirming Corollary 8.6.3.

The next result is a Wielandt-Hoffman theorem for singular values:

Theorem 8.6.4 *If A and $A + E$ are in $\mathbb{R}^{m \times n}$ with $m \geq n$, then*

$$
\sum_{k=1}^{n} (\sigma_k(A + E) - \sigma_k(A))^2 \leq \| E \|_F^2 .
$$

Proof. Apply Theorem 8.1.4 to $\begin{bmatrix} 0 & A^T \\ A & 0 \end{bmatrix}$ and $\begin{bmatrix} 0 & (A+E)^T \\ A+E & 0 \end{bmatrix}$. $\square$

Example 8.6.3 If

$$
A = \begin{bmatrix} 1 & 4 \\ 2 & 5 \\ 3 & 6 \end{bmatrix} \quad \text{and} \quad A + E = \begin{bmatrix} 1 & 4 \\ 2 & 5 \\ 3 & 6.01 \end{bmatrix}
$$

then

$$
\sum_{k=1}^{2} (\sigma_k(A + E) - \sigma_k(A))^2 = .472 \times 10^{-4} \leq 10^{-4} = \| E \|_F^2 .
$$

See Example 8.6.1.

For $A \in \mathbb{R}^{m \times n}$ we say that the k-dimensional subspaces $S \subseteq \mathbb{R}^n$ and
$T \subseteq \mathbb{R}^m$ form a *singular subspace pair* if $x \in S$ and $y \in T$ imply $Ax \in T$
and $A^T y \in S$. The following result is concerned with the perturbation of
singular subspace pairs.

Theorem 8.6.5 *Let $A, E \in \mathbb{R}^{m \times n}$ with $m \geq n$ be given and suppose that*
$V \in \mathbb{R}^{n \times n}$ *and* $U \in \mathbb{R}^{m \times m}$ *are orthogonal. Assume that*

$$
V = \begin{bmatrix} V_1 & V_2 \end{bmatrix} \qquad\qquad U = \begin{bmatrix} U_1 & U_2 \end{bmatrix}
$$
$$
\quad\ r \quad\ n-r \qquad\qquad\qquad r \quad\ m-r
$$

and that $\mathrm{ran}(V_1)$ *and* $\mathrm{ran}(U_1)$ *form a singular subspace pair for A. Let*

$$
U^H A V = \begin{bmatrix} A_{11} & 0 \\ 0 & A_{22} \end{bmatrix} \quad \begin{matrix} r \\ m-r \end{matrix}
$$
$$
\qquad\qquad\quad r \quad\ \ n-r
$$

$$U^H E V = \begin{bmatrix} E_{11} & E_{12} \\ E_{21} & E_{22} \end{bmatrix} \begin{matrix} r \\ m-r \end{matrix}$$
$$\begin{matrix} r & n-r \end{matrix}$$

and assume that

$$\delta = \min_{\substack{\sigma \in \sigma(A_{11}) \\ \gamma \in \sigma(A_{22})}} |\sigma - \gamma| > 0.$$

If

$$\| E \|_F \le \frac{\delta}{4},$$

then there exist matrices $P \in \mathbb{R}^{(n-r) \times r}$ *and* $Q \in \mathbb{R}^{(m-r) \times r}$ *satisfying*

$$\left\| \begin{bmatrix} Q \\ P \end{bmatrix} \right\|_F \le 4 \frac{\| E \|_F}{\delta}$$

such that $\mathrm{ran}(V_1 + V_2 Q)$ *and* $\mathrm{ran}(U_1 + U_2 P)$ *is a singular subspace pair for* $A + E$.

Proof. See Stewart (1973), Theorem 6.4. $\square$

Roughly speaking, the theorem says that $O(\epsilon)$ changes in A can alter a singular subspace by an amount ϵ/δ, where δ measures the separation of the relevant singular values.

Example 8.6.4 The matrix $A = \mathrm{diag}(2.000, 1.001, .999) \in \mathbb{R}^{4 \times 3}$ has singular subspace pairs $(\mathrm{span}\{v_i\}, \mathrm{span}\{u_i\})$ for $i = 1, 2, 3$ where $v_i = e_i^{(3)}$ and $u_i = e_i^{(4)}$. Suppose

$$A + E = \begin{bmatrix} 2.000 & .010 & .010 \\ .010 & 1.001 & .010 \\ .010 & .010 & .999 \\ .010 & .010 & .010 \end{bmatrix}$$

The corresponding columns of the matrices

$$\hat{U} = [\, \hat{u}_1 \, \hat{u}_2 \, \hat{u}_3 \,] = \begin{bmatrix} .9999 & -.0144 & .0007 \\ .0101 & .7415 & .6708 \\ .0101 & .6707 & -.7616 \\ .0051 & .0138 & -.0007 \end{bmatrix}$$

$$\hat{V} = [\, \hat{v}_1 \, \hat{v}_2 \, \hat{v}_3 \,] = \begin{bmatrix} .9999 & -.0143 & .0007 \\ .0101 & .7416 & .6708 \\ .0101 & .6707 & -.7416 \end{bmatrix}$$

define singular subspace pairs for $A+E$. Note that the pair $\{\mathrm{span}\{\hat{v}_i\}, \mathrm{span}\{\hat{u}_i\}\}$, is close to $\{\mathrm{span}\{v_i\}, \mathrm{span}\{u_i\}\}$ for $i = 1$ but not for $i = 2$ or 3. On the other hand, the singular subspace pair $\{\mathrm{span}\{\hat{v}_2, \hat{v}_3\}, \mathrm{span}\{\hat{u}_2, \hat{u}_3\}\}$ is close to $\{\mathrm{span}\{v_2, v_3\}, \mathrm{span}\{u_2, u_3\}\}$.

8.6.2   The SVD Algorithm

We now show how a variant of the QR algorithm can be used to compute the SVD of an $A \in \mathbb{R}^{m \times n}$ with $m \geq n$. At first glance, this appears straightforward. Equation (8.6.1) suggests that we

- form $C = A^T A$,

- use the symmetric QR algorithm to compute $V_1^T C V_1 = \mathrm{diag}(\sigma_i^2)$,

- apply QR with column pivoting to AV_1 obtaining $U^T(AV_1)\Pi = R$.

Since R has orthogonal columns, it follows that $U^T A(V_1\Pi)$ is diagonal. However, as we saw in Example 5.3.2, the formation of $A^T A$ can lead to a loss of information. The situation is not quite so bad here, since the original A is used to compute U.

A preferable method for computing the SVD is described in Golub and Kahan (1965). Their technique finds U and V simultaneously by *implicitly* applying the symmetric QR algorithm to $A^T A$. The first step is to reduce A to upper bidiagonal form using Algorithm 5.4.2:

$$
U_B^T A V_B = \begin{bmatrix} B \\ 0 \end{bmatrix} \qquad B = \begin{bmatrix} d_1 & f_1 & & \cdots & & 0 \\ 0 & d_2 & \ddots & & & \vdots \\ & & \ddots & \ddots & \ddots & \\ \vdots & & & \ddots & \ddots & f_{n-1} \\ 0 & \cdots & & & 0 & d_n \end{bmatrix} \in \mathbb{R}^{n \times n}.
$$

The remaining problem is thus to compute the SVD of B. To this end, consider applying an implicit-shift QR step (Algorithm 8.3.2) to the tridiagonal matrix $T = B^T B$:

- Compute the eigenvalue λ of

$$
T(m{:}n, m{:}n) = \begin{bmatrix} d_m^2 + f_{m-1}^2 & d_m f_m \\ d_m f_m & d_n^2 + f_m^2 \end{bmatrix} \qquad m = n - 1
$$

that is closer to $d_n^2 + f_m^2$.

- Compute $c_1 = \cos(\theta_1)$ and $s_1 = \sin(\theta_1)$ such that

$$
\begin{bmatrix} c_1 & s_1 \\ -s_1 & c_1 \end{bmatrix}^T \begin{bmatrix} d_1^2 - \lambda \\ d_1 f_1 \end{bmatrix} = \begin{bmatrix} \times \\ 0 \end{bmatrix}
$$

and set $G_1 = G(1, 2, \theta_1)$.

- Compute Givens rotations $G_2, \ldots, G_{n-1}$ so that if $Q = G_1 \cdots G_{n-1}$ then $Q^T T Q$ is tridiagonal and $Q e_1 = G_1 e_1$.

Note that these calculations require the explicit formation of $B^T B$, which, as we have seen, is unwise from the numerical standpoint.

Suppose instead that we apply the Givens rotation G_1 above to B directly. Illustrating with the $n = 6$ case this gives

$$
B \leftarrow B G_1 = \begin{bmatrix}
\times & \times & 0 & 0 & 0 & 0 \\
+ & \times & \times & 0 & 0 & 0 \\
0 & 0 & \times & \times & 0 & 0 \\
0 & 0 & 0 & \times & \times & 0 \\
0 & 0 & 0 & 0 & \times & \times \\
0 & 0 & 0 & 0 & 0 & \times
\end{bmatrix}.
$$

We then can determine Givens rotations $U_1, V_2, U_2, \ldots, V_{n-1}$, and U_{n-1} to chase the unwanted nonzero element down the bidiagonal:

$$
B \leftarrow U_1^T B = \begin{bmatrix}
\times & \times & + & 0 & 0 & 0 \\
0 & \times & \times & 0 & 0 & 0 \\
0 & 0 & \times & \times & 0 & 0 \\
0 & 0 & 0 & \times & \times & 0 \\
0 & 0 & 0 & 0 & \times & \times \\
0 & 0 & 0 & 0 & 0 & \times
\end{bmatrix}
$$

$$
B \leftarrow B V_2 = \begin{bmatrix}
\times & \times & 0 & 0 & 0 & 0 \\
0 & \times & \times & 0 & 0 & 0 \\
0 & + & \times & \times & 0 & 0 \\
0 & 0 & 0 & \times & \times & 0 \\
0 & 0 & 0 & 0 & \times & \times \\
0 & 0 & 0 & 0 & 0 & \times
\end{bmatrix}
$$

$$
B \leftarrow U_2^T B = \begin{bmatrix}
\times & \times & 0 & 0 & 0 & 0 \\
0 & \times & \times & + & 0 & 0 \\
0 & 0 & \times & \times & 0 & 0 \\
0 & 0 & 0 & \times & \times & 0 \\
0 & 0 & 0 & 0 & \times & \times \\
0 & 0 & 0 & 0 & 0 & \times
\end{bmatrix}
$$

and so on. The process terminates with a new bidiagonal $\bar{B}$ that is related to B as follows:

$$
\bar{B} = (U_{n-1}^T \cdots U_1^T) B (G_1 V_2 \cdots V_{n-1}) = \bar{U}^T B \bar{V}.
$$

Since each V_i has the form $V_i = G(i, i+1, \theta_i)$ where $i = 2{:}n - 1$, it follows that $\bar{V}e_1 = Qe_1$. By the implicit Q theorem we can assert that $\bar{V}$ and Q are essentially the same. Thus, we can implicitly effect the transition from T to $\bar{T} = \bar{B}^T \bar{B}$ by working directly on the bidiagonal matrix B.

Of course, for these claims to hold it is necessary that the underlying tridiagonal matrices be unreduced. Since the subdiagonal entries of $B^T B$ are of the form d_{i-1}, f_i, it is clear that we must search the bidiagonal band for zeros. If $f_k = 0$ for some k, then

$$B = \begin{bmatrix} B_1 & 0 \\ 0 & B_2 \end{bmatrix} \begin{matrix} k \\ n - k \end{matrix}$$
$$\begin{matrix} k & n-k \end{matrix}$$

and the original SVD problem decouples into two smaller problems involving the matrices B_1 and B_2. If $d_k = 0$ for some $k < n$, then premultiplication by a sequence of Givens transformations can zero f_k. For example, if $n = 6$ and $k = 3$, then by rotating in row planes (3,4), (3,5), and (3,6) we can zero the entire third row:

$$B = \begin{bmatrix} \times & \times & 0 & 0 & 0 & 0 \\ 0 & \times & \times & 0 & 0 & 0 \\ 0 & 0 & 0 & \times & 0 & 0 \\ 0 & 0 & 0 & \times & \times & 0 \\ 0 & 0 & 0 & 0 & \times & \times \\ 0 & 0 & 0 & 0 & 0 & \times \end{bmatrix} \xrightarrow{(3,4)} \begin{bmatrix} \times & \times & 0 & 0 & 0 & 0 \\ 0 & \times & \times & 0 & 0 & 0 \\ 0 & 0 & 0 & 0 & + & 0 \\ 0 & 0 & 0 & \times & \times & 0 \\ 0 & 0 & 0 & 0 & \times & \times \\ 0 & 0 & 0 & 0 & 0 & \times \end{bmatrix}$$

$$\xrightarrow{(3,5)} \begin{bmatrix} \times & \times & 0 & 0 & 0 & 0 \\ 0 & \times & \times & 0 & 0 & 0 \\ 0 & 0 & 0 & 0 & 0 & + \\ 0 & 0 & 0 & \times & \times & 0 \\ 0 & 0 & 0 & 0 & \times & \times \\ 0 & 0 & 0 & 0 & 0 & \times \end{bmatrix} \xrightarrow{(3,6)} \begin{bmatrix} \times & \times & 0 & 0 & 0 & 0 \\ 0 & \times & \times & 0 & 0 & 0 \\ 0 & 0 & 0 & 0 & 0 & 0 \\ 0 & 0 & 0 & \times & \times & 0 \\ 0 & 0 & 0 & 0 & \times & \times \\ 0 & 0 & 0 & 0 & 0 & \times \end{bmatrix}$$

If $d_n = 0$, then the last column can be zeroed with a series of column rotations in planes $(n - 1, n)$, $(n - 2, n), \ldots, (1, n)$. Thus, we can decouple if $f_1 \cdots f_{n-1} = 0$ or $d_1 \cdots d_n = 0$.

Algorithm 8.6.1 (Golub-Kahan SVD Step) Given a bidiagonal matrix $B \in \mathbb{R}^{m \times n}$ having no zeros on its diagonal or superdiagonal, the following algorithm overwrites B with the bidiagonal matrix $\bar{B} = \bar{U}^T B \bar{V}$ where $\bar{U}$ and $\bar{V}$ are orthogonal and $\bar{V}$ is essentially the orthogonal matrix that would be obtained by applying Algorithm 8.3.2 to $T = B^T B$.

Let μ be the eigenvalue of the trailing 2-by-2 submatrix of $T = B^T B$
   that is closer to t_{nn}.
$y = t_{11} - \mu$
$z = t_{12}$
for $k = 1{:}n - 1$
   Determine $c = \cos(\theta)$ and $s = \sin(\theta)$ such that
   $$[\, y \quad z \,] \begin{bmatrix} c & s \\ -s & c \end{bmatrix} = [\, * \quad 0 \,]$$
   $B = BG(k, k + 1, \theta)$
   $y = b_{kk}; \; z = b_{k+1,k}$
   Determine $c = \cos(\theta)$ and $s = \sin(\theta)$ such that
   $$\begin{bmatrix} c & s \\ -s & c \end{bmatrix}^T \begin{bmatrix} y \\ z \end{bmatrix} = \begin{bmatrix} * \\ 0 \end{bmatrix}$$
   $B = G(k, k + 1, \theta)^T B$
   if $k < n - 1$
      $y = b_{k,k+1}; \; z = b_{k,k+2}$
   end
end

An efficient implementation of this algorithm would store B's diagonal and
superdiagonal in vectors $a(1{:}n)$ and $f(1{:}n - 1)$ respectively and would require $30n$ flops and $2n$ square roots. Accumulating U requires $6mn$ flops.
Accumulating V requires $6n^2$ flops.

   Typically, after a few of the above SVD iterations, the superdiagonal
entry f_{n-1} becomes negligible. Criteria for smallness within B's band are
usually of the form

$$|f_i| \;\leq\; \epsilon(\, |d_i| + |d_{i+1}| \,)$$
$$|d_i| \;\leq\; \epsilon \, \| \, B \, \|$$

where ϵ is a small multiple of the unit roundoff and $\| \cdot \|$ is some compu-
tationally convenient norm.

   Combining Algorithm 5.4.2 (bidiagonalization), Algorithm 8.6.1, and
the decoupling calculations mentioned earlier gives

Algorithm 8.6.2 (The SVD Algorithm) Given $A \in \mathbb{R}^{m \times n}$ $(m \geq n)$ and
ϵ, a small multiple of the unit roundoff, the following algorithm overwrites
A with $U^T A V = D + E$, where $U \in \mathbb{R}^{m \times n}$ is orthogonal, $V \in \mathbb{R}^{n \times n}$ is
orthogonal, $D \in \mathbb{R}^{m \times n}$ is diagonal, and E satisfies $\| \, E \, \|_2 \approx \mathbf{u} \| \, A \, \|_2$.

   Use Algorithm 5.4.2 to compute the bidiagonalization

$$\begin{bmatrix} B \\ 0 \end{bmatrix} \leftarrow (U_1 \cdots U_n)^T A (V_1 \cdots V_{n-2})$$

until $q = n$

Set $b_{i,i+1}$ to zero if $|b_{i,i+1}| \le \epsilon(|b_{ii}| + |b_{i+1,i+1}|)$
    for any $i = 1{:}n - 1$.

Find the largest q and the smallest p such that if

$$
B = \begin{bmatrix} B_{11} & 0 & 0 \\ 0 & B_{22} & 0 \\ 0 & 0 & B_{33} \end{bmatrix} \begin{matrix} p \\ n-p-q \\ q \end{matrix}
$$
$$
\begin{matrix} p & n-p-q & q \end{matrix}
$$

then B_{33} is diagonal and B_{22} has nonzero superdiagonal.

if $q < n$

    if any diagonal entry in B_{22} is zero, then zero
        the superdiagonal entry in the same row.

    else

        Apply Algorithm 8.3.1 to B_{22},
        $B = \text{diag}(I_p, U, I_{q+m-n})^T B \text{diag}(I_p, V, I_q)$

    end

    end

end

The amount of work required by this algorithm and its numerical properties are discussed in §5.4.5 and §5.5.8.

Example 8.6.5 If Algorithm 8.6.2 is applied to

$$
A = \begin{bmatrix} 1 & 1 & 0 & 0 \\ 0 & 2 & 1 & 0 \\ 0 & 0 & 3 & 1 \\ 0 & 0 & 0 & 4 \end{bmatrix}
$$

then the superdiagonal elements converge to zero as follows:

Iteration	$O(|a_{21}|)$	$O(|a_{32}|)$	$O(|a_{43}|)$
1	10^0	10^0	10^0
2	10^0	10^0	10^0
3	10^0	10^0	10^0
4	10^0	10^{-1}	10^{-2}
5	10^0	10^{-1}	10^{-8}
6	10^0	10^{-1}	10^{-27}
7	10^0	10^{-1}	converg.
8	10^0	10^{-4}	
9	10^{-1}	10^{-14}	
10	10^{-1}	converg.	
11	10^{-4}		
12	10^{-12}		
13	converg.		

Observe the cubic-like convergence.

8.6.3 Jacobi SVD Procedures

It is straightforward to adapt the Jacobi procedures of §8.4 to the SVD problem. Instead of solving a sequence of 2-by-2 symmetric eigenproblems, we solve a sequence of 2-by-2 SVD problems. Thus, for a given index pair (p, q) we compute a pair of rotations such that

$$\begin{bmatrix} c_1 & s_1 \\ -s_1 & c_1 \end{bmatrix}^T \begin{bmatrix} a_{pp} & a_{pq} \\ a_{qp} & a_{qq} \end{bmatrix} \begin{bmatrix} c_2 & s_2 \\ -s_2 & c_2 \end{bmatrix} = \begin{bmatrix} d_p & 0 \\ 0 & d_q \end{bmatrix}.$$

See P8.6.8. The resulting algorithm is referred to as *two-sided* because each update involves a pre- and post-multiplication.

A *one-sided* Jacobi algorithm involves a sequence of pairwise column orthogonalizations. For a given index pair (p, q) a Jacobi rotation $J(p, q, \theta)$ is determined so that columns p and q of $AJ(p.q, \theta)$ are orthogonal to each other. See P8.6.8. Note that this corresponds to zeroing the (p, q) and (q, p) entries in $A^T A$. Once AV has sufficiently orthogonal columns, the rest of the SVD (U and Σ) follows from column scaling: $AV = U\Sigma$.

Problems

P8.6.1 Show that if $B \in \mathbf{R}^{n \times n}$ is an upper bidiagonal matrix having a repeated singular value, then B must have a zero on its diagonal or superdiagonal.

P8.6.2 Give formulae for the eigenvectors of $\begin{bmatrix} 0 & A^T \\ A & 0 \end{bmatrix}$ in terms of the singular vectors of $A \in \mathbf{R}^{m \times n}$ where $m \geq n$.

P8.6.3 Give an algorithm for reducing a complex matrix A to *real* bidiagonal form using complex Householder transformations.

P8.6.4 Relate the singular values and vectors of $A = B + iC$ ($B, C \in \mathbf{R}^{m \times n}$) to those of $\begin{bmatrix} B & -C \\ C & B \end{bmatrix}$.

P8.6.5 Complete the proof of Theorem 8.6.1.

P8.6.6 Assume that $n = 2m$ and that $S \in \mathbf{R}^{n \times n}$ is skew-symmetric and tridiagonal. Show that there exists a permutation $P \in \mathbf{R}^{n \times n}$ such that $P^T SP$ has the following form:

$$P^T SP = \begin{bmatrix} 0 & -B^T \\ B & 0 \end{bmatrix} \begin{matrix} m \\ m \end{matrix}$$
$$\qquad\qquad\quad m \quad\ m$$

Describe B. Show how to compute the eigenvalues and eigenvectors of S via the SVD of B. Repeat for the case $n = 2m + 1$.

P8.6.7 (a) Let

$$C = \begin{bmatrix} w & \tilde{} \\ y & z \end{bmatrix}$$

be real. Give a stable algorithm for computing c and s with $c^2 + s^2 = 1$ such that

$$B = \begin{bmatrix} c & s \\ -s & c \end{bmatrix} C$$

is symmetric. (b) Combine (a) with the Jacobi trigonometric calculations in the text to obtain a stable algorithm for computing the SVD of C. (c) Part (b) can be used to develop a Jacobi-like algorithm for computing the SVD of $A \in \mathbb{R}^{n \times n}$. For a given (p,q) with $p < q$, Jacobi transformations $J(p,q,\theta_1)$ and $J(p,q,\theta_2)$ are determined such that if

$$B = J(p,q,\theta_1)^T A J(p,q,\theta_2),$$

then $b_{pq} = b_{qp} = 0$. Show

$$\text{off}(B)^2 = \text{off}(A)^2 - b_{pq}^2 - b_{qp}^2.$$

How might p and q be determined? How could the algorithm be adapted to handle the case when $A \in \mathbb{R}^{m \times n}$ with $m > n$?

P8.6.8 Let x and y be in $\mathbb{R}^m$ and define the orthogonal matrix Q by

$$Q = \begin{bmatrix} c & s \\ -s & c \end{bmatrix}.$$

Give a stable algorithm for computing c and s such that the columns of $[x,y]Q$ are orthogonal to each other.

P8.6.9 Suppose $B \in \mathbb{R}^{n \times n}$ is upper bidiagonal with $b_{nn} = 0$. Show how to construct orthogonal U and V (product of Givens rotations) so that $U^T BV$ is upper bidiagonal with a zero nth column.

P8.6.10 Suppose $B \in \mathbb{R}^{n \times n}$ is upper bidiagonal with diagonal entries $d(1{:}n)$ and superdiagonal entries $f(1{:}n-1)$. State and prove a singular value version of Theorem 8.5.1.

Notes and References for Sec. 8.6

The mathematical properties of the SVD are discussed in Stewart and Sun (1990) as well as

A.R. Amir-Moez (1965). *Extremal Properties of Linear Transformations and Geometry of Unitary Spaces*, Texas Tech University Mathematics Series, no. 243, Lubbock, Texas.

G.W. Stewart (1973). "Error and Perturbation Bounds for Subspaces Associated with Certain Eigenvalue Problems," *SIAM Review 15*, 727–64.

P.A. Wedin (1972). "Perturbation Bounds in Connection with the Singular Value Decomposition," *BIT 12*, 99–111.

G.W. Stewart (1979). "A Note on the Perturbation of Singular Values," *Lin. Alg. and Its Applic. 28*, 213–16.

G.W. Stewart (1984). "A Second Order Perturbation Expansion for Small Singular Values," *Lin. Alg. and Its Applic. 56*, 231–236.

R.J. Vaccaro (1994). "A Second-Order Perturbation Expansion for the SVD," *SIAM J. Matrix Anal. Applic. 15*, 661–671.

The idea of adapting the symmetric QR algorithm to compute the SVD first appeared in

G.H. Golub and W. Kahan (1965). "Calculating the Singular Values and Pseudo-Inverse of a Matrix," *SIAM J. Num. Anal. Ser. B 2*, 205–24.

and then came some early implementations:

P.A. Businger and G.H. Golub (1969). "Algorithm 358: Singular Value Decomposition of a Complex Matrix," *Comm. Assoc. Comp. Mach. 12*, 564–65.

G.H. Golub and C. Reinsch (1970). "Singular Value Decomposition and Least Squares Solutions," *Numer. Math. 14*, 403–20. See also Wilkinson and Reinsch (1971, 134–51).

Interesting algorithmic developments associated with the SVD appear in

J.J.M. Cuppen (1983). "The Singular Value Decomposition in Product Form," *SIAM J. Sci. and Stat. Comp. 4*, 216–222.

J.J. Dongarra (1983). "Improving the Accuracy of Computed Singular Values," *SIAM J. Sci. and Stat. Comp. 4*, 712–719.

S. Van Huffel, J. Vandewalle, and A. Haegemans (1987). "An Efficient and Reliable Algorithm for Computing the Singular Subspace of a Matrix Associated with its Smallest Singular Values," *J. Comp. and Appl. Math. 19*, 313–330.

P. Deift, J. Demmel, L.-C. Li, and C. Tomei (1991). "The Bidiagonal Singular Value Decomposition and Hamiltonian Mechanics," *SIAM J. Num. Anal. 28*, 1463–1516.

R. Mathias and G.W. Stewart (1993). "A Block QR Algorithm and the Singular Value Decomposition," *Lin. Alg. and Its Applic. 182*, 91–100.

Å. Björck, E. Grimme, and P. Van Dooren (1994). "An Implicit Shift Bidiagonalization Algorithm for Ill-Posed Problems," *BIT 34*, 510–534.

The Polar decomposition of a matrix can be computed immediately from its SVD. However, special algorithms have been developed just for this purpose.

N.J. Higham (1986). "Computing the Polar Decomposition—with Applications," *SIAM J. Sci. and Stat. Comp. 7*, 1160–1174.

N.J. Higham and P. Papadimitriou (1994). "A Parallel Algorithm for Computing the Polar Decomposition," *Parallel Comp. 20*, 1161–1173.

Jacobi methods for the SVD fall into two categories. The two-sided Jacobi algorithms repeatedly perform the update $A \leftarrow U^T A V$ producing a sequence of iterates that are increasingly diagonal.

E.G. Kogbetliantz (1955). "Solution of Linear Equations by Diagonalization of Coefficient Matrix," *Quart. Appl. Math. 13*, 123–132.

G.E. Forsythe and P. Henrici (1960). "The Cyclic Jacobi Method for Computing the Principal Values of a Complex Matrix," *Trans. Amer. Math. Soc. 94*, 1–23.

C.C. Paige and P. Van Dooren (1986). "On the Quadratic Convergence of Kogbetliantz's Algorithm for Computing the Singular Value Decomposition," *Lin. Alg. and Its Applic. 77*, 301–313.

J.P. Charlier and P. Van Dooren (1987). "On Kogbetliantz's SVD Algorithm in the Presence of Clusters," *Lin. Alg. and Its Applic. 95*, 135–160.

Z. Bai (1988). "Note on the Quadratic Convergence of Kogbetliantz's Algorithm for Computing the Singular Value Decomposition," *Lin. Alg. and Its Applic. 104*, 131–140.

J.P. Charlier, M. Vanbegin, P. Van Dooren (1988). "On Efficient Implementation of Kogbetliantz's Algorithm for Computing the Singular Value Decomposition," *Numer. Math. 52*, 279–300.

K.V. Fernando (1989). "Linear Convergence of the Row Cyclic Jacobi and Kogbetliantz methods," *Numer. Math. 56*, 73–92.

The one-sided Jacobi SVD procedures repeatedly perform the update $A \leftarrow AV$ producing a sequence of iterates with columns that are increasingly orthogonal.

J.C. Nash (1975). "A One-Sided Tranformation Method for the Singular Value Decomposition and Algebraic Eigenproblem," *Comp. J. 18*, 74–76.

P.C. Hansen (1988). "Reducing the Number of Sweeps in Hestenes Method," in *Singular Value Decomposition and Signal Processing*, ed. E.F. Deprettere, North Holland.

K. Veselić and V. Hari (1989). "A Note on a One-Sided Jacobi Algorithm," *Numer. Math. 56*, 627–633.

Numerous parallel implementations have been developed.

F.T. Luk (1980). "Computing the Singular Value Decomposition on the ILLIAC IV," *ACM Trans. Math. Soft. 6*, 524–39.

R.P. Brent and F.T. Luk (1985). "The Solution of Singular Value and Symmetric Eigenvalue Problems on Multiprocessor Arrays," *SIAM J. Sci. and Stat. Comp. 6*, 69–84.

R.P. Brent, F.T. Luk, and C. Van Loan (1985). "Computation of the Singular Value Decomposition Using Mesh Connected Processors," *J. VLSI Computer Systems 1*, 242–270.

F.T. Luk (1986). "A Triangular Processor Array for Computing Singular Values," *Lin. Alg. and Its Applic. 77*, 259–274.

M. Berry and A. Sameh (1986). "Multiprocessor Jacobi Algorithms for Dense Symmetric Eigenvalue and Singular Value Decompositions," in *Proc. International Conference on Parallel Processing*, 433–440.

R. Schreiber (1986). "Solving Eigenvalue and Singular Value Problems on an Undersized Systolic Array," *SIAM J. Sci. and Stat. Comp. 7*, 441–451.

C.H. Bischof and C. Van Loan (1986). "Computing the SVD on a Ring of Array Processors," in *Large Scale Eigenvalue Problems*, eds. J. Cullum and R. Willoughby, North Holland, 51-66.

C.H. Bischof (1987). "The Two-Sided Block Jacobi Method on Hypercube Architectures," in *Hypercube Multiprocessors*, ed. M.T. Heath, SIAM Press, Philadelphia.

C.H. Bischof (1989). "Computing the Singular Value Decomposition on a Distributed System of Vector Processors," *Parallel Computing 11*, 171–186.

S. Van Huffel and H. Park (1994). "Parallel Tri- and Bidiagonalization of Bordered Bidiagonal Matrices," *Parallel Computing 20*, 1107–1128.

B. Lang (1996). "Parallel Reduction of Banded Matrices to Bidiagonal Form," *Parallel Computing 22*, 1–18.

The divide and conquer algorithms devised for for the symmetric eigenproblem have SVD analogs:

E.R. Jessup and D.C. Sorensen (1994). "A Parallel Algorithm for Computing the Singular Value Decomposition of a Matrix," *SIAM J. Matrix Anal. Appl. 15*, 530–548.

M. Gu and S.C. Eisenstat (1995). "A Divide-and-Conquer Algorithm for the Bidiagonal SVD," *SIAM J. Matrix Anal. Appl. 16*, 79–92.

Careful analyses of the SVD calculation include

J.W. Demmel and W. Kahan (1990). "Accurate Singular Values of Bidiagonal Matrices," *SIAM J. Sci. and Stat. Comp. 11*, 873–912.

K.V. Fernando and B.N. Parlett (1994). "Accurate Singular Values and Differential qd Algorithms," *Numer. Math. 67*, 191–230.

S. Chandrasekaren and I.C.F. Ipsen (1994). "Backward Errors for Eigenvalue and Singular Value Decompositions," *Numer. Math. 68*, 215–223.

High accuracy SVD calculation and connections among the Cholesky, Schur, and singular value computations are discussed in

J.W. Demmel and K. Veselič (1992). "Jacobi's Method is More Accurate than QR," *SIAM J. Matrix Anal. Appl. 13*, 1204–1245.

R. Mathias (1995). "Accurate Eigensystem Computations by Jacobi Methods," *SIAM J. Matrix Anal. Appl. 16*, 977–1003.

8.7 Some Generalized Eigenvalue Problems

Given a symmetric matrix $A \in \mathbb{R}^{n \times n}$ and a symmetric positive definite $B \in \mathbb{R}^{n \times n}$, we consider the problem of finding a nonzero vector x and a scalar λ so $Ax = \lambda Bx$. This is the *symmetric-definite generalized eigenproblem*. The scalar λ can be thought of as a *generalized eigenvalue*. As λ varies, $A - \lambda B$ defines a *pencil* and our job is to determine

$$\lambda(A, B) = \{ \lambda \mid \det(A - \lambda I) = 0 \}.$$

A symmetric-definite generalized eigenproblem can be transformed to an equivalent problem with a congruence transformation:

$$A - \lambda B \text{ is singular} \quad \Leftrightarrow \quad (X^T A X) - \lambda(X^T B X) \text{ is singular}$$

Thus, if X is nonsingular, then $\lambda(A, B) = \lambda(X^T A X, X^T B X)$.

In this section we present various structure-preserving procedures that solve such eigenproblems through the careful selection of X. The related generalized singular value decomposition problem is also discussed.

8.7.1 Mathematical Background

We seek is a stable, efficient algorithm that computes X such that $X^T A X$ and $X^T B X$ are both in "canonical form." The obvious form to aim for is diagonal form.

Theorem 8.7.1 *Suppose A and B are n-by-n symmetric matrices, and define $C(\mu)$ by*

$$C(\mu) = \mu A + (1 - \mu)B \qquad \mu \in \mathbb{R}. \tag{8.7.1}$$

If there exists a $\mu \in [0, 1]$ such that $C(\mu)$ is non-negative definite and

$$\text{null}(C(\mu)) = \text{null}(A) \cap \text{null}(B)$$

then there exists a nonsingular X such that both $X^T A X$ and $X^T B X$ are diagonal.

Proof. Let $\mu \in [0, 1]$ be chosen so that $C(\mu)$ is non-negative definite with the property that $\text{null}(C(\mu)) = \text{null}(A) \cap \text{null}(B)$. Let

$$Q_1^T C(\mu) Q_1 = \begin{bmatrix} D & 0 \\ 0 & 0 \end{bmatrix} \qquad D = \text{diag}(d_1, \ldots, d_k), \ d_i > 0$$

be the Schur decomposition of $C(\mu)$ and define $X_1 = Q_1 \text{diag}(D^{-1/2}, I_{n-k})$. If $A_1 = X_1^T A X_1$, $B_1 = X_1^T B X_1$, and $C_1 = X_1^T C(\mu) X_1$, then

$$C_1 = \begin{bmatrix} I_k & 0 \\ 0 & 0 \end{bmatrix} = \mu A_1 + (1 - \mu) B_1.$$

Since span$\{e_{k+1},\ldots,e_n\}$ = null(C_1) = null$(A_1)\cap$null(B_1) it follows that A_1 and B_1 have the following block structure:

$$A_1 = \begin{bmatrix} A_{11} & 0 \\ 0 & 0 \end{bmatrix} \begin{matrix} k \\ n-k \end{matrix} \qquad B_1 = \begin{bmatrix} B_{11} & 0 \\ 0 & 0 \end{bmatrix} \begin{matrix} k \\ n-k \end{matrix} \quad .$$
$$\begin{matrix} k & n-k \end{matrix} \qquad\qquad\qquad \begin{matrix} k & n-k \end{matrix}$$

Moreover $I_k = \mu A_{11} + (1-\mu)B_{11}$.

Suppose $\mu \neq 0$. It then follows that if $Z^T B_{11} Z = \text{diag}(b_1,\ldots,b_k)$ is the Schur decomposition of B_{11} and we set $X = X_1\text{diag}(Z,I_{n-k})$ then

$$X^T BX = \text{diag}(b_1,\ldots,b_k,0,\ldots,0) \equiv D_B$$

and

$$\begin{aligned} X^T AX &= \frac{1}{\mu}X^T \left(C(\mu)-(1-\mu)B\right)X \\ &= \frac{1}{\mu}\left(\begin{bmatrix} I_k & 0 \\ 0 & 0 \end{bmatrix} - (1-\mu)D_B\right) \equiv D_A . \end{aligned}$$

On the other hand, if $\mu = 0$, then let $Z^T A_{11} Z = \text{diag}(a_1,\ldots,a_k)$ be the Schur decomposition of A_{11} and set $X = X_1\text{diag}(Z,I_{n-k})$. It is easy to verify that in this case as well, both $X^T AX$ and $X^T BX$ are diagonal. $\square$

Frequently, the conditions in Theorem 8.7.1 are satisfied because either A or B is positive definite.

Corollary 8.7.2 *If $A - \lambda B \in \mathbb{R}^{n\times n}$ is symmetric-definite, then there exists a nonsingular $X = [\,x_1,\ldots,x_n\,]$ such that*

$$X^T AX = \text{diag}(a_1,\ldots,a_n) \quad and \quad X^T BX = \text{diag}(b_1,\ldots,b_n) .$$

Moreover, $Ax_i = \lambda_i Bx_i$ for $i = 1{:}n$ where $\lambda_i = a_i/b_i$.

Proof. By setting $\mu = 0$ in Theorem 8.7.1 we see that symmetric-definite pencils can be simultaneously diagonalized. The rest of the corollary is easily verified. $\square$

Example 8.7.1 If

$$A = \begin{bmatrix} 229 & 163 \\ 163 & 116 \end{bmatrix} \quad and \quad B = \begin{bmatrix} 81 & 59 \\ 59 & 43 \end{bmatrix}$$

then $A - \lambda B$ is symmetric-definite and $\lambda(A,B) = \{5,-1/2\}$. If

$$X = \begin{bmatrix} 3 & -5 \\ -4 & 7 \end{bmatrix}$$

then $X^T AX = \text{diag}(5, -1)$ and $X^T BX = \text{diag}(1, 2)$.

Stewart (1979) has worked out a perturbation theory for symmetric pencils $A - \lambda B$ that satisfy

$$c(A, B) = \min_{\|x\|_2 = 1} (x^T Ax)^2 + (x^T Bx)^2 > 0 \qquad (8.7.2)$$

The scalar $c(A, B)$ is called the *Crawford number* of the pencil $A - \lambda B$.

Theorem 8.7.3 *Suppose $A - \lambda B$ is an n-by-n symmetric-definite pencil with eigenvalues*

$$\lambda_1 \geq \lambda_2 \geq \cdots \geq \lambda_n.$$

Suppose E_A and E_B are symmetric n-by-n matrices that satisfy

$$\epsilon^2 = \| E_A \|_2^2 + \| E_B \|_2^2 < c(A, B).$$

Then $(A + E_A) - \lambda(B + E_B)$ is symmetric-definite with eigenvalues

$$\mu_1 \geq \cdots \geq \mu_n$$

that satisfy

$$|\arctan(\lambda_i) - \arctan(\mu_i)| \leq \arctan(\epsilon/c(A, B))$$

for $i = 1{:}n$.

Proof. See Stewart (1979). □

8.7.2 Methods for the Symmetric-Definite Problem

Turning to algorithmic matters, we first present a method for solving the symmetric-definite problem that utilizes both the Cholesky factorization and the symmetric QR algorithm.

Algorithm 8.7.1 Given $A = A^T \in \mathbb{R}^{n \times n}$ and $B = B^T \in \mathbb{R}^{n \times n}$ with B positive definite, the following algorithm computes a nonsingular X such that $X^T BX = I_n$ and $X^T AX = \text{diag}(a_1, \ldots, a_n)$.

> Compute the Cholesky factorization $B = GG^T$
>   using Algorithm 4.2.2.
> Compute $C = G^{-1}AG^{-T}$.
> Use the symmetric QR algorithm to compute the Schur
>   decomposition $Q^T CQ = \text{diag}(a_1, \ldots, a_n)$.
> Set $X = G^{-T}Q$.

This algorithm requires about $14n^3$ flops. In a practical implementation, A can be overwritten by the matrix C. See Martin and Wilkinson (1968c) for details. Note that

$$\lambda(A, B) = \lambda(A, GG^T) = \lambda(G^{-1}AG^{-T}, I) = \lambda(C) = \{a_1, \ldots, a_n\}.$$

If $\hat{a}_i$ is a computed eigenvalue obtained by Algorithm 8.7.1, then it can be shown that $\hat{a}_i \in \lambda(G^{-1}AG^{-T} + E_i)$, where $\| E_i \|_2 \approx \mathbf{u}\| A \|_2\| B^{-1} \|_2$. Thus, if B is ill-conditioned, then $\hat{a}_i$ may be severely contaminated with roundoff error even if a_i is a well-conditioned generalized eigenvalue. The problem, of course, is that in this case, the matrix $C = G^{-1}AG^{-T}$ can have some very large entries if B, and hence G, is ill-conditioned. This difficulty can sometimes be overcome by replacing the matrix G in Algorithm 8.7.1 with $VD^{-1/2}$ where $V^T BV = D$ is the Schur decomposition of B. If the diagonal entries of D are ordered from smallest to largest, then the large entries in C are concentrated in the upper left-hand corner. The small eigenvalues of C can then be computed without excessive roundoff error contamination (or so the heuristic goes). For further discussion, consult Wilkinson (1965, pp.337–38).

Example 8.7.2 If

$$A = \begin{bmatrix} 1 & 2 & 3 \\ 2 & 4 & 5 \\ 3 & 5 & 6 \end{bmatrix} \quad \text{and} \quad G = \begin{bmatrix} .001 & 0 & 0 \\ 1 & .001 & 0 \\ 2 & 1 & .001 \end{bmatrix}$$

and $B = GG^T$, then the two smallest eigenvalues of $A - \lambda B$ are

$$a_1 = -0.619402940600584 \qquad a_2 = 1.627440079051887.$$

If 17-digit floating point arithmetic is used, then these eigenvalues are computed to full machine precision when the symmetric QR algorithm is applied to $fl(D^{-1/2}V^T AVD^{-1/2})$, where $B = VDV^T$ is the Schur decomposition of B. On the other hand, if Algorithm 8.7.1 is applied, then

$$\hat{a}_1 = -0.619373517376444 \qquad \hat{a}_2 = 1.627516601905228.$$

The reason for obtaining only four correct significant digits is that $\kappa_2(B) \approx 10^{18}$.

The condition of the matrix X in Algorithm 8.7.1 can sometimes be improved by replacing B with a suitable convex combination of A and B. The connection between the eigenvalues of the modified pencil and those of the original are detailed in the proof of Theorem 8.7.1.

Other difficulties concerning Algorithm 8.7.1 revolve around the fact that $G^{-1}AG^{-T}$ is generally full even when A and B are sparse. This is a serious problem, since many of the symmetric-definite problems arising in practice are large and sparse.

Crawford (1973) has shown how to implement Algorithm 8.7.1 effectively when A and B are banded. Aside from this case, however, the simultaneous diagonalization approach is impractical for the large, sparse symmetric-definite problem.

An alternative idea is to extend the Rayleigh quotient iteration (8.4.4) as follows:

x_0 given with $\| x_0 \|_2 = 1$
for $k = 0, 1, \ldots$
$\quad \mu_k = x_k^T A x_k / x_k^T B x_k$ (8.7.3)
$\quad$ Solve $(A - \mu_k B) z_{k+1} = B x_k$ for z_{k+1}.
$\quad x_{k+1} = z_{k+1} / \| z_{k+1} \|_2$
end

The mathematical basis for this iteration is that

$$\lambda = \frac{x^T A x}{x^T B x} \qquad (8.7.4)$$

minimizes

$$f(\lambda) = \| A x - \lambda B x \|_B \qquad (8.7.5)$$

where $\| \cdot \|_B$ is defined by $\| z \|_B^2 = z^T B^{-1} z$. The mathematical properties of (8.7.3) are similar to those of (8.4.4). Its applicability depends on whether or not systems of the form $(A - \mu B) z = x$ can be readily solved. A similar comment pertains to the following generalized orthogonal iteration:

$Q_0 \in \mathbb{R}^{n \times p}$ given with $Q_0^T Q_0 = I_p$
for $k = 1, 2, \ldots$
$\quad$ Solve $B Z_k = A Q_{k-1}$ for Z_k.   (8.7.6)
$\quad Z_k = Q_k R_k$ (QR factorization)
end

This is mathematically equivalent to (7.3.4) with A replaced by $B^{-1} A$. Its practicality depends on how easy it is to solve linear systems of the form $B z = y$.

Sometimes A and B are so large that neither (8.7.3) nor (8.7.6) can be invoked. In this situation, one can resort to any of a number of gradient and coordinate relaxation algorithms. See Stewart (1976) for an extensive guide to the literature.

8.7.3   The Generalized Singular Value Problem

We conclude with some remarks about symmetric pencils that have the form $A^T A - \lambda B^T B$ where $A \in \mathbb{R}^{m \times n}$ and $B \in \mathbb{R}^{p \times n}$. This pencil underlies the *generalized singular value decomposition* (GSVD), a decomposition that is useful in several constrained least squares problems. (Cf. §12.1.) Note that by Theorem 8.7.1 there exists a nonsingular $X \in \mathbb{R}^{n \times n}$ such that $X^T (A^T A) X$ and $X^T (B^T B) X$ are both diagonal. The value of the GSVD is that these diagonalizations can be achieved without forming $A^T A$ and $B^T B$.

Theorem 8.7.4 (Generalized Singular Value Decomposition) *If we have $A \in \mathbb{R}^{m \times n}$ with $m \geq n$ and $B \in \mathbb{R}^{p \times n}$, then there exist orthogonal $U \in \mathbb{R}^{m \times m}$ and $V \in \mathbb{R}^{p \times p}$ and an invertible $X \in \mathbb{R}^{n \times n}$ such that*

$$U^T A X = C = \operatorname{diag}(c_1, \dots, c_n) \qquad c_i \geq 0$$

and

$$V^T B X = S = \operatorname{diag}(s_1, \dots, s_q) \qquad s_i \geq 0$$

where $q = \min(p, n)$.

Proof. The proof of this decomposition appears in Van Loan (1976). We present a more constructive proof along the lines of Paige and Saunders (1981). For clarity we assume that $\operatorname{null}(A) \cap \operatorname{null}(B) = \{0\}$ and $p \geq n$. We leave it to the reader to extend the proof so that it covers theses cases.

Let

$$\begin{bmatrix} A \\ B \end{bmatrix} = \begin{bmatrix} Q_1 \\ Q_2 \end{bmatrix} R \tag{8.7.6}$$

be a QR factorization with $Q_1 \in \mathbb{R}^{m \times n}$, $Q_2 \in \mathbb{R}^{p \times n}$, and $R \in \mathbb{R}^{n \times n}$. Paige and Saunders show that the SVD's of Q_1 and Q_2 are related in the sense that

$$Q_1 = UCW^T \qquad Q_2 = VSW^T \tag{8.7.7}$$

Here, U, V, and W are orthogonal, $C = \operatorname{diag}(c_i)$ with $0 \leq c_1 \leq \cdots \leq c_n$, $S = \operatorname{diag}(s_i)$ with $s_1 \geq \cdots \geq s_n$, and $C^T C + S^T S = I_n$. The decomposition (8.7.7) is a variant of the CS decomposition in §2.6 and from it we conclude that $A = Q_1 R = UC(W^T R)$ and $B = Q_2 R = VS(W^T R)$. The theorem follows by setting $X = (W^T R)^{-1}$, $D_A = C$, and $D_B = S$. The invertibility of R follows from our assumption that $\operatorname{null}(A) \cap \operatorname{null}(B) = \{0\}$. $\square$

The elements of the set $\sigma(A, B) \equiv \{ c_1/s_1, \dots, c_n/s_q \}$ are referred to as the *generalized singular values* of A and B. Note that $\sigma \in \sigma(A, B)$ implies that $\sigma^2 \in \lambda(A^T A, B^T B)$. The theorem is a generalization of the SVD in that if $B = I_n$, then $\sigma(A, B) = \sigma(A)$.

Our proof of the GSVD is of practical importance since Stewart (1983) and Van Loan (1985) have shown how to stably compute the CS decomposition. The only tricky part is the inversion of $W^T R$ to get X. Note that the columns of $X = [\, x_1, \dots, x_n \,]$ satisfy

$$s_i^2 A^T A x_i = c_i^2 B^T B x_i \qquad i = 1{:}n$$

and so if $s_i \neq 0$ then $A^T A x_i = \sigma_i^2 B^T B x_i$ where $\sigma_i = c_i/s_i$. Thus, the x_i are aptly termed the *generalized singular vectors* of the pair (A, B).

In several applications an orthonormal basis for some designated generalized singular vector subspace space $\text{span}\{x_{i_1}, \ldots, x_{i_k}\}$ is required. We show how this can be accomplished without any matrix inversions or cross products:

- Compute the QR factorization

$$\begin{bmatrix} A \\ B \end{bmatrix} = \begin{bmatrix} Q_1 \\ Q_2 \end{bmatrix} R.$$

- Compute the CS decomposition

$$Q_1 = UCW^T \qquad Q_2 = VSW^T$$

and order the diagonals of C and S so that

$$\{ c_1/s_1, \ldots, c_k/s_k \} = \{ c_{i_1}/s_{i_1}, \ldots, c_{i_k}/s_{i_k} \}.$$

- Compute orthogonal Z and upper triangular T so $TZ = W^T R$. (See P8.7.5.) Note that if $X^{-1} = W^T R = TZ$, then $X = Z^T T^{-1}$ and so the first k rows of Z are an orthonormal basis for $\text{span}\{x_1, \ldots, x_k\}$.

Problems

P8.7.1 Suppose $A \in \mathbf{R}^{n \times n}$ is symmetric and $G \in \mathbf{R}^{n \times n}$ is lower triangular and nonsingular. Give an efficient algorithm for computing $C = G^{-1}AG^{-T}$.

P8.7.2 Suppose $A \in \mathbf{R}^{n \times n}$ is symmetric and $B \in \mathbf{R}^{n \times n}$ is symmetric positive definite. Give an algorithm for computing the eigenvalues of AB that uses the Cholesky factorization and the symmetric QR algorithm.

P8.7.3 Show that if C is real and diagonalizable, then there exist symmetric matrices A and B, B nonsingular, such that $C = AB^{-1}$. This shows that symmetric pencils $A - \lambda B$ are essentially general.

P8.7.4 Show how to convert an $Ax = \lambda Bx$ problem into a generalized singular value problem if A and B are both symmetric and non-negative definite.

P8.7.5 Given $Y \in \mathbf{R}^{n \times n}$ show how to compute Householder matrices $H_2, \ldots, H_n$ so that $Y H_n \cdots H_2 = T$ is upper triangular. Hint: H_k zeros out the kth row.

P8.7.6 Suppose

$$\begin{bmatrix} 0 & A \\ A^T & 0 \end{bmatrix} \begin{bmatrix} y \\ z \end{bmatrix} = \lambda \begin{bmatrix} B_1 & 0 \\ 0 & B_2 \end{bmatrix} \begin{bmatrix} y \\ z \end{bmatrix}$$

where $A \in \mathbf{R}^{m \times n}$, $B_1 \in \mathbf{R}^{m \times m}$, and $B_2 \in \mathbf{R}^{n \times n}$. Assume that B_1 and B_2 are positive definite with Cholesky triangles G_1 and G_2 respectively. Relate the generalized eigenvalues of this problem to the singular values of $G_1^{-1}AG_2^{-T}$.

P8.7.7 Suppose A and B are both symmetric positive definite. Show how to compute $\lambda(A, B)$ and the corresponding eigenvectors using the Cholesky factorization and CS decomposition.

Notes and References for Sec. 8.7

An excellent survey of computational methods for symmetric-definite pencils is given in

G.W. Stewart (1976). "A Bibliographical Tour of the Large Sparse Generalized Eigenvalue Problem," in *Sparse Matrix Computations* , ed., J.R. Bunch and D.J. Rose, Academic Press, New York.

Some papers of particular interest include

R.S. Martin and J.H. Wilkinson (1968c). "Reduction of a Symmetric Eigenproblem $Ax = \lambda Bx$ and Related Problems to Standard Form," *Numer. Math. 11*, 99–110.

G. Peters and J.H. Wilkinson (1969). "Eigenvalues of $Ax = \lambda Bx$ with Band Symmetric A and B," *Comp. J. 12*, 398-404.

G. Fix and R. Heiberger (1972). "An Algorithm for the Ill-Conditioned Generalized Eigenvalue Problem," *SIAM J. Num. Anal. 9*, 78–88.

C.R. Crawford (1973). "Reduction of a Band Symmetric Generalized Eigenvalue Problem," *Comm. ACM 16*, 41–44.

A. Ruhe (1974). "SOR Methods for the Eigenvalue Problem with Large Sparse Matrices," *Math. Comp.* 28, 695–710.

C.R. Crawford (1976). "A Stable Generalized Eigenvalue Problem," *SIAM J. Num. Anal. 13*, 854–60.

A. Bunse-Gerstner (1984). "An Algorithm for the Symmetric Generalized Eigenvalue Problem," *Lin. Alg. and Its Applic. 58*, 43–68.

C.R. Crawford (1986). "Algorithm 646 PDFIND: A Routine to Find a Positive Definite Linear Combination of Two Real Symmetric Matrices," *ACM Trans. Math. Soft. 12*, 278–282.

C.R. Crawford and Y.S. Moon (1983). "Finding a Positive Definite Linear Combination of Two Hermitian Matrices," *Lin. Alg. and Its Applic. 51*, 37–48.

W. Shougen and Z. Shuqin (1991). "An Algorithm for $Ax = \lambda Bx$ with Symmetric and Positive Definite A and B," *SIAM J. Matrix Anal. Appl. 12*, 654–660.

K. Li and T-Y. Li (1993). "A Homotopy Algorithm for a Symmetric Generalized Eigenproblem," *Numerical Algorithms 4*, 167–195.

K. Li, T-Y. Li, and Z. Zeng (1994). "An Algorithm for the Generalized Symmetric Tridiagonal Eigenvalue Problem," *Numerical Algorithms 8*, 269–291.

H. Zhang and W.F. Moss (1994). "Using Parallel Banded Linear System Solvers in Generalized Eigenvalue Problems," *Parallel Computing 20*, 1089–1106

The simultaneous reduction of two symmetric matrices to diagonal form is discussed in

A. Berman and A. Ben-Israel (1971). "A Note on Pencils of Hermitian or Symmetric Matrices," *SIAM J. Applic. Math. 21*, 51–54.

F. Uhlig (1973). "Simultaneous Block Diagonalization of Two Real Symmetric Matrices," *Lin. Alg. and Its Applic. 7*, 281–89.

F. Uhlig (1976). "A Canonical Form for a Pair of Real Symmetric Matrices That Generate a Nonsingular Pencil," *Lin. Alg. and Its Applic. 14*, 189–210.

K.N. Majinder (1979). "Linear Combinations of Hermitian and Real Symmetric Matrices," *Lin. Alg. and Its Applic. 25*, 95–105.

The perturbation theory that we presented for the symmetric-definite problem was taken from

G.W. Stewart (1979). "Perturbation Bounds for the Definite Generalized Eigenvalue Problem," *Lin. Alg. and Its Applic. 23*, 69–86.

See also

L. Elsner and J. Guang Sun (1982). "Perturbation Theorems for the Generalized Eigenvalue Problem,; *Lin. Alg. and its Applic. 48*, 341-357.

J. Guang Sun (1982). "A Note on Stewart's Theorem for Definite Matrix Pairs," *Lin. Alg. and Its Applic. 48*, 331–339.

J. Guang Sun (1983). "Perturbation Analysis for the Generalized Singular Value Problem," *SIAM J. Numer. Anal. 20*, 611–625.

C.C. Paige (1984). "A Note on a Result of Sun J.-Guang: Sensitivity of the CS and GSV Decompositions," *SIAM J. Numer. Anal. 21*, 186–191.

The generalized SVD and some of its applications are discussed in

C.F. Van Loan (1976). "Generalizing the Singular Value Decomposition," *SIAM J. Num. Anal. 13*, 76–83.

C.C. Paige and M. Saunders (1981). "Towards A Generalized Singular Value Decomposition," *SIAM J. Num. Anal. 18*, 398–405.

B. Kågström (1985). "The Generalized Singular Value Decomposition and the General $A - \lambda B$ Problem," *BIT 24*, 568-583.

Stable methods for computing the CS and generalized singular value decompositions are described in

G.W. Stewart (1983). "A Method for Computing the Generalized Singular Value Decomposition," in *Matrix Pencils* , ed. B. Kågström and A. Ruhe, Springer-Verlag, New York, pp. 207–20.

C.F. Van Loan (1985). "Computing the CS and Generalized Singular Value Decomposition," *Numer. Math. 46*, 479–492.

M.T. Heath, A.J. Laub, C.C. Paige, and R.C. Ward (1986). "Computing the SVD of a Product of Two Matrices," *SIAM J. Sci. and Stat. Comp. 7*, 1147-1159.

C.C. Paige (1986). "Computing the Generalized Singular Value Decomposition," *SIAM J. Sci. and Stat. Comp. 7*, 1126–1146.

L.M. Ewerbring and F.T. Luk (1989). "Canonical Correlations and Generalized SVD: Applications and New Algorithms," *J. Comput. Appl. Math. 27*, 37–52.

J. Erxiong (1990). "An Algorithm for Finding Generalized Eigenpairs of a Symmetric Definite Matrix Pencil," *Lin.Alg. and Its Applic. 132*, 65–91.

P.C. Hansen (1990). "Relations Between SVD and GSVD of Discrete Regularization Problems in Standard and General Form," *Lin.Alg. and Its Applic. 141*, 165–176.

H. Zha (1991). "The Restricted Singular Value Decomposition of Matrix Triplets," *SIAM J. Matrix Anal. Appl. 12*, 172–194.

B. De Moor and G.H. Golub (1991). "The Restricted Singular Value Decomposition: Properties and Applications," *SIAM J. Matrix Anal. Appl. 12*, 401–425.

V. Hari (1991). "On Pairs of Almost Diagonal Matrices," *Lin. Alg. and Its Applic. 148*, 193–223.

B. De Moor and P. Van Dooren (1992). "Generalizing the Singular Value and QR Decompositions," *SIAM J. Matrix Anal. Appl. 13*, 993–1014.

H. Zha (1992). "A Numerical Algorithm for Computing the Restricted Singular Value Decomposition of Matrix Triplets," *Lin.Alg. and Its Applic. 168*, 1–25.

R-C. Li (1993). "Bounds on Perturbations of Generalized Singular Values and of Associated Subspaces," *SIAM J. Matrix Anal. Appl. 14*, 195–234.

K. Veselić (1993). "A Jacobi Eigenreduction Algorithm for Definite Matrix Pairs," *Numer. Math. 64*, 241–268.

Z. Bai and H. Zha (1993). "A New Preprocessing Algorithm for the Computation of the Generalized Singular Value Decomposition," *SIAM J. Sci. Comp. 14*, 1007–1012.

L. Kaufman (1993). "An Algorithm for the Banded Symmetric Generalized Matrix Eigenvalue Problem," *SIAM J. Matrix Anal. Appl. 14*, 372–389.

G.E. Adams, A.W. Bojanczyk, and F.T. Luk (1994). "Computing the PSVD of Two 2×2 Triangular Matrices," *SIAM J. Matrix Anal. Appl. 15*, 366–382.

Z. Drmač (1994). *The Generalized Singular Value Problem*, Ph.D. Thesis, FernUniversitat, Hagen, Germany.

R-C. Li (1994). "On Eigenvalue Variations of Rayleigh Quotient Matrix Pencils of a Definite Pencil," *Lin. Alg. and Its Applic. 208/209*, 471–483.

Chapter 9

Lanczos Methods

In this chapter we develop the Lanczos method, a technique that can be used to solve certain large, sparse, symmetric eigenproblems $Ax = \lambda x$. The method involves partial tridiagonalizations of the given matrix A. However, unlike the Householder approach, no intermediate, full submatrices are generated. Equally important, information about A's extremal eigenvalues tends to emerge long before the tridiagonalization is complete. This makes the Lanczos algorithm particularly useful in situations where a few of A's largest or smallest eigenvalues are desired.

The derivation and exact arithmetic attributes of the method are presented in §9.1. The key aspects of the Kaniel-Paige theory are detailed. This theory explains the extraordinary convergence properties of the Lanczos process. Unfortunately, roundoff errors make the Lanczos method somewhat difficult to use in practice. The central problem is a loss of orthogonality among the Lanczos vectors that the iteration produces. There are several ways to cope with this as we discuss §9.2.

In §9.3 we show how the "Lanczos idea" can be applied to solve an assortment of singular value, least squares, and linear equations problems. Of particular interest is the development of the conjugate gradient method for symmetric positive definite linear systems. The Lanczos-conjugate gradient connection is explored further in the next chapter. In §9.4 we discuss the Arnoldi iteration which is based on the Hessenberg decomposition and a version of the Lanczos process that can (sometimes) be used to tridiagonalize unsymmetric matrices.

Before You Begin

Chapters 5 and 8 are required for §9.1-9.3 and Chapter 7 is needed for §9.4. Within this chapter there are the following dependencies:

$$\S 9.1 \quad \rightarrow \quad \S 9.2 \quad \rightarrow \quad \S 9.3$$
$$\downarrow$$
$$\S 9.4$$

A wide range of Lanczos papers are collected in Brown, Chu, Ellison, and Plemmons (1994). Other complementary references include Parlett (1980), Saad (1992), and Chatelin (1993). The two volume work by Cullum and Willoughby (1985a,1985b) includes both analysis and software.

9.1   Derivation and Convergence Properties

Suppose $A \in \mathbb{R}^{n \times n}$ is large, sparse, and symmetric and assume that a few of its largest and/or smallest eigenvalues are desired. This problem can be solved by a method attributed to Lanczos (1950). The method generates a sequence of tridiagonal matrices T_k with the property that the extremal eigenvalues of $T_k \in \mathbb{R}^{k \times k}$ are progressively better estimates of A's extremal eigenvalues. In this section, we derive the technique and investigate its exact arithmetic properties. Throughout the section $\lambda_i(\cdot)$ designates the ith largest eigenvalue.

9.1.1   Krylov Subspaces

The derivation of the Lanczos algorithm can proceed in several ways. So that its remarkable convergence properties do not come as a complete surprise, we prefer to lead into the technique by considering the optimization of the Rayleigh quotient

$$r(x) \;=\; \frac{x^T A x}{x^T x} \qquad x \neq 0.$$

Recall from Theorem 8.1.2 that the maximum and minimum values of $r(x)$ are $\lambda_1(A)$ and $\lambda_n(A)$, respectively. Suppose $\{q_i\} \subseteq \mathbb{R}^n$ is a sequence of orthonormal vectors and define the scalars M_k and m_k by

$$M_k = \lambda_1(Q_k^T A Q_k) = \max_{y \neq 0} \frac{y^T (Q_k^T A Q_k) y}{y^T y} = \max_{\|y\|_2 = 1} r(Q_k y) \leq \lambda_1(A)$$

$$m_k = \lambda_k(Q_k^T A Q_k) = \min_{y \neq 0} \frac{y^T (Q_k^T A Q_k) y}{y^T y} = \min_{\|y\|_2 = 1} r(Q_k y) \geq \lambda_n(A)$$

where $Q_k = [\, q_1, \ldots, q_k \,]$. The Lanczos algorithm can be derived by considering how to generate the q_k so that M_k and m_k are increasingly better estimates of $\lambda_1(A)$ and $\lambda_n(A)$.

Suppose $u_k \in \mathrm{span}\{q_1, \ldots, q_k\}$ is such that $M_k = r(u_k)$. Since $r(x)$ increases most rapidly in the direction of the gradient

$$\nabla r(x) = \frac{2}{x^T x}(Ax - r(x)x),$$

we can ensure that $M_{k+1} > M_k$ if q_{k+1} is determined so

$$\nabla r(u_k) \in \mathrm{span}\{q_1, \ldots, q_{k+1}\}. \tag{9.1.1}$$

(This assumes $\nabla r(u_k) \neq 0$.) Likewise, if $v_k \in \mathrm{span}\{q_1, \ldots, q_k\}$ satisfies $r(v_k) = m_k$, then it makes sense to require

$$\nabla r(v_k) \in \mathrm{span}\{q_1, \ldots, q_{k+1}\} \tag{9.1.2}$$

since $r(x)$ decreases most rapidly in the direction of $-\nabla r(x)$.

At first glance, the task of finding a single q_{k+1} that satisfies these two requirements appears impossible. However, since $\nabla r(x) \in \mathrm{span}\{x, Ax\}$, it is clear that (9.1.1) and (9.1.2) can be simultaneously satisfied if

$$\mathrm{span}\{q_1, \ldots, q_k\} = \mathrm{span}\{q_1, Aq_1, \ldots, A^{k-1}q_1\}$$

and we choose q_{k+1} so

$$\mathrm{span}\{q_1, \ldots, q_{k+1}\} = \mathrm{span}\{q_1, Aq_1, \ldots, A^{k-1}q_1, A^k q_1\}.$$

Thus, we are led to the problem of computing orthonormal bases for the *Krylov subspaces*

$$\mathcal{K}(A, q_1, k) = \mathrm{span}\{q_1, Aq_1, \ldots, A^{k-1}q_1\}.$$

These are just the range spaces of the Krylov matrices

$$K(A, q_1, n) = [\, q_1, Aq_1, A^2 q_1, \ldots, A^{n-1} q_1 \,].$$

presented in §8.3.2.

9.1.2   Tridiagonalization

In order to find this basis efficiently we exploit the connection between the tridiagonalization of A and the QR factorization of $K(A, q_1, n)$. Recall that if $Q^T A Q = T$ is tridiagonal with $Q e_1 = q_1$, then

$$K(A, q_1, n) = Q [\, e_1, T e_1, T^2 e_1, \ldots, T^{n-1} e_1 \,]$$

is the QR factorization of $K(A, q_1, n)$ where $e_1 = I_n(:,1)$. Thus the q_k can effectively be generated by tridiagonalizing A with an orthogonal matrix whose first column is q_1.

Householder tridiagonalization, discussed in §8.3.1, can be adapted for this purpose. However, this approach is impractical if A is large and sparse because Householder similarity transformations tend to destroy sparsity. As a result, unacceptably large, dense matrices arise during the reduction.

Loss of sparsity can sometimes be controlled by using Givens rather than Householder transformations. See Duff and Reid (1976). However, any method that computes T by successively updating A is not useful in the majority of cases when A is sparse.

This suggests that we try to compute the elements of the tridiagonal matrix $T = Q^T A Q$ directly. Setting $Q = [\, q_1, \ldots, q_n \,]$ and

$$
T = \begin{bmatrix}
\alpha_1 & \beta_1 & & \cdots & & 0 \\
\beta_1 & \alpha_2 & \ddots & & & \vdots \\
 & \ddots & \ddots & \ddots & & \\
\vdots & & \ddots & \ddots & & \beta_{n-1} \\
0 & \cdots & & & \beta_{n-1} & \alpha_n
\end{bmatrix}
$$

and equating columns in $AQ = QT$, we find

$$
Aq_k = \beta_{k-1} q_{k-1} + \alpha_k q_k + \beta_k q_{k+1} \qquad \beta_0 q_0 \equiv 0
$$

for $k = 1{:}n - 1$. The orthonormality of the q_i implies $\alpha_k = q_k^T A q_k$. Moreover, if $r_k = (A - \alpha_k I)q_k - \beta_{k-1} q_{k-1}$ is nonzero, then $q_{k+1} = r_k/\beta_k$ where $\beta_k = \pm \| r_k \|_2$. If $r_k = 0$, then the iteration breaks down but (as we shall see) not without the acquisition of valuable invariant subspace information. So by properly sequencing the above formulae we obtain the *Lanczos iteration*:

$$
\begin{aligned}
& r_0 = q_1; \ \beta_0 = 1; \ q_0 = 0; \ k = 0 \\
& \textbf{while } (\beta_k \neq 0) \\
& \qquad q_{k+1} = r_k/\beta_k; \ k = k + 1; \ \alpha_k = q_k^T A q_k \\
& \qquad r_k = (A - \alpha_k I)q_k - \beta_{k-1} q_{k-1}; \ \beta_k = \| r_k \|_2 \\
& \textbf{end}
\end{aligned}
\qquad (9.1.3)
$$

There is no loss of generality in choosing the β_k to be positive. The q_k are called *Lanczos vectors*.

9.1.3   Termination and Error Bounds

The iteration halts before complete tridiagonalization if q_1 is contained in a proper invariant subspace. This is one of several mathematical properties of the method that we summarize in the following theorem.

Theorem 9.1.1 *Let $A \in \mathbb{R}^{n \times n}$ be symmetric and assume $q_1 \in \mathbb{R}^n$ has unit 2-norm. Then the Lanczos iteration (9.1.3) runs until $k = m$, where $m = \text{rank}(K(A, q_1, n))$ Moreover, for $k = 1{:}m$ we have*

$$AQ_k = Q_k T_k + r_k e_k^T \qquad (9.1.4)$$

where

$$T_k = \begin{bmatrix} \alpha_1 & \beta_1 & \cdots & & 0 \\ \beta_1 & \alpha_2 & \ddots & & \vdots \\ & \ddots & \ddots & \ddots & \\ \vdots & & \ddots & \ddots & \beta_{k-1} \\ 0 & \cdots & & \beta_{k-1} & \alpha_k \end{bmatrix}$$

and $Q_k = [\, q_1, \ldots, q_k \,]$ has orthonormal columns that span $\mathcal{K}(A, q_1, k)$.

Proof. The proof is by induction on k. Suppose the iteration has produced $Q_k = [\, q_1, \ldots, q_k \,]$ such that $\text{ran}(Q_k) = \mathcal{K}(A, q_1, k)$ and $Q_k^T Q_k = I_k$. It is easy to see from (9.1.3) that (9.1.4) holds. Thus, $Q_k^T A Q_k = T_k + Q_k^T r_k e_k^T$. Since $\alpha_i = q_i^T A q_i$ for $i = 1{:}k$ and

$$q_{i+1}^T A q_i = q_{i+1}^T (A q_i - \alpha_i q_i - \beta_{i-1} q_{i-1}) = q_{i+1}^T (\beta_i q_{i+1}) = \beta_i$$

for $i = 1{:}k-1$, we have $Q_k^T A Q_k = T_k$. Consequently, $Q_k^T r_k = 0$.
If $r_k \neq 0$, then $q_{k+1} = r_k / \| \, r_k \, \|_2$ is orthogonal to $q_1, \ldots, q_k$ and

$$q_{k+1} \in \text{span}\{A q_k, q_k, q_{k-1}\} \subseteq \mathcal{K}(A, q_1, k+1).$$

Thus, $Q_{k+1}^T Q_{k+1} = I_{k+1}$ and $\text{ran}(Q_{k+1}) = \mathcal{K}(A, q_1, k+1)$. On the other hand, if $r_k = 0$, then $A Q_k = Q_k T_k$. This says that $\text{ran}(Q_k) = \mathcal{K}(A, q_1, k)$ is invariant. From this we conclude that $k = m = \text{rank}(K(A, q_1, n))$. $\square$

Encountering a zero β_k in the Lanczos iteration is a welcome event in that it signals the computation of an exact invariant subspace. However, an exact zero or even a small β_k is a rarity in practice. Nevertheless, the extremal eigenvalues of T_k turn out to be surprisingly good approximations to A's extremal eigenvalues. Consequently, other explanations for the convergence of T_k's eigenvalues must be sought. The following result is a step in this direction.

Theorem 9.1.2 *Suppose that k steps of the Lanczos algorithm have been performed and that $S_k^T T_k S_k = \text{diag}(\theta_1, \ldots, \theta_k)$ is the Schur decomposition of the tridiagonal matrix T_k. If $Y_k = [\, y_1, \ldots, y_k \,] = Q_k S_k \in \mathbb{R}^{n \times k}$, then for $i = 1{:}k$ we have $\| \, A y_i - \theta_i y_i \, \|_2 = |\beta_k| \, |s_{ki}|$ where $S_k = (s_{pq})$.*

Proof. Post-multiplying (9.1.4) by S_k gives

$$A Y_k = Y_k \text{diag}(\theta_1, \ldots, \theta_k) + r_k e_k^T S_k,$$

and so $Ay_i = \theta_i y_i + r_k(e_k^T Se_i)$. The proof is complete by taking norms and recalling that $\| r_k \|_2 = |\beta_k|$. $\square$

The theorem provides computable error bounds for T_k's eigenvalues:

$$\min_{\mu \in \lambda(A)} |\theta_i - \mu| \leq |\beta_k| |s_{ki}| \qquad i = 1{:}k$$

Note that in the terminology of Theorem 8.1.15, the (θ_i, y_i) are Ritz pairs for the subspace $\text{ran}(Q_k)$.

Another way that T_k can be used to provide estimates of A's eigenvalues is described in Golub (1974) and involves the judicious construction of a rank-one matrix E such that $\text{ran}(Q_k)$ is invariant for $A + E$. In particular, if we use the Lanczos method to compute $AQ_k = Q_k T_k + r_k e_k^T$ and set $E = \tau ww^T$, where $\tau = \pm 1$ and $w = aq_k + br_k$, then it can be shown that

$$(A + E)Q_k = Q_k(T_k + \tau a^2 e_k e_k^T) + (1 + \tau ab)r_k e_k^T.$$

If $0 = 1 + \tau ab$, then the eigenvalues of $\bar{T}_k = T_k + \tau a^2 e_k e_k^T$, a tridiagonal matrix, are also eigenvalues of $A + E$. Using Theorem 8.1.8 it can be shown that the interval $[\lambda_i(\tilde{T}_k), \lambda_{i-1}(\tilde{T}_k)]$ contains an eigenvalue of A for $i = 2{:}k$.

These bracketing intervals depend on the choice of τa^2. Suppose we have an approximate eigenvalue of λ of A. One possibility is to choose τa^2 so that $\det(T_k - \lambda I_k) = (\alpha_2 + \tau a^2 - \lambda)p_{k-1}(\lambda) - \beta_{k-1}^2 p_{k-2}(\lambda) = 0$ where the polynomials $p_i(x) = \det(T_i - xI_i)$ can be evaluated at λ using the three-term recurrence (8.5.2). (This assumes that $p_{k-1}(\lambda) \neq 0$.) Eigenvalue estimation in this spirit is discussed in Lehmann (1963) and Householder (1968).

9.1.4   The Kaniel-Paige Convergence Theory

The preceding discussion indicates how eigenvalue estimates can be obtained via the Lanczos algorithm, but it reveals nothing about rate of convergence. Results of this variety constitute what is known as the *Kaniel-Paige theory*, a sample of which follows.

Theorem 9.1.3 *Let A be an n-by-n symmetric matrix with eigenvalues $\lambda_1 \geq \cdots \geq \lambda_n$ and corresponding orthonormal eigenvectors $z_1, \ldots, z_n$. If $\theta_1 \geq \cdots \geq \theta_k$ are the eigenvalues of the matrix T_k obtained after k steps of the Lanczos iteration, then*

$$\lambda_1 \geq \theta_1 \geq \lambda_1 - \frac{(\lambda_1 - \lambda_n)\tan(\phi_1)^2}{(c_{k-1}(1 + 2\rho_1))^2}$$

where $\cos(\phi_1) = |q_1^T z_1|$, $\rho_1 = (\lambda_1 - \lambda_2)/(\lambda_2 - \lambda_n)$, and $c_{k-1}(x)$ is the Chebyshev polynomial of degree $k - 1$.

Proof. From Theorem 8.1.2, we have

$$\theta_1 = \max_{y \neq 0} \frac{y^T T_k y}{y^T y} = \max_{y \neq 0} \frac{(Q_k y)^T A (Q_k y)}{(Q_k y)^T (Q_k y)} = \max_{0 \neq w \in \mathcal{K}(A, q_1, k)} \frac{w^T A w}{w^T w}$$

Since λ_1 is the maximum of $w^T A w / w^T w$ over all nonzero w, it follows that $\lambda_1 \geq \theta_1$. To obtain the lower bound for θ_1, note that

$$\theta_1 = \max_{p \in \mathcal{P}_{k-1}} \frac{q_1^T p(A) A p(A) q_1}{q_1^T p(A)^2 q_1}$$

where $\mathcal{P}_{k-1}$ is the set of $k-1$ degree polynomials. If $q_1 = \sum_{i=1}^{n} d_i z_i$ then

$$\frac{q_1^T p(A) A p(A) q_1}{q_1^T p(A)^2 q_1} = \frac{\displaystyle\sum_{i=1}^{n} d_i^2 p(\lambda_i)^2 \lambda_i}{\displaystyle\sum_{i=1}^{n} d_i^2 p(\lambda_i)^2}$$

$$\geq \lambda_1 - (\lambda_1 - \lambda_n) \frac{\displaystyle\sum_{i=2}^{n} d_i^2 p(\lambda_i)^2}{d_1^2 p(\lambda_1)^2 + \displaystyle\sum_{i=2}^{n} d_i^2 p(\lambda_i)^2} .$$

We can make the lower bound tight by selecting a polynomial $p(x)$ that is large at $x = \lambda_1$ in comparison to its value at the remaining eigenvalues. One way of doing this is to set

$$p(x) = c_{k-1}\left(-1 + 2\frac{x - \lambda_n}{\lambda_2 - \lambda_n}\right)$$

where $c_{k-1}(z)$ is the $(k-1)$-st Chebyshev polynomial generated via the recursion

$$c_k(z) = 2z c_{k-1}(z) - c_{k-2}(z) \qquad c_0 = 1, \ c_1 = z.$$

These polynomials are bounded by unity on $[-1, 1]$, but grow very rapidly outside this interval. By defining $p(x)$ this way it follows that $|p(\lambda_i)|$ is bounded by unity for $i = 2{:}n$, while $p(\lambda_1) = c_{k-1}(1 + 2\rho_1)$. Thus,

$$\theta_1 \geq \lambda_1 - (\lambda_1 - \lambda_n) \frac{1 - d_1^2}{d_1^2} \frac{1}{c_{k-1}(1 + 2\rho_1)^2} .$$

The desired lower bound is obtained by noting that $\tan(\phi_1)^2 = (1 - d_1^2)/d_1^2$. $\square$

An analogous result pertaining to θ_k follows immediately from this theorem:

Corollary 9.1.4 *Using the same notation as the theorem,*

$$\lambda_n \leq \theta_k \leq \lambda_n + \frac{(\lambda_1 - \lambda_n)\tan(\phi_n)^2}{c_{k-1}(1 + 2\rho_n)^2}$$

where $\rho_n = (\lambda_{n-1} - \lambda_n)/(\lambda_1 - \lambda_{n-1})$ *and* $\cos(\phi_n) = q_n^T z_n$.

Proof. Apply Theorem 9.1.3 with A replaced by $-A$. $\square$

9.1.5   The Power Method Versus the Lanczos Method

It is worthwhile to compare θ_1 with the corresponding power method estimate of λ_1. (See §8.2.1.) For clarity, assume $\lambda_1 \geq \cdots \geq \lambda_n \geq 0$. After $k-1$ power method steps applied to q_1, a vector is obtained in the direction of

$$v = A^{k-1}q_1 = \sum_{i=1}^n c_i \lambda_i^{k-1} z_i$$

along with an eigenvalue estimate

$$\gamma_1 = \frac{v^T A v}{v^T v}.$$

Using the proof and notation of Theorem 9.1.3, it is easy to show that

$$\lambda_1 \geq \gamma_1 \geq \lambda_1 - (\lambda_1 - \lambda_n)\tan(\phi_1)^2 \left(\frac{\lambda_2}{\lambda_1}\right)^{2k-1}. \qquad (9.1.5)$$

(Hint: Set $p(x) = x^{k-1}$ in the proof.) Thus, we can compare the quality of the lower bounds for θ_1 and γ_1 by comparing

$$L_{k-1} \equiv 1 \Big/ \left[c_{k-1}\left(2\frac{\lambda_1}{\lambda_2} - 1\right)\right]^2 \geq 1 \Big/ \left[c_{k-1}(1 + 2\rho_1)\right]^2$$

and

$$R_{k-1} = \left(\frac{\lambda_2}{\lambda_1}\right)^{2(k-1)}.$$

This is done in following table for representative values of k and λ_2/λ_1.

The superiority of the Lanczos estimate is self-evident. This should be no surprise, since θ_1 is the maximum of $r(x) = x^T A x / x^T x$ over *all* of $\mathcal{K}(A, q_1, k)$, while $\gamma_1 = r(v)$ for a particular v in $\mathcal{K}(A, q_1, k)$, namely $v = A^{k-1}q_1$.

λ_1/λ_2	$k=5$	$k=10$	$k=15$	$k=20$	$k=25$
1.50	$\dfrac{1.1\times10^{-4}}{3.9\times10^{-2}}$	$\dfrac{2.0\times10^{-10}}{6.8\times10^{-4}}$	$\dfrac{3.9\times10^{-16}}{1.2\times10^{-5}}$	$\dfrac{7.4\times10^{-22}}{2.0\times10^{-7}}$	$\dfrac{1.4\times10^{-27}}{3.5\times10^{-9}}$
1.10	$\dfrac{2.7\times10^{-2}}{4.7\times10^{-1}}$	$\dfrac{5.5\times10^{-5}}{1.8\times10^{-1}}$	$\dfrac{1.1\times10^{-7}}{6.9\times10^{-2}}$	$\dfrac{2.1\times10^{-10}}{2.7\times10^{-2}}$	$\dfrac{4.2\times10^{-13}}{1.0\times10^{-2}}$
1.01	$\dfrac{5.6\times10^{-1}}{9.2\times10^{-1}}$	$\dfrac{1.0\times10^{-1}}{8.4\times10^{-1}}$	$\dfrac{1.5\times10^{-2}}{7.6\times10^{-1}}$	$\dfrac{2.0\times10^{-3}}{6.9\times10^{-1}}$	$\dfrac{2.8\times10^{-4}}{6.2\times10^{-1}}$

TABLE 9.1.1 L_{k-1}/R_{k-1}

9.1.6   Convergence of Interior Eigenvalues

We conclude with some remarks about error bounds for T_k's interior eigenvalues. The key idea in the proof of Theorem 9.1.3 is the use of the translated Chebyshev polynomial. With this polynomial we amplified the component of q_1 in the direction z_1. A similar idea can be used to obtain bounds for an interior Ritz value θ_i. However, the bounds are not as satisfactory because the "amplifying polynomial" has the form $q(x)\Pi_{i=1}^{k-1}(x - \lambda_i)$, where $q(x)$ is the $(k-1)$ degree of the Chebyshev polynomial on the interval $[\lambda_{i+1}, \lambda_n]$. For details, see Kaniel (1966), Paige (1971), or Saad (1980).

Problems

P9.1.1 Suppose $A \in \mathbf{R}^{n\times n}$ is skew-symmetric. Derive a Lanczos-like algorithm for computing a skew-symmetric tridiagonal matrix T_m such that $AQ_m = Q_mT_m$, where $Q_m^TQ_m = I_m$.

P9.1.2 Let $A \in \mathbf{R}^{n\times n}$ be symmetric and define $r(x) = x^TAx/x^Tx$. Suppose $S \subseteq \mathbf{R}^n$ is a subspace with the property that $x \in S$ implies $\nabla r(x) \in S$. Show that S is invariant for A.

P9.1.3 Show that if a symmetric matrix $A \in \mathbf{R}^{n\times n}$ has a multiple eigenvalue, then the Lanczos iteration terminates prematurely.

P9.1.4 Show that the index m in Theorem 9.1.1 is the dimension of the smallest invariant subspace for A that contains q_1.

P9.1.5 Let $A \in \mathbf{R}^{n\times n}$ be symmetric and consider the problem of determining an orthonormal sequence $q_1, q_2, \ldots$ with the property that once $Q_k =[\, q_1, \ldots, q_k\,]$ is known, q_{k+1} is chosen so as to minimize $\mu_k = \| (I - Q_{k+1}Q_{k+1}^T)AQ_k \|_F$. Show that if $\mathrm{span}\{q_1, \ldots, q_k\} = \mathcal{K}(A, q_1, k)$, then it is possible to choose q_{k+1} so $\mu_k = 0$. Explain how this optimization problem leads to the Lanczos iteration.

P9.1.6 Suppose $A \in \mathbf{R}^{n\times n}$ is symmetric and that we wish to compute its largest eigenvalue. Let η be an approximate eigenvector and set

$$\alpha = \frac{\eta^T A\eta}{\eta^T\eta}$$
$$z = A\eta - \alpha\eta.$$

(a) Show that the interval $[\alpha - \delta, \alpha + \delta]$ must contain an eigenvalue of A where $\delta = \| z \|_2 / \| \eta \|_2$. (b) Consider the new approximation $\bar{\eta} = a\eta + bz$ and show how to determine the scalars a and b so that

$$\bar{\alpha} = \frac{\bar{\eta}^T A \bar{\eta}}{\bar{\eta}^T \bar{\eta}}$$

is maximized. (c) Relate the above computations to the first two steps of the Lanczos process.

Notes and References for Sec. 9.1

The classic reference for the Lanczos method is

C. Lanczos (1950). "An Iteration Method for the Solution of the Eigenvalue Problem of Linear Differential and Integral Operators," *J. Res. Nat. Bur. Stand. 45*, 255–82.

Although the convergence of the Ritz values is alluded to this paper, for more details we refer the reader to

S. Kaniel (1966). "Estimates for Some Computational Techniques in Linear Algebra," *Math. Comp. 20*, 369–78.
C.C. Paige (1971). "The Computation of Eigenvalues and Eigenvectors of Very Large Sparse Matrices," Ph.D. thesis, London University.
Y. Saad (1980). "On the Rates of Convergence of the Lanczos and the Block Lanczos Methods," *SIAM J. Num. Anal.17*, 687–706.

The connections between the Lanczos algorithm, orthogonal polynomials, and the theory of moments are discussed in

N.J. Lehmann (1963). "Optimale Eigenwerteinschliessungen," *Numer. Math. 5*, 246–72.
A.S. Householder (1968). "Moments and characteristic Roots II," *Numer. Math. 11*, 126–28.
G.H. Golub (1974). "Some Uses of the Lanczos Algorithm in Numerical Linear Algebra," *in Topics in Numerical Analysis*, ed., J.J.H. Miller, Academic Press, New York.

We motivated our discussion of the Lanczos algorithm by discussing the inevitability of fill-in when Householder or Givens transformations are used to tridiagonalize. Actually, fill-in can sometimes be kept to an acceptable level if care is exercised. See

I.S. Duff (1974). "Pivot Selection and Row Ordering in Givens Reduction on Sparse Matrices," *Computing 13*, 239–48.
I.S. Duff and J.K. Reid (1976). "A Comparison of Some Methods for the Solution of Sparse Over-Determined Systems of Linear Equations," *J. Inst. Maths. Applic. 17*, 267–80.
L. Kaufman (1979). "Application of Dense Householder Transformations to a Sparse Matrix," *ACM Trans. Math. Soft. 5*, 442–50.

9.2   Practical Lanczos Procedures

Rounding errors greatly affect the behavior of the Lanczos iteration. The basic difficulty is caused by loss of orthogonality among the Lanczos vectors, a phenomenon that muddies the issue of termination and complicates the relationship between A's eigenvalues and those of the tridiagonal matrices T_k. This troublesome feature, coupled with the advent of Householder's perfectly stable method of tridiagonalization, explains why the Lanczos algorithm was disregarded by numerical analysts during the 1950's and 1960's. However, interest in the method was rejuvenated with the development of the Kaniel-Paige theory and because the pressure to solve large, sparse eigenproblems increased with increased computer power. With many fewer than n iterations typically required to get good approximate extremal eigenvalues, the Lanczos method became attractive as a sparse matrix technique rather than as a competitor of the Householder approach.

Successful implementations of the Lanczos iteration involve much more than a simple encoding of (9.1.3). In this section we outline some of the practical ideas that have been proposed to make Lanczos procedure viable in practice.

9.2.1   Exact Arithmetic Implementation

With careful overwriting in (9.1.3) and exploitation of the formula

$$\alpha_k = q_k^T(Aq_k - \beta_{k-1}q_{k-1}),$$

the whole Lanczos process can be implemented with just two n-vectors of storage.

Algorithm 9.2.1. (**The Lanczos Algorithm**) Given a symmetric $A \in \mathbb{R}^{n \times n}$ and $w \in \mathbb{R}^n$ having unit 2-norm, the following algorithm computes a k-by-k symmetric tridiagonal matrix T_k with the property that $\lambda(T_k) \subset \lambda(A)$. It assumes the existence of a function **A.mult**(w) that returns the matrix-vector product Aw. The diagonal and subdiagonal elements of T_k are stored in $\alpha(1{:}k)$ and $\beta(1{:}k-1)$ respectively.

$$v(1{:}n) = 0; \ \beta_0 = 1; \ k = 0$$
$$\textbf{while } \beta_k \neq 0$$
$$\qquad \textbf{if } k \neq 0$$
$$\qquad\qquad \textbf{for } i = 1{:}n$$
$$\qquad\qquad\qquad t = w_i; \ w_i = v_i/\beta_k; \ v_i = -\beta_k t$$
$$\qquad\qquad \textbf{end}$$
$$\qquad \textbf{end}$$
$$\qquad v = v + \textbf{A.mult}(w)$$
$$\qquad k = k+1; \ \alpha_k = w^T v; \ v = v - \alpha_k w; \ \beta_k = \| v \|_2$$
$$\textbf{end}$$

Note that A is not altered during the entire process. Only a procedure **A.mult**$(\cdot)$ for computing matrix-vector products involving A need be supplied. If A has an average of about i nonzeros per row, then approximately $(2i + 8)n$ flops are involved in a single Lanczos step.

Upon termination the eigenvalues of T_k can be found using the symmetric tridiagonal QR algorithm or any of the special methods of §8.5, such as bisection.

The Lanczos vectors are generated in the n-vector w. If they are desired for later use, then special arrangements must be made for their storage. In the typical sparse matrix setting they could be stored on a disk or some other secondary storage device until required.

9.2.2   Roundoff Properties

The development of a practical, easy-to-use Lanczos procedure requires an appreciation of the fundamental error analyses of Paige (1971, 1976, 1980). An examination of his results is the best way to motivate the several modified Lanczos procedures of this section.

After j steps of the algorithm we obtain the matrix of computed Lanczos vectors $\hat{Q}_k = [\hat{q}_1, \ldots, \hat{q}_k]$ and the associated tridiagonal matrix

$$
\hat{T}_k = \begin{bmatrix}
\hat{\alpha}_1 & \hat{\beta}_1 & \cdots & & 0 \\
\hat{\beta}_1 & \hat{\alpha}_2 & \ddots & & \vdots \\
& \ddots & \ddots & \ddots & \\
\vdots & & \ddots & \ddots & \hat{\beta}_{k-1} \\
0 & \cdots & & \hat{\beta}_{k-1} & \hat{\alpha}_k
\end{bmatrix} .
$$

Paige (1971, 1976) shows that if $\hat{r}_k$ is the computed analog of r_k, then

$$
A\hat{Q}_k = \hat{Q}_k\hat{T}_k + \hat{r}_k e_k^T + E_k \tag{9.2.1}
$$

where

$$
\| E_k \|_2 \approx \mathbf{u}\| A \|_2 . \tag{9.2.2}
$$

This indicates that the important equation $AQ_k = Q_kT_k + r_ke_k^T$ is satisfied to working precision.

Unfortunately, the picture is much less rosy with respect to the orthogonality among the $\hat{q}_i$. (Normality is not an issue. The computed Lanczos vectors essentially have unit length.) If $\hat{\beta}_k = fl(\| \hat{r}_k \|_2)$ and we compute $\hat{q}_{k+1} = fl(\hat{r}_k/\hat{\beta}_k)$, then a simple analysis shows that $\hat{\beta}_k\hat{q}_{k+1} \approx \hat{r}_k + w_k$ where $\| w_k \|_2 \approx \mathbf{u}\| \hat{r}_k \|_2 \approx \mathbf{u}\| A \|_2$. Thus, we may conclude that

$$
|\hat{q}_{k+1}^T\hat{q}_i| \approx \frac{|\hat{r}_k^T\hat{q}_i| + \mathbf{u}\| A \|_2}{|\hat{\beta}_k|}
$$

for $i = 1{:}k$. In other words, significant departures from orthogonality can be expected when $\hat{\beta}_k$ is small, *even* in the ideal situation where $\hat{r}_k^T\hat{Q}_k$ is zero. A small $\hat{\beta}_k$ implies cancellation in the computation of $\hat{r}_k$. We stress that loss of orthogonality is due to this cancellation and is not the result of the gradual accumulation of roundoff error.

Example 9.2.1 The matrix

$$A = \begin{bmatrix} 2.64 & -.48 \\ -.48 & 2.36 \end{bmatrix}$$

has eigenvalues $\lambda_1 = 3$ and $\lambda_2 = 2$. If the Lanczos algorithm is applied to this matrix with $q_1 = [.810, -.586]^T$ and three-digit floating point arithmetic is performed, then $\hat{q}_2 = [.707, .707]^T$. Loss of orthogonality occurs because span$\{q_1\}$ is almost invariant for A. (The vector $x = [.8, -.6]^T$ is the eigenvector affiliated with λ_1.)

Further details of the Paige analysis are given shortly. Suffice it to say now that loss of orthogonality always occurs in practice and with it, an apparent deterioration in the quality of $\hat{T}_k$'s eigenvalues. This can be quantified by combining (9.2.1) with Theorem 8.1.16. In particular, if in that theorem we set $F_1 = \hat{r}_k e_k^T + E_k$, $X_1 = \hat{Q}_k$, $S = \hat{T}_k$, and assume that

$$\tau = \| \hat{Q}_k^T \hat{Q}_k - I_k \|_2$$

satisfies $\tau < 1$, then there exist eigenvalues $\mu_1, \ldots, \mu_k \in \lambda(A)$ such that

$$|\mu_i - \lambda_i(T_k)| \leq \sqrt{2}\,(\| \hat{r}_k \|_2 + \| E_k \|_2 + \tau(2 + \tau)\| A \|_2)$$

for $i = 1{:}k$. An obvious way to control the τ factor is to orthogonalize each newly computed Lanczos vector against its predecessors. This leads directly to our first "practical" Lanczos procedure.

9.2.3 Lanczos with Complete Reorthogonalization

Let $r_0, \ldots, r_{k-1} \in \mathbb{R}^n$ be given and suppose that Householder matrices $H_0, \ldots, H_{k-1}$ have been computed such that $(H_0 \cdots H_{k-1})^T [r_0, \ldots, r_{k-1}]$ is upper triangular. Let $[q_1, \ldots, q_k]$ denote the first k columns of the Householder product $(H_0 \cdots H_{k-1})$. Now suppose that we are given a vector $r_k \in \mathbb{R}^n$ and wish to compute a unit vector q_{k+1} in the direction of

$$w = r_k - \sum_{i=1}^{k}(q_i^T r_k)q_i \in \text{span}\{q_1, \ldots, q_k\}^{\perp}.$$

If a Householder matrix H_k is determined so $(H_0 \cdots H_k)^T [r_0, \ldots, r_k]$ is upper triangular, then it follows that column $(k + 1)$ of $H_0 \cdots H_k$ is the desired unit vector.

If we incorporate these Householder computations into the Lanczos process, then we can produce Lanczos vectors that are orthogonal to machine precision:

$r_0 = q_1$ (given unit vector)
Determine Householder H_0 so $H_0 r_0 = e_1$.
$\alpha_1 = q_1^T A q_1$
for $k = 1{:}n - 1$
$\qquad r_k = (A - \alpha_k I)q_k - \beta_{k-1}q_{k-1}$ $\qquad (\beta_0 q_0 \equiv 0)$ $\qquad\qquad$ (9.2.3)
$\qquad w = (H_{k-1} \cdots H_0)r_k$
$\qquad$ Determine Householder H_k so $H_k w = (w_1, \ldots, w_k, \beta_k, 0, \ldots, 0)^T$.
$\qquad q_{k+1} = H_0 \cdots H_k e_{k+1}$; $\alpha_{k+1} = q_{k+1}^T A q_{k+1}$
end

This is an example of a *complete reorthorgonalization* Lanczos scheme. A thorough analysis may be found in Paige (1970). The idea of using Householder matrices to enforce orthogonality appears in Golub, Underwood, and Wilkinson (1972).

That the computed $\hat{q}_i$ in (9.2.3) are orthogonal to working precision follows from the roundoff properties of Householder matrices. Note that by virtue of the definition of q_{k+1} , it makes no difference if $\beta_k = 0$. For this reason, the algorithm may safely run until $k = n - 1$. (However, in practice one would terminate for a much smaller value of k.)

Of course, in any implementation of (9.2.3), one stores the Householder vectors v_k and never explicitly forms the corresponding P_k. Since we have $H_k(1{:}k, 1{:}k) = I_k$ there is no need to compute the first k components of $w = (H_{k-1} \cdots H_0)r_k$, for in exact arithmetic these components would be zero.

Unfortunately, these economies make but a small dent in the computational overhead associated with complete reorthogonalization. The Householder calculations increase the work in the kth Lanczos step by $O(kn)$ flops. Moreover, to compute q_{k+1}, the Householder vectors associated with $H_0, \ldots, H_k$ must be accessed. For large n and k, this usually implies a prohibitive amount of data transfer.

Thus, there is a high price associated with complete reorthogonalization. Fortunately, there are more effective courses of action to take, but these demand that we look more closely at how orthogonality is lost.

9.2.4   Selective Orthogonalization

A remarkable, ironic consequence of the Paige (1971) error analysis is that loss of orthogonality goes hand in hand with convergence of a Ritz pair. To be precise, suppose the symmetric QR algorithm is applied to $\hat{T}_k$ and renders computed Ritz values $\hat{\theta}_1, \ldots, \hat{\theta}_k$ and a nearly orthogonal matrix of eigenvectors $\hat{S}_k = (\hat{s}_{pq})$. If $\hat{Y}_k = [\hat{y}_1, \ldots, \hat{y}_k] = fl(\hat{Q}_k \hat{S}_k)$, then it can be shown that for $i = 1{:}k$ we have

$$|\hat{q}_{k+1}^T \hat{y}_i| \approx \frac{\mathbf{u} \| A \|_2}{|\hat{\beta}_k| |\hat{s}_{ki}|} \qquad\qquad (9.2.4)$$

and

$$\| A\hat{y}_i - \hat{\theta}_i \hat{y}_i \|_2 \approx |\hat{\beta}_k| |\hat{s}_{ki}| . \qquad (9.2.5)$$

That is, the most recently computed Lanczos vector $\hat{q}_{k+1}$ tends to have a nontrivial and unwanted component in the direction of any converged Ritz vector. Consequently, instead of orthogonalizing $\hat{q}_{k+1}$ against all of the previously computed Lanczos vectors, we can achieve the same effect by orthogonalizing it against the much smaller set of converged Ritz vectors.

The practical aspects of enforcing orthogonality in this way are discussed in Parlett and Scott (1979). In their scheme, known as *selective orthogonalization,* a computed Ritz pair $(\hat{\theta}, \hat{y})$ is called "good" if it satisfies

$$\| A\hat{y} - \hat{\theta}\hat{y} \|_2 \approx \sqrt{\mathbf{u}}\| A \|_2 .$$

As soon as $\hat{q}_{k+1}$ is computed, it is orthogonalized against each good Ritz vector. This is much less costly than complete reorthogonalization, since there are usually many fewer good Ritz vectors than Lanczos vectors.

One way to implement selective orthogonalization is to diagonalize $\hat{T}_k$ at each step and then examine the $\hat{s}_{ki}$ in light of (9.2.4) and (9.2.5). A much more efficient approach is to estimate the loss-of-orthogonality measure $\| I_k - \hat{Q}_k^T \hat{Q}_k \|_2$ using the following result:

Lemma 9.2.1 *Suppose* $S_+ = [\, S \; d \,]$ *where* $S \in \mathbb{R}^{n \times k}$ *and* $d \in \mathbb{R}^n$. *If* S *satisfies* $\| I_k - S^T S \|_2 \le \mu$ *and* $|1 - d^T d| \le \delta$ *then* $\| I_{k+1} - S_+^T S_+ \|_2 \le \mu_+$ *where*

$$\mu_+ = \frac{1}{2} \left(\mu + \delta + \sqrt{(\mu - \delta)^2 + 4\| S^T d \|_2^2} \, \right)$$

Proof. See Kahan and Parlett (1974) or Parlett and Scott (1979). $\square$

Thus, if we have a bound for $\| I_k - \hat{Q}_k^T \hat{Q}_k \|_2$ we can generate a bound for $\| I_{k+1} - \hat{Q}_{k+1}^T \hat{Q}_{k+1} \|_2$ by applying the lemma with $S = \hat{Q}_k$ and $d = \hat{q}_{k+1}$. (In this case $\delta \approx \mathbf{u}$ and we assume that $\hat{q}_{k+1}$ has been orthogonalized against the set of currently good Ritz vectors.) It is possible to estimate the norm of $\hat{Q}_k^T \hat{q}_{k+1}$ from a simple recurrence that spares one the need for accessing $\hat{q}_1, \ldots, \hat{q}_k$. See Kahan and Parlett (1974) or Parlett and Scott (1979). The overhead is minimal, and when the bounds signal loss of orthogonality, it is time to contemplate the enlargement of the set of good Ritz vectors. Then and only then is $\hat{T}_k$ diagonalized.

9.2.5   The Ghost Eigenvalue Problem

Considerable effort has been spent in trying to develop a workable Lanczos procedure that does not involve any kind of orthogonality enforcement. Research in this direction focuses on the problem of "ghost" or "spurious" eigenvalues. These are multiple eigenvalues of $\hat{T}_k$ that correspond to simple eigenvalues of A. They arise because the iteration essentially restarts itself when orthogonality to a converged Ritz vector is lost. (By way of analogy, consider what would happen during orthogonal iteration §8.2.8 if we "forgot" to orthogonalize.)

The problem of identifying ghost eigenvalues and coping with their presence is discussed in Cullum and Willoughby (1979) and Parlett and Reid (1981). It is a particularly pressing problem in those applications where all of A's eigenvalues are desired, for then the above orthogonalization procedures are too expensive to implement.

Difficulties with the Lanczos iteration can be expected even if A has a genuinely multiple eigenvalue. This follows because the $\hat{T}_k$ are unreduced, and unreduced tridiagonal matrices cannot have multiple eigenvalues. Our next practical Lanczos procedure attempts to circumvent this difficulty.

9.2.6   Block Lanczos

Just as the simple power method has a block analog in simultaneous iteration, so does the Lanczos algorithm have a block version. Suppose $n = rp$ and consider the decomposition

$$Q^T A Q = \bar{T} = \begin{bmatrix} M_1 & B_1^T & \cdots & & 0 \\ B_1 & M_2 & \ddots & & \vdots \\ & \ddots & \ddots & \ddots & \\ \vdots & & \ddots & \ddots & B_{r-1}^T \\ 0 & \cdots & & B_{r-1} & M_r \end{bmatrix} \qquad (9.2.6)$$

where

$$Q = [X_1, \ldots, X_r] \qquad X_i \in \mathbb{R}^{n \times p}$$

is orthogonal, each $M_i \in \mathbb{R}^{p \times p}$, and each $B_i \in \mathbb{R}^{p \times p}$ is upper triangular. Comparing blocks in $AQ = Q\bar{T}$ shows that

$$AX_k = X_{k-1}B_{k-1}^T + X_k M_k + X_{k+1}B_k \qquad X_0 B_0 \equiv 0$$

for $k = 1{:}r - 1$. From the orthogonality of Q we have

$$M_k = X_k^T A X_k$$

for $k = 1{:}r$. Moreover, if we define

$$R_k = AX_k - X_k M_k - X_{k-1}B_{k-1}^T \in \mathbb{R}^{n \times p}$$

then $X_{k+1}B_k = R_k$ is a QR factorization of R_k. These observations suggest that the block tridiagonal matrix $\bar{T}$ in (9.2.6) can be generated as follows:

$X_1 \in \mathbb{R}^{p \times p}$ given with $X_1^T X_1 = I_p$.

$M_1 = X_1^T A X_1$

for $k = 1{:}r - 1$ (9.2.7)

$\quad R_k = A X_k - X_k M_k - X_{k-1} B_{k-1}^T \qquad (X_0 B_0^T \equiv 0)$

$\quad X_{k+1} B_k = R_k \qquad$ (QR factorization of R_k)

$\quad M_{k+1} = X_{k+1}^T A X_{k+1}$

end

At the beginning of the kth pass through the loop we have

$$A[\,X_1,\dots,X_k\,] = [\,X_1,\dots,X_k\,]\bar{T}_k + R_k[\,0,\dots,0,I_p\,] \qquad (9.2.8)$$

where

$$\bar{T}_k = \begin{bmatrix} M_1 & B_1^T & \cdots & & 0 \\ B_1 & M_2 & \ddots & & \vdots \\ & \ddots & \ddots & \ddots & \\ \vdots & & \ddots & \ddots & B_{k-1}^T \\ 0 & \cdots & & B_{k-1} & M_k \end{bmatrix}.$$

Using an argument similar to the one used in the proof of Theorem 9.1.1, we can show that the X_k are mutually orthogonal provided none of the R_k are rank-deficient. However if $\text{rank}(R_k) < p$ for some k, then it is possible to choose the columns of X_{k+1} such that $X_{k+1}^T X_i = 0$, for $i = 1{:}k$. See Golub and Underwood (1977).

Because $\bar{T}_k$ has bandwidth p, it can be efficiently reduced to tridiagonal form using an algorithm of Schwartz (1968). Once tridiagonal form is achieved, the Ritz values can be obtained via the symmetric QR algorithm.

In order to intelligently decide when to use block Lanczos, it is necessary to understand how the block dimension affects convergence of the Ritz values. The following generalization of Theorem 9.1.3 sheds light on this issue.

Theorem 9.2.2 *Let A by an n-by-n symmetric matrix with eigenvalues $\lambda_1 \geq \cdots \geq \lambda_n$ and corresponding orthonormal eigenvectors $z_1,\dots,z_n$. Let $\mu_1 \geq \cdots \geq \mu_p$ be the p largest eigenvalues of the matrix $\bar{T}_k$ obtained after k steps of the block Lanczos iteration (9.2.7). If $Z_1 = [\,z_1,\dots,z_p\,]$ and $\cos(\theta_p) = \sigma_p(Z_1^T X_1) > 0$, then for $i = 1{:}p$, $\lambda_i \geq \mu_i \geq \lambda_i - \epsilon_i^2$ where*

$$\epsilon_i^2 = \frac{(\lambda_1 - \lambda_i)\tan^2(\theta_p)}{\left[c_{k-1}\left(\frac{1+\gamma_i}{1-\gamma_i}\right)\right]^2} \qquad \gamma_i = \frac{\lambda_i - \lambda_{p+1}}{\lambda_i - \lambda_n}$$

and $c_{k-1}(z)$ is the Chebyshev polynomial of degree $k - 1$.

Proof. See Underwood (1975). □

Analogous inequalities can be obtained for $\bar{T}_k$'s smallest eigenvalues by applying the theorem with A replaced by $-A$.

Based on Theorem 9.2.2 and scrutiny of the block Lanczos iteration (9.2.7) we may conclude that:

- the error bound for the Ritz values improve with increased p.

- the amount of work required to compute $\bar{T}_k$'s eigenvalues is proportional to p^2.

- the block dimension should be at least as large as the largest multiplicity of any sought-after eigenvalue.

How to determine block dimension in the face of these tradeoffs is discussed in detail by Scott (1979).

Loss of orthogonality also plagues the block Lanczos algorithm. However, all of the orthogonality enforcement schemes described above can be extended to the block setting.

9.2.7   s-Step Lanczos

The block Lanczos algorithm (9.2.7) can be used in an iterative fashion to calculate selected eigenalues of A. To fix ideas, suppose we wish to calculate the p largest eigenvalues. If $X_1 \in \mathbb{R}^{n \times p}$ is a given matrix having orthonormal columns, we may proceed as follows:

> **until** $\| AX_1 - X_1\bar{T}_s \|_F$ is small enough
>> Generate $X_2, \ldots, X_s \in \mathbb{R}^{n \times p}$ via the block Lanczos algorithm.
>> Form $\bar{T}_s = [X_1, \ldots, X_s]^T A [X_1, \ldots, X_s]$, an sp-by-sp, p-diagonal matrix.
>> Compute an orthogonal matrix $U = [u_1, \ldots, u_{sp}]$ such that $U^T \bar{T}_s U = \text{diag}(\theta_1, \ldots, \theta_{sp})$ with $\theta_1 \geq \cdots \geq \theta_{sp}$.
>> Set $X_1 = [X_1, \ldots, X_s][u_1, \ldots, u_p]$.
> **end**

This is the block analog of the *s-step Lanczos algorithm* , which has been extensively analyzed by Cullum and Donath (1974) and Underwood (1975).

The same idea can also be used to compute several of A's smallest eigenvalues or a mixture of both large and small eigenvalues. See Cullum (1978). The choice of the parameters s and p depends upon storage constraints as well as upon the factors we mentioned above in our discussion of block dimension. The block dimension p may be diminished as the good Ritz vectors emerge. However this demands that orthogonality to the converged vectors be enforced. See Cullum and Donath (1974).

Problems

P9.2.1 Prove Lemma 9.2.1.

P9.2.2 If $\text{rank}(R_k) < p$ in (9.2.7), does it follow that $\text{range}([\,X_1,\ldots,X_k\,])$ contains an eigenvector of A?

Notes and References for Sec. 9.2

Of the several computational variants of the Lanczos Method, Algorithm 9.2.1 is the most stable. For details, see

C.C. Paige (1972). "Computational Variants of the Lanczos Method for the Eigenproblem," *J. Inst. Math. Applic. 10*, 373–81.

Other practical details associated with the implementation of the Lanczos procedure are discussed in

D.S. Scott (1979). "How to Make the Lanczos Algorithm Converge Slowly," *Math. Comp. 33*, 239–47.

B.N. Parlett, H. Simon, and L.M. Stringer (1982). "On Estimating the Largest Eigenvalue with the Lanczos Algorithm," *Math. Comp. 38*, 153–166.

B.N. Parlett and B. Nour-Omid (1985). "The Use of a Refined Error Bound When Updating Eigenvalues of Tridiagonals," *Lin. Alg. and Its Applic. 68*, 179–220.

J. Kuczyński and H. Woźniakowski (1992). "Estimating the Largest Eigenvalue by the Power and Lanczos Algorithms with a Random Start," *SIAM J. Matrix Anal. Appl. 13*, 1094–1122.

The behavior of the Lanczos method in the presence of roundoff error was originally reported in

C.C. Paige (1971). "The Computation of Eigenvalues and Eigenvectors of Very Large Sparse Matrices," Ph.D. thesis, University of London.

Important follow-up papers include

C.C. Paige (1976). "Error Analysis of the Lanczos Algorithm for Tridiagonalizing Symmetric Matrix," *J. Inst. Math. Applic. 18*, 341–49.

C.C. Paige (1980). "Accuracy and Effectiveness of the Lanczos Algorithm for the Symmetric Eigenproblem," *Lin. Alg. and Its Applic. 34*, 235–58.

For a discussion about various reorthogonalization schemes, see

C.C. Paige (1970). "Practical Use of the Symmetric Lanczos Process with Reorthogonalization," *BIT 10*, 183–95.

G.H. Golub, R. Underwood, and J.H. Wilkinson (1972). "The Lanczos Algorithm for the Symmetric $Ax = \lambda Bx$ Problem," Report STAN-CS-72-270, Department of Computer Science, Stanford University, Stanford, California.

B.N. Parlett and D.S. Scott (1979). "The Lanczos Algorithm with Selective Orthogonalization," *Math. Comp. 33*, 217–38.

H. Simon (1984). "Analysis of the Symmetric Lanczos Algorithm with Reorthogonalization Methods," *Lin. Alg. and Its Applic. 61*, 101–132.

Without any reorthogonalization it is necessary either to monitor the loss of orthogonality and quit at the appropriate instant or else to devise some scheme that will aid in the distinction between the ghost eigenvalues and the actual eigenvalues. See

W. Kahan and B.N. Parlett (1976). "How Far Should You Go with the Lanczos Process?" in *Sparse Matrix Computations*, ed. J. Bunch and D. Rose, Academic Press, New York, pp. 131-44.
J. Cullum and R.A. Willoughby (1979). "Lanczos and the Computation in Specified Intervals of the Spectrum of Large, Sparse Real Symmetric Matrices, in *Sparse Matrix Proc.* , 1978, ed. I.S. Duff and G.W. Stewart, SIAM Publications, Philadelphia, PA.
B.N. Parlett and J.K. Reid (1981). "Tracking the Progress of the Lanczos Algorithm for Large Symmetric Eigenproblems," *IMA J. Num. Anal. 1*, 135–55.
D. Calvetti, L. Reichel, and D.C. Sorensen (1994). "An Implicitly Restarted Lanczos Method for Large Symmetric Eigenvalue Problems," *ETNA 2*, 1–21.

The block Lanczos algorithm is discussed in

J. Cullum and W.E. Donath (1974). "A Block Lanczos Algorithm for Computing the q Algebraically Largest Eigenvalues and a Corresponding Eigenspace of Large Sparse Real Symmetric Matrices," *Proc. of the 1974 IEEE Conf. on Decision and Control*, Phoenix, Arizona, pp. 505–9.
R. Underwood (1975). "An Iterative Block Lanczos Method for the Solution of Large Sparse Symmetric Eigenproblems," Report STAN-CS-75-495, Department of Computer Science, Stanford University, Stanford, California.
G.H. Golub and R. Underwood (1977). "The Block Lanczos Method for Computing Eigenvalues," in *Mathematical Software III* , ed. J. Rice, Academic Press, New York, pp. 364–77.
J. Cullum (1978). "The Simultaneous Computation of a Few of the Algebraically Largest and Smallest Eigenvalues of a Large Sparse Symmetric Matrix," *BIT 18*, 265–75.
A. Ruhe (1979). "Implementation Aspects of Band Lanczos Algorithms for Computation of Eigenvalues of Large Sparse Symmetric Matrices," *Math. Comp. 33*, 680-87.

The block Lanczos algorithm generates a symmetric band matrix whose eigenvalues can be computed in any of several ways. One approach is described in

H.R. Schwartz (1968). "Tridiagonalization of a Symmetric Band Matrix," *Numer. Math. 12*, 231–41. See also Wilkinson and Reinsch (1971, 273–83).

In some applications it is necessary to obtain estimates of interior eigenvalues. The Lanczos algorithm, however, tends to find the extreme eigenvalues first. The following papers deal with this issue:

A.K. Cline, G.H. Golub, and G.W. Platzman (1976). "Calculation of Normal Modes of Oceans Using a Lanczos Method," in *Sparse Matrix Computations* , ed. J.R. Bunch and D.J. Rose, Academic Press, New York, pp. 409–26.
T. Ericsson and A. Ruhe (1980). "The Spectral Transformation Lanczos Method for the Numerical Solution of Large Sparse Generalized Symmetric Eigenvalue Problems," *Math. Comp. 35*, 1251–68.
R.G. Grimes, J.G. Lewis, and H.D. Simon (1994). "A Shifted Block Lanczos Algorithm for Solving Sparse Symmetric Generalized Eigenproblems," *SIAM J. Matrix Anal. Appl. 15*, 228–272.

9.3   Applications to $Ax = b$ and Least Squares

In this section we briefly show how the Lanczos iteration can be embellished to solve large sparse linear equation and least squares problems. For further details, we recommend Saunders (1995).

9.3.1   Symmetric Positive Definite Systems

Suppose $A \in \mathbb{R}^{n \times n}$ is symmetric and positive definite and consider the functional $\phi(x)$ defined by

$$\phi(x) = \frac{1}{2} x^T A x - x^T b$$

where $b \in \mathbb{R}^n$. Since $\nabla \phi(x) = Ax - b$, it follows that $x = A^{-1}b$ is the unique minimizer of ϕ. Hence, an approximate minimizer of ϕ can be regarded as an approximate solution to $Ax = b$.

Suppose $x_0 \in \mathbb{R}^n$ is an initial guess. One way to produce a vector sequence $\{x_k\}$ that converges to x is to generate a sequence of orthonormal vectors $\{q_k\}$ and to let x_k minimize ϕ over the set

$$x_0 + \text{span}\{q_1, \ldots, q_k\} = \{x_0 + a_1 q_1 + \cdots + a_k q_k : a_k \in \mathbb{R}\}$$

for $k = 1{:}n$. If $Q_k = [\, q_1, \ldots, q_k \,]$, then this just means choosing $y \in \mathbb{R}^k$ such that

$$
\begin{aligned}
\phi(x_0 + Q_k y) &= \frac{1}{2}(x_0 + Q_k y)^T A(x_0 + Q_k y) - (x_0 + Q_k y)^T b \\
&= \frac{1}{2} y^T (Q_k^T A Q_k) y - y^T Q_k^T (b - A x_0) + \phi(x_0)
\end{aligned}
$$

is minimized. By looking at the gradient of this expression with respect to y we see that

$$x_k = x_0 + Q_k y_k \qquad\qquad (9.3.1)$$

where

$$(Q_k^T A Q_k) y_k = Q_k^T (b - A x_0). \qquad\qquad (9.3.2)$$

When $k = n$ the minimization is over all of $\mathbb{R}^n$ and so $A x_n = b$.

For large sparse A it is necessary to overcome two hurdles in order to make this an effective solution process:

- the linear system (9.3.2) must be "easily" solved.

- we must be able to compute x_k *without* having to refer to $q_1, \ldots, q_k$ explicitly as (9.3.1) suggests. Otherwise there would be an excessive amount of data movement.

We show that both of these requirements are met if the q_k are Lanczos vectors.

After k steps of the Lanczos algorithm we obtain the factorization

$$AQ_k = Q_k T_k + r_k e_k^T \qquad (9.3.3)$$

where

$$T_k = Q_k^T A Q_k = \begin{bmatrix} \alpha_1 & \beta_1 & \cdots & & 0 \\ \beta_1 & \alpha_2 & \ddots & & \vdots \\ & \ddots & \ddots & \ddots & \\ \vdots & & \ddots & \ddots & \beta_{k-1} \\ 0 & \cdots & & \beta_{k-1} & \alpha_k \end{bmatrix}. \qquad (9.3.4)$$

With this approach (9.3.2) becomes a symmetric positive definite tridiagonal system which may be solved via the LDL^T factorization. (See Algorithm 4.3.6.) In particular, by setting

$$L_k = \begin{bmatrix} 1 & 0 & 0 & 0 \\ \mu_1 & 1 & 0 & 0 \\ \vdots & \ddots & \ddots & \vdots \\ 0 & \cdots & \mu_{k-1} & 1 \end{bmatrix} \quad \text{and} \quad D_k = \begin{bmatrix} d_1 & 0 & \cdots & 0 \\ 0 & d_2 & & \vdots \\ \vdots & & \ddots & 0 \\ 0 & \cdots & 0 & d_k \end{bmatrix}$$

we find by comparing entries in

$$T_k = L_k D_k L_k^T \qquad (9.3.5)$$

that

$d_1 = \alpha_1$
for $i = 2:k$
$\qquad \mu_{i-1} = \beta_{i-1}/d_{i-1}$
$\qquad d_i = \alpha_i - \beta_{i-1}\mu_{i-1}$
end

Note that we need only calculate the quantities

$$\begin{aligned} \mu_{k-1} &= \beta_{k-1}/d_{k-1} \\ d_k &= \alpha_k - \beta_{k-1}\mu_{k-1} \end{aligned} \qquad (9.3.6)$$

in order to obtain L_k and D_k from L_{k-1} and D_{k-1}.

As we mentioned, it is critical to be able to compute x_k in (9.3.1) efficiently. To this end we define $C_k \in \mathbb{R}^{n \times k}$ and $p_k \in \mathbb{R}^k$ by the equations

$$\begin{aligned} C_k L_k^T &= Q_k \\ L_k D_k p_k &= Q_k^T(b - Ax_0) \end{aligned} \qquad (9.3.7)$$

and observe that

$$x_k = x_0 + Q_k T_k^{-1} Q_k^T b = x_0 + Q_k (L_k D_k L_k^T)^{-1} Q_k^T b = x_0 + C_k p_k.$$

Let $C_k = [\, c_1, \ldots, c_k \,]$ be a column partitioning. It follows from (9.3.7) that

$$[\, c_1, \; \mu_1 c_1 + c_2, \; \cdots, \; \mu_{k-1} c_{k-1} + c_k \,] = [\, q_1, \ldots, q_k \,]$$

and therefore $C_k = [\, C_{k-1}, c_k \,]$ where

$$c_k = q_k - \mu_{k-1} c_{k-1}.$$

Also observe that if we set $p_k = [\, \rho_1, \ldots, \rho_k \,]^T$ in $L_k D_k p_k = Q_k^T b$, then that equation becomes

$$\left[\begin{array}{c|c} L_{k-1} D_{k-1} & 0 \\ \hline 0 \cdots 0 \; \mu_{k-1} d_{k-1} & d_k \end{array} \right] \left[\begin{array}{c} \rho_1 \\ \rho_2 \\ \vdots \\ \rho_{k-1} \\ \rho_k \end{array} \right] = \left[\begin{array}{c} q_1^T b \\ q_2^T b \\ \vdots \\ q_{k-1}^T b \\ q_k^T b \end{array} \right].$$

Since $L_{k-1} D_{k-1} p_{k-1} = Q_{k-1}^T b$, it follows that

$$p_k = \left[\begin{array}{c} p_{k-1} \\ \rho_k \end{array} \right]$$

where

$$\rho_k = \left(q_k^T b - \mu_{k-1} d_{k-1} \rho_{k-1} \right) / d_k$$

and thus,

$$x_k = x_0 + C_k p_k = x_0 + C_{k-1} p_{k-1} + \rho_k c_k = x_{k-1} + \rho_k c_k.$$

This is precisely the kind of recursive formula for x_k that we need. Together with (9.3.6) and (9.3.7) it enables us to make the transition from $(q_{k-1}, c_{k-1}, x_{k-1})$ to (q_k, c_k, x_k) with a minimal work and storage.

A further simplification results if we set q_1 to be a unit vector in the direction of the initial residual $b - Ax_0$. With this choice for a Lanczos starting vector, $q_i^T (b - Ax_0) = 0$ for $i \geq 2$. It follows from (9.3.3) that

$$\begin{aligned} b - Ax_k &= b - A(x_0 + Q_k y_k) = (b - Ax_0) - (Q_k T_k + r_k e_k^T) y_k \\ &= (b - Ax_0) - Q_k Q_k^T (b - Ax_0) - r_k e_k^T y = -r_k e_k^T y. \end{aligned}$$

Thus, if $\beta_k = \|\,r_k\,\|_2 = 0$ in the Lanczos iteration, then $Ax_k = b$. Moreover, $\|\,Ax_k - b\,\|_2 = \beta_k |e_k^T y_k|$ and so estimates of the current residual can be obtained as a by-product of the iteration. Overall, we have the following procedure.

Algorithm 9.3.1 If $A \in \mathbb{R}^{n \times n}$ is symmetric positive definite, $b \in \mathbb{R}^n$, and $x_0 \in \mathbb{R}^n$ is an initial guess $(Ax_0 \approx b)$, then this algorithm computes the solution to $Ax = b$.

$$r_0 = b - Ax_0$$
$$\beta_0 = \|\,r_0\,\|_2$$
$$q_0 = 0$$
$$k = 0$$
while $\beta_k \neq 0$
$$q_{k+1} = r_k / \beta_k$$
$$k = k + 1$$
$$\alpha_k = q_k^T A q_k$$
$$r_k = (A - \alpha_k I)q_k - \beta_{k-1}q_{k-1}$$
$$\beta_k = \|\,r_k\,\|_2$$
    if $k = 1$
$$d_1 = \alpha_1$$
$$c_1 = q_1$$
$$\rho_1 = \beta_0 / \alpha_1$$
$$x_1 = \rho_1 q_1$$
    else
$$\mu_{k-1} = \beta_{k-1} / d_{k-1}$$
$$d_k = \alpha_k - \beta_{k-1}\mu_{k-1}$$
$$c_k = q_k - \mu_{k-1}c_{k-1}$$
$$\rho_k = -\mu_{k-1}d_{k-1}\rho_{k-1} / d_k$$
$$x_k = x_{k-1} + \rho_k c_k$$
    end
end
$$x = x_k$$

This algorithm requires one matrix-vector multiplication and a couple of saxpy operations per iteration. The numerical behavior of Algorithm 9.3.1 is discussed in the next chapter, where it is rederived and identified as the widely known method of *conjugate gradients*.

9.3.2 Symmetric Indefinite Systems

A key feature in the above development is the idea of computing the LDL^T factorization of the tridiagonal matrices T_k. Unfortunately, this is potentially unstable if A, and consequently T_k, is not positive definite. A way around this difficulty proposed by Paige and Saunders (1975) is to develop the recursion for x_k via an "LQ" factorization of T_k. In particular, at the kth step of the iteration, we have Givens rotations $J_1, \ldots, J_{k-1}$ such that

$$
T_k J_1 \cdots J_{k-1} \;=\; L_k \;=\;
\begin{bmatrix}
d_1 & 0 & 0 & \cdots & \cdots & \cdots & 0 \\
e_1 & d_2 & 0 & \cdots & \cdots & \cdots & 0 \\
f_1 & e_2 & d_3 & \cdots & \cdots & \cdots & 0 \\
\vdots & \ddots & \ddots & \ddots & & & \\
& & & \ddots & \ddots & \ddots & 0 \\
0 & 0 & 0 & \cdots & f_{k-2} & e_{k-1} & \bar{d}_k
\end{bmatrix} .
$$

Note that with this factorization x_k is given by

$$
x_k \;=\; x_0 + Q_k y_k \;=\; Q_k T_k^{-1} Q_k^T b \;=\; W_k s_k
$$

where

$$
W_k \;=\; Q_k J_1 \cdots J_{k-1} \in \mathbb{R}^{n \times k}
$$

and $s_k \in \mathbb{R}^k$ solves

$$
L_k s_k \;=\; Q_k^T b.
$$

Scrutiny of these equations enables one to develop a formula for computing x_k from x_{k-1} and an easily computed multiple of w_k, the last column of W_k. This defines the SYMMLQ method set forth in Paige and Saunders (1975).

A different idea is to notice from (9.3.3) and the definition $\beta_k q_{k+1} = r_k$ that

$$
AQ_k = Q_k T_k + \beta_k q_{k+1} e_k^T = Q_{k+1} H_k
$$

where

$$
H_k = \begin{bmatrix} T_k \\ \beta_k e_k^T \end{bmatrix} .
$$

This $(k+1)$-by-k matrix is upper Hessenberg and figures in the MINRES method of Paige and Saunders (1975). In this technique x_k minimizes $\| Ax - b \|_2$ over the set $x_0 + \text{span}\{q_1, \ldots, q_k\}$. Note that

$$
\begin{aligned}
\| A(x_0 + Q_k y) - b \|_2 &= \| AQ_k y - (b - Ax_0) \|_2 \\
&= \| Q_{k+1} H_k y - (b - Ax_0) \|_2 = \| H_k y - \beta_0 e_1 \|_2
\end{aligned}
$$

where it is assumed that $q_1 = (b - Ax_0)/\beta_0$ is a unit vector. As in SYMMLQ, it is possible to develop recursions that permit the efficient computation of x_k from its predecessor x_{k-1}. The QR factorization of H_k is involved.

The behavior of the conjugate gradient method is detailed in the next chapter. The convergence of SYMMLQ and MINRES is more complicated and is discussed in Paige, Parlett, and Van Der Vorst (1995).

9.3.3 Bidiagonalization and the SVD

Suppose $U^T AV = B$ represents the bidiagonalization of $A \in \mathbb{R}^{m \times n}$ with

$$
\begin{aligned}
U &= [\, u_1, \ldots, u_m \,] & U^T U &= I_m \\
V &= [\, v_1, \ldots, v_n \,] & V^T V &= I_n
\end{aligned}
$$

and

$$
B = \begin{bmatrix}
\alpha_1 & \beta_1 & & \cdots & & 0 \\
0 & \alpha_2 & \ddots & & & \vdots \\
& & \ddots & \ddots & \ddots & \\
& & & \ddots & \ddots & \beta_{n-1} \\
\vdots & & & & \ddots & \\
0 & \cdots & & 0 & & \alpha_n
\end{bmatrix}. \tag{9.3.8}
$$

Recall from §5.4.3 that this factorization may be computed using Householder transformations and that it serves as a front end for the SVD algorithm.

Unfortunately, if A is large and sparse, then we can expect large, dense submatrices to arise during the Householder bidiagonalization. Consequently, it would be nice to develop a means for computing B directly without any orthogonal updates of the matrix A.

Proceeding just as we did in §9.1.2 we compare columns in the equations $AV = UB$ and $A^T U = VB^T$ for $k = 1:n$ and obtain

$$
\begin{aligned}
Av_k &= \alpha_k u_k + \beta_{k-1} u_{k-1} & \beta_0 u_0 &\equiv 0 \\
A^T u_k &= \alpha_k v_k + \beta_k v_{k+1} & \beta_n v_{n+1} &\equiv 0
\end{aligned} \tag{9.3.9}
$$

Defining

$$
\begin{aligned}
r_k &= Av_k - \beta_{k-1} u_{k-1} \\
p_k &= A^T u_k - \alpha_k v_k
\end{aligned}
$$

we may conclude from orthonormality that $\alpha_k = \pm \| r_k \|_2$, $u_k = r_k / \alpha_k$, $\beta_k = \pm \| p_k \|_2$, and $v_{k+1} = p_k / \beta_k$. Properly sequenced, these equations define the Lanczos method for bidiagonalizing a rectangular matrix:

$$
\begin{aligned}
&v_1 = \text{given unit 2-norm } n\text{-vector} \\
&p_0 = v_1; \ \beta_0 = 1; \ k = 0; \ u_0 = 0 \\
&\textbf{while } \beta_k \neq 0 \\
&\quad v_{k+1} = p_k / \beta_k \\
&\quad k = k + 1 \\
&\quad r_k = Av_k - \beta_{k-1} u_{k-1} \\
&\quad \alpha_k = \| r_k \|_2 \\
&\quad u_k = r_k / \alpha_k \\
&\quad p_k = A^T u_k - \alpha_k v_k \\
&\quad \beta_k = \| p_k \|_2 \\
&\textbf{end}
\end{aligned} \tag{9.3.10}
$$

If $\text{rank}(A) = n$, then we can guarantee that no zero α_k arise. Indeed, if $\alpha_k = 0$ then $\text{span}\{Av_1, \ldots, Av_k\} \subset \text{span}\{u_1, \ldots, u_{k-1}\}$ which implies rank deficiency.

If $\beta_k = 0$, then it is not hard to verify that

$$A[\, v_1, \ldots, v_k \,] = [\, u_1, \ldots, u_k \,] B_k$$

$$A^T[\, u_1, \ldots, u_k \,] = [\, v_1, \ldots, v_k \,] B_k^T$$

where $B_k = B(1{:}k, 1{:}k)$ and B is prescribed by (9.3.8). Thus, the v vectors and the u vectors are singular vectors and $\sigma(B_k) \subset \sigma(A)$. Lanczos bidiagonalization is discussed in Paige (1974). See also Cullum and Willoughby (1985a, 1985b). It is essentially equivalent to applying the Lanczos tridiagonalization scheme to the symmetric matrix

$$C = \begin{bmatrix} 0 & A \\ A^T & 0 \end{bmatrix}.$$

We showed that $\lambda_i(C) = \sigma_i(A) = -\lambda_{n+m-i+1}(C)$ for $i = 1{:}n$ at the beginning of §8.6. Because of this, it is not surprising that the large singular values of the bidiagonal matrix tend to be very good approximations to the large singular values of A. The small singular values of A correspond to the interior eigenvalues of C and are not so well approximated. The equivalent of the Kaniel-Paige theory for the Lanczos bidiagonalization may be found in Luk (1978) as well as in Golub, Luk, and Overton (1981). The analytic, algorithmic, and numerical developments of the previous two sections all carry over naturally to the bidiagonalization.

9.3.4   Least Squares

The full-rank LS problem min $\| Ax - b \|_2$ can be solved via the bidiagonalization. In particular,

$$x_{LS} = Vy_{LS} = \sum_{i=1}^{n} y_i v_i$$

where $y = [y_1, \ldots, y_n]^T$ solves the system $By = [u_1^T b, \ldots, u_n^T b]^T$. Note that because B is *upper* bidiagonal, we cannot solve for y until the bidiagonalization is complete. Moreover, we are required to save the vectors $v_1, \ldots, v_n$, an unhappy circumstance if n is large.

The development of a sparse least squares algorithm based on the bidiagonalization can be accomplished more favorably if A is reduced to lower bidiagonal form

$$U^T A V = B = \begin{bmatrix} \begin{array}{ccccc} \alpha_1 & 0 & \cdots & \cdots & 0 \\ \beta_1 & \alpha_2 & \cdots & \cdots & \vdots \\ \vdots & \ddots & \ddots & & \vdots \\ \vdots & & \ddots & \ddots & 0 \\ 0 & \cdots & & \ddots & \alpha_n \\ 0 & \cdots & \cdots & 0 & \beta_n \end{array} \\ \hline 0 \end{bmatrix}$$

where $V = [\, v_1, \ldots, v_n \,]$ and $U = [\, u_1, \ldots, u_m \,]$ are orthogonal. Comparing columns in the equations $A^T U = V B^T$ and $AV = UB$ we obtain

$$\begin{aligned} A^T u_k &= \beta_{k-1} v_{k-1} + \alpha_k v_k & \beta_0 v_0 &\equiv 0 \\ A v_k &= \alpha_k u_k + \beta_k u_{k+1} & \end{aligned}$$

It is straightforward to develop a Lanczos procedure from these equations and the resulting algorithm is very similar to (9.3.10), only u_1 is the starting vector.

Define the matrices $V_k = [\, v_1, \ldots, v_k \,]$, $U_k = [\, u_1, \ldots, u_k \,]$, and $B_k = B(1{:}k{+}1, 1{:}k)$ and observe that $AV_k = U_{k+1} B_k$. Our goal is to compute x_k, the minimizer of $\|\, Ax - b \,\|_2$ over all vectors of the form $x = x_0 + V_k y$, where $y \in \mathbb{R}^k$ and $x_0 \in \mathbb{R}^n$ is an initial guess. If $u_1 = (b - Ax_0)/\|\, b - Ax_0 \,\|_2$, then

$$A(x_0 + V_k y) - b = U_{k+1} B_k y - \beta_1 U_{k+1} e_1 = U_{k+1}(B_k y - \beta_1 e_1)$$

where $e_1 = I_m(:, 1)$. It follows that if y_k solves the $(k+1)$-by-k lower bidiagonal LS problem

$$\min \|\, B_{k+1} y - \beta_1 e_1 \,\|_2$$

then $x_k = x_0 + V_k y_k$. Since B_k is lower bidiagonal, it is easy to compute Givens rotations $J_1, \ldots, J_k$ such that

$$J_k \cdots J_1 B_k = \begin{bmatrix} R_k \\ 0 \end{bmatrix} \begin{array}{c} k \\ 1 \end{array}$$

is upper bidiagonal. If

$$J_k \cdots J_1 U_{k+1}^T b = \begin{bmatrix} d_k \\ u \end{bmatrix} \begin{array}{c} k \\ 1 \end{array},$$

then it follows that $x_k = x_0 + V_k y_k = W_k d_k$ where $W_k = V_k R_k^{-1}$. Paige and Saunders (1982a) show how x_k can be obtained from x_{k-1} via a simple recursion that involves the last column of W_k. The net result is a sparse LS algorithm referred to as LSQR that requires only a few n-vectors of storage to implement.

Problems

P9.3.1 Modify Algorithm 9.3.1 so that it implements the indefinite symmetric solver outlined in §9.3.2.

P9.3.2 How many vector workspaces are required to implement efficiently (9.3.10)?

P9.3.3 Suppose A is rank deficient and $\alpha_k = 0$ in (9.3.10). How could u_k be obtained so that the iteration could continue?

P9.3.4 Work out the lower bidiagonal version of (9.3.10) and detail the least square solver sketched in §9.3.4.

Notes and References for Sec. 9.3

Much of the material in this section has been distilled from the following papers:

C.C. Paige (1974). "Bidiagonalization of Matrices and Solution of Linear Equations," *SIAM J. Num. Anal. 11*, 197–209.
C.C. Paige and M.A. Saunders (1975). "Solution of Sparse Indefinite Systems of Linear Equations," *SIAM J. Num. Anal. 12*, 617–29.
C.C. Paige and M.A. Saunders (1982a). "LSQR: An Algorithm for Sparse Linear Equations and Sparse Least Squares," *ACM Trans. Math. Soft. 8*, 43–71.
C.C. Paige and M.A. Saunders (1982b). "Algorithm 583 LSQR: Sparse Linear Equations and Least Squares Problems," *ACM Trans. Math. Soft. 8*, 195–209.
M.A. Sanders (1995). "Solution of Sparse Rectangular Systems," *BIT 35*, 588–604.

See also Cullum and Willoughby (19855a,1985b) and

O. Widlund (1978). "A Lanczos Method for a Class of Nonsymmetric Systems of Linear Equations," *SIAM J. Numer. Anal. 15*, 801–12.
B.N. Parlett (1980). "A New Look at the Lanczos Algorithm for Solving Symmetric Systems of Linear Equations," *Lin. Alg. and Its Applic. 29*, 323–46.
G.H. Golub, F.T. Luk, and M. Overton (1981). "A Block Lanczos Method for Computing the Singular Values and Corresponding Singular Vectors of a Matrix," *ACM Trans. Math. Soft. 7*, 149–69.
J. Cullum, R.A. Willoughby, and M. Lake (1983). "A Lanczos Algorithm for Computing Singular Values and Vectors of Large Matrices," *SIAM J. Sci. and Stat. Comp. 4*, 197–215.
Y. Saad (1987). "On the Lanczos Method for Solving Symmetric Systems with Several Right Hand Sides," *Math. Comp. 48*, 651–662.
M. Berry and G.H. Golub (1991). "Estimating the Largest Singular Values of Large Sparse Matrices via Modified Moments," *Numerical Algorithms 1*, 353–374.
C.C. Paige, B.N. Parlett, and H.A. Van Der Vorst (1995). "Approximate Solutions and Eigenvalue Bounds from Krylov Subspaces," *Numer. Linear Algebra with Applic. 2*, 115–134.

9.4 Arnoldi and Unsymmetric Lanczos

If A is not symmetric, then the orthogonal tridiagonalization $Q^T A Q = T$ does not exist in general. There are two ways to proceed. The Arnoldi approach involves the column-by-column generation of an orthogonal Q such that $Q^T A Q = H$ is the Hessenberg reduction of §7.4. The unsymmetric Lanczos approach computes the columns of $Q = [q_1, \ldots, q_n]$ and $P = [p_1, \ldots, p_n]$ so that $P^T A Q = T$ is tridiagonal and $P^T Q = I_n$. Both methods are interesting as large sparse unsymmetric eigenvalue solvers and both can be adapted for sparse unsymmetric $Ax = b$ solving. (See §10.4.)

9.4.1 The Basic Arnoldi Iteration

One way to extend the Lanczos process to unsymmetric matrices is due to Arnoldi (1951) and revolves around the Hessenberg reduction $Q^T A Q = H$. In particular, if $Q = [q_1, \ldots, q_n]$ and we compare columns in $AQ = QH$, then

$$Aq_k = \sum_{i=1}^{k+1} h_{ik} q_i \qquad 1 \le k \le n-1.$$

Isolating the last term in the summation gives

$$h_{k+1,k} q_{k+1} = Aq_k - \sum_{i=1}^{k} h_{ik} q_i : \equiv r_k$$

where $h_{ik} = q_i^T A q_k$ for $i = 1{:}k$. It follows that if $r_k \ne 0$, then q_{k+1} is specified by

$$q_{k+1} = r_k / h_{k+1,k}$$

where $h_{k+1,k} = \| r_k \|_2$. These equations define the *Arnoldi process* and in strict analogy to the symmetric Lanczos process (9.1.3) we obtain :

$$
\begin{aligned}
&r_0 = q_1 \\
&h_{10} = 1 \\
&k = 0 \\
&\textbf{while } (h_{k+1,k} \ne 0) \\
&\qquad q_{k+1} = r_k / h_{k+1,k} \\
&\qquad k = k + 1 \\
&\qquad r_k = Aq_k \\
&\qquad \textbf{for } i = 1{:}k \\
&\qquad\qquad h_{ik} = q_i^T w \\
&\qquad\qquad r_k = r_k - h_{ik} q_i \\
&\qquad \textbf{end} \\
&\qquad h_{k+1,k} = \| r_k \|_2 \\
&\textbf{end}
\end{aligned}
\qquad (9.4.1)
$$

We assume that q_1 is a given unit 2-norm starting vector. The q_k are called the *Arnoldi vectors* and they define an orthonormal basis for the Krylov subspace $\mathcal{K}(A, q_1, k)$:

$$\text{span}\{q_1, \ldots, q_k\} = \text{span}\{q_1, Aq_1, \ldots, A^{k-1}q_1\}. \qquad (9.4.2)$$

The situation after k steps is summarized by the *k-step Arnoldi factorization*

$$AQ_k = Q_k H_k + r_k e_k^T \qquad (9.4.3)$$

where $Q_k = [q_1, \ldots, q_k]$, $e_k = I_k(:, k)$, and

$$H_k = \begin{bmatrix} h_{11} & h_{12} & \cdots & \cdots & h_{1k} \\ h_{21} & h_{22} & \cdots & \cdots & h_{2k} \\ 0 & h_{32} & \ddots & & \vdots \\ \vdots & & \ddots & \ddots & \vdots \\ 0 & \cdots & \cdots & h_{k,k-1} & h_{kk} \end{bmatrix}.$$

If $r_k = 0$, then the columns of Q_k define an invariant subspace and $\lambda(H_k) \subseteq \lambda(A)$. Otherwise, the focus is on how to extract information about A's eigensystem from the Hessenberg matrix H_k and the matrix Q_k of Arnoldi vectors.

If $y \in \mathbb{R}^k$ is a unit 2-norm eigenvector for H_k and $H_k y = \lambda y$, then from (9.4.3)

$$(A - \lambda I)x = (e_k^T y)r_k$$

where $x = Q_k y$. We call λ a *Ritz value* and x the corresponding *Ritz vector*. The size of $|e_k^T y| \| r_k \|_2$ can be used to obtain error bounds, although the relevant perturbation theorems are not as routine to apply as in the symmetric case.

Some numerical properties of the Arnoldi iteration are discussed in Wilkinson (1965, pp.382). As with the symmetric Lanczos iteration, loss of orthogonality among the q_i is an issue. But two other features of (9.4.1) must be addressed before a practical Arnoldi eigensolver can be obtained:

- The Arnoldi vectors $q_1, \ldots, q_k$ are referenced in step k and the computation of $H_k(1{:}k, k)$ involves $O(kn)$ flops. Thus, there is a steep penalty associated with the generation of long Arnoldi sequences.

- The eigenvalues of H_k do not approximate the eigenvalues of A in the style of Kaniel and Paige. This is in contrast to the symmetric case where information about A's extremal eigenvalues emerges quickly . With Arnoldi, the early extraction of eigenvalue information depends crucially on the choice of q_1.

These realities suggest a framework in which we use Arnoldi with repeated, carefully chosen restarts and a controlled iteration maximum. (Recall the s-step Lanczos process of §9.2.7.)

9.4.2 Arnoldi with Restarting

Consider running Arnoldi for m steps and then restarting the process with a vector q_+ chosen from the span of the Arnoldi vectors $q_1, \ldots, q_m$. Because of the Krylov connection (9.4.2), q_+ has the form

$$q_+ = p(A)q_1$$

for some polynomial of degree $m - 1$. If $Av_i = \lambda_i v_i$ for $i = 1{:}n$ and q_1 has the eigenvector expansion

$$q_1 = a_1 v_1 + \cdots + a_n v_n,$$

then

$$q_+ = a_1 p(\lambda_1) v_1 + \cdots + a_n p(\lambda_n) v_n.$$

Note that $\mathcal{K}(A, q_+, m)$ is rich in eigenvectors that are emphasized by $p(\lambda)$. That is, if $p(\lambda_{wanted})$ is large compared to $p(\lambda_{unwanted})$, then the Krylov space $\mathcal{K}(A, q_+, m)$ will have much better approximations to the eigenvector x_{wanted} than to the eigenvector $x_{unwanted}$. (It is possible to couch this argument in terms of Schur invariant subspaces rather than in terms of particular eigenvectors.)

Thus the act of picking a good restart vector q_+ from $\mathcal{K}(A, q_1, m)$ is the act of picking a polynomial "filter" that tunes out unwanted portions of the spectrum. Various heuristics for doing this have been developed based on computed Ritz vectors. See Saad (1980, 1984, 1992).

We describe a method due to Sorensen (1992) that determines the restart vector implicitly using the QR iteration with shifts. The restart occurs after every m steps and we assume that $m > j$ where j is the number of sought-after eigenvalues. The choice of the Arnoldi length parameter m depends on the problem dimension n, the effect of orthogonality loss, and system storage constraints.

After m steps we have the Arnoldi factorization

$$AQ_c = Q_c H_c + r_c e_m^T$$

where $Q_c \in \mathbb{R}^{n \times m}$ has orthonormal columns, $H_c \in \mathbb{R}^{m \times m}$ is upper Hessenberg, and $Q_c^T r_c = 0$. The subscript "c" stands for "current." The QR iteration with shifts is then applied to H_c:

$$
\begin{aligned}
&H^{(1)} = H_c \\
&\textbf{for } i = 1{:}p \\
&\qquad H^{(i)} - \mu_i I = V_i R_i \\
&\qquad H^{(i+1)} = R_i V_i + \mu_i I \\
&\textbf{end} \\
&H_+ = H^{(p+1)}
\end{aligned}
$$

Here $p = m - j$ and it is assumed that the implicitly shifted QR process of
§7.5.5 is applied. The selection of the shifts will be discussed shortly.
    The orthogonal matrix $V = V_1 \cdots V_p$ has three crucial properties:

(1) $H_+ = V^T H_c V$. This is because $V_i^T H^{(i)} V_i = H^{(i+1)}$.

(2) $[V]_{mi} = 0$ for $i = 1{:}j - 1$. This is because each V_i is upper Hessenberg
    and so $V \in \mathbb{R}^{m \times m}$ has lower bandwidth $p = m - j$.

(3) The first column of V has the form

$$V e_1 = \alpha (H_c - \mu_p I)(H_c - \mu_{p-1} I) \cdots (H_c - \mu_1 I) e_1 \qquad (9.4.4)$$

    where α is a scalar.

To be convinced of property (3), consider the $p = 2$ case:

$$
\begin{aligned}
V R_2 R_1 &= V_1 (V_2 R_2) R_1 = V_1 (H^{(2)} - \mu_2 I) R_1 \\
&= V_1 (V_1^T H^{(1)} V_1 - \mu_2 I) R_1 = (H^{(1)} - \mu_2 I) V_1 R_1 \\
&= (H^{(1)} - \mu_2 I)(H^{(1)} - \mu_1 I) = (H_c - \mu_2 I)(H_c - \mu_1 I).
\end{aligned}
$$

Since $R_2 R_1$ is upper triangular, the first column of $V = V_1 V_2$ is a multiple
of $(H_c - \mu_2 I)(H_c - \mu_1 I)$.
    We now show how to restart the Arnoldi process using the matrix V to
implicitly select the new starting vector. From (1) we obtain the following
transformation of (9.4.3):

$$AQ_+ = Q_+ H_+ + r_c e_m^T V$$

where $Q_+ = Q_c V$. This is *not* a new length-m Arnoldi factorization because
$e_m^T V$ is not a multiple of e_m^T. However, in view of property (2),

$$AQ_+(:, 1{:}j) = Q_+(:, 1{:}j) H_+(1{:}j, 1{:}j) + v_{mj} r_c e_j^T \qquad (9.4.5)$$

is a length-j Arnoldi factorization. By "jumping into" the basic Arnoldi
iteration at step $j+1$ and performing p steps, we can extend (9.4.5) to a new
length-m Arnoldi factorization. Moreover, using property (3) the associated
starting vector $q_1^{(new)} = Q_+(:, 1)$ has the following characterization:

$$
\begin{aligned}
Q_+(:, 1) = Q_c V e_1 &= \alpha Q_c (H_c - \mu_p I) \cdots (H_c - \mu_1 I) e_1 \\
&= \alpha (A - \mu_p I) \cdots (A_c - \mu_1 I) Q_c e_1 \qquad (9.4.6)
\end{aligned}
$$

The last equation follows from the identity

$$(A - \mu I) Q_c = Q_c (H_c - \mu I) + r e_m^T$$

and the fact that $e_m^T f(H_c) e_1 = 0$ for any polynomial $f(\cdot)$ of degree $p - 1$ or
less.

Thus, $q_1^{(new)} = p(A)q_1$ where $p(\lambda)$ is the polynomial

$$p(\lambda) = (\lambda - \mu_1)(\lambda - \mu_2) \cdots (\lambda - \mu_p).$$

This shows that the shifts are the zeros of the filtering polynomial. One interesting choice for the shifts is to compute $\lambda(H_c)$ and to identify the eigenvalues of interest $\tilde{\lambda}_1, \ldots, \tilde{\lambda}_j$:

$$\lambda(H_c) = \{\tilde{\lambda}_1, \ldots, \tilde{\lambda}_j\} \cup \{\tilde{\lambda}_{j+1}, \ldots, \tilde{\lambda}_m\}.$$

Setting $\mu_i = \tilde{\lambda}_{i+j}$ for $i = 1{:}p$ is one way of generating a filter polynomial that de-emphasizes the unwanted portion of the spectrum.

We have just presented the rudiments of the *implicitly restarted Arnoldi method*. It has many attractive attributes. For implementation details and further analysis, see Lehoucq and Sorensen (1996) and Morgan (1996).

9.4.3 Unsymmetric Lanczos Tridiagonalization

Another way to extend the symmetric Lanczos process is to reduce A to tridiagonal form using a general similarity transformation. Suppose $A \in \mathbb{R}^{n \times n}$ and that a nonsingular matrix Q exists so

$$Q^{-1}AQ = T = \begin{bmatrix} \alpha_1 & \gamma_1 & & \cdots & & 0 \\ \beta_1 & \alpha_2 & \ddots & & & \vdots \\ & \ddots & \ddots & \ddots & & \\ \vdots & & \ddots & \ddots & \ddots & \gamma_{n-1} \\ 0 & \cdots & & & \beta_{n-1} & \alpha_n \end{bmatrix}.$$

With the column partitionings

$$\begin{aligned} Q &= [\, q_1, \ldots, q_n \,] \\ Q^{-T} = P &= [\, p_1, \ldots, p_n \,] \end{aligned}$$

we find upon comparing columns in $AQ = QT$ and $A^T P = PT^T$ that

$$\begin{aligned} Aq_k &= \gamma_{k-1}q_{k-1} + \alpha_k q_k + \beta_k q_{k+1} & \gamma_0 q_0 &\equiv 0 \\ A^T p_k &= \beta_{k-1}p_{k-1} + \alpha_k p_k + \gamma_k p_{k+1} & \beta_0 p_0 &\equiv 0 \end{aligned}$$

for $k = 1{:}n-1$. These equations together with the *biorthogonality condition* $P^T Q = I_n$ imply

$$\alpha_k = p_k^T A q_k$$

and

$$\begin{aligned} \beta_k q_{k+1} \equiv r_k &= (A - \alpha_k I)q_k - \gamma_{k-1}q_{k-1} \\ \gamma_k p_{k+1} \equiv s_k &= (A - \alpha_k I)^T p_k - \beta_{k-1}p_{k-1}. \end{aligned}$$

There is some flexibility in choosing the scale factors β_k and γ_k. Note that

$$1 = p_{k+1}^T q_{k+1} = (s_k/\gamma_k)^T (r_k/\beta_k).$$

It follows that once β_k is specified γ_k is given by

$$\gamma_k = s_k^T r_k/\beta_k.$$

With the "canonical" choice $\beta_k = \| r_k \|_2$ we obtain

> q_1, p_1 given unit 2-norm vectors with $p_1^T q_1 \neq 0$.
> $k = 0$
> $q_0 = 0; \; r_0 = q_1$
> $p_0 = 0; \; s_0 = p_1$
> **while** $(r_k \neq 0) \wedge (s_k \neq 0) \wedge (s_k^T r_k \neq 0)$
> $\quad \beta_k = \| r_k \|_2$
> $\quad \gamma_k = s_k^T r_k/\beta_k$
> $\quad q_{k+1} = r_k/\beta_k$
> $\quad p_{k+1} = s_k/\gamma_k$
> $\quad k = k+1$ (9.4.7)
> $\quad \alpha_k = p_k^T A q_k$
> $\quad r_k = (A - \alpha_k I)q_k - \gamma_{k-1}q_{k-1}$
> $\quad s_k = (A - \alpha_k I)^T p_k - \beta_{k-1}p_{k-1}$
> **end**

If

$$T_k = \begin{bmatrix} \alpha_1 & \gamma_1 & & \cdots & & 0 \\ \beta_1 & \alpha_2 & \ddots & & & \vdots \\ & \ddots & \ddots & \ddots & & \\ & & \ddots & \ddots & \ddots & \\ \vdots & & & \ddots & \ddots & \gamma_{k-1} \\ 0 & \cdots & & & \beta_{k-1} & \alpha_k \end{bmatrix},$$

then the situation at the bottom of the loop is summarized by the equations

$$A[\,q_1,\ldots,q_k\,] = [\,q_1,\ldots,q_k\,]T_k + r_k e_k^T \qquad (9.4.8)$$
$$A^T[\,p_1,\ldots,p_k\,] = [\,p_1,\ldots,p_k\,]T_k^T + s_k e_k^T. \qquad (9.4.9)$$

If $r_k = 0$, then the iteration terminates and span$\{q_1,\ldots,q_k\}$ is an invariant subspace for A. If $s_k = 0$, then the iteration also terminates and span$\{p_1,\ldots,p_k\}$ is an invariant subspace for A^T. However, if neither of these conditions are true and $s_k^T r_k = 0$, then the tridiagonalization process ends without any invariant subspace information. This is called *serious breakdown*. See Wilkinson (1965, p.389) for an early discussion of the matter.

9.4.4 The Look-Ahead Idea

It is interesting to look the serious breakdown issue in the block version of (9.4.7). For clarity assume that $A \in \mathbb{R}^{n \times n}$ with $n = rp$. Consider the factorization

$$P^T A Q = \begin{bmatrix} M_1 & C_1^T & \cdots & & 0 \\ B_1 & M_2 & \ddots & & \vdots \\ & \ddots & \ddots & \ddots & \\ \vdots & & \ddots & \ddots & C_{r-1}^T \\ 0 & \cdots & & B_{r-1} & M_r \end{bmatrix} \qquad (9.4.10)$$

where all the blocks are p-by-p. Let $Q = [\, Q_1, \ldots, Q_r \,]$ and $P = [\, P_1, \ldots, P_r \,]$ be conformable partitionings of Q and P. Comparing block columns in the equations $AQ = QT$ and $A^T P = PT^T$ we obtain

$$\begin{aligned} Q_{k+1} B_k &= A Q_k - Q_k M_k - Q_{k-1} C_{k-1}^T &\equiv R_k \\ P_{k+1} C_k &= A^T P_k - P_k M_k^T - P_{k-1} B_{k-1}^T &\equiv S_k \end{aligned}$$

Note that $M_k = P_k^T A Q_k$. If $S_k^T R_k \in \mathbb{R}^{p \times p}$ is nonsingular and we compute $B_k, C_k \in \mathbb{R}^{p \times p}$ so that

$$C_k^T B_k = S_k^T R_k,$$

then

$$\begin{aligned} Q_{k+1} &= R_k B_k^{-1} & (9.4.11) \\ P_{k+1} &= S_k C_k^{-1} & (9.4.12) \end{aligned}$$

satisfy $P_{k+1}^T Q_{k+1} = I_p$. Serious breakdown in this setting is associated with having a singular $S_k^T R_k$.

One way of solving the serious breakdown problem in (9.4.7) is to go after a factorization of the form (9.4.10) in which the block sizes are dynamically determined. Roughly speaking, in this approach matrices Q_{k+1} and P_{k+1} are built up column by column with special recursions that culminate in the production of a nonsingular $P_{k+1}^T Q_{k+1}$. The computations are arranged so that the biorthogonality conditions $P_i^T Q_{k+1} = 0$ and $Q_i^T P_{k+1} = 0$ hold for $i = 1{:}k$.

A method of this form belongs to the family of *look-ahead Lanczos* methods. The length of a look-ahead step is the width of the Q_{k+1} and P_{k+1} that it produces. If that width is one, a conventional block Lanczos step may be taken. Length-2 look-ahead steps are discussed in Parlett, Taylor and Liu (1985). The notion of *incurable breakdown* is also presented by these authors. Freund, Gutknecht, and Nachtigal (1993) cover the general case along with a host of implementation details. Floating point considerations require the handling of "near" serious breakdown. In practice, each M_k that is 2-by-2 or larger corresponds to an instance of near serious breakdown.

Problems

P9.4.1 Prove that the Arnoldi vectors in (9.4.1) are mutually orthogonal.

P9.4.2 Prove (9.4.4).

P9.4.3 Prove (9.4.6).

P9.4.4 Give an example of a starting vector for which the unsymmetric Lanczos iteration (9.4.7) breaks down without rendering any invariant subspace information. Use

$$A = \begin{bmatrix} 1 & 6 & 2 \\ 3 & 0 & 2 \\ 1 & 3 & 5 \end{bmatrix}.$$

P9.4.5 Suppose $H \in \mathbf{R}^{n \times n}$ is upper Hessenberg. Discuss the computation of a unit upper triangular matrix U such that $HU = UT$ where T is tridiagonal.

P9.4.6 Show that the QR algorithm for eigenvalues does not preserve tridiagonal structure in the unsymmetric case.

Notes and References for Sec. 9.4

References for the Arnoldi iteration and its practical implementation include Saad (1992) and

W.E. Arnoldi (1951). "The Principle of Minimized Iterations in the Solution of the Matrix Eigenvalue Problem," *Quarterly of Applied Mathematics 9*, 17–29.

Y. Saad (1980). "Variations of Arnoldi's Method for Computing Eigenelements of Large Unsymmetric Matrices.," *Lin. Alg. and Its Applic. 34*, 269–295.

Y. Saad (1984). "Chebyshev Acceleration Techniques for Solving Nonsymmetric Eigenvalue Problems," *Math. Comp. 42*, 567–588.

D.C. Sorensen (1992). "Implicit Application of Polynomial Filters in a k-Step Arnoldi Method," *SIAM J. Matrix Anal. Appl. 13*, 357–385.

D.C. Sorensen (1995). "Implicitly Restarted Arnoldi/Lanczos Methods for Large Scale Eigenvalue Calculations," in *Proceedings of the ICASE/LaRC Workshop on Parallel Numerical Algorithms, May 23–25, 1994*, D.E. Keyes, A. Sameh, and V. Venkatakrishnan (eds), Kluwer.

R.B, Lehoucq (1995). "Analysis and Implementation of an Implicitly Restarted Arnoldi Iteration," Ph.D. thesis, Rice University, Houston Texas.

R.B. Lehoucq (1996). "Restarting an Arnoldi Reduction," Report MCS-P591-0496, Argonne National Laboratory, Argonne Illinois.

R.B. Lehoucq and D.C. Sorensen (1996). "Deflation Techniques for an Implicitly Restarted Iteration," *SIAM J. Matrix Analysis and Applic,* to appear.

R.B. Morgan (1996). "On Restarting the Arnoldi Method for Large Nonsymmetric Eigenvalue Problems," *Math Comp 65*, 1213–1230.

Related papers include

A. Ruhe (1984). "Rational Krylov Algorithms for Eigenvalue Computation," *Lin. Alg. and Its Applic. 58*, 391–405.

A. Ruhe (1994). "Rational Krylov Algorithms for Nonsymmetric Eigenvalue Problems II. Matrix Pairs," *Lin. Alg. and Its Applic. 197*, 283–295.

A. Ruhe (1994). "The Rational Krylov Algorithm for Nonsymmetric Eigenvalue Problems III: Complex Shifts for Real Matrices," *BIT 34*,165–176.

T. Huckle (1994). "The Arnoldi Method for Normal Matrices," *SIAM J. Matrix Anal. Appl. 15*, 479–489.

C.C. Paige, B.N. Parlett,and H.A. Van Der Vorst (1995). "Approximate Solutions and Eigenvalue Bounds from Krylov Subspaces," *Numer. Linear Algebra with Applic. 2*, 115–134.

K.C. Toh and L.N. Trefethen (1996). "Calculation of Pseudospectra by the Arnoldi Iteration," *SIAM J. Sci. Comp. 17*, 1–15.

The unsymmetric Lanczos process and related look ahead ideas are nicely presented in

B.N. Parlett, D. Taylor, and Z. Liu (1985). "A Look-Ahead Lanczos Algorithm for Unsymmetric Matrices," *Math. Comp. 44*, 105–124.

R.W. Freund, M. Gutknecht, and N. Nachtigal (1993). "An Implementation of the Look-Ahead Lanczos Algorithm for Non-Hermitian Matrices," *SIAM J. Sci. and Stat.Comp. 14*, 137–158.

See also

Y. Saad (1982). "The Lanczos Biorthogonalization Algorithm and Other Oblique Projection Methods for Solving Large Unsymmetric Eigenproblems," *SIAM J. Numer. Anal. 19*, 485–506.

G.A. Geist (1991). "Reduction of a General Matrix to Tridiagonal Form," *SIAM J. Matrix Anal. Appl. 12*, 362–373.

C. Brezinski, M. Zaglia, and H. Sadok (1991). "Avoiding Breakdown and Near Breakdown in Lanczos Tyoe Algorithms," *Numer. Alg. 1*, 261–284.

S.K. Kim and A.T. Chronopoulos (1991). "A Class of Lanczos-Like Algorithms Implemented on Parallel Computers," *Parallel Comput. 17*, 763–778.

B.N. Parlett (1992). "Reduction to Tridiagonal Form and Minimal Realizations," *SIAM J. Matrix Anal. Appl. 13*, 567–593.

M. Gutknecht (1992). "A Completed Theory of the Unsymmetric Lanczos Process and Related Algorithms, Part I," *SIAM J. Matrix Anal. Appl. 13*, 594–639.

M. Gutknecht (1994). "A Completed Theory of the Unsymmetric Lanczos Process and Related Algorithms, Part II," *SIAM J. Matrix Anal. Appl. 15*, 15–58.

Z. Bai (1994). "Error Analysis of the Lanczos Algorithm for Nonsymmetric Eigenvalue Problem," *Math. Comp. 62*, 209–226.

T. Huckle (1995). "Low-Rank Modification of the Unsymmetric Lanczos Algorithm," *Math.Comp. 64*, 1577–1588.

Z. Jia (1995). "The Convergence of Generalized Lanczos Methods for Large Unsymmetric Eigenproblems," *SIAM J. Matrix Anal. Applic 16*, 543–562.

M.T. Chu, R.E. Funderlic, and G.H. Golub (1995). "A Rank-One Reduction Formula and Its Applications to Matrix Factorizations," *SIAM Review 37*, 512–530.

Other papers include

H.A. Van der Vorst (1982). "A Generalized Lanczos Scheme," *Math. Comp. 39*, 559–562.

D. Boley and G.H. Golub (1984). "The Lanczos-Arnoldi Algorithm and Controllability," *Syst. Control Lett. 4*, 317–324.

Chapter 10

Iterative Methods for Linear Systems

We concluded the previous chapter by showing how the Lanczos iteration could be used to solve various linear equation and least squares problems. The methods developed were suitable for large sparse problems because they did not require the factorization of the underlying matrix. In this section, we continue the discussion of linear equation solvers that have this property.

The first section is a brisk review of the classical iterations: Jacobi, Gauss-Seidel, SOR, Chebyshev semi-iterative, and so on. Our treatment of these methods is brief because our principal aim in this chapter is to highlight the method of conjugate gradients. In §10.2, we carefully develop this important technique in a natural way from the method of steepest descent. Recall that the conjugate gradient method has already been introduced via the Lanczos iteration in §9.3. The reason for deriving the method again is to motivate some of its practical variants, which are the subject of §10.3. Extensions to unsymmetric problems are treated in §10.4.

We warn the reader of an inconsistency in the notation of this chapter In §10.1, methods are developed at the "(i,j) level" necessitating the use of superscripts: $x_i^{(k)}$ denotes the i-th component of a vector $x^{(k)}$. In the other

508

sections, however, algorithmic developments can proceed without explicit mention of vector/matrix entries. Hence, in §10.2-§10.4 we dispense with superscripts and denote vector sequences by $\{x_k\}$.

Before You Begin

Chapter 1, §§2.1-2.5, and §2.7, Chapter 3, and §§4.1-4.3 are assumed. Other dependencies include:

$$
\begin{array}{ccccccc}
 & & & & & \text{Chapter 9} & \\
 & & & & & \downarrow & \\
\S 10.1 & \rightarrow & \S 10.2 & \rightarrow & \S 10.3 & \rightarrow & \S 10..4 \\
 & & & & \uparrow & & \\
 & & & & \S 7.4 & &
\end{array}
$$

Texts devoted to iterative solvers include Varga (1962), Young (1971), Hageman and Young (1981), and Axelsson (1994). The software "templates" volume by Barrett *et al* (1993) is particularly useful. The direct (non-iterative) solution of large sparse systems is sometimes preferred. See George and Liu (1981) and Duff, Erisman, and Reid (1986).

10.1   The Standard Iterations

The linear equation solvers in Chapters 3 and 4 involve the factorization of the coefficient matrix A. Methods of this type are called *direct methods*. Direct methods can be impractical if A is large and sparse, because the sought-after factors can be dense. An exception to this occurs when A is banded (cf. §4.3). Yet in many band matrix problems even the band itself is sparse making algorithms such as band Cholesky difficult to implement.

One reason for the great interest in sparse linear equation solvers is the importance of being able to obtain numerical solutions to partial differential equations. Indeed, researchers in computational PDE's have been responsible for many of the sparse matrix techniques that are presently in general use.

Roughly speaking, there are two approaches to the sparse $Ax = b$ problem. One is to pick an appropriate direct method and adapt it to exploit A's sparsity. Typical adaptation strategies involve the intelligent use of data structures and special pivoting strategies that minimize fill-in.

In contrast to the direct methods are the *iterative methods*. These methods generate a sequence of approximate solutions $\{x^{(k)}\}$ and essentially involve the matrix A only in the context of matrix-vector multiplication. The evaluation of an iterative method invariably focuses on how quickly the iterates $x^{(k)}$ converge. In this section, we present some basic iterative methods, discuss their practical implementation, and prove a few representative theorems concerned with their behavior.

10.1.1   The Jacobi and Gauss-Seidel Iterations

Perhaps the simplest iterative scheme is the *Jacobi iteration*. It is defined for matrices that have nonzero diagonal elements. The method can be motivated by rewriting the 3-by-3 system $Ax = b$ as follows:

$$
\begin{aligned}
x_1 &= (b_1 - a_{12}x_2 - a_{13}x_3)/a_{11} \\
x_2 &= (b_2 - a_{21}x_1 - a_{23}x_3)/a_{22} \\
x_3 &= (b_3 - a_{31}x_1 - a_{32}x_2)/a_{33}
\end{aligned}
$$

Suppose $x^{(k)}$ is an approximation to $x = A^{-1}b$. A natural way to generate a new approximation $x^{(k+1)}$ is to compute

$$
\begin{aligned}
x_1^{(k+1)} &= (b_1 - a_{12}x_2^{(k)} - a_{13}x_3^{(k)})/a_{11} \\
x_2^{(k+1)} &= (b_2 - a_{21}x_1^{(k)} - a_{23}x_3^{(k)})/a_{22} \qquad (10.1.1) \\
x_3^{(k+1)} &= (b_3 - a_{31}x_1^{(k)} - a_{32}x_2^{(k)})/a_{33}
\end{aligned}
$$

This defines the Jacobi iteration for the case $n = 3$. For general n we have

> **for** $i = 1{:}n$
> $$
> x_i^{(k+1)} = \left(b_i - \sum_{j=1}^{i-1} a_{ij}x_j^{(k)} - \sum_{j=i+1}^{n} a_{ij}x_j^{(k)} \right) \Bigg/ a_{ii} \qquad (10.1.2)
> $$
> **end**

Note that in the Jacobi iteration one does not use the most recently available information when computing $x_i^{(k+1)}$. For example, $x_1^{(k)}$ is used in the calculation of $x_2^{(k+1)}$ even though component $x_1^{(k+1)}$ is known. If we revise the Jacobi iteration so that we always use the most current estimate of the exact x_i then we obtain

> **for** $i = 1{:}n$
> $$
> x_i^{(k+1)} = \left(b_i - \sum_{j=1}^{i-1} a_{ij}x_j^{(k+1)} - \sum_{j=i+1}^{n} a_{ij}x_j^{(k)} \right) \Bigg/ a_{ii} \qquad (10.1.3)
> $$
> **end**

This defines what is called the *Gauss-Seidel iteration*.

For both the Jacobi and Gauss-Seidel iterations, the transition from $x^{(k)}$ to $x^{(k+1)}$ can be succinctly described in terms of the matrices L, D, and U defined by:

$$
L = \begin{bmatrix}
0 & 0 & \cdots & \cdots & 0 \\
a_{21} & 0 & \cdots & & \vdots \\
a_{31} & a_{32} & \ddots & & 0 \\
\vdots & & & 0 & 0 \\
a_{n1} & a_{n2} & \cdots & a_{n,n-1} & 0
\end{bmatrix}
$$

$$D = \text{diag}(a_{11}, \ldots, a_{nn}) \qquad (10.1.4)$$

$$U = \begin{bmatrix} 0 & a_{12} & \cdots & \cdots & a_{1n} \\ 0 & 0 & \cdots & & \vdots \\ 0 & 0 & \ddots & & a_{n-2,n} \\ \vdots & & & & a_{n-1,n} \\ 0 & 0 & \cdots & 0 & 0 \end{bmatrix}$$

In particular, the Jacobi step has the form $M_J x^{(k+1)} = N_J x^{(k)} + b$ where $M_J = D$ and $N_J = -(L+U)$. On the other hand, Gauss-Seidel is defined by $M_G x^{(k+1)} = N_G x^{(k)} + b$ with $M_G = (D+L)$ and $N_G = -U$.

10.1.2   Splittings and Convergence

The Jacobi and Gauss-Seidel procedures are typical members of a large family of iterations that have the form

$$M x^{(k+1)} = N x^{(k)} + b \qquad (10.1.5)$$

where $A = M - N$ is a *splitting* of the matrix A. For the iteration (10.1.5) to be practical, it must be "easy" to solve a linear system with M as the matrix. Note that for Jacobi and Gauss-Seidel, M is diagonal and lower triangular respectively.

Whether or not (10.1.5) converges to $x = A^{-1}b$ depends upon the eigenvalues of $M^{-1}N$. In particular, if the *spectral radius* of an n-by-n matrix G is defined by

$$\rho(G) = \max\{ |\lambda| : \lambda \in \lambda(G) \},$$

then it is the size of $\rho(M^{-1}N)$ that is critical to the success of (10.1.5).

Theorem 10.1.1 *Suppose $b \in \mathbb{R}^n$ and $A = M - N \in \mathbb{R}^{n \times n}$ is nonsingular. If M is nonsingular and the spectral radius of $M^{-1}N$ satisfies the inequality $\rho(M^{-1}N) < 1$, then the iterates $x^{(k)}$ defined by $M x^{(k+1)} = N x^{(k)} + b$ converge to $x = A^{-1}b$ for any starting vector $x^{(0)}$.*

Proof. Let $e^{(k)} = x^{(k)} - x$ denote the error in the kth iterate. Since $Mx = Nx + b$ it follows that $M(x^{(k+1)} - x) = N(x^{(k)} - x)$, and thus, the error in $x^{(k+1)}$ is given by $e^{(k+1)} = M^{-1}Ne^{(k)} = (M^{-1}N)^{k+1}e^{(0)}$. From Lemma 7.3.2 we know that $(M^{-1}N)^k \to 0$ iff $\rho(M^{-1}N) < 1$. $\square$

This result is central to the study of iterative methods where algorithmic development typically proceeds along the following lines:

- A splitting $A = M - N$ is proposed where linear systems of the form $Mz = d$ are "easy" to solve.

- Classes of matrices are identified for which the iteration matrix $G = M^{-1}N$ satisfies $\rho(G) < 1$.

- Further results about $\rho(G)$ are established to gain intuition about how the error $e^{(k)}$ tends to zero.

For example, consider the Jacobi iteration, $Dx^{(k+1)} = -(L+U)x^{(k)} + b$. One condition that guarantees $\rho(M_J^{-1}N_J) < 1$ is strict diagonal dominance. Indeed, if A has that property (defined in §3.4.10), then

$$\rho(M_J^{-1}N_J) \leq \| D^{-1}(L+U) \|_\infty = \max_{1 \leq i \leq n} \sum_{\substack{j=1 \\ j \neq i}}^n \left| \frac{a_{ij}}{a_{ii}} \right| < 1$$

Usually, the "more dominant" the diagonal the more rapid the convergence but there are counterexamples. See P10.1.7.

A more complicated spectral radius argument is needed to show that Gauss-Seidel converges for symmetric positive definite A.

Theorem 10.1.2 *If $A \in \mathbb{R}^{n \times n}$ is symmetric and positive definite, then the Gauss-Seidel iteration (10.1.3) converges for any $x^{(0)}$.*

Proof. Write $A = L + D + L^T$ where $D = \text{diag}(a_{ii})$ and L is strictly lower triangular. In light of Theorem 10.1.1 our task is to show that the matrix $G = -(D+L)^{-1}L^T$ has eigenvalues that are inside the unit circle. Since D is positive definite we have $G_1 \equiv D^{1/2}GD^{-1/2} = -(I + L_1)^{-1}L_1^T$, where $L_1 = D^{-1/2}LD^{-1/2}$. Since G and G_1 have the same eigenvalues, we must verify that $\rho(G_1) < 1$. If $G_1 x = \lambda x$ with $x^H x = 1$, then we have $-L_1^T x = \lambda(I + L_1)x$ and thus, $-x^H L_1^T x = \lambda(1 + x^H L_1 x)$. Letting $a + bi = x^H L_1 x$ we have

$$|\lambda|^2 = \left| \frac{-a + bi}{1 + a + bi} \right|^2 = \frac{a^2 + b^2}{1 + 2a + a^2 + b^2} .$$

However, since $D^{-1/2}AD^{-1/2} = I + L_1 + L_1^T$ is positive definite, it is not hard to show that $0 < 1 + x^H L_1 x + x^H L_1^T x = 1 + 2a$ implying $|\lambda| < 1$. $\square$

This result is frequently applicable because many of the matrices that arise from discretized elliptic PDE's are symmetric positive definite. Numerous other results of this flavor appear in the literature.

10.1.3   Practical Implementation of Gauss-Seidel

We now focus on some practical details associated with the Gauss-Seidel iteration. With overwriting the Gauss-Seidel step (10.1.3) is particularly simple to implement:

for $i = 1{:}n$

$$x_i = \left(b_i - \sum_{j=1}^{i-1} a_{ij}x_j - \sum_{j=i+1}^{n} a_{ij}x_j \right) \bigg/ a_{ii}$$

end

This computation requires about twice as many flops as there are nonzero entries in the matrix A. It makes no sense to be more precise about the work involved because the actual implementation depends greatly upon the structure of the problem at hand.

In order to stress this point we consider the application of (10.1.3) to the NM-by-NM block tridiagonal system

$$\begin{bmatrix} T & -I_N & \cdots & 0 \\ -I_N & T & \ddots & \vdots \\ & \ddots & \ddots & \ddots \\ \vdots & & \ddots & \ddots & -I_N \\ 0 & \cdots & & -I_N & T \end{bmatrix} \begin{bmatrix} g_1 \\ g_2 \\ \vdots \\ \vdots \\ g_M \end{bmatrix} = \begin{bmatrix} f_1 \\ f_2 \\ \vdots \\ \vdots \\ f_M \end{bmatrix} \qquad (10.1.6)$$

where

$$T = \begin{bmatrix} 4 & -1 & \cdots & 0 \\ -1 & 4 & \ddots & \vdots \\ & \ddots & \ddots & \ddots \\ \vdots & & \ddots & \ddots & -1 \\ 0 & \cdots & & -1 & 4 \end{bmatrix} , \; g_j = \begin{bmatrix} G(1,j) \\ G(2,j) \\ \vdots \\ \vdots \\ G(N,j) \end{bmatrix} , \; f_j = \begin{bmatrix} F(1,j) \\ F(2,j) \\ \vdots \\ \vdots \\ F(N,j) \end{bmatrix} .$$

This problem arises when the Poisson equation is discretized on a rectangle. It is easy to show that the matrix A is positive definite.

With the convention that $G(i, j) = 0$ whenever $i \in \{0, N + 1\}$ or $j \in \{0, M + 1\}$ we see that with overwriting the Gauss-Seidel step takes on the form:

for $j = 1{:}M$
    for $i = 1{:}N$
        $G(i,j) = (F(i,j) + G(i-1,j) + G(i+1,j)+$
                    $G(i,j-1) + G(i,j+1))/4$
    end
end

Note that in this problem no storage is required for the matrix A.

10.1.4   Successive Over-Relaxation

The Gauss-Seidel iteration is very attractive because of its simplicity. Unfortunately, if the spectral radius of $M_G^{-1}N_G$ is close to unity, then it may be prohibitively slow because the error tends to zero like $\rho(M_G^{-1}N_G)^k$. To rectify this, let $\omega \in \mathbb{R}$ and consider the following modification of the Gauss-Seidel step:

> **for** $i = 1{:}n$
>
> $$x_i^{(k+1)} = \omega \left(b_i - \sum_{j=1}^{i-1} a_{ij} x_j^{(k+1)} - \sum_{j=i+1}^{n} a_{ij} x_j^{(k)} \right) \Big/ a_{ii}$$
>
> $$+\; (1-\omega) x_i^{(k)} \qquad\qquad (10.1.7)$$
>
> **end**

This defines the method of *successive over-relaxation* (SOR). Using (10.1.4) we see that in matrix terms, the SOR step is given by

$$M_\omega x^{(k+1)} = N_\omega x^{(k)} + \omega b \qquad\qquad (10.1.8)$$

where $M_\omega = D + \omega L$ and $N_\omega = (1-\omega)D - \omega U$. For a few structured (but important) problems such as (10.1.6), the value of the *relaxation parameter* ω that minimizes $\rho(M_\omega^{-1}N_\omega)$ is known. Moreover, a significant reduction in $\rho(M_1^{-1}N_1) = \rho(M_G^{-1}N_G)$ can result. In more complicated problems, however, it may be necessary to perform a fairly sophisticated eigenvalue analysis in order to determine an appropriate ω.

10.1.5   The Chebyshev Semi-Iterative Method

Another way to accelerate the convergence of an iterative method makes use of Chebyshev polynomials. Suppose $x^{(1)}, \ldots, x^{(k)}$ have been generated via the iteration $M x^{(j+1)} = N x^{(j)} + b$ and that we wish to determine coefficients $\nu_j(k)$, $j = 0{:}k$ such that

$$y^{(k)} = \sum_{j=0}^{k} \nu_j(k) x^{(j)} \qquad\qquad (10.1.9)$$

represents an improvement over $x^{(k)}$. If $x^{(0)} = \cdots = x^{(k)} = x$, then it is reasonable to insist that $y^{(k)} = x$. Hence, we require

$$\sum_{j=0}^{k} \nu_j(k) = 1. \qquad\qquad (10.1.10)$$

Subject to this constraint, we consider how to choose the $\nu_j(k)$ so that the error in $y^{(k)}$ is minimized.

Recalling from the proof of Theorem 10.1.1 that $x^{(k)} - x = (M^{-1}N)^k e^{(0)}$ where $e^{(0)} = x^{(0)} - x$, we see that

$$y^{(k)} - x \; = \; \sum_{j=0}^{k} \nu_j(k)(x^{(j)} - x) \; = \; \sum_{j=0}^{k} \nu_j(k)(M^{-1}N)^j e^{(0)} .$$

Working in the 2-norm we therefore obtain

$$\| \, y^{(k)} - x \, \|_2 \; \leq \; \| \, p_k(G) \, \|_2 \, \| \, e^{(0)} \, \|_2 \qquad\qquad (10.1.11)$$

where $G = M^{-1}N$ and

$$p_k(z) \; = \; \sum_{j=0}^{k} \nu_j(k) z^j .$$

Note that the condition (10.1.10) implies $p_k(1) = 1$.

At this point we assume that G is symmetric with eigenvalues λ_i that satisfy $-1 < \alpha \leq \lambda_n \leq \cdots \leq \lambda_1 \leq \beta < 1$. It follows that

$$\| \, p_k(G) \, \|_2 \; = \; \max_{\lambda_i \in \lambda(A)} \; |p_k(\lambda_i)| \; \leq \; \max_{\alpha \leq \lambda \leq \beta} \; |p_k(\lambda)| .$$

Thus, to make the norm of $p_k(G)$ small, we need a polynomial $p_k(z)$ that is small on $[\alpha, \beta]$ subject to the constraint that $p_k(1) = 1$.

Consider the Chebyshev polynomials $c_j(z)$ generated by the recursion $c_j(z) = 2zc_{j-1}(z) - c_{j-2}(z)$ where $c_0(z) = 1$ and $c_1(z) = z$. These polynomials satisfy $|c_j(z)| \leq 1$ on [-1, 1] but grow rapidly off this interval. As a consequence, the polynomial

$$p_k(z) \; = \; \frac{c_k \left(-1 + 2\dfrac{z - \alpha}{\beta - \alpha} \right)}{c_k(\mu)}$$

where

$$\mu \; = \; -1 + 2\frac{1 - \alpha}{\beta - \alpha} \; = \; 1 + 2\frac{1 - \beta}{\beta - \alpha}$$

satisfies $p_k(1) = 1$ and tends to be small on $[\alpha, \beta]$. From the definition of $p_k(z)$ and equation (10.1.11) we see

$$\| \, y^{(k)} - x \, \|_2 \; \leq \; \frac{\| \, x - x^{(0)} \, \|_2}{|c_k(\mu)|} .$$

Thus, the larger μ is, the greater the acceleration of convergence.

In order for the above to be a practical acceleration procedure, we need a more efficient method for calculating $y^{(k)}$ than (10.1.9). We have been tacitly assuming that n is large and thus the retrieval of $x^{(0)}, \dots, x^{(k)}$ for large k would be inconvenient or even impossible.

Fortunately, it is possible to derive a three-term recurrence among the $y^{(k)}$ by exploiting the three-term recurrence among the Chebyshev polynomials. In particular, it can be shown that if

$$\omega_{k+1} = 2\frac{2 - \beta - \alpha}{\beta - \alpha}\frac{c_k(\mu)}{c_{k+1}(\mu)}$$

then

$$y^{(k+1)} = \omega_{k+1}(y^{(k)} - y^{(k-1)} + \gamma z^{(k)}) + y^{(k-1)}$$

$$Mz^{(k)} = b - Ay^{(k)} \qquad\qquad (10.1.12)$$

$$\gamma = 2/(2 - \alpha - \beta),$$

where $y^{(0)} = x^{(0)}$ and $y^{(1)} = x^{(1)}$. We refer to this scheme as the Chebyshev semi-iterative method associated with $My^{(k+1)} = Ny^{(k)} + b$. For the acceleration to be effective we need good lower and upper bounds α and β. As in SOR, these parameters may be difficult to ascertain except in a few structured problems.

Chebyshev semi-iterative methods are extensively analyzed in Varga (1962, chapter 5), as well as in Golub and Varga (1961).

10.1.6   Symmetric SOR

In deriving the Chebyshev acceleration we assumed that the iteration matrix $G = M^{-1}N$ was symmetric. Thus, our simple analysis does not apply to the unsymmetric SOR iteration matrix $M_\omega^{-1}N_\omega$. However, it is possible to symmetrize the SOR method making it amenable to Chebyshev acceleration. The idea is to couple SOR with the *backward SOR* scheme

for $i = n: -1:1$

$$x_i^{(k+1)} = \omega\left(b_i - \sum_{j=1}^{i-1} a_{ij}x_j^{(k+1)} - \sum_{j=i+1}^{n} a_{ij}x_j^{(k)}\right) \Big/ a_{ii}$$

$$+ (1 - \omega)x_i^{(k)} \qquad\qquad (10.1.13)$$

end

This iteration is obtained by updating the unknowns in reverse order in (10.1.7). Backward SOR can be described in matrix terms using (10.1.4). In particular, we have $\tilde{M}_\omega x^{(k+1)} = \tilde{N}_\omega x^{(k)} + \omega b$ where

$$\tilde{M}_\omega = D + \omega U \qquad \text{and} \qquad \tilde{N}_\omega = (1 - \omega)D - \omega L. \qquad (10.1.14)$$

If A is symmetric ($U = L^T$), then $\tilde{M}_\omega = M_\omega^T$ and $\tilde{N}_\omega = N_\omega^T$, and we have the iteration

$$M_\omega x^{(k+1/2)} = N_\omega x^{(k)} + \omega b$$

$$M_\omega^T x^{(k+1)} = N_\omega^T x^{(k+1/2)} + \omega b. \qquad (10.1.15)$$

It is clear that $G = M_\omega^{-T} N_\omega^T M_\omega^{-1} N_\omega$ is the iteration matrix for this method. From the definitions of M_ω and N_ω it follows that

$$G = M^{-1}N \equiv (M_\omega D^{-1} M_\omega^T)^{-1}(N_\omega^T D^{-1} N_\omega). \qquad (10.1.16)$$

If D has positive diagonal entries and $KK^T = (N_\omega^T D^{-1} N_\omega)$ is the Cholesky factorization, then $K^T G K^{-T} = K^T (M_\omega D^{-1} M_\omega^T)^{-1} K$. Thus, G is similar to a symmetric matrix and has real eigenvalues.

The iteration (10.1.15) is called the *symmetric successive over-relaxation* (SSOR) method. It is frequently used in conjunction with the Chebyshev semi-iterative acceleration.

Problems

P10.1.1 Show that the Jacobi iteration can be written in the form $x^{(k+1)} = x^{(k)} + Hr^{(k)}$ where $r^{(k)} = b - Ax^{(k)}$. Repeat for the Gauss-Seidel iteration.

P10.1.2 Show that if A is strictly diagonally dominant, then the Gauss-Seidel iteration converges.

P10.1.3 Show that the Jacobi iteration converges for 2-by-2 symmetric positive definite systems.

P10.1.4 Show that if $A = M - N$ is singular, then we can never have $\rho(M^{-1}N) < 1$ even if M is nonsingular.

P10.1.5 Prove (10.1.16).

P10.1.6 Prove the converse of Theorem 10.1.1. In other words, show that if the iteration $Mx^{(k+1)} = Nx^{(k)} + b$ always converges, then $\rho(M^{-1}N) < 1$.

P10.1.7 (Supplied by R.S. Varga) Suppose that

$$A_1 = \begin{bmatrix} 1 & -1/2 \\ -1/2 & 1 \end{bmatrix} \qquad A_2 = \begin{bmatrix} 1 & -3/4 \\ -1/12 & 1 \end{bmatrix}.$$

Let J_1 and J_2 be the associated Jacobi iteration matrices. Show that $\rho(J_1) > \rho(J_2)$ thereby refuting the claim that greater diagonal dominance implies more rapid Jacobi convergence.

P10.1.8 The Chebyshev algorithm is defined in terms of parameters

$$\omega_{k+1} = \frac{2c_k(1/\rho)}{\rho c_{k+1}(1/\rho)}$$

where $c_k(\lambda) = \cosh[k \cosh^{-1}(\lambda)]$ with $\lambda > 1$. (a) Show that $1 < \omega_k < 2$ for $k > 1$ whenever $0 < \rho < 1$. (b) Verify that $\omega_{k+1} < \omega_k$. (c) Determine $\lim \omega_k$ as $k \to \infty$.

P10.1.9 Consider the 2-by-2 matrix

$$A = \begin{bmatrix} 1 & \rho \\ -\rho & 1 \end{bmatrix}.$$

(a) Under what conditions will Gauss-Seidel converge with this matrix? (b) For what range of ω will the SOR method converge? What is the optimal choice for this parameter? (c) Repeat (a) and (b) for the matrix

$$A = \begin{bmatrix} I_n & S \\ -S^T & I_n \end{bmatrix}$$

where $S \in \mathbf{R}^{n \times n}$. Hint: Use the SVD of S.

P10.1.10 We want to investigate the solution of $Au = f$ where $A \neq A^T$. For a model problem, consider the finite difference approximation to

$$-u'' + \sigma u' = 0 \qquad 0 < x < 1$$

where $u(0) = 10$ and $u(1) = 10 \exp^\sigma$. This leads to the difference equation

$$-u_{i-1} + 2u_i - u_{i+1} + R(u_{i+1} - u_{i-1}) = 0 \qquad i = 1{:}n$$

where $R = \sigma h/2$, $u_0 = 10$, and $u_{n+1} = 10 \exp^\sigma$. The number R should be less than 1. What is the convergence rate for the iteration $Mu^{(k+1)} = Nu^{(k)} + f$ where $M = (A + A^T)/2$ and $N = (A^T - A)/2$?

P10.1.11 Consider the iteration

$$y^{(k+1)} = \omega(By^{(k)} + d - y^{(k-1)}) + y^{(k-1)}$$

where B has Schur decomposition $Q^T BQ = \text{diag}(\lambda_1, \ldots, \lambda_n)$ with $\lambda_1 \geq \cdots \geq \lambda_n$. Assume that $x = Bx + d$. (a) Derive an equation for $e^{(k)} = y^{(k)} - x$. (b) Assume $y^{(1)} = By^{(0)} + d$. Show that $e^{(k)} = p_k(B)e^{(0)}$ where p_k is an even polynomial if k is even and an odd polynomial if k is odd. (c) Write $f^{(k)} = Q^T e^{(k)}$. Derive a difference equation for $f_j^{(k)}$ for $j = 1{:}n$. Try to specify the exact solution for general $f_j^{(0)}$ and $f_j^{(1)}$. (d) Show how to determine an optimal ω.

Notes and References for Sec. 10.1

As we mentioned, Young (1971) has the most comprehensive treatment of the SOR method. The object of "SOR theory" is to guide the user in choosing the relaxation parameter ω. In this setting, the ordering of equations and unknowns is critical. See

M.J.M. Bernal and J.H. Verner (1968). "On Generalizing of the Theory of Consistent Orderings for Successive Over-Relaxation Methods," *Numer. Math. 12*, 215–22.

D.M. Young (1970). "Convergence Properties of the Symmetric and Unsymmetric Over-Relaxation Methods," *Math. Comp. 24*, 793–807.

D.M. Young (1972). "Generalization of Property A and Consistent Ordering," *SIAM J. Num. Anal. 9*, 454–63.

R.A. Nicolaides (1974). "On a Geometrical Aspect of SOR and the Theory of Consistent Ordering for Positive Definite Matrices," *Numer. Math. 12*, 99–104.

L. Adams and H. Jordan (1986). "Is SOR Color-Blind?" *SIAM J. Sci. Stat. Comp. 7*, 490–506.

M. Eiermann and R.S. Varga (1993). "Is the Optimal ω Best for the SOR Iteration Method," *Lin. Alg. and Its Applic. 182*, 257–277.

An analysis of the Chebyshev semi-iterative method appears in

G.H. Golub and R.S. Varga (1961). "Chebychev Semi-Iterative Methods, Successive Over-Relaxation Iterative Methods, and Second-Order Richardson Iterative Methods, Parts I and II," *Numer. Math. 3*, 147–56, 157–68.

This work is premised on the assumption that the underlying iteration matrix has real eigenvalues. How to proceed when this is not the case is discussed in

T.A. Manteuffel (1977). "The Tchebychev Iteration for Nonsymmetric Linear Systems," *Numer. Math. 28*, 307–27.

M. Eiermann and W. Niethammer (1983). "On the Construction of Semi-iterative Methods," *SIAM J. Numer. Anal. 20*, 1153–1160.

W. Niethammer and R.S. Varga (1983). "The Analysis of k-step Iterative Methods for Linear Systems from Summability Theory," *Numer. Math. 41*, 177–206.

G.H. Golub and M. Overton (1988). "The Convergence of Inexact Chebychev and Richardson Iterative Methods for Solving Linear Systems," *Numer. Math. 53*, 571–594.

D. Calvetti, G.H. Golub, and L. Reichel (1994). "An Adaptive Chebyshev Iterative Method for Nonsymmetric Linear Systems Based on Modified Moments," *Numer. Math. 67*, 21–40.

Other unsymmetric methods include

M. Eiermann, W. Niethammer, and R.S. Varga (1992). "Acceleration of Relaxation Methods for Non-Hermitian Linear Systems," *SIAM J. Matrix Anal. Appl. 13*, 979–991.

H. Elman and G.H. Golub (1990). "Iterative Methods for Cyclically Reduced Non-Self-Adjoint Linear Systems I," *Math. Comp. 54*, 671–700.

H. Elman and G.H. Golub (1990). "Iterative Methods for Cyclically Reduced Non-Self-Adjoint Linear Systems II," *Math. Comp. 56*, 215–242.

R. Bramley and A. Sameh (1992). "Row Projection Methods for Large Nonsymmetric Linear Systems," *SIAM J. Sci. Statist. Comput. 13*, 168–193.

Sometimes it is possible to "symmetrize" an iterative method, thereby simplifying the acceleration process, since all the relevant eigenvalues are real. This is the case for the SSOR method discussed in

J.W. Sheldon (1955). "On the Numerical Solution of Elliptic Difference Equations," *Math. Tables Aids Comp. 9*, 101–12.

The parallel implementation of the classical iterations has received some attention. See

D.J. Evans (1984). "Parallel SOR Iterative Methods," *Parallel Computing 1*, 3–18.

N. Patel and H. Jordan (1984). "A Parallelized Point Rowwise Successive Over-Relaxation Method on a Multiprocessor," *Parallel Computing 1*, 207–222.

R.J. Plemmons (1986). "A Parallel Block Iterative Scheme Applied to Computations in Structural Analysis," *SIAM J. Alg. and Disc. Methods 7*, 337–347.

C. Kamath and A. Sameh (1989). "A Projection Method for Solving Nonsymmetric Linear Systems on Multiprocessors," *Parallel Computing 9*, 291–312.

We have seen that the condition $\kappa(A)$ is an important issue when direct methods are applied to $Ax = b$. However, the condition of the system also has a bearing on iterative method performance. See

M. Arioli and F. Romani (1985). "Relations Between Condition Numbers and the Convergence of the Jacobi Method for Real Positive Definite Matrices," *Numer. Math. 46*, 31–42.

M. Arioli, I.S. Duff, and D. Ruiz (1992). "Stopping Criteria for Iterative Solvers," *SIAM J. Matrix Anal. Appl. 13*, 138–144.

Iterative methods for singular systems are discussed in

A. Dax (1990). "The Convergence of Linear Stationary Iterative Processes for Solving Singular Unstructured Systems of Linear Equations," *SIAM Review 32*, 611–635.

Finally, the effect of rounding errors on the methods of this section are treated in

H. Wozniakowski (1978). "Roundoff-Error Analysis of Iterations for Large Linear Systems," *Numer. Math. 30*, 301–314.

P.A. Knight (1993). "Error Analysis of Stationary Iteration and Associated Problems," Ph.D. thesis, Department of Mathematics, University of Manchester, England.

10.2 The Conjugate Gradient Method

A difficulty associated with the SOR, Chebyshev semi-iterative, and related methods is that they depend upon parameters that are sometimes hard to choose properly. For example, the Chebyshev acceleration scheme needs good estimates of the largest and smallest eigenvalue of the underlying iteration matrix $M^{-1}N$. Unless this matrix is sufficiently structured, it may be analytically impossible and/or computationally expensive to do this.

In this section, we present a method without this difficulty for the symmetric positive definite $Ax = b$ problem, the well-known Hestenes-Stiefel conjugate gradient method. We derived this method in §9.3.1 from the Lanczos algorithm. The derivation now is from a different point of view and it will set the stage for various important generalizations in §10.3 and §10.4.

10.2.1 Steepest Descent

The starting point in the derivation is to consider how we might go about minimizing the function

$$\phi(x) = \frac{1}{2}x^T A x - x^T b$$

where $b \in \mathbb{R}^n$ and $A \in \mathbb{R}^{n \times n}$ is assumed to be positive definite and symmetric. The minimum value of $\phi(x)$ is $-b^T A^{-1} b / 2$, achieved by setting $x = A^{-1}b$. Thus, minimizing ϕ and solving $Ax = b$ are equivalent problems if A is symmetric positive definite.

One of the simplest strategies for minimizing ϕ is the *method of steepest descent*. At a current point x_c the function ϕ decreases most rapidly in the direction of the negative gradient: $-\nabla \phi(x_c) = b - Ax_c$. We call

$$r_c = b - Ax_c$$

the *residual* of x_c. If the residual is nonzero, then there exists a positive α such that $\phi(x_c + \alpha r_c) < \phi(x_c)$. In the method of steepest descent (with exact line search) we set $\alpha = r_c^T r_c / r_c^T A r_c$ thereby minimizing

$$\phi(x_c + \alpha r_c) = \phi(x_c) - \alpha r_c^T r_c + \frac{1}{2} \alpha^2 r_c^T A r_c.$$

This gives

$$
\begin{aligned}
&x_0 = \text{initial guess} \\
&r_0 = b - A x_0 \\
&k = 0 \\
&\text{while } r_k \neq 0 \\
&\quad k = k + 1 \\
&\quad \alpha_k = r_{k-1}^T r_{k-1} / r_{k-1}^T A r_{k-1} \\
&\quad x_k = x_{k-1} + \alpha_k r_{k-1} \\
&\quad r_k = b - A x_k \\
&\text{end}
\end{aligned}
\qquad (10.2.1)
$$

It can be shown that

$$\left(\phi(x_k) + \frac{1}{2} b^T A^{-1} b \right) \leq \left(1 - \frac{1}{\kappa_2(A)} \right) \left(\phi(x_{k-1}) + \frac{1}{2} b^T A^{-1} b \right) \quad (10.2.2)$$

which implies global convergence. Unfortunately, the rate of convergence may be prohibitively slow if the condition $\kappa_2(A) = \lambda_1(A)/\lambda_n(A)$ is large. Geometrically this means that the level curves of ϕ are very elongated hyperellipsoids and minimization corresponds to finding the lowest point in a relatively flat, steep-sided valley. In steepest descent, we are forced to traverse back and forth *across* the valley rather than *down* the valley. Stated another way, the gradient directions that arise during the iteration are not different enough.

10.2.2 General Search Directions

To avoid the pitfalls of steepest descent, we consider the successive min-imization of ϕ along a set of directions $\{p_1, p_2, \ldots\}$ that do not neces-sarily correspond to the residuals $\{r_0, r_1, \ldots\}$. It is easy to show that $\phi(x_{k-1} + \alpha p_k)$ is minimized by setting

$$\alpha = \alpha_k = p_k^T r_{k-1} / p_k^T A p_k.$$

With this choice it can be shown that

$$\phi(x_{k-1} + \alpha_k p_k) = \phi(x_{k-1}) - \frac{1}{2} \frac{(p_k^T r_{k-1})^2}{p_k^T A p_k}. \qquad (10.2.3)$$

To ensure a reduction in the size of ϕ we insist that p_k *not* be orthogonal to r_{k-1}. This leads to the following framework:

$x_0 =$ initial guess
$r_0 = b - Ax_0$
$k = 0$
while $r_k \neq 0$
    $k = k + 1$ (10.2.4)
    Choose a direction p_k such that $p_k^T r_{k-1} \neq 0$.
    $\alpha_k = p_k^T r_{k-1}/p_k^T A p_k$
    $x_k = x_{k-1} + \alpha_k p_k$
    $r_k = b - A x_k$
end

Note that

$$x_k \in x_0 + \text{span}\{p_1, \ldots, p_k\} \equiv \{x_0 + \gamma_1 p_1 + \cdots + \gamma_k p_k : \gamma_i \in \mathbb{R}\}.$$

Our goal is to choose the search directions in a way that guarantees convergence without the shortcomings of steepest descent.

10.2.3   A-Conjugate Search Directions

If the search directions are linearly independent *and* x_k solves the problem

$$\min_{x \in x_0 + \text{span}\{p_1, \ldots, p_k\}} \phi(x) \tag{10.2.5}$$

for $k = 1, 2, \ldots$, then convergence is guaranteed in at most n steps. This is because x_n minimizes ϕ over $\mathbb{R}^n$ and therefore satisfies $Ax_n = b$.

However, for this to be a viable approach the search directions must have the property that it is "easy" to compute x_k given x_{k-1}. Let us see what this says about the determination of p_k. If

$$x_k = x_0 + P_{k-1} y + \alpha p_k$$

where $P_{k-1} = [\, p_1, \ldots, p_{k-1} \,]$, $y \in \mathbb{R}^{k-1}$, and $\alpha \in \mathbb{R}$, then

$$\phi(x_k) = \phi(x_0 + P_{k-1} y) + \alpha y^T P_{k-1}^T A p_k + \frac{\alpha^2}{2} p_k^T A p_k - \alpha p_k^T r_0.$$

If $p_k \in \text{span}\{Ap_1, \ldots, Ap_{k-1}\}^{\perp}$, then the cross term $\alpha y^T P_{k-1}^T A p_k$ is zero and the search for the minimizing x_k splits into a pair of uncoupled minimizations, one for y and one for α:

$$\min_{x_k \in x_0 + \text{span}\{p_1, \ldots, p_k\}} \phi(x_k) = \min_{y, \alpha} \phi(x_0 + P_{k-1} y + \alpha p_k)$$

$$= \min_{y, \alpha} \left(\phi(x_0 + P_{k-1} y) + \frac{\alpha^2}{2} p_k^T A p_k - \alpha p_k^T r_0 \right)$$

$$= \quad \min_{y} \; \phi(x_0 + P_{k-1}y) \; + \quad \min_{\alpha} \left(\frac{\alpha^2}{2} p_k^T A p_k - \alpha p_k^T r_0 \right).$$

Note that if y_{k-1} solves the first min problem then $x_{k-1} = x_0 + P_{k-1}y_{k-1}$ minimizes ϕ over $x_0 + \text{span}\{p_1, \ldots, p_{k-1}\}$. The solution to the α min problem is given by $\alpha_k = p_k^T r_0 / p_k^T A p_k$. Note that because of A-conjugacy,

$$
\begin{aligned}
p_k^T r_{k-1} &= p_k^T (b - Ax_{k-1}) \\
&= p_k^T (b - A(x_0 + P_{k-1}y_{k-1})) = p_k^T r_0.
\end{aligned}
$$

With these results it follows that $x_k = x_{k-1} + \alpha_k p_k$ and we obtain the following instance of (10.2.4):

> x_0 = initial guess
> $k = 0$
> $r_0 = b - Ax_0$
> while $r_k \neq 0$
> $\quad k = k + 1$
> $\quad$ Choose $p_k \in \text{span}\{Ap_1, \ldots, Ap_{k-1}\}^\perp$ so $p_k^T r_{k-1} \neq 0$. (10.2.6)
> $\quad \alpha_k = p_k^T r_{k-1} / p_k^T A p_k$
> $\quad x_k = x_{k-1} + \alpha_k p_k$
> $\quad r_k = b - Ax_k$
> end

The following lemma shows that it is possible to find the search directions with the required properties.

Lemma 10.2.1 *If $r_{k-1} \neq 0$, then there exists a $p_k \in \text{span}\{Ap_1, \ldots, Ap_{k-1}\}^\perp$ such that $p_k^T r_{k-1} \neq 0$.*

Proof. For the case $k = 1$, set $p_1 = r_0$. If $k > 1$, then since $r_{k-1} \neq 0$ it follows that

$$
\begin{aligned}
A^{-1}b \notin x_0 + \text{span}\{p_1, \ldots, p_{k-1}\} \quad &\Rightarrow \quad b \notin Ax_0 + \text{span}\{Ap_1, \ldots, Ap_{k-1}\} \\
&\Rightarrow \quad r_0 \notin \text{span}\{Ap_1, \ldots, Ap_{k-1}\}.
\end{aligned}
$$

Thus there exists a $p \in \text{span}\{Ap_1, \ldots, Ap_{k-1}\}^\perp$ such that $p^T r_0 \neq 0$. But $x_{k-1} \in x_0 + \text{span}\{p_1, \ldots, p_{k-1}\}$ and so $r_{k-1} \in r_0 + \text{span}\{Ap_1, \ldots, Ap_{k-1}\}$. It follows that $p^T r_{k-1} = p^T r_0 \neq 0$. $\square$

The search directions in (10.2.6) are said to be *A-conjugate* because $p_i^T A p_j = 0$ for all $i \neq j$. Note that if $P_k = [p_1, \ldots, p_k]$ is the matrix of these vectors, then

$$P_k^T A P_k = \text{diag}(p_1^T A p_1, \ldots, p_k^T A p_k)$$

is nonsingular since A is positive definite and the search directions are nonzero. It follows that P_k has full column rank. This guarantees convergence in (10.2.6) in at most n steps because x_n (if we get that far) minimizes $\phi(x)$ over $\text{ran}(P_n) = \mathbb{R}^n$.

10.2.4   Choosing a Best Search Direction

A way to combine the positive aspects of steepest descent and A-conjugate searching is to choose p_k in (10.2.6) to be the closest vector to r_{k-1} that is A-conjugate to $p_1, \dots, p_{k-1}$. This defines "version zero" of the *method of conjugate gradients*:

$$
\begin{aligned}
&x_0 = \text{initial guess} \\
&k = 0 \\
&r_0 = b - Ax_0 \\
&\textbf{while } r_k \neq 0 \\
&\qquad k = k + 1 \\
&\qquad \textbf{if } k = 1 \\
&\qquad\qquad p_1 = r_0 \\
&\qquad \textbf{else} \\
&\qquad\qquad \text{Let } p_k \text{ minimize } \| \, p - r_{k-1} \, \|_2 \text{ over all vectors} \\
&\qquad\qquad\qquad p \in \text{span}\{Ap_1, \dots, Ap_{k-1}\}^{\perp} \\
&\qquad \textbf{end} \\
&\qquad \alpha_k = p_k^T r_{k-1} / p_k^T A p_k \\
&\qquad x_k = x_{k-1} + \alpha_k p_k \\
&\qquad r_k = b - A x_k \\
&\textbf{end} \\
&x = x_k
\end{aligned}
\tag{10.2.7}
$$

To make this an effective sparse $Ax = b$ solver, we need an efficient method for computing p_k. A considerable amount of analysis is required to develop the final recursions. The first step is to show that p_k is the minimum residual of a certain least squares problem.

Lemma 10.2.2 *For $k \geq 2$ the vectors p_k generated by (10.2.7) satisfy*

$$
p_k = r_{k-1} - A P_{k-1} z_{k-1},
$$

where $P_{k-1} = [p_1, \dots, p_{k-1}]$ and z_{k-1} solves $\displaystyle \min_{z \in \mathbb{R}^{k-1}} \| \, r_{k-1} - A P_{k-1} z \, \|_2$.

Proof. Suppose z_{k-1} solves the above LS problem and let p be the associated minimum residual:

$$
p = r_{k-1} - A P_{k-1} z_{k-1}.
$$

It follows that $p^T A P_{k-1} = 0$. Moreover, $p = [I - (AP_{k-1})(AP_{k-1})^{+}]r_{k-1}$ is the orthogonal projection of r_{k-1} into $\text{ran}(AP_{k-1})^{\perp}$ and so it is the closest vector in $\text{ran}(AP_{k-1})^{\perp}$ to r_{k-1}. Thus, $p = p_k$. $\square$

With this result we can establish a number of important relationships between the residuals r_k, the search directions p_k, and the Krylov subspaces

$$\mathcal{K}(r_0, A, k) = \text{span}\{r_0, Ar_0, \ldots, A^{k-1}r_0\}.$$

Theorem 10.2.3 *After k iterations in (10.2.7) we have*

$$r_k = r_{k-1} - \alpha_k A p_k \qquad (10.2.8)$$
$$P_k^T r_k = 0 \qquad (10.2.9)$$
$$\text{span}\{p_1, \ldots, p_k\} = \text{span}\{r_0, \ldots, r_{k-1}\} = \mathcal{K}(r_0, A, k) \quad (10.2.10)$$

and the residuals $r_0, \ldots, r_k$ *are mutually orthogonal.*

Proof. Equation (10.2.8) follows by applying A to both sides of $x_k = x_{k-1} + \alpha_k p_k$ and using the definition of the residual.

To prove (10.2.9), we recall that $x_k = x_0 + P_k y_k$ where y_k is the minimizer of

$$\phi(x_0 + P_k y) = \phi(x_0) + \frac{1}{2} y^T (P_k^T A P_k) y - y^T P_k (b - A x_0).$$

But this means that y_k solves the linear system $(P_k^T A P_k) y = P_k^T (b - A x_0)$. Thus

$$0 = P_k^T (b - A x_0) - P_k^T A P_k y_k = P_k^T (b - A(x_0 + P_k y_k)) = P_k^T r_k.$$

To prove (10.2.10) we note from (10.2.8) that

$$\{A p_1, \ldots, A p_{k-1}\} \subseteq \text{span}\{r_0, \ldots, r_{k-1}\}$$

and so from Lemma 10.2.2,

$$p_k = r_{k-1} - [A p_1, \ldots, A p_{k-1}] z_{k-1} \in \text{span}\{r_0, \ldots, r_{k-1}\}$$

It follows that

$$[p_1, \ldots, p_k] = [r_0, \ldots, r_{k-1}] T$$

for some upper triangular T. Since the search directions are independent, T is nonsingular. This shows

$$\text{span}\{p_1, \ldots, p_k\} = \text{span}\{r_0, \ldots, r_{k-1}\}.$$

Using (10.2.8) we see that

$$r_k \in \text{span}\{r_{k-1}, A p_k\} \subseteq \text{span}\{r_{k-1}, A r_0, \ldots, A r_{k-1}\}.$$

The Krylov space connection in (10.2.10) follows from this by induction.

Finally, to establish the mutual orthogonality of the residuals, we note from (10.2.9) that r_k is orthogonal to any vector in the range of P_k. But from (10.2.10) this subspace contains $r_0, \ldots, r_{k-1}$. □

Using these facts we next show that p_k is a simple linear combination of its predecessor p_{k-1} and the "current" residual r_{k-1}.

Corollary 10.2.4 *The residuals and search directions in (10.2.7) have the property that $p_k \in \text{span}\{p_{k-1}, r_{k-1}\}$ for $k \geq 2$.*

Proof. If $k = 2$, then from (10.2.10) $p_2 \in \text{span}\{r_0, r_1\}$. But $p_1 = r_0$ and so p_2 is a linear combination of p_1 and r_1.

If $k > 2$, then partition the vector z_{k-1} of Lemma 10.2.2 as

$$z_{k-1} = \begin{bmatrix} w \\ \mu \end{bmatrix} \begin{matrix} k-2 \\ 1 \end{matrix}.$$

Using the identity $r_{k-1} = r_{k-2} - \alpha_{k-1} A p_{k-1}$, we see that

$$\begin{aligned} p_k &= r_{k-1} - AP_{k-1}z_{k-1} = r_{k-1} - AP_{k-2}w - \mu A p_{k-1} \\ &= \left(1 + \frac{\mu}{\alpha_{k-1}}\right) r_{k-1} + s_{k-1} \end{aligned}$$

where

$$\begin{aligned} s_{k-1} &= -\frac{\mu}{\alpha_{k-1}} r_{k-2} - AP_{k-2}w \\ &\in \text{span}\{r_{k-2}, AP_{k-2}w\} \\ &\subseteq \text{span}\{r_{k-2}, Ap_1, \ldots, Ap_{k-2}\} \\ &\subseteq \text{span}\{r_1, \ldots, r_{k-2}\} \end{aligned}$$

Because the r_i are mutually orthogonal, it follows that s_{k-1} and r_{k-1} are orthogonal to each other. Thus, the least squares problem of Lemma 10.2.2 boils down to choosing μ and w such that

$$\| p_k \|_2^2 = \left(1 + \frac{\mu}{\alpha_{k-1}}\right)^2 \| r_{k-1} \|_2^2 + \| s_{k-1} \|_2^2$$

is minimum. Since the 2-norm of $r_{k-2} - AP_{k-2}z$ is minimized by z_{k-2} giving residual p_{k-1}, it follows that s_{k-1} is a multiple of p_{k-1}. Consequently, $p_k \in \text{span}\{r_{k-1}, p_{k-1}\}$. $\square$

We are now set to derive a very simple expression for p_k. Without loss of generality we may assume from Corollary 10.2.4 that

$$p_k = r_{k-1} + \beta_k p_{k-1}.$$

Since $p_{k-1}^T A p_k = 0$ it follows that

$$\beta_k = -\frac{p_{k-1}^T A r_{k-1}}{p_{k-1}^T A p_{k-1}}$$

This leads us to "version 1" of the conjugate gradient method:

$x_0 = $ initial guess
$k = 0$
$r_0 = b - Ax_0$
while $r_k \neq 0$
    $k = k + 1$
    if $k = 1$
        $p_1 = r_0$
    else
$$\beta_k = -p_{k-1}^T A r_{k-1}/p_{k-1}^T A p_{k-1}$$
$$p_k = r_{k-1} + \beta_k p_{k-1} \qquad (10.2.11)$$
    end
    $\alpha_k = p_k^T r_{k-1}/p_k^T A p_k$
    $x_k = x_{k-1} + \alpha_k p_k$
    $r_k = b - Ax_k$
end
$x = x_k$

In this implementation, the method requires three separate matrix-vector multiplications per step. However, by computing residuals recursively via $r_k = r_{k-1} - \alpha_k A p_k$ and substituting

$$r_{k-1}^T r_{k-1} = -\alpha_{k-1} r_{k-1}^T A p_{k-1} \qquad (10.2.12)$$

and

$$r_{k-2}^T r_{k-2} = \alpha_{k-1} p_{k-1}^T A p_{k-1} \qquad (10.2.13)$$

into the formula for β_k, we obtain the following more efficient version:

Algorithm 10.2.1 [Conjugate Gradients] If $A \in \mathbb{R}^{n \times n}$ is symmetric positive definite, $b \in \mathbb{R}^n$, and $x_0 \in \mathbb{R}^n$ is an initial guess $(Ax_0 \approx b)$, then the following algorithm computes $x \in \mathbb{R}^n$ so $Ax = b$.

$k = 0$
$r_0 = b - Ax_0$
while $r_k \neq 0$
    $k = k + 1$
    if $k = 1$
        $p_1 = r_0$
    else
$$\beta_k = r_{k-1}^T r_{k-1}/r_{k-2}^T r_{k-2}$$
$$p_k = r_{k-1} + \beta_k p_{k-1}$$
    end
    $\alpha_k = r_{k-1}^T r_{k-1}/p_k^T A p_k$
    $x_k = x_{k-1} + \alpha_k p_k$
    $r_k = r_{k-1} - \alpha_k A p_k$
end
$x = x_k$

This procedure is essentially the form of the conjugate gradient algorithm that appears in the original paper by Hestenes and Stiefel (1952). Note that only one matrix-vector multiplication is required per iteration.

10.2.5   The Lanczos Connection

In §9.3.1 we derived the conjugate gradient method from the Lanczos algorithm. Now let us look at the connections between these two algorithms in the reverse direction by "deriving" the Lanczos process from conjugate gradients. Define the matrix of residuals $R_k \in \mathbb{R}^{n \times k}$ by

$$R_k = [r_0, \ldots, r_{k-1}]$$

and the upper bidiagonal matrix $B_k \in \mathbb{R}^{k \times k}$ by

$$B_k = \begin{bmatrix} 1 & -\beta_2 & 0 & \cdots & 0 \\ 0 & 1 & -\beta_3 & & \vdots \\ & \ddots & \ddots & \ddots & 0 \\ \vdots & & \ddots & \ddots & -\beta_k \\ 0 & \cdots & & 0 & 1 \end{bmatrix}.$$

From the equations $p_i = r_{i-1} + \beta_i p_{i-1}$, $i = 2{:}k$, and $p_1 = r_0$ it follows that $R_k = P_k B_k$. Since the columns of $P_k = [p_1, \ldots, p_k]$ are A-conjugate, we see that $R_k^T A R_k = B_k^T \operatorname{diag}(p_1^T A p_1, \ldots, p_k^T A p_k) B_k$ is tridiagonal. From (10.2.10) it follows that if

$$\Delta = \operatorname{diag}(\rho_0, \ldots, \rho_{k-1}) \qquad \rho_i = \| r_i \|_2$$

then the columns of $R_k \Delta^{-1}$ form an orthonormal basis for the subspace span$\{r_0, A r_0, \ldots, A^{k-1} r_0\}$. Consequently, the columns of this matrix are essentially the Lanczos vectors of Algorithm 9.3.1, i.e.,

$$q_i = \pm r_{i-1}/\rho_{i-1} \qquad i = 1{:}k.$$

Moreover, the tridiagonal matrix associated with these Lanczos vectors is given by

$$T_k = \Delta^{-1} B_k^T \operatorname{diag}(p_i^T A p_i) B_k \Delta^{-1}. \tag{10.2.14}$$

The diagonal and subdiagonal of this matrix involve quantities that are readily available during the conjugate gradient iteration. Thus, we can obtain good estimates of A's extremal eigenvalues (and condition number) as we generate the x_k in Algorithm 10.2.1.

10.2.6 Some Practical Details

The termination criteria in Algorithm 10.2.1 is unrealistic. Rounding errors lead to a loss of orthogonality among the residuals and finite termination is not mathematically guaranteed. Moreover, when the conjugate gradient method is applied, n is usually so big that $O(n)$ iterations represents an unacceptable amount of work. As a consequence of these observations, it is customary to regard the method as a genuinely iterative technique with termination based upon an iteration maximum k_{max} and the residual norm. This leads to the following practical version of Algorithm 10.2.1:

$$
\begin{aligned}
&x = \text{initial guess} \\
&k = 0 \\
&r = b - Ax_0 \\
&\rho_0 = \| r \|_2^2 \\
&\textbf{while } (\sqrt{\rho_k} > \epsilon \| b \|_2) \wedge (k < k_{max}) \\
&\quad k = k + 1 \\
&\quad \textbf{if } k = 1 \\
&\quad\quad p = r \\
&\quad \textbf{else} \\
&\quad\quad \beta_k = \rho_{k-1}/\rho_{k-2} \\
&\quad\quad p = r + \beta_k p \\
&\quad \textbf{end} \\
&\quad w = Ap \\
&\quad \alpha_k = \rho_{k-1}/p^T w \\
&\quad x = x + \alpha_k p \\
&\quad r = r - \alpha_k w \\
&\quad \rho_k = \| r \|_2^2 \\
&\textbf{end}
\end{aligned}
\tag{10.2.16}
$$

This algorithm requires one matrix-vector multiplication and 10n flops per iteration. Notice that just four n-vectors of storage are essential: x, r, p, and w. The subscripting of the scalars is not necessary and is only done here to facilitate comparison with Algorithm 10.2.1.

It is also possible to base the termination criteria on heuristic estimates of the error $A^{-1}r_k$ by approximating $\| A^{-1} \|_2$ with the reciprocal of the smallest eigenvalue of the tridiagonal matrix T_k given in (10.2.14).

The idea of regarding conjugate gradients as an iterative method began with Reid (1971). The iterative point of view is useful but then the *rate* of convergence is central to the method's success.

10.2.7 Convergence Properties

We conclude this section by examining the convergence of the conjugate gradient iterates $\{x_k\}$. Two results are given and they both say that the method performs well when A is near the identity either in the sense of a low rank perturbation or in the sense of norm.

Theorem 10.2.5 *If $A = I + B$ is an n-by-n symmetric positive definite matrix and rank$(B) = r$ then Algorithm 10.2.1 converges in at most $r + 1$ steps.*

Proof. The dimension of

$$\text{span}\{r_0, Ar_0, \ldots, A^{k-1}r_0\} = \text{span}\{r_0, Br_0, \ldots, B^{k-1}r_0\}$$

cannot exceed $r + 1$. Since $p_1, \ldots, p_k$ span this subspace and are independent, the iteration cannot progress beyond $r + 1$ steps. $\square$

An important metatheorem follows from this result:

- If A is close to a rank r correction to the identity, then Algorithm 10.2.1 almost converges after $r + 1$ steps.

We show how this heuristic can be exploited in the next section.

An error bound of a different flavor can be obtained in terms of the A-norm which we define as follows:

$$\| w \|_A = \sqrt{w^T A w}.$$

Theorem 10.2.6 *Suppose $A \in \mathbb{R}^{n \times n}$ is symmetric positive definite and $b \in \mathbb{R}$. If Algorithm 10.2.1 produces iterates $\{x_k\}$ and $\kappa = \kappa_2(A)$ then*

$$\| x - x_k \|_A \le 2\| x - x_0 \|_A \left(\frac{\sqrt{\kappa} - 1}{\sqrt{\kappa} + 1} \right)^k.$$

Proof. See Luenberger (1973, p.187). $\square$

The accuracy of the $\{x_k\}$ is often much better than this theorem predicts. However, a heuristic version of Theorem 10.2.6 turns out to be very useful:

- The conjugate gradient method converges very fast in the A-norm if $\kappa_2(A) \approx 1$.

In the next section we show how we can sometimes convert a given $Ax = b$ problem into a related $\tilde{A}\tilde{x} = \tilde{b}$ problem with $\tilde{A}$ being close to the identity.

Problems

P10.2.1 Verify that the residuals in (10.2.1) satisfy $r_i^T r_j = 0$ whenever $j = i + 1$.

P10.2.2 Verify (10.2.2).

P10.2.3 Verify (10.2.3).

P10.2.4 Verify (10.2.12) and (10.2.13).

P10.2.5 Give formula for the entries of the tridiagonal matrix T_k in (10.2.14).

P10.2.6 Compare the work and storage requirements associated with the practical implementation of Algorithms 9.3.1 and 10.2.1.

P10.2.7 Show that if $A \in \mathbb{R}^{n \times n}$ is symmetric positive definite and has k distinct eigenvalues, then the conjugate gradient method does not require more than $k + 1$ steps to converge.

P10.2.8 Use Theorem 10.2.6 to verify that

$$\| x_k - A^{-1}b \|_2 \leq 2\sqrt{\kappa} \left(\frac{\sqrt{\kappa} - 1}{\sqrt{\kappa} + 1} \right)^k \| x_0 - A^{-1}b \|_2.$$

Notes and References for Sec. 10.2

The conjugate gradient method is a member of a larger class of methods that are referred to as *conjugate direction* algorithms. In a conjugate direction algorithm the search directions are all *B*-conjugate for some suitably chosen matrix *B*. A discussion of these methods appears in

J.E. Dennis Jr. and K. Turner (1987). "Generalized Conjugate Directions," *Lin. Alg. and Its Applic. 88/89*, 187–209.

G.W. Stewart (1973). "Conjugate Direction Methods for Solving Systems of Linear Equations," *Numer. Math. 21*, 284–97.

Some historical and unifying perspectives are offered in

G. Golub and D. O'Leary (1989). "Some History of the Conjugate Gradient and Lanczos Methods," *SIAM Review 31*, 50–102.

M.R. Hestenes (1990). "Conjugacy and Gradients," in *A History of Scientific Computing*, Addison-Wesley, Reading, MA.

S. Ashby, T.A. Manteuffel, and P.E. Saylor (1992). "A Taxonomy for Conjugate Gradient Methods," *SIAM J. Numer. Anal. 27*, 1542–1568.

The classic reference for the conjugate gradient method is

M.R. Hestenes and E. Stiefel (1952). "Methods of Conjugate Gradients for Solving Linear Systems," *J. Res. Nat. Bur. Stand. 49*, 409–36.

An exact arithmetic analysis of the method may be found in chapter 2 of

M.R. Hestenes (1980). *Conjugate Direction Methods in Optimization*, Springer-Verlag, Berlin.

See also

O. Axelsson (1977). "Solution of Linear Systems of Equations: Iterative Methods," in *Sparse Matrix Techniques: Copenhagen, 1976*, ed. V.A. Barker, Springer-Verlag, Berlin.

For a discussion of conjugate gradient convergence behavior, see

D. G. Luenberger (1973). *Introduction to Linear and Nonlinear Programming*, Addison-Wesley, New York.

A. van der Sluis and H.A. Van Der Vorst (1986). "The Rate of Convergence of Conjugate Gradients," *Numer. Math. 48*, 543–560.

The idea of using the conjugate gradient method as an iterative method was first discussed in

J.K. Reid (1971). " On the Method of Conjugate Gradients for the Solution of Large Sparse Systems of Linear Equations," in *Large Sparse Sets of Linear Equations* , ed. J.K. Reid, Academic Press, New York, pp. 231–54.

Several authors have attempted to explain the algorithm's behavior in finite precision arithmetic. See

H. Wozniakowski (1980). "Roundoff Error Analysis of a New Class of Conjugate Gradient Algorithms," *Lin. Alg. and Its Applic. 29,*
A. Greenbaum and Z. Strakos (1992). "Predicting the Behavior of Finite Precision lanczos and Conjugate Gradient Computations," *SIAM J. Matrix Ana. Applic. 13,* 121–137.

See also the analysis in

G.W. Stewart (1975). "The Convergence of the Method of Conjugate Gradients at Isolated Extreme Points in the Spectrum," *Numer. Math. 24,* 85–93.
A. Jennings (1977). "Influence of the Eigenvalue Spectrum on the Convergence Rate of the Conjugate Gradient Method," *J. Inst. Math. Applic. 20,* 61–72.
J. Cullum and R. Willoughby (1980). "The Lanczos Phenomena: An Interpretation Based on Conjugate Gradient Optimization," *Lin. Alg. and Its Applic. 29,* 63–90.

Finally, we mention that the method can be used to compute an eigenvector of a large sparse symmetric matrix:

A. Ruhe and T. Wiberg (1972). "The Method of Conjugate Gradients Used in Inverse Iteration," *BIT 12,* 543–54.

10.3   Preconditioned Conjugate Gradients

We concluded the previous section by observing that the method of conjugate gradients works well on matrices that are either well conditioned or have just a few distinct eigenvalues. (The latter being the case when A is a lower rank perturbation of the identity.) In this section we show how to *precondition* a linear system so that the matrix of coefficients assumes one of these nice forms. Our treatment is quite brief and informal. Golub and Meurant (1983) and Axelsson (1985) have more comprehensive expositions.

10.3.1   Derivation

Consider the n-by-n symmetric positive definite linear system $Ax = b$. The idea behind preconditioned conjugate gradients is to apply the "regular" conjugate gradient method to the transformed system

$$\tilde{A}\tilde{x} = \tilde{b}, \qquad (10.3.1)$$

where $\tilde{A} = C^{-1}AC^{-1}$, $\tilde{x} = Cx$, $\tilde{b} = C^{-1}b$, and C is symmetric positive definite. In view of our remarks in §10.2.8, we should try to choose C

so that $\tilde{A}$ is well conditioned or a matrix with clustered eigenvalues. For reasons that will soon emerge, the matrix C^2 must also be "simple."

If we apply Algorithm 10.2.1 to (10.3.1), then we obtain the iteration

$$
\begin{aligned}
&k = 0\\
&\tilde{x}_0 = \text{initial guess } (\tilde{A}\tilde{x}_0 \approx \tilde{b})\\
&\tilde{r}_0 = \tilde{b} - \tilde{A}\tilde{x}_0\\
&\textbf{while } \tilde{r}_k \neq 0\\
&\qquad k = k + 1\\
&\qquad \textbf{if } k = 1\\
&\qquad\qquad \tilde{p}_1 = \tilde{r}_0\\
&\qquad \textbf{else}\\
&\qquad\qquad \beta_k = \tilde{r}_{k-1}^T \tilde{r}_{k-1} / \tilde{r}_{k-2}^T \tilde{r}_{k-2}\\
&\qquad\qquad \tilde{p}_k = \tilde{r}_{k-1} + \beta_k \tilde{p}_{k-1}\\
&\qquad \textbf{end}\\
&\qquad \alpha_k = \tilde{r}_{k-1}^T \tilde{r}_{k-1} / \tilde{p}_k^T C^{-1} A C^{-1} \tilde{p}_k\\
&\qquad \tilde{x}_k = \tilde{x}_{k-1} + \alpha_k \tilde{p}_k\\
&\qquad \tilde{r}_k = \tilde{r}_{k-1} - \alpha_k C^{-1} A C^{-1} \tilde{p}_k\\
&\textbf{end}\\
&\tilde{x} = \tilde{x}_k
\end{aligned}
\tag{10.3.2}
$$

Here, $\tilde{x}_k$ should be regarded as an approximation to $\tilde{x}$ and $\tilde{r}_k$ is the residual in the transformed coordinates, i.e., $\tilde{r}_k = \tilde{b} - \tilde{A}\tilde{x}_k$. Of course, once we have $\tilde{x}$ then we can obtain x via the equation $x = C^{-1}\tilde{x}$. However, it is possible to avoid explicit reference to the matrix C^{-1} by defining $\tilde{p}_k = Cp_k$, $\tilde{x}_k = Cx_k$, and $\tilde{r}_k = C^{-1}r_k$. Indeed, if we substitute these definitions into (10.3.2) and recall that $\tilde{b} = C^{-1}b$ and $\tilde{x} = Cx$, then we obtain

$$
\begin{aligned}
&k = 0\\
&x_0 = \text{initial guess } (Ax_0 \approx b)\\
&r_0 = b - Ax_0\\
&\textbf{while } C^{-1}r_k \neq 0\\
&\qquad k = k + 1\\
&\qquad \textbf{if } k = 1\\
&\qquad\qquad Cp_1 = C^{-1}r_0\\
&\qquad \textbf{else}\\
&\qquad\qquad \beta_k = (C^{-1}r_{k-1})^T(C^{-1}r_{k-1})/(C^{-1}r_{k-2})^T(C^{-1}r_{k-2})\\
&\qquad\qquad Cp_k = C^{-1}r_{k-1} + \beta_k Cp_{k-1}\\
&\qquad \textbf{end}\\
&\qquad \alpha_k = (C^{-1}r_{k-1})^T(C^{-1}r_{k-1})/(Cp_k)^T(C^{-1}AC^{-1})(Cp_k)\\
&\qquad Cx_k = Cx_{k-1} + \alpha_k Cp_k\\
&\qquad C^{-1}r_k = C^{-1}r_{k-1} - \alpha_k(C^{-1}AC^{-1})Cp_k\\
&\textbf{end}\\
&Cx = Cx_k
\end{aligned}
\tag{10.3.3}
$$

If we define the *preconditioner* M by $M = C^2$ (also positive definite) and let z_k be the solution of the system $Mz_k = r_k$ then (10.3.3) simplifies to

Algorithm 10.3.1 [Preconditioned Conjugate Gradients] Given a symmetric positive definite $A \in \mathbb{R}^{n \times n}$, $b \in \mathbb{R}^n$, a symmetric positive definite preconditioner M, and an initial guess x_0 ($Ax_0 \approx b$), the following algorithm solves the linear system $Ax = b$.

$k = 0$
$r_0 = b - Ax_0$
while ($r_k \neq 0$)
 Solve $Mz_k = r_k$.
 $k = k + 1$
 if $k = 1$
 $p_1 = z_0$
 else
 $\beta_k = r_{k-1}^T z_{k-1} / r_{k-2}^T z_{k-2}$
 $p_k = z_{k-1} + \beta_k p_{k-1}$
 end
 $\alpha_k = r_{k-1}^T z_{k-1} / p_k^T A p_k$
 $x_k = x_{k-1} + \alpha_k p_k$
 $r_k = r_{k-1} - \alpha_k A p_k$
end
$x = x_k$

A number of important observations should be made about this procedure:

- It can be shown that the residuals and search directions satisfy

$$r_j^T M^{-1} r_i = 0 \qquad i \neq j \qquad (10.3.4)$$

$$p_j^T (C^{-1} A C^{-1}) p_i = 0 \qquad i \neq j \qquad (10.3.5)$$

- The denominators $r_{k-2}^T z_{k-2} = z_{k-2}^T M z_{k-2}$ never vanish because M is positive definite.

- Although the transformation C figured heavily in the derivation of the algorithm, its action is only felt through the preconditioner $M = C^2$.

- For Algorithm 10.3.1 to be an effective sparse matrix technique, linear systems of the form $Mz = r$ must be easily solved *and* convergence must be rapid.

The choice of a good preconditioner can have a dramatic effect upon the rate of convergence. Some of the possibilities are now discussed.

10.3.2   Incomplete Cholesky Preconditioners

One of the most important preconditioning strategies involves computing an *incomplete Cholesky factorization* of A. The idea behind this approach is to calculate a lower triangular matrix H with the property that H has some tractable sparsity structure and is somehow "close" to A's exact Cholesky factor G. The preconditioner is then taken to be $M = HH^T$. To appreciate this choice consider the following facts:

- There exists a unique symmetric positive definite matrix C such that $M = C^2$.

- There exists an orthogonal Q such that $C = QH^T$, i.e., H^T is the upper triangular factor of a QR factorization of C.

We therefore obtain the heuristic

$$\begin{aligned} \tilde{A} &= C^{-1}AC^{-1} = C^{-T}AC^{-1} \\ &= (HQ^T)^{-1}A(QH^T)^{-1} = Q(H^{-1}GG^TH^{-T})Q^T \approx I \end{aligned} \tag{10.3.6}$$

Thus, the better H approximates G the smaller the condition of $\tilde{A}$, and the better the performance of Algorithm 10.3.1.

An easy but effective way to determine such a simple H that approximates G is to step through the Cholesky reduction setting h_{ij} to zero if the corresponding a_{ij} is zero. Pursuing this with the outer product version of Cholesky we obtain

$$\begin{aligned} &\textbf{for } k = 1{:}n \\ &\quad A(k,k) = \sqrt{A(k,k)} \\ &\quad \textbf{for } i = k+1{:}n \\ &\quad\quad \textbf{if } A(i,k) \neq 0 \\ &\quad\quad\quad A(i,k) = A(i,k)/A(k,k) \\ &\quad\quad \textbf{end} \\ &\quad \textbf{end} \\ &\quad \textbf{for } j = k+1{:}n \\ &\quad\quad \textbf{for } i = j{:}n \\ &\quad\quad\quad \textbf{if } A(i,j) \neq 0 \\ &\quad\quad\quad\quad A(i,j) = A(i,j) - A(i,k)A(j,k) \\ &\quad\quad\quad \textbf{end} \\ &\quad\quad \textbf{end} \\ &\quad \textbf{end} \\ &\textbf{end} \end{aligned} \tag{10.3.7}$$

In practice, the matrix A and its incomplete Cholesky factor H would be stored in an appropriate data structure and the looping in the above algorithm would take on a very special appearance.

Unfortunately, (10.3.7) is not always stable. Classes of positive definite matrices for which incomplete Cholesky is stable are identified in Manteuffel (1979). See also Elman (1986).

10.3.3   Incomplete Block Preconditioners

As with just about everything else in this book, the incomplete factorization ideas outlined in the previous subsection have a block analog. We illustrate this by looking at the *incomplete block Cholesky factorization* of the symmetric, positive definite, block tridiagonal matrix

$$
A \; = \; \begin{bmatrix} A_1 & E_1^T & 0 \\ E_1 & A_2 & E_2^T \\ 0 & E_2 & A_3 \end{bmatrix} .
$$

For purposes of illustration, we assume that the A_i are tridiagonal and the E_i are diagonal. Matrices with this structure arise from the standard 5-point discretization of self-adjoint elliptic partial differential equations over a two-dimensional domain.

The 3-by-3 case is sufficiently general. Our discussion is based upon Concus, Golub, and Meurant (1985). Let

$$
G \; = \; \begin{bmatrix} G_1 & 0 & 0 \\ F_1 & G_2 & 0 \\ 0 & F_2 & G_3 \end{bmatrix}
$$

be the exact block Cholesky factor of A. Although G is sparse as a block matrix, the individual blocks are dense with the exception of G_1. This can be seen from the required computations:

$$
\begin{aligned}
G_1 G_1^T &= B_1 \equiv A_1 \\
F_1 &= E_1 G_1^{-1} \\
G_2 G_2^T &= B_2 \equiv A_2 - F_1 F_1^T = A_2 - E_1 B_1^{-1} E_1^T \\
F_2 &= E_2 G_2^{-1} \\
G_3 G_3^T &= B_3 \equiv A_3 - F_2 F_2^T = A_3 - E_2 B_2^{-1} E_2^T
\end{aligned}
$$

We therefore seek an approximate block Cholesky factor of the form

$$
\tilde{G} \; = \; \begin{bmatrix} \tilde{G}_1 & 0 & 0 \\ \tilde{F}_1 & \tilde{G}_2 & 0 \\ 0 & \tilde{F}_2 & \tilde{G}_3 \end{bmatrix}
$$

so that we can easily solve systems that involve the preconditioner $M = \tilde{G}\tilde{G}^T$. This involves the imposition of sparsity on $\tilde{G}$'s blocks and here is a reasonable approach given that the A_i are tridiagonal and the E_i are diagonal:

$$
\begin{aligned}
\tilde{G}_1 \tilde{G}_1^T &= \tilde{B}_1 \equiv A_1 \\
\tilde{F}_1 &= E_1 \tilde{G}_1^{-1} \\
\tilde{G}_2 \tilde{G}_2^T &= \tilde{B}_2 \equiv A_2 - E_1 \Lambda_1 E_1^T, \quad \Lambda_1 \text{ (tridiagonal)} \approx \tilde{B}_1^{-1} \\
\tilde{F}_2 &= E_2 \tilde{G}_2^{-1} \\
\tilde{G}_3 \tilde{G}_3^T &= \tilde{B}_3 \equiv A_3 - E_2 \Lambda_2 E_2^T, \quad \Lambda_2 \text{ (tridiagonal)} \approx \tilde{B}_2^{-1}
\end{aligned}
$$

Note that all the $\tilde{B}_i$ are tridiagonal. Clearly, the Λ_i must be carefully chosen to ensure that the $\tilde{B}_i$ are also symmetric and positive definite. It then follows that the $\tilde{G}_i$ are lower bidiagonal. The $\tilde{F}_i$ are full, but they need not be explicitly formed. For example, in the course of solving the system $Mz = r$ we must solve a system of the form

$$
\begin{bmatrix} \tilde{G}_1 & 0 & 0 \\ \tilde{F}_1 & \tilde{G}_2 & 0 \\ 0 & \tilde{F}_2 & \tilde{G}_3 \end{bmatrix} \begin{bmatrix} w_1 \\ w_2 \\ w_3 \end{bmatrix} = \begin{bmatrix} r_1 \\ r_2 \\ r_3 \end{bmatrix}.
$$

Forward elimination can be used to carry out matrix-vector products that involve the $\tilde{F}_i = E_i \tilde{G}_i^{-1}$:

$$
\begin{aligned}
\tilde{G}_1 w_1 &= r_1 \\
\tilde{G}_2 w_2 &= r_2 - \tilde{F}_1 w_1 = r_2 - E_1 \tilde{G}_1^{-1} w_1 \\
\tilde{G}_3 w_3 &= r_3 - \tilde{F}_2 w_2 = r_3 - E_2 \tilde{G}_2^{-1} w_2
\end{aligned}
$$

The choice of Λ_i is delicate as the resulting $\tilde{B}_i$ must be positive definite. As we have organized the computation, the central issue is how to approximate the inverse of an m-by-m symmetric, positive definite, tridiagonal matrix $T = (t_{ij})$ with a symmetric tridiagonal matrix Λ. There are several reasonable approaches:

- Set $\Lambda = \operatorname{diag}(1/t_{11}, \ldots, 1/t_{nn})$.

- Take Λ to be the tridiagonal part of T^{-1}. This can be efficiently computed since there exist $u, v \in \mathbb{R}^m$ such that the lower triangular part of T^{-1} is the lower triangular part of uv^T. See Asplund(1959).

- Set $\Lambda = U^T U$ where U is the lower bidiagonal portion of G^{-1} where $T = GG^T$ is the Cholesky factorization. This can be found in $O(m)$ flops.

For a discussion of these approximations and what they imply about the associated preconditioners, see Concus, Golub, and Meurant (1985).

10.3.4   Domain Decomposition Ideas

The numerical solution of elliptic partial differential equations often leads to linear systems of the form

$$
\begin{bmatrix}
A_1 & \cdots & & \cdots & B_1 \\
\vdots & A_2 & & & B_2 \\
& & \ddots & & \vdots \\
\vdots & & & A_p & B_p \\
B_1^T & B_2^T & \cdots & B_p^T & Q
\end{bmatrix}
\begin{bmatrix}
x_1 \\ x_2 \\ \vdots \\ x_p \\ z
\end{bmatrix}
=
\begin{bmatrix}
d_1 \\ d_2 \\ \vdots \\ d_p \\ f
\end{bmatrix}
\qquad (10.3.8)
$$

if the unknowns are properly sequenced. See Meurant (1984). Here, the A_i are symmetric positive definite, the B_i are sparse, and the last block column is generally much narrower than the others.

An example with $p = 2$ serves to connect (10.3.8) and its block structure with the underlying problem geometry and the chosen *domain decomposition*. Suppose we are to solve Poisson's equation on the following domain:

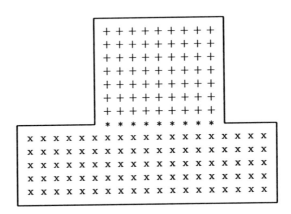

With the usual discretization, an unknown at a mesh point is coupled only to its "north", "east", "south", and "west" neighbor. There are three "types" of variables: those interior to the top subdomain (aggregated in the subvector x_1 and associated with the "+" mesh points), those interior to the bottom subdomain (aggregated in the subvector x_2 and associated with the "x" mesh points), and those on the interface between the two subdomains (aggregated in the subvector z and associated with the "*" mesh points). Note that the interior unknowns of one subdomain are not coupled to the interior unknowns of another subdomain, which accounts for the zero blocks in (10.3.8). Also observe that the number of interface unknowns is typically small compared to the overall number of unknowns.

Now let us explore the preconditioning possibilities associated with (10.3.8). We continue with the $p = 2$ case for simplicity. If we set

$$
M = L \begin{bmatrix} M_1^{-1} & 0 & 0 \\ 0 & M_2^{-1} & 0 \\ 0 & 0 & S^{-1} \end{bmatrix} L^T
$$

where

$$
L = \begin{bmatrix} M_1 & 0 & 0 \\ 0 & M_2 & 0 \\ B_1^T & B_2^T & S \end{bmatrix}
$$

then

$$
M = \begin{bmatrix} M_1 & 0 & B_1 \\ 0 & M_2 & B_2 \\ B_1^T & B_2^T & S_* \end{bmatrix} \tag{10.3.9}
$$

with $S_* = S + B_1^T M_1^{-1} B_1 + B_2^T M_2^{-1} B_2$. Let us consider how we might choose the block parameters M_1, M_2, and S so as to produce an effective preconditioner.

If we compare (10.3.9) with the $p = 2$ version of (10.3.8) we see that it makes sense for M_i to approximate A_i and for S_* to approximate Q. The latter is achieved if $S \approx Q - B_1^T M_1^{-1} B_1 - B_2^T M_2^{-1} B_2$. There are several approaches to selecting S and they all address the fact that we cannot form the dense matrices $B_i M_i^{-1} B_i^T$. For example, as discussed in the previous subsection, tridiagonal approximations of the M_i^{-1} could be used. See Meurant (1989).

If the subdomains are sufficiently regular and it is feasible to solve linear systems that involve the A_i exactly (say by using a fast Poisson solver), then we can set $M_i = A_i$. It follows that $M = A + E$ where the rank$(E) = m$ with m being the number of interface unknowns. Thus, the preconditioned conjugate gradient algorithm would theoretically converge in $m + 1$ steps.

Regardless of the approximations that must be incorporated in the process, we see that there are significant opportunities for parallelism because the subdomain problems are decoupled. Indeed, the number of subdomains p is usually a function of both the problem geometry and the number of processors that are available for the computation.

10.3.5 Polynomial Preconditioners

The vector z defined by the preconditioner system $Mz = r$ should be thought of as an approximate solution to $Az = r$ insofar as M is an approximation of A. One way to obtain such an approximate solution is to apply p steps of a stationary method $M_1 z^{(k+1)} = N_1 z^{(k)} + r$, $z^{(0)} = 0$. It follows that if $G = M_1^{-1} N_1$ then

$$z \equiv z^{(p)} = (I + G + \cdots G^{p-1}) M_1^{-1} r.$$

Thus, if $M^{-1} = (I + G + \cdots G^{p-1}) M_1^{-1}$ then $Mz = r$ and we can think of M as a preconditioner. Of course, it is important that M be symmetric positive definite and this constrains the choice of M_1, N_1, and p. Because M is a polynomial in G it is referred to as a *polynomial preconditioner*. This type of preconditioner is attractive from the vector/parallel point of view and has therefore attracted considerable attention.

10.3.6   Another Perspective

The polynomial preconditioner discussion points to an important connection between the classical iterations and the preconditioned conjugate gradient algorithm. Many iterative methods have as their basic step

$$x_k = x_{k-2} + \omega_k (\gamma_k z_{k-1} + x_{k-1} - x_{k-2}) \qquad (10.3.10)$$

where $M z_{k-1} = r_{k-1} = b - A x_{k-1}$. For example, if we set $\omega_k = 1$, and $\gamma_k = 1$, then

$$x_k = M^{-1}(b - A x_{k-1}) + x_{k-1},$$

i.e., $M x_k = N x_{k-1} + b$, where $A = M - N$. Thus, the Jacobi, Gauss-Seidel, SOR, and SSOR methods of §10.1 have the form (10.3.10). So also does the Chebyshev semi-iterative method (10.1.12).

Following Concus, Golub, and O'Leary (1976), it is also possible to organize Algorithm 10.3.1 with a central step of the form (10.3.10):

$$x_{-1} = 0; \; x_0 = \text{initial guess}; \; k = 0; \; r_0 = b - A x_0$$
$$\textbf{while } r_k \neq 0$$
$$\qquad k = k + 1$$
$$\qquad \text{Solve } M z_{k-1} = r_{k-1} \text{ for } z_{k-1}.$$
$$\qquad \gamma_{k-1} = z_{k-1}^T M z_{k-1} / z_{k-1}^T A z_{k-1}$$
$$\qquad \textbf{if } k = 1 \qquad\qquad\qquad\qquad\qquad\qquad\qquad (10.3.11)$$
$$\qquad\qquad \omega_1 = 1$$
$$\qquad \textbf{else}$$
$$\qquad\qquad \omega_k = \left(1 - \frac{\gamma_{k-1}}{\gamma_{k-2}} \frac{z_{k-1}^T M z_{k-1}}{z_{k-2}^T M z_{k-2}} \frac{1}{\omega_{k-1}} \right)^{-1}$$
$$\qquad \textbf{end}$$
$$\qquad x_k = x_{k-2} + \omega_k (\gamma_{k-1} z_{k-1} + x_{k-1} - x_{k-2})$$
$$\qquad r_k = b - A x_k$$
$$\textbf{end}$$
$$x = x_n$$

Thus, we can think of the scalars ω_k and γ_k in (10.3.11) as acceleration parameters that can be chosen to speed the convergence of the iteration $Mx_k = Nx_{k-1} + b$. Hence, any iterative method based on the splitting $A = M - N$ can be accelerated by the conjugate gradient algorithm as long as M (the preconditioner) is symmetric and positive definite.

Problems

P10.3.1 Detail an incomplete factorization procedure that is based on gaxpy Cholesky, i.e., Algorithm 4.2.1.

P10.3.2 How many n-vectors of storage is required by a practical implementation of Algorithm 10.3.1? Ignore workspaces that may be required when $Mz = r$ is solved.

Notes and References for Sec. 10.3

Our discussion of the preconditioned conjugate gradient is drawn from several sources including

P. Concus, G.H. Golub, and D.P. O'Leary (1976). " A Generalized Conjugate Gradient Method for the Numerical Solution of Elliptic Partial Differential Equations," in *Sparse Matrix Computations* , ed. J.R. Bunch and D.J. Rose, Academic Press, New York.

G.H. Golub and G. Meurant (1983). *Résolution Numérique des Grandes Systèmes Linéaires,* Collection de la Direction des Etudes et Recherches de l'Electricité de France, vol. 49, Eyolles, Paris.

O. Axelsson (1985). "A Survey of Preconditioned Iterative Methods for Linear Systems of Equations," *BIT 25,* 166–187.

P. Concus, G.H. Golub, and G. Meurant (1985). "Block Preconditioning for the Conjugate Gradient Method," *SIAM J. Sci. and Stat. Comp. 6,* 220–252.

O. Axelsson and G. Lindskog (1986). "On the Rate of Convergence of the Preconditioned Conjugate Gradient Method," *Numer. Math. 48,* 499–523.

Incomplete factorization ideas are detailed in

J.A. Meijerink and H.A. Van der vorst (1977). "An Iterative Solution Method for Linear Equation Systems of Which the Coefficient Matrix is a Symmetric M-Matrix," *Math. Comp. 31,* 148–62.

T.A. Mantueffel (1979). " Shifted Incomplete Cholesky Factorization," in *Sparse Matrix Proceedings* , 1978, ed. I.S. Duff and G.W. Stewart, SIAM Publications, Philadelphia, PA.

T.F. Chan, K.R. Jackson, and B. Zhu (1983). "Alternating Direction Incomplete Factorizations," *SIAM J. Numer. Anal. 20,* 239–257.

G. Roderigue and D. Wolitzer (1984). "Preconditioning by Incomplete Block Cyclic Reduction," *Math. Comp. 42,* 549–566.

O. Axelsson (1985). "Incomplete Block Matrix Factorization Preconditioning Methods. The Ultimate Answer?", *J. Comput. Appl. Math. 12&13,* 3–18.

O. Axelsson (1986). "A General Incomplete Block Matrix Factorization Method," *Lin. Alg. Appl. 74,* 179–190.

H. Elman (1986). "A Stability Analysis of Incomplete LU Factorization," *Math. Comp. 47,* 191–218.

T. Chan (1991). "Fourier Analysis of Relaxed Incomplete Factorization Preconditioners," *SIAM J. Sci. Statist. Comput. 12,* 668–680.

Y. Notay (1992). "On the Robustness of Modified Incomplete Factorization Methods,"
   J. Comput. Math. 40, 121–141.

For information on domain decomposition and other "pde driven" preconditioning ideas,
see

J.H. Bramble, J.E. Pasciak, and A.H. Schatz (1986). "The construction of Preconditioners for Elliptic Problems by Substructuring I," *Math. Comp. 47,* 103–134.
J.H. Bramble, J.E. Pasciak, and A.H. Schatz (1986). "The construction of Preconditioners for Elliptic Problems by Substructuring II," *Math. Comp. 49,* 1–17.
G. Meurant (1989). "Domain Decomposition Methods for Partial Differential Equations
   on Parallel Computers," to appear *Int'l J. Supercomputing Applications.*
W.D. Gropp and D.E. Keyes (1992). "Domain Decomposition with Local Mesh Refine-
   ment," *SIAM J. Sci. Statist. Comput. 13,* 967–993.
D.E. Keyes, T.F. Chan, G. Meurant, J.S. Scroggs, and R.G. Voigt (eds) (1992). *Do-
   main Decomposition Methods for Partial Differential Equations,* SIAM Publications,
   Philadelphia, PA.
M. Mu. (1995). "A New family of Preconditioners for Domain Decomposition," *SIAM
   J. Sci. Comp. 16,* 289–306.

Various aspects of polynomial preconditioners are discussed in

O.G. Johnson, C.A. Micchelli, and G. Paul (1983). "Polynomial Preconditioners for
   Conjugate Gradient Calculations," *SIAM J. Numer. Anal. 20,* 362–376.
S.C. Eisenstat (1984). "Efficient Implementation of a Class of Preconditioned Conjugate
   Gradient Methods," *SIAM J. Sci. and Stat. Computing 2,* 1–4.
Y. Saad (1985). "Practical Use of Polynomial Preconditionings for the Conjugate Gra-
   dient Method," *SIAM J. Sci. and Stat. Comp. 6,* 865–882.
L. Adams (1985). "m-step Preconditioned Congugate Gradient Methods," *SIAM J. Sci.
   and Stat. Comp. 6,* 452–463.
S.F. Ashby (1987). "Polynomial Preconditioning for Conjugate Gradient Methods,"
   Ph.D. Thesis, Dept. of Computer Science, University of Illinois.
S. Ashby, T. Manteuffel, and P. Saylor (1989). "Adaptive Polynomial Preconditioning
   for Hermitian Indefinite Linear Systems," *BIT 29,* 583–609.
R.W. Freund (1990). "On Conjugate Gradient Type Methods and Polynomial Pre-
   conditioners for a Class of Complex Non-Hermitian Matrices," *Numer. Math. 57,*
   285–312.
S. Ashby, T. Manteuffel, and J. Otto (1992). "A Comparison of Adaptive Chebyshev
   and Least Squares Polynomial Preconditioning for Hermitian Positive Definite Linear
   Systems," *SIAM J. Sci. Stat. Comp. 13,* 1–29.

Numerous vector/parallel implementations of the cg method have been developed. See

P.F. Dubois, A. Greenbaum, and G.H. Rodrigue (1979). "Approximating the Inverse
   of a Matrix for Use on Iterative Algorithms on Vector Processors," *Computing 22,*
   257–268.
H.A. Van der Vorst (1982). "A Vectorizable Variant of Some ICCG Methods," *SIAM J.
   Sci. and Stat. Comp. 3,* 350–356.
G. Meurant (1984). "The Block Preconditioned Conjugate Gradient Method on Vector
   Computers," *BIT 24,* 623–633.
T. Jordan (1984). "Conjugate Gradient Preconditioners for Vector and Parallel Pro-
   cessors," in G. Birkoff and A. Schoenstadt (eds), *Proceedings of the Conference on
   Elliptic Problem Solvers,* Academic Press, NY.
H.A. Van der Vorst (1986). "The Performance of Fortran Implementations for Precon-
   ditioned Conjugate Gradients on Vector Computers," *Parallel Computing 3,* 49–58.
M.K. Seager (1986). "Parallelizing Conjugate Gradient for the Cray X-MP," *Parallel
   Computing 3,* 35–47.

O. Axelsson and B. Polman (1986). "On Approximate Factorization Methods for Block Matrices Suitable for Vector and Parallel Processors," *Lin. Alg. and Its Applic.* 77, 3–26.

D.P. O'Leary (1987). "Parallel Implementation of the Block Conjugate Gradient Algorithm," *Parallel Computers 5*, 127–140.

R. Melhem(1987). "Toward Efficient Implementation of Preconditioned Conjugate Gradient Methods on Vector Supercomputers," *Int'l J. Supercomputing Applications 1*, 70–98.

E.L. Poole and J.M. Ortega (1987). "Multicolor ICCG Methods for Vector Computers," *SIAM J. Numer. Anal. 24*, 1394–1418.

C.C. Ashcraft and R. Grimes (1988). "On Vectorizing Incomplete Factorization and SSOR Preconditioners," *SIAM J. Sci. and Stat. Comp. 9*, 122–151.

U. Meier and A. Sameh (1988). "The Behavior of Conjugate Gradient Agorithms on a Multivector Processor with a Hierarchical Memory," *J. Comput. Appl. Math. 24*, 13–32.

W.D. Gropp and D.E. Keyes (1988). "Complexity of Parallel Implementation of Domain Decomposition Techniques for Elliptic Partial Differential Equations," *SIAM J. Sci. and Stat. Comp. 9*, 312–326.

H. Van Der Vorst (1989). "High Performance Preconditioning," *SIAM J. Sci. and Stat. Comp. 10*, 1174–1185.

H. Elman (1989). "Approximate Schur Complement Preconditioners on Serial and Parallel Computers," *SIAM J. Sci. Stat. Comput. 10*, 581–605.

O. Axelsson and V. Eijkhout (1989). "Vectorizable Preconditioners for Elliptic Difference Equations in Three Space Dimensions," *J. Comput. Appl. Math. 27*, 299–321.

S.L. Johnsson and K. Mathur (1989). "Experience with the Conjugate Gradient Method for Stress Analysis on a Data Parallel Supercomputer," *International Journal on Numerical Methods in Engineering 27*, 523–546.

L. Mansfield (1991). "Damped Jacobi Preconditioning and Coarse Grid Deflation for Conjugate Gradient Iteration on Parallel Computers," *SIAM J. Sci. and Stat. Comp. 12*, 1314–1323.

V. Eijkhout (1991). "Analysis of Parallel Incomplete Point Factorizations," *Lin. Alg. and Its Applic. 154–156*, 723–740.

S. Doi (1991). "On Parallelism and Convergence of Incomplete LU Factorizations," *Appl. Numer. Math. 7*, 417–436.

Preconditioners for large Toeplitz systems are discussed in

G. Strang (1986). "A Proposal for Toeplitz Matrix Calculations," *Stud. Appl. Math. 74*, 171–176.

T.F. Chan (1988). "An Optimal Circulant Preconditioner for Toeplitz Systems," *SIAM. J. Sci. Stat. Comp. 9*, 766–771.

R.H. Chan (1989). "The Spectrum of a Family of Circulant Preconditioned Toeplitz Systems," *SIAM J. Num. Anal. 26*, 503–506.

R.H. Chan (1991). "Preconditioners for Toeplitz Systems with Nonnegative Generating Functions," *IMA J. Num. Anal. 11*, 333–345.

T. Huckle (1992). "Circulant and Skewcirculant Matrices for Solving Toeplitz Matrix Problems," *SIAM J. Matrix Anal. Appl. 13*, 767–777.

T. Huckle (1992). "A Note on Skew-Circulant Preconditioners for Elliptic Problems," *Numerical Algorithms 2*, 279–286.

R.H. Chan, J.G. Nagy, and R.J. Plemmons (1993). "FFT based Preconditioners for Toeplitz Block Least Squares Problems," *SIAM J. Num. Anal. 30*, 1740–1768.

M. Hanke and J.G. Nagy (1994). "Toeplitz Approximate Inverse Preconditioner for Banded Toeplitz Matrices," *Numerical Algorithms 7*, 183–199.

R.H. Chan, J.G. Nagy, and R.J. Plemmons (1994). "Circulant Preconditioned Toeplitz Least Squares Iterations," *SIAM J. Matrix Anal. Appl. 15*, 80–97.

T.F. Chan and J.A. Olkin (1994). "Circulant Preconditioners for Toeplitz Block Matrices," *Numerical Algorithms 6*, 89–101.

Finally, we offer an assortment of references concerned with the practical application of the cg method:

J.K. Reid (1972). "The Use of Conjugate Gradients for Systems of Linear Equations Possessing Property A," *SIAM J. Num. Anal. 9*, 325–32.

D.P. O'Leary (1980). "The Block Conjugate Gradient Algorithm and Related Methods," *Lin. Alg. and Its Applic. 29*, 293–322.

R.C. Chin, T.A. Manteuffel, and J. de Pillis (1984). "ADI as a Preconditioning for Solving the Convection-Diffusion Equation," *SIAM J. Sci. and Stat. Comp. 5*, 281–299.

I. Duff and G. Meurant (1989). "The Effect of Ordering on Preconditioned Conjugate Gradients," *BIT 29*, 635–657.

A. Greenbaum and G. Rodrigue (1989). "Optimal Preconditioners of a Given Sparsity Pattern," *BIT 29*, 610–634.

O. Axelsson and P. Vassilevski (1989). "Algebraic Multilevel Preconditioning Methods I," *Numer. Math. 56*, 157–177.

O. Axelsson and P. Vassilevski (1990). "Algebraic Multilevel Preconditioning Methods II," *SIAM J. Numer. Anal. 27*. 1569–1590.

M. Hanke and M. Neumann (1990). "Preconditionings and Splittings for Rectangular Systems," *Numer. Math. 57*, 85–96.

A. Greenbaum (1992). "Diagonal Scalings of the Laplacian as Preconditioners for Other Elliptic Differential Operators," *SIAM J. Matrix Anal. Appl. 13*, 826–846.

P.E. Gill, W. Murray, D.B. Ponceleón, and M.A. Saunders (1992). "Preconditioners for Indefinite Systems Arising in Optimization," *SIAM J. Matrix Anal. Appl. 13*, 292–311.

G. Meurant (1992). "A Review on the Inverse of Symmetric Tridiagonal and Block Tridiagonal Matrices," *SIAM J. Matrix Anal. Appl. 13*, 707–728.

S. Holmgren and K. Otto (1992). "Iterative Solution Methods and Preconditioners for Block-Tridiagonal Systems of Equations," *SIAM J. Matrix Anal. Appl. 13*, 863–886.

S.A. Vavasis (1992). "Preconditioning for Boundary Integral Equations," *SIAM J. Matrix Anal. Appl. 13*, 905–925.

P. Joly and G. Meurant (1993). "Complex Conjugate Gradient Methods," *Numerical Algorithms 4*, 379–406.

X.-C. Cai and O. Widlund (1993). "Multiplicative Schwarz Algorithms for Some Nonsymmetric and Indefinite Problems," *SIAM J. Numer. Anal. 30*, 936–952.

10.4   Other Krylov Subspace Methods

The conjugate gradient method presented over the previous two sections is applicable to symmetric positive definite systems. The MINRES and SYMMLQ variants developed in §9.3.2 in connection with the symmetric Lanczos process can handle symmetric indefinite systems. Now we push the generalizations even further in pursuit of iterative methods that are applicable to unsymmetric systems.

The discussion is patterned after the survey article by Freund, Golub, and Nachtigal (1992) and Chapter 9 of Golub and Ortega (1993). We focus on cg-type algorithms that involve optimization over Krylov spaces.

Bear in mind that there is a large gap between our algorithmic specifications and production software. A good place to build an appreciation for this point is the *Templates* book by Barrett *et al* (1993). The book by Saad (1996) is also highly recommended

10.4.1 Normal Equation Approaches

The method of normal equations for the least squares problem is appealing because it allows us to use simple "Cholesky technology" instead of more complicated methods that involve orthogonalization. Likewise, in the unsymmetric $Ax = b$ problem it is tempting to solve the equivalent symmetric positive definite system

$$A^T Ax = A^T b$$

using existing conjugate gradient technology. Indeed, if we make the substitution $A \leftarrow A^T A$ in Algorithm 10.2.1 and note that a normal equation residual $A^T b - A^T Ax_k$ is A^T times the "true" residual $b - Ax_k$, then we obtain the Conjugate Gradient Normal Equation Residual method:

Algorithm 10.4.1 [CGNR] If $A \in \mathbb{R}^{n \times n}$ is nonsingular, $b \in \mathbb{R}^n$, and $x_0 \in \mathbb{R}^n$ is an initial guess ($Ax_0 \approx b$), then the following algorithm computes $x \in \mathbb{R}^n$ so $Ax = b$.

$k = 0$
$r_0 = b - Ax_0$
while $r_k \neq 0$
$\quad k = k + 1$
$\quad$ if $k = 1$
$\quad\quad p_1 = A^T r_0$
$\quad$ else
$\quad\quad \beta_k = (A^T r_{k-1})^T (A^T r_{k-1}) / (A^T r_{k-2})^T (A^T r_{k-2})$
$\quad\quad p_k = A^T r_{k-1} + \beta_k p_{k-1}$
$\quad$ end
$\quad \alpha_k = (A^T r_{k-1})^T (A^T r_{k-1}) / (Ap_k)^T (Ap_k)$
$\quad x_k = x_{k-1} + \alpha_k p_k$
$\quad r_k = r_{k-1} - \alpha_k Ap_k$
end
$x = x_k$

Another way to make an unsymmetric $Ax = b$ problem "cg-friendly" is to work with the system

$$AA^T y = b \qquad x = A^T y.$$

In "y space" the cg algorithm takes on the following form:

$k = 0$
$y_0 = $ initial guess $(AA^T y_0 = b)$
$r_0 = b - AA^T y_0$
while $r_k \neq 0$
    $k = k + 1$
    if $k = 1$
        $p_1 = r_0$
    else
        $\beta_k = r_{k-1}^T r_{k-1} / r_{k-2}^T r_{k-2}$
        $p_k = r_{k-1} + \beta_k p_{k-1}$
    end
    $\alpha_k = r_{k-1}^T r_{k-1} / p_k^T AA^T p_k$
    $y_k = y_{k-1} + \alpha_k p_k$
    $r_k = r_{k-1} - \alpha_k AA^T p_k$
end
$y = y_k$

Making the substitutions $x_k \leftarrow A^T y_k$ and $p_k \leftarrow A^T p_k$ and simplifying we obtain the $\underline{C}$onjugate $\underline{G}$radient $\underline{N}$ormal Equation $\underline{E}$rror method:

Algorithm 10.4.2 [CGNE]   If $A \in \mathbb{R}^{n \times n}$ is nonsingular, $b \in \mathbb{R}^n$, and $x_0 \in \mathbb{R}^n$ is an initial guess $(Ax_0 \approx b)$, then the following algorithm computes $x \in \mathbb{R}^n$ so $Ax = b$.

$k = 0$
$r_0 = b - Ax_0$
while $r_k \neq 0$
    $k = k + 1$
    if $k = 1$
        $p_1 = A^T r_0$
    else
        $\beta_k = r_{k-1}^T r_{k-1} / r_{k-2}^T r_{k-2}$
        $p_k = A^T r_{k-1} + \beta_k p_{k-1}$
    end
    $\alpha_k = r_{k-1}^T r_{k-1} / p_k^T p_k$
    $x_k = x_{k-1} + \alpha_k p_k$
    $r_k = r_{k-1} - \alpha_k A p_k$
end
$x = x_k$

In general these two normal equation approaches are handicapped by the squaring of the condition number. (Recall Theorem 10.2.6.) However, there are some occasions where they are effective and we refer the reader to Freund, Golub and Nachtigal (1991).

10.4.2 A Note on Objective Functions

Based on what we know about the cg method, the CGNR iterate x_k minimizes

$$\phi_1(x) = \frac{1}{2} x^T (A^T A) x - x^T A^T b$$

over the set

$$S_k^{(CGNR)} = x_0 + \mathcal{K}(A^T A, r_0, k).$$

It is easy to show that

$$\frac{1}{2} \| b - Ax \|_2^2 = \phi_1(x) + \frac{1}{2} b^T b$$

and so x_k minimizes the residual $\| b - Ax \|_2$ over $S_k^{(CGNR)}$. The "R" in "CGNR" is there because of the residual-based optimization.

On the other hand, the CGNE (implicit) iterate y_k minimizes

$$\phi_2(y) = \frac{1}{2} y^T (AA^T) y - y^T b$$

over the set $y_0 + \mathcal{K}(AA^T, b - AA^T y_0, k)$. With the change of variable $x = A^T y$ it can be shown that x_k minimizes

$$\frac{1}{2} x^T x - x^T A^{-1} b = \frac{1}{2} \| x - A^{-1} b \|_2^2 + \frac{1}{2} \| A^{-1} b \|_2^2$$

over

$$S_k^{(CGNE)} = x_0 + \mathcal{K}(A^T A, A^T r_0, k). \tag{10.4.1}$$

Thus CGNE minimizes the error at each step and that explains the "E" in "CGNE".

10.4.3 The Conjugate Residual Method

Recall that if A is symmetric positive definite, then it has a symmetric positive definite square root $A^{1/2}$. (See §4.2.10.) Note that in this case $Ax = b$ and

$$A^{1/2} x = A^{-1/2} b$$

are equivalent and that the former is the normal equation version of the latter. If we apply CGNR to this square root system and simplify the results, then we obtain

Algorithm 10.2.3 [Conjugate Residuals] If $A \in \mathbb{R}^{n \times n}$ is symmetric positive definite, $b \in \mathbb{R}^n$, and $x_0 \in \mathbb{R}^n$ is an initial guess ($Ax_0 \approx b$), then the following algorithm computes $x \in \mathbb{R}^n$ so $Ax = b$.

$$k = 0$$
$$r_0 = b - Ax_0$$
$$\text{while } r_k \neq 0$$
$$\qquad k = k + 1$$
$$\qquad \text{if } k = 1$$
$$\qquad\qquad p_1 = r_0$$
$$\qquad \text{else}$$
$$\qquad\qquad \beta_k = r_{k-1}^T A r_{k-1} / r_{k-2}^T A r_{k-2}$$
$$\qquad\qquad A p_k = A r_{k-1} + \beta_k A p_{k-1}$$
$$\qquad \text{end}$$
$$\qquad \alpha_k = r_{k-1}^T A r_{k-1} / (A p_k)^T (A p_k)$$
$$\qquad x_k = x_{k-1} + \alpha_k p_k$$
$$\qquad r_k = r_{k-1} - \alpha_k A p_k$$
$$\text{end}$$
$$x = x_k$$

It follows from our comments about CGNR that $\| A^{-1/2}(b - Ax) \|_2$ is minimized over the set $x_0 + \mathcal{K}(A, r_0, k)$ during the kth iteration

10.4.4   GMRES

In §9.3.2 we briefly discussed the Lanczos-based MINRES method for symmetric, possibly indefinite, $Ax = b$ problems. In that method the iterate x_k minimizes $\| b - Ax \|_2$ over the set

$$\mathcal{S}_k = x_0 + \text{span}\{r_0, Ar_0, \ldots, A^{k-1}r_0\} = x_0 + \mathcal{K}(A, r_0, k) \qquad (10.4.2)$$

The key idea behind the algorithm is to express x_k in terms of the Lanczos vectors $q_1, q_2, \ldots, q_k$ which span $\mathcal{K}(A, r_0, k)$ if q_1 is a multiple of the initial residual $r_0 = b - Ax_0$.

In the <u>G</u>eneralized <u>M</u>inimum <u>Res</u>idual (GMRES) method of Saad and Schultz (1986) the same approach is taken except that the iterates are expressed in terms of Arnoldi vectors instead of Lanczos vectors in order to handle unsymmetric A. After k steps of the Arnoldi iteration (9.4.1) we have the factorization

$$AQ_k = Q_{k+1}\tilde{H}_k \qquad (10.4.3)$$

where the columns of $Q_{k+1} = [\, Q_k \; q_{k+1} \,]$ are the orthonormal Arnoldi vectors and

$$\tilde{H}_k = \begin{bmatrix} h_{11} & h_{12} & \cdots & & \cdots & h_{1k} \\ h_{21} & h_{22} & \cdots & & \cdots & h_{2k} \\ 0 & \ddots & \ddots & & & \vdots \\ \vdots & & \ddots & \ddots & & \vdots \\ 0 & \cdots & \cdots & h_{k,k-1} & & h_{kk} \\ 0 & \cdots & \cdots & & 0 & h_{k+1,k} \end{bmatrix} \in \mathbb{R}^{k+1 \times k}$$

is upper Hessenberg. In the kth step of GMRES, $\| b - Ax_k \|_2$ is minimized subject to the constraint that x_k has the form $x_k = x_0 + Q_k y_k$ for some $y_k \in \mathbb{R}^k$. If $q_1 = \rho_0 r_0$ where $\rho_0 = 1/\| r_0 \|_2$, then it follows that

$$
\begin{aligned}
\| b - A(x_0 + Q_k y_k) \|_2 &= \| r_0 - AQ_k y_k \|_2 \\
&= \| r_0 - Q_{k+1} \tilde{H}_k y_k \|_2 \\
&= \| \rho_0 e_1 - \tilde{H}_k y_k \|_2 .
\end{aligned}
$$

Thus, y_k is the solution to a $(k+1)$-by-k least squares problem and the GMRES iterate is given by $x_k = x_0 + Q_k y_k$.

Algorithm 10.4.4 [GMRES] If $A \in \mathbb{R}^{n \times n}$ is nonsingular, $b \in \mathbb{R}^n$, and $x_0 \in \mathbb{R}^n$ is an initial guess ($Ax_0 \approx b$), then the following algorithm computes $x \in \mathbb{R}^n$ so $Ax = b$.

$r_0 = b - Ax_0$
$h_{10} = \| r_0 \|_2$
$k = 0$
while ($h_{k+1,k} > 0$)
$\quad q_{k+1} = r_k/h_{k+1,k}$
$\quad k = k + 1$
$\quad r_k = Aq_k$
$\quad$ for $i = 1{:}k$
$\quad\quad h_{ik} = q_i^T w$
$\quad\quad r_k = r_k - h_{ik} q_i$
$\quad$ end
$\quad h_{k+1,k} = \| r_k \|_2$
$\quad x_k = x_0 + Q_k y_k$ where $\| \rho_0 e_1 - \tilde{H}_k y_k \|_2 = \min$
end
$x = x_k$

It is easy to verify that

$$
\| b - Ax_k \|_2 = h_{k+1,k}
$$

The upper Hessenberg least square problem can be efficiently solved using Givens rotations. In practice there is no need to form x_k until one is happy with its residual.

The main problem with "unlimited GMRES" is that the kth iteration involves $O(kn)$ flops. Thus like Arnoldi, a practical GMRES implementation requires a restart strategy to avoid excessive amounts of computation and memory traffic. For example, if at most m steps are tolerable, then x_m can be used as the initial vector for the next GMRES sequence.

10.4.5 Preconditioning

Preconditioning is the other key to making GMRES effective. Analogous to the development of the preconditioned conjugate gradient method in §10.3, we obtain a nonsingular matrix $M = M_1 M_2$ that approximates A in some sense and then apply GMRES to the system $\tilde{A}\tilde{x} = \tilde{b}$ where $\tilde{A} = M_1^{-1} A M_2^{-1}$, $\tilde{b} = M_1^{-1}b$, and $\tilde{x} = M_2 x$. If we write down the GMRES iteration for the tilde system and manipulate the equations to restore the original variables, then the resulting iteration requires the solution of linear systems that involve the preconditioner M. Thus, the act of finding a good preconditioner $M = M_1 M_2$ is the act of making $\tilde{A} = M_1^{-1} A M_2^{-1}$ look as much as possible like the identity subject to the constraint that linear systems with M are easy to solve.

10.4.6 The Biconjugate Gradient Method

Just as Arnoldi underwrites GMRES, the unsymmetric Lanczos process underwrites the Biconjugate gradient (BiCG) method. The starting point in the development of BiCG is to go back to the Lanczos derivation of the conjugate gradient method in §9.3.1. In terms of Lanczos vectors, the kth cg iterate is given by $x_k = x_0 + Q_k y_k$ where Q_k is the matrix of Lanczos vectors, $T_k = Q_k^T A Q_k$ is tridiagonal, and y_k solves $T_k y_k = Q_k^T r_0$. Note that

$$Q_k^T(b - Ax_k) = Q_k^T(r_0 - AQ_k y_k) = 0.$$

Thus, we can characterize x_k by insisting that it come from $x_0 + \mathcal{K}(A, r_0, k)$ and that it produce a residual that is orthogonal to a given subspace, say $\mathcal{K}(A, r_0, k)$.

In the unsymmetric case we can extend this notion by producing a sequence of iterates $\{x_k\}$ with the property that x_k belongs to $x_0 + \mathcal{K}(A, r_0, k)$ and produces a residual that is orthogonal to $\mathcal{K}(A^T, s_0, k)$ for some $s_0 \in \mathbb{R}^n$. Simplifications occur if the unsymmetric Lanczos process is used to generate bases for the two involved Krylov spaces. In particular, after k steps of the unsymmetric Lanczos algorithm (9.4.7) we have $Q_k, P_k \in \mathbb{R}^{n \times k}$ such that $P_k^T Q_k = I_k$ and a tridiagonal matrix $T_k = P_k^T A Q_k$ such that

$$
\begin{aligned}
AQ_k &= Q_k T_k + r_k e_k^T & P_k^T r_k &= 0 \\
A^T P_k &= P_k T_k^T + s_k e_k^T & Q_k^T s_k &= 0
\end{aligned}
\tag{10.4.4}
$$

In BiCG we set $x_k = x_0 + Q_k y_k$ where $T_k y_k = Q_k^T r_0$. Note that the *Galerkin condition*

$$P_k^T(b - Ax_k) = P_k^T(r_0 - AQ_k y_k) = 0$$

holds.

As might be expected, it is possible to develop recursions so that x_k can be computed as a simple combination of x_{k-1} and q_{k-1}, instead of as a linear combination of all the previous q-vectors.

The BiCG method is subject to serious breakdown because of its dependence on the unsymmetric Lanczos process. However, by relying on a look-ahead Lanczos procedure it is possible to overcome some of these difficulties.

10.4.7   QMR

Another iteration that runs off of the unsymmetric Lanczos process is the quasi-minimum residual (QMR) method of Freund and Nachtigal (1991). As in BiCG the kth iterate has the form $x_k = x_0 + Q_k y_k$. It is easy to show that after k steps in (9.4.7) we have the factorization

$$AQ_k = Q_{k+1}\tilde{T}_k$$

where $\tilde{T}_k \in \mathbb{R}^{k+1 \times k}$ is tridiagonal. It follows that if $q_1 = \rho(b - Ax_0)$, then

$$
\begin{aligned}
b - Ax_k &= b - A(x_0 + Q_k y_k) \\
&= r_0 - AQ_k y_k \\
&= r_0 - Q_{k+1}\tilde{T}_k y_k \\
&= Q_{k+1}(\rho e_1 - \tilde{T}_k y_k).
\end{aligned}
$$

If y_k is chosen to minimize the 2-norm of this vector, then in exact arithmetic $x_0 + Q_k y_k$ defines the GMRES iterate. In QMR, y_k is chosen to minimize $\| \rho e_1 - \tilde{T}_k y_k \|_2$.

10.4.8   Summary

The methods that we have presented do not submit to a linear ranking. The choice of a technique is complicated and depends on a host of factors. A particularly cogent assessment of the major algorithms is given in Barrett *et al* (1993).

Problems

P10.4.1 Analogous to (10.2.16), develop efficient implementations of the CGNR, CGNE, Conjugate residual methods.

P10.4.2 Establish the mathematical equivalence of the CGNR and the LSQR method outlined in §9.3.4.

P10.4.3 Prove (10.4.3).

P10.4.4 Develop an efficient preconditioned GMRES implementation. Proceeding as we did in §10.3 for preconditioned conjugate gradient method. (See (10.3.2) and (10.3.3) in particular.)

P10.4.5 Prove that the GMRES least squares problem has full rank.

Notes and References for Sec. 10.4

The following papers serve as excellent introductions to the world of unsymmetric iteration:

S. Eisenstat, H. Elman, and M. Schultz (1983). "Variational Iterative Methods for Nonsymmetric Systems of Equations," *SIAM J. Num. Anal. 20*, 345–357.

R.W. Freund, G.H. Golub, and N. Nachtigal (1992). "Iterative Solution of Linear Systems," *Acta Numerica 1*, 57–100.

N. Nachtigal, S. Reddy, and L. Trefethen (1992). "How Fast Are Nonsymmetric Matrix Iterations," *SIAM J. Matrix Anal. Appl. 13*, 778–795.

A. Greenbaum and L.N. Trefethen (1994). "GMRES/CR and Arnoldi/Lanczos as Matrix Approximation Problems," *SIAM J. Sci. Comp. 15*, 359–368.

Krylov space methods and analysis are featured in the following papers:

W.E. Arnoldi (1951). "The Principle of Minimized Iterations in the Solution of the Matrix Eigenvalue Problem," *Quart. Appl. Math. 9*, 17–29.

Y. Saad (1981). "Krylov Subspace Methods for Solving Large Unsymmetric Linear Systems," *Math. Comp. 37*, 105–126.

Y. Saad (1984). "Practical Use of Some Krylov Subspace Methods for Solving Indefinite and Nonsymmetric Linear Systems," *SIAM J. Sci. and Stat. Comp. 5*, 203–228.

Y. Saad (1989). "Krylov Subspace Methods on Supercomputers," *SIAM J. Sci. and Stat. Comp. 10*, 1200–1322.

C.-M. Huang and D.P. O'Leary (1993). "A Krylov Multisplitting Algorithm for Solving Linear Systems of Equations," *Lin. Alg. and Its Applic. 194*, 9–29.

C.C. Paige, B.N. Parlett, and H.A. Van Der Vorst (1995). "Approximate Solutions and Eigenvalue Bounds from Krylov Subspaces," *Numer. Linear Algebra with Applic. 2*, 115–134.

References for the GMRES method include

Y. Saad and M. Schultz (1986). "GMRES: A Generalized Minimal Residual Algorithm for Solving Nonsymmetric Linear Systems," *SIAM J. Scientific and Stat. Comp. 7*, 856–869.

H.F. Walker (1988). "Implementation of the GMRES Method Using Householder Transformations," *SIAM J. Sci. Stat. Comp. 9*, 152–163.

C. Vuik and H.A. van der Vorst (1992). "A Comparison of Some GMRES-like Methods," *Lin. Alg. and Its Applic. 160*, 131–162.

N. Nachtigal, L. Reichel, and L. Trefethen (1992). "A Hybrid GMRES Algorithm for Nonsymmetric Linear Systems," *SIAM J. Matrix Anal. Appl. 13*, 796–825.

Y. Saad (1993). "A Flexible Inner-Outer Preconditioned GMRES Algorithm," *SIAM J. Sci. Comput. 14*, 461–469.

Z. Bai, D. Hu, and L. Reichel (1994). "A Newton Basis GMRES Implementation," *IMA J. Num. Anal. 14*, 563–581.

R.B. Morgan (1995). "A Restarted GMRES Method Augmented with Eigenvectors," *SIAM J. Matrix Anal. Applic. 16*, 1154–1171.

Preconditioning ideas for unsymmetric problems are discussed in the following papers:

Y. Saad (1988). "Preconditioning Techniques for Indefinite and Nonsymmetric Linear Systems," *J. Comput. Appl. Math. 24*, 89–105.

L. Yu. Kolotilina and A. Yu. Yeremin (1993). "Factorized Sparse Approximate Inverse Preconditioning I: Theory," *SIAM J. Matrix Anal. Applic. 14*, 45–58.

I.E. Kaporin (1994). "New Convergence Results and Preconditioning Strategies for the Conjugate Gradient Method," *Num. Lin. Alg. Applic. 1*, 179–210.

L. Yu. Kolotilina and A. Yu. Yeremin (1995). "Factorized Sparse Approximate Inverse Preconditioning II: Solution of 3D FE Systems on Massively Parallel Computers," *Intern. J. High Speed Comput. 7*, 191–215.

H. Elman (1996). "Fast Nonsymmetric Iterations and Preconditioning for Navier-Stokes Equations," *SIAM J. Sci. Comput. 17*, 33–46.

M. Benzi, C.D. Meyer, and M. Tuma (1996). "A Sparse Approximate Inverse Preconditioner for the Conjugate Gradient Method," *SIAM J. Sci. Comput. 17*, to appear.

Some representative papers concerned with the development of nonsymmetric conjugate
gradient procedures include

D.M. Young and K.C. Jea (1980). "Generalized Conjugate Gradient Acceleration of
Nonsymmetrizable Iterative Methods," *Lin. Alg. and Its Applic. 34*, 159–94.
O. Axelsson (1980). "Conjugate Gradient Type Methods for Unsymmetric and Incon-
sistent Systems of Linear Equations," *Lin. Alg. and Its Applic. 29*, 1–16.
K.C. Jea and D.M. Young (1983). "On the Simplification of Generalized Conjugate
Gradient Methods for Nonsymmetrizable Linear Systems," *Lin. Alg. and Its Applic.
52/53*, 399–417.
V. Faber and T. Manteuffel (1984). "Necessary and Sufficient Conditions for the Exis-
tence of a Conjugate Gradient Method," *SIAM J. Numer. Anal. 21* 352–362.
Y. Saad and M. Schultz (1985). "Conjugate Gradient-Like Algorithms for Solving Non-
symmetric Linear Systems," *Math. Comp. 44*, 417–424.
H.A. Van der Vorst (1986). "An Iterative Solution Method for Solving $f(A)x = b$ Using
Krylov Subspace Information Obtained for the Symmetric Positive Definite Matrix
A," *J. Comp. and App. Math. 18*, 249–263.
M.A. Saunders, H.D. Simon, and E.L. Yip (1988). "Two Conjugate Gradient-Type
Methods for Unsymmetric Linear Equations," *SIAM J. Num. Anal. 25*, 927–940.
R. Freund (1992). "Conjugate Gradient-Type Methods for Linear Systems with Complex
Symmetric Coefficient Matrices," *SIAM J. Sci. Statist. Comput. 13*, 425–448.

More Lanczos-based solvers are discussed in

Y. Saad (1982). "The Lanczos Biorthogonalization Algorithm and Other Oblique Pro-
jection Methods for Solving Large Unsymmetric Systems," *SIAM J. Numer. Anal.
19*, 485–506.
Y. Saad (1987). "On the Lanczos Method for Solving Symmetric Systems with Several
Right Hand Sides," *Math. Comp. 48*, 651–662.
C. Brezinski and H. Sadok (1991). "Avoiding Breakdown in the CGS Algorithm," *Nu-
mer. Alg. 1*, 199–206.
C. Brezinski, M. Zaglia, and H. Sadok (1992). "A Breakdown Free Lanczos Type Algo-
rithm for Solving Linear Systems," *Numer. Math. 63*, 29–38.
S.K. Kim and A.T. Chronopoulos (1991). "A Class of Lanczos-Like Algorithms Imple-
mented on Parallel Computers," *Parallel Comput. 17*, 763–778.
W. Joubert (1992). "Lanczos Methods for the Solution of Nonsymmetric Systems of
Linear Equations," *SIAM J. Matrix Anal. Appl. 13*, 926–943.
R.W. Freund, M. Gutknecht, and N. Nachtigal (1993). "An Implementation of the
Look-Ahead Lanczos Algorithm for Non-Hermitian Matrices," *SIAM J. Sci. and
Stat.Comp. 14*, 137–158.

The QMR method is detailed in the following papers

R.W. Freund and N. Nachtigal (1991). "QMR: A Quasi-Minimal Residual Method for
Non-Hermitian Linear Systems," *Numer. Math. 60*, 315–339.
R.W. Freund (1993). "A Transpose-Free Quasi-Minimum Residual Algorithm for Non-
hermitian Linear System," *SIAM J. Sci. Comput. 14*, 470–482.
R.W. Freund and N.M. Nachtigal (1994). "An Implementation of the QMR Method
Based on Coupled Two-term Recurrences," *SIAM J. Sci. Comp. 15*, 313–337.

The residuals in BiCG tend to display erratic behavior prompting the development of
stabilizing techniques:

H. van der Vorst (1992). "BiCGSTAB: A Fast and Smoothly Converging Variant of the Bi-CG for the Solution of Nonsymmetric Linear Systems," *SIAM J. Sci. and Stat. Comp. 13*, 631–644.

M. Gutknecht (1993). "Variants of BiCBSTAB for Matrices with Complex Spectrum," *SIAM J. Sci. and Stat. Comp. 14*, 1020–1033.

G.L.G. Sleijpen and D.R. Fokkema (1993). "BICGSTAB(ℓ) for Linear Equations Involving Unsymmetric Matrices with Complex Spectrum," *Electronic Transactions on Numerical Analysis 1*, 11–32.

C. Brezinski and M. Redivo-Zaglia (1995). "Look-Ahead in BiCGSTAB and Other Product-Type Methods for Linear Systems," *BIT 35*, 169–201.

In some applications it is awkward to produce matrix-vector product code for both Ax and $A^T x$. Transpose free methods are popular in this context. See

P. Sonneveld (1989). "CGS, A Fast Lanczos-Type Solver for Nonsymmetric Linear Systems," *SIAM J. Sci. and Stat. Comp. 10*, 36–52.

G. Radicati di Brozolo and Y. Robert (1989). "Parallel Conjugate Gradient-like Algorithms for Solving Sparse Nonsymmetric Linear Systems on a Vector Multiprocessor," *Parallel Computing 11*, 233–240.

C. Brezinski and M. Redivo-Zaglia (1994). "Treatment of Near-Breakdown in the CGS Algorithms," *Numerical Algorithms 7*, 33–73.

E.M. Kasenally (1995). "GMBACK: A Generalized Minimum Backward Error Algorithm for Nonsymmetric Linear Systems," *SIAM J. Sci. Comp. 16*, 698–719.

C.C. Paige, B.N. Parlett, and H.A. van der Vorst (1995). "Approximate Solutions and Eigenvalue Bounds from Krylov Subspaces," *Num. Lin. Alg. with Applic. 2*, 115–133.

M. Hochbruck and Ch. Lubich (1996), "On Krylov Subspace Approximations to the Matrix Exponential Operator," *SIAM J. Numer. Anal.*, to appear.

M. Hochbruck and Ch. Lubich (1996), "Error Analysis of Krylov Method in a Nutshell," *SIAM J. Sci. Comput.*, to appear.

Connections between the pseudoinverse of a rectangular matrix A and the conjugate gradient method applied to $A^T A$ are pointed out in the paper

M. Hestenes (1975). "Pseudoinverses and Conjugate Gradients," *CACM 18*, 40–43.

Chapter 11

Functions of Matrices

Computing a function $f(A)$ of an n-by-n matrix A is a frequently occurring problem in control theory and other application areas. Roughly speaking, if the scalar function $f(z)$ is defined on $\lambda(A)$, then $f(A)$ is defined by substituting "A" for "z" in the "formula" for $f(z)$. For example, if $f(z) = (1 + z)/(1 - z)$ and $1 \notin \lambda(A)$, then $f(A) = (I + A)(I - A)^{-1}$.

The computations get particularly interesting when the function f is transcendental. One approach in this more complicated situation is to compute an eigenvalue decomposition $A = YBY^{-1}$ and use the formula $f(A) = Yf(B)Y^{-1}$. If B is sufficiently simple, then it is often possible to calculate $f(B)$ directly. This is illustrated in §11.1 for the Jordan and Schur decompositions. Not surprisingly, reliance on the latter decomposition results in a more stable $f(A)$ procedure.

Another class of methods for the matrix function problem is to approximate the desired function $f(A)$ with an easy-to-calculate function $g(A)$. For example, g might be a truncated Taylor series approximate to f. Error bounds associated with the approximation of matrix functions are given in §11.2.

In the last section we discuss the special and very important problem of computing the matrix exponential e^A.

Before You Begin

Chapters 1, 2, 3, 7 and 8 are assumed. Within this chapter there are the following dependencies:

$$\S 11.1 \quad \rightarrow \quad \S 11.2 \quad \rightarrow \quad \S 11.3$$

Complementary references include Mirsky (1955), Gantmacher (1959), Bellman (1969), and Horn and Johnson (1991). Some Matlab functions important to this chapter are expm, expm1, expm2, expm3, logm, sqrtm, and funm.

11.1 Eigenvalue Methods

Given an n-by-n matrix A and a scalar function $f(z)$, there are several ways to define the *matrix function* $f(A)$. A very informal definition might be to substitute "A" for "z" in the formula for $f(z)$. For example, if $p(z) = 1 + z$ and $r(z) = (1 - (z/2))^{-1}(1 + (z/2))$ for $z \neq 2$, then it is certainly reasonable to define $p(A)$ and $r(A)$ by

$$p(A) = I + A$$

and

$$r(A) = \left(I - \frac{A}{2}\right)^{-1}\left(I + \frac{A}{2}\right) \qquad 2 \notin \lambda(A).$$

"A-for-z" substitution also works for transcendental functions, i.e.,

$$e^A = \sum_{k=0}^{\infty} \frac{A^k}{k!}.$$

To make subsequent algorithmic developments precise, however, we need a more precise definition of $f(A)$.

11.1.1 A Definition

There are many ways to establish rigorously the notion of a matrix function. See Rinehart (1955). Perhaps the most elegant approach is in terms of a line integral. Suppose $f(z)$ is analytic inside on a closed contour Γ which encircles $\lambda(A)$. We define $f(A)$ to be the matrix

$$f(A) = \frac{1}{2\pi i} \oint_\Gamma f(z)(zI - A)^{-1}dz. \qquad (11.1.1)$$

This definition is immediately recognized as a matrix version of the Cauchy integral theorem. The integral is defined on an element-by-element basis:

$$f(A) = (f_{kj}) \quad \Longrightarrow \quad f_{kj} = \frac{1}{2\pi i} \oint_\Gamma f(z)e_k^T(zI - A)^{-1}e_j dz.$$

Notice that the entries of $(zI - A)^{-1}$ are analytic on Γ and that $f(A)$ is defined whenever $f(z)$ is analytic in a neighborhood of $\lambda(A)$.

11.1.2   The Jordan Characterization

Although fairly useless from the computational point of view, the definition
(11.1.1) can be used to derive more practical characterizations of $f(A)$. For
example, if $f(A)$ is defined and

$$A = XBX^{-1} = X\text{diag}(B_1,\ldots,B_p)X^{-1}, \qquad B_i \in \mathbb{C}^{n_i \times n_i}$$

then it is easy to verify that

$$f(A) = Xf(B)X^{-1} = X\text{diag}(f(B_1),\ldots,f(B_p))X^{-1}. \qquad (11.1.2)$$

For the case when the B_i are Jordan blocks we obtain the following:

Theorem 11.1.1 *Let $X^{-1}AX = \text{diag}(J_1,\ldots,J_p)$ be the Jordan canonical
form (JCF) of $A \in \mathbb{C}^{n \times n}$ with*

$$J_i = \begin{bmatrix} \lambda_i & 1 & \cdots & \cdots & 0 \\ 0 & \lambda_i & 1 & \cdots & \vdots \\ \vdots & \ddots & \ddots & \ddots & \vdots \\ \vdots & \vdots & \ddots & \ddots & 1 \\ 0 & \cdots & \cdots & 0 & \lambda_i \end{bmatrix}$$

being an m_i-by-m_i Jordan block. If $f(z)$ is analytic on an open set containing $\lambda(A)$, then

$$f(A) = X\text{diag}(f(J_1),\ldots,f(J_p))X^{-1}$$

where

$$f(J_i) = \begin{bmatrix} f(\lambda_i) & f^{(1)}(\lambda_i) & \cdots & \cdots & \dfrac{f^{(m_i-1)}(\lambda_i)}{(m_i-1)!} \\ 0 & f(\lambda_i) & \ddots & \cdots & \vdots \\ \vdots & \vdots & \ddots & \ddots & \vdots \\ \vdots & \vdots & \vdots & \ddots & f^{(1)}(\lambda_i) \\ 0 & \cdots & \cdots & \cdots & f(\lambda_i) \end{bmatrix}.$$

Proof. In view of the remarks preceding the statement of the theorem, it
suffices to examine $f(G)$ where

$$G = \lambda I + E \qquad E = (\delta_{i,j-1})$$

is a q-by-q Jordan block. Suppose $(zI - G)$ is nonsingular. Since

$$(zI - G)^{-1} = \sum_{k=0}^{q-1} \frac{E^k}{(z - \lambda)^{k+1}}$$

it follows from Cauchy's integral theorem that

$$f(G) = \sum_{k=0}^{q-1} \left[\frac{1}{2\pi i} \oint_\Gamma \frac{f(z)}{(z-\lambda)^{k+1}} dz \right] E^k = \sum_{k=0}^{q-1} \frac{f^{(k)}(\lambda)}{k!} E^k .$$

The theorem follows from the observation that $E^k = (\delta_{i,j-k})$. $\square$

Corollary 11.1.2 *If $A \in \mathbb{C}^{n \times n}$, $A = X\mathrm{diag}(\lambda_1, \dots, \lambda_n)X^{-1}$, and $f(A)$ is defined, then*

$$f(A) = X\mathrm{diag}(f(\lambda_1), \dots, f(\lambda_n))X^{-1}.$$

Proof. The Jordan blocks are all 1-by-1. $\square$

These results illustrate the close connection between $f(A)$ and the eigensystem of A. Unfortunately, the JCF approach to the matrix function problem has dubious computational merit unless A is diagonalizable with a well-conditioned matrix of eigenvectors. Indeed, rounding errors of order $\mathbf{u}\kappa_2(X)$ can be expected to contaminate the computed result, since a linear system involving the matrix X must be solved. The following example suggests that ill-conditioned similarity transformations should be avoided when computing a function of a matrix.

Example 11.1.1 If

$$A = \begin{bmatrix} 1 + 10^{-5} & 1 \\ 0 & 1 - 10^{-5} \end{bmatrix},$$

then any matrix of eigenvectors is a column scaled version of

$$X = \begin{bmatrix} 1 & -1 \\ 0 & 2(1 - 10^{-5}) \end{bmatrix}$$

and has a 2-norm condition number of order 10^5. Using a computer with machine precision $\mathbf{u} \approx 10^{-7}$ we find

$$fl[X^{-1}\mathrm{diag}(\exp(1 + 10^{-5}), \exp(1 - 10^{-5}))X] = \begin{bmatrix} 2.718307 & 2.750000 \\ 0.000000 & 2.718254 \end{bmatrix}$$

while

$$e^A = \begin{bmatrix} 2.718309 & 2.718282 \\ 0.000000 & 2.718255 \end{bmatrix} .$$

11.1.3   A Schur Decomposition Approach

Some of the difficulties associated with the Jordan approach to the matrix function problem can be circumvented by relying upon the Schur decomposition. If $A = QTQ^H$ is the Schur decomposition of A, then

$$f(A) = Qf(T)Q^H.$$

For this to be effective, we need an algorithm for computing functions of upper triangular matrices. Unfortunately, an explicit expression for $f(T)$ is very complicated as the following theorem shows.

Theorem 11.1.3 *Let $T = (t_{ij})$ be an n-by-n upper triangular matrix with $\lambda_i = t_{ii}$ and assume $f(T)$ is defined. If $f(T) = (f_{ij})$, then $f_{ij} = 0$ if $i > j$, $f_{ij} = f(\lambda_i)$ for $i = j$, and for all $i < j$ we have*

$$f_{ij} = \sum_{(s_0,\ldots,s_k) \in S_{ij}} t_{s_0,s_1} t_{s_1,s_2} \cdots t_{s_{k-1},s_k} f[\lambda_{s_0},\ldots,\lambda_{s_k}],$$

where S_{ij} is the set of all strictly increasing sequences of integers that start at i and end at j and $f[\lambda_{s_0},\ldots,\lambda_{s_k}]$ is the kth order divided difference of f at $\{\lambda_{s_0},\ldots,\lambda_{s_k}\}$.

Proof. See Descloux (1963), Davis (1973), or Van Loan (1975). □

Computing $f(T)$ via Theorem 11.1.3 would require $O(2^n)$ flops. Fortunately, Parlett (1974) has derived an elegant recursive method for determining the strictly upper triangular portion of the matrix $F = f(T)$. It requires only $2n^3/3$ flops and can be derived from the following commutivity result:

$$FT = TF. \tag{11.1.3}$$

Indeed, by comparing (i,j) entries in this equation, we find

$$\sum_{k=i}^{j} f_{ik} t_{kj} = \sum_{k=i}^{j} t_{ik} f_{kj} \qquad j > i$$

and thus, if t_{ii} and t_{jj} are distinct,

$$f_{ij} = t_{ij} \frac{f_{jj} - f_{ii}}{t_{jj} - t_{ii}} + \sum_{k=i+1}^{j-1} \frac{t_{ik} f_{kj} - f_{ik} t_{kj}}{t_{jj} - t_{ii}}. \tag{11.1.4}$$

From this we conclude that f_{ij} is a linear combination of its neighbors to its left and below in the matrix F. For example, the entry f_{25} depends upon f_{22}, f_{23}, f_{24}, f_{55}, f_{45}, and f_{35}. Because of this, the entire upper triangular portion of F can be computed one superdiagonal at a time beginning with the diagonal, $f(t_{11}),\ldots, f(t_{nn})$. The complete procedure is as follows:

Algorithm 11.1.1 This algorithm computes the matrix function $F = f(T)$ where T is upper triangular with distinct eigenvalues and f is defined on $\lambda(T)$.

    for $i = 1{:}n$
        $f_{ii} = f(t_{ii})$
    end

for $p = 1{:}n - 1$
    for $i = 1{:}n - p$
        $j = i + p$
        $s = t_{ij}(f_{jj} - f_{ii})$
        for $k = i + 1{:}j - 1$
            $s = s + t_{ik}f_{kj} - f_{ik}t_{kj}$
        end
        $f_{ij} = s/(t_{jj} - t_{ii})$
    end
end

This algorithm requires $2n^3/3$ flops. Assuming that $T = QAQ^H$ is the Schur form of A, $f(A) = QFQ^H$ where $F = f(T)$. Clearly, most of the work in computing $f(A)$ by this approach is in the computation of the Schur decomposition, unless f is extremely expensive to evaluate.

Example 11.1.2 If
$$T = \begin{bmatrix} 1 & 2 & 3 \\ 0 & 3 & 4 \\ 0 & 0 & 5 \end{bmatrix}$$
and $f(z) = (1 + z)/z$ then $F = (f_{ij}) = f(T)$ is defined by

$$\begin{aligned} f_{11} &= (1+1)/1 = 2 \\ f_{22} &= (1+3)/3 = 4/3 \\ f_{33} &= (1+5)/5 = 6/5 \\ f_{12} &= t_{12}(f_{22} - f_{11})/(t_{22} - t_{11}) = -2/3 \\ f_{23} &= t_{23}(f_{33} - f_{22})/(t_{33} - t_{22}) = -4/15 \\ f_{13} &= [t_{13}(f_{33} - f_{11}) + (t_{12}f_{23} - f_{12}t_{23})]/(t_{33} - t_{11}) = -1/15. \end{aligned}$$

11.1.4   A Block Schur Approach

If A has close or multiple eigenvalues, then Algorithm 11.1.1 leads to poor results. In this case, it is advisable to use a block version of Algorithm 11.1.1. We outline such a procedure due to Parlett (1974a). The first step is to choose Q in the Schur decomposition such that close or multiple eigenvalues are clustered in blocks $T_{11}, \ldots, T_{pp}$ along the diagonal of T. In particular, we must compute a partitioning

$$T = \begin{bmatrix} T_{11} & T_{12} & \cdots & T_{1p} \\ 0 & T_{22} & \cdots & T_{2p} \\ \vdots & \vdots & \ddots & \vdots \\ 0 & 0 & \cdots & T_{pp} \end{bmatrix} \qquad F = \begin{bmatrix} F_{11} & F_{12} & \cdots & F_{1p} \\ 0 & F_{22} & \cdots & F_{2p} \\ \vdots & \vdots & \ddots & \vdots \\ 0 & 0 & \cdots & F_{pp} \end{bmatrix}$$

where $\lambda(T_{ii}) \cap \lambda(T_{jj}) \neq \emptyset$, $i \neq j$. The actual determination of the block sizes can be done using the methods of §7.6.

Next, we compute the submatrices $F_{ii} = f(T_{ii})$ for $i = 1{:}p$. Since the eigenvalues of T_{ii} are presumably close, these calculations require special methods. (Some possibilities are discussed in the next two sections.) Once the diagonal blocks of F are known, the blocks in the strict upper triangle of F can be found recursively, as in the scalar case. To derive the governing equations, we equate (i, j) blocks in $FT = TF$ for $i < j$ and obtain the following generalization of (11.1.4):

$$ F_{ij}T_{jj} - T_{ii}F_{ij} = T_{ij}F_{jj} - F_{ii}T_{ij} + \sum_{k=i+1}^{j-1} (T_{ik}F_{kj} - F_{ik}T_{kj}) . \qquad (11.1.5) $$

This is a linear system whose unknowns are the elements of the block F_{ij} and whose right-hand side is "known" if we compute the F_{ij} one block super-diagonal at a time. We can solve (11.1.5) using the Bartels-Stewart algorithm (Algorithm 7.6.2).

The block Schur approach described here is useful when computing real functions of real matrices. After computing the real Schur form $A = QTQ^T$, the block algorithm can be invoked in order to handle the 2-by-2 bumps along the diagonal of T.

Problems

P11.1.1 Using the definition (11.1.1) show that (a) $Af(A) = f(A)A$, (b) $f(A)$ is upper triangular if A is upper triangular, and (c) $f(A)$ is Hermitian if A is Hermitian.

P11.1.2 Rewrite Algorithm 11.1.1 so that $f(T)$ is computed column by column.

P11.1.3 Suppose $A = X\mathrm{diag}(\lambda_i)X^{-1}$ where $X = [\, x_1, \ldots, x_n \,]$ and $X^{-1} = [\, y_1, \ldots, y_n \,]^H$. Show that if $f(A)$ is defined, then

$$ f(A) = \sum_{k=1}^{n} f(\lambda_i)x_i y_i^H . $$

P11.1.4 Show that

$$ T = \begin{bmatrix} T_{11} & T_{12} \\ 0 & T_{22} \end{bmatrix} \begin{matrix} p \\ q \end{matrix} \qquad \Rightarrow \qquad f(T) = \begin{bmatrix} F_{11} & F_{12} \\ 0 & F_{22} \end{bmatrix} \begin{matrix} p \\ q \end{matrix} $$
$$ \begin{matrix} p & q \end{matrix} \qquad\qquad\qquad\qquad\qquad \begin{matrix} p & q \end{matrix} $$

where $F_{11} = f(T_{11})$ and $F_{22} = f(T_{22})$. Assume $f(T)$ is defined.

Notes and References for Sec. 11.1

The contour integral representation of $f(A)$ given in the text is useful in functional analysis because of its generality. See

N. Dunford and J. Schwartz (1958). *Linear Operators, Part I*, Interscience, New York.

As we discussed, other definitions of $f(A)$ are possible. However, for the matrix functions typically encountered in practice, all these definitions are equivalent. See

R.F. Rinehart (1955). "The Equivalence of Definitions of a Matric Function," *Amer. Math. Monthly 62*, 395–414.

Various aspects of the Jordan representation are detailed in

J.S. Frame (1964). "Matrix Functions and Applications, Part II," *IEEE Spectrum 1 (April)*, 102–8.
J.S. Frame (1964). "Matrix Functions and Applications, Part IV," *IEEE Spectrum 1 (June)*, 123–31.

The following are concerned with the Schur decomposition and its relationship to the $f(A)$ problem:

D. Davis (1973). "Explicit Functional Calculus," *Lin. Alg. and Its Applic. 6*, 193–99.
J. Descloux (1963). "Bounds for the Spectral Norm of Functions of Matrices," *Numer. Math. 5*, 185–90.
C.F. Van Loan (1975). "A Study of the Matrix Exponential," Numerical Analysis Report No. 10, Dept. of Maths., University of Manchester, England.

Algorithm 11.1.1 and the various computational difficulties that arise when it is applied to a matrix having close or repeated eigenvalues are discussed in

B.N. Parlett (1976). "A Recurrence Among the Elements of Functions of Triangular Matrices," *Lin. Alg. and Its Applic. 14*, 117–21.

A compromise between the Jordan and Schur approaches to the $f(A)$ problem results if A is reduced to block diagonal form as described in §7.6.3. See

B. Kågström (1977). "Numerical Computation of Matrix Functions," Department of Information Processing Report UMINF-58.77, University of Ümea, Sweden.

The sensitivity of matrix functions to perturbation is discussed in

C.S. Kenney and A.J. Laub (1989). "Condition Estimates for Matrix Functions," *SIAM J. Matrix Anal. Appl. 10*, 191–209.
C.S. Kenney and A.J. Laub (1994). "Small-Sample Statistical Condition Estimates for General Matrix Functions," *SIAM J. Sci. Comp. 15*, 36–61.

A theme in this chapter is that if A is nonnormal, then there is more to computing $f(A)$ than just computing $f(z)$ on $\lambda(A)$. The pseudo-eigenvalue concept is a way of understanding this phenomena. See

L.N. Trefethen (1992). "Pseudospectra of Matrices," in *Numerical Analysis 1991*, D.F. Griffiths and G.A. Watson (eds), Longman Scientific & Technical, Harlow, Essex, UK.

More details are offered in §11.3.4.

11.2 Approximation Methods

We now consider a class of methods for computing matrix functions which at first glance do not appear to involve eigenvalues. These techniques are based on the idea that if $g(z)$ approximates $f(z)$ on $\lambda(A)$, then $f(A)$ approximates $g(A)$, e.g.,

$$e^A \approx I + A + \frac{A^2}{2!} + \cdots + \frac{A^q}{q!}.$$

We begin by bounding $\| f(A) - g(A) \|$ using the Jordan and Schur matrix function representations. We follow this discussion with some comments on the evaluation of matrix polynomials.

11.2.1   A Jordan Analysis

The Jordan representation of matrix functions (Theorem 11.1.1) can be used to bound the error in an approximant $g(A)$ of $f(A)$.

Theorem 11.2.1 *Let* $X^{-1}AX = \mathrm{diag}(J_1, \ldots, J_p)$ *be the JCF of* $A \in \mathbb{C}^{n \times n}$ *with*

$$
J_i = \begin{bmatrix}
\lambda_i & 1 & \cdots & \cdots & 0 \\
0 & \lambda_i & 1 & \vdots & \vdots \\
\vdots & \vdots & \ddots & \ddots & \vdots \\
\vdots & \vdots & \vdots & \ddots & 1 \\
0 & \cdots & \cdots & \cdots & \lambda_i
\end{bmatrix}
$$

being an m_i-*by*-m_i *Jordan block. If* $f(z)$ *and* $g(z)$ *are analytic on an open set containing* $\lambda(A)$, *then*

$$
\| f(A) - g(A) \|_2 \le \kappa_2(X) \max_{\substack{1 \le i \le p \\ 0 \le r \le m_i - 1}} m_i \frac{\left| f^{(r)}(\lambda_i) - g^{(r)}(\lambda_i) \right|}{r!}.
$$

Proof. Defining $h(z) = f(z) - g(z)$ we have

$$
\| f(A) - g(A) \|_2 = \| X \mathrm{diag}(h(J_1), \ldots, h(J_p)) X^{-1} \|_2
$$

$$
\le \kappa_2(X) \max_{1 \le i \le p} \| h(J_i) \|_2 .
$$

Using Theorem 11.1.1 and equation (2.3.8) we conclude that

$$
\| h(J_i) \|_2 \le m_i \max_{0 \le r \le m_i - 1} \frac{\left| h^{(r)}(\lambda_i) \right|}{r!}
$$

thereby proving the theorem. $\square$

11.2.2   A Schur Analysis

If we rely on the Schur instead of the Jordan decomposition we obtain an alternative bound.

Theorem 11.2.2 *Let $Q^H A Q = T = \text{diag}(\lambda_i) + N$ be the Schur decomposition of $A \in \mathbb{C}^{n \times n}$, with N being the strictly upper triangular portion of T. If $f(z)$ and $g(z)$ are analytic on a closed convex set Ω whose interior contains $\lambda(A)$, then*

$$\| f(A) - g(A) \|_F \leq \sum_{r=0}^{n-1} \delta_r \frac{\| \, |N|^r \, \|_F}{r!}$$

where

$$\delta_r = \sup_{z \in \Omega} \left| f^{(r)}(z) - g^{(r)}(z) \right| .$$

Proof. Let $h(z) = f(z) - g(z)$ and set $H = (h_{ij}) = h(A)$. Let $S_{ij}^{(r)}$ denote the set of strictly increasing integer sequences $(s_0, \ldots, s_r)$ with the property that $s_0 = i$ and $s_r = j$. Notice that

$$S_{ij} = \bigcup_{r=1}^{j-i} S_{ij}^{(r)}$$

and so from Theorem 11.1.3, we obtain the following for all $i < j$:

$$h_{ij} = \sum_{r=1}^{j-1} \sum_{s \in S_{ij}^{(r)}} n_{s_0, s_1} n_{s_1, s_2} \cdots n_{s_{r-1}, s_r} h[\lambda_{s_0}, \ldots, \lambda_{s_r}].$$

Now since Ω is convex and h analytic, we have

$$|h[\lambda_{s_0}, \ldots, \lambda_{s_r}]| \leq \sup_{z \in \Omega} \frac{\left| h^{(r)}(z) \right|}{r!} = \frac{\delta_r}{r!} . \tag{11.2.1}$$

Furthermore if $|N|^r = (n_{ij}^{(r)})$ for $r \geq 1$, then it can be shown that

$$n_{ij}^{(r)} = \begin{cases} 0 & j < i + r \\[2mm] \displaystyle\sum_{s \in S_{ij}^{(r)}} \left| n_{s_0, s_1} n_{s_1, s_2} \cdots n_{s_{r-1}, s_r} \right| & j \geq i + r \end{cases} \tag{11.2.2}$$

The theorem now follows by taking absolute values in the expression for h_{ij} and then using (11.2.1) and (11.2.2). $\square$

The bounds in the above theorems suggest that there is more to approximating $f(A)$ than just approximating $f(z)$ on the spectrum of A. In particular, we see that if the eigensystem of A is ill-conditioned and/or A's departure from normality is large, then the discrepancy between $f(A)$ and $g(A)$ may be considerably larger than the maximum of $|f(z) - g(z)|$ on $\lambda(A)$. Thus, even though approximation methods avoid eigenvalue computations, they appear to be influenced by the structure of A's eigensystem, a point that we pursue further in the next section.

Example 11.2.1 Suppose

$$A = \begin{bmatrix} -.01 & 1 & 1 \\ 0 & 0 & 1 \\ 0 & 0 & .01 \end{bmatrix}.$$

If $f(z) = e^z$ and $g(z) = 1 + z + z^2/2$, then $\| f(A) - g(A) \| \approx 10^{-5}$ in either the Frobenius norm or the 2-norm. Since $\kappa_2(X) \approx 10^7$, the error predicted by Theorem 11.2.1 is $O(1)$, rather pessimistic. On the other hand, the error predicted by the Schur decomposition approach is $O(10^{-2})$.

11.2.3   Taylor Approximants

A popular way of approximating a matrix function such as e^A is through the truncation of its Taylor series. The conditions under which a matrix function $f(A)$ has a Taylor series representation are easily established.

Theorem 11.2.3 If $f(z)$ has a power series representation

$$f(z) = \sum_{k=0}^{\infty} c_k z^k$$

on an open disk containing $\lambda(A)$, then

$$f(A) = \sum_{k=0}^{\infty} c_k A^k .$$

Proof. We prove the theorem for the case when A is diagonalizable. In P11.2.1, we give a hint as to how to proceed without this assumption. Suppose $X^{-1}AX = D = \operatorname{diag}(\lambda_1,\ldots,\lambda_n)$. Using Corollary 11.1.2, we have

$$\begin{aligned} f(A) &= X\operatorname{diag}\left(f(\lambda_1),\ldots,f(\lambda_n)\right) X^{-1} \\ &= X\operatorname{diag}\left(\sum_{k=0}^{\infty} c_k\lambda_1^k,\ldots,\sum_{k=0}^{\infty} c_k\lambda_n^k \right) X^{-1} \\ &= X\left(\sum_{k=0}^{\infty} c_k D^k \right) X^{-1} = \sum_{k=0}^{\infty} c_k(XDX^{-1})^k = \sum_{k=0}^{\infty} c_k A^k. \ \square \end{aligned}$$

Several important transcendental matrix functions have particularly simple
series representations:

$$\log(I - A) = \sum_{k=1}^{\infty} \frac{A^k}{k} \qquad |\lambda| < 1, \ \lambda \in \lambda(A)$$

$$\sin(A) = \sum_{k=0}^{\infty} (-1)^k \frac{A^{2k+1}}{(2k+1)!}$$

$$\cos(A) = \sum_{k=0}^{\infty} (-1)^k \frac{A^{2k}}{(2k)!} \ .$$

The following theorem bounds the errors that arise when matrix functions
such as these are approximated via truncated Taylor series.

Theorem 11.2.4 *If $f(z)$ has the Taylor series*

$$f(z) = \sum_{k=0}^{\infty} \alpha_k z^k$$

on an open disk containing the eigenvalues of $A \in \mathbb{C}^{n \times n}$, then

$$\| f(A) - \sum_{k=0}^{q} \alpha_k A^k \|_2 \leq \frac{n}{(q+1)!} \max_{0 \leq s \leq 1} \| A^{q+1} f^{(q+1)}(As) \|_2 \ .$$

Proof. Define the matrix $E(s)$ by

$$f(As) = \sum_{k=0}^{q} \alpha_k (As)^k + E(s) \qquad 0 \leq s \leq 1. \tag{11.2.3}$$

If $f_{ij}(s)$ is the (i,j) entry of $f(As)$, then it is necessarily analytic and so

$$f_{ij}(s) = \left(\sum_{k=0}^{q} \frac{f_{ij}^{(k)}(0)}{k!} s^k \right) + \frac{f_{ij}^{(q+1)}(\varepsilon_{ij})}{(q+1)!} s^{q+1} \tag{11.2.4}$$

where ε_{ij} satisfies $0 \leq \varepsilon_{ij} \leq s \leq 1$.

By comparing powers of s in (11.2.3) and (11.2.4) we conclude that
$e_{ij}(s)$, the (i,j) entry of $E(s)$, has the form

$$e_{ij}(s) = \frac{f_{ij}^{(q+1)}(\varepsilon_{ij})}{(q+1)!} s^{q+1} \ .$$

Now $f_{ij}^{(q-1)}(s)$ is the (i, j) entry of $A^{q+1} f^{(q+1)}(As)$ and therefore

$$|e_{ij}(s)| \leq \max_{0 \leq s \leq 1} \frac{f_{ij}^{(q+1)}(s)}{(q+1)!} \leq \max_{0 \leq s \leq 1} \frac{\| A^{q+1} f^{(q+1)}(As) \|_2}{(q+1)!}.$$

The theorem now follows by applying (2.3.8). $\square$

Example 11.2.2 If

$$A = \begin{bmatrix} -49 & 24 \\ -64 & 31 \end{bmatrix},$$

then

$$e^A = \begin{bmatrix} -0.735759 & .0551819 \\ -1.471518 & 1.103638 \end{bmatrix}.$$

For $q = 59$, Theorem 11.2.4 predicts that

$$\| e^A - \sum_{k=0}^{q} \frac{A^k}{k!} \|_2 \leq \frac{n}{(q+1)!} \max_{0 \leq s \leq 1} \| A^{q+1} e^{As} \|_2 \leq 10^{-60}.$$

However, if $u \approx 10^{-7}$, then we find

$$fl\left(\sum_{k=0}^{59} \frac{A^k}{k!}\right) = \begin{bmatrix} -22.25880 & -1.4322766 \\ -61.49931 & -3.474280 \end{bmatrix}.$$

The problem is that some of the partial sums have large elements. For example, $I + \cdots + A^{17}/17!$ has entries of order 10^7. Since the machine precision is approximately 10^{-7}, rounding errors larger than the norm of the solution are sustained.

Example 11.2.2 highlights a shortcoming of truncated Taylor series approximation: It tends to be worthwhile only near the origin. The problem can sometimes be circumvented through a change of scale. For example, by repeated application of the *double angle* formulae:

$$\cos(2A) = 2\cos(A)^2 - I \qquad \sin(2A) = 2\sin(A)\cos(A)$$

it is possible to "build up" the sine and cosine of a matrix from suitably truncated Taylor series approximates:

$S_0 = $ Taylor approximate to $\sin(A/2^k)$
$C_0 = $ Taylor approximate to $\cos(A/2^k)$
for $j = 1:k$
    $S_j = 2S_{j-1}C_{j-1}$
    $C_j = 2C_{j-1}^2 - I$
end

Here k is a positive integer chosen so that, say, $\| A \|_\infty \approx 2^k$. See Serbin and Blalock (1979).

11.2.4   Evaluating Matrix Polynomials

Since the approximation of transcendental matrix functions so often involves the evaluation of polynomials, it is worthwhile to look at the details of computing

$$p(A) = b_0 I + b_1 A + \cdots + b_q A^q$$

where the scalars $b_0, \ldots, b_q \in \mathbb{R}$ are given. The most obvious approach is to invoke Horner's scheme:

Algorithm 11.2.1 Given a matrix A and $b(0{:}q)$, the following algorithm computes $F = b_q A^q + \cdots + b_1 A + b_0 I$.

$F = b_q A + b_{q-1} I$
for $k = q - 2{:} - 1{:}0$
　　$F = AF + b_k I$
end

This requires $q - 1$ matrix multiplications. However, unlike the scalar case, this summation process is not optimal. To see why, suppose $q = 9$ and observe that

$$\begin{aligned} p(A) = {}& A^3(A^3(b_9 A^3 + (b_8 A^2 + b_7 A + b_6 I)) \\ & + (b_5 A^2 + b_4 A + b_3 I)) + b_2 A^2 + b_1 A + b_0 I. \end{aligned}$$

Thus, $F = p(A)$ can be evaluated with only four matrix multiplies:

$$\begin{aligned} A_2 &= A^2 \\ A_3 &= A A_2 \\ F_1 &= b_9 A_3 + b_8 A_2 + b_7 A + b_6 I \\ F_2 &= A_3 F_1 + b_5 A_2 + b_4 A + b_3 I \\ F &= A_3 F_2 + b_2 A_2 + b_1 A + b_0 I. \end{aligned}$$

In general, if s is any integer satisfying $1 \le s \le \sqrt{q}$ then

$$p(A) = \sum_{k=0}^{r} B_k (A^s)^k \qquad r = \text{floor}(q/s) \qquad (11.2.5)$$

where

$$B_k = \begin{cases} b_{sk+s-1} A^{s-1} + \cdots + b_{sk+1} A + b_{sk} I & k = 0{:}r - 1 \\ b_q A^{q-sr} + \cdots + b_{sr+1} A + b_k I & k = r. \end{cases}$$

Once $A^2, \ldots, A^s$ are computed, Horner's rule can be applied to (11.2.5) and the net result is that $p(A)$ can be computed with $s + r - 1$ matrix multiplies. By choosing $s = \text{floor}(\sqrt{q})$, the number of matrix multiplies is approximately minimized. This technique is discussed in Paterson and Stockmeyer (1973). Van Loan (1978) shows how the procedure can be implemented without storage arrays for $A^2, \ldots, A^s$.

11.2.5   Computing Powers of a Matrix

The problem of raising a matrix to a given power deserves special mention. Suppose it is required to compute A^{13}. Noting that $A^4 = (A^2)^2$, $A^8 = (A^4)^2$ and $A^{13} = A^8 A^4 A$, we see that this can be accomplished with just 5 matrix multiplications. In general we have

Algorithm 11.2.2 (Binary Powering)   Given a positive integer s and $A \in \mathbb{R}^{n \times n}$, the following algorithm computes $F = A^s$ where s is a positive integer and $A \in \mathbb{R}^{n \times n}$.

Let $s = \displaystyle\sum_{k=0}^{t} \beta_k 2^k$ be the binary expansion of s with $\beta_t \neq 0$.
$Z = A;\ q = 0$
while $\beta_q = 0$
$\qquad Z = Z^2;\ q = q + 1$
end
$F = Z$
for $k = q + 1{:}t$
$\qquad Z = Z^2$
$\qquad$ **if** $\beta_k \neq 0$
$\qquad\qquad F = FZ$
$\qquad$ **end**
end

This algorithm requires at most $2 \,\text{floor}[\log_2(s)]$ matrix multiplies. If s is a power of 2, then only $\log_2(s)$ matrix multiplies are needed.

11.2.6   Integrating Matrix Functions

We conclude this section with some remarks on the integration of matrix functions. Suppose $f(At)$ is defined for all $t \in [a, b]$ and that we wish to compute

$$F = \int_a^b f(At)\,dt.$$

As in (11.1.1) the integration is on an element-by-element basis.

Ordinary quadrature rules can be applied to F. For example, with Simpson's rule, we have

$$F \approx \tilde{F} = \frac{h}{3} \sum_{k=0}^{m} w_k f(A(a + kh)) \qquad (11.2.6)$$

where m is even, $h = (b - a)/m$ and

$$w_k = \begin{cases} 1 & k = 0, m \\ 4 & k \text{ odd} \\ 2 & k \text{ even}, k \neq 0, m . \end{cases}$$

If $(d^4/dz^4)f(zt) = f^{(4)}(zt)$ is continuous for $t \in [a, b]$ and if $f^{(4)}(At)$ is defined on this same interval, then it can be shown that $\tilde{F} = F + E$ where

$$\| E \|_2 \leq \frac{nh^4(b - a)}{180} \max_{a \leq t \leq b} \| f^{(4)}(At) \|_2 . \qquad (11.2.7)$$

Let f_{ij} and e_{ij} denote the (i, j) entries of F and E, respectively. Under the above assumptions we can apply the standard error bounds for Simpson's rule and obtain

$$|e_{ij}| \leq \frac{h^4(b - a)}{180} \max_{a \leq t \leq b} |e_i^T f^{(4)}(At)e_j| .$$

The inequality (11.2.7) now follows since $\| E \|_2 \leq n \max |e_{ij}|$ and

$$\max_{a \leq t \leq b} |e_i^T f^{(4)}(At)e_j| \leq \max_{a \leq t \leq b} \| f^{(4)}(At) \|_2 .$$

Of course, in the practical application of (11.2.6), the function evaluations $f(A(a + kh))$ normally have to be approximated. Thus, the overall error involves the error in approximating $f(A(a+kh))$ as well as the Simpson rule error.

Problems

P11.2.1 (a) Suppose $G = \lambda I + E$ is a p-by-p Jordan block, where $E = (\delta_{i,j-1})$. Show that

$$(\lambda I + E)^k = \sum_{j=0}^{\min\{p-1,k\}} \binom{k}{j} \lambda^{k-j} E^j .$$

(b) Use (a) and Theorem 11.1.1 to prove Theorem 11.2.3.

P11.2.2 Verify (11.2.2).

P11.2.3 Show that if $\| A \|_2 < 1$, then $\log(I + A)$ exists and satisfies the bound

$\| \log(I + A) \|_2 \leq \| A \|_2/(1 - \| A \|_2)$.

P11.2.4   Let A by an n-by-n symmetric positive definite matrix. (a) Show that there exists a unique symmetric positive definite X such that $A = X^2$. (b) Show that if $X_0 = I$ and $X_{k+1} = (X_k + AX_k^{-1})/2$ then $X_k \rightarrow \sqrt{A}$ quadratically where $\sqrt{A}$ denotes the matrix X in part (a).

P11.2.5   Specialize Algorithm 11.2.1 to the case when A is symmetric. Repeat for the case when A is upper triangular. In both instances, give the associated flop counts.

P11.2.6   Show that $X(t) = C_1 \cos(t\sqrt{A}) + C_2\sqrt{A^{-1}} \sin(t\sqrt{A})$ solves the initial value problem $\ddot{X}(t) = -AX(t)$, $X(0) = C_1$, $\dot{X}(0) = C_2$. Assume that A is symmetric positive definite.

P11.2.7   Using Theorem 11.2.4, bound the error in the approximations:

$$\sin(A) \approx \sum_{k=0}^{q}(-1)^k \frac{A^{2k+1}}{(2k+1)!} \qquad \cos(A) \approx \sum_{k=0}^{q}(-1)^k \frac{A^{2k}}{(2k)!} .$$

P11.2.8   Suppose $A \in \mathbf{R}^{n \times n}$ is nonsingular and $X_0 \in \mathbf{R}^{n \times n}$ is given. The iteration defined by

$$X_{k+1} = X_k(2I - AX_k)$$

is the matrix analog of Newton's method applied to the function $f(x) = a - (1/x)$. Use the SVD to analyze this iteration. Do the iterates converge to A^{-1}? Discuss the choice of X_0.

Notes and References for Sec. 11.2

The optimality of Horner's rule for polynomial evaluation is discussed in

D. Knuth (1981). *The Art of Computer Programming , vol. 2. Seminumerical Algorithms* , 2nd ed., Addison-Wesley, Reading, Massachusetts.

M.S. Paterson and L.J. Stockmeyer (1973). "On the Number of Nonscalar Multiplications Necessary to Evaluate Polynomials," *SIAM J. Comp. 2*, 60–66.

The Horner evaluation of matrix polynomials is analyzed in

C.F. Van Loan (1978). "A Note on the Evaluation of Matrix Polynomials," *IEEE Trans. Auto. Cont. AC-24*, 320–21.

Other aspects of matrix function computation are discussed in

N.J. Higham and P.A. Knight (1995). "Matrix Powers in Finite Precision Arithmetic," *SIAM J. Matrix Anal. Appl. 16*, 343–358.

R. Mathias (1993). "Approximation of Matrix-Valued Functions," *SIAM J. Matrix Anal. Appl. 14*, 1061–1063.

S. Friedland (1991). "Revisiting Matrix Squaring," *Lin. Alg. and Its Applic. 154-156*, 59–63.

H. Bolz and W. Niethammer (1988). "On the Evaluation of Matrix Functions Given by Power Series," *SIAM J. Matrix Anal. Appl. 9*, 202–209.

The Newton and Language representations for $f(A)$ and their relationship to other matrix function definitions is discussed in

R.F. Rinehart (1955). "The Equivalence of Definitions of a Matric Function," *Amer. Math. Monthly 62*, 395–414.

The "double angle" method for computing the cosine of matrix is analyzed in

S. Serbin and S. Blalock (1979). "An Algorithm for Computing the Matrix Cosine," *SIAM J. Sci. Stat. Comp. 1*, 198–204.

The square root is a particularly important matrix function. See §4.2.10. Several approaches are possible:

Å. Björck and S. Hammarling (1983). "A Schur Method for the Square Root of a Matrix," *Lin. Alg. and Its Applic. 52/53*, 127–140.
N.J. Higham (1986). "Newton's Method for the Matrix Square Root," *Math. Comp. 46*, 537–550.
N.J. Higham (1987). "Computing Real Square Roots of a Real Matrix," *Lin. Alg. and Its Applic. 88/89*, 405–430.

11.3 The Matrix Exponential

One of the most frequently computed matrix functions is the exponential

$$e^{At} = \sum_{k=0}^{\infty} \frac{(At)^k}{k!} .$$

Numerous algorithms for computing e^{At} have been proposed, but most of them are of dubious numerical quality, as is pointed out in the survey article by Moler and Van Loan (1978). In order to illustrate what the computational difficulties are, we present a "scaling and squaring" method based upon Padé approximation. A brief analysis of the method follows that involves some e^{At} perturbation theory and comments about the shortcomings of eigenanalysis in settings where non-normality prevails.

11.3.1 A Padé Approximation Method

Following the discussion in §11.2, if $g(z) \approx e^z$, then $g(A) \approx e^A$. A very useful class of approximants for this purpose are the Padé functions defined by

$$R_{pq}(z) = D_{pq}(z)^{-1} N_{pq}(z),$$

where

$$N_{pq}(z) = \sum_{k=0}^{p} \frac{(p+q-k)!p!}{(p+q)!k!(p-k)!} z^k$$

and

$$D_{pq}(z) = \sum_{k=0}^{q} \frac{(p+q-k)!q!}{(p+q)!k!(q-k)!} (-z)^k .$$

Notice that $R_{p0}(z) = 1 + z + \cdots + z^p/p!$ is the pth order Taylor polynomial.

Unfortunately, the Padé approximants are good only near the origin, as the following identity reveals:

$$e^A = R_{pq}(A) + \frac{(-1)^q}{(p+q)!} A^{p+q+1} D_{pq}(A)^{-1} \int_0^1 u^p (1-u)^q e^{A(1-u)} du. \quad (11.3.1)$$

However, this problem can be overcome by exploiting the fact that $e^A = (e^{A/m})^m$. In particular, we can scale A by m such that $F_{pq} = R_{pq}(A/m)$ is a suitably accurate approximation to $e^{A/m}$. We then compute F_{pq}^m using Algorithm 11.2.2. If m is a power of two, then this amounts to repeated squaring and so is very efficient. The success of the overall procedure depends on the accuracy of the approximant

$$F_{pq} = \left(R_{pq}\left(\frac{A}{2^j}\right) \right)^{2^j}.$$

In Moler and Van Loan (1978) it is shown that if

$$\frac{\| A \|_\infty}{2^j} \leq \frac{1}{2},$$

then there exists an $E \in \mathbb{R}^{n \times n}$ such that

$$
\begin{aligned}
F_{pq} &= e^{A+E} \\
AE &= EA \\
\| E \|_\infty &\leq \varepsilon(p,q)\| A \|_\infty \\
\varepsilon(p,q) &= 2^{3-(p+q)} \frac{p!q!}{(p+q)!(p+q+1)!}.
\end{aligned}
$$

These results form the basis of an effective e^A procedure with error control. Using the above formulae it is easy to establish the inequality:

$$\frac{\| e^A - F_{pq} \|_\infty}{\| e^A \|_\infty} \leq \varepsilon(p,q)\| A \|_\infty e^{\varepsilon(p,q)\| A \|_\infty}.$$

The parameters p and q can be determined according to some relative error tolerance. Note that since F_{pq} requires about $j + \max(p,q)$ matrix multiplies it makes sense to set $p = q$ as this choice minimizes $\varepsilon(p,q)$ for a given amount of work. Encapsulating these ideas we obtain

Algorithm 11.3.1 Given $\delta > 0$ and $A \in \mathbb{R}^{n \times n}$, the following algorithm computes $F = e^{A+E}$ where $\| E \|_\infty \leq \delta \| A \|_\infty$

$j = \max(0, 1 + \text{floor}(\log_2(\| A \|_\infty)))$
$A = A/2^j$
Let q be the smallest non-negative integer such that $\varepsilon(q,q) \leq \delta$.

$D = I; \ N = I; \ X = I; \ c = 1$
for $k = 1{:}q$
$\quad c = c(q - k + 1)/[(2q - k + 1)k]$
$\quad X = AX; \ N = N + cX; \ D = D + (-1)^k cX$
end
Solve $DF = N$ for F using Gaussian elimination.
for $k = 1{:}j$
$\quad F = F^2$
end

This algorithm requires about $2(q + j + 1/3)n^3$flops. The roundoff error properties of have essentially been analyzed by Ward (1977).

The special Horner techniques of §11.2 can be applied to quicken the computation of $D = D_{qq}(A)$ and $N = N_{qq}(A)$. For example, if $q = 8$ we have $N_{qq}(A) = U + AV$ and $D_{qq}(A) = U - AV$ where

$$U = c_0 I + c_2 A^2 + (c_4 I + c_6 A^2 + c_8 A^4)A^4$$

and

$$V = c_1 I + c_3 A^2 + (c_5 I + c_7 A^2)A^4.$$

Clearly, N and D can be found in 5 matrix multiplies rather than the 7 required by Algorithm 11.3.1.

11.3.2 Perturbation Theory

Is Algorithm 11.3.1 stable in the presence of roundoff error? To answer this question we need to understand the sensitivity of the matrix exponential to perturbations in A. The starting point in the discussion is the initial value problem

$$\dot{X}(t) = AX(t) \qquad X(0) = I$$

where $A, X(t) \in \mathbb{R}^{n \times n}$. This has the unique solution $X(t) = e^{At}$, a characterization of the matrix exponential that can be used to establish the identity

$$e^{(A+E)t} - e^{At} = \int_0^t e^{A(t-s)} E e^{(A+E)s} ds \ .$$

From this it follows that

$$\frac{\| \, e^{(A+E)t} - e^{At} \, \|_2}{\| \, e^{At} \, \|_2} \ \leq \ \frac{\| \, E \, \|_2}{\| \, e^{At} \, \|_2} \int_0^t \| \, e^{A(t-s)} \, \|_2 \, \| \, e^{(A+E)s} \, \|_2 ds \ .$$

Further simplifications result if we bound the norms of the exponentials that appear in the integrand. One way of doing this is through the Schur decomposition. If $Q^H A Q = \text{diag}(\lambda_i) + N$ is the Schur decomposition of $A \in \mathbb{C}^{n \times n}$, then it can be shown that

$$\| \, e^{At} \, \|_2 \ \leq \ e^{\alpha(A)t M_S(t)}, \tag{11.3.2}$$

where

$$\alpha(A) = \max\{\text{Re}(\lambda) : \lambda \in \lambda(A)\} \qquad (11.3.3)$$

and

$$M_S(t) = \sum_{k=0}^{n-1} \frac{\| Nt \|_2^k}{k!}.$$

The quantity $\alpha(A)$ is called the *spectral abscissa* and with a little manipulation it can be shown that

$$\frac{\| e^{(A+E)t} - e^{At} \|_2}{\| e^{At} \|_2} \leq t \| E \|_2 M_S(t)^2 \exp(t M_S(t) \| E \|_2).$$

Notice that $M_S(t) \equiv 1$ if and only if A is normal, suggesting that the matrix exponential problem is "well behaved" if A is normal. This observation is confirmed by the behavior of the *matrix exponential condition number* $\nu(A, t)$, defined by

$$\nu(A, t) = \max_{\| E \|_2 \leq 1} \left\| \int_0^t e^{A(t-s)} E e^{As} ds \right\|_2 \frac{\| A \|_2}{\| e^{At} \|_2}.$$

This quantity, discussed in Van Loan (1977), measures the sensitivity of the map $A \to e^{At}$ in that for a given t, there is a matrix E for which

$$\frac{\| e^{(A+E)t} - e^{At} \|_2}{\| e^{At} \|_2} \approx \nu(A, t) \frac{\| E \|_2}{\| A \|_2}.$$

Thus, if $\nu(A, t)$ is large, small changes in A can induce relatively large changes in e^{At}. Unfortunately, it is difficult to characterize precisely those A for which $\nu(A, t)$ is large. (This is in contrast to the linear equation problem $Ax = b$, where the ill-conditioned A are neatly described in terms of SVD.) One thing we can say, however, is that $\nu(A, t) \geq t \| A \|_2$, with equality holding for all non-negative t if and only if A is normal.

Dwelling a little more on the effect of non-normality, we know from the analysis of §11.2 that approximating e^{At} involves more than just approximating e^{zt} on $\lambda(A)$. Another clue that eigenvalues do not "tell the whole story" in the e^{At} problem has to do with the inability of the spectral abscissa (11.3.3) to predict the size of $\| e^{At} \|_2$ as a function of time. If A is normal, then

$$\| e^{At} \|_2 = e^{\alpha(A)t} \qquad (11.3.4)$$

Thus, there is uniform decay if the eigenvalues of A are in the open left half plane. But if A is non-normal, then e^{At} can grow before decay "sets in." The 2-by-2 example

$$A = \begin{bmatrix} -1 & M \\ 0 & -1 \end{bmatrix} \Leftrightarrow e^{At} = e^{-t} \begin{bmatrix} 1 & tM \\ 0 & 1 \end{bmatrix}$$

plainly illustrates this point.

11.3.3   Some Stability Issues

With this discussion we are ready to begin thinking about the stability of Algorithm 11.3.1. A potential difficulty arises during the squaring process if A is a matrix whose exponential grows before it decays. If

$$G = R_{qq}\left(\frac{A}{2^j}\right) \approx e^{A/2^j},$$

then it can be shown that rounding errors of order

$$\gamma = \mathbf{u}\| G^2 \|_2 \| G^4 \|_2 \| G^8 \|_2 \cdots \| G^{2^{j-1}} \|_2$$

can be expected to contaminate the computed G^{2^j}. If $\| e^{At} \|_2$ has a substantial initial growth, then it may be the case that

$$\gamma \gg \mathbf{u}\| G^{2^j} \|_2 \approx \mathbf{u}\| e^A \|_2$$

thus ruling out the possibility of small relative errors.

If A is normal, then so is the matrix G and therefore $\| G^m \|_2 = \| G \|_2^m$ for all positive integers m. Thus, $\gamma \approx \mathbf{u}\| G^{2^j} \|_2 \approx \mathbf{u}\| e^A \|_2$ and so the initial growth problems disappear. The algorithm can essentially be guaranteed to produce small relative error when A is normal. On the other hand, it is more difficult to draw conclusions about the method when A is non-normal because the connection between $\nu(A,t)$ and the initial growth phenomena is unclear. However, numerical experiments suggest that Algorithm 11.3.1 fails to produce a relatively accurate e^A only when $v(A,1)$ is correspondingly large.

11.3.4   Eigenvalues and Pseudo-Eigenvalues

We closed §7.1 with a comment that the eigenvalues of a matrix are generally not good "informers" when it comes to measuring nearness to singularity, unless the matrix is normal. It is the singular values that shed light on $Ax = b$ sensitivity. Our discussion of the matrix exponential is another warning to the same effect. The spectrum of a non-normal A does not completely describe e^{At} behavior.

In many applications, the eigenvalues of a matrix "say something" about an underlying phenomenon that is being modeled. If the eigenvalues are extremely sensitive to perturbation, then what they say can be misleading. This has prompted the development of the idea of pseudospectra. For $\epsilon \geq 0$, the ϵ-*pseudospectrum* of a matrix A is a subset of the complex plane defined by

$$\lambda_\epsilon(A) = \left\{ z \in \mathbb{C} : \| (zI - A)^{-1} \|_2 \geq \frac{1}{\epsilon} \right\} \qquad (11.3.5)$$

Qualitatively, z is a pseudo-eigenvalue of A if $zI - A$ is sufficiently close to singular. By convention we set $\lambda_0(A) = \lambda(A)$. Here are some pseudospectra properties:

1. If $\epsilon_1 \le \epsilon_2$, then $\lambda_{\epsilon_1}(A) \subseteq \lambda_{\epsilon_2}(A)$.

2. $\lambda_\epsilon(A) = \{z \in \mathbb{C} : \sigma_{min}(zI - A) \le \epsilon\}$.

3. $\lambda_\epsilon(A) = \{z \in \mathbb{C} : z \in \lambda(A + E), \text{for some } E \text{ with } \| E \|_2 \le \epsilon\}$.

Plotting the pseudospectra of a non-normal matrix A can provide insight into behavior. Here "behavior" can mean anything from the mathematical behavior of an iteration to solve $Ax = b$ to the physical behavior predicted by a model that involves A. See Higham and Trefethen (1993), Nachtigal, Reddy, and Trefethen (1992), and Trefethen, Trefethen, Reddy, and Driscoll (1993).

Problems

P11.3.1 Show that $e^{(A+B)t} = e^{At}e^{Bt}$ for all t if and only if $AB = BA$. (Hint: Express both sides as a power series in t and compare the coefficient of t.)

P11.3.2 Suppose that A is skew-symmetric. Show that both e^A and the (1,1) Padé approximate $R_{11}(A)$ are orthogonal. Are there any other values of p and q for which $R_{pq}(A)$ is orthogonal?

P11.3.3 Show that if A is nonsingular, then there exists a matrix X such that $A = e^X$. Is X unique?

P11.3.4 Show that if

$$\exp\left(\begin{bmatrix} -A^T & P \\ 0 & A \end{bmatrix} z\right) = \begin{bmatrix} F_{11} & F_{12} \\ 0 & F_{22} \end{bmatrix} \begin{matrix} n \\ n \end{matrix}$$
$$\begin{matrix} n & n \end{matrix}$$

then

$$F_{11}^T F_{12} = \int_0^z e^{A^T t} P e^{At} dt.$$

P11.3.5 Give an algorithm for computing e^A when $A = uv^T$, $u, v \in \mathbb{R}^n$.

P11.3.6 Suppose $A \in \mathbb{R}^{n \times n}$ and that $v \in \mathbb{R}^n$ has unit 2-norm. Define the function $\phi(t) = \| e^{At} v \|_2^2 / 2$ and show that

$$\dot{\phi}(t) \le \mu(A)\phi(t)$$

where $\mu(A) = \lambda_1((A + A^T)/2)$. Conclude that $\| e^{At} \|_2 \le e^{\mu(A)t}$ where $t \ge 0$.

P11.3.7 Prove the three pseudospectra properties given in the text.

Notes and References for Sec. 11.3

Much of what appears in this section and an extensive bibliography may be found in the following survey article:

C.B. Moler and C.F. Van Loan (1978). "Nineteen Dubious Ways to Compute the Exponential of a Matrix," *SIAM Review 20*, 801–36.

Scaling and squaring with Padé approximants (Algorithm 11.3.1) and a careful implementation of Parlett's Schur decomposition method (Algorithm 11.1.1) were found to be among the less dubious of the nineteen methods scrutinized. Various aspects of Padé approximation of the matrix exponential are discussed in

W. Fair and Y. Luke (1970). "Padé Approximations to the Operator Exponential,"
    Numer. Math. 14, 379–82.
C.F. Van Loan (1977). "On the Limitation and Application of Padé Approximation to
    the Matrix Exponential," in Padé and Rational Approximation, ed. E.B. Saff and
    R.S. Varga, Academic Press, New York.
R.C. Ward (1977). "Numerical Computation of the Matrix Exponential with Accuracy
    Estimate," SIAM J. Num. Anal. 14, 600–14.
A. Wragg (1973). "Computation of the Exponential of a Matrix I: Theoretical Consid-
    erations," J. Inst. Math. Applic. 11, 369–75.
A. Wragg (1975). "Computation of the Exponential of a Matrix II: Practical Consider-
    ations," J. Inst. Math. Applic. 15, 273–78.

A proof of equation (11.3.1) for the scalar case appears in

R.S. Varga (1961). "On Higher-Order Stable Implicit Methods for Solving Parabolic
    Partial Differential Equations," J. Math. Phys. 40, 220–31.

There are many applications in control theory calling for the computation of the matrix exponential. In the linear optimal regular problem, for example, various integrals
involving the matrix exponential are required. See

J. Johnson and C.L. Phillips (1971). "An Algorithm for the Computation of the Integral
    of the State Transition Matrix," IEEE Trans. Auto. Cont. AC-16, 204–5.
C.F. Van Loan (1978). "Computing Integrals Involving the Matrix Exponential," IEEE
    Trans. Auto. Cont. AC-23, 395–404.

An understanding of the map $A \rightarrow \exp(At)$ and its sensitivity is helpful when assessing
the performance of algorithms for computing the matrix exponential. Work in this direction includes

B. Kågström (1977). "Bounds and Perturbation Bounds for the Matrix Exponential,"
    BIT 17, 39–57.
C.F. Van Loan (1977). "The Sensitivity of the Matrix Exponential," SIAM J. Num.
    Anal. 14, 971–81.
R. Mathias (1992). "Evaluating the Frechet Derivative of the Matrix Exponential,"
    Numer. Math. 63, 213–226.

The computation of a logarithm of a matrix is an important area demanding much more
work. These calculations arise in various "system identification" problems. See

B. Singer and S. Spilerman (1976). "The Representation of Social Processes by Markov
    Models," Amer. J. Sociology 82, 1–54.
B.W. Helton (1968). "Logarithms of Matrices," Proc. Amer. Math. Soc. 19, 733–36.

For pointers into the pseudospectra literature we recommend

L.N. Trefethen (1992). "Pseudospecta of Matrices," in Numerical Analysis 1991, D.F.
    Griffiths and G.A. Watson (eds), Longman Scientific and Technical, Harlow, Essex,
    UK, 234–262.
D.J. Higham and L.N. Trefethen (1993). "Stiffness of ODES," BIT 33, 285–303.
L.N. Trefethen, A.E. Trefethen, S.C. Reddy, and T.A. Driscoll (1993). "Hydrodynamic
    Stability Without Eigenvalues," Science 261, 578–584.

as well as Chaitin-Chatelin and Frayssé (1996, chapter 10).

Chapter 12

Special Topics

In this final chapter we discuss an assortment of problems that represent important applications of the singular value, QR, and Schur decompositions. We first consider least squares minimization with constraints. Two types of constraints are considered in §12.1, quadratic inequality and linear equality. The next two sections are also concerned with variations on the standard LS problem. In §12.2 we consider how the vector of observations b might be approximated by some subset of A's columns, a course of action that is sometimes appropriate if A is rank-deficient. In §12.3 we consider a variation of ordinary regression known as total least squares that has appeal when A is contaminated with error. More applications of the SVD are considered in §12.4, where various subspace calculations are considered. In §12.5 we investigate the updating of orthogonal factorizations when the matrix A undergoes a low-rank perturbation. Some variations of the basic eigenvalue problem are discussed in §12.6.

Before You Begin

Because of the topical nature of this chapter, it doesn't make sense to have a chapter-wide, before-you-begin advisory. Instead, each section will begin with pointers to earlier portions of the book, and, if appropriate, pointers to LAPACK and other texts.

12.1   Constrained Least Squares

In the least squares setting it is sometimes natural to minimize $\| Ax - b \|_2$
over a proper subset of $\mathbb{R}^n$. For example, we may wish to predict b as best
we can with Ax subject to the constraint that x is a unit vector. Or, perhaps
the solution defines a fitting function $f(t)$ which is to have prescribed values
at a finite number of points. This can lead to an equality constrained least
squares problem. In this section we show how these problems can be solved
using the QR factorization and the SVD.

Chapter 5 and §8.7 should be understood before reading this section.
LAPACK connections include:

LAPACK: Tools for Generalized/Constrained LS Problems	
_GGLSE	Solves the equality constrained LS problem
_GGQRF	Computes the generalized QR factorization of a matrix pair
_GGRQF	Computes the generalized RQ factorization of a matrix pair
_GGSVP	Converts the GSVD problem to triangular form
_TGSJA	Computes the GSVD of a pair of triangular matrices

Complementary references include Lawson and Hanson (1974) and Björck
(1996).

12.1.1   The Problem LSQI

Least squares minimization with a quadratic inequality constraint—the
LSQI problem—is a technique that can be used whenever the solution to
the ordinary LS problem needs to be *regularized*. A simple LSQI problem
that arises when attempting to fit a function to noisy data is

$$\text{minimize } \| Ax - b \|_2 \quad \text{subject to } \| Bx \|_2 \leq \alpha \qquad (12.1.1)$$

where $A \in \mathbb{R}^{m \times n}$, $b \in \mathbb{R}^m$, $B \in \mathbb{R}^{n \times n}$ (nonsingular), and $\alpha \geq 0$. The constraint defines a hyperellipsoid in $\mathbb{R}^n$ and is usually chosen to damp out
excessive oscillation in the fitting function. This can be done, for example,
if B is a discretized second derivative operator.

More generally, we have the problem

$$\text{minimize } \| Ax - b \|_2 \quad \text{subject to } \| Bx - d \|_2 \leq \alpha \qquad (12.1.2)$$

where $A \in \mathbb{R}^{m \times n}$ ($m \geq n$), $b \in \mathbb{R}^m$, $B \in \mathbb{R}^{p \times n}$, $d \in \mathbb{R}^p$, and $\alpha \geq 0$. The
generalized singular value decomposition of §8.7.3 sheds light on the solv-
ability of (12.1.2). Indeed, if

$$
\begin{aligned}
U^T A X &= \text{diag}(\alpha_1, \ldots, \alpha_n) \qquad U^T U = I_m \\
V^T B X &= \text{diag}(\beta_1, \ldots, \beta_q) \qquad V^T V = I_p, \quad q = \min\{p, n\}
\end{aligned}
\qquad (12.1.3)
$$

is the generalized singular value decomposition of A and B, then (12.1.2) transforms to

$$\text{minimize } \| D_A y - \tilde{b} \|_2 \quad \text{subject to } \| D_B y - \tilde{d} \|_2 \leq \alpha$$

where $\tilde{b} = U^T b$, $\tilde{d} = V^T d$, and $y = X^{-1} x$. The simple form of the objective function

$$\| D_A y - \tilde{b} \|_2^2 = \sum_{i=1}^{n} (\alpha_i y_i - \tilde{b}_i)^2 + \sum_{i=n+1}^{m} \tilde{b}_i^2 \tag{12.1.4}$$

and the constraint equation

$$\| D_B y - \tilde{d} \|_2^2 = \sum_{i=1}^{r} (\beta_i y_i - \tilde{d}_i)^2 + \sum_{i=r+1}^{p} \tilde{d}_i^2 \leq \alpha^2 \tag{12.1.5}$$

facilitate the analysis of the LSQI problem. Here, $r = \text{rank}(B)$ and we assume that $\beta_{r+1} = \cdots = \beta_q = 0$.

To begin with, the problem has a solution if and only if

$$\sum_{i=r+1}^{p} \tilde{d}_i^2 \leq \alpha^2 .$$

If we have equality in this expression then consideration of (12.1.4) and (12.1.5) shows that the vector defined by

$$y_i = \begin{cases} \tilde{d}_i/\beta_i & i = 1{:}r \\ \tilde{b}_i/\alpha_i & i = r+1{:}n, \alpha_i \neq 0 \\ 0 & i = r+1{:}n, \alpha_i = 0 \end{cases} \tag{12.1.6}$$

solves the LSQI problem. Otherwise

$$\sum_{i=r+1}^{p} \tilde{d}_i^2 < \alpha^2. \tag{12.1.7}$$

and we have more alternatives to pursue. The vector $y \in \mathbb{R}^n$, defined by

$$y_i = \begin{cases} \tilde{b}_i/\alpha_i & \alpha_i \neq 0 \\ \tilde{d}_i/\beta_i & \alpha_i = 0 \end{cases} \quad i = 1{:}n$$

is a minimizer of $\| D_A y - \tilde{b} \|_2$. If this vector is also feasible, then we have a solution to (12.1.2). (This is not necessarily the solution of minimum 2-norm, however.) We therefore assume that

$$\sum_{\substack{i=1 \\ \alpha_i \neq 0}}^{q} \left(\beta_i \frac{\tilde{b}_i}{\alpha_i} - \tilde{d}_i \right)^2 + \sum_{i=q+1}^{p} \tilde{d}_i^2 > \alpha^2. \tag{12.1.8}$$

This implies that the solution to the LSQI problem occurs on the boundary of the feasible set. Thus, our remaining goal is to

$$\text{minimize } \| D_A y - \tilde{b} \|_2 \quad \text{subject to } \| D_B y - \tilde{d} \|_2 = \alpha .$$

To solve this problem, we use the method of Lagrange multipliers. Defining

$$h(\lambda, y) = \| D_A y - \tilde{b} \|_2^2 + \lambda \left(\| D_B y - \tilde{d} \|_2^2 - \alpha^2 \right)$$

we see that the equations $0 = \partial h / \partial y_i$, $i = 1{:}n$, lead to the linear system

$$(D_A^T D_A + \lambda D_B^T D_B) y = D_A^T \tilde{b} + \lambda D_B^T \tilde{d}.$$

Assuming that the matrix of coefficients is nonsingular, this has a solution $y(\lambda)$ where

$$y_i(\lambda) = \begin{cases} \dfrac{\alpha_i \tilde{b}_i + \lambda \beta_i \tilde{d}_i}{\alpha_i^2 + \lambda \beta_i^2} & i = 1{:}q \\[2mm] \tilde{b}_i / \alpha_i & i = q+1{:}n \end{cases}$$

To determine the Lagrange parameter we define

$$\phi(\lambda) \equiv \| D_B y(\lambda) - \tilde{d} \|_2^2 = \sum_{i=1}^{r} \left(\alpha_i \frac{\beta_i \tilde{b}_i - \alpha_i \tilde{d}_i}{\alpha_i^2 + \lambda \beta_i^2} \right)^2 + \sum_{i=r+1}^{p} \tilde{d}_i^2$$

and seek a solution to $\phi(\lambda) = \alpha^2$. Equations of this type are referred to as *secular equations* and we encountered them earlier in §8.5.3. From (12.1.8) we see that $\phi(0) > \alpha^2$. Now $\phi(\lambda)$ is monotone decreasing for $\lambda > 0$, and (12.1.8) therefore implies the existence of a unique positive λ^* for which $\phi(\lambda^*) = \alpha^2$. It is easy to show that this is the desired root. It can be found through the application of any standard root-finding technique, such as Newton's method. The solution of the original LSQI problem is then $x = Xy(\lambda^*)$.

12.1.2   LS Minimization Over a Sphere

For the important case of minimization over a sphere ($B = I_n$, $d = 0$), we have the following procedure:

Algorithm 12.1.1   Given $A \in \mathbb{R}^{m \times n}$ with $m \geq n$, $b \in \mathbb{R}^m$, and $\alpha > 0$, the following algorithm computes a vector $x \in \mathbb{R}^n$ such that $\| Ax - b \|_2$ is minimum, subject to the constraint that $\| x \|_2 \leq \alpha$.

> Compute the SVD $A = U \Sigma V^T$, save $V = [\, v_1, \ldots, v_n \,]$, and
>   form $b = U^T b$.
> $r = \text{rank}(A)$

if $\displaystyle\sum_{i=1}^{r}\left(\frac{b_i}{\sigma_i}\right)^2 > \alpha^2$

   Find λ^* such that $\displaystyle\sum_{i=1}^{r}\left(\frac{\sigma_i b_i}{\sigma_i^2 + \lambda^*}\right)^2 = \alpha^2.$

   $\displaystyle x = \sum_{i=1}^{r}\left(\frac{\sigma_i b_i}{\sigma_i^2 + \lambda^*}\right)v_i$

else

   $\displaystyle x = \sum_{i=1}^{r}\left(\frac{b_i}{\sigma_i}\right)v_i$

end

The SVD is the dominant computation in this algorithm.

Example 12.1.1 The secular equation for the problem

$$\min_{\|x\|_2 = 1} \left\| \begin{bmatrix} 2 & 0 \\ 0 & 1 \\ 0 & 0 \end{bmatrix} \begin{bmatrix} x_1 \\ x_2 \end{bmatrix} - \begin{bmatrix} 4 \\ 2 \\ 3 \end{bmatrix} \right\|_2$$

is given by

$$\left(\frac{8}{\lambda+4}\right)^2 + \left(\frac{2}{\lambda+1}\right)^2 = 1.$$

For this problem we find $\lambda^* = 4.57132$ and $x = [.93334 \ .35898]^T$.

12.1.3 Ridge Regression

The problem solved by Algorithm 12.1.1 is equivalent to the Lagrange multiplier problem of determining $\lambda > 0$ such that

$$(A^T A + \lambda I)x = A^T b \tag{12.1.9}$$

and $\| x \|_2 = \alpha$. This equation is precisely the normal equation formulation for the *ridge regression* problem

$$\min_{x} \left\| \begin{bmatrix} A \\ \sqrt{\lambda}I \end{bmatrix} x - \begin{bmatrix} b \\ 0 \end{bmatrix} \right\|_2^2 = \min_{x} \| Ax - b \|_2^2 + \lambda\| x \|_2^2 .$$

In the general ridge regression problem one has some criteria for selecting the ridge parameter λ, e.g., $\| x(\lambda) \|_2 = \alpha$ for some given α. We describe a λ-selection procedure that is discussed in Golub, Heath, and Wahba (1979).
   Set $D_k = I - e_k e_k^T = \text{diag}(1,\ldots,1,0,1,\ldots,1) \in \mathbb{R}^{m\times m}$ and let $x_k(\lambda)$ solve

$$\min_{x} \| D_k(Ax - b) \|_2^2 + \lambda\| x \|_2^2 . \tag{12.1.10}$$

Thus, $x_k(\lambda)$ is the solution to the ridge regression problem with the kth row of A and kth component of b deleted, i.e., the kth experiment is ignored. Now consider choosing λ so as to minimize the *cross-validation weighted square error* $C(\lambda)$ defined by

$$C(\lambda) \;=\; \frac{1}{m} \sum_{k=1}^{m} w_k (a_k^T x_k(\lambda) - b_k)^2 \,.$$

Here, $w_1, \ldots, w_m$ are non-negative weights and a_k^T is the kth row of A. Noting that

$$\| Ax_k(\lambda) - b \|_2^2 \;=\; \| D_k(Ax_k(\lambda) - b) \|_2^2 \;+\; (a_k^T x_k(\lambda) - b_k)^2$$

we see that $\left(a_k^T x_k(\lambda) - b_k\right)^2$ is the increase in the sum of squares resulting when the kth row is "reinstated." Minimizing $C(\lambda)$ is tantamount to choosing λ such that the final model is not overly dependent on any one experiment.

A more rigorous analysis can make this statement precise and also suggest a method for minimizing $C(\lambda)$. Assuming that $\lambda > 0$, an algebraic manipulation shows that

$$x_k(\lambda) \;=\; x(\lambda) \;+\; \frac{a_k^T x(\lambda) - b_k}{1 - z_k^T a_k} z_k \qquad\qquad (12.1.11)$$

where $z_k = (A^T A + \lambda I)^{-1} a_k$ and $x(\lambda) = (A^T A + \lambda I)^{-1} A^T b$. Applying $-a_k^T$ to (12.1.11) and then adding b_k to each side of the resulting equation gives

$$b_k - a_k^T x_k(\lambda) \;=\; \frac{e_k^T (I - A(A^T A + \lambda I)^{-1} A^T) b}{e_k^T (I - A(A^T A + \lambda I)^{-1} A^T) e_k} \,. \qquad\qquad (12.1.12)$$

Noting that the residual $r = (r_1, \ldots, r_m)^T = b - Ax(\lambda)$ is given by the formula $r = [I - A(A^T A + \lambda I)^{-1} A^T] b$, we see that

$$C(\lambda) \;=\; \frac{1}{m} \sum_{k=1}^{m} w_k \left(\frac{r_k}{\partial r_k / \partial b_k} \right)^2 .$$

The quotient $r_k/(\partial r_k/\partial b_k)$ may be regarded as an inverse measure of the "impact" of the kth observation b_k on the model. When $\partial r_k/\partial b_k$ is small, this says that the error in the model's prediction of b_k is somewhat independent of b_k. The tendency for this to be true is lessened by basing the model on the λ^* that minimizes $C(\lambda)$.

The actual determination of λ^* is simplified by computing the SVD of A. Indeed, if $U^T A V = \mathrm{diag}(\sigma_1, \ldots, \sigma_n)$ with $\sigma_1 \geq \ldots \geq \sigma_n$ and $\tilde{b} = U^T b$, then it can be shown from (12.1.12) that

$$C(\lambda) = \frac{1}{m} \sum_{k=1}^{m} w_k \left[\frac{\tilde{b}_k - \sum_{j=1}^{r} u_{kj} \tilde{b}_j \left(\frac{\sigma_j^2}{\sigma_j^2 + \lambda} \right)}{1 - \sum_{j=1}^{r} u_{kj}^2 \left(\frac{\sigma_j^2}{\sigma_j^2 + \lambda} \right)} \right]^2 .$$

The minimization of this expression is discussed in Golub, Heath, and Wahba (1979).

12.1.4 Equality Constrained Least Squares

We conclude the section by considering the least squares problem with linear equality constraints:

$$\min_{Bx=d} \| Ax - b \|_2 \qquad (12.1.13)$$

Here $A \in \mathbb{R}^{m \times n}$, $B \in \mathbb{R}^{p \times n}$, $b \in \mathbb{R}^m$, $d \in \mathbb{R}^p$, and $\text{rank}(B) = p$. We refer to (12.1.13) as the *LSE* problem. By setting $\alpha = 0$ in (12.1.2) we see that the LSE problem is a special case of the LSQI problem. However, it is simpler to approach the LSE problem directly rather than through Lagrange multipliers.

Assume for clarity that both A and B have full rank. Let

$$Q^T B^T = \begin{bmatrix} R \\ 0 \end{bmatrix} \begin{matrix} p \\ n-p \end{matrix}$$

be the QR factorization of B^T and set

$$AQ = [\ A_1 \quad A_2\ ] \qquad Q^T x = \begin{bmatrix} y \\ z \end{bmatrix} \begin{matrix} p \\ n-p \end{matrix} .$$
$$p \quad\ \ n-p$$

It is clear that with these transformations (12.1.13) becomes

$$\min_{R^T y=d} \| A_1 y + A_2 z - b \|_2 .$$

Thus, y is determined from the constraint equation $R^T y = d$ and the vector z is obtained by solving the unconstrained *LS* problem

$$\min_{z} \| A_2 z - (b - A_1 y) \|_2 .$$

Combining the above, we see that $x = Q \begin{bmatrix} y \\ z \end{bmatrix}$ solves (12.1.13).

Algorithm 12.1.2 Suppose $A \in \mathbb{R}^{m \times n}$, $B \in \mathbb{R}^{p \times n}$, $b \in \mathbb{R}^m$, and $d \in \mathbb{R}^p$. If $\text{rank}(A) = n$ and $\text{rank}(B) = p$, then the following algorithm minimizes $\| Ax - b \|_2$ subject to the constraint $Bx = d$.

> $B^T = QR$ (QR factorization)
> Solve $R(1{:}p, 1{:}p)^T y = d$ for y.
> $A = AQ$
> Find z so $\| A(:, p+1{:}n)z - (b - A(:, 1{:}p)y) \|_2$ is minimized.
> $x = Q(:, 1{:}p)y + Q(:, p+1{:}n)z$

Note that this approach to the LSE problem involves two factorizations and a matrix multiplication.

12.1.5   The Method of Weighting

An interesting way to obtain an approximate solution to (12.1.13) is to solve the unconstrained LS problem

$$\min_x \; \left\| \begin{bmatrix} A \\ \lambda B \end{bmatrix} x - \begin{bmatrix} b \\ \lambda d \end{bmatrix} \right\|_2 \tag{12.1.14}$$

for large λ. The generalized singular value decomposition of §8.7.3 sheds light on the quality of the approximation. Let

$$U^T A X = \text{diag}(\alpha_1, \ldots, \alpha_n) = D_A \in \mathbb{R}^{m \times n}$$
$$V^T B X = \text{diag}(\beta_1, \ldots, \beta_p) = D_B \in \mathbb{R}^{p \times n}$$

be the GSVD of (A, B) and assume that both matrices have full rank for clarity. If $U = [\, u_1, \ldots, u_m \,]$, $V = [\, v_1, \ldots, v_p \,]$, and $X = [\, x_1, \ldots, x_n \,]$, then it is easy to show that

$$x = \sum_{i=1}^{p} \frac{v_i^T d}{\beta_i} x_i + \sum_{i=p+1}^{n} \frac{u_i^T b}{\alpha_i} x_i \tag{12.1.15}$$

is the exact solution to (12.1.13), while

$$x(\lambda) = \sum_{i=1}^{p} \frac{\alpha_i u_i^T b + \lambda^2 \beta_i^2 v_i^T d}{\alpha_i^2 + \lambda^2 \beta_i^2} x_i + \sum_{i=p+1}^{n} \frac{u_i^T b}{\alpha_i} x_i \tag{12.1.16}$$

solves (12.1.14). Since

$$x(\lambda) - x = \sum_{i=1}^{p} \frac{\alpha_i (\beta_i u_i^T b - \alpha_i v_i^T d)}{\beta_i (\alpha_i^2 + \lambda^2 \beta_i^2)} x_i \tag{12.1.17}$$

it follows that $x(\lambda) \to x$ as $\lambda \to \infty$.

The appeal of this approach to the LSE problem is that no special subroutines are required: an ordinary LS solver will do. However, for large values of λ numerical problems can arise and it is necessary to take precautions. See Powell and Reid (1968) and Van Loan (1982a).

Example 12.1.2 The problem

$$\min_{x_1 = x_2} \left\| \begin{bmatrix} 1 & 2 \\ 3 & 4 \\ 5 & 6 \end{bmatrix} \begin{bmatrix} x_1 \\ x_2 \end{bmatrix} - \begin{bmatrix} 7 \\ 1 \\ 3 \end{bmatrix} \right\|_2$$

has solution $x = [.3407821 , .3407821]^T$. This can be approximated by solving

$$\min \left\| \begin{bmatrix} 1 & 2 \\ 3 & 4 \\ 5 & 6 \\ 1000 & -1000 \end{bmatrix} \begin{bmatrix} x_1 \\ x_2 \end{bmatrix} - \begin{bmatrix} 7 \\ 1 \\ 3 \\ 0 \end{bmatrix} \right\|_2$$

which has solution $x = [.3407810 , .3407829]^T$.

Problems

P 12.1.1 (a) Show that if $\text{null}(A) \cap \text{null}(B) \neq \{0\}$, then (12.1.2) cannot have a unique solution. (b) Give an example which shows that the converse is not true. (Hint: $A^+ b$ feasible.)

P12.1.2 Let $p_0(x), \ldots, p_n(x)$ be given polynomials and $(x_0, y_0), \ldots, (x_m, y_m)$ a given set of coordinate pairs with $x_i \in [a, b]$. It is desired to find a polynomial $p(x) = \sum_{k=0}^{n} a_k p_k(x)$ such that $\sum_{i=0}^{m} (p(x_i) - y_i)^2$ is minimized subject to the constraint that

$$\int_a^b [p''(x)]^2 dx \approx h \sum_{i=0}^{N} \left(\frac{p(z_{i-1}) - 2p(z_i) + p(z_{i+1})}{h^2} \right)^2 \leq \alpha^2$$

where $z_i = a + ih$ and $b = a + Nh$. Show that this leads to an LSQI problem of the form (12.1.1).

P12.1.3 Suppose $Y = [y_1, \ldots, y_k] \in \mathbf{R}^{m \times k}$ has the property that

$$Y^T Y = \text{diag}(d_1^2, \ldots, d_k^2) \qquad d_1 \geq d_2 \geq \cdots \geq d_k > 0 .$$

Show that if $Y = QR$ is the QR factorization of Y, then R is diagonal with $|r_{ii}| = d_i$.

P12.1.4 (a) Show that if $(A^T A + \lambda I)x = A^T b$, $\lambda > 0$, and $\| x \|_2 = \alpha$, then $z = (Ax - b)/\lambda$ solves the *dual equations* $(AA^T + \lambda I)z = -b$ with $\| A^T z \|_2 = \alpha$. (b) Show that if $(AA^T + \lambda I)z = -b$, $\| A^T z \|_2 = \alpha$, then $x = -A^T z$ satisfies $(A^T A + \lambda I)x = A^T b$, $\| x \|_2 = \alpha$.

P12.1.5 Suppose A is the m-by-1 matrix of ones and let $b \in \mathbf{R}^m$. Show that the cross-validation technique with unit weights prescribes an optimal λ given by

$$\lambda = \left(\left(\frac{\tilde{b}}{s} \right)^2 - \frac{1}{m} \right)^{-1}$$

where $\tilde{b}^T = (b_1 + \cdots + b_m)/m$ and $s = \sum_{i=1}^{m} (b_i - \tilde{b})^2/(m-1)$.

P12.1.6 Establish equations (12.1.15), (12.1.16), and (12.1.17).

P12.1.7 Develop an SVD version of Algorithm 12.1.2 that can handle rank deficiency in A and B.

P12.1.8 Suppose

$$A = \left[\begin{array}{c} A_1 \\ A_2 \end{array} \right]$$

where $A_1 \in \mathbf{R}^{n \times n}$ is nonsingular and $A_2 \in \mathbf{R}^{(m-n) \times n}$. Show that

$$\sigma_{min}(A) \geq \sqrt{1 + \sigma_{min}(A_2 A_1^{-1})^2}\, \sigma_{min}(A_1)\,.$$

P12.1.9 Consider the problem

$$\min_{\substack{x^T B x = \beta^2 \\ x^T C x = \gamma^2}} \| Ax - b \|_2 \qquad A \in \mathbf{R}^{m \times n},\ b \in \mathbf{R}^m,\ B, C \in \mathbf{R}^{n \times n}$$

Assume that B and C are positive definite and that $Z \in \mathbf{R}^{n \times n}$ is a nonsingular matrix with the property that $Z^T B Z = \mathrm{diag}(\lambda_1, \ldots, \lambda_n)$ and $Z^T C Z = I_n$. Assume that $\lambda_1 \geq \cdots \geq \lambda_n$. (a) Show that the the set of feasible x is empty unless $\lambda_n \leq \beta^2/\gamma^2 \leq \lambda_1$. (b) Using Z, show how the two constraint problem can be converted to a single constraint problem of the form

$$\min_{y^T W y = \beta^2 - \lambda_n \gamma^2} \| \bar{A} x - b \|_2$$

where $W = \mathrm{diag}(\lambda_1, \ldots, \lambda_n) - \lambda_n I$.

P12.1.10 Suppose $p \geq m \geq n$ and that $A \in \mathbf{R}^{m \times n}$ and $B \in \mathbf{R}^{m \times p}$ Show how to compute orthogonal $Q \in \mathbf{R}^{m \times m}$ and orthogonal $V \in \mathbf{R}^{n \times n}$ so that

$$Q^T A = \left[\begin{array}{c} R \\ 0 \end{array} \right] \qquad Q^T B V = [0\,,\, S]$$

where $R \in \mathbf{R}^{n \times n}$ and $S \in \mathbf{R}^{m \times m}$ are upper triangular.

P12.1.11 Suppose $r \in \mathbf{R}^m$, $y \in \mathbf{R}^n$, and $\delta > 0$. Show how to solve the problem

$$\min_{\substack{E \in \mathbf{R}^{m \times n} \\ \| E \|_F \leq \delta}} \| Ey - r \|_2$$

Repeat with "min" replaced by "max".

Notes and References for Sec. 12.1

Roughly speaking, regularization is a technique for transforming a poorly conditioned problem into a stable one. Quadratically constrained least squares is an important example. See

L. Eldén (1977). "Algorithms for the Regularization of Ill-Conditioned Least Squares Problems," *BIT 17*, 134–45.

References for cross-validation include

G.H. Golub, M. Heath, and G. Wahba (1979). "Generalized Cross-Validation as a Method for Choosing a Good Ridge Parameter," *Technometrics 21*, 215–23.

L. Eldén (1985). "A Note on the Computation of the Generalized Cross-Validation Function for Ill-Conditioned Least Squares Problems," *BIT 24*, 467–472.

The LSQI problem is discussed in

G.E. Forsythe and G.H. Golub (1965). "On the Stationary Values of a Second-Degree Polynomial on the Unit Sphere," *SIAM J. App. Math. 14,* 1050–68.

L. Eldén (1980). "Perturbation Theory for the Least Squares Problem with Linear Equality Constraints," *SIAM J. Num. Anal. 17,* 338–50.

W. Gander (1981). "Least Squares with a Quadratic Constraint," *Numer. Math. 36,* 291–307.

L. Eldén (1983). "A Weighted Pseudoinverse, Generalized Singular Values, and Constrained Least Squares Problems," *BIT 22 ,* 487–502.

G.W. Stewart (1984). "On the Asymptotic Behavior of Scaled Singular Value and QR Decompositions," *Math. Comp. 43,* 483–490.

G.H. Golub and U. von Matt (1991). "Quadratically Constrained Least Squares and Quadratic Problems," *Numer. Math. 59,* 561–580.

T.F. Chan, J.A. Olkin, and D. Cooley (1992). "Solving Quadratically Constrained Least Squares Using Black Box Solvers," *BIT 32,* 481–495.

Other computational aspects of the LSQI problem involve updating and the handling of banded and sparse problems. See

K. Schittkowski and J. Stoer (1979). "A Factorization Method for the Solution of Constrained Linear Least Squares Problems Allowing for Subsequent Data changes," *Numer. Math. 31,* 431–463.

D.P. O'Leary and J.A. Simmons (1981). "A Bidiagonalization-Regularization Procedure for Large Scale Discretizations of Ill-Posed Problems," *SIAM J. Sci. and Stat. Comp. 2,* 474–489.

Å. Björck (1984). "A General Updating Algorithm for Constrained Linear Least Squares Problems," *SIAM J. Sci. and Stat. Comp. 5,* 394–402.

L. Eldén (1984). "An Algorithm for the Regularization of Ill-Conditioned, Banded Least Squares Problems," *SIAM J. Sci. and Stat. Comp. 5,* 237–254.

Various aspects of the LSE problem are discussed and analyzed in

M.J.D. Powell and J.K. Reid (1968). "On Applying Householder's Method to Linear Least Squares Problems," *Proc. IFIP Congress,* pp. 122–26.

C. Van Loan (1985). "On the Method of Weighting for Equality Constrained Least Squares Problems," *SIAM J. Numer. Anal. 22,* 851–864.

J.L. Barlow, N.K. Nichols, and R.J. Plemmons (1988). "Iterative Methods for Equality Constrained Least Squares Problems," SIAM J. Sci. and Stat. Comp. 9, 892–906.

J.L. Barlow (1988). "Error Analysis and Implementation Aspects of Deferred Correction for Equality Constrained Least-Squares Problems," *SIAM J. Num. Anal. 25,* 1340–1358.

J.L. Barlow and S.L. Handy (1988). "The Direct Solution of Weighted and Equality Constrained Least-Squares Problems," *SIAM J. Sci. Stat. Comp. 9,* 704–716.

J.L. Barlow and U.B. Vemulapati (1992). "A Note on Deferred Correction for Equality Constrained Least Squares Problems," *SIAM J. Num. Anal. 29,* 249–256.

M. Wei (1992). "Perturbation Theory for the Rank-Deficient Equality Constrained Least Squares Problem," *SIAM J. Num. Anal. 29,* 1462–1481.

M. Wei (1992). "Algebraic Properties of the Rank-Deficient Equality-Constrained and Weighted Least Squares Problems," *Lin. Alg. and Its Applic. 161,* 27–44.

M. Gulliksson and P-Å. Wedin (1992). "Modifying the QR-Decomposition to Constrained and Weighted Linear Least Squares," *SIAM J. Matrix Anal. Appl. 13,* 1298–1313.

Å. Björck and C.C. Paige (1994). "Solution of Augmented Linear Systems Using Orthogonal Factorizations," *BIT 34,* 1–24.

M. Gulliksson (1994). "Iterative Refinement for Constrained and Weighted Linear Least Squares," *BIT 34,* 239–253.

M. Gulliksson (1995). "Backward Error Analysis for the Constrained and Weighted
   Linear Least Squares Problem When Using the Weighted QR Factorization," *SIAM
   J. Matrix. Anal. Appl. 13*, 675–687.

Generalized factorizations have an important bearing on generalized least squares problems.

C.C. Paige (1985). "The General Linear Model and the Generalized Singular Value
   Decomposition," *Lin. Alg. and Its Applic. 70*, 269–284.
C.C. Paige (1990). "Some Aspects of Generalized QR Factorization," in *Reliable Nu-
   merical Computations*, M. Cox and S. Hammarling (eds), Clarendon Press, Oxford.
E. Anderson, Z. Bai, and J. Dongarra (1992). "Generalized QR Factorization and Its
   Applications," *Lin. Alg. and Its Applic. 162/163/164*, 243–271.

12.2   Subset Selection Using the SVD

As described in §5.5, the rank-deficient LS problem min $\| Ax - b \|_2$ can be
approached by approximating the minimum norm solution

$$x_{LS} = \sum_{i=1}^{r} \frac{u_i^T b}{\sigma_i} v_i \qquad r = \text{rank}(A)$$

with

$$x_{\tilde{r}} = \sum_{i=1}^{\tilde{r}} \frac{u_i^T b}{\sigma_i} v_i \qquad \tilde{r} \le r$$

where

$$A = U\Sigma V^T = \sum_{i=1}^{r} \sigma_i u_i v_i^T \tag{12.2.1}$$

is the SVD of A and $\tilde{r}$ is some numerically determined estimate of r. Note
that $x_{\tilde{r}}$ minimizes $\| A_{\tilde{r}} x - b \|_2$ where

$$A_{\tilde{r}} = \sum_{i=1}^{\tilde{r}} \sigma_i u_i v_i^T$$

is the closest matrix to A that has rank $\tilde{r}$. See Theorem 2.5.3.

Replacing A by $A_{\tilde{r}}$ in the LS problem amounts to filtering the small
singular values and can make a great deal of sense in those situations where
A is derived from noisy data. In other applications, however, rank deficiency
implies redundancy among the factors that comprise the underlying model.
In this case, the model-builder may not be interested in a predictor such
as $A_{\tilde{r}} x_{\tilde{r}}$ that involves all n redundant factors. Instead, a predictor Ay may
be sought where y has at most $\tilde{r}$ nonzero components. The position of the
nonzero entries determines which columns of A, i.e., which factors in the
model, are to be used in approximating the observation vector b. How to
pick these columns is the problem of *subset selection* and is the subject of
this section.

The contents of this section depends heavily upon §2.6 and Chapter 5.

12.2.1 QR with Column Pivoting

QR with column pivoting can be regarded as a method for selecting an independent subset of A's columns from which b might be predicted. Suppose we apply Algorithm 5.4.1 to $A \in \mathbb{R}^{m \times n}$ and compute an orthogonal Q and a permutation Π such that $R = Q^T A \Pi$ is upper triangular. If $R(1{:}\tilde{r}, 1{:}\tilde{r})z = \tilde{b}(1{:}\tilde{r})$ where $\tilde{b} = Q^T b$ and we set

$$y = \Pi \begin{bmatrix} z \\ 0 \end{bmatrix},$$

then Ay is an approximate LS predictor of b that involves the first $\tilde{r}$ columns of $A\Pi$.

12.2.2 Using the SVD

Although QR with column pivoting is a fairly reliable way to handle near rank deficiency, the SVD is sometimes preferable for reasons discussed in §5.5. We therefore describe an SVD-based subset selection procedure due to Golub, Klema, and Stewart (1976) that proceeds as follows:

- Compute the SVD $A = U\Sigma V^T$ and use it to determine a rank estimate $\tilde{r}$.

- Calculate a permutation matrix P such that the columns of the matrix $B_1 \in \mathbb{R}^{m \times \tilde{r}}$ in $AP = [\, B_1\ B_2\,]$ are "sufficiently independent."

- Predict b with the vector Ay where $y = P \begin{bmatrix} z \\ 0 \end{bmatrix}$ and $z \in \mathbb{R}^{\tilde{r}}$ minimizes $\| B_1 z - b \|_2$

The second step is key. Since

$$\min_{z \in \mathbb{R}^{\tilde{r}}} \| B_1 z - b \|_2 = \| Ay - b \|_2 \geq \min_{x \in \mathbb{R}^n} \| Ax - b \|_2$$

it can be argued that the permutation P should be chosen to make the residual $(I - B_1 B_1^+)b$ as small as possible. Unfortunately, such a solution procedure can be unstable. For example, if

$$A = \begin{bmatrix} 1 & 1 & 0 \\ 1 & 1+\epsilon & 1 \\ 0 & 0 & 1 \end{bmatrix}, \qquad b = \begin{bmatrix} 1 \\ -1 \\ 0 \end{bmatrix},$$

$\tilde{r} = 2$, and $P = I$, then $\min \| B_1 z - b \|_2 = 0$, but $\| B_1^+ b \|_2 = O(1/\epsilon)$. On the other hand, any proper subset involving the third column of A is strongly independent but renders a much worse residual.

This example shows that there can be a trade-off between the independence of the chosen columns and the norm of the residual that they render. How to proceed in the face of this trade-off requires additional mathematical machinery in the form of useful bounds on $\sigma_{\tilde{r}}(B_1)$, the smallest singular value of B_1.

Theorem 12.2.1 *Let the SVD of $A \in \mathbb{R}^{m \times n}$ be given by (12.2.1), and define the matrix $B_1 \in \mathbb{R}^{m \times \tilde{r}}$, $\tilde{r} \leq \mathrm{rank}(A)$, by*

$$AP = \begin{array}{c} [\ B_1 \quad B_2 \ ] \\ \ \tilde{r} \quad\ \ n - \tilde{r} \end{array}$$

where $P \in \mathbb{R}^{n \times n}$ is a permutation. If

$$P^T V = \begin{array}{c} \left[\begin{array}{cc} \tilde{V}_{11} & \tilde{V}_{12} \\ \tilde{V}_{21} & \tilde{V}_{22} \end{array} \right] & \begin{array}{c} \tilde{r} \\ n - \tilde{r} \end{array} \\ \ \tilde{r} \quad n - \tilde{r} \end{array} \qquad (12.2.2)$$

and $\tilde{V}_{11}$ is nonsingular, then

$$\frac{\sigma_{\tilde{r}}(A)}{\| \tilde{V}_{11}^{-1} \|_2} \leq \sigma_{\tilde{r}}(B_1) \leq \sigma_{\tilde{r}}(A).$$

Proof. The upper bound follows from the minimax characterization of singular values given in §8.6.1

To establish the lower bound, partition the diagonal matrix of singular values as follows:

$$\Sigma = \begin{array}{c} \left[\begin{array}{cc} \Sigma_1 & 0 \\ 0 & \Sigma_2 \end{array} \right] & \begin{array}{c} \tilde{r} \\ m - \tilde{r} \end{array} \\ \ \tilde{r} \quad n - \tilde{r} \end{array} \ .$$

If $w \in \mathbb{R}^{\tilde{r}}$ is a unit vector with the property that $\| B_1 w \|_2 = \sigma_{\tilde{r}}(B_1)$, then

$$\begin{aligned} \sigma_{\tilde{r}}(B_1)^2 &= \| B_1 w \|_2^2 = \left\| U \Sigma V^T P \left[\begin{array}{c} w \\ 0 \end{array} \right] \right\|_2^2 \\ &= \| \Sigma_1 \tilde{V}_{11}^T w \|_2^2 + \| \Sigma_2 \tilde{V}_{12}^T w \|_2^2 \ . \end{aligned}$$

The theorem now follows because $\| \Sigma_1 \tilde{V}_{11}^T w \|_2 \geq \sigma_{\tilde{r}}(A)/\| \tilde{V}_{11}^{-1} \|_2$. $\square$

This result suggests that in the interest of obtaining a sufficiently independent subset of columns, we choose the permutation P such that the resulting $\tilde{V}_{11}$ submatrix is as well-conditioned as possible. A heuristic solution to this problem can be obtained by computing the QR with column-pivoting factorization of the matrix $[\ V_{11}^T \ V_{21}^T \ ]$, where

$$V = \begin{array}{c} \left[\begin{array}{cc} V_{11} & V_{12} \\ V_{21} & V_{22} \end{array} \right] & \begin{array}{c} \tilde{r} \\ n - \tilde{r} \end{array} \\ \ \tilde{r} \quad n - \tilde{r} \end{array}$$

is a partitioning of the matrix V in (12.2.1). In particular, if we apply QR with column pivoting (Algorithm 5.4.1) to compute

$$Q^T [\, V_{11}^T \; V_{21}^T \,] P \; = \; [\; R_{11} \quad R_{12} \;]$$
$$\phantom{Q^T [\, V_{11}^T \; V_{21}^T \,] P \; = \; [\;} \tilde{r} \qquad n - \tilde{r}$$

where Q is orthogonal, P is a permutation matrix, and R_{11} is upper triangular, then (12.2.2) implies:

$$\left[\begin{array}{c} \tilde{V}_{11} \\ \tilde{V}_{21} \end{array} \right] = P^T \left[\begin{array}{c} V_{11} \\ V_{21} \end{array} \right] = \left[\begin{array}{c} R_{11}^T Q^T \\ R_{12}^T Q^T \end{array} \right].$$

Note that R_{11} is nonsingular and that $\| \tilde{V}_{11}^{-1} \|_2 = \| R_{11}^{-1} \|_2$. Heuristically, column pivoting tends to produce a well-conditioned R_{11}, and so the overall process tends to produce a well-conditoned $\tilde{V}_{11}$. Thus we obtain

Algorithm 12.2.1   Given $A \in \mathbb{R}^{m \times n}$ and $b \in \mathbb{R}^m$ the following algorithm computes a permutation P, a rank estimate $\tilde{r}$, and a vector $z \in \mathbb{R}^{\tilde{r}}$ such that the first $\tilde{r}$ columns of $B = AP$ are independent and such that $\| B(:, 1{:}\tilde{r}) z - b \|_2$ is minimized.

> Compute the SVD $U^T A V = \mathrm{diag}(\sigma_1, \dots, \sigma_n)$ and save V.
> Determine $\tilde{r} \le \mathrm{rank}(A)$.
> Apply QR with column pivoting: $Q^T V(:, 1{:}\tilde{r})^T P = [\, R_{11} \; R_{12} \,]$ and set
> $\qquad AP = [\, B_1 \; B_2 \,]$ with $B_1 \in \mathbb{R}^{m \times \tilde{r}}$ and $B_2 \in \mathbb{R}^{m \times (n - \tilde{r})}$.
> Determine $z \in \mathbb{R}^{\tilde{r}}$ such that $\| b - B_1 z \|_2 = \min$.

Example 12.2.1 Let

$$A = \left[\begin{array}{ccc} 3 & 4 & 1 \\ 7 & 4 & -3 \\ 2 & 5 & 3 \\ -1 & 4 & 5 \end{array} \right], \qquad b = \left[\begin{array}{c} 1 \\ 1 \\ 1 \\ 1 \end{array} \right],$$

$\mathrm{rank}(A) = 2$, and $x_{LS} = [\, .0815 \; .1545 \; .0730 \,]^T$. If Algorithm 12.2.1 is applied, we find $P = [\, e_2 \; e_1 \; e_3 \,]$ and solution $x = [\, .0845 \; .2275 \; .0000 \,]^T$. Note: $\| b - A x_{LS} \|_2 \approx \| b - Ax \|_2 = .1966$.

12.2.3   More on Column Independence vs. Residual

We return to the discussion of the trade-off between column independence and norm of the residual. In particular, to assess the above method of subset selection, we need to examine the residual of the vector y that it produces $r_y = b - Ay = b - B_1 z = (I - B_1 B_1^+) b$. Here, $B_1 = B(:, 1{:}\tilde{r})$ with

$B = AP$. To this end, it is appropriate to compare r_y with $r_{x_{\tilde{r}}} = b - Ax_{\tilde{r}}$ since we are regarding A as a rank-$\tilde{r}$ matrix and since $x_{\tilde{r}}$ solves the nearest rank-$\tilde{r}$ LS problem, namely, min $\| A_{\tilde{r}}x - b \|_2$.

Theorem 12.2.2 *If* r_y *and* $r_{x_{\tilde{r}}}$ *are defined as above and if* $\tilde{V}_{11}$ *is the leading* r-by-r *principal submatrix of* $P^T V$, *then*

$$\| r_{x_{\tilde{r}}} - r_y \|_2 \leq \frac{\sigma_{\tilde{r}+1}(A)}{\sigma_{\tilde{r}}(A)} \| \tilde{V}_{11}^{-1} \|_2 \| b \|_2$$

Proof. Note that $r_{x_{\tilde{r}}} = (I - U_1 U_1^T)b$ and $r_y = (I - Q_1 Q_1^T)b$ where

$$U = \begin{bmatrix} U_1 & U_2 \end{bmatrix}$$
$$\quad\quad \tilde{r} \quad m - \tilde{r}$$

is a partitioning of the matrix U in (12.2.1) and where $Q_1 = B_1(B_1^T B_1)^{-1/2}$. Using Theorem 2.6.1 we obtain

$$\| r_{x_{\tilde{r}}} - r_y \|_2 \leq \| U_1 U_1^T - Q_1 Q_1^T \|_2 \| b \|_2 = \| U_2^T Q_1 \|_2 \| b \|_2$$

while Theorem 12.2.1 permits us to conclude that

$$\| U_2^T Q_1 \|_2 \leq \| U_2^T B_1 \|_2 \| (B_1^T B_1)^{-1/2} \|_2 \leq \sigma_{\tilde{r}+1}(A)\frac{1}{\sigma_{\tilde{r}}(B_1)}$$

$$\leq \frac{\sigma_{\tilde{r}+1}(A)}{\sigma_{\tilde{r}}(A)} \| \tilde{V}_{11}^{-1} \|_2. \quad \square$$

Noting that

$$\| r_{x_{\tilde{r}}} - r_y \|_2 = \left\| B_1 y - \sum_{i=1}^{r}(u_i^T b)u_i \right\|_2$$

we see that Theorem 12.2.2 sheds light on how well $B_1 y$ can predict the "stable" component of b, i.e., $U_1^T b$. Any attempt to approximate $U_2^T b$ can lead to a large norm solution. Moreover, the theorem says that if $\sigma_{\tilde{r}+1}(A) \ll \sigma_{\tilde{r}}(A)$, then any reasonably independent subset of columns produces essentially the same-sized residual. On the other hand, if there is no well-defined gap in the singular values, then the determination of $\tilde{r}$ becomes difficult and the entire subset selection problem more complicated.

Problems

P12.2.1 Suppose $A \in \mathbf{R}^{m \times n}$ and that $\| u^T A \|_2 = \sigma$ with $u^T u = 1$. Show that if $u^T(Ax - b) = 0$ for $x \in \mathbf{R}^n$ and $b \in \mathbf{R}^m$, then $\| x \|_2 \geq |u^T b|/\sigma$.

P12.2.2 Show that if $B_1 \in \mathbf{R}^{m \times k}$ is comprised of k columns from $A \in \mathbf{R}^{m \times n}$ then $\sigma_k(B_1) \leq \sigma_k(A)$.

P12.2.3 In equation (12.2.2) we know that the matrix

$$P^T V = \begin{bmatrix} \tilde{V}_{11} & \tilde{V}_{12} \\ \tilde{V}_{21} & \tilde{V}_{22} \end{bmatrix} \begin{matrix} \tilde{r} \\ n - \tilde{r} \end{matrix}$$
$$\quad\quad\; \tilde{r} \quad\; n - \tilde{r}$$

is orthogonal. Thus, $\| \tilde{V}_{11}^{-1} \|_2 = \| \tilde{V}_{22}^{-1} \|_2$ from the CS decomposition (Theorem 2.6.3). Show how to compute P by applying the QR with column pivoting algorithm to $[\tilde{V}_{22}^T \; \tilde{V}_{12}^T]$. (For $\tilde{r} > n/2$, this procedure would be more economical than the technique discussed in the text.) Incorporate this observation in Algorithm 12.2.1.

Notes and References for Sec. 12.2

The material in this section is derived from

G.H. Golub, V. Klema and G.W. Stewart (1976). "Rank Degeneracy and Least Squares Problems," Technical Report TR-456, Department of Computer Science, University of Maryland, College Park, MD.

A subset selection procedure based upon the total least squares fitting technique of §12.3 is given in

S. Van Huffel and J. Vandewalle (1987). "Subset Selection Using the Total Least Squares Approach in Collinearity Problems with Errors in the Variables," *Lin. Alg. and Its Applic. 88/89,* 695–714.

The literature on subset selection is vast and we refer the reader to

H. Hotelling (1957). "The Relations of the Newer Multivariate Statistical Methods to Factor Analysis," *Brit. J. Stat. Psych. 10,* 69–79.

12.3 Total Least Squares

The problem of minimizing $\| D(Ax - b) \|_2$ where $A \in \mathbb{R}^{m \times n}$, and $D = \text{diag}(d_1, \ldots, d_m)$ is nonsingular can be recast as follows:

$$\min_{b+r \,\in\, \text{range}(A)} \quad \| Dr \|_2 \qquad r \in \mathbb{R}^m . \tag{12.3.1}$$

In this problem, there is a tacit assumption that the errors are confined to the "observation" b. When error is also present in the "data" A, then it may be more natural to consider the problem

$$\min_{b+r \,\in\, \text{range}(A+E)} \quad \| D[\, E \,,\, r\,]T \|_F \qquad E \in \mathbb{R}^{m \times n}, \; r \in \mathbb{R}^m \tag{12.3.2}$$

where $D = \text{diag}(d_1, \ldots, d_m)$ and $T = \text{diag}(t_1, \ldots, t_{n+1})$ are nonsingular. This problem, discussed in Golub and Van Loan (1980), is referred to as the *total least squares* (TLS) problem.

If a minimizing $[E_0 \,,\, r_0]$ can be found for (12.3.2), then any x satisfying $(A + E_0)x = b + r_0$ is called a TLS solution. However, it should be realized that (12.3.2) may fail to have a solution altogether. For example, if

$$A = \begin{bmatrix} 1 & 0 \\ 0 & 0 \\ 0 & 0 \end{bmatrix}, \; b = \begin{bmatrix} 1 \\ 1 \\ 1 \end{bmatrix}, \; D = I_3, \; T = I_3, \text{ and } E_\epsilon = \begin{bmatrix} 0 & 0 \\ 0 & \epsilon \\ 0 & \epsilon \end{bmatrix}$$

then for all $\epsilon > 0$, $b \in \text{ran}(A + E_\epsilon)$. However, there is no smallest value of $\|\,[\,E\,,\,r\,]\,\|_F$ for which $b + r \in \text{ran}(A + E)$.

A generalization of (12.3.2) results if we allow multiple right-hand sides. In particular, if $B \in \mathbb{R}^{m \times k}$, then we have the problem

$$\min_{\text{range}(B+R) \subseteq \text{range}(A+E)} \|\, D[\,E\,,\,R\,]T\,\|_F \qquad (12.3.3)$$

where $E \in \mathbb{R}^{m \times n}$ and $R \in \mathbb{R}^{m \times k}$ and the matrices $D = \text{diag}(d_1, \ldots, d_m)$ and $T = \text{diag}(t_1, \ldots, t_{n+k})$ are nonsingular. If $[\,E_0\,,\,R_0\,]$ solves (12.3.3), then any $X \in \mathbb{R}^{n \times k}$ that satisfies $(A + E_0)X = (B + R_0)$ is said to be a TLS solution to (12.3.3).

In this section we discuss some of the mathematical properties of the total least squares problem and show how it can be solved using the SVD. Chapter 5 is the only prerequisite. A very detailed treatment of the TLS problem is given in the monograph by Van Huffel and Vanderwalle (1991).

12.3.1   Mathematical Background

The following theorem gives conditions for the uniqueness and existence of a TLS solution to the multiple right-hand side problem.

Theorem 12.3.1 *Let A, B, D, and T be as above and assume $m \geq n + k$. Let*

$$C = D[\,A\,,\,B\,]T = [\begin{matrix} C_1 & C_2 \end{matrix}]$$
$$\qquad\qquad\quad n \quad\ k$$

have SVD $U^T C V = \text{diag}(\sigma_1, \ldots, \sigma_{n+k}) = \Sigma$ where U, V, and Σ are partitioned as follows:

$$U = [\begin{matrix} U_1 & U_2 \end{matrix}] \qquad V = \begin{bmatrix} V_{11} & V_{12} \\ V_{21} & V_{22} \end{bmatrix} \begin{matrix} n \\ k \end{matrix}$$
$$\quad\ \ n \quad\ k \qquad\qquad\qquad\ n \quad\ \ k$$

$$\Sigma = \begin{bmatrix} \Sigma_1 & 0 \\ 0 & \Sigma_2 \end{bmatrix} \begin{matrix} n \\ k \end{matrix}$$
$$\qquad\quad n \quad\ k$$

If $\sigma_n(C_1) > \sigma_{n+1}(C)$, then the matrix $[\,E_0\,,\,R_0\,]$ defined by

$$D[\,E_0\,,\,R_0\,]T = -U_2\Sigma_2[\,V_{12}^T\,,\,V_{22}^T\,] \qquad (12.3.4)$$

solves (12.3.3). If $T_1 = \text{diag}(t_1, \ldots, t_n)$ and $T_2 = \text{diag}(t_{n+1}, \ldots, t_{n+k})$ then the matrix

$$X_{TLS} = -T_1 V_{12} V_{22}^{-1} T_2^{-1}$$

exists and is the unique solution to $(A + E_0)X = B + R_0$.

Proof. We first establish two results that follow from the assumption $\sigma_n(C_1) > \sigma_{n+1}(C)$. From the equation $CV = U\Sigma$ we have $C_1V_{12}+C_2V_{22} = U_2\Sigma_2$. We wish to show that V_{22} is nonsingular. Suppose $V_{22}x = 0$ for some unit 2-norm x. It follows from $V_{12}^TV_{12} + V_{22}^TV_{22} = I$ that $\| V_{12}x \|_2 = 1$. But then

$$\sigma_{n+1}(C) \geq \| U_2\Sigma_2 x \|_2 = \| C_1V_{12}x \|_2 \geq \sigma_n(C_1) ,$$

a contradiction. Thus, the submatix V_{22} is nonsingular.

The other fact that follows from $\sigma_n(C_1) > \sigma_{n+1}(C)$ concerns the strict separation of $\sigma_n(C)$ and $\sigma_{n+1}(C)$. From Corollary 8.3.3, we have $\sigma_n(C) \geq \sigma_n(C_1)$ and so $\sigma_n(C) \geq \sigma_n(C_1) > \sigma_{n+1}(C)$.

Now we are set to prove the theorem. If $\text{ran}(B + R) \subset \text{ran}(A + E)$, then there is an X (n-by-k) so $(A + E)X = B + R$, i.e.,

$$\{ D[\, A\,,\, B\,]T + D[\, E\,,\, R\,]T \}T^{-1} \begin{bmatrix} X \\ -I_k \end{bmatrix} = 0\,. \qquad (12.3.5)$$

Thus, the matrix in curly brackets has, at most, rank n. By following the argument in Theorem 2.5.3, it can be shown that

$$\| D[\, E\,,\, R\,]T \|_F \geq \sum_{i=n+1}^{n+k} \sigma_i(C)^2$$

and that the lower bound is realized by setting $[\, E\,,\, R\,] = [\, E_0\,,\, R_0\,]$. The inequality $\sigma_n(C) > \sigma_{n+1}(C)$ ensures that $[E_0\,,\, R_0]$ is the unique minimizer. The null space of

$$\{D[\, A\,,\, B\,]T + D[\, E_0\,,\, R_0\,]T\} = U_1\Sigma_1[\, V_{11}^T \; V_{21}^T\,]$$

is the range of $\begin{bmatrix} V_{12} \\ V_{22} \end{bmatrix}$. Thus, from (12.3.5)

$$T^{-1} \begin{bmatrix} X \\ -I_k \end{bmatrix} = \begin{bmatrix} V_{12} \\ V_{22} \end{bmatrix} S$$

for some k-by-k matrix S. From the equations $T_1^{-1}X = V_{12}S$ and $-T_2^{-1} = V_{22}S$ we see that $S = -V_{22}^{-1}T_2^{-1}$. Thus, we must have

$$X = T_1V_{12}S = -T_1V_{12}V_{22}^{-1}T_2^{-1} = X_{TLS}. \;\square$$

If $\sigma_n(C) = \sigma_{n+1}(C)$, then the TLS problem may still have a solution, although it may not be unique. In this case, it may be desirable to single out a "minimal norm" solution. To this end, consider the τ-norm defined on $\mathbb{R}^{n\times k}$ by $\| Z \|_\tau = \| T_1^{-1}ZT_2 \|_2$. If X is given by (12.3.5), then from the CS decomposition (Theorem 2.6.3) we have

$$\| X \|_\tau^2 = \| V_{12}V_{22}^{-1} \|_2^2 = \frac{1 - \sigma_k(V_{22})^2}{\sigma_k(V_{22})^2}.$$

This suggests choosing V in Theorem 12.3.1 so that $\sigma_k(V_{22})$ is maximized.

12.3.2 Computations for the k=1 Case

We show how to maximize V_{22} in the important $k = 1$ case. Suppose the singular values of C satisfy $\sigma_{n-p} > \sigma_{n-p+1} = \cdots = \sigma_{n+1}$ and let $V = [\, v_1, \ldots, v_{n+1} \,]$ be a column partitioning of V. If $\tilde{Q}$ is a Householder matrix such that

$$
V(:, n+1-p{:}n+1)\tilde{Q} \;=\; \begin{bmatrix} W & z \\ 0 & \alpha \end{bmatrix} \begin{matrix} n \\ 1 \end{matrix}
$$
$$
\phantom{V(:, n+1-p{:}n+1)\tilde{Q} \;=\;} \begin{matrix} p & 1 \end{matrix}
$$

then $\begin{bmatrix} z \\ \alpha \end{bmatrix}$ has the largest $(n + 1)$-st component of all the vectors in span$\{v_{n+1-p}, \ldots, v_{n+1}\}$. If $\alpha = 0$, the TLS problem has no solution. Otherwise $x_{TLS} = -T_1 z/(t_{n+1}\alpha)$. Moreover,

$$
\begin{bmatrix} I_{n-1} & 0 \\ 0 & Q \end{bmatrix} U^T(D[\, A, b\,]T)V \begin{bmatrix} I_{n-p} & 0 \\ 0 & \tilde{Q} \end{bmatrix} = \Sigma
$$

and so

$$
D[\, E_0, r_0\,]T \;=\; -D[\, A, b\,]T \begin{bmatrix} z \\ \alpha \end{bmatrix} [\, z^T\ \alpha\,].
$$

Overall, we have the following algorithm:

Algorithm 12.3.1 Given $A \in \mathbb{R}^{m \times n}$ $(m > n)$, $b \in \mathbb{R}^m$, and nonsingular $D = \text{diag}(d_1, \ldots, d_m)$ and $T = \text{diag}(t_1, \ldots, t_{n+1})$, the following algorithm computes (if possible) a vector $x_{TLS} \in \mathbb{R}^n$ such that $(A + E_0)x = (b + r_0)$ and $\| D[\, E_0, r_0\,]T \|_F$ is minimal.

> Compute the SVD $U^T(D[\, A, b\,]T)V = \text{diag}(\sigma_1, \ldots, \sigma_{n+1})$. Save V.
> Determine p such that $\sigma_1 \geq \cdots \geq \sigma_{n-p} > \sigma_{n-p+1} = \cdots = \sigma_{n+1}$.
> Compute a Householder matrix P such that if $\tilde{V} = VP$, then
> $\quad \tilde{V}(n + 1, n - p + 1{:}n) = 0$
>
> if $\tilde{v}_{n+1,n+1} \neq 0$
> $\quad$ for $i = 1{:}n$
> $\quad\quad x_i = -t_i \tilde{v}_{i,n+1}/(t_{n+1}\tilde{v}_{n+1,n+1})$
> $\quad$ end
> end

This algorithm requires about $2mn^2 + 12n^3$ flops and most of these are associated with the SVD computation.

Example 12.3.1 The TLS problem $\displaystyle\min_{(a+e)x=b+r} \| [\, e, r\,] \|_F$ where $a = [1, 2, 3, 4]^T$ and $b = [2.01, 3.99, 5.80, 8.30]^T$ has solution $x_{TLS} = 2.0212$, $e = [-.0045, -.0209, -.1048, .0855]^T$, and $r = [.0022, .0103, .0519, -.0423]^T$. Note that for this data $x_{LS} = 2.0197$.

12.3.3 Geometric Interpretation

It can be shown that the TLS solution x_{TLS} minimizes

$$\psi(x) = \sum_{i=1}^{m} d_i^2 \frac{|a_i^T x - b_i|^2}{x^T T_1^{-2} x + t_{n+1}^{-2}}$$

where a_i^T is ith row of A and b_i is the ith component of b. A geometrical interpretation of the TLS problem is made possible by this observation. Indeed,

$$\frac{|a_i^T x - b_i|^2}{x^T T_1^{-2} x + t_{n+1}^{-2}}$$

is the square of the distance from $\begin{bmatrix} a_i \\ b_i \end{bmatrix} \in \mathbb{R}^{n+1}$ to the nearest point in the subspace

$$P_x = \left\{ \begin{bmatrix} a \\ b \end{bmatrix} : a \in \mathbb{R}^n,\ b \in \mathbb{R},\ b = x^T a \right\}$$

where distance in $\mathbb{R}^{n+1}$ is measured by the norm $\| z \| = \| Tz \|_2$. A great deal has been written about this kind of fitting. See Pearson (1901) and Madansky (1959).

Problems

P12.3.1 Consider the TLS problem (12.3.2) with nonsingular D and T. (a) Show that if rank$(A) < n$, then (12.3.2) has a solution if and only if $b \in$ ran(A). (b) Show that if rank$(A) = n$, then (12.3.2) has no solution if $A^T D^2 b = 0$ and $|t_{n+1}|\| Db \|_2 \geq \sigma_n(DAT_1)$ where $T_1 = \text{diag}(t_1, \ldots, t_n)$.

P12.3.2 Show that if $C = D[\, A,\, b\,]T = [\, A_1,\, d\,]$ and $\sigma_n(C) > \sigma_{n+1}(C)$, then the TLS solution x satisfies $(A_1^T A_1 - \sigma_{n+1}(C)^2 I)x = A_1^T d$.

P12.3.3 Show how to solve (12.3.2) with the added constraint that the first p columns of the minimizing E are zero.

Notes and References for Sec. 12.3

This section is based upon

G.H. Golub and C.F. Van Loan (1980). "An Analysis of the Total Least Squares Problem," *SIAM J. Num. Anal.* 17, 883–93.

The bearing of the SVD on the TLS problem is set forth in

G.H. Golub and C. Reinsch (1970). "Singular Value Decomposition and Least Squares Solutions," *Numer. Math. 14*, 403–420.

G.H. Golub (1973). "Some Modified Matrix Eigenvalue Problems," *SIAM Review 15*, 318–334.

The most detailed study of the TLS problem is

S. Van Huffel and J. Vandewalle (1991). *The Total Least Squares Problem: Computational Aspects and Analysis*, SIAM Publications, Philadelphia.

If some of the columns of A are known exactly then it is sensible to force the TLS perturbation matrix E to be zero in the same columns. Aspects of this constrained TLS problem are discussed in

J.W. Demmel (1987). "The Smallest Perturbation of a Submatrix which Lowers the Rank and Constrained Total Least Squares Problems, *SIAM J. Numer. Anal. 24*, 199–206.

S. Van Huffel and J. Vandewalle (1988). "The Partial Total Least Squares Algorithm," *J. Comp. and App. Math. 21*, 333–342.

S. Van Huffel and J. Vandewalle (1988). "Analysis and Solution of the Nongeneric Total Least Squares Problem," *SIAM J. Matrix Anal. Appl. 9*, 360–372.

S. Van Huffel and J. Vandewalle (1989). "Analysis and Properties of the Generalized Total Least Squares Problem $AX \approx B$ When Some or All Columns in A are Subject to Error," *SIAM J. Matrix Anal. Appl. 10*, 294–315.

S. Van Huffel and H. Zha (1991). "The Restricted Total Least Squares Problem: Formulation, Algorithm, and Properties," *SIAM J. Matrix Anal. Appl. 12*, 292–309.

S. Van Huffel (1992). "On the Significance of Nongeneric Total Least Squares Problems," *SIAM J. Matrix Anal. Appl. 13*, 20–35.

M. Wei (1992). "The Analysis for the Total Least Squares Problem with More than One Solution," *SIAM J. Matrix Anal. Appl. 13*, 746–763.

S. Van Huffel and H. Zha (1993). "An Efficient Total Least Squares Algorithm Based On a Rank-Revealing Two-Sided Orthogonal Decomposition," *Numerical Algorithms 4*, 101–133.

C.C. Paige and M. Wei (1993). "Analysis of the Generalized Total Least Squares Problem $AX = B$ when Some of the Columns are Free of Error," *Numer. Math. 65*, 177–202.

R.D. Fierro and J.R. Bunch (1994). "Collinearity and Total Least Squares," *SIAM J. Matrix Anal. Appl. 15*, 1167–1181.

Other references concerned with least squares fitting when there are errors in the data matrix include

K. Pearson (1901). "On Lines and Planes of Closest Fit to Points in Space," *Phil. Mag. 2*, 559–72.

A. Wald (1940). "The Fitting of Straight Lines if Both Variables are Subject to Error," *Annals of Mathematical Statistics 11*, 284–300.

A. Madansky (1959). "The Fitting of Straight Lines When Both Variables Are Subject to Error," *J. Amer. Stat. Assoc. 54*, 173–205.

I. Linnik (1961). *Method of Least Squares and Principles of the Theory of Observations*, Pergamon Press, New York.

W.G. Cochrane (1968). "Errors of Measurement in Statistics," *Technometrics 10*, 637–66.

R.F. Gunst, J.T. Webster, and R.L. Mason (1976). "A Comparison of Least Squares and Latent Root Regression Estimators," *Technometrics 18*, 75–83.

G.W. Stewart (1977c). "Sensitivity Coefficients for the Effects of Errors in the Independent Variables in a Linear Regression," Technical Report TR-571, Department of Computer Science, University of Maryland, College Park, MD.

A. Van der Sluis and G.W. Veltkamp (1979). "Restoring Rank and Consistency by Orthogonal Projection," *Lin. Alg. and Its Applic. 28*, 257–78.

12.4   Computing Subspaces with the SVD

It is sometimes necessary to investigate the relationship between two given subspaces. How close are they? Do they intersect? Can one be "rotated" into the other? And so on. In this section we show how questions like these can be answered using the singular value decomposition. Knowledge of Chapter 5 and §8.6 are assumed.

12.4.1   Rotation of Subspaces

Suppose $A \in \mathbb{R}^{m \times p}$ is a data matrix obtained by performing a certain set of experiments. If the same set of experiments is performed again, then a different data matrix, $B \in \mathbb{R}^{m \times p}$, is obtained. In the *orthogonal Procrustes problem* the possibility that B can be rotated into A is explored by solving the following problem:

$$\text{minimize } \| A - BQ \|_F \qquad \text{subject to } Q^T Q = I_p. \qquad (12.4.1)$$

Recall that the trace of a matrix is the sum of its diagonal entries and thus, $\text{tr}(C^T C) = \| C \|_F^2$. It follows that if $Q \in \mathbb{R}^{p \times p}$ is orthogonal, then

$$\| A - BQ \|_F^2 = \text{tr}(A^T A) + \text{tr}(B^T B) - 2\,\text{tr}(Q^T B^T A).$$

Thus, (12.4.1) is equivalent to the problem of maximizing $\text{tr}(Q^T B^T A)$.

The maximizing Q can be found by calculating the SVD of $B^T A$. Indeed, if $U^T(B^T A)V = \Sigma = \text{diag}(\sigma_1, \ldots, \sigma_p)$ is the SVD of this matrix and we define the orthogonal matrix Z by $Z = V^T Q^T U$, then

$$\text{tr}(Q^T B^T A) = \text{tr}(Q^T U \Sigma V^T) = \text{tr}(Z\Sigma) = \sum_{i=1}^{p} z_{ii}\sigma_i \leq \sum_{i=1}^{p} \sigma_i.$$

Clearly, the upper bound is attained by setting $Q = UV^T$ for then $Z = I_p$. This gives the following algorithm:

Algorithm 12.4.1 Given A and B in $\mathbb{R}^{m \times p}$, the following algorithm finds an orthogonal $Q \in \mathbb{R}^{p \times p}$ such that $\| A - BQ \|_F$ is minimum.

$\quad C = B^T A$
$\quad$ Compute the SVD $U^T C V = \Sigma$. Save U and V.
$\quad Q = UV^T.$

The solution matrix Q is the orthogonal polar factor of $B^T A$. See §4.2.10.

Example 12.4.1

$$Q = \begin{bmatrix} .9999 & -.0126 \\ .0126 & .9999 \end{bmatrix} \quad \text{minimizes} \quad \left\| \begin{bmatrix} 1 & 2 \\ 3 & 4 \\ 5 & 6 \\ 7 & 8 \end{bmatrix} Q - \begin{bmatrix} 1.2 & 2.1 \\ 2.9 & 4.3 \\ 5.2 & 6.1 \\ 6.8 & 8.1 \end{bmatrix} \right\|_F.$$

12.4.2   Intersection of Null Spaces

Let $A \in \mathbb{R}^{m \times n}$ and $B \in \mathbb{R}^{p \times n}$ be given, and consider the problem of finding an orthonormal basis for $\text{null}(A) \cap \text{null}(B)$. One approach is to compute the null space of the matrix

$$C = \begin{bmatrix} A \\ B \end{bmatrix}$$

since $Cx = 0 \Leftrightarrow x \in \text{null}(A) \cap \text{null}(B)$. However, a more economical procedure results if we exploit the following theorem.

Theorem 12.4.1 *Suppose $A \in \mathbb{R}^{m \times n}$ and let $\{z_1, \ldots, z_t\}$ be an orthonormal basis for $\text{null}(A)$. Define $Z = [z_1, \ldots, z_t]$ and let $\{w_1, \ldots, w_q\}$ be an orthonormal basis for $\text{null}(BZ)$ where $B \in \mathbb{R}^{p \times n}$. If $W = [w_1, \ldots, w_q]$, then the columns of ZW form an orthonormal basis for $\text{null}(A) \cap \text{null}(B)$.*

Proof. Since $AZ = 0$ and $(BZ)W = 0$, we clearly have $\text{ran}(ZW) \subset \text{null}(A) \cap \text{null}(B)$. Now suppose x is in both $\text{null}(A)$ and $\text{null}(B)$. It follows that $x = Za$ for some $0 \neq a \in \mathbb{R}^t$. But since $0 = Bx = BZa$, we must have $a = Wb$ for some $b \in \mathbb{R}^q$. Thus, $x = ZWb \in \text{ran}(ZW)$. $\square$

When the SVD is used to compute the orthonormal bases in this theorem we obtain the following procedure:

Algorithm 12.4.2   Given $A \in \mathbb{R}^{m \times n}$ and $B \in \mathbb{R}^{p \times n}$, the following algorithm computes and integer s and a matrix $Y = [y_1, \ldots, y_s]$ having orthonormal columns which span $\text{null}(A) \cap \text{null}(B)$. If the intersection is trivial then $s = 0$.

> Compute the SVD $U_A^T A V_A = \text{diag}(\sigma_i)$. Save V_A and set
> $\quad\quad r = \text{rank}(A)$.
> **if** $r < n$
> $\quad\quad C = BV_A(:, r + 1{:}n)$
> $\quad\quad$ Compute the SVD $U_C^T C V_C = \text{diag}(\gamma_i)$. Save V_C and set
> $\quad\quad\quad\quad q = \text{rank}(C)$.
> $\quad\quad$ **if** $q < n - r$
> $\quad\quad\quad\quad s = n - r - q$
> $\quad\quad\quad\quad Y = V_A(:, r + 1{:}n)V_C(:, q + 1{:}n - r)$
> $\quad\quad$ **else**
> $\quad\quad\quad\quad s = 0$
> $\quad\quad$ **end**
> **else**
> $\quad\quad s = 0$
> **end**

The amount of work required by this algorithm depends upon the relative sizes of m, n, p, and r.

We mention that a practical implementation of this algorithm requires a means for deciding when a computed singular value $\hat{\sigma}_i$ is negligible. The use of a tolerance δ for this purpose (e.g. $\hat{\sigma}_i < \delta \Rightarrow \hat{\sigma}_i = 0$) implies that the columns of the computed $\hat{Y}$ "almost" define a common null space of A and B in the sense that $\| A\hat{Y} \|_2 \approx \| B\hat{Y} \|_2 \approx \delta$.

Example 12.4.2 If

$$
A = \begin{bmatrix} 1 & -1 & 1 \\ 1 & -1 & 1 \\ 1 & -1 & 1 \end{bmatrix} \quad \text{and} \quad B = \begin{bmatrix} 4 & 2 & 0 \\ 2 & 1 & 0 \\ 6 & 3 & 0 \end{bmatrix}
$$

then $\mathrm{null}(A) \cap \mathrm{null}(B) = \mathrm{span}\{x\}$, where $x = [1 \ -2 \ -3]^T$. Applying Algorithm 12.4.2 we find

$$
V_{2A}V_{2C} = \begin{bmatrix} -.8165 & .0000 \\ -.4082 & .7071 \\ .4082 & .7071 \end{bmatrix} \begin{bmatrix} -.3273 \\ -.9449 \end{bmatrix} \approx \begin{bmatrix} .2673 \\ -.5345 \\ -.8018 \end{bmatrix} \approx .2673 \begin{bmatrix} 1 \\ -2 \\ -3 \end{bmatrix} .
$$

12.4.3   Angles Between Subspaces

Let F and G be subspaces in $\mathbb{R}^m$ whose dimensions satisfy

$$
p = \dim(F) \geq \dim(G) = q \geq 1.
$$

The *principal angles* $\theta_1, \ldots, \theta_q \in [0, \pi/2]$ between F and G are defined recursively by

$$
\cos(\theta_k) = \max_{u \in F} \max_{v \in G} u^T v = u_k^T v_k
$$

subject to:

$$
\begin{aligned}
\| u \| = \| v \| &= 1 \\
u^T u_i &= 0 \qquad i = 1{:}k-1 \\
v^T v_i &= 0 \qquad i = 1{:}k-1.
\end{aligned}
$$

Note that the principal angles satisfy $0 \leq \theta_1 \leq \cdots \leq \theta_q \leq \pi/2$. The vectors $\{u_1, \ldots, u_q\}$ and $\{v_1, \ldots, v_q\}$ are called the *principal vectors* between the subspaces F and G.

Principal angles and vectors arise in many important statistical applications. The largest principal angle is related to the notion of distance between equidimensional subspaces that we discussed in §2.6.3 If $p = q$ then $\mathrm{dist}(F, G) = \sqrt{1 - \cos(\theta_p)^2} = \sin(\theta_p)$.

If the columns of $Q_F \in \mathbb{R}^{m \times p}$ and $Q_G \in \mathbb{R}^{m \times q}$ define orthonormal bases for F and G respectively, then

$$
\max_{\substack{u \in F \\ \|u\|_2 = 1}} \max_{\substack{v \in G \\ \|v\|_2 = 1}} u^T v = \max_{\substack{y \in \mathbb{R}^p \\ \|y\|_2 = 1}} \max_{\substack{z \in \mathbb{R}^q \\ \|z\|_2 = 1}} y^T (Q_F^T Q_G) z
$$

From the minimax characterization of singular values given in Theorem 8.6.1 it follows that if $Y^T(Q_F^T Q_G)Z = \text{diag}(\sigma_1, \ldots, \sigma_q)$ is the SVD of $Q_F^T Q_G$, then we may define the u_k, v_k, and θ_k by

$$
\begin{aligned}
[\, u_1, \ldots, u_p \,] &= Q_F Y \\
[\, v_1, \ldots, v_q \,] &= Q_G Z \\
\cos(\theta_k) &= \sigma_k \qquad k = 1{:}q
\end{aligned}
$$

Typically, the spaces F and G are defined as the ranges of given matrices $A \in \mathbb{R}^{m \times p}$ and $B \in \mathbb{R}^{m \times q}$. In this case the desired orthonormal bases can be obtained by computing the QR factorizations of these two matrices.

Algorithm 12.4.3 Given $A \in \mathbb{R}^{m \times p}$ and $B \in \mathbb{R}^{m \times q}$ ($p \geq q$) each with linearly independent columns, the following algorithm computes the orthogonal matrices $U = [\, u_1, \ldots, u_q \,]$ and $V = [\, v_1, \ldots, v_q \,]$ and $\cos(\theta_1), \ldots \cos(\theta_q)$ such that the θ_k are the principal angles between $\text{ran}(A)$ and $\text{ran}(B)$ and u_k and v_k are the associated principal vectors.

Use Algorithm 5.2.1 to compute the QR factorizations

$$
\begin{aligned}
A &= Q_A R_A & Q_A^T Q_A &= I_p, & R_A &\in \mathbb{R}^{p \times p} \\
B &= Q_B R_B & Q_B^T Q_B &= I_q, & R_B &\in \mathbb{R}^{q \times q}
\end{aligned}
$$

$C = Q_A^T Q_B$
Compute the SVD $Y^T C Z = \text{diag}(\cos(\theta_k))$.
$Q_A Y(:, 1{:}q) = [\, u_1, \ldots, u_q \,]$
$Q_B Z = [\, v_1, \ldots, v_q \,]$

This algorithm requires about $4m(q^2 + 2p^2) + 2pq(m + q) + 12q^3$ flops.

The idea of using the SVD to compute the principal angles and vectors is due to Björck and Golub (1973). The problem of rank deficiency in A and B is also treated in this paper.

12.4.4   Intersection of Subspaces

Algorithm 12.4.3 can also be used to compute an orthonormal basis for $\text{ran}(A) \cap \text{ran}(B)$ where $A \in \mathbb{R}^{m \times p}$ and $B \in \mathbb{R}^{m \times q}$

Theorem 12.4.2 Let $\{\cos(\theta_k), u_k, v_k\}_{k=1}^q$ be defined by Algorithm 12.4.3. If the index s is defined by $1 = \cos(\theta_1) = \cdots = \cos(\theta_s) > \cos(\theta_{s+1})$, then we have

$$
\text{ran}(A) \cap \text{ran}(B) = \text{span}\{u_1, \ldots, u_s\} = \text{span}\{v_1, \ldots, v_s\} \, .
$$

Proof. The proof follows from the observation that if $\cos(\theta_k) = 1$, then $u_k = v_k$. $\square$

With inexact arithmetic, it is necessary to compute the approximate multiplicity of the unit cosines in Algorithm 12.4.3.

Example 12.4.3 If

$$A = \begin{bmatrix} 1 & 2 \\ 3 & 4 \\ 5 & 6 \end{bmatrix} \quad \text{and} \quad B = \begin{bmatrix} 1 & 5 \\ 3 & 7 \\ 5 & -1 \end{bmatrix}$$

then the cosines of the principal angles between ran(A) and ran(B) are 1.000 and .856.

Problems

P12.4.1 Show that if A and B are m-by-p matrices, with $p \le m$, then

$$\min_{Q^T Q = I_p} \| A - BQ \|_F^2 = \sum_{i=1}^{p} (\sigma_i(A)^2 - 2\sigma_i(B^T A) + \sigma_i(B)^2).$$

P12.4.2 Extend Algorithm 12.4.2 so that it can compute an orthonormal basis for $\text{null}(A_1) \cap \cdots \cap \text{null}(A_s)$.

P12.4.3 Extend Algorithm 12.4.3 to handle the case when A and B are rank deficient.

P12.4.4 Relate the principal angles and vectors between ran(A) and ran(B) to the eigenvalues and eigenvectors of the generalized eigenvalue problem

$$\begin{bmatrix} 0 & A^T B \\ B^T A & 0 \end{bmatrix} \begin{bmatrix} y \\ z \end{bmatrix} = \sigma \begin{bmatrix} A^T A & 0 \\ 0 & B^T B \end{bmatrix} \begin{bmatrix} y \\ z \end{bmatrix}.$$

P12.4.5 Suppose $A, B \in \mathbf{R}^{m \times n}$ and that A has full column rank. Show how to compute a symmetric matrix $X \in \mathbf{R}^{n \times n}$ that minimizes $\| AX - B \|_F$. Hint: Compute the SVD of A.

Notes and References for Sec. 12.4

The problem of minimizing $\| A - BQ \|_F$ over all orthogonal matrices arises in psychometrics. See

B. Green (1952). "The Orthogonal Approximation of an Oblique Structure in Factor Analysis," *Psychometrika 17*, 429–40.

P. Schonemann (1966). "A Generalized Solution of the Orthogonal Procrustes Problem," *Psychometrika 31*, 1–10.

I.Y. Bar-Itzhack (1975). "Iterative Optimal Orthogonalization of the Strapdown Matrix," *IEEE Trans. Aerospace and Electronic Systems 11*, 30–37.

R.J. Hanson and M.J. Norris (1981). "Analysis of Measurements Based on the Singular Value Decomposition," *SIAM J. Sci. and Stat. Comp. 2*, 363–374.

H. Park (1991). "A Parallel Algorithm for the Unbalanced Orthogonal Procrustes Problem," *Parallel Computing 17*, 913–923.

When $B = I$, this problem amounts to finding the closest orthogonal matrix to A. This is equivalent to the polar decomposition problem of §4.2.10. See

A. Björck and C. Bowie (1971). "An Iterative Algorithm for Computing the Best Estimate of an Orthogonal Matrix," *SIAM J. Num. Anal. 8*, 358–64.

N.J. Higham (1986). "Computing the Polar Decomposition—with Applications," *SIAM J. Sci. and Stat. Comp. 7*, 1160–1174.

If A is reasonably close to being orthogonal itself, then Björck and Bowie's technique is more efficient than the SVD algorithm.

The problem of minimizing $\| AX - B \|_F$ subject to the constraint that X is symmetric is studied in

N.J. Higham (1988). "The Symmetric Procrustes Problem," *BIT 28*, 133–43.

Using the SVD to solve the canonical correlation problem is discussed in

A. Björck and G.H. Golub (1973). "Numerical Methods for Computing Angles Between Linear Subspaces," *Math. Comp. 27*, 579–94.

G.H. Golub and H. Zha (1994). "Perturbation Analysis of the Canonical Correlations of Matrix Pairs," *Lin. Alg. and Its Applic. 210*, 3–28.

The SVD has other roles to play in statistical computation.

S.J. Hammarling (1985). "The Singular Value Decomposition in Multivariate Statistics," *ACM SIGNUM Newsletter 20*, 2-25.

12.5   Updating Matrix Factorizations

In many applications it is necessary to re-factor a given matrix $A \in \mathbb{R}^{m \times n}$ after it has been altered in some minimal sense. For example, given that we have the QR factorization of A, we may need to calculate the QR factorization of a matrix A that is obtained by (a) adding a general rank-one matrix to A, (b) appending a row (or column) to A, or (c) deleting a row (or column) from A. In this section we show that in situations like these, it is much more efficient to "update" A's QR factorization than to generate it from scratch. We also show how to update the null space of a matrix after it has been augmented with an additional row.

Before beginning, we mention that there are also techniques for updating the factorizations $PA = LU$, $A = GG^T$, and $A = LDL^T$. Updating these factorizations, however, can be quite delicate because of pivoting requirements and because when we tamper with a positive definite matrix the result may not be positive definite. See Gill, Golub, Murray, and Saunders (1974) and Stewart (1979). Along these lines we briefly discuss hyperbolic transformations and their use in the Cholesky downdating problem.

Familiarity with §3.5, §4.1, §5.1, §5.2, §5.4, and §5.5 is required. Complementary reading includes Gill, Murray, and Wright (1991).

12.5.1   Rank-One Changes

Suppose we have the QR factorization $QR = B \in \mathbb{R}^{m \times n}$ and that we need to compute the QR factorization $B + uv^T = Q_1 R_1$ where $u, v \in \mathbb{R}^n$ are given. Observe that

$$B + uv^T = Q(R + wv^T) \qquad (12.5.1)$$

where $w = Q^T u$. Suppose that we compute rotations $J_{n-1}, \dots, J_2, J_1$ such that

$$J_1^T \cdots J_{n-1}^T w = \pm \| w \|_2 e_1 .$$

Here, each J_k is a rotation in planes k and $k+1$. (For details, see Algorithm 5.1.3.) If these same Givens rotations are applied to R, it can be shown that

$$H = J_1^T \cdots J_{n-1}^T R \qquad (12.5.2)$$

is upper Hessenberg. For example, in the $n = 4$ case we start with

$$R = \begin{bmatrix} \times & \times & \times & \times \\ 0 & \times & \times & \times \\ 0 & 0 & \times & \times \\ 0 & 0 & 0 & \times \end{bmatrix} \qquad w = \begin{bmatrix} \times \\ \times \\ \times \\ \times \end{bmatrix}$$

and then update as follows:

$$R = J_3^T R = \begin{bmatrix} \times & \times & \times & \times \\ 0 & \times & \times & \times \\ 0 & 0 & \times & \times \\ 0 & 0 & \times & \times \end{bmatrix} \qquad w = J_3^T w = \begin{bmatrix} \times \\ \times \\ \times \\ 0 \end{bmatrix}$$

$$R = J_2^T R = \begin{bmatrix} \times & \times & \times & \times \\ 0 & \times & \times & \times \\ 0 & \times & \times & \times \\ 0 & 0 & \times & \times \end{bmatrix} \qquad w = J_2^T w = \begin{bmatrix} \times \\ \times \\ 0 \\ 0 \end{bmatrix}$$

$$H = J_1^T R = \begin{bmatrix} \times & \times & \times & \times \\ \times & \times & \times & \times \\ 0 & \times & \times & \times \\ 0 & 0 & \times & \times \end{bmatrix} \qquad w = J_1^T w = \begin{bmatrix} \times \\ 0 \\ 0 \\ 0 \end{bmatrix}$$

Consequently,

$$(J_1^T \cdots J_{n-1}^T)(R + wv^T) = H \pm \| w \|_2 e_1 v^T = H_1 \qquad (12.5.3)$$

is also upper Hessenberg.

In Algorithm 5.2.3, we show how to compute the QR factorization of an upper Hessenberg matrix in $O(n^2)$ flops. In particular, we can find Givens rotations G_k, $k = 1{:}n - 1$ such that

$$G_{n-1}^T \cdots G_1^T H_1 = R_1 \qquad (12.5.4)$$

is upper triangular. Combining (12.5.1) through (12.5.4) we obtain the QR factorization $B + uv^T = Q_1 R_1$ where

$$Q_1 = Q J_{n-1} \cdots J_1 G_1 \cdots G_{n-1}.$$

A careful assessment of the work reveals that about $26n^2$ flops are required. The vector $w = Q^T u$ requires $2n^2$ flops. Computing H and accumulating the J_k into Q. involves $12n^2$ flops. Finally, computing R_1 and multiplying the G_k into Q involves $12n^2$ flops.

The technique readily extends to the case when B is rectangular. It can also be generalized to compute the QR factorization of $B + UV^T$ where rank$(UV^T) = p > 1$.

12.5.2   Appending or Deleting a Column

Assume that we have the QR factorization

$$QR = A = [\, a_1, \ldots, a_n \,] \qquad a_i \in \mathbb{R}^m \qquad (12.5.5)$$

and partition the upper triangular matrix $R \in \mathbb{R}^{m \times n}$ as follows:

$$R = \begin{bmatrix} R_{11} & v & R_{13} \\ 0 & r_{kk} & w^T \\ 0 & 0 & R_{33} \\ k-1 & 1 & n-k \end{bmatrix} \begin{matrix} k-1 \\ 1 \\ m-k \end{matrix} \;\; .$$

Now suppose that we want to compute the QR factorization of

$$\tilde{A} = [\, a_1, \ldots, a_{k-1}, a_{k+1}, \ldots, a_n \,] \in \mathbb{R}^{m \times (n-1)} \, .$$

Note that $\tilde{A}$ is just A with its kth column deleted and that

$$Q^T \tilde{A} = \begin{bmatrix} R_{11} & R_{13} \\ 0 & w^T \\ 0 & R_{33} \end{bmatrix} = H$$

is upper Hessenberg, e.g.,

$$H = \begin{bmatrix} \times & \times & \times & \times & \times \\ 0 & \times & \times & \times & \times \\ 0 & 0 & \times & \times & \times \\ 0 & 0 & \times & \times & \times \\ 0 & 0 & 0 & \times & \times \\ 0 & 0 & 0 & 0 & \times \\ 0 & 0 & 0 & 0 & 0 \end{bmatrix} \qquad m = 7, \; n = 6, \; k = 3$$

Clearly, the unwanted subdiagonal elements $h_{k+1,k}, \ldots, h_{n,n-1}$ can be zeroed by a sequence of Givens rotations: $G_{n-1}^T \cdots G_k^T H = R_1$. Here, G_i is a rotation in planes i and $i+1$ for $i = k{:}n - 1$. Thus, if $Q_1 = QG_k \cdots G_{n-1}$ then $\tilde{A} = Q_1 R_1$ is the QR factorization of $\tilde{A}$.

The above update procedure can be executed in $O(n^2)$ flops and is very useful in certain least squares problems. For example, one may wish to examine the significance of the kth factor in the underlying model by deleting the kth column of the corresponding data matrix and solving the resulting LS problem.

In a similar vein, it is useful to be able to compute efficiently the solution to the LS problem after a column has been appended to A. Suppose we have the QR factorization (12.5.5) and now wish to compute the QR factorization of

$$\tilde{A} = [\, a_1, \ldots, a_k, z, a_{k+1}, \ldots, a_n \,]$$

where $z \in \mathbb{R}^m$ is given. Note that if $w = Q^T z$ then

$$Q^T A = [\, Q^T a_1, \ldots, Q^T a_k, w, Q^T a_{k+1}, \ldots, Q^T a_n \,] = \tilde{A}$$

is upper triangular except for the presence of a "spike" in its $k+1$-st column, e.g.,

$$\tilde{A} = \begin{bmatrix} \times & \times & \times & \times & \times & \times \\ 0 & \times & \times & \times & \times & \times \\ 0 & 0 & \times & \times & \times & \times \\ 0 & 0 & 0 & \times & \times & \times \\ 0 & 0 & 0 & \times & 0 & \times \\ 0 & 0 & 0 & \times & 0 & 0 \\ 0 & 0 & 0 & \times & 0 & 0 \end{bmatrix} \qquad m = 7,\ n = 5,\ k = 3$$

It is possible to determine Givens rotations $J_{m-1}, \ldots, J_{k+1}$ so that

$$J_{k+1}^T \cdots J_{m-1}^T w = \begin{bmatrix} w_1 \\ \vdots \\ w_{k+1} \\ 0 \\ \vdots \\ 0 \end{bmatrix}$$

with $J_{k+1}^T \cdots J_{m-1}^T \tilde{A} = \tilde{R}$ upper triangular. We illustrate this by continuing with the above example:

$$H = J_6^T \tilde{A} = \begin{bmatrix} \times & \times & \times & \times & \times & \times \\ 0 & \times & \times & \times & \times & \times \\ 0 & 0 & \times & \times & \times & \times \\ 0 & 0 & 0 & \times & \times & \times \\ 0 & 0 & 0 & \times & 0 & \times \\ 0 & 0 & 0 & \times & 0 & 0 \\ 0 & 0 & 0 & 0 & 0 & 0 \end{bmatrix}$$

$$H = J_5^T H = \begin{bmatrix} \times & \times & \times & \times & \times & \times \\ 0 & \times & \times & \times & \times & \times \\ 0 & 0 & \times & \times & \times & \times \\ 0 & 0 & 0 & \times & \times & \times \\ 0 & 0 & 0 & \times & 0 & \times \\ 0 & 0 & 0 & 0 & 0 & \times \\ 0 & 0 & 0 & 0 & 0 & 0 \end{bmatrix}$$

$$H = J_4^T H = \begin{bmatrix} \times & \times & \times & \times & \times & \times \\ 0 & \times & \times & \times & \times & \times \\ 0 & 0 & \times & \times & \times & \times \\ 0 & 0 & 0 & \times & \times & \times \\ 0 & 0 & 0 & 0 & \times & \times \\ 0 & 0 & 0 & 0 & 0 & \times \\ 0 & 0 & 0 & 0 & 0 & 0 \end{bmatrix}$$

This update requires $O(mn)$ flops.

12.5.3   Appending or Deleting a Row

Suppose we have the QR factorization $QR = A \in \mathbb{R}^{m \times n}$ and now wish to obtain the QR factorization of

$$\tilde{A} = \begin{bmatrix} w^T \\ A \end{bmatrix}$$

where $w \in \mathbb{R}^n$. Note that

$$\text{diag}(1, Q^T)\tilde{A} = \begin{bmatrix} w^T \\ R \end{bmatrix} = H$$

is upper Hessenberg. Thus, Givens rotations $J_1, \ldots, J_n$ could be determined so $J_n^T \cdots J_1^T H = R_1$ is upper triangular. It follows that

$$\tilde{A} = Q_1 R_1$$

is the desired QR factorization, where $Q_1 = \text{diag}(1, Q)J_1 \cdots J_n$.

No essential complications result if the new row is added between rows k and $k+1$ of A. We merely apply the above with A replaced by PA and Q replaced by PQ where

$$P = \begin{bmatrix} 0 & I_{m-k} \\ I_k & 0 \end{bmatrix}.$$

Upon completion $\text{diag}(1, P^T)Q_1$ is the desired orthogonal factor.

Lastly, we consider how to update the QR factorization $QR = A \in \mathbb{R}^{m \times n}$ when the first row of A is deleted. In particular, we wish to compute the QR factorization of the submatrix A_1 in

$$A = \begin{bmatrix} z^T \\ A_1 \end{bmatrix} \begin{matrix} 1 \\ m-1 \end{matrix}$$

(The procedure is similar when an arbitrary row is deleted.) Let q^T be the first row of Q and compute Givens rotations $G_1, \ldots, G_{m-1}$ such that

$$G_1^T \cdots G_{m-1}^T q = \alpha e_1,$$

where $\alpha = \pm 1$. Note that

$$H = G_1^T \cdots G_{m-1}^T R = \begin{bmatrix} v^T \\ R_1 \end{bmatrix} \begin{matrix} 1 \\ m-1 \end{matrix}$$

is upper Hessenberg and that

$$Q G_{m-1} \cdots G_1 = \begin{bmatrix} \alpha & 0 \\ 0 & Q_1 \end{bmatrix}$$

where $Q_1 \in \mathbb{R}^{(m-1) \times (m-1)}$ is orthogonal. Thus,

$$A = \begin{bmatrix} z^T \\ A_1 \end{bmatrix} = (Q G_{m-1} \cdots G_1)(G_1^T \cdots G_{m-1}^T R) = \begin{bmatrix} \alpha & 0 \\ 0 & Q_1 \end{bmatrix} \begin{bmatrix} v^T \\ R_1 \end{bmatrix}$$

from which we conclude that $A_1 = Q_1 R_1$ is the desired QR factorization.

12.5.4   Hyperbolic Transformation Methods

Recall that the "R" in $A = QR$ is the transposed Cholesky factor in $A^T A = GG^T$. Thus, there is a close connection between the QR modifications just discussed and analogous modifications of the Cholesky factorization. We illustrate this with the *Cholesky downdating problem* which corresponds to the removal of an A-row in QR. In the Cholesky downdating problem we have the Cholesky factorization

$$GG^T = A^T A = \begin{bmatrix} z^T \\ A_1 \end{bmatrix}^T \begin{bmatrix} z^T \\ A_1 \end{bmatrix} \qquad (12.5.6)$$

where $A \in \mathbb{R}^{m \times n}$ with $m > n$ and $z \in \mathbb{R}^n$. Our task is to find a lower triangular G_1 such that $G_1 G_1^T = A_1^T A_1$. There are several approaches to this interesting and important problem. Simply because it is an opportunity to introduce some new ideas, we present a downdating procedure that relies on *hyperbolic transformations*.

We start with a definition. $H \in \mathbb{R}^{m \times m}$ is *pseudo-orthogonal* with respect to the *signature matrix* $S = \text{diag}(\pm 1) \in \mathbb{R}^{m \times m}$ if $H^T S H = S$. Now from (12.5.6) we have $A^T A = A_1^T A_1 + z z^T = G G^T$ and so

$$A_1^T A_1 = A^T A - z z^T = G G^T - z z^T = [\, G \, z \,] \begin{bmatrix} I_n & 0 \\ 0 & -1 \end{bmatrix} \begin{bmatrix} G^T \\ z^T \end{bmatrix}.$$

Define the signature matrix

$$S = \begin{bmatrix} I_n & 0 \\ 0 & -1 \end{bmatrix} \tag{12.5.7}$$

and suppose that we can find $H \in \mathbb{R}^{(n+1) \times (n+1)}$ such that $H^T S H = S$ with the property that

$$H \begin{bmatrix} G^T \\ z^T \end{bmatrix} = \begin{bmatrix} G_1^T \\ 0 \end{bmatrix} \tag{12.5.8}$$

is upper triangular. It follows that

$$A_1^T A_1 = [\, G \, z \,] H^T S H \begin{bmatrix} G^T \\ z^T \end{bmatrix} = [G_1 \, 0] S \begin{bmatrix} G_1 \\ 0 \end{bmatrix} = G_1 G_1^T$$

is the sought after Cholesky factorization.

We now show how to construct the hyperbolic transformation H in (12.5.8) using *hyperbolic rotations*. A 2-by-2 hyperbolic rotation has the form

$$H = \begin{bmatrix} \cosh(\theta) & -\sinh(\theta) \\ -\sinh(\theta) & \cosh(\theta) \end{bmatrix} = \begin{bmatrix} c & -s \\ -s & c \end{bmatrix}.$$

Note that if $H \in \mathbb{R}^{2 \times 2}$ is a hyperbolic rotation then $H^T S H = S$ where $S = \text{diag}(-1,1)$. Paralleling our Givens rotations developments, let us see how hyperbolic rotations can be used for zeroing. From

$$\begin{bmatrix} c & -s \\ -s & c \end{bmatrix} \begin{bmatrix} x_1 \\ x_2 \end{bmatrix} = \begin{bmatrix} r \\ 0 \end{bmatrix} \qquad c^2 - s^2 = 1$$

we obtain the equation $c x_2 = s x_1$. Note that there is no solution to this equation if $x_1 = x_2 \neq 0$, a clue that hyperbolic rotations are not as numerically solid as their Givens rotation counterparts. If $x_1 \neq x_2$ then it is possible to compute the cosh-sinh pair:

> if $x_2 = 0$
> $s = 0;\ c = 1$
> else
>   if $|x_2| < |x_1|$
> $\tau = x_2/x_1;\ c = 1/\sqrt{1 - \tau^2};\ s = c\tau$
>   elseif $|x_1| < |x_2|$
> $\tau = x_1/x_2;\ s = 1/\sqrt{1 - \tau^2};\ c = s\tau$
>   end
> end

$$\tag{12.5.9}$$

Observe that the norm of the hyperbolic rotation produced by this algorithm gets large as x_1 gets close to x_2.

Now any matrix $H = H(p, n + 1, \theta) \in \mathbb{R}^{(n+1)\times(n+1)}$ that is the identity everywhere except $h_{p,p} = h_{n+1,n+1} = \cosh(\theta)$ and $h_{p,n+1} = h_{n+1,p} = -\sinh(\theta)$ satisfies $H^T S H = S$ where S is prescribed in (12.5.7). Using (12.5.9), we attempt to generate hyperbolic rotations $H_k = H(1, k, \theta_k)$ for $k = 2{:}n + 1$ so that

$$H_n \cdots H_1 \begin{bmatrix} G^T \\ z^T \end{bmatrix} = \begin{bmatrix} \tilde{G}^T \\ 0 \end{bmatrix}.$$

This turns out to be possible if A has full column rank. Hyperbolic rotation H_k zeros entry $(k + 1, k)$. In other words, if A has full column rank, then it can be shown that each call to (12.5.9) results in a cosh-sinh pair. See Alexander, Pan, and Plemmons (1988).

12.5.5   Updating the ULV Decomposition

Suppose $A \in \mathbb{R}^{m\times n}$ is rank deficient and that we have a basis for its null space. If we add a row to A,

$$\tilde{A} = \begin{bmatrix} A \\ z^T \end{bmatrix},$$

then how easy is it to compute a null basis for $\tilde{A}$? When a sequence of such update problems are involved the issue is one of *tracking* the null space. Subspace tracking arises in a number of real-time signal processing applications.

Working with the SVD is awkward in this context because $O(n^3)$ flops are required to recompute the SVD of a matrix that has undergone a unit rank perturbation. On the other hand, Stewart (1993) has shown that the null space updating problem becomes $O(n^2)$ per step if we properly couple the ideas of condition estimation of §3.5.4 and complete orthogonal decomposition. Recall from §5.4.2 that a complete orthogonal decomposition is two-sided and reveals the rank of the underlying matrix,

$$U^T A V = \begin{bmatrix} T_{11} & 0 \\ 0 & 0 \end{bmatrix}, \qquad T_{11} \in \mathbb{R}^{r\times r}, \ r = \operatorname{rank}(A).$$

A pair of QR factorizations (one with column pivoting) can be used to compute this. In this case $T_{11} = L$ is lower triangular in exact arithmetic. But with noise and roundoff we instead compute

$$U^T A V = \begin{bmatrix} L & 0 \\ H & E \\ 0 & 0 \end{bmatrix} \qquad (12.5.10)$$

where $L \in \mathbb{R}^{r \times r}$ and $E \in \mathbb{R}^{(n-r) \times (n-r)}$ are lower triangular and H and E are "small" compared to $\sigma_{min}(L)$. In this case we refer to (12.5.10) as a *rank-revealing ULV decomposition.*[1] Note that if

$$V = \begin{bmatrix} V_1 & V_2 \end{bmatrix} \qquad\qquad U = \begin{bmatrix} U_1 & U_2 \end{bmatrix}$$
$$ r \quad n-r \qquad\qquad\qquad r \quad m-r$$

then the columns of V_2 define an approximate null space:

$$\| A V_2 \|_2 = \| U_2 E \|_2 \leq \| E \|_2 .$$

Our goal is to produce a rank-revealing ULV decomposition for the row-appended matrix $\tilde{A}$. To be more specific, our aim is to show how to produce updates of L, E, H, V, and (possibly) the rank in $O(n^2)$ flops. Note that

$$\begin{bmatrix} U & 0 \\ 0 & 1 \end{bmatrix}^T \begin{bmatrix} A \\ z^T \end{bmatrix} V = \begin{bmatrix} L & 0 \\ H & E \\ 0 & 0 \\ w^T & y^T \end{bmatrix}.$$

By permuting the bottom row up "underneath" H and E we see that the challenge is to compute a rank-revealing ULV decomposition of

$$\begin{bmatrix} L & 0 \\ H & E \\ w^T & y^T \end{bmatrix} = \begin{bmatrix} \ell & 0 & 0 & 0 & 0 & 0 & 0 \\ \ell & \ell & 0 & 0 & 0 & 0 & 0 \\ \ell & \ell & \ell & 0 & 0 & 0 & 0 \\ \ell & \ell & \ell & \ell & 0 & 0 & 0 \\ h & h & h & h & e & 0 & 0 \\ h & h & h & h & e & e & 0 \\ h & h & h & h & e & e & e \\ w & w & w & w & y & y & y \end{bmatrix} \qquad (12.5.11)$$

in $O(n^2)$ flops. Here and in the sequel, we set $r = 4$ and $n = 7$ to illustrate the main ideas. Bear in mind that the h and e entries are small and that

---

[1]Dual to this is the URV decomposition in which the rank-revealing form is upper triangular. There are updating situations that sometimes favor the manipulation of this form instead of ULV.

we have deduced that the numerical rank is four. In practice, this involves comparisons with a small tolerance as discussed in §5.5.7.

Using zeroing techniques similar to those presented in §12.5.3, the bottom row can be zeroed with a sequence of row rotations giving

$$
\left[\frac{\tilde{L}}{0} \right] =
\left[
\begin{array}{cccc|ccc}
\times & 0 & 0 & 0 & 0 & 0 & 0 \\
\times & \times & 0 & 0 & 0 & 0 & 0 \\
\times & \times & \times & 0 & 0 & 0 & 0 \\
\times & \times & \times & \times & 0 & 0 & 0 \\
\times & \times & \times & \times & \times & 0 & 0 \\
\times & \times & \times & \times & \times & \times & 0 \\
\times & \times & \times & \times & \times & \times & \times \\
0 & 0 & 0 & 0 & 0 & 0 & 0
\end{array}
\right] .
$$

Because this zeroing process intermingles the (presumably large) entries of the bottom row with the entries from each of the other rows, the triangular form typically is *not* rank revealing. However, we can restore the rank-revealing structure with a combination of condition estimation and zero-chasing with rotations. Let us assume that with the added row, the new null space has dimension two.

With a reliable condition estimator we produce a unit 2-norm vector p such that

$$
\| \, p^T \tilde{L} \, \|_2 \approx \sigma_{min}(\tilde{L}).
$$

See §3.5.4. Rotations $\{U_{i,i+1}\}_{i=1}^{6}$ can be found such that

$$
U_{67}^T U_{56}^T U_{45}^T U_{34}^T U_{23}^T U_{12}^T p = e_8 = I_8(:,8).
$$

The matrix

$$
H = U_{67}^T U_{56}^T U_{45}^T U_{34}^T U_{23}^T U_{12}^T \tilde{L}
$$

is lower Hessenberg and can be restored to a lower triangular form L_+ by a sequence of column rotations:

$$
L_+ = H V_{12} V_{23} V_{34} V_{45} V_{56} V_{67}.
$$

It follows that

$$
e_8^T L_+ = \left(e_8^T H \right) V_{12} V_{23} V_{34} V_{45} V_{56} V_{67} = \left(p^T \tilde{L} \right) V_{12} V_{23} V_{34} V_{45} V_{56} V_{67}
$$

has approximate norm $\sigma_{min}(\tilde{L})$. Thus, we obtain a lower triangular matrix of the form

CHAPTER 12. SPECIAL TOPICS

$$
\begin{bmatrix}
\times & 0 & 0 & 0 & 0 & 0 & 0 \\
\times & \times & 0 & 0 & 0 & 0 & 0 \\
\times & \times & \times & 0 & 0 & 0 & 0 \\
\times & \times & \times & \times & 0 & 0 & 0 \\
\times & \times & \times & \times & \times & 0 & 0 \\
\times & \times & \times & \times & \times & \times & 0 \\
\hline
h & h & h & h & h & h & e
\end{bmatrix}
$$

with small h's and e. We can repeat the condition estimation and zero chasing on the leading 6-by-6 portion thereby producing (perhaps) another row of small numbers:

$$
\begin{bmatrix}
\times & 0 & 0 & 0 & 0 & 0 & 0 \\
\times & \times & 0 & 0 & 0 & 0 & 0 \\
\times & \times & \times & 0 & 0 & 0 & 0 \\
\times & \times & \times & \times & 0 & 0 & 0 \\
\times & \times & \times & \times & \times & 0 & 0 \\
\hline
h & h & h & h & h & e & 0 \\
h & h & h & h & h & e & e
\end{bmatrix} .
$$

(If not, then the revealed rank is 6.) Continuing in this way, we can restore any lower triangular matrix to rank-revealing form.

In the event that the y vector in (12.5.11) is small, we can reach rank-revealing form by a different, more efficient route. We start with a sequence of left and right Givens rotations to zero all but the first component of y:

$$
\begin{array}{c}
V_{67} \\
\longrightarrow
\end{array}
\begin{bmatrix}
\ell & 0 & 0 & 0 & 0 & 0 & 0 \\
\ell & \ell & 0 & 0 & 0 & 0 & 0 \\
\ell & \ell & \ell & 0 & 0 & 0 & 0 \\
\ell & \ell & \ell & \ell & 0 & 0 & 0 \\
\hline
h & h & h & h & e & 0 & 0 \\
h & h & h & h & e & e & e \\
h & h & h & h & e & e & e \\
\hline
x & x & x & x & y & y & 0
\end{bmatrix}
\quad
\begin{array}{c}
U_{67} \\
\longrightarrow
\end{array}
\begin{bmatrix}
\ell & 0 & 0 & 0 & 0 & 0 & 0 \\
\ell & \ell & 0 & 0 & 0 & 0 & 0 \\
\ell & \ell & \ell & 0 & 0 & 0 & 0 \\
\ell & \ell & \ell & \ell & 0 & 0 & 0 \\
\hline
h & h & h & h & e & 0 & 0 \\
h & h & h & h & e & e & 0 \\
h & h & h & h & e & e & e \\
\hline
x & x & x & x & y & y & 0
\end{bmatrix}
$$

$$
\begin{array}{c}
V_{56} \\
\longrightarrow
\end{array}
\begin{bmatrix}
\ell & 0 & 0 & 0 & 0 & 0 & 0 \\
\ell & \ell & 0 & 0 & 0 & 0 & 0 \\
\ell & \ell & \ell & 0 & 0 & 0 & 0 \\
\ell & \ell & \ell & \ell & 0 & 0 & 0 \\
\hline
h & h & h & h & e & e & 0 \\
h & h & h & h & e & e & 0 \\
h & h & h & h & e & e & e \\
\hline
x & x & x & x & y_* & 0 & 0
\end{bmatrix}
\quad
\begin{array}{c}
U_{56} \\
\longrightarrow
\end{array}
\begin{bmatrix}
\ell & 0 & 0 & 0 & 0 & 0 & 0 \\
\ell & \ell & 0 & 0 & 0 & 0 & 0 \\
\ell & \ell & \ell & 0 & 0 & 0 & 0 \\
\ell & \ell & \ell & \ell & 0 & 0 & 0 \\
\hline
h & h & h & h & e & 0 & 0 \\
h & h & h & h & e & e & 0 \\
h & h & h & h & e & e & e \\
\hline
x & x & x & x & y_* & 0 & 0
\end{bmatrix}
$$

Here, "U_{ij}" means a rotation of rows i and j and "V_{ij}" means a rotation of columns i and j. It is important to observe that there is no intermingling of small and large numbers during this process. The h's and e's are still small.

Following this, we produce a sequence of rotations that transform the matrix to

$$
\begin{bmatrix}
\ell & 0 & 0 & 0 & 0 & 0 & 0 \\
\ell & \ell & 0 & 0 & 0 & 0 & 0 \\
\ell & \ell & \ell & 0 & 0 & 0 & 0 \\
\ell & \ell & \ell & \ell & 0 & 0 & 0 \\
\hline
h & h & h & h & e & 0 & 0 \\
h & h & h & h & e & e & 0 \\
h & h & h & h & e & e & e \\
y & y & y & y & y & 0 & 0
\end{bmatrix}
\tag{12.5.12}
$$

where all the y's are small:

$$
\begin{array}{cc}
U_{48} \\ \longrightarrow
\end{array}
\begin{bmatrix}
\ell & 0 & 0 & 0 & 0 & 0 & 0 \\
\ell & \ell & 0 & 0 & 0 & 0 & 0 \\
\ell & \ell & \ell & 0 & 0 & 0 & 0 \\
\ell & \ell & \ell & \ell & \mu & 0 & 0 \\
\hline
h & h & h & h & e & 0 & 0 \\
h & h & h & h & e & e & e \\
h & h & h & h & e & e & e \\
x & x & x & 0 & y & 0 & 0
\end{bmatrix}
\qquad
\begin{array}{cc}
U_{38} \\ \longrightarrow
\end{array}
\begin{bmatrix}
\ell & 0 & 0 & 0 & 0 & 0 & 0 \\
\ell & \ell & 0 & 0 & 0 & 0 & 0 \\
\ell & \ell & \ell & 0 & \mu & 0 & 0 \\
\ell & \ell & \ell & \ell & \mu & 0 & 0 \\
\hline
h & h & h & h & e & 0 & 0 \\
h & h & h & h & e & e & 0 \\
h & h & h & h & e & e & e \\
x & x & 0 & 0 & y & 0 & 0
\end{bmatrix}
$$

$$
\begin{array}{cc}
U_{28} \\ \longrightarrow
\end{array}
\begin{bmatrix}
\ell & 0 & 0 & 0 & 0 & 0 & 0 \\
\ell & \ell & 0 & 0 & \mu & 0 & 0 \\
\ell & \ell & \ell & 0 & \mu & 0 & 0 \\
\ell & \ell & \ell & \ell & \mu & 0 & 0 \\
\hline
h & h & h & h & e & e & 0 \\
h & h & h & h & e & e & 0 \\
h & h & h & h & e & e & e \\
x & 0 & 0 & 0 & y & 0 & 0
\end{bmatrix}
\qquad
\begin{array}{cc}
U_{18} \\ \longrightarrow
\end{array}
\begin{bmatrix}
\ell & 0 & 0 & 0 & \mu & 0 & 0 \\
\ell & \ell & 0 & 0 & \mu & 0 & 0 \\
\ell & \ell & \ell & 0 & \mu & 0 & 0 \\
\ell & \ell & \ell & \ell & \mu & 0 & 0 \\
\hline
h & h & h & h & e & 0 & 0 \\
h & h & h & h & e & e & 0 \\
h & h & h & h & e & e & e \\
0 & 0 & 0 & 0 & y_{**} & 0 & 0
\end{bmatrix}
$$

Note that y_{**} is small because of 2-norm preservation. Column rotations in planes (1,5), (2,5), (3,5), and (4,5) can remove the μ's:

$$
V_{15} \longrightarrow
\left[
\begin{array}{cccc|ccc}
\ell & 0 & 0 & 0 & 0 & 0 & 0 \\
\ell & \ell & 0 & 0 & \mu & 0 & 0 \\
\ell & \ell & \ell & 0 & \mu & 0 & 0 \\
\ell & \ell & \ell & \ell & \mu & 0 & 0 \\
h & h & h & h & e & 0 & 0 \\
h & h & h & h & e & e & 0 \\
h & h & h & h & e & e & e \\
y & 0 & 0 & 0 & y & 0 & 0
\end{array}
\right]
\qquad
V_{25} \longrightarrow
\left[
\begin{array}{cccc|ccc}
\ell & 0 & 0 & 0 & 0 & 0 & 0 \\
\ell & \ell & 0 & 0 & 0 & 0 & 0 \\
\ell & \ell & \ell & 0 & \mu & 0 & 0 \\
\ell & \ell & \ell & \ell & \mu & 0 & 0 \\
h & h & h & h & e & 0 & 0 \\
h & h & h & h & e & e & 0 \\
h & h & h & h & e & e & e \\
y & y & 0 & 0 & y & 0 & 0
\end{array}
\right]
$$

$$
V_{35} \longrightarrow
\left[
\begin{array}{cccc|ccc}
\ell & 0 & 0 & 0 & 0 & 0 & 0 \\
\ell & \ell & 0 & 0 & 0 & 0 & 0 \\
\ell & \ell & \ell & 0 & 0 & 0 & 0 \\
\ell & \ell & \ell & \ell & \mu & 0 & 0 \\
h & h & h & h & e & 0 & 0 \\
h & h & h & h & e & e & 0 \\
h & h & h & h & e & e & e \\
y & y & y & 0 & y & 0 & 0
\end{array}
\right]
\qquad
V_{45} \longrightarrow
\left[
\begin{array}{cccc|ccc}
\ell & 0 & 0 & 0 & 0 & 0 & 0 \\
\ell & \ell & 0 & 0 & 0 & 0 & 0 \\
\ell & \ell & \ell & 0 & 0 & 0 & 0 \\
\ell & \ell & \ell & \ell & 0 & 0 & 0 \\
h & h & h & h & e & 0 & 0 \\
h & h & h & h & e & e & 0 \\
h & h & h & h & e & e & e \\
y & y & y & y & y & 0 & 0
\end{array}
\right]
$$

thus producing the structure displayed in (12.5.12). All the y's are small and thus a sequence of row rotations $U_{57}, U_{47}, \ldots, U_{17}$, can be constructed to clean out the bottom row giving the rank-revealed form

$$
\left[
\begin{array}{cccc|ccc}
\ell & 0 & 0 & 0 & 0 & 0 & 0 \\
\ell & \ell & 0 & 0 & 0 & 0 & 0 \\
\ell & \ell & \ell & 0 & 0 & 0 & 0 \\
\ell & \ell & \ell & \ell & 0 & 0 & 0 \\
h & h & h & h & e & 0 & 0 \\
h & h & h & h & e & e & 0 \\
h & h & h & h & e & e & e \\
0 & 0 & 0 & 0 & 0 & 0 & 0
\end{array}
\right].
$$

Problems

P12.5.1 Suppose we have the QR factorization for $A \in \mathbf{R}^{m \times n}$ and now wish to minimize $\| (A + uv^T)x - b \|_2$ where $u, b \in \mathbf{R}^m$ and $v \in \mathbf{R}^n$ are given. Give an algorithm for solving this problem that requires $O(mn)$ flops. Assume that Q must be updated.

P12.5.2 Suppose we have the QR factorization $QR = A \in \mathbf{R}^{m \times n}$. Give an algorithm for computing the QR factorization of the matrix A obtained by deleting the kth row of A. Your algorithm should require $O(mn)$ flops.

P12.5.3 Suppose $T \in \mathbf{R}^{n \times n}$ is tridiagonal and symmetric and that $v \in \mathbf{R}^n$. Show how the Lanczos algorithm can be used (in principle) to compute an orthogonal $Q \in \mathbf{R}^{n \times n}$ in $O(n^2)$ flops such that $Q^T (T + vv^T)Q = \tilde{T}$ is also tridiagonal.

P12.5.4 Suppose

$$
A = \left[\begin{array}{c} c^T \\ B \end{array} \right] \qquad c \in \mathbf{R}^n,\ B \in \mathbf{R}^{(m-1) \times n}
$$

has full column rank and $m > n$. Using the Sherman-Morrison-Woodbury formula show that

$$\frac{1}{\sigma_{min}(B)} \leq \frac{1}{\sigma_{min}(A)} + \frac{\| (A^T A)^{-1} c \|_2^2}{1 - c^T (A^T A)^{-1} c}.$$

P12.5.5   As a function of x_1 and x_2, what is the 2-norm of the hyperbolic rotation produced by (12.5.9)?

P12.5.6   Show that the hyperbolic reduction in §12.5.4 does not breakdown if A has full column rank.

P12.5.7   Assume

$$A = \begin{bmatrix} R & H \\ 0 & E \end{bmatrix}$$

where R and E are square with

$$\rho = \frac{\| E \|_2}{\sigma_{min}(R)} < 1.$$

Show that if

$$Q = \begin{bmatrix} Q_{11} & Q_{12} \\ Q_{21} & Q_{22} \end{bmatrix}$$

is orthogonal and

$$\begin{bmatrix} R & H \\ 0 & E \end{bmatrix} \begin{bmatrix} Q_{11} & Q_{12} \\ Q_{21} & Q_{22} \end{bmatrix} = \begin{bmatrix} R_1 & 0 \\ H_1 & E_1 \end{bmatrix},$$

then $\| H_1 \|_2 \leq \rho \| H \|_2$.

Notes and References for Sec. 12.5

Numerous aspects of the updating problem are presented in

P.E. Gill, G.H. Golub, W. Murray, and M.A. Saunders (1974). "Methods for Modifying Matrix Factorizations," *Math. Comp. 28*, 505-35.

Applications in the area of optimization are covered in

R.H. Bartels (1971). "A Stabilization of the Simplex Method," *Numer. Math. 16*, 414-434.

P.E. Gill, W. Murray, and M.A. Saunders (1975). "Methods for Computing and Modifying the LDV Factors of a Matrix," *Math. Comp. 29*, 1051-77.

D. Goldfarb (1976). "Factored Variable Metric Methods for Unconstrained Optimization," *Math. Comp. 30*, 796-811.

J.E. Dennis and R.B. Schnabel (1983). *Numerical Methods for Unconstrained Optimization and Nonlinear Equations*, Prentice-Hall, Englewood Cliffs, NJ.

W.W. Hager (1989). "Updating the Inverse of a Matrix," *SIAM Review 31*, 221-239.

S.K. Eldersveld and M.A. Saunders (1992). "A Block-LU Update for Large-Scale Linear Programming," *SIAM J. Matrix Anal. Appl. 13*, 191-201.

Updating issues in the least squares setting are discussed in

J. Daniel, W.B. Gragg, L. Kaufman, and G.W. Stewart (1976). "Reorthogonaization and Stable Algorithms for Updating the Gram-Schmidt QR Factorization," *Math. Comp. 30*, 772-95.

S. Qiao (1988). "Recursive Least Squares Algorithm for Linear Prediction Problems," *SIAM J. Matrix Anal. Appl. 9*, 323-328.

Å. Björck, H. Park, and L. Eldén (1994). "Accurate Downdating of Least Squares Solutions," *SIAM J. Matrix Anal. Appl. 15*, 549-568.

S.J. Olszanskyj, J.M. Lebak, and A.W. Bojanczyk (1994). "Rank-k Modification Methods for Recursive Least Squares Problems," *Numerical Algorithms 7*, 325–354.

L. Eldén and H. Park (1994). "Block Downdating of Least Squares Solutions," *SIAM J. Matrix Anal. Appl. 15*, 1018–1034.

Another important topic concerns the updating of condition estimates:

W.R. Ferng, G.H. Golub, and R.J. Plemmons (1991). "Adaptive Lanczos Methods for Recursive Condition Estimation," *Numerical Algorithms 1*, 1-20.

G. Shroff and C.H. Bischof (1992). "Adaptive Condition Estimation for Rank-One Updates of QR Factorizations," *SIAM J. Matrix Anal. Appl. 13*, 1264–1278.

D.J. Pierce and R.J. Plemmons (1992). "Fast Adaptive Condition Estimation," *SIAM J. Matrix Anal. Appl. 13*, 274–291.

Hyperbolic transformations are discussed in

G.H. Golub (1969). "Matrix Decompositions and Statistical Computation," in *Statistical Computation*, ed., R.C. Milton and J.A. Nelder, Academic Press, New York, pp. 365–97.

C.M. Rader and A.O. Steinhardt (1988). "Hyperbolic Householder Transforms," *SIAM J. Matrix Anal. Appl. 9*, 269–290.

S.T. Alexander, C.T. Pan, and R.J. Plemmons (1988). "Analysis of a Recursive Least Squares Hyperbolic Rotation Algorithm for Signal Processing," *Lin. Alg. and Its Applic. 98*, 3–40.

G. Cybenko and M. Berry (1990). "Hyperbolic Householder Algorithms for Factoring Structured Matrices," *SIAM J. Matrix Anal. Appl. 11*, 499–520.

A.W. Bojanczyk, R. Onn, and A.O. Steinhardt (1993). "Existence of the Hyperbolic Singular Value Decomposition," *Lin. Alg. and Its Applic. 185*, 21–30.

Cholesky update issues have also attracted a lot of attention.

G.W. Stewart (1979). "The Effects of Rounding Error on an Algorithm for Downdating a Cholesky Factorization," *J. Inst. Math. Applic. 23*, 203–13.

A.W. Bojanczyk, R.P. Brent, P. Van Dooren, and F.R. de Hoog (1987). "A Note on Downdating the Cholesky Factorization," *SIAM J. Sci. and Stat. Comp. 8*, 210–221.

C.S. Henkel, M.T. Heath, and R.J. Plemmons (1988). "Cholesky Downdating on a Hypercube," in G. Fox (1988), 1592–1598.

C.-T. Pan (1993). "A Perturbation Analysis of the Problem of Downdating a Cholesky Factorization," *Lin. Alg. and Its Applic. 183*, 103–115.

L. Eldén and H. Park (1994). "Perturbation Analysis for Block Downdating of a Cholesky Decomposition," *Numer. Math. 68*, 457–468.

Updating and downdating the ULV and URV decompositions and related topics are covered in

C.H. Bischof and G.M. Shroff (1992). "On Updating Signal Subspaces," *IEEE Trans. Signal Proc. 40*, 96–105.

G.W. Stewart (1992). "An Updating Algorithm for Subspace Tracking," *IEEE Trans. Signal Proc. 40*, 1535–1541.

G.W. Stewart (1993). "Updating a Rank-Revealing ULV Decomposition," *SIAM J. Matrix Anal. Appl. 14*, 494–499.

G.W. Stewart (1994). "Updating URV Decompositions in Parallel," *Parallel Computing 20*, 151–172.

H. Park and L. Eldén (1995). "Downdating the Rank-Revealing URV Decomposition," *SIAM J. Matrix Anal. Appl. 16*, 138–155.

Finally, we mention the following paper concerned with SVD updating:

M. Moonen, P. Van Dooren, and J. Vandewalle (1992). "A Singular Value Decomposition Updating Algorithm," *SIAM J. Matrix Anal. Appl. 13*, 1015–1038.

12.6   Modified/Structured Eigenproblems

In this section we treat an array of constrained, inverse, and structured eigenvalue problems. Although the examples are not related, collectively they show how certain special eigenproblems can be solved using the basic factorization ideas presented in earlier chapters.

The dependence of this section upon earlier portions of the book is as follows:

§§5.1, 5.2, 8.1, 8.3	→ §12.6.1
§§8.1, 8.3, 9.1	→ §12.6.2
§§4.7, 8.1	→ §12.6.3
§§5.1, 5.2, 5.4, 7.4, 8.1, 8.2, 8.3, 8.6	→ §12.6.4

12.6.1   A Constrained Eigenvalue Problem

Let $A \in \mathbb{R}^{n \times n}$ be symmetric. The gradient of $r(x) = x^T A x / x^T x$ is zero if and only if x is an eigenvector of A. Thus the stationary values of $r(x)$ are therefore the eigenvalues of A.

In certain applications it is necessary to find the stationary values of $r(x)$ subject to the constraint $C^T x = 0$ where $C \in \mathbb{R}^{n \times p}$ with $n \geq p$. Suppose

$$Q^T C Z = \begin{bmatrix} S & 0 \\ 0 & 0 \end{bmatrix} \begin{matrix} r \\ n-r \end{matrix} \qquad r = \text{rank}(C)$$
$$\begin{matrix} r & p-r \end{matrix}$$

is a complete orthogonal decomposition of C. Define $B \in \mathbb{R}^{n \times n}$ by

$$Q^T A Q = B = \begin{bmatrix} B_{11} & B_{12} \\ B_{21} & B_{22} \end{bmatrix} \begin{matrix} r \\ n-r \end{matrix}$$
$$\begin{matrix} r & n-r \end{matrix}$$

and set

$$y = Q^T x = \begin{bmatrix} u \\ v \end{bmatrix} \begin{matrix} r \\ n-r \end{matrix} \quad .$$

Since $C^T x = 0$ transforms to $S^T u = 0$, the original problem becomes one of finding the stationary values of $r(y) = y^T B y / y^T y$ subject to the constraint that $u = 0$. But this amounts merely to finding the stationary values (eigenvalues) of the $(n-r)$-by-$(n-r)$ symmetric matrix B_{22}.

12.6.2 Two Inverse Eigenvalue Problems

Consider the $r = 1$ case in the previous subsection. Let $\tilde{\lambda}_1 \leq \ldots \leq \tilde{\lambda}_{n-1}$ be the stationary values of $x^T A x / x^T x$ subject to the constraint $c^T x = 0$. From Theorem 8.1.7, it is easy to show that these stationary values interlace the eigenvalues λ_i of A:

$$\lambda_n \leq \tilde{\lambda}_{n-1} \leq \lambda_{n-1} \leq \cdots \leq \lambda_2 \leq \tilde{\lambda}_1 \leq \lambda_1.$$

Now suppose that A has distinct eigenvalues and that we are *given* the values $\tilde{\lambda}_1, \ldots, \tilde{\lambda}_{n-1}$ that satisfy

$$\lambda_n < \tilde{\lambda}_{n-1} < \lambda_{n-1} < \cdots < \lambda_2 < \tilde{\lambda}_1 < \lambda_1.$$

We seek to determine a unit vector $c \in \mathbb{R}^n$ such that the $\tilde{\lambda}_i$ are the stationary values of $x^T A x$ subject to $x^T x = 1$ and $c^T x = 0$.

In order to determine the properties that c must have, we use the method of Lagrange multipliers. Equating the gradient of

$$\phi(x, \lambda, \mu) = x^T A x - \lambda(x^T x - 1) + 2\mu x^T c$$

to zero we obtain the important equation $(A - \lambda I)x = -\mu c$. Thus, $A - \lambda I$ is nonsingular and so $x = -\mu(A - \lambda I)^{-1}c$. Applying c^T to both sides of this equation and substituting the eigenvalue decomposition $Q^T A Q = \text{diag}(\lambda_i)$ we obtain

$$0 = \sum_{i=1}^{n} \frac{d_i^2}{\lambda_i - \lambda}$$

where $d = Q^T c$, i.e.,

$$p(\lambda) \equiv \sum_{i=1}^{n} d_i^2 \prod_{\substack{j=1 \\ j \neq i}}^{n} (\lambda_j - \lambda) = 0.$$

Notice that $1 = \| c \|_2^2 = \| d \|_2 = d_1^2 + \cdots + d_n^2$ is the coefficient of $(-\lambda)^{n-1}$. Since $p(\lambda)$ is a polynomial having zeroes $\tilde{\lambda}_1, \ldots, \tilde{\lambda}_{n-1}$ we must have

$$p(\lambda) = \prod_{j=1}^{n-1} (\tilde{\lambda}_j - \lambda).$$

It follows from these two formulas for $p(\lambda)$ that

$$d_k^2 = \frac{\displaystyle\prod_{j=1}^{n-1} (\tilde{\lambda}_j - \lambda_k)}{\displaystyle\prod_{\substack{j=1 \\ j \neq k}}^{n-1} (\lambda_j - \lambda_k)} \qquad k = 1{:}n. \tag{12.6.1}$$

This determines each d_k up to its sign. Thus there are 2^n different solutions $c = Qd$ to the original problem.

A related inverse eigenvalue problem involves finding a tridiagonal matrix

$$T = \begin{bmatrix} \alpha_1 & \beta_1 & & \cdots & & 0 \\ \beta_1 & \alpha_2 & \ddots & & & \vdots \\ & \ddots & \ddots & \ddots & & \\ \vdots & & \ddots & \ddots & \beta_{n-1} \\ 0 & \cdots & & \beta_{n-1} & \alpha_n \end{bmatrix}$$

such that T has prescribed eigenvalues $\{\lambda_1, \ldots, \lambda_n\}$ and $T(2{:}n, 2{:}n)$ has prescribed eigenvalues $\{\tilde{\lambda}_1, \ldots, \tilde{\lambda}_{n-1}\}$ with

$$\lambda_1 > \tilde{\lambda}_1 > \lambda_2 > \cdots > \lambda_{n-1} > \tilde{\lambda}_{n-1} > \lambda_n .$$

We show how to compute the tridiagonal T via the Lanczos process. Note that the $\tilde{\lambda}_i$ are the stationary values of

$$\phi(y) = \frac{y^T \Lambda y}{y^T y}$$

subject to $d^T y = 0$ where $\Lambda = \text{diag}(\lambda_1, \ldots, \lambda_n)$ and d is specified by (12.6.1). If we apply the Lanczos iteration (9.1.3) with $A = \Lambda$ and $q_1 = d$, then it produces an orthogonal matrix Q and a tridiagonal matrix T such that $Q^T \Lambda Q = T$. With the definition $x = Q^T y$, it follows that the $\tilde{\lambda}_i$ are the stationary values of

$$\psi(x) = \frac{x^T T x}{x^T x}$$

subject to $e_1^T x = 0$. But these are precisely the eigenvalues of $T(2{:}n, 2{:}n)$!

12.6.3   A Toeplitz Eigenproblem

Assume that

$$T = \begin{bmatrix} 1 & r^T \\ r & G \end{bmatrix}$$

is symmetric, positive definite, and Toeplitz with $r \in \mathbb{R}^{n-1}$. Our goal is to compute the smallest eigenvalue $\lambda_{min}(T)$ of T given that

$$\lambda_{min}(T) < \lambda_{min}(G).$$

This problem is considered in Cybenko and Van Loan (1986) and has applications in signal processing.

Suppose

$$\begin{bmatrix} 1 & r^T \\ r & G \end{bmatrix} \begin{bmatrix} \alpha \\ y \end{bmatrix} = \lambda \begin{bmatrix} \alpha \\ y \end{bmatrix},$$

i.e.,

$$\alpha + r^T y = \lambda \alpha$$
$$\alpha r + G y = \lambda y.$$

If $\lambda \notin \lambda(G)$, then $y = -\alpha(G - \lambda I)^{-1}r$, $\alpha \neq 0$, and

$$\alpha + r^T \left[-\alpha(G - \lambda I)^{-1}r \right] = \lambda \alpha.$$

Thus, λ is a zero of the rational function

$$f(\lambda) = 1 - \lambda - r^T(G - \lambda I)^{-1}r.$$

We have dealt with similar functions in §8.5 and §12.1. In this case, f always has a negative slope

$$f'(\lambda) = -1 - \| (G - \lambda I)^{-1}r \|_2^2 \leq -1.$$

If $\lambda < \lambda_{min}(G)$, then it also has a negative second derivative:

$$f''(\lambda) = -2r^T(G - \lambda I)^{-3}r \leq 0.$$

Using these facts it can be shown that if

$$\lambda_{min}(T) \leq \lambda^{(0)} < \lambda_{min}(G), \qquad (12.6.2)$$

then the Newton iteration

$$\lambda^{(k+1)} = \lambda^{(k)} - \frac{f(\lambda^{(k)})}{f'(\lambda^{(k)})} \qquad (12.6.3)$$

converges to $\lambda_{min}(T)$ monotonically from the right. Note that

$$\lambda^{(k+1)} = \lambda^{(k)} + \frac{1 + r^T w - \lambda^{(k)}}{1 + w^T w}$$

where w solves the "shifted" Yule-Walker system

$$(G - \lambda^{(k)}I)w = -r.$$

Since, $\lambda^{(k)} < \lambda_{min}(G)$, this system is positive definite and Algorithm 4.7.1 is applicable if we simply apply it to the normalized Toeplitz matrix $(G - \lambda^{(k)}I)/(1 - \lambda^{(k)})$.

A starting value that satisfies (12.6.2) can be obtained by examining the Durbin algorithm when it is applied to $T_\lambda = (T - \lambda I)/(1 - \lambda)$. For this matrix the "$r$" vector is $r/(1 - \lambda)$ and so the Durbin algorithm (4.7.1) transforms to

$$r = r/(1 - \lambda)$$
$$y^{(1)} = -r_1$$
for $k = 1{:}n - 1$

$$\beta_k = 1 + [r^{(k)}]^T y^{(k)}$$ $\qquad$ (12.6.4)
$$\alpha_k = -(r_{k+1} + r^{(k)^T} E_k y^{(k)})/\beta_k$$
$$z^{(k)} = y^{(k)} + \alpha_k E_k y^{(k)}$$

$$y^{(k+1)} = \left[\begin{array}{c} z^{(k)} \\ \alpha_k \end{array} \right]$$

end

From the discussion in §4.7.2 we know that $\beta_1, \ldots, \beta_k > 0$ implies that $T_\lambda(1{:}k + 1, 1{:}k + 1)$ is positive definite. Hence, a suitably modified (12.6.4) can be used to compute $m(\lambda)$, the largest index m such that $\beta_1, \ldots, \beta_m$ are all positive but that $\beta_{m+1} \leq 0$. Note that if $m(\lambda) = n - 2$, then (12.6.2) holds. This suggests the following bisection scheme:

Choose L and R so $L \leq \lambda_{min}(T) < \lambda_{min}(G) \leq R$.
Until $m = n - 2$
$\qquad \lambda = (L + R)/2$
$\qquad m = m(\lambda)$
$\qquad$ if $m < n - 2$ $\qquad$ (12.6.5)
$\qquad\qquad R = \lambda$
$\qquad$ end
$\qquad$ if $m = n - 1$
$\qquad\qquad L = \lambda$
$\qquad$ end
end

The bracketing interval $[L, R]$ always contains a λ such that $m(\lambda) = n - 2$ and so the current λ has this property upon termination.

There are several possible choices for a starting interval. One idea is to set $L = 0$ and $R = 1 - |r_1|$ since

$$0 < \lambda_{min}(T) < \lambda_{min}(G) \leq \lambda_{min}\left(\left[\begin{array}{cc} 1 & r_1 \\ r_1 & 1 \end{array} \right] \right) = 1 - |r_1|$$

where the upper bound follows from Theorem 8.1.7.

Note that the iterations in (12.6.4) and (12.6.5) involve at most $O(n^2)$ flops. A heuristic argument that $O(\log n)$ iterations are required is given in Cybenko and Van Loan (1986).

12.6.4   An Orthogonal Matrix Eigenproblem

Computing the eigenvalues and eigenvectors of a real orthogonal matrix $A \in \mathbb{R}^{n \times n}$ is a problem that arises in signal processing, see Cybenko (1985).

The eigenvalues of A are on the unit circle and moreover,

$$\cos(\theta) \pm i \sin(\theta) \in \lambda(A) \quad \Leftrightarrow \quad \cos(\theta) \in \lambda\left(\frac{A + A^{-1}}{2}\right) = \lambda\left(\frac{A + A^T}{2}\right).$$

This suggests computing $\text{Re}(\lambda(A))$ via the Schur decomposition

$$Q^T \left(\frac{A + A^T}{2}\right) Q = \text{diag}(\cos(\theta_1), \ldots, \cos(\theta_n))$$

and then computing $\text{Im}(\lambda(A))$ with the formula $s = \sqrt{1 - c^2}$. Unfortunately, if $|c| \approx 1$, then this formula does not produce an accurate sine because of floating point cancellation. We could work with the skew-symmetric matrix $(A - A^T)/2$ to get the "small sine" eigenvalues, but then we are talking about a method that requires a pair of full Schur decomposition problems and the approach begins to lose its appeal.

A way around these difficulties that involves an interesting SVD application is proposed by Ammar, Gragg, and Reichel (1986). We present just the eigenvalue portion of their algorithm. The derivation is instructive because it involves practically every decomposition that we have studied.

The first step is to orthogonally reduce A to upper Hessenberg form, $Q^T A Q = H$. (Frequently, A is already in Hessenberg form.) Without loss of generality, we may assume that H is unreduced with positive subdiagonal elements.

If n is odd, then it must have a real eigenvalue because the eigenvalues of a real matrix come in complex conjugate pairs. In this case it is possible to deflate the problem with $O(n)$ work to size $n - 1$ by carefully working with the eigenvector equation $Hx = x$ (or $Hx = -x$). See Gragg (1986). Thus, we may assume that n is even.

For $1 \leq k \leq n - 1$, define the reflection $G_k \in \mathbb{R}^{n \times n}$ by

$$G_k = G_k(\phi_k) = \begin{bmatrix} I_{k-1} & 0 & 0 & 0 \\ 0 & -c_k & s_k & 0 \\ 0 & s_k & c_k & 0 \\ 0 & 0 & 0 & I_{n-k-1} \end{bmatrix}$$

where $c_k = \cos(\phi_k)$, $s_k = \sin(\phi_k)$, and $0 < \phi_k < \pi$. It is possible to determine $G_1, \ldots, G_{n-1}$ such that

$$H = (G_1 \cdots G_{n-1}) \, \text{diag}(1, \ldots, 1, -c_n)$$

where $c_n = \pm 1$. This is just the QR decomposition of H. The sines $s_1, \ldots, s_{n-1}$ are the subdiagonal entries of H. The "R" matrix is diagonal because it is orthogonal and triangular. Since the determinant of a reflection is -1, $\det(H) = c_n$. This quantity is the product of H's eigenvalues and so if $c_n = -1$, then $\{-1, 1\} \subseteq \lambda(H)$. In this situation it is also possible to deflate.

So altogether we may assume that n is even and

$$H = G_1(\phi_1) \cdots G_{n-1}(\phi_{n-1})G_n(\phi_n)$$

where $G_n = G_n(\phi_n) = \text{diag}(1, \ldots, 1, -c_n)$ and $c_n = 1$. Designate the sought after eigenvalues by

$$\lambda(H) = \{\, \cos(\theta_k) \pm i \cdot \sin(\theta_k) \,\}_{k=1}^m \qquad (12.6.4)$$

where $m = n/2$.

The cosines $c_1, \ldots, c_n$ are called the *Schur parameters* and as we mentioned, the corresponding sines are the subdiagonal entries of H. Using these numbers it is possible to construct *explicitly* a pair of bidiagonal matrices $B_C, B_S \in \mathbb{R}^{n \times n}$ with the property that

$$\sigma(B_C(1{:}m, 1{:}m)) = \{\cos(\theta_1/2), \ldots, \cos(\theta_m/2)\} \qquad (12.6.5)$$
$$\sigma(B_S(1{:}m, 1{:}m)) = \{\sin(\theta_1/2), \ldots, \sin(\theta_m/2)\} \qquad (12.6.6)$$

The singular values of $B_C(1{:}m, 1{:}m)$ and $B_S(1{:}m, 1{:}m)$ can be computed using the bidiagonal SVD algorithm. The angle θ_k can be accurately computed from $\sin(\theta_k/2)$ if $0 < \theta_k \le \pi/2$ and accurately computed from $\cos(\theta_k/2)$ if $\pi/2 \le \theta_k < \pi$. The construction of B_C and B_S is based on three facts:

1. H is similar to

$$\tilde{H} = H_o H_e$$

where H_o and H_e are the odd and even reflection products

$$H_o = G_1 G_3 \cdots G_{n-1}$$
$$H_e = G_2 G_4 \cdots G_n.$$

These matrices are block diagonal with 2-by-2 and 1-by-1 blocks, i.e.,

$$H_o = \text{diag}(R(\phi_1), R(\phi_3), \ldots, R(\phi_{n-1})) \qquad (12.6.7)$$
$$H_e = \text{diag}(1, R(\phi_2), R(\phi_4), \ldots, R(\phi_{n-2}), -1) \qquad (12.6.8)$$

where

$$R(\phi) = \begin{bmatrix} -\cos(\phi) & \sin(\phi) \\ \sin(\phi) & \cos(\phi) \end{bmatrix}. \qquad (12.6.9)$$

2. The eigenvalues of the symmetric tridiagonal matrices

$$C = \frac{H_o + H_e}{2} \quad \text{and} \quad S = \frac{H_o - H_e}{2} \qquad (12.6.10)$$

are given by

$$\lambda(C) = \{\, \pm\cos(\theta_1/2), \ldots, \pm\cos(\theta_m/2) \,\} \qquad (12.6.11)$$
$$\lambda(S) = \{\, \pm\sin(\theta_1/2), \ldots, \pm\sin(\theta_m/2) \,\}. \qquad (12.6.12)$$

3. It is possible to construct bidiagonalizations

$$U_C^T C V_C = B_C \quad \text{and} \quad U_S^T S V_S = B_S$$

that satisfy (12.6.5) and (12.6.6). The transformations U_C, V_C, U_S, and V_S are products of known reflections G_k and simple permutations.

We begin the verification of these three facts by showing that H is similar to $H_o H_e$. The $n = 8$ case is sufficient for this purpose. Define the orthogonal matrix P by

$$P = F_7 F_5 F_3 \quad \text{where} \quad \begin{cases} F_3 = G_3 G_4 G_5 G_6 G_7 G_8 \\ F_5 = G_5 G_6 G_7 G_8 \\ F_7 = G_7 G_8. \end{cases}$$

Since reflections are symmetric and $G_i G_j = G_j G_i$ if $|i - j| \geq 2$, we see that

$$\begin{aligned} F_3 H F_3^T &= (G_3 G_4 G_5 G_6 G_7 G_8)(G_1 G_2 G_3 G_4 G_5 G_6 G_7 G_8)(G_3 G_4 G_5 G_6 G_7 G_8)^T \\ &= (G_3 G_4 G_5 G_6 G_7 G_8) G_1 G_2 \\ &= G_1 G_3 G_2 G_4 G_5 G_6 G_7 G_8, \end{aligned}$$

$$\begin{aligned} F_5 (F_3 H F_3^T) F_5^T &= (G_5 G_6 G_7 G_8)(G_1 G_3 G_2 G_4 G_5 G_6 G_7 G_8)(G_5 G_6 G_7 G_8)^T, \\ &= (G_5 G_6 G_7 G_8) G_1 G_3 G_2 G_4 \\ &= G_1 G_3 G_5 G_2 G_4 G_6 G_7 G_8 \end{aligned}$$

$$\begin{aligned} P H P^T &= F_7 (F_5 F_3 H F_3^T F_5^T) F_7^T \\ &= (G_7 G_8)(G_1 G_3 G_5 G_2 G_4 G_6 G_7 G_8)(G_7 G_8)^T \\ &= (G_1 G_3 G_5 G_7)(G_2 G_4 G_6 G_8) \equiv H_o H_e. \end{aligned}$$

The second of the three facts that we need to establish relates the eigenvalues of $\tilde{H} = H_o H_e$ to the eigenvalues of the C and S matrices defined in (12.6.10). It follows from (12.6.7) and (12.6.8) that these matrices are symmetric, tridiagonal, and unreduced, e.g.,

$$C = \frac{1}{2} \begin{bmatrix} -c_1 & s_1 & 0 & 0 \\ s_1 & c_1 - c_2 & s_2 & 0 \\ 0 & s_2 & c_2 - c_3 & s_3 \\ 0 & 0 & s_3 & c_3 - c_4 \end{bmatrix}$$

$$S = \frac{1}{2} \begin{bmatrix} -c_1 & s_1 & 0 & 0 \\ s_1 & c_1 + c_2 & -s_2 & 0 \\ 0 & -s_2 & -c_2 - c_3 & s_3 \\ 0 & 0 & s_3 & c_3 + c_4 \end{bmatrix}.$$

By working with the definitions it is easy to verify that

$$\frac{\tilde{H} + \tilde{H}^T}{2} = \frac{H_o H_e + (H_o H_e)^{-1}}{2} = \frac{H_o H_e + H_e H_o}{2} = 2C^2 - I$$

and

$$\frac{\tilde{H} + \tilde{H}^T}{2i} = \frac{H_o H_e - (H_o H_e)^{-1}}{2i} = \frac{H_o H_e - H_e H_o}{2i} = -2iSC.$$

This shows that $\text{Re}(\lambda(\tilde{H})) = \lambda(2C^2 - I)$ and $\text{Im}(\lambda(\tilde{H})) = \lambda(-2iCS)$ thereby establishing (12.6.11) and (12.6.12).

Instead of thinking of these half-angle cosines and sines as eigenvalues of n-by-n matrices, it is more efficient to think of them as singular values of m-by-m matrices. This brings us to the bidiagonalization of C and S. The orthogonal equivalence transformations that carry out this task are based upon the Schur decompositions of H_o and H_e. A 2-by-2 reflection $R(\phi)$ defined by (12.6.9) has eigenvalues 1 and -1 and the following Schur decomposition:

$$R(\phi/2)R(\phi)R(\phi/2) = \begin{bmatrix} 1 & 0 \\ 0 & -1 \end{bmatrix}.$$

Thus, if

$$\begin{aligned} Q_o &= \text{diag}(R(\phi_1/2), R(\phi_3/2), \ldots, R(\phi_{n-1}/2)) \\ Q_e &= \text{diag}(1, R(\phi_2/2), R(\phi_4/2), \ldots, R(\phi_{n-2}/2), -1) \end{aligned}$$

then from (12.6.7) and (12.6.8) H_o and H_e have the following Schur decompositions:

$$\begin{aligned} Q_o H_o Q_o &= D_o = \text{diag}(1, -1, 1, -1, \cdots, 1, -1) \\ Q_e H_e Q_e &= D_e = \text{diag}(1, 1, -1, 1, -1, \cdots, 1, -1, -1). \end{aligned}$$

The matrices

$$\begin{aligned} C^{(1)} &= Q_o C Q_e = \frac{1}{2} Q_o \left(H_o + H_e \right) Q_e = \frac{1}{2} \left(D_o(Q_o Q_e) + (Q_o Q_e) D_e \right) \\ S^{(1)} &= Q_o S Q_e = \frac{1}{2} Q_o \left(H_o - H_e \right) Q_e = \frac{1}{2} \left(D_o(Q_o Q_e) - (Q_o Q_e) D_e \right) \end{aligned}$$

have the same singular values as C and S respectively. To analyze their structure we first note that $Q_0 Q_e$ is banded:

$$Q_o Q_e = \begin{bmatrix} \times & \times & \times & 0 & 0 & 0 & 0 & 0 \\ \times & \times & \times & 0 & 0 & 0 & 0 & 0 \\ 0 & \times & \times & \times & \times & 0 & 0 & 0 \\ 0 & \times & \times & \times & \times & 0 & 0 & 0 \\ 0 & 0 & 0 & \times & \times & \times & \times & 0 \\ 0 & 0 & 0 & \times & \times & \times & \times & 0 \\ 0 & 0 & 0 & 0 & 0 & \times & \times & \times \\ 0 & 0 & 0 & 0 & 0 & \times & \times & \times \end{bmatrix}.$$

(The main ideas from this point on are amply communicated with $n = 8$ examples.) If $D_o(i,i)$ and $D_e(j,j)$ have the *opposite* sign, then $C_{ij}^{(1)} = 0$ from which we conclude that $C^{(1)}$ has the form

$$
C^{(1)} = Q_o C Q_e \;=\;
\begin{bmatrix}
a_0 & b_1 & 0 & 0 & 0 & 0 & 0 & 0 \\
0 & 0 & b_2 & 0 & 0 & 0 & 0 & 0 \\
0 & a_2 & 0 & b_3 & 0 & 0 & 0 & 0 \\
0 & 0 & a_3 & 0 & b_4 & 0 & 0 & 0 \\
0 & 0 & 0 & a_4 & 0 & b_5 & 0 & 0 \\
0 & 0 & 0 & 0 & a_5 & 0 & b_6 & 0 \\
0 & 0 & 0 & 0 & 0 & a_6 & 0 & 0 \\
0 & 0 & 0 & 0 & 0 & 0 & a_7 & b_8
\end{bmatrix}.
$$

Analogously, if $D_o(i,i)$ and $D_e(j,j)$ have the *same* sign, then $S_{ij}^{(1)} = 0$ from which we conclude that $S^{(1)}$ has the form

$$
S^{(1)} = Q_o S Q_e \;=\;
\begin{bmatrix}
0 & 0 & f_1 & 0 & 0 & 0 & 0 & 0 \\
e_2 & d_2 & 0 & 0 & 0 & 0 & 0 & 0 \\
0 & 0 & d_3 & 0 & f_3 & 0 & 0 & 0 \\
0 & e_4 & 0 & d_4 & 0 & 0 & 0 & 0 \\
0 & 0 & 0 & 0 & d_5 & 0 & f_5 & 0 \\
0 & 0 & 0 & e_6 & 0 & d_6 & 0 & 0 \\
0 & 0 & 0 & 0 & 0 & 0 & d_7 & f_7 \\
0 & 0 & 0 & 0 & 0 & 0 & e_8 & 0
\end{bmatrix}.
$$

Row/column permutations of these matrices result in bidiagonal forms:

$$
B_C \;=\; C^{(1)}([\,1\,3\,5\,7\,2\,4\,6\,8\,],[\,1\,2\,4\,6\,3\,5\,7\,8\,])
$$

$$
=\;
\left[
\begin{array}{cccc|cccc}
a_0 & b_1 & 0 & 0 & 0 & 0 & 0 & 0 \\
0 & a_2 & b_3 & 0 & 0 & 0 & 0 & 0 \\
0 & 0 & a_4 & b_5 & 0 & 0 & 0 & 0 \\
0 & 0 & 0 & a_6 & 0 & 0 & 0 & 0 \\
\hline
0 & 0 & 0 & 0 & b_2 & 0 & 0 & 0 \\
0 & 0 & 0 & 0 & a_3 & b_4 & 0 & 0 \\
0 & 0 & 0 & 0 & 0 & a_5 & b_6 & 0 \\
0 & 0 & 0 & 0 & 0 & 0 & a_7 & b_8
\end{array}
\right]
$$

$$B_S = S^{(1)}([\,2\,4\,6\,8\,1\,3\,5\,7\,],[\,1\,2\,4\,6\,3\,5\,7\,8\,])$$

$$= \begin{bmatrix} e_2 & d_2 & 0 & 0 & 0 & 0 & 0 & 0 \\ 0 & e_4 & d_4 & 0 & 0 & 0 & 0 & 0 \\ 0 & 0 & e_6 & d_6 & 0 & 0 & 0 & 0 \\ 0 & 0 & 0 & e_8 & 0 & 0 & 0 & 0 \\ 0 & 0 & 0 & 0 & f_1 & 0 & 0 & 0 \\ 0 & 0 & 0 & 0 & d_3 & f_3 & 0 & 0 \\ 0 & 0 & 0 & 0 & 0 & d_5 & f_5 & 0 \\ 0 & 0 & 0 & 0 & 0 & 0 & d_7 & f_7 \end{bmatrix}.$$

It is not hard to verify that a's, b's, d's, e's, and f's are all nonzero and this implies that the singular values of $B_C(1{:}m, 1{:}m)$ and $B_S(1{:}m, 1{:}m)$ are distinct. Since

$$\sigma(C) = \sigma(B_C) = \{\, \cos(\theta_1/2), \cos(\theta_1/2), \ldots, \cos(\theta_m/2), \cos(\theta_m/2) \,\}$$
$$\sigma(S) = \sigma(B_S) = \{\, \sin(\theta_1/2), \sin(\theta_1/2), \ldots, \sin(\theta_m/2), \sin(\theta_m/2) \,\}$$

we have verified (12.6.5) and (12.6.6).

Problems

P12.6.1 Let $A \in \mathbf{R}^{m \times n}$ and consider the problem of finding the stationary values of

$$R(x, y) = \frac{y^T A x}{\| y \|_2 \| x \|_2} \qquad y \in \mathbf{R}^m, x \in \mathbf{R}^n$$

subject to the constraints

$$C^T x = 0 \quad C \in \mathbf{R}^{n \times p} \quad n \geq p$$
$$D^T y = 0 \quad D \in \mathbf{R}^{m \times q} \quad m \geq q$$

Show how to solve this problem by first computing complete orthogonal decompositions of C and D and then computing the SVD of a certain submatrix of a transformed A.

P12.6.2 Suppose $A \in \mathbf{R}^{m \times n}$ and $B \in \mathbf{R}^{p \times n}$. Assume that rank$(A) = n$ and rank$(B) = p$. Using the methods of this section, show how to solve

$$\min_{Bx=0} \frac{\| b - Ax \|_2^2}{\| x \|_2^2 + 1} = \min_{Bx=0} \frac{\left\| [A, b] \begin{bmatrix} x \\ -1 \end{bmatrix} \right\|_2^2}{\left\| \begin{bmatrix} x \\ -1 \end{bmatrix} \right\|_2^2}$$

Show that this is a constrained TLS problem. Is there always a solution?

P12.6.3 Suppose $A \in \mathbf{R}^{n \times n}$ is symmetric and that $B \in \mathbf{R}^{p \times n}$ has rank p. Let $d \in \mathbf{R}^p$. Show how to solve the problem of minimizing $x^T A x$ subject to the constraints $\| x \|_2 = 1$ and $Bx = d$. Indicate when a solution fails to exist.

P12.6.4 Assume that $A \in \mathbf{R}^{n \times n}$ is symmetric, large, and sparse and that $C \in \mathbf{R}^{n \times p}$ is also large and sparse. How can the Lanczos process be used to find the stationary values of

$$r(x) = \frac{x^T A x}{x^T x}$$

subject to the constraint $C^T x = 0$? Assume that a sparse QR factorization $C = QR$ is available.

P12.6.5 Relate the eigenvalues and eigenvectors of

$$A = \begin{bmatrix} 0 & A_1 & 0 & 0 \\ 0 & 0 & A_2 & 0 \\ 0 & 0 & 0 & A_3 \\ A_4 & 0 & 0 & 0 \end{bmatrix}.$$

to the eigenvalues and eigenvectors of $\tilde{A} = A_1 A_2 A_3 A_4$. Assume that the diagonal blocks in A are square.

P12.6.6 Prove that if (12.6.2) holds, then (12.6.3) converges to $\lambda_{min}(T)$ montonically from the right.

P12.6.7 Recall from §4.7 that it is possible to compute the inverse of a symmetric positive definite Toeplitz matrix in $O(n^2)$ flops. Use this fact to obtain an initial bracketing interval for (12.6.5) that is based on $\| T^{-1} \|_\infty$ and $\| G^{-1} \|_\infty$.

P12.6.8 A matrix $A \in \mathbf{R}^{n \times n}$ is *centrosymmetric* if it is symmetric and persymmetric, i.e., $A = E_n A E_n$ where $E_n = I_n(:, n: -1:1)$. Show that if $n = 2m$ and Q is the orthogonal matrix

$$Q = \frac{1}{\sqrt{2}} \begin{bmatrix} I_m & I_m \\ E_m & -E_m \end{bmatrix},$$

then

$$Q^T AQ = \begin{bmatrix} A_{11} + A_{12} E_m & 0 \\ 0 & A_{11} - A_{12} E_m \end{bmatrix}$$

where $A_{11} = A(1{:}m, 1{:}m)$ and $A_{12} = A(1{:}m, m+1{:}n)$. Show that if $n = 2m$, then the Schur decomposition of a centrosymmetric matrix can be computed with one-fourth the flops that it takes to compute the Schur decomposition of a symmetric matrix, assuming that the QR algorithm is used in both cases. Repeat the problem if $n = 2m + 1$.

P12.6.9 Suppose $F, G \in \mathbf{R}^{n \times n}$ are symmetric and that

$$Q = \begin{bmatrix} Q_1 & Q_2 \\ p & n-p \end{bmatrix}$$

is an n-by-n orthogonal matrix. Show how to compute Q and p so that

$$f(Q, p) = \text{tr}(Q_1^T F Q_1) + \text{tr}(Q_2^T G Q_2)$$

is maximized. Hint: $\text{tr}(Q_1^T F Q_1) + \text{tr}(Q_2^T G Q_2) = \text{tr}(Q_1^T (F - G) Q_1) + \text{tr}(G)$.

P12.6.10 Suppose $A \in \mathbf{R}^{n \times n}$ is given and consider the problem of minimizing $\| A - S \|_F$ over all symmetric positive semidefinite matrices S that have rank r or less. Show that

$$S = \sum_{i=1}^{\min\{k,r\}} \lambda_i q_i q_i^T$$

solves this problem where

$$\frac{A + A^T}{2} = Q\text{diag}(\lambda_1, \dots, \lambda_n) Q^T$$

is the Schur decomposition of A's symmetric part, $Q = [\, q_1, \dots, q_n \,]$, and

$$\lambda_1 \geq \cdots \geq \lambda_k > 0 \geq \lambda_{k+1} \geq \cdots \geq \lambda_n.$$

P12.6.11 Verify for general n (even) that H is similar to $H_o H_e$ where these matrices are defined in §12.6.4.

P12.6.12 Verify that the bidiagonal matrices $B_C(1{:}m, 1{:}m)$ and $B_S(1{:}m, 1{:}m)$ in §12.6.4

have nonzero entries on their diagonal and superdiagonal and specify their value.

P12.6.13 A real $2n$-by-$2n$ matrix of the form

$$M = \left[\begin{array}{cc} A & G \\ F & -A^T \end{array} \right]$$

is *Hamiltonian* if $A \in \mathbf{R}^{n \times n}$ and $F, G \in \mathbf{R}^{n \times n}$ are symmetric. Equivalently, if the orthogonal matrix J is defined by

$$J = \left[\begin{array}{cc} 0 & I_n \\ -I_n & 0 \end{array} \right],$$

then $M \in \mathbf{R}^{2n \times 2n}$ is Hamiltonian if and only if $J^T M J = -M^T$. (a) Show that the eigenvalues of a Hamiltonian matrix come in plus-minus pairs. (b) A matrix $S \in \mathbf{R}^{2n \times 2n}$ is *symplectic* if $J^T S J = -S^{-T}$. Show that if S is symplectic and M is Hamiltonian, then $S^{-1} M S$ is also Hamiltonian. (c) Show that if $Q \in \mathbf{R}^{2n \times 2n}$ is orthogonal *and* symplectic, then

$$Q = \left[\begin{array}{cc} Q_1 & Q_2 \\ -Q_2 & Q_1 \end{array} \right]$$

where $Q_1^T Q_1 + Q_2^T Q_2 = I_n$ and $Q_2^T Q_1$ is symmetric. Thus, a Givens rotation of the form $G(i, i+n, \theta)$ is orthogonal symplectic as is the direct sum of n-by-n Householders. (d) Show how to compute a symplectic orthogonal U such that

$$U^T M U = \left[\begin{array}{cc} H & R \\ D & -H^T \end{array} \right]$$

where H is upper Hessenberg and D is diagonal.

Notes and References for Sec. 12.6

The inverse eigenvalue problems discussed in this §12.6.1 and §12.6.2 appear in the following survey articles:

G.H. Golub (1973). "Some Modified Matrix Eigenvalue Problems," *SIAM Review 15*, 318–44.

D. Boley and G.H. Golub (1987). "A Survey of Matrix Inverse Eigenvalue Problems," *Inverse Problems 3*, 595–622.

References for the stationary value problem include

G.E. Forsythe and G.H. Golub (1965). "On the Stationary Values of a Second-Degree Polynomial on the Unit Sphere," *SIAM J. App. Math. 13*, 1050–68.

G.H. Golub and R. Underwood (1970). "Stationary Values of the Ratio of Quadratic Forms Subject to Linear Constraints," *Z. Angew. Math. Phys. 21*, 318–26.

S. Leon (1994). "Maximizing Bilinear Forms Subject to Linear Constraints," *Lin. Alg. and Its Applic. 210*, 49–58.

An algorithm for minimizing $x^T A x$ where x satisfies $Bx = d$ and $\| x \|_2 = 1$ is presented in

W. Gander, G.H. Golub, and U. von Matt (1991). "A Constrained Eigenvalue Problem," in *Numerical Linear Algebra, Digital Signal Processing, and Parallel Algorithms*, G.H. Golub and P. Van Dooren (eds), Springer-Verlag, Berlin.

Selected papers that discuss a range of inverse eigenvalue problems include

G.H. Golub and J.H. Welsch (1969). "Calculation of Gauss Quadrature Rules," *Math. Comp. 23*, 221–30.

S. Friedland (1975). "On Inverse Multiplicative Eigenvalue Problems for Matrices," *Lin. Alg. and Its Applic. 12*, 127–38.

D.L. Boley and G.H. Golub (1978). "The Matrix Inverse Eigenvalue Problem for Periodic Jacobi Matrices," in *Proc. Fourth Symposium on Basic Problems of Numerical Mathematics*, Prague, pp. 63–76.

W.E. Ferguson (1980). "The Construction of Jacobi and Periodic Jacobi Matrices with Prescribed Spectra," *Math. Comp. 35*, 1203–1220.

J. Kautsky and G.H. Golub (1983). "On the Calculation of Jacobi Matrices," *Lin. Alg. and Its Applic. 52/53*, 439–456.

D. Boley and G.H. Golub (1984). "A Modified Method for Restructuring Periodic Jacobi Matrices," *Math. Comp. 42*, 143–150.

W.B. Gragg and W.J. Harrod (1984). "The Numerically Stable Reconstruction of Jacobi Matrices from Spectral Data," *Numer. Math. 44*, 317–336.

S. Friedland, J. Nocedal, and M.L. Overton (1987). "The Formulation and Analysis of Numerical Methods for Inverse Eigenvalue Problems," *SIAM J. Numer. Anal. 24*, 634–667.

M.T. Chu (1992). "Numerical Methods for Inverse Singular Value Problems," *SIAM J. Num. Anal. 29*, 885–903.

G. Ammar and G. He (1995). "On an Inverse Eigenvalue Problem for Unitary Matrices," *Lin. Alg. and Its Applic. 218*, 263–271.

H. Zha and Z. Zhang (1995). "A Note on Constructing a Symmetric Matrix with Specified Diagonal Entries and Eigenvalues," *BIT 35*, 448–451.

Various Toeplitz eigenvalue computations are presented in

G. Cybenko and C. Van Loan (1986). "Computing the Minimum Eigenvalue of a Symmetric Positive Definite Toeplitz Matrix," *SIAM J. Sci. and Stat. Comp. 7*, 123–131.

W.F. Trench (1989). "Numerical Solution of the Eigenvalue Problem for Hermitian Toeplitz Matrices," *SIAM J. Matrix Anal. Appl. 10*, 135–146.

L. Reichel and L.N. Trefethen (1992). "Eigenvalues and Pseudo-eigenvalues of Toeplitz Matrices," *Lin. Alg. and Its Applic. 162/163/164*, 153–186.

S.L. Handy and J.L. Barlow (1994). "Numerical Solution of the Eigenproblem for Banded, Symmetric Toeplitz Matrices," *SIAM J. Matrix Anal. Appl. 15*, 205–214.

Unitary/orthogonal eigenvalue problems are treated in

H. Rutishauser (1966). "Bestimmung der Eigenwerte Orthogonaler Matrizen," *Numer. Math. 9*, 104–108.

P.J. Eberlein and C.P. Huang (1975). "Global Convergence of the QR Algorithm for Unitary Matrices with Some Results for Normal Matrices," *SIAM J. Numer. Anal. 12*, 421–453.

G. Cybenko (1985). "Computing Pisarenko Frequency Estimates," in *Proceedings of the Princeton Conference on Information Science and Systems*, Dept. of Electrical Engineering, Princeton University.

W. B. Gragg (1986). "The QR Algorithm for Unitary Hessenberg Matrices," *J. Comp. Appl. Math. 16*, 1–8.

G.S. Ammar, W.B. Gragg, and L. Reichel (1985). "On the Eigenproblem for Orthogonal Matrices," *Proc. IEEE Conference on Decision and Control*, 1963–1966.

W.B. Gragg and L. Reichel (1990). "A Divide and Conquer Method for Unitary and Orthogonal Eigenproblems," *Numer. Math. 57*, 695–718.

Hamiltonian eigenproblems (see P12.6.13) occur throughout optimal control theory and are very important.

C.C. Paige and C. Van Loan (1981). "A Schur Decomposition for Hamiltonian Matrices," *Lin. Alg. and Its Applic. 41*, 11–32.

C. Van Loan (1984). "A Symplectic Method for Approximating All the Eigenvalues of a Hamiltonian Matrix," *Lin. Alg. and Its Applic. 61*, 233–252.

R. Byers (1986) "A Hamiltonian QR Algorithm," *SIAM J. Sci. and Stat. Comp.* 7, 212–229.

V. Mehrmann (1988). "A Symplectic Orthogonal Method for Single Input or Single Output Discrete Time Optimal Quadratic Control Problems," *SIAM J. Matrix Anal. Appl.* 9, 221–247.

G. Ammar and V. Mehrmann (1991). "On Hamiltonian and Symplectic Hessenberg Forms," *Lin.Alg. and Its Application 149*, 55–72.

A. Bunse-Gerstner, R. Byers, and V. Mehrmann (1992). "A Chart of Numerical Methods for Structured Eigenvalue Problems," *SIAM J. Matrix Anal. Appl. 13*, 419–453.

Other papers on modified/structured eigenvalue problems include

A. Bunse-Gerstner and W.B. Gragg (1988). "Singular Value Decompositions of Complex Symmetric Matrices," *J. Comp. Applic. Math. 21*, 41–54.

R. Byers (1988). "A Bisection Method for Measuring the Distance of a Stable Matrix to the Unstable Matrices," *SIAM J. Sci. Stat. Comp.* 9, 875–881.

J.W. Demmel and W. Gragg (1993). "On Computing Accurate Singular Values and Eigenvalues of Matrices with Acyclic Graphs," *Lin. Alg. and Its Applic. 185*, 203–217.

A. Bunse-Gerstner, R. Byers, and V. Mehrmann (1993). "Numerical Methods for Simultaneous Diagonalization," *SIAM J. Matrix Anal. Appl. 14*, 927–949.

Bibliography

J.O. Aasen (1971). "On the Reduction of a Symmetric Matrix to Tridiagonal Form," *BIT 11*, 233–42.

N.N. Abdelmalek (1971). "Roundoff Error Analysis for Gram-Schmidt Method and Solution of Linear Least Squares Problems," *BIT 11*, 345–68.

G.E. Adams, A.W. Bojanczyk, and F.T. Luk (1994). "Computing the PSVD of Two 2×2 Triangular Matrices," *SIAM J. Matrix Anal. Appl. 15*, 366–382.

L. Adams (1985). "m-step Preconditioned Congugate Gradient Methods," *SIAM J. Sci. and Stat. Comp. 6*, 452–463.

L. Adams and P. Arbenz (1994). "Towards a Divide and Conquer Algorithm for the Real Nonsymmetric Eigenvalue Problem," *SIAM J. Matrix Anal. Appl. 15*, 1333–1353.

L. Adams and T. Crockett (1984). "Modeling Algorithm Execution Time on Processor Arrays," *Computer 17*, 38–43.

L. Adams and H. Jordan (1986). "Is SOR Color-Blind?" *SIAM J. Sci. Stat. Comp. 7*, 490–506.

S.T. Alexander, C.T. Pan, and R.J. Plemmons (1988). "Analysis of a Recursive Least Squares Hyperbolic Rotation Algorithm for Signal Processing," *Lin. Alg. and Its Applic. 98*, 3–40.

E.L. Allgower (1973). "Exact Inverses of Certain Band Matrices," *Numer. Math. 21*, 279–84.

A.R. Amir-Moez (1965). *Extremal Properties of Linear Transformations and Geometry of Unitary Spaces*, Texas Tech University Mathematics Series, no. 243, Lubbock, Texas.

G.S. Ammar and W.B. Gragg (1988). "Superfast Solution of Real Positive Definite Toeplitz Systems," *SIAM J. Matrix Anal. Appl. 9*, 61–76.

G.S. Ammar, W.B. Gragg, and L. Reichel (1985). "On the Eigenproblem for Orthogonal Matrices," *Proc. IEEE Conference on Decision and Control*, 1963–1966.

G.S. Ammar and G. He (1995). "On an Inverse Eigenvalue Problem for Unitary Matrices," *Lin. Alg. and Its Applic. 218*, 263–271.

G.S. Ammar and V. Mehrmann (1991). "On Hamiltonian and Symplectic Hessenberg Forms," *Lin.Alg. and Its Applic. 149*, 55–72.

P. Amodio and L. Brugnano (1995). "The Parallel QR Factorization Algorithm for Tridiagonal Linear Systems," *Parallel Computing 21*, 1097–1110.

C. Ancourt, F. Coelho, F. Irigoin, and R. Keryell (1993). "A Linear Algebra Framework for Static HPF Code Distribution," *Proceedings of the 4th Workshop on Compilers for Parallel Computers*, Delft, The Netherlands.

A.A. Anda and H. Park (1994). "Fast Plane Rotations with Dynamic Scaling," *SIAM J. Matrix Anal. Appl. 15*, 162–174.

E. Anderson, Z. Bai, C. Bischof, J. Demmel, J. Dongarra, J. DuCroz, A. Greenbaum, S. Hammarling, A. McKenney, S. Ostrouchov, and D. Sorensen (1995). *LAPACK Users' Guide, Release 2.0, 2nd ed.*, SIAM Publications, Philadelphia, PA.

E. Anderson, Z. Bai, and J. Dongarra (1992). "Generalized QR Factorization and Its Application," *Lin. Alg. and Its Applic. 162/163/164*, 243–271.

N. Anderson and I. Karasalo (1975). "On Computing Bounds for the Least Singular Value of a Triangular Matrix," *BIT 15*, 1–4.

P. Anderson and G. Loizou (1973). "On the Quadratic Convergence of an Algorithm Which Diagonalizes a Complex Symmetric Matrix," *J. Inst. Math. Applic. 12*, 261–71.

P. Anderson and G. Loizou (1976). "A Jacobi-Type Method for Complex Symmetric Matrices (Handbook)," *Numer. Math. 25*, 347–63.

H.C. Andrews and J. Kane (1970). "Kronecker Matrices, Computer Implementation, and Generalized Spectra," *J. Assoc. Comput. Mach. 17*, 260–268.

P. Arbenz, W. Gander, and G.H. Golub (1988). "Restricted Rank Modification of the Symmetric Eigenvalue Problem: Theoretical Considerations," *Lin. Alg. and Its Applic. 104*, 75–95.

P. Arbenz and G.H. Golub (1988). "On the Spectral Decomposition of Hermitian Matrices Subject to Indefinite Low Rank Perturbations with Applications," *SIAM J. Matrix Anal. Appl. 9*, 40–58.

T.A. Arias, A. Edelman, and S. Smith (1996). "Conjugate Gradient and Newton's Method on the Grassman and Stiefel Manifolds," to appear in *SIAM J. Matrix Anal. Appl.*

M. Arioli, J. Demmel, and I. Duff (1989). "Solving Sparse Linear Systems with Sparse Backward Error," *SIAM J. Matrix Anal. Appl. 10*, 165–190.

M. Arioli, I.S. Duff, and D. Ruiz (1992). "Stopping Criteria for Iterative Solvers," *SIAM J. Matrix Anal. Appl. 13*, 138–144.

M. Arioli and A. Laratta (1985). "Error Analysis of an Algorithm for Solving an Underdetermined System," *Numer. Math. 46*, 255–268.

M. Arioli and F. Romani (1985). "Relations Between Condition Numbers and the Convergence of the Jacobi Method for Real Positive Definite Matrices," *Numer. Math. 46*, 31–42.

W.F. Arnold and A.J. Laub (1984). "Generalized Eigenproblem Algorithms and Software for Algebraic Riccati Equations," *Proc. IEEE 72*, 1746–1754.

W.E. Arnoldi (1951). "The Principle of Minimized Iterations in the Solution of the Matrix Eigenvalue Problem," *Quarterly of Applied Mathematics 9*, 17–29.

S. Ashby (1987). "Polynomial Preconditioning for Conjugate Gradient Methods," Ph.D. Thesis, Dept. of Computer Science, University of Illinois.

S. Ashby, T. Manteuffel, and J. Otto (1992). "A Comparison of Adaptive Chebyshev and Least Squares Polynomial Preconditioning for Hermitian Positive Definite Linear Systems," *SIAM J. Sci. Stat. Comp. 13*, 1–29.

S. Ashby, T. Manteuffel, and P. Saylor (1989). "Adaptive Polynomial Preconditioning for Hermitian Indefinite Linear Systems," *BIT 29*, 583–609.

S. Ashby, T. Manteuffel, and P. Saylor (1990). "A Taxonomy of Conjugate Gradient Methods," *SIAM J. Num. Anal. 27*, 1542–1568.

C.C. Ashcraft and R. Grimes (1988). "On Vectorizing Incomplete Factorization and SSOR Preconditioners," *SIAM J. Sci. and Stat. Comp. 9*, 122–151.

G. Auchmuty (1991). "A Posteriori Error Estimates for Linear Equations," *Numer. Math. 61*, 1–6.

O. Axelsson (1977). "Solution of Linear Systems of Equations: Iterative Methods," in *Sparse Matrix Techniques: Copenhagen*, 1976, ed. V.A. Barker, Springer-Verlag, Berlin.

O. Axelsson (1980). "Conjugate Gradient Type Methods for Unsymmetric and Inconsistent Systems of Linear Equations," *Lin. Alg. and Its Applic. 29*, 1–16.

O. Axelsson (1985). "Incomplete Block Matrix Factorization Preconditioning Methods. The Ultimate Answer?", *J. Comput. Appl. Math. 12&13*, 3–18.

O. Axelsson (1985). "A Survey of Preconditioned Iterative Methods for Linear Systems of Equations," *BIT 25*, 166–187.

O. Axelsson (1986). "A General Incomplete Block Matrix Factorization Method," *Lin. Alg. Appl. 74*, 179–190.

O. Axelsson, ed. (1989). "Preconditioned Conjugate Gradient Methods," *BIT 29:4*.

O. Axelsson (1994). *Iterative Solution Methods*, Cambridge University Press.

O. Axelsson and V. Eijkhout (1989). "Vectorizable Preconditioners for Elliptic Difference Equations in Three Space Dimensions," *J. Comput. Appl. Math. 27*, 299–321.

O. Axelsson and G. Lindskog (1986). "On the Rate of Convergence of the Preconditioned Conjugate Gradient Method," *Numer. Math. 48*, 499–523.

O. Axelsson and B. Polman (1986). "On Approximate Factorization Methods for Block Matrices Suitable for Vector and Parallel Processors," *Lin. Alg. and Its Applic. 77*, 3–26.

O. Axelsson and P. Vassilevski (1989). "Algebraic Multilevel Preconditioning Methods I," *Numer. Math. 56*, 157–177.

O. Axelsson and P. Vassilevski (1990). "Algebraic Multilevel Preconditioning Methods II," *SIAM J. Numer. Anal. 27*. 1569–1590.

Z. Bai (1988). "Note on the Quadratic Convergence of Kogbetliantz's Algorithm for Computing the Singular Value Decomposition," *Lin. Alg. and Its Applic. 104*, 131–140.

Z. Bai (1994). "Error Analysis of the Lanczos Algorithm for Nonsymmetric Eigenvalue Problem," *Math. Comp. 62*, 209–226.

Z. Bai and J.W. Demmel (1989). "On a Block Implementation of Hessenberg Multishift QR Iteration," *Int'l J. of High Speed Comput. 1*, 97–112.

Z. Bai and J.W. Demmel (1993). "On Swapping Diagonal Blocks in Real Schur Form," *Lin. Alg. and Its Applic. 186*, 73–95.

Z. Bai, J.W. Demmel, and A. McKenney (1993). "On Computing Condition Numbers for the Nonsymmetric Eigenproblem," *ACM Trans. Math. Soft. 19*, 202–223.

Z. Bai, D. Hu, and L. Reichel (1994). "A Newton Basis GMRES Implementation," *IMA J. Num. Anal. 14*, 563–581.

Z. Bai and H. Zha (1993). "A New Preprocessing Algorithm for the Computation of the Generalized Singular Value Decomposition," *SIAM J. Sci. Comp. 14*, 1007–1012.

D.H. Bailey (1988). "Extra High Speed Matrix Multiplication on the Cray-2," *SIAM J. Sci. and Stat. Comp. 9*, 603–607.

D.H. Bailey (1993). "Algorithm 719: Multiprecision Translation and Execution of FORTRAN Programs," *ACM Trans. Math. Soft. 19*, 288–319.

D.H. Bailey, K.Lee, and H.D. Simon (1991). "Using Strassen's Algorithm to Accelerate the Solution of Linear Systems," *J. Supercomputing 4*, 357–371.

D.H. Bailey, H.D. Simon, J. T. Barton, M.J. Fouts (1989). "Floating Point Arithmetic in Future Supercomputers," *Int'l J. Supercomputing Appl. 3*, 86–90.

I.Y. Bar-Itzhack (1975). "Iterative Optimal Orthogonalization of the Strapdown Matrix," *IEEE Trans. Aerospace and Electronic Systems 11*, 30–37.

J.L. Barlow (1986). "On the Smallest Positive Singular Value of an M-Matrix with Applications to Ergodic Markov Chains," *SIAM J. Alg. and Disc. Struct. 7*, 414–424.

J.L. Barlow (1988). "Error Analysis and Implementation Aspects of Deferred Correction for Equality Constrained Least-Squares Problems," *SIAM J. Num. Anal. 25*, 1340–1358.

J.L. Barlow (1993). "Error Analysis of Update Methods for the Symmetric Eigenvalue Problem," *SIAM J. Matrix Anal. Appl. 14*, 598–618.

J.L. Barlow and J. Demmel (1990). "Computing Accurate Eigensystems of Scaled Diagonally Dominant Matrices," *SIAM J. Numer. Anal. 27*, 762–791.

J.L. Barlow and S.L. Handy (1988). "The Direct Solution of Weighted and Equality Constrained Least-Squares Problems," *SIAM J. Sci. Stat. Comp. 9*, 704–716.

J.L. Barlow, M.M. Monahemi, and D.P. O'Leary (1992). "Constrained Matrix Sylvester Equations," *SIAM J. Matrix Anal. Appl. 13*, 1–9.

J.L. Barlow, N.K. Nichols, and R.J. Plemmons (1988). "Iterative Methods for Equality Constrained Least Squares Problems," SIAM J. Sci. and Stat. Comp. 9, 892–906.

J.L. Barlow and U.B. Vemulapati (1992). "Rank Detection Methods for Sparse Matrices," *SIAM J. Matrix. Anal. Appl. 13*, 1279–1297.

J.L. Barlow and U.B. Vemulapati (1992). "A Note on Deferred Correction for Equality Constrained Least Squares Problems," *SIAM J. Num. Anal. 29*, 249–256.

S. Barnett and C. Storey (1968). "Some Applications of the Lyapunov Matrix Equation," *J. Inst. Math. Applic. 4*, 33–42.

R. Barrett, M. Berry, T.F. Chan, J. Demmel, J. Donato, J. Dongarra, V. Eijkhout, R. Pozo, C. Romine, H. van der Vorst (1993). *Templates for the Solution of Linear Systems: Building Blocks for Iterative Methods*, SIAM Publications, Philadelphia, PA.

A. Barrlund (1991). "Perturbation Bounds for the LDL^T and LU Decompositions," *BIT 31*, 358–363.

A. Barrlund (1994). "Perturbation Bounds for the Generalized QR Factorization," *Lin. Alg. and Its Applic. 207*, 251–271.

R.H. Bartels (1971). "A Stabilization of the Simplex Method," *Numer. Math. 16*, 414-434.

R.H. Bartels, A.R. Conn, and C. Charalambous (1978). "On Cline's Direct Method for Solving Overdetermined Linear Systems in the L_∞ Sense," *SIAM J. Num. Anal. 15*, 255–70.

R.H. Bartels and G.W. Stewart (1972). "Solution of the Equation $AX + XB = C$," *Comm. ACM 15*, 820–26.

S.G. Bartels and D.J. Higham (1992). "The Structured Sensitivity of Vandermonde-Like Systems," *Numer. Math. 62*, 17–34.

W. Barth, R.S. Martin, and J.H. Wilkinson (1967). "Calculation of the Eigenvalues of a Symmetric Tridiagonal Matrix by the Method of Bisection," *Numer. Math. 9*, 386–93. See also Wilkinson and Reinsch (1971, 249–256).

V. Barwell and J.A. George (1976). "A Comparison of Algorithms for Solving Symmetric Indefinite Systems of Linear Equations," *ACM Trans. Math. Soft. 2*, 242–51.

K.J. Bathe and E.L. Wilson (1973). "Solution Methods for Eigenvalue Problems in Structural Mechanics," *Int. J. Numer. Meth. Eng. 6*, 213-26.

S. Batterson (1994). "Convergence of the Francis Shifted QR Algorithm on Normal Matrices," *Lin. Alg. and Its Applic. 207*, 181–195.

S. Batterson and J. Smillie (1989). "The Dynamics of Rayleigh Quotient Iteration," *SIAM J. Num. Anal. 26*, 624–636.

D. Bau, I. Kodukula, V. Kotlyar, K. Pingali, and P. Stodghil (1993). "Solving Alignment Using Elementary Linear Algebra," in *Proceedings of the 7th International Workshop on Languages and Compilers for Parallel Computing*, Lecture Notes in Computer Science 892, Springer-Verlag, New York, 46–60.

F.L. Bauer (1963). "Optimally Scaled Matrices," *Numer. Math. 5*, 73–87.

F.L. Bauer (1965). "Elimination with Weighted Row Combinations for Solving Linear Equations and Least Squares Problems," *Numer. Math. 7*, 338–52. See also Wilkinson and Reinsch (1971, 119–33).

F.L. Bauer and C.T. Fike (1960). "Norms and Exclusion Theorems," *Numer. Math. 2*, 123–44.

F.L. Bauer and C. Reinsch (1968). "Rational QR Transformations with Newton Shift for Symmetric Tridiagonal Matrices," *Numer. Math. 11*, 264–72. See also Wilkinson and Reinsch (1971, pp.257–65).

F.L. Bauer and C. Reinsch (1971). "Inversion of Positive Definite Matrices by the Gauss-Jordan Method," in *Handbook for Automatic Computation Vol. 2, Linear Algebra*, J.H. Wilkinson and C. Reinsch, eds. Springer-Verlag, New York, 45–49.

C. Bavely and G.W. Stewart (1979). "An Algorithm for Computing Reducing Subspaces by Block Diagonalization," *SIAM J. Num. Anal. 16*, 359–67.

C. Beattie and D.W. Fox (1989). "Localization Criteria and Containment for Rayleigh Quotient Iteration," *SIAM J. Matrix Anal. Appl. 10*, 80–93.

R. Beauwens and P. de Groen, eds. (1992). *Iterative Methods in Linear Algebra*, Elsevier (North-Holland), Amsterdam.

T. Beelen and P. Van Dooren (1988). "An Improved Algorithm for the Computation of Kronecker's Canonical Form of a Singular Pencil," *Lin. Alg. and Its Applic. 105*, 9–65.

R. Bellman (1970). *Introduction to Matrix Analysis*, 2nd ed., McGraw-Hill, New York.

E. Beltrami (1873). "Sulle Funzioni Bilineari," *Gionale di Mathematiche 11*, 98–106.

A. Berman and A. Ben-Israel (1971). "A Note on Pencils of Hermitian or Symmetric Matrices," *SIAM J. Applic. Math. 21*, 51–54.

A. Berman and R.J. Plemmons (1979). *Nonnegative Matrices in the Mathematical Sciences*, Academic Press, New York.

M.J.M. Bernal and J.H. Verner (1968). "On Generalizing of the Theory of Consistent Orderings for Successive Over-Relaxation Methods," *Numer. Math. 12*, 215–22.

J. Berntsen (1989). "Communication Efficient Matrix Multiplication on Hypercubes," *Parallel Computing 12*, 335–342.

M.W. Berry, J.J. Dongarra, and Y. Kim (1995). "A Parallel Algorithm for the Reduction of a Nonsymmetric Matrix to Block Upper Hessenberg Form," *Parallel Computing 21*, 1189–1211.

M.W. Berry and G.H. Golub (1991). "Estimating the Largest Singular Values of Large Sparse Matrices via Modified Moments," *Numerical Algorithms 1*, 353–374.

M.W. Berry and A. Sameh (1986). "Multiprocessor Jacobi Algorithms for Dense Symmetric Eigenvalue and Singular Value Decompositions," in *Proc. International Conference on Parallel Processing*, 433–440.

D.P. Bertsekas and J. N. Tsitsiklis (1989). *Parallel and Distributed Computation: Numerical Methods*, Prentice Hall, Englewood Cliffs, NJ.

R. Bevilacqua, B. Codenotti, and F. Romani (1988). "Parallel Solution of Block Tridiagonal Linear Systems," *Lin.Alg. and Its Applic. 104*, 39–57.

C.H. Bischof (1987). "The Two-Sided Block Jacobi Method on Hypercube Architectures," in *Hypercube Multiprocessors*, ed. M.T. Heath, SIAM Publications, Philadelphia, PA.

C.H. Bischof (1988). "QR Factorization Algorithms for Coarse Grain Distributed Systems," PhD Thesis, Dept. of Computer Science, Cornell University, Ithaca, NY.

C.H. Bischof (1989). "Computing the Singular Value Decomposition on a Distributed System of Vector Processors," *Parallel Computing 11*, 171–186.

C.H. Bischof (1990). "Incremental Condition Estimation," *SIAM J. Matrix Anal. Appl. 11*, 644–659.

C.H. Bischof (1990). "Incremental Condition Estimation for Sparse Matrices," *SIAM J. Matrix Anal. Appl. 11*, 312–322.

C.H. Bischof and P.C. Hansen (1992). "A Block Algorithm for Computing Rank-Revealing QR Factorizations," *Numerical Algorithms 2*, 371-392.

C.H. Bischof and G.M. Shroff (1992). "On Updating Signal Subspaces," *IEEE Trans. Signal Proc. 40*, 96–105.

C.H. Bischof and C. Van Loan (1986). "Computing the SVD on a Ring of Array Processors," in *Large Scale Eigenvalue Problems*, eds. J. Cullum and R. Willoughby, North Holland, 51-66.

C.H. Bischof and C. Van Loan (1987). "The WY Representation for Products of Householder Matrices," *SIAM J. Sci. and Stat. Comp. 8*, s2–s13.

Å. Björck (1967). "Iterative Refinement of Linear Least Squares Solutions I," *BIT 7*, 257–78.

A. Björck (1967). "Solving Linear Least Squares Problems by Gram-Schmidt Orthogonalization," *BIT 7*, 1–21.

Å. Björck (1968). "Iterative Refinement of Linear Least Squares Solutions II," *BIT 8*, 8–30.

Å. Björck (1984). "A General Updating Algorithm for Constrained Linear Least Squares Problems," *SIAM J. Sci. and Stat. Comp. 5*, 394–402.

Å. Björck (1987). "Stability Analysis of the Method of Seminormal Equations for Linear Least Squares Problems," *Linear Alg. and Its Applic. 88/89*, 31–48.

Å. Björck (1991). "Component-wise Perturbation Analysis and Error Bounds for Linear Least Squares Solutions," *BIT 31*, 238–244.

Å. Björck (1992). "Pivoting and Stability in the Augmented System Method," *Proceedings of the 14th Dundee Conference*, D.F. Griffiths and G.A. Watson (eds), Longman Scientific and Technical, Essex, U.K.

A. Björck (1994). "Numerics of Gram-Schmidt Orthogonalization," *Lin. Alg. and Its Applic. 197/198*, 297–316.

Å. Björck (1996). *Numerical Methods for Least Squares Problems*, SIAM Publications, Philadelphia, PA.

A. Björck and C. Bowie (1971). "An Iterative Algorithm for Computing the Best Estimate of an Orthogonal Matrix," *SIAM J. Num. Anal. 8*, 358–64.

Å. Björck and I.S. Duff (1980). "A Direct Method for the Solution of Sparse Linear Least Squares Problems," *Lin. Alg. and Its Applic. 34*, 43–67.

A. Björck and T. Elfving (1973). "Algorithms for Confluent Vandermonde Systems," *Numer. Math. 21*, 130–37.

Å. Björck and G.H. Golub (1967). "Iterative Refinement of Linear Least Squares Solutions by Householder Transformation," *BIT 7*, 322–37.

A. Björck and G.H. Golub (1973). "Numerical Methods for Computing Angles Between Linear Subspaces," *Math. Comp. 27*, 579–94.

Å. Björck, E. Grimme, and P. Van Dooren (1994). "An Implicit Shift Bidiagonalization Algorithm for Ill-Posed Problems," *BIT 34*, 510–534.

Å. Björck and S. Hammarling (1983). "A Schur Method for the Square Root of a Matrix," *Lin. Alg. and Its Applic. 52/53*, 127–140.

P. Bjørstad, F. Manne, T.Sørevik, and M. Vajteršic (1992). "Efficient Matrix Multiplication on SIMD Computers," *SIAM J. Matrix Anal. Appl. 13*, 386–401.

Å. Björck and C.C. Paige (1992). "Loss and Recapture of Orthogonality in the Modified Gram-Schmidt Algorithm," *SIAM J. Matrix Anal. Appl. 13*, 176–190.

Å. Björck and C.C. Paige (1994). "Solution of Augmented Linear Systems Using Orthogonal Factorizations," *BIT 34*, 1–24.

Å. Björck, H. Park, and L. Eldén (1994). "Accurate Downdating of Least Squares Solutions," *SIAM J. Matrix Anal. Appl. 15*, 549–568.

A. Björck and V. Pereyra (1970). "Solution of Vandermonde Systems of Equations," *Math. Comp. 24*, 893–903.

Å. Björck, R.J. Plemmons, and H. Schneider, eds. (1981). *Large-Scale Matrix Problems*, North-Holland, New York.

J.M. Blue (1978). "A Portable FORTRAN Program to Find the Euclidean Norm of a Vector," *ACM Trans. Math. Soft. 4*, 15–23.

E. Bodewig (1959). *Matrix Calculus*, North Holland, Amsterdam.

Z. Bohte (1975). "Bounds for Rounding Errors in the Gaussian Elimination for Band Systems," *J. Inst. Math. Applic. 16*, 133–42.

A.W. Bojanczyk, R.P. Brent, and F.R. de Hoog (1986). "QR Factorization of Toeplitz Matrices," *Numer. Math. 49*, 81–94.

A.W. Bojanczyk, R.P. Brent, F.R. de Hoog, and D.R. Sweet (1995). "On the Stability of the Bareiss and Related Toeplitz Factorization Algorithms," *SIAM J. Matrix Anal. Appl. 16*, 40–57.

A.W. Bojanczyk, R.P. Brent, P. Van Dooren, and F.R. de Hoog (1987). "A Note on Downdating the Cholesky Factorization," *SIAM J. Sci. and Stat. Comp. 8*, 210–221.

A.W. Bojanczyk and G. Cybenko, eds. (1995). *Linear Algebra for Signal Processing*, IMA Volumes in Mathematics and Its Applications, Springer-Verlag, New York.

A.W. Bojanczyk, R. Onn, and A.O. Steinhardt (1993). "Existence of the Hyperbolic Singular Value Decomposition," *Lin. Alg. and Its Applic. 185*, 21–30.

D.L. Boley and G.H. Golub (1978). "The Matrix Inverse Eigenvalue Problem for Periodic Jacobi Matrices," in *Proc. Fourth Symposium on Basic Problems of Numerical Mathematics*, Prague, pp. 63–76.

D.L. Boley and G.H. Golub (1984). "A Modified Method for Restructuring Periodic Jacobi Matrices," *Math. Comp. 42*, 143–150.

D.L. Boley and G.H. Golub (1984). "The Lanczos-Arnoldi Algorithm and Controllability," *Syst. Control Lett. 4*, 317–324.

D.L. Boley and G.H. Golub (1987). "A Survey of Matrix Inverse Eigenvalue Problems," *Inverse Problems 3*, 595–622.

H. Bolz and W. Niethammer (1988). "On the Evaluation of Matrix Functions Given by Power Series," *SIAM J. Matrix Anal. Appl. 9*, 202–209.

S. Bondeli and W. Gander (1994). "Cyclic Reduction for Special Tridiagonal Systems," *SIAM J. Matrix Anal. Appl. 15*, 321–330.

J. Boothroyd and P.J. Eberlein (1968). "Solution to the Eigenproblem by a Norm-Reducing Jacobi-Type Method (Handbook)," *Numer. Math. 11*, 1–12. See also Wilkinson and Reinsch (1971, pp.327–38).

H.J. Bowdler, R.S. Martin, G. Peters, and J.H. Wilkinson (1966). "Solution of Real and Complex Systems of Linear Equations," *Numer. Math. 8*, 217–234. (See also Wilkinson and Reinsch (1971, 93–110).

H.J. Bowdler, R.S. Martin, C. Reinsch, and J.H. Wilkinson (1968). "The QR and QL Algorithms for Symmetric Matrices," *Numer. Math. 11*, 293–306. See also Wilkinson and Reinsch (1971, pp.227–240).

J.H. Bramble, J.E. Pasciak, and A.H. Schatz (1986). "The construction of Preconditioners for Elliptic Problems by Substructuring I," *Math. Comp. 47*, 103–134.

J.H. Bramble, J.E. Pasciak, and A.H. Schatz (1986). "The construction of Preconditioners for Elliptic Problems by Substructuring II," *Math. Comp. 49*, 1–17.

R. Bramley and A. Sameh (1992). "Row Projection Methods for Large Nonsymmetric Linear Systems," *SIAM J. Sci. Statist. Comput. 13*, 168–193.

R.P. Brent (1970). "Error Analysis of Algorithms for Matrix Multiplication and Triangular Decomposition Using Winograd's Identity," *Numer. Math. 16*, 145–156.

R.P. Brent (1978). "A Fortran Multiple Precision Arithmetic Package," *ACM Trans. Math. Soft. 4*, 57–70.

R.P. Brent (1978). "Algorithm 524 MP, a Fortran Multiple Precision Arithmetic Package," *ACM Trans. Math. Soft. 4*, 71–81.

R.P. Brent and F.T. Luk (1982) "Computing the Cholesky Factorization Using a Systolic Architecture," *Proc. 6th Australian Computer Science Conf.* 295–302.

R.P. Brent and F.T. Luk (1985). "The Solution of Singular Value and Symmetric Eigenvalue Problems on Multiprocessor Arrays," *SIAM J. Sci. and Stat. Comp. 6*, 69–84.

R.P. Brent, F.T. Luk, and C. Van Loan (1985). "Computation of the Singular Value Decomposition Using Mesh Connected Processors," *J. VLSI Computer Systems 1*, 242–270.

C. Brezinski and M. Redivo-Zaglia (1994). "Treatment of Near-Breakdown in the CGS Algorithms," *Numer. Alg. 7*, 33–73.

C. Brezinski and M. Redivo-Zaglia (1995). "Look-Ahead in BiCGSTAB and Other Product-Type Methods for Linear Systems," *BIT 35*, 169–201.

C. Brezinski and H. Sadok (1991). "Avoiding Breakdown in the CGS Algorithm," *Numer. Alg. 1*, 199–206.

C. Brezinski, M. Zaglia, and H. Sadok (1991). "Avoiding Breakdown and Near Breakdown in Lanczos Tyoe Algorithms," *Numer. Alg. 1*, 261–284.

C. Brezinski, M. Zaglia, and H. Sadok (1992). "A Breakdown Free Lanczos Type Algorithm for Solving Linear Systems," *Numer. Math. 63*, 29–38.

K.W. Brodlie and M.J.D. Powell (1975). "On the Convergence of Cyclic Jacobi Methods," *J. Inst. Math. Applic. 15*, 279–87.

J.D. Brown, M.T. Chu, D.C. Ellison, and R.J. Plemmons, eds. (1994). *Proceedings of the Cornelius Lanczos International Centenary Conference*, SIAM Publications, Philadelphia, PA.

C.G. Broyden (1973). "Some Condition Number Bounds for the Gaussian Elimination Process," *J. Inst. Math. Applic. 12*, 273–86.

A. Buckley (1974). "A Note on Matrices $A = I + H$, H Skew-Symmetric," *Z. Angew. Math. Mech. 54*, 125–26.

A. Buckley (1977). "On the Solution of Certain Skew-Symmetric Linear Systems," *SIAM J. Num. Anal. 14*, 566–70.

J.R. Bunch (1971). "Analysis of the Diagonal Pivoting Method," *SIAM J. Num. Anal. 8*, 656–680.

J.R. Bunch (1971). "Equilibration of Symmetric Matrices in the Max-Norm," *J. ACM* *18*, 566–72.

J.R. Bunch (1974). "Partial Pivoting Strategies for Symmetric Matrices," *SIAM J. Num. Anal. 11*, 521–528.

J.R. Bunch (1976). "Block Methods for Solving Sparse Linear Systems," in *Sparse Matrix Computations*, J.R. Bunch and D.J. Rose (eds), Academic Press, New York.

J.R. Bunch (1982). "A Note on the Stable Decomposition of Skew Symmetric Matrices," *Math. Comp. 158*, 475–480.

J.R. Bunch (1985). "Stability of Methods for Solving Toeplitz Systems of Equations," *SIAM J. Sci. Stat. Comp. 6*, 349–364.

J.R. Bunch, J.W. Demmel, and C.F. Van Loan (1989). "The Strong Stability of Algorithms for Solving Symmetric Linear Systems," *SIAM J. Matrix Anal. Appl. 10*, 494–499.

J.R. Bunch and L. Kaufman (1977). "Some Stable Methods for Calculating Inertia and Solving Symmetric Linear Systems," *Math. Comp. 31*, 162–79.

J.R. Bunch, L. Kaufman, and B.N. Parlett (1976). "Decomposition of a Symmetric Matrix," *Numer. Math. 27*, 95–109.

J.R. Bunch, C.P. Nielsen, and D.C. Sorensen (1978). "Rank-One Modification of the Symmetric Eigenproblem," *Numer. Math. 31*, 31–48.

J.R. Bunch and B.N. Parlett (1971). "Direct Methods for Solving Symmetric Indefinite Systems of Linear Equations," *SIAM J. Num. Anal. 8*, 639–55.

J.R. Bunch and D.J. Rose, eds. (1976). *Sparse Matrix Computations*, Academic Press, New York.

O. Buneman (1969). "A Compact Non-Iterative Poisson Solver," Report 294, Stanford University Institute for Plasma Research, Stanford, California.

A. Bunse-Gerstner (1984). "An Algorithm for the Symmetric Generalized Eigenvalue Problem," *Lin. Alg. and Its Applic. 58*, 43–68.

A. Bunse-Gerstner, R. Byers, and V. Mehrmann (1992). "A Chart of Numerical Methods for Structured Eigenvalue Problems," *SIAM J. Matrix Anal. Appl. 13*, 419–453.

A. Bunse-Gerstner, R. Byers, and V. Mehrmann (1993). "Numerical Methods for Simultaneous Diagonalization," *SIAM J. Matrix Anal. Appl. 14*, 927–949.

A. Bunse-Gerstner and W.B. Gragg (1988). "Singular Value Decompositions of Complex Symmetric Matrices," *J. Comp. Applic. Math. 21*, 41–54.

J.V. Burke and M.L. Overton (1992). "Stable Perturbations of Nonsymmetric Matrices," *Lin.Alg. and Its Applic. 171*, 249–273.

P.A. Businger (1968). "Matrices Which Can be Optimally Scaled," *Numer. Math. 12*, 346–48.

P.A. Businger (1969). "Reducing a Matrix to Hessenberg Form," *Math. Comp. 23*, 819–21.

P.A. Businger (1971). "Monitoring the Numerical Stability of Gaussian Elimination," *Numer. Math. 16*, 360–61.

P.A. Businger (1971). "Numerically Stable Deflation of Hessenberg and Symmetric Tridiagonal Matrices,*BIT 11*, 262–70.

P.A. Businger and G.H. Golub (1965). "Linear Least Squares Solutions by Householder Transformations," *Numer. Math. 7*, 269–76. See also Wilkinson and Reinsch (1971,111–18).

P.A. Businger and G.H. Golub (1969). "Algorithm 358: Singular Value Decomposition of a Complex Matrix," *Comm. Assoc. Comp. Mach. 12*, 564–65.

B.L. Buzbee (1986) "A Strategy for Vectorization," *Parallel Computing 3*, 187-192.

B.L. Buzbee and F.W. Dorr (1974). "The Direct Solution of the Biharmonic Equation on Rectangular Regions and the Poisson Equation on Irregular Regions,"*SIAM J. Num. Anal. 11*, 753–63.

B.L. Buzbee, F.W. Dorr, J.A. George, and G.H. Golub (1971). "The Direct Solution of the Discrete Poisson Equation on Irregular Regions,"*SIAM J. Num. Anal. 8*, 722–36.

B.L. Buzbee, G.H. Golub, and C.W. Nielson (1970). "On Direct Methods for Solving Poisson's Equations," *SIAM J. Num. Anal. 7*, 627-56.

R. Byers (1984). "A Linpack-Style Condition Estimator for the Equation $AX - XB^T = C$," *IEEE Trans. Auto. Cont. AC-29*, 926–928.

R. Byers (1986) "A Hamiltonian QR Algorithm," *SIAM J. Sci. and Stat. Comp. 7*, 212–229.

R. Byers (1988). "A Bisection Method for Measuring the Distance of a Stable Matrix to the Unstable Matrices," *SIAM J. Sci. Stat. Comp. 9*, 875–881.

R. Byers and S.G. Nash (1987). "On the Singular Vectors of the Lyapunov Operator," *SIAM J. Alg. and Disc. Methods 8*, 59–66.

X.-C. Cai and O. Widlund (1993). "Multiplicative Schwarz Algorithms for Some Nonsymmetric and Indefinite Problems," *SIAM J. Numer. Anal. 30*, 936–952.

D. Calvetti, G.H. Golub, and L. Reichel (1994). "An Adaptive Chebyshev Iterative Method for Nonsymmetric Linear Systems Based on Modified Moments," *Numer. Math. 67*, 21–40.

D. Calvetti and L. Reichel (1992). "A Chebychev-Vandermonde Solver," *Lin. Alg. and Its Applic. 172*, 219–229.

D. Calvetti and L. Reichel (1993). "Fast Inversion of Vandermonde-Like Matrices Involving Orthogonal Polynomials," *BIT 33*, 473–484.

D. Calvetti, L. Reichel, and D.C. Sorensen (1994). "An Implicitly Restarted Lanczos Method for Large Symmetric Eigenvalue Problems," *ETNA 2*, 1–21.

L.E. Cannon (1969). *A Cellular Computer to Implement the Kalman Filter Algorithm*, Ph.D. Thesis, Montana State University.

R. Carter (1991). "Y-MP Floating Point and Cholesky Factorization," *Int'l J. High Speed Computing 3*, 215–222.

F. Chaitin-Chatelin and V. Frayseé (1996). *Lectures on Finite Precision Computations*, SIAM Publications, Philadelphia, PA.

R.H. Chan (1989). "The Spectrum of a Family of Circulant Preconditioned Toeplitz Systems," *SIAM J. Num. Anal. 26*, 503–506.

R.H. Chan (1991). "Preconditioners for Toeplitz Systems with Nonnegative Generating Functions," *IMA J. Num. Anal. 11*, 333–345.

R.H. Chan, J.G. Nagy, and R.J. Plemmons (1993). "FFT based Preconditioners for Toeplitz Block Least Squares Problems," *SIAM J. Num. Anal. 30*, 1740–1768.

R.H. Chan, J.G. Nagy, and R.J. Plemmons (1994). "Circulant Preconditioned Toeplitz Least Squares Iterations," *SIAM J. Matrix Anal. Appl. 15*, 80–97.

S.P. Chan and B.N. Parlett (1977). "Algorithm 517: A Program for Computing the Condition Numbers of Matrix Eigenvalues Without Computing Eigenvectors," *ACM Trans. Math. Soft. 3*, 186–203.

T.F. Chan (1982). "An Improved Algorithm for Computing the Singular Value Decomposition," *ACM Trans. Math. Soft. 8*, 72–83.

T.F. Chan (1984). "Deflated Decomposition Solutions of Nearly Singular Systems," *SIAM J. Num. Anal. 21*, 738–754.

T.F. Chan (1985). "On the Existence and Computation of LU Factorizations with small pivots," *Math. Comp. 42*, 535–548.

T.F. Chan (1987). "Rank Revealing QR Factorizations," *Lin. Alg. and Its Applic. 88/89*, 67–82.

T.F. Chan (1988). "An Optimal Circulant Preconditioner for Toeplitz Systems," *SIAM. J. Sci. Stat. Comp. 9*, 766–771.

T.F. Chan (1991). "Fourier Analysis of Relaxed Incomplete Factorization Preconditioners," *SIAM J. Sci. Statist. Comput. 12*, 668–680.

T.F. Chan and P. Hansen (1992). "A Look-Ahead Levinson Algorithm for Indefinite Toeplitz Systems," *SIAM J. Matrix Anal. Appl. 13*, 490–506.

T.F. Chan and P. Hansen (1992). "Some Applications of the Rank Revealing QR Factorization," *SIAM J. Sci. and Stat. Comp. 13*, 727–741.

T.F. Chan, K.R. Jackson, and B. Zhu (1983). "Alternating Direction Incomplete Factorizations," *SIAM J. Numer. Anal. 20*, 239–257.

T.F. Chan and J.A. Olkin (1994). "Circulant Preconditioners for Toeplitz Block Matrices," *Numerical Algorithms 6*, 89–101.

T.F. Chan, J.A. Olkin, and D. Cooley (1992). "Solving Quadratically Constrained Least Squares Using Black Box Solvers," *BIT 32*, 481–495.

S. Chandrasekaren and I.C.F. Ipsen (1994). "On Rank-Revealing Factorizations," *SIAM J. Matrix Anal. Appl. 15*, 592–622.

S. Chandrasekaren and I.C.F. Ipsen (1994). "Backward Errors for Eigenvalue and Singular Value Decompositions," *Numer. Math. 68*, 215–223.

S. Chandrasekaren and I.C.F. Ipsen (1995). "On the Sensitivity of Solution Components in Linear Systems of Equations," *SIAM J. Matrix Anal. Appl. 16*, 93–112.

H.Y. Chang and M. Salama (1988). "A Parallel Householder Tridiagonalization Strategy Using Scattered Square Decomposition," *Parallel Computing 6*, 297-312.

J.P. Charlier, M. Vanbegin, P. Van Dooren (1988). "On Efficient Implementation of Kogbetliantz's Algorithm for Computing the Singular Value Decomposition," *Numer. Math. 52*, 279–300.

J.P. Charlier and P. Van Dooren (1987). "On Kogbetliantz's SVD Algorithm in the Presence of Clusters," *Lin. Alg. and Its Applic. 95*, 135–160.

B.A. Chartres and J.C. Geuder (1967). "Computable Error Bounds for Direct Solution of Linear Equations," *J. ACM 14*, 63–71.

F. Chatelin (1993). *Eigenvalues of Matrices*, John Wiley and Sons, New York.

S. Chen, J. Dongarra, and C. Hsuing (1984). "Multiprocessing Linear Algebra Algorithms on the Cray X-MP-2: Experiences with Small Granularity," *J. Parallel and Distributed Computing 1*, 22–31.

S. Chen, D. Kuck, and A. Sameh (1978). "Practical Parallel Band Triangular Systems Solvers," *ACM Trans. Math. Soft. 4*, 270–277.

K.H. Cheng and S. Sahni (1987). "VLSI Systems for Band Matrix Multiplication," *Parallel Computing 4*, 239–258.

R.C. Chin, T.A. Manteuffel, and J. de Pillis (1984). "ADI as a Preconditioning for Solving the Convection-Diffusion Equation," *SIAM J. Sci. and Stat. Comp. 5*, 281–299.

J. Choi, J.J. Dongarra, and D.W. Walker (1995). "Parallel Matrix Transpose Algorithms on Distributed Memory Concurrent Computers," *Parallel Computing 21*, 1387–1406.

M.T. Chu (1992). "Numerical Methods for Inverse Singular Value Problems," *SIAM J. Num. Anal. 29*, 885–903.

M.T. Chu, R.E. Funderlic, and G.H. Golub (1995). "A Rank-One Reduction Formula and Its Applications to Matrix Factorizations," *SIAM Review 37*, 512–530.

P.G. Ciarlet (1989). *Introduction to Numerical Linear Algebra and Optimisation*, Cambridge University Press.

A.K. Cline (1973). "An Elimination Method for the Solution of Linear Least Squares Problems," *SIAM J. Num. Anal. 10*, 283–89.

A.K. Cline (1976). "A Descent Method for the Uniform Solution to Overdetermined Systems of Equations," *SIAM J. Num. Anal. 13*, 293–309.

A.K. Cline, A.R. Conn, and C. Van Loan (1982). "Generalizing the LINPACK Condition Estimator," in *Numerical Analysis* , ed., J.P. Hennart, Lecture Notes in Mathematics no. 909, Springer-Verlag, New York.

A.K. Cline, G.H. Golub, and G.W. Platzman (1976). "Calculation of Normal Modes of Oceans Using a Lanczos Method," in *Sparse Matrix Computations* , ed. J.R. Bunch and D.J. Rose, Academic Press, New York, pp. 409–26.

A.K. Cline, C.B. Moler, G.W. Stewart, and J.H. Wilkinson (1979). "An Estimate for the Condition Number of a Matrix," *SIAM J. Num. Anal. 16*, 368-75.

A.K. Cline and R.K. Rew (1983). "A Set of Counter examples to Three Condition Number Estimators," *SIAM J. Sci. and Stat. Comp. 4*, 602–611.

R.E. Cline and R.J. Plemmons (1976). "L_2-Solutions to Underdetermined Linear Systems," *SIAM Review 18*, 92-106.

M. Clint and A. Jennings (1970). "The Evaluation of Eigenvalues and Eigenvectors of Real Symmetric Matrices by Simultaneous Iteration," *Comp. J. 13*, 76–80.

M. Clint and A. Jennings (1971). "A Simultaneous Iteration Method for the Unsymmetric Eigenvalue Problem," *J. Inst. Math. Applic. 8*, 111–21.

W.G. Cochrane (1968). "Errors of Measurement in Statistics," *Technometrics 10,* 637–66.

W.J. Cody (1988). "ALGORITHM 665 MACHAR: A Subroutine to Dynamically Determine Machine Parameters," *ACM Trans. Math. Soft. 14,* 303–311.

A.M. Cohen (1974). "A Note on Pivot Size in Gaussian Elimination," *Lin. Alg. and Its Applic. 8,* 361–68.

T.F. Coleman and Y. Li (1992). "A Globally and Quadratically Convergent Affine Scaling Method for Linear L_1 Problems," *Mathematical Programming, 56, Series A,* 189–222.

T.F. Coleman and D.C. Sorensen (1984). "A Note on the Computation of an Orthonormal Basis for the Null Space of a Matrix," *Mathematical Programming 29,* 234–242.

T.F. Coleman and C.F. Van Loan (1988). *Handbook for Matrix Computations,* SIAM Publications, Philadelphia, PA.

L. Colombet, Ph. Michallon, and D. Trystram (1996). "Parallel Matrix-Vector Product on Rings with a Minimum of Communication," *Parallel Computing 22,* 289–310.

P. Concus and G.H. Golub (1973). "Use of Fast Direct Methods for the Efficient Numerical Solution of Nonseparable Elliptic Equations," *SIAM J. Num. Anal. 10,* 1103–20.

P. Concus, G.H. Golub, and G. Meurant (1985). "Block Preconditioning for the Conjugate Gradient Method," *SIAM J. Sci. and Stat. Comp. 6,* 220–252.

P. Concus, G.H. Golub, and D.P. O'Leary (1976). " A Generalized Conjugate Gradient Method for the Numerical Solution of Elliptic Partial Differential Equations," in *Sparse Matrix Computations* , ed. J.R. Bunch and D.J. Rose, Academic Press, New York.

K. Connolly, J.J. Dongarra, D. Sorensen, and J. Patterson (1988). "Programming Methodology and Performance Issues for Advanced Computer Architectures," *Parallel Computing 5,* 41-58.

J.M. Conroy (1989). "A Note on the Parallel Cholesky Factorization of Wide Banded Matrices," *Parallel Computing 10,* 239–246.

S.D. Conte and C. de Boor (1980). *Elementary Numerical Analysis: An Algorithmic Approach,* 3rd ed., McGraw-Hill, New York.

J.E. Cope and B.W. Rust (1979). "Bounds on solutions of systems with accurate data," *SIAM J. Num. Anal. 16,* 950–63.

M. Costnard, M. Marrakchi, and Y. Robert (1988). "Parallel Gaussian Elimination on an MIMD Computer," *Parallel Computing 6,* 275–296.

M. Costnard, J.M. Muller, and Y. Robert (1986). "Parallel QR Decomposition of a Rectangular Matrix," *Numer. Math. 48,* 239–250.

M. Costnard and D. Trystram (1995). *Parallel Algorithms and Architectures,* International Thomson Computer Press, New York.

R.W. Cottle (1974). "Manifestations of the Schur Complement," *Lin. Alg. and Its Applic. 8,* 189-211.

M.G. Cox (1981). "The Least Squares Solution of Overdetermined Linear Equations having Band or Augmented Band Structure," *IMA J. Num. Anal. 1,* 3–22.

C.R. Crawford (1973). "Reduction of a Band Symmetric Generalized Eigenvalue Problem," *Comm. ACM 16,* 41–44.

C.R. Crawford (1976). "A Stable Generalized Eigenvalue Problem," *SIAM J. Num. Anal. 13,* 854–60.

C.R. Crawford (1986). "Algorithm 646 PDFIND: A Routine to Find a Positive Definite Linear Combination of Two Real Symmetric Matrices," *ACM Trans. Math. Soft. 12,* 278–282.

C.R. Crawford and Y.S. Moon (1983). "Finding a Positive Definite Linear Combination of Two Hermitian Matrices," *Lin. Alg. and Its Applic. 51,* 37–48.

S. Crivelli and E.R. Jessup (1995). "The Cost of Eigenvalue Computation on Distributed Memory MIMD Computers," *Parallel Computing 21,* 401–422.

C.W. Cryer (1968). "Pivot Size in Gaussian Elimination," *Numer. Math. 12,* 335–45.

J. Cullum (1978). "The Simultaneous Computation of a Few of the Algebraically Largest and Smallest Eigenvalues of a Large Sparse Symmetric Matrix," *BIT 18*, 265–75.

J. Cullum and W.E. Donath (1974). "A Block Lanczos Algorithm for Computing the q Algebraically Largest Eigenvalues and a Corresponding Eigenspace of Large Sparse Real Symmetric Matrices," *Proc. of the 1974 IEEE Conf. on Decision and Control*, Phoenix, Arizona, pp. 505–9.

J. Cullum and R.A. Willoughby (1977). "The Equivalence of the Lanczos and the Conjugate Gradient Algorithms," IBM Research Report RE-6903.

J. Cullum and R.A. Willoughby (1979). "Lanczos and the Computation in Specified Intervals of the Spectrum of Large, Sparse Real Symmetric Matrices, in *Sparse Matrix Proc.* , 1978, ed. I.S. Duff and G.W. Stewart, SIAM Publications, Philadelphia, PA.

J. Cullum and R. Willoughby (1980). "The Lanczos Phenomena: An Interpretation Based on Conjugate Gradient Optimization," *Lin. Alg. and Its Applic. 29*, 63–90.

J. Cullum and R.A. Willoughby (1985). *Lanczos Algorithms for Large Symmetric Eigenvalue Computations, Vol. I Theory*, Birkhaüser, Boston.

J. Cullum and R.A. Willoughby (1985). *Lanczos Algorithms for Large Symmetric Eigenvalue Computations, Vol. II Programs*, Birkhaüser, Boston.

J. Cullum and R.A. Willoughby, eds. (1986). *Large Scale Eigenvalue Problems*, North-Holland, Amsterdam.

J. Cullum, R.A. Willoughby, and M. Lake (1983). "A Lanczos Algorithm for Computing Singular Values and Vectors of Large Matrices," *SIAM J. Sci. and Stat. Comp. 4*, 197–215.

J.J.M. Cuppen (1981). "A Divide and Conquer Method for the Symmetric Eigenproblem," *Numer. Math. 36*, 177–95.

J.J.M. Cuppen (1983). "The Singular Value Decomposition in Product Form," *SIAM J. Sci. and Stat. Comp. 4*, 216–222.

J.J.M. Cuppen (1984). "On Updating Triangular Products of Householder Matrices," *Numer. Math. 45*, 403-410.

E. Cuthill (1972). "Several Strategies for Reducing the Bandwidth of Matrices," in *Sparse Matrices and Their Applications*, ed. D.J. Rose and R.A. Willoughby, Plenum Press, New York.

G. Cybenko (1978). "Error Analysis of Some Signal Processing Algorithms," Ph.D. thesis, Princeton University.

G. Cybenko (1980). "The Numerical Stability of the Levinson-Durbin Algorithm for Toeplitz Systems of Equations," *SIAM J. Sci. and Stat. Comp. 1*, 303-19.

G. Cybenko (1984). "The Numerical Stability of the Lattice Algorithm for Least Squares Linear Prediction Problems," *BIT 24*, 441–455.

G. Cybenko (1985). "Computing Pisarenko Frequency Estimates," in *Proceedings of the Princeton Conference on Information Science and Systems*, Dept. of Electrical Engineering, Princeton University.

G. Cybenko and M. Berry (1990). "Hyperbolic Householder Algorithms for Factoring Structured Matrices," *SIAM J. Matrix Anal. Appl. 11*, 499–520.

G. Cybenko and C. Van Loan (1986). "Computing the Minimum Eigenvalue of a Symmetric Positive Definite Toeplitz Matrix," *SIAM J. Sci. and Stat. Comp. 7*, 123–131.

K. Dackland, E. Elmroth, and B. Kågström (1992). "Parallel Block Factorizations on the Shared Memory Multiprocessor IBM 3090 VF/600J," *International J. Supercomputer Applications, 6*, 69–97.

J. Daniel, W.B. Gragg, L.Kaufman, and G.W. Stewart (1976). "Reorthogonalization and Stable Algorithms for Updating the Gram-Schmidt QR Factorization," *Math. Comp. 30*, 772–795.

B. Danloy (1976). "On the Choice of Signs for Householder Matrices," *J. Comp. Appl. Math. 2*, 67–69.

B.N. Datta (1989). "Parallel and Large-Scale Matrix Computations in Control: Some Ideas," *Lin. Alg. and Its Applic. 121*, 243–264.

B.N. Datta (1995). *Numerical Linear Algebra and Applications*. Brooks/Cole Publishing Company, Pacific Grove, California.

B.N. Datta, C.R. Johnson. M.A. Kaashoek, R. Plemmons, and E.D. Sontag, eds. (1988), *Linear Algebra in Signals, Systems, and Control*, SIAM Publications, Philadelphia, PA.

K. Datta (1988). "The Matrix Equation $XA - BX = R$ and Its Applications," *Lin. Alg. and Its Applic. 109*, 91–105.

C. Davis and W.M. Kahan (1970). "The Rotation of Eigenvectors by a Perturbation, III," *SIAM J. Num. Anal. 7*, 1–46.

D. Davis (1973). "Explicit Functional Calculus," *Lin. Alg. and Its Applic. 6*, 193–99.

G.J. Davis (1986). "Column LU Pivoting on a Hypercube Multiprocessor," *SIAM J. Alg. and Disc. Methods 7*, 538–550.

J. Day and B. Peterson (1988). "Growth in Gaussian Elimination," *Amer. Math. Monthly 95*, 489–513.

A. Dax (1990). "The Convergence of Linear Stationary Iterative Processes for Solving Singular Unstructured Systems of Linear Equations," *SIAM Review 32*, 611–635.

C. de Boor (1979). "Efficient Computer Manipulation of Tensor Products," *ACM Trans. Math. Soft. 5*, 173–182.

C. de Boor and A. Pinkus (1977). "A Backward Error Analysis for Totally Positive Linear Systems," *Numer. Math. 27*, 485-90.

T. Dehn, M. Eiermann, K. Giebermann, and V. Sperling (1995). "Structured Sparse Matrix Vector Multiplication on Massively Parallel SIMD Architectures," *Parallel Computing 21*, 1867–1894.

P. Deift, J. Demmel, L.-C. Li, and C. Tomei (1991). "The Bidiagonal Singular Value Decomposition and Hamiltonian Mechanics," *SIAM J. Num. Anal. 28*, 1463–1516.

P. Deift, T. Nanda, and C. Tomei (1983). "Ordinary Differential Equations and the Symmetric Eigenvalue Problem," *SIAM J. Numer. Anal. 20*, 1–22.

T. Dekker and W. Hoffman (1989). "Rehabilitation of the Gauss-Jordan Algorithm," *Numer. Math. 54*, 591–599.

T.J. Dekker and J.F. Traub (1971). "The Shifted QR Algorithm for Hermitian Matrices," *Lin. Alg. and Its Applic. 4*, 137–54.

J.M. Delosme and I.C.F. Ipsen (1986). "Parallel Solution of Symmetric Positive Definite Systems with Hyperbolic Rotations," *Lin. Alg. and Its Applic. 77*, 75–112.

C.J. Demeure (1989). "Fast QR Factorization of Vandermonde Matrices," *Lin. Alg. and Its Applic. 122/123/124*, 165–194.

J.W. Demmel (1983). "A Numerical Analyst's Jordan Canonical Form," Ph.D. Thesis, Berkeley.

J.W. Demmel (1983). "The Condition Number of Equivalence Transformations that Block Diagonalize Matrix Pencils," *SIAM J. Numer. Anal. 20*, 599–610.

J.W. Demmel (1984). "Underflow and the Reliability of Numerical Software," *SIAM J. Sci. and Stat. Comp. 5*, 887–919.

J.W. Demmel (1987). "Three Methods for Refining Estimates of Invariant Subspaces," *Computing 38*, 43–57.

J.W. Demmel (1987). "On the Distance to the Nearest Ill-Posed Problem," *Numer. Math. 51*, 251–289.

J.W. Demmel (1987). "A Counterexample for two Conjectures About Stability," *IEEE Trans. Auto. Cont. AC-32*, 340–342.

J.W. Demmel (1987). "The smallest perturbation of a submatrix which lowers the rank and constrained total least squares problems, *SIAM J. Numer. Anal. 24*, 199–206.

J.W. Demmel (1988). "The Probability that a Numerical Analysis Problem is Difficult," *Math. Comp. 50*, 449–480.

J.W. Demmel (1992). "The Componentwise Distance to the Nearest Singular Matrix," *SIAM J. Matrix Anal. Appl. 13*, 10–19.

J.W. Demmel (1996). *Numerical Linear Algebra*, SIAM Publications, Philadelphia, PA.

J.W. Demmel and W. Gragg (1993). "On Computing Accurate Singular Values and Eigenvalues of Matrices with Acyclic Graphs," *Lin. Alg. and Its Applic. 185*, 203–217.

J.W. Demmel, M.T. Heath, and H.A. Van Der Vorst (1993) "Parallel Numerical Linear Algebra," in *Acta Numerica 1993*, Cambridge University Press.

J.W. Demmel and N.J. Higham (1992). "Stability of Block Algorithms with Fast Level-3 BLAS," *ACM Trans. Math. Soft. 18*, 274–291.

J.W. Demmel and N.J. Higham (1993). "Improved Error Bounds for Underdetermined System Solvers," *SIAM J. Matrix Anal. Appl. 14*, 1–14.

J.W. Demmel, N.J. Higham, and R.S. Schreiber (1995). "Stability of Block LU Factorization," *Numer. Lin. Alg. with Applic. 2*, 173–190.

J.W. Demmel and B. Kågström (1986). "Stably Computing the Kronecker Structure and Reducing Subspaces of Singular Pencils $A - \lambda B$ for Uncertain Data," in *Large Scale Eigenvalue Problems*, J. Cullum and R.A. Willoughby (eds), North-Holland, Amsterdam.

J.W. Demmel and B. Kågström (1987). "Computing Stable Eigendecompositions of Matrix Pencils," *Linear Alg. and Its Applic 88/89*, 139–186.

J.W. Demmel and B. Kågström (1988). "Accurate Solutions of Ill-Posed Problems in Control Theory," *SIAM J. Matrix Anal. Appl.* 126–145.

J.W. Demmel and W. Kahan (1990). "Accurate Singular Values of Bidiagonal Matrices," *SIAM J. Sci. and Stat. Comp. 11*, 873–912.

J.W. Demmel and K. Veselič (1992). "Jacobi's Method is More Accurate than QR," *SIAM J. Matrix Anal. Appl. 13*, 1204–1245.

B. De Moor and G.H. Golub (1991). "The Restricted Singular Value Decomposition: Properties and Applications," *SIAM J. Matrix Anal. Appl. 12*, 401–425.

B. De Moor and P. Van Dooren (1992). "Generalizing the Singular Value and QR Decompositions," *SIAM J. Matrix Anal. Appl. 13*, 993–1014.

J.E. Dennis and R.B. Schnabel (1983). *Numerical Methods for Unconstrained Optimization and Nonlinear Equations*, Prentice-Hall, Englewood Cliffs, NJ.

J.E. Dennis Jr. and K. Turner (1987). "Generalized Conjugate Directions," *Lin. Alg. and Its Applic. 88/89*, 187–209.

E.F. Deprettere, ed. (1988). *SVD and Signal Processing*. Elsevier, Amsterdam.

J. Descloux (1963). "Bounds for the Spectral Norm of Functions of Matrices," *Numer. Math. 5*, 185–90.

M.A. Diamond and D.L.V. Ferreira (1976). "On a Cyclic Reduction Method for the Solution of Poisson's Equation," *SIAM J. Num. Anal. 13*, 54–70.

S. Doi (1991). "On Parallelism and Convergence of Incomplete LU Factorizations," *Appl. Numer. Math. 7*, 417–436.

J.J. Dongarra (1983). "Improving the Accuracy of Computed Singular Values," *SIAM J. Sci. and Stat. Comp. 4*, 712–719.

J.J. Dongarra, J.R. Bunch, C.B. Moler, and G.W. Stewart (1979). *LINPACK Users Guide*, SIAM Publications, Philadelphia, PA.

J.J. Dongarra, J. Du Croz, I.S. Duff, and S.J. Hammarling (1990). "A Set of Level 3 Basic Linear Algebra Subprograms," *ACM Trans. Math. Soft. 16*, 1–17.

J.J. Dongarra, J. Du Croz, I.S. Duff, and S.J. Hammarling (1990). "Algorithm 679. A Set of Level 3 Basic Linear Algebra Subprograms: Model Implementation and Test Programs," *ACM Trans. Math. Soft. 16*, 18–28.

J.J. Dongarra, J. Du Croz, S. Hammarling, and R.J. Hanson (1988). "An Extended Set of Fortran Basic Linear Algebra Subprograms," *ACM Trans. Math. Soft. 14*, 1–17.

J.J. Dongarra, J. Du Croz, S. Hammarling, and R.J. Hanson (1988). "Algorithm 656 An Extended Set of Fortran Basic Linear Algebra Subprograms: Model Implementation and Test Programs," *ACM Trans. Math. Soft. 14*, 18–32.

J.J. Dongarra, I. Duff, P. Gaffney, and S. McKee, eds. (1989), *Vector and Parallel Computing*, Ellis Horwood, Chichester, England.

J.J. Dongarra, I. Duff, D. Sorensen, and H. van der Vorst (1990). *Solving Linear Systems on Vector and Shared Memory Computers*, SIAM Publications, Philadelphia, PA.

J.J. Dongarra and S. Eisenstat (1984). "Squeezing the Most Out of an Algorithm in Cray Fortran," *ACM Trans. Math. Soft. 10*, 221-230.

J.J. Dongarra, F.G. Gustavson, and A. Karp (1984). "Implementing Linear Algebra Algorithms for Dense Matrices on a Vector Pipeline Machine," *SIAM Review 26*, 91–112.

J.J. Dongarra, S. Hammarling, and D.C. Sorensen (1989). "Block Reduction of Matrices to Condensed Forms for Eigenvalue Computations," *JACM 27*, 215–227.

J.J. Dongarra, S. Hammarling, and J.H. Wilkinson (1992). "Numerical Considerations in Computing Invariant Subspaces," *SIAM J. Matrix Anal. Appl. 13*, 145–161.

J.J. Dongarra and T. Hewitt (1986). "Implementing Dense Linear Algebra Algorithms Using Multitasking on the Cray X-MP-4 (or Approaching the Gigaflop)," *SIAM J. Sci. and Stat. Comp. 7*, 347–350.

J.J. Dongarra and A. Hinds (1979). "Unrolling Loops in Fortran," *Software Practice and Experience 9*, 219–229.

J.J. Dongarra and R.E. Hiromoto (1984). "A Collection of Parallel Linear Equation Routines for the Denelcor HEP,' *Parallel Computing 1*, 133–142.

J.J. Dongarra, L. Kaufman, and S. Hammarling (1986). "Squeezing the Most Out of Eigenvalue Solvers on High Performance Computers," *Lin. Alg. and Its Applic. 77*, 113–136.

J.J. Dongarra, C.B. Moler, and J.H. Wilkinson (1983). "Improving the Accuracy of Computed Eigenvalues and Eigenvectors," *SIAM J. Numer. Anal. 20*, 23–46.

J.J. Dongarra and A.H. Sameh (1984). "On Some Parallel Banded System Solvers," *Parallel Computing 1*, 223–235.

J.J. Dongarra, A. Sameh, and D. Sorensen (1986). "Implementation of Some Concurrent Algorithms for Matrix Factorization," *Parallel Computing 3*, 25–34.

J.J. Dongarra and D.C. Sorensen (1986). "Linear Algebra on High Performance Computers," *Appl. Math. and Comp. 20*, 57–88.

J.J. Dongarra and D.C. Sorensen (1987). "A Portable Environment for Developing Parallel Programs," *Parallel Computing 5*, 175–186.

J.J. Dongarra and D.C. Sorensen (1987). "A Fully Parallel Algorithm for the Symmetric Eigenvalue Problem," *SIAM J. Sci. and Stat. Comp. 8*, S139–S154.

J.J. Dongarra and D. Walker (1995). "Software Libraries for Linear Algebra Computations on High Performance Computers," *SIAM Review 37*, 151–180.

F.W. Dorr (1970). "The Direct Solution of the Discrete Poisson Equation on a Rectangle," *SIAM Review 12*, 248–63.

F.W. Dorr (1973). "The Direct Solution of the Discrete Poisson Equation in $O(n^2)$ Operations," *SIAM Review 15*, 412–415.

C.C. Douglas, M. Heroux, G. Slishman, and R.M. Smith (1994). "GEMMW: A Portable Level 3 BLAS Winograd Variant of Strassen's Matrix-Matrix Multiply Algorithm," *J. Comput. Phys. 110*, 1–10.

Z. Drmač (1994). *The Generalized Singular Value Problem*, Ph.D. Thesis, FernUniversitat, Hagen, Germany.

Z. Dramăc, M. Omladič, and K. Veselič (1994). "On the Perturbation of the Cholesky Factorization," *SIAM J. Matrix Anal. Appl. 15*,1319–1332.

P.F. Dubois, A. Greenbaum, and G.H. Rodrigue (1979). "Approximating the Inverse of a Matrix for Use on Iterative Algorithms on Vector Processors," *Computing 22*, 257–268.

A.A. Dubrulle (1970). "A Short Note on the Implicit QL Algorithm for Symmetric Tridiagonal Matrices," *Numer. Math. 15*, 450.

A.A. Dubrulle and G.H. Golub (1994). "A Multishift QR Iteration Without Computation of the Shifts," *Numerical Algorithms 7*, 173–181.

A.A. Dubrulle, R.S. Martin, and J.H. Wilkinson (1968). "The Implicit QL Algorithm," *Numer. Math. 12*, 377-83. see also Wilkinson and Reinsch (1971, pp.241–48).

J.J. Du Croz and N.J. Higham (1992). "Stability of Methods for Matrix Inversion," *IMA J. Num. Anal. 12*, 1–19.

I.S. Duff (1974). "Pivot Selection and Row Ordering in Givens Reduction on Sparse Matrices," *Computing 13*, 239–48.

I.S. Duff (1977). "A Survey of Sparse Matrix Research," *Proc. IEEE 65*, 500–535.

I.S. Duff, ed. (1981). *Sparse Matrices and Their Uses*, Academic Press, New York.

I.S. Duff, A.M. Erisman, and J.K. Reid (1986). *Direct Methods for Sparse Matrices*, Oxford University Press.

I.S. Duff, N.I.M. Gould, J.K. Reid, J.A. Scott, and K. Turner (1991). "The Factorization of Sparse Indefinite Matrices," *IMA J. Num. Anal. 11*, 181–204.

I.S. Duff and G. Meurant (1989). "The Effect of Ordering on Preconditioned Conjugate Gradients," *BIT 29*, 635–657.

I.S. Duff and J.K. Reid (1975). "On the Reduction of Sparse Matrices to Condensed Forms by Similarity Transformations," *J. Inst. Math. Applic. 15*, 217–24.

I.S. Duff and J.K. Reid (1976). "A Comparison of Some Methods for the Solution of Sparse Over-Determined Systems of Linear Equations," *J. Inst. Math. Applic. 17*, 267–80.

I.S. Duff and G.W. Stewart, eds. (1979). *Sparse Matrix Proceedings, 1978*, SIAM Publications, Philadelphia, PA.

N. Dunford and J. Schwartz (1958). *Linear Operators, Part I*, Interscience, New York.

J. Durbin (1960). "The Fitting of Time Series Models," *Rev. Inst. Int. Stat. 28* 233–43.

P.J. Eberlein (1965). "On Measures of Non-Normality for Matrices," *Amer. Math. Soc. Monthly 72*, 995–96.

P.J. Eberlein (1970). "Solution to the Complex Eigenproblem by a Norm-Reducing Jacobi-type Method," *Numer. Math. 14*, 232–45. See also Wilkinson and Reinsch (1971, pp.404–17).

P.J. Eberlein (1971). "On the Diagonalization of Complex Symmetric Matrices," *J. Inst. Math. Applic. 7*, 377–83.

P.J. Eberlein (1987). "On Using the Jacobi Method on a Hypercube," in *Hypercube Multiprocessors*, ed. M.T. Heath, SIAM Publications, Philadelphia, PA.

P.J. Eberlein and C.P. Huang (1975). "Global Convergence of the QR Algorithm for Unitary Matrices with Some Results for Normal Matrices," *SIAM J. Numer. Anal. 12*, 421–453.

C. Eckart and G. Young (1939). "A Principal Axis Transformation for Non-Hermitian Matrices," *Bull. Amer. Math. Soc. 45*, 118–21.

A. Edelman (1992). "The Complete Pivoting Conjecture for Gaussian Elimination is False," *The Mathematica Journal 2*, 58–61.

A. Edelman (1993). "Large Dense Numerical Linear Algebra in 1993: The Parallel Computing Influence," *Int'l J. Supercomputer Appl. 7*, 113–128.

A. Edelman, E. Elmroth, and B. Kågström (1996). "A Geometric Approach to Perturbation Theory of Matrices and Matrix Pencils," *SIAM J. Matrix Anal.*, to appear.

A. Edelman and W. Mascarenhas (1995). "On the Complete Pivoting Conjecture for a Hadamard Matrix of Order 12," *Linear and Multilinear Algebra 38*, 181–185.

A. Edelman and H. Murakami (1995). "Polynomial Roots from Companion Matrix Eigenvalues," *Math. Comp. 64*, 763–776.

M. Eiermann and W. Niethammer (1983). "On the Construction of Semi-iterative Methods," *SIAM J. Numer. Anal. 20*, 1153–1160.

M. Eiermann, W. Niethammer, and R.S. Varga (1992). "Acceleration of Relaxation Methods for Non-Hermitian Linear Systems," *SIAM J. Matrix Anal. Appl. 13*, 979–991.

M. Eiermann and R.S. Varga (1993). "Is the Optimal ω Best for the SOR Iteration Method," *Lin. Alg. and Its Applic. 182*, 257–277.

V. Eijkhout (1991). "Analysis of Parallel Incomplete Point Factorizations," *Lin. Alg. and Its Applic. 154–156*, 723–740.

S.C. Eisenstat (1984). "Efficient Implementation of a Class of Preconditioned Conjugate Gradient Methods," *SIAM J. Sci. and Stat. Computing 2*, 1–4.

S.C. Eisenstat, H. Elman, and M. Schultz (1983). "Variational Iterative Methods for Nonsymmetric Systems of Equations," *SIAM J. Num. Anal. 20*, 345–357.

S.C. Eisenstat, M.T. Heath, C.S. Henkel, and C.H. Romine (1988). "Modified Cyclic Algorithms for Solving Triangular Systems on Distributed Memory Multiprocessors," *SIAM J. Sci. and Stat. Comp. 9*, 589–600.

L. Eldén (1977). "Algorithms for the Regularization of Ill-Conditioned Least Squares Problems," *BIT 17*, 134–45.

L. Eldén (1980). "Perturbation Theory for the Least Squares Problem with Linear Equality Constraints," *SIAM J. Num. Anal. 17*, 338–50.

L. Eldén (1983). "A Weighted Pseudoinverse, Generalized Singular Values, and Constrained Least Squares Problems," *BIT 22* , 487–502.

L. Eldén (1984). "An Algorithm for the Regularization of Ill-Conditioned, Banded Least Squares Problems," *SIAM J. Sci. and Stat. Comp. 5*, 237–254.

L. Eldén (1985). "A Note on the Computation of the Generalized Cross-Validation Function for Ill-Conditioned Least Squares Problems," *BIT 24*, 467–472.

L. Eldén and H. Park (1994). "Perturbation Analysis for Block Downdating of a Cholesky Decomposition," *Numer. Math. 68*, 457–468.

L. Eldén and H. Park (1994). "Block Downdating of Least Squares Solutions," *SIAM J. Matrix Anal. Appl. 15*, 1018–1034.

L. Eldén and R. Schreiber (1986). "An Application of Systolic Arrays to Linear Discrete Ill-Posed Problems," *SIAM J. Sci. and Stat. Comp. 7*, 892–903.

H. Elman (1986). "A Stability Analysis of Incomplete LU Factorization," *Math. Comp. 47*, 191–218.

H. Elman (1989). "Approximate Schur Complement Preconditioners on Serial and Parallel Computers," *SIAM J. Sci. Stat. Comput. 10*, 581–605.

H. Elman (1996). "Fast Nonsymmetric Iterations and Preconditioning for Navier-Stokes Equations," *SIAM J. Sci. Comput. 17*, 33–46.

H. Elman and G.H. Golub (1990). "Iterative Methods for Cyclically Reduced Non-Self-Adjoint Linear Systems I," *Math. Comp. 54*, 671–700.

H. Elman and G.H. Golub (1990). "Iterative Methods for Cyclically Reduced Non-Self-Adjoint Linear Systems II," *Math. Comp. 56*, 215–242.

E. Elmroth and B. Kågström (1996). "The Set of 2-by-3 Matrix Pencils–Kronecker Structure and their Transitions under Perturbations," *SIAM J. Matrix Anal.*, to appear.

L. Elsner and J.-G. Sun (1982). "Perturbation Theorems for the Generalized Eigenvalue Problem,; *Lin. Alg. and its Applic. 48*, 341-357.

W. Enright (1979). "On the Efficient and Reliable Numerical Solution of Large Linear Systems of O.D.E.'s," *IEEE Trans. Auto. Cont.* AC-24, 905–8.

W. Enright and S. Serbin (1978). "A Note on the Efficient Solution of Matrix Pencil Systems," *BIT 18*, 276–81.

I. Erdelyi (1967). "On the Matrix Equation $Ax = \lambda Bx$," *J. Math. Anal. and Applic. 17*, 119–32.

T. Ericsson and A. Ruhe (1980). "The Spectral Transformation Lanczos Method for the Numerical Solution of Large Sparse Generalized Symmetric Eigenvalue Problems," *Math. Comp. 35*, 1251–68.

A.M. Erisman and J.K. Reid (1974). "Monitoring the Stability of the Triangular Factorization of a Sparse Matrix," *Numer. Math. 22*, 183–86.

J. Erxiong (1990). "An Algorithm for Finding Generalized Eigenpairs of a Symmetric Definite Matrix Pencil," *Lin.Alg. and Its Applic. 132*, 65–91.

J. Erxiong (1992). "A Note on the Double-Shift QL Algorithm," *Lin.Alg. and Its Applic. 171*,.121–132.

D.J. Evans (1984). "Parallel SOR Iterative Methods," *Parallel Computing 1*, 3–18.

D.J. Evans and R. Dunbar (1983). "The Parallel Solution of Triangular Systems of Equations," *IEEE Trans. Comp.* C-32, 201–204.

L.M. Ewerbring and F.T. Luk (1989). "Canonical Correlations and Generalized SVD: Applications and New Algorithms," *J. Comput. Appl. Math. 27*, 37–52.

V. Faber and T. Manteuffel (1984). "Necessary and Sufficient Conditions for the Existence of a Conjugate Gradient Method," *SIAM J. Numer. Anal. 21* 352–362.

V.N. Faddeeva (1959). *Computational Methods of Linear Algebra*, Dover, New York.

V. Fadeeva and D. Fadeev (1977). "Parallel Computations in Linear Algebra," *Kibernetica 6*, 28–40.

W. Fair and Y. Luke (1970). "Padé Approximations to the Operator Exponential," *Numer. Math. 14*, 379–82.

R.W. Farebrother (1987). *Linear Least Squares Computations*, Marcel Dekker, New York.

D.G. Feingold and R.S. Varga (1962). "Block Diagonally Dominant Matrices and Generalizations of the Gershgorin Circle Theorem," *Pacific J. Math. 12*, 1241–50.

T. Fenner and G. Loizou (1974). "Some New Bounds on the Condition Numbers of Optimally Scaled Matrices," *J. ACM 21*, 514–24.

W.E. Ferguson (1980). "The Construction of Jacobi and Periodic Jacobi Matrices with Prescribed Spectra," *Math. Comp. 35*, 1203–1220.

K.V. Fernando (1989). "Linear Convergence of the Row Cyclic Jacobi and Kogbetliantz methods," *Numer. Math. 56*, 73–92.

K.V. Fernando and B.N. Parlett (1994). "Accurate Singular Values and Differential qd Algorithms," *Numer. Math. 67*, 191–230.

W.R. Ferng, G.H. Golub, and R.J. Plemmons (1991). "Adaptive Lanczos Methods for Recursive Condition Estimation," *Numerical Algorithms 1*, 1-20.

R.D. Fierro and J.R. Bunch (1994). "Collinearity and Total Least Squares," *SIAM J. Matrix Anal. Appl. 15*, 1167–1181.

R.D. Fierro and P.C. Hansen (1995). "Accuracy of TSVD Solutions Computed from Rank-Revealing Decompositions," *Numer. Math. 70*, 453–472.

C. Fischer and R.A. Usmani (1969). "Properties of Some Tridiagonal Matrices and Their Application to Boundary Value Problems," *SIAM J. Num. Anal. 6*, 127–42.

G. Fix and R. Heiberger (1972). "An Algorithm for the Ill-Conditioned Generalized Eigenvalue Problem," *SIAM J. Num. Anal. 9*, 78–88.

U. Flaschka, W-W. Li, and J-L. Wu (1992). "A KQZ Algorithm for Solving Linear-Response Eigenvalue Equations," *Lin. Alg. and Its Applic. 165*, 93–123.

R. Fletcher (1976). "Factorizing Symmetric Indefinite Matrices," *Lin. Alg. and Its Applic. 14*, 257–72.

A. Forsgren (1995). "On Linear Least-Squares Problems with Diagonally Dominant Weight Matrices," Technical Report TRITA-MAT-1995-OS2, Department of Mathematics, Royal Institute of Technology, S-100 44, Stockholm, Sweden.

G.E. Forsythe (1960). "Crout with Pivoting," *Comm. ACM 3*, 507–8.

G.E. Forsythe and G.H. Golub (1965). "On the Stationary Values of a Second-Degree Polynomial on the Unit Sphere," *SIAM J. App. Math. 13*, 1050–68.

G.E. Forsythe and P. Henrici (1960). "The Cyclic Jacobi Method for Computing the Principal Values of a Complex Matrix," *Trans. Amer. Math. Soc. 94*, 1–23.

G.E. Forsythe and C. Moler (1967). *Computer Solution of Linear Algebraic Systems*, Prentice-Hall, Englewood Cliffs, NJ.

L.V. Foster (1986). "Rank and Null Space Calculations Using Matrix Decomposition without Column Interchanges," *Lin. Alg. and Its Applic. 74*, 47–71.

L.V. Foster (1994). "Gaussian Elimination with Partial Pivoting Can Fail in Practice," *SIAM J. Matrix Anal. Appl. 15*, 1354–1362.

R. Fourer (1984). "Staircase Matrices and Systems," *SIAM Review 26*, 1–71.

L. Fox (1964). *An Introduction to Numerical Linear Algebra*, Oxford University Press, Oxford, England.

G.C. Fox, ed. (1988). *The Third Conference on Hypercube Concurrent Computers and Applications, Vol. II – Applications*, ACM Press, New York.

G.C. Fox, M. A. Johnson, G. A. Lyzenga, S. W. Otto, J. K. Salmon and D. W. Walker (1988). *Solving Problems on Concurrent Processors, Volume 1*. Prentice Hall, Englewood Cliffs, NJ.

G.C. Fox, S.W. Otto, and A.J. Hey (1987). "Matrix Algorithms on a Hypercube I: Matrix Multiplication," *Parallel Computing 4*, 17–31.

G.C. Fox, R.D. Williams, and P. C. Messina (1994). *Parallel Computing Works!*, Morgan Kaufmann, San Francisco.

J.S. Frame (1964). "Matrix Functions and Applications, Part II," *IEEE Spectrum 1 (April)*, 102–8.

J.S. Frame (1964). "Matrix Functions and Applications, Part IV," *IEEE Spectrum 1 (June)*, 123–31.

J.G.F. Francis (1961). "The QR Transformation: A Unitary Analogue to the LR Transformation, Parts I and II" *Comp. J. 4*, 265–72, 332–45.

J.N. Franklin (1968). *Matrix Theory* Prentice Hall, Englewood Cliffs, NJ.

T. L. Freeman and C. Phillips (1992). *Parallel Numerical Algorithms*, Prentice Hall, New York.

R.W. Freund (1990). "On Conjugate Gradient Type Methods and Polynomial Preconditioners for a Class of Complex Non-Hermitian Matrices," *Numer. Math. 57*, 285–312.

R.W. Freund (1992). "Conjugate Gradient-Type Methods for Linear Systems with Complex Symmetric Coefficient Matrices," *SIAM J. Sci. Statist. Comput. 13*, 425–448.

R.W. Freund (1993). "A Transpose-Free Quasi-Minimum Residual Algorithm for Non-hermitian Linear System," *SIAM J. Sci. Comput. 14*, 470–482.

R.W. Freund and N. Nachtigal (1991). "QMR: A Quasi-Minimal Residual Method for Non-Hermitian Linear Systems," *Numer. Math. 60*, 315–339.

R.W. Freund and N.M. Nachtigal (1994). "An Implementation of the QMR Method Based on Coupled Two-term Recurrences," *SIAM J. Sci. Comp. 15*, 313–337.

R.W. Freund, G.H. Golub, and N. Nachtigal (1992). "Iterative Solution of Linear Systems," *Acta Numerica 1*, 57–100.

R.W. Freund, M. Gutknecht, and N. Nachtigal (1993). "An Implementation of the Look-Ahead Lanczos Algorithm for Non-Hermitian Matrices," *SIAM J. Sci. and Stat.Comp. 14*, 137–158.

R.W. Freund and H. Zha (1993). "A Look-Ahead Algorithm for the Solution of General Hankel Systems," *Numer. Math. 64*, 295–322.

S. Friedland (1975). "On Inverse Multiplicative Eigenvalue Problems for Matrices," *Lin. Alg. and Its Applic. 12*, 127–38.

S. Friedland (1991). "Revisiting Matrix Squaring," *Lin. Alg. and Its Applic. 154-156*, 59–63.

S. Friedland, J. Nocedal, and M.L. Overton (1987). "The Formulation and Analysis of Numerical Methods for Inverse Eigenvalue Problems," *SIAM J. Numer. Anal. 24*, 634–667.

C.E. Froberg (1965). "On Triangularization of Complex Matrices by Two Dimensional Unitary Tranformations," *BIT 5*, 230–34.

R.E. Funderlic and A. Geist (1986). "Torus Data Flow for Parallel Computation of Missized Matrix Problems," *Lin. Alg. and Its Applic. 77*, 149–164.

G. Galimberti and V. Pereyra (1970). "Numerical Differentiation and the Solution of Multidimensional Vandermonde Systems," *Math. Comp. 24*, 357–64.

G. Galimberti and V. Pereyra (1971). "Solving Confluent Vandermonde Systems of Hermitian Type," *Numer. Math. 18*, 44–60.

K.A Gallivan, M. Heath, E. Ng, J. Ortega, B. Peyton, R. Plemmons, C. Romine, A. Sameh, and B. Voigt (1990), *Parallel Algorithms for Matrix Computations*, SIAM Publications, Philadelphia, PA.

K.A. Gallivan, W. Jalby, and U. Meier (1987). "The Use of BLAS3 in Linear Algebra on a Parallel Processor with a Hierarchical Memory," *SIAM J. Sci. and Stat. Comp. 8*, 1079–1084.

K.A. Gallivan, W. Jalby, U. Meier, and A.H. Sameh (1988). "Impact of Hierarchical Memory Systems on Linear Algebra Algorithm Design," *Int'l J. Supercomputer Applic. 2*, 12-48.

K.A. Gallivan, R.J. Plemmons, and A.H. Sameh (1990). "Parallel Algorithms for Dense Linear Algebra Computations," *SIAM Review 32*, 54–135.

E. Gallopoulos and Y. Saad (1989). "A Parallel Block Cyclic Reduction Algorithm for the Fast Solution of Elliptic Equations," *Parallel Computing 10*, 143–160.

W. Gander (1981). "Least Squares with a Quadratic Constraint," *Numer. Math. 36*, 291–307.

W. Gander, G.H. Golub, and U. von Matt (1991). "A Constrained Eigenvalue Problem," in *Numerical Linear Algebra, Digital Signal Processing, and Parallel Algorithms,* G.H. Golub and P. Van Dooren (eds), Springer-Verlag, Berlin.

D. Gannon and J. Van Rosendale (1984). "On the Impact of Communication Complexity on the Design of Parallel Numerical Algorithms," *IEEE Trans. Comp. C-33,* 1180–1194.

F.R. Gantmacher (1959). *The Theory of Matrices, vols. 1 and 2,* Chelsea, New York.

B.S. Garbow, J.M. Boyle, J.J. Dongarra, and C.B. Moler (1972). *Matrix Eigensystem Routines: EISPACK Guide Extension,* Lecture Notes in Computer Science, Volume 51, Springer-Verlag, New York.

J. Gardiner, M.R. Wette, A.J. Laub, J.J. Amato, and C.B. Moler (1992). "Algorithm 705: A FORTRAN-77 Software Package for Solving the Sylvester Matrix Equation $AXB^T + CXD^T = E$," *ACM Trans. Math. Soft. 18,* 232–238.

W. Gautschi (1975). "Norm Estimates for Inverses of Vandermonde Matrices," *Numer. Math. 23,* 337–47.

W. Gautschi (1975). "Optimally Conditioned Vandermonde Matrices," *Numer. Math. 24,* 1–12.

G.A. Geist (1991). "Reduction of a General Matrix to Tridiagonal Form," *SIAM J. Matrix Anal. Appl. 12,* 362–373.

G.A. Geist and M.T. Heath (1986). "Matrix Factorization on a Hypercube," in M.T. Heath (ed) (1986). *Proceedings of First SIAM Conference on Hypercube Multiprocessors,* SIAM Publications, Philadelphia, PA.

G.A. Geist and C.H. Romine (1988). "LU Factorization Algorithms on Distributed Memory Multiprocessor Architectures," *SIAM J. Sci. and Stat. Comp. 9,* 639–649.

W.M. Gentleman (1973). "Least Squares Computations by Givens Transformations without Square Roots," *J. Inst. Math. Appl. 12,* 329–36.

W.M. Gentleman (1973). "Error Analysis of QR Decompositions by Givens Transformations," *Lin. Alg. and Its Applic. 10,* 189–97.

W.M. Gentleman and H.T. Kung (1981). "Matrix Triangularization by Systolic Arrays," SPIE Proceedings, Vol. 298, 19–26.

J.A. George (1973). "Nested Dissection of a Regular Finite Element Mesh," *SIAM J. Num. Anal. 10,* 345–63.

J.A. George (1974). "On Block Elimination for Sparse Linear Systems," *SIAM J. Num. Anal. 11,* 585–603.

J.A. George and M.T. Heath (1980). "Solution of Sparse Linear Least Squares Problems Using Givens Rotations," *Lin. Alg. and Its Applic. 34,* 69–83.

A. George, M.T. Heath, and J. Liu (1986). "Parallel Cholesky Factorization on a Shared Memory Multiprocessor," *Lin. Alg. and Its Applic. 77,* 165–187.

A. George and J. W-H. Liu (1981). *Computer Solution of Large Sparse Positive Definite Systems.* Prentice-Hall Inc., Englewood Cliffs, New Jersey.

A.R. Ghavimi and A.J. Laub (1995). "Residual Bounds for Discrete-Time Lyapunov Equations," *IEEE Trans. Auto. Cont. 40,* 1244–1249.

N.E. Gibbs and W.G. Poole, Jr. (1974). "Tridiagonalization by Permutations," *Comm. ACM 17,* 20–24.

N.E. Gibbs, W.G. Poole, Jr., and P.K. Stockmeyer (1976). "An Algorithm for Reducing the Bandwidth and Profile of a Sparse Matrix," *SIAM J. Num. Anal. 13,* 236–50.

N.E. Gibbs, W.G. Poole, Jr., and P.K. Stockmeyer (1976). "A Comparison of Several Bandwidth and Profile Reduction Algorithms," *ACM Trans. Math. Soft. 2,* 322–30.

P.E. Gill, G.H. Golub, W. Murray, and M.A. Saunders (1974). "Methods for Modifying Matrix Factorizations," *Math. Comp. 28,* 505-35.

P.E. Gill and W. Murray (1976). "The Orthogonal Factorization of a Large Sparse Matrix," in *Sparse Matrix Computations,* ed. J.R. Bunch and D.J. Rose, Academic Press, New York, pp. 177-200.

P.E. Gill, W. Murray, D.B. Ponceleón, and M.A. Saunders (1992). "Preconditioners for Indefinite Systems Arising in Optimization," *SIAM J. Matrix Anal. Appl. 13,* 292–311.

P.E. Gill, W. Murray, and M.A. Saunders (1975). "Methods for Computing and Modifying the LDV Factors of a Matrix," *Math. Comp. 29*, 1051–77.

P.E. Gill, W. Murray, and M.H. Wright (1991). *Numerical Linear Algebra and Optimization, Vol. 1*, Addison-Wesley, Reading, MA.

W. Givens (1958). "Computation of Plane Unitary Rotations Transforming a General Matrix to Triangular Form," *SIAM J. App. Math. 6*, 26–50.

J. Gluchowska and A. Smoktunowicz (1990). "Solving the Linear Least Squares Problem with Very High Relative Accuracy," *Computing 45*, 345–354.

I.C. Gohberg and M.G. Krein (1969). *Introduction to the Theory of Linear Non-Self Adjoint Operators* , Amer. Math. Soc., Providence, R.I.

I.C. Gohberg, P. Lancaster, and L. Rodman (1986). *Invariant Subspaces of Matrices With Applications*, John Wiley and Sons, New York.

D. Goldberg (1991). "What Every Computer Scientist Should Know About Floating Point Arithmetic," *ACM Surveys 23*, 5–48.

D. Goldfarb (1976). "Factored Variable Metric Methods for Unconstrained Optimization," *Math. Comp. 30*, 796–811.

H.H. Goldstine and L.P. Horowitz (1959). "A Procedure for the Diagonalization of Normal Matrices," *J. Assoc. Comp. Mach. 6*, 176–95.

G.H. Golub (1965). "Numerical Methods for Solving Linear Least Squares Problems," *Numer. Math. 7*, 206–16.

G.H. Golub (1969). "Matrix Decompositions and Statistical Computation," in *Statistical Computation* , ed. R.C. Milton and J.A. Nelder, Academic Press, New York, pp. 365–97.

G.H. Golub (1973). "Some Modified Matrix Eigenvalue Problems," *SIAM Review 15*, 318–334.

G.H. Golub (1974). "Some Uses of the Lanczos Algorithm in Numerical Linear Algebra," in *Topics in Numerical Analysis*, ed., J.J.H. Miller, Academic Press, New York.

G.H. Golub, M. Heath, and G. Wahba (1979). "Generalized Cross-Validation as a Method for Choosing a Good Ridge Parameter," *Technometrics 21*, 215–23.

G.H. Golub, A. Hoffman, and G.W. Stewart (1988). "A Generalization of the Eckart-Young-Mirsky Approximation Theorem." *Lin. Alg. and Its Applic. 88/89*, 317–328.

G.H. Golub and W. Kahan (1965). "Calculating the Singular Values and Pseudo-Inverse of a Matrix," *SIAM J. Num. Anal. 2*, 205–24.

G.H. Golub, V. Klema and G.W. Stewart (1976). "Rank Degeneracy and Least Squares Problems," Technical Report TR-456, Department of Computer Science, University of Maryland, College Park, MD.

G.H. Golub, F.T. Luk, and M. Overton (1981). "A Block Lanczos Method for Computing the Singular Values and Corresponding Singular Vectors of a Matrix," *ACM Trans. Math. Soft. 7*, 149–69.

G.H. Golub and G. Meurant (1983). *Résolution Numérique des Grandes Systèmes Linéaires*, Collection de la Direction des Etudes et Recherches de l'Electricité de France, vol. 49, Eyolles, Paris.

G.H. Golub and C.D. Meyer (1986). "Using the QR Factorization and Group Inversion to Compute, Differentiate, and estimate the Sensitivity of Stationary Probabilities for Markov Chains," *SIAM J. Alg. and Dis. Methods, 7*, 273–281.

G.H. Golub, S. Nash, and C. Van Loan (1979). "A Hessenberg-Schur Method for the Matrix Problem $AX + XB = C$," *IEEE Trans. Auto. Cont. AC-24*, 909–13.

G.H. Golub and D. O'Leary (1989). "Some History of the Conjugate Gradient and Lanczos Methods," *SIAM Review 31*, 50–102.

G.H. Golub and J.M. Ortega (1993). *Scientific Computing: An Introduction with Parallel Computing*, Academic Press, Boston.

G.H. Golub and M. Overton (1988). "The Convergence of Inexact Chebychev and Richardson Iterative Methods for Solving Linear Systems," *Numer. Math. 53*, 571–594.

G.H. Golub and V. Pereyra (1973). "The Differentiation of Pseudo-Inverses and Nonlinear Least Squares Problems Whose Variables Separate," *SIAM J. Num. Anal. 10*, 413–32.

G.H. Golub and V. Pereyra (1976). "Differentiation of Pseudo-Inverses, Separable Nonlinear Least Squares Problems and Other Tales," in *Generalized Inverses and Applications* , ed. M.Z. Nashed, Academic Press, New York, pp. 303–24.

G.H. Golub and C. Reinsch (1970). "Singular Value Decomposition and Least Squares Solutions," *Numer. Math. 14*, 403–20. See also Wilkinson and Reinsch (1971, 134–51).

G.H. Golub and W.P. Tang (1981). "The Block Decomposition of a Vandermonde Matrix and Its Applications,"*BIT 21*, 505–17.

G.H. Golub and R. Underwood (1977). "The Block Lanczos Method for Computing Eigenvalues," in *Mathematical Software III* , ed. J. Rice, Academic Press, New York, pp. 364–77.

G.H. Golub, R. Underwood, and J.H. Wilkinson (1972). "The Lanczos Algorithm for the Symmetric $Ax = \lambda Bx$ Problem," Report STAN-CS-72-270, Department of Computer Science, Stanford University, Stanford, California.

G.H. Golub and P. Van Dooren, eds. (1991). *Numerical Linear Algebra, Digital Signal Processing, and Parallel Algorithms.*. Springer-Verlag, Berlin.

G.H. Golub and C.F. Van Loan (1979). "Unsymmetric Positive Definite Linear Systems," *Lin. Alg. and Its Applic. 28*, 85–98.

G.H. Golub and C.F. Van Loan (1980). "An Analysis of the Total Least Squares Problem," *SIAM J. Num. Anal. 17*, 883–93.

G.H. Golub and J.M. Varah (1974). "On a Characterization of the Best L_2-Scaling of a Matrix," *SIAM J. Num. Anal. 11*, 472–79.

G.H. Golub and R.S. Varga (1961). "Chebychev Semi-Iterative Methods, Successive Over-Relaxation Iterative Methods, and Second-Order Richardson Iterative Methods, Parts I and II," *Numer. Math. 3*, 147–56, 157–68.

G.H. Golub and J.H. Welsch (1969). "Calculation of Gauss Quadrature Rules," *Math. Comp. 23*, 221–30.

G.H. Golub and J.H. Wilkinson (1966). "Note on Iterative Refinement of Least Squares Solutions," *Numer. Math. 9*, 139–48.

G.H. Golub and J.H. Wilkinson (1976). "Ill-Conditioned Eigensystems and the Computation of the Jordan Canonical Form," *SIAM Review 18*, 578–619.

G.H. Golub and H. Zha (1994). "Perturbation Analysis of the Canonical Correlations of Matrix Pairs," *Lin. Alg. and Its Applic. 210*, 3–28.

N. Gould (1991). "On Growth in Gaussian Elimination with Complete Pivoting," *SIAM J. Matrix Anal. Appl. 12*, 354–361.

R.J. Goult, R.F. Hoskins, J.A. Milner and M.J. Pratt (1974). *Computational Methods in Linear Algebra*, John Wiley and Sons, New York.

A.R. Gourlay (1970). "Generalization of Elementary Hermitian Matrices," *Comp. J. 13*, 411–12.

A.R. Gourlay and G.A. Watson (1973). *Computational Methods for Matrix Eigenproblems*, John Wiley & Sons, New York.

W. Govaerts (1991). "Stable Solvers and Block Elimination for Bordered Systems," *SIAM J. Matrix Anal. Appl. 12*, 469–483.

W. Govaerts and J.D. Pryce (1990). "Block Elimination with One Iterative Refinement Solves Bordered Linear Systems Accurately," *BIT 30*, 490–507.

W. Govaerts and J.D. Pryce (1993). "Mixed Block Elimination for Linear Systems with Wider Borders," *IMA J. Num. Anal. 13*, 161–180.

W. B. Gragg (1986). "The QR Algorithm for Unitary Hessenberg Matrices," *J. Comp. Appl. Math. 16*, 1–8.

W.B. Gragg and W.J. Harrod (1984). "The Numerically Stable Reconstruction of Jacobi Matrices from Spectral Data," *Numer. Math. 44*, 317–336.

W.B. Gragg and L. Reichel (1990). "A Divide and Conquer Method for Unitary and Orthogonal Eigenproblems," *Numer. Math. 57*, 695–718.

A. Graham (1981). *Kronecker Products and Matrix Calculus with Applications*, Ellis Horwood Ltd., Chichester, England.

B. Green (1952). "The Orthogonal Approximation of an Oblique Structure in Factor Analysis," *Psychometrika 17*, 429–40.

A. Greenbaum (1992). "Diagonal Scalings of the Laplacian as Preconditioners for Other Elliptic Differential Operators," *SIAM J. Matrix Anal. Appl. 13*, 826–846.

A. Greenbaum and G. Rodrigue (1989). "Optimal Preconditioners of a Given Sparsity Pattern," *BIT 29*, 610–634.

A. Greenbaum and Z. Strakos (1992). "Predicting the Behavior of Finite Precision Lanczos and Conjugate Gradient Computations," *SIAM J. Matrix Anal. Appl. 13*, 121–137.

A. Greenbaum and L.N. Trefethen (1994). "GMRES/CR and Arnoldi/Lanczos as Matrix Approximation Problems," *SIAM J. Sci. Comp. 15*, 359–368.

J. Greenstadt (1955). "A Method for Finding Roots of Arbitrary Matrices," *Math. Tables and Other Aids to Comp. 9*, 47–52.

R.G. Grimes and J.G. Lewis (1981). "Condition Number Estimation for Sparse Matrices," *SIAM J. Sci. and Stat. Comp. 2*, 384–88.

R.G. Grimes, J.G. Lewis, and H.D. Simon (1994). "A Shifted Block Lanczos Algorithm for Solving Sparse Symmetric Generalized Eigenproblems," *SIAM J. Matrix Anal. Appl. 15*, 228–272.

W.D. Gropp and D.E. Keyes (1988). "Complexity of Parallel Implementation of Domain Decomposition Techniques for Elliptic Partial Differential Equations," *SIAM J. Sci. and Stat. Comp. 9*, 312–326.

W.D. Gropp and D.E. Keyes (1992). "Domain Decomposition with Local Mesh Refinement," *SIAM J. Sci. Statist. Comput. 13*, 967–993.

M. Gu and S.C. Eisenstat (1995). "A Divide-and-Conquer Algorithm for the Bidiagonal SVD," *SIAM J. Matrix Anal. Appl. 16*, 79–92.

M. Gu and S.C. Eisenstat (1995). "A Divide-and-Conquer Algorithm for the Symmetric Tridiagonal Eigenproblem," *SIAM J. Matrix Anal. Appl. 16*, 172–191.

M. Gulliksson (1994). "Iterative Refinement for Constrained and Weighted Linear Least Squares," *BIT 34*, 239–253.

M. Gulliksson (1995). "Backward Error Analysis for the Constrained and Weighted Linear Least Squares Problem When Using the Weighted QR Factorization," *SIAM J. Matrix. Anal. Appl. 13*, 675–687.

M. Gulliksson and P-Å. Wedin (1992). "Modifying the QR-Decomposition to Constrained and Weighted Linear Least Squares," *SIAM J. Matrix Anal. Appl. 13*, 1298–1313.

R.F. Gunst, J.T. Webster, and R.L. Mason (1976). "A Comparison of Least Squares and Latent Root Regression Estimators," *Technometrics 18*, 75–83.

K.K. Gupta (1972). "Solution of Eigenvalue Problems by Sturm Sequence Method," *Int. J. Numer. Meth. Eng. 4*, 379–404.

M. Gutknecht (1992). "A Completed Theory of the Unsymmetric Lanczos Process and Related Algorithms, Part I," *SIAM J. Matrix Anal. Appl. 13*, 594–639.

M. Gutknecht (1993). "Variants of BiCBSTAB for Matrices with Complex Spectrum," *SIAM J. Sci. and Stat. Comp. 14*, 1020–1033.

M. Gutknecht (1994). "A Completed Theory of the Unsymmetric Lanczos Process and Related Algorithms, Part II," *SIAM J. Matrix Anal. Appl. 15*, 15–58.

W. Hackbusch (1994). *Iterative Solution of Large Sparse Systems of Equations*, Springer-Verlag, New York.

D. Hacon (1993). "Jacobi's Method for Skew-Symmetric Matrices," *SIAM J. Matrix Anal. Appl. 14*, 619–628.

L.A. Hageman and D.M. Young (1981). *Applied Iterative Methods*, Academic Press, New York.

W.W. Hager (1984). "Condition Estimates," *SIAM J. Sci. and Stat. Comp. 5*, 311–316.

W.W. Hager (1988). *Applied Numerical Linear Algebra*, Prentice-Hall, Englewood Cliffs, NJ.

S.J. Hammarling (1974). "A Note on Modifications to the Givens Plane Rotation," *J. Inst. Math. Appl. 13*, 215–18.

S.J. Hammarling (1985). "The Singular Value Decomposition in Multivariate Statistics," *ACM SIGNUM Newsletter 20*, 2-25.

S.L. Handy and J.L. Barlow (1994). "Numerical Solution of the Eigenproblem for Banded, Symmetric Toeplitz Matrices," *SIAM J. Matrix Anal. Appl. 15*, 205–214.

M. Hanke and J.G. Nagy (1994). "Toeplitz Approximate Inverse Preconditioner for Banded Toeplitz Matrices," *Numerical Algorithms 7*, 183–199.

M. Hanke and M. Neumann (1990). "Preconditionings and Splittings for Rectangular Systems," *Numer. Math. 57*, 85–96.

E.R. Hansen (1962). "On Quasicyclic Jacobi Methods," *ACM J. 9*, 118–35.

E.R. Hansen (1963). "On Cyclic Jacobi Methods," *SIAM J. Appl. Math. 11*, 448–59.

P.C. Hansen (1987). "The Truncated SVD as a Method for Regularization," *BIT 27*, 534–553.

P.C. Hansen (1988). "Reducing the Number of Sweeps in Hestenes Method," in *Singular Value Decomposition and Signal Processing*, ed. E.F. Deprettere, North Holland.

P.C. Hansen (1990). "Relations Between SVD and GSVD of Discrete Regularization Problems in Standard and General Form," *Lin.Alg. and Its Applic. 141*, 165–176.

P.C. Hansen and H. Gesmar (1993). "Fast Orthogonal Decomposition of Rank-Deficient Toeplitz Matrices," *Numerical Algorithms 4*, 151–166.

R.J. Hanson and C.L. Lawson (1969). "Extensions and Applications of the Householder Algorithm for Solving Linear Least Square Problems," *Math. Comp. 23*, 787–812.

V. Hari (1982). "On the Global Convergence of the Eberlein Method for Real Matrices," *Numer. Math. 39*, 361–370.

V. Hari (1991). "On Pairs of Almost Diagonal Matrices," *Lin. Alg. and Its Applic. 148*, 193–223.

M.T. Heath, ed. (1986). *Proceedings of First SIAM Conference on Hypercube Multiprocessors*, SIAM Publications, Philadelphia, PA.

M.T. Heath, ed. (1987). *Hypercube Multiprocessors*, SIAM Publications, Philadelphia, PA.

M.T. Heath (1997). *Scientific Computing: An Introductory Survey*, McGraw-Hill, New York.

M.T. Heath, A.J. Laub, C.C. Paige, and R.C. Ward (1986). "Computing the SVD of a Product of Two Matrices," *SIAM J. Sci. and Stat. Comp. 7*, 1147-1159.

M.T. Heath, E. Ng, and B.W. Peyton (1991). "Parallel Algorithms for Sparse Linear Systems," *SIAM Review 33*, 420–460.

M.T. Heath and C.H. Romine (1988). "Parallel Solution of Triangular Systems on Distributed Memory Multiprocessors," *SIAM J. Sci. and Stat. Comp. 9*, 558–588.

M. Hegland (1991). "On the Parallel Solution of Tridiagonal Systems by Wrap-Around Partitioning and Incomplete LU Factorization," *Numer. Math. 59*, 453–472.

G. Heinig and P. Jankowski (1990). "Parallel and Superfast Algorithms for Hankel Systems of Equations," *Numer. Math. 58*, 109–127.

D.E. Heller (1976). "Some Aspects of the Cyclic Reduction Algorithm for Block Tridiagonal Linear Systems," *SIAM J. Num. Anal. 13*, 484–96.

D.E. Heller (1978). "A Survey of Parallel Algorithms in Numerical Linear Algebra," *SIAM Review 20*, 740-777.

D.E. Heller and I.C.F. Ipsen (1983). "Systolic Networks for Orthogonal Decompositions," *SIAM J. Sci. and Stat. Comp. 4*, 261–269.

B.W. Helton (1968). "Logarithms of Matrices," *Proc. Amer. Math. Soc. 19*, 733–36.

H.V. Henderson and S.R. Searle (1981). "The Vec-Permutation Matrix, The Vec Operator, and Kronecker Products: A Review," *Linear and Multilinear Algebra 9*, 271–288.

B. Hendrickson and D. Womble (1994). "The Torus-Wrap Mapping for Dense Matrix Calculations on Massively Parallel Computers," *SIAM J. Sci. Comput. 15*, 1201–1226.

C.S. Henkel, M.T. Heath, and R.J. Plemmons (1988). "Cholesky Downdating on a Hypercube," in G. Fox (1988), 1592–1598.

P. Henrici (1958). "On the Speed of Convergence of Cyclic and Quasicyclic Jacobi Methods for Computing the Eigenvalues of Hermitian Matrices," *SIAM J. Appl. Math.* 6, 144–62.

P. Henrici (1962). "Bounds for Iterates, Inverses, Spectral Variation and Fields of Values of Non-normal Matrices," *Numer. Math.* 4, 24–40.

P. Henrici and K. Zimmermann (1968). "An Estimate for the Norms of Certain Cyclic Jacobi Operators," *Lin. Alg. and Its Applic.* 1, 489–501.

M.R. Hestenes (1980). *Conjugate Direction Methods in Optimization*, Springer-Verlag, Berlin.

M.R. Hestenes (1990). "Conjugacy and Gradients," in *A History of Scientific Computing*, Addison-Wesley, Reading, MA.

M.R. Hestenes and E. Stiefel (1952). "Methods of Conjugate Gradients for Solving Linear Systems," *J. Res. Nat. Bur. Stand.* 49, 409–36.

G. Hewer and C. Kenney (1988). "The Sensitivity of the Stable Lyapunov Equation," *SIAM J. Control Optim* 26, 321–344.

D.J. Higham (1995). "Condition Numbers and Their Condition Numbers," *Lin. Alg. and Its Applic.* 214, 193–213.

D.J. Higham and N.J. Higham (1992). "Componentwise Perturbation Theory for Linear Systems with Multiple Right-Hand Sides," *Lin. Alg. and Its Applic.* 174, 111–129.

D.J. Higham and N.J. Higham (1992). "Backward Error and Condition of Structured Linear Systems," *SIAM J. Matrix Anal. Appl.* 13, 162–175.

D.J. Higham and L.N. Trefethen (1993). "Stiffness of ODES," *BIT 33*, 285–303.

N.J. Higham (1985). "Nearness Problems in Numerical Linear Algebra," PhD Thesis, University of Manchester, England.

N.J. Higham (1986). "Newton's Method for the Matrix Square Root," *Math. Comp.* 46, 537–550.

N.J. Higham (1986). "Computing the Polar Decomposition—with Applications," *SIAM J. Sci. and Stat. Comp.* 7, 1160–1174.

N.J. Higham (1986). "Efficient Algorithms for computing the condition number of a tridiagonal matrix," *SIAM J. Sci. and Stat. Comp.* 7, 150–165.

N.J. Higham (1987). "A Survey of Condition Number Estimation for Triangular Matrices," *SIAM Review 29*, 575–596.

N.J. Higham (1987). "Error Analysis of the Björck-Pereyra Algorithms for Solving Vandermonde Systems," *Numer. Math. 50*, 613–632.

N.J. Higham (1987). "Computing Real Square Roots of a Real Matrix," *Lin. Alg. and Its Applic. 88/89*, 405–430.

N.J. Higham (1988). "Fast Solution of Vandermonde-like Systems Involving Orthogonal Polynomials," *IMA J. Num. Anal.* 8, 473–486.

N.J. Higham (1988). "Computing a Nearest Symmetric Positive Semidefinite Matrix," *Lin. Alg. and Its Applic. 103*, 103–118.

N.J. Higham (1988). "The Symmetric Procrustes Problem," *BIT 28*, 133–43.

N.J. Higham (1988). "FORTRAN Codes for Estimating the One-Norm of a Real or Complex Matrix with Applications to Condition Estimation (Algorithm 674)," *ACM Trans. Math. Soft. 14*, 381–396.

N.J. Higham (1989). "Matrix Nearness Problems and Applications," in *Applications of Matrix Theory*, M.J.C. Gover and S. Barnett (eds), Oxford University Press, Oxford UK, 1–27.

N.J. Higham (1989). "The Accuracy of Solutions to Triangular Systems," *SIAM J. Num. Anal. 26*, 1252–1265.

N.J. Higham (1990). "Bounding the Error in Gaussian Elimination for Tridiagonal Systems," *SIAM J. Matrix Anal. Appl. 11*, 521–530.

N.J. Higham (1990). "Stability Analysis of Algorithms for Solving Confluent Vandermonde-like Systems," *SIAM J. Matrix Anal. Appl. 11*, 23–41.

N.J. Higham (1990). "Analysis of the Cholesky Decomposition of a Semidefinite Matrix," in *Reliable Numerical Computation*, M.G. Cox and S.J. Hammarling (eds), Oxford University Press, Oxford, UK, 161–185.

N.J. Higham (1990). "Exploiting Fast Matrix Multiplication within the Level 3 BLAS," *ACM Trans. Math. Soft. 16*, 352–368.

N.J. Higham (1991). "Iterative Refinement Enhances the Stability of QR Factorization Methods for Solving Linear Equations," *BIT 31*, 447–468.

N.J. Higham (1992). "Stability of a Method for Multiplying Complex Matrices with Three Real Matrix Multiplications," *SIAM J. Matrix Anal. Appl. 13*, 681–687.

N.J. Higham (1992). "Estimating the Matrix p-Norm," *Numer. Math. 62*, 539–556.

N.J. Higham (1993). "Optimization by Direct Search in Matrix Computations," *SIAM J. Matrix Anal. Appl. 14*, 317–333.

N.J. Higham (1993). "Perturbation Theory and Backward Error for AX - XB = C," *BIT 33*, 124–136.

N.J. Higham (1994). "The Matrix Sign Decomposition and Its Relation to the Polar Decomposition," *Lin. Alg. and Its Applic. 212/213*, 3–20.

N.J. Higham (1994). "A Survey of Componentwise Perturbation Theory in Numerical Linear Algebra," in *Mathematics of Computation 1943-1993: A Half Century of Computational Mathematics*, W. Gautschi (ed.), Volume 48 of *Proceedings of Symposia in Applied Mathematics*, American Mathematical Society, Providence, Rhode Island.

N.J. Higham (1995). "Stability of Parallel Triangular System Solvers," *SIAM J. Sci. Comp. 16*, 400–413.

N.J. Higham (1996). *Accuracy and Stability of Numerical Algorithms*, SIAM Publications, Philadelphia, PA.

N.J. Higham and D.J. Higham (1989). "Large Growth Factors in Gaussian Elimination with Pivoting," *SIAM J. Matrix Anal. Appl. 10*, 155–164.

N.J. Higham and P.A. Knight (1995). "Matrix Powers in Finite Precision Arithmetic," *SIAM J. Matrix Anal. Appl. 16*, 343–358.

N.J. Higham and P. Papadimitriou (1994). "A Parallel Algorithm for Computing the Polar Decomposition," *Parallel Comp. 20*, 1161–1173.

R.W. Hockney (1965). "A Fast Direct Solution of Poisson's Equation Using Fourier Analysis, "*J. ACM 12*, 95-113.

R.W. Hockney and C.R. Jesshope (1988). *Parallel Computers 2*, Adam Hilger, Bristol and Philadelphia.

W. Hoffman and B.N. Parlett (1978). "A New Proof of Global Convergence for the Tridiagonal QL Algorithm," *SIAM J. Num. Anal. 15*, 929–37.

S. Holmgren and K. Otto (1992). "Iterative Solution Methods and Preconditioners for Block-Tridiagonal Systems of Equations," *SIAM J. Matrix Anal. Appl. 13*, 863–886.

H. Hotelling (1957). "The Relations of the Newer Multivariate Statistical Methods to Factor Analysis," *Brit. J. Stat. Psych. 10*, 69–79.

P.D. Hough and S.A. Vavasis (1996). "Complete Orthogonal Decomposition for Weighted Least Squares," *SIAM J. Matrix Anal. Appl.*, to appear.

A.S. Householder (1958). "Unitary Triangularization of a Nonsymmetric Matrix," *J. ACM. 5*, 339–42.

A.S. Householder (1964). *The Theory of Matrices in Numerical Analysis* , Dover Publications, New York.

A.S. Householder (1968). "Moments and characteristic Roots II," *Numer. Math. 11*, 126–28.

R. Horn and C. Johnson (1985). *Matrix Analysis*, Cambridge University Press, New York.

R. Horn and C. Johnson (1991). *Topics in Matrix Analysis*, Cambridge University Press, New York.

C.P. Huang (1975). "A Jacobi-Type Method for Triangularizing an Arbitrary Matrix," *SIAM J. Num. Anal. 12*, 566–70.

C.P. Huang (1981). "On the Convergence of the QR Algorithm with Origin Shifts for Normal Matrices," *IMA J. Num. Anal. 1*, 127–33.

C.-M. Huang and D.P. O'Leary (1993). "A Krylov Multisplitting Algorithm for Solving Linear Systems of Equations," *Lin. Alg. and Its Applic. 194*, 9–29.

T. Huckle (1992). "Circulant and Skewcirculant Matrices for Solving Toeplitz Matrix Problems," *SIAM J. Matrix Anal. Appl. 13*, 767–777.

T. Huckle (1992). "A Note on Skew-Circulant Preconditioners for Elliptic Problems," *Numerical Algorithms 2*, 279–286.

T. Huckle (1994). "The Arnoldi Method for Normal Matrices," *SIAM J. Matrix Anal. Appl. 15*, 479–489.

T. Huckle (1995). "Low-Rank Modification of the Unsymmetric Lanczos Algorithm," *Math.Comp. 64*, 1577–1588.

T.E. Hull and J.R. Swensen (1966). "Tests of Probabilistic Models for Propagation of Roundoff Errors," *Comm. ACM. 9*, 108–13.

T-M. Hwang, W-W. Lin, and E.K. Yang (1992). "Rank-Revealing LU Factorizations," *Lin. Alg. and Its Applic. 175*, 115–141.

Y. Ikebe (1979). "On Inverses of Hessenberg Matrices," *Lin. Alg. and Its Applic. 24*, 93–97.

I.C.F. Ipsen, Y. Saad, and M. Schultz (1986). "Dense Linear Systems on a Ring of Processors," *Lin. Alg. and Its Applic. 77*, 205–239.

C.G.J. Jacobi (1846). "Uber ein Leichtes Verfahren Die in der Theorie der Sacularstroungen Vorkommendern Gleichungen Numerisch Aufzulosen," *Crelle's J. 30*, 51–94.

P. Jacobson, B. Kågström, and M. Rannar (1992). "Algorithm Development for Distributed Memory Multicomputers Using Conlab," *Scientific Programming, 1*, 185–203.

H.J. Jagadish and T. Kailath (1989). "A Family of New Efficient Arrays for Matrix Multiplication," *IEEE Trans. Comput. 38*, 149–155.

W. Jalby and B. Philippe (1991). "Stability Analysis and Improvement of the Block Gram-Schmidt Algorithm," *SIAM J. Sci. Stat. Comp. 12*, 1058–1073.

M. Jankowski and M. Wozniakowski (1977). "Iterative Refinement Implies Numerical Stability," *BIT 17*, 303–311.

K.C. Jea and D.M. Young (1983). "On the Simplification of Generalized Conjugate Gradient Methods for Nonsymmetrizable Linear Systems," *Lin. Alg. and Its Applic. 52/53*, 399–417.

A. Jennings (1977). "Influence of the Eigenvalue Spectrum on the Convergence Rate of the Conjugate Gradient Method," *J. Inst. Math. Applic. 20*, 61–72.

A. Jennings (1977). *Matrix Computation for Engineers and Scientists*, John Wiley and Sons, New York.

A. Jennings and J.J. McKeowen (1992). *Matrix Computation (2nd ed)*, John Wiley and Sons, New York.

A. Jennings and D.R.L. Orr (1971). "Application of the Simultaneous Iteration Method to Undamped Vibration Problems," *Inst. J. Numer. Math. Eng. 3*, 13–24.

A. Jennings and M.R. Osborne (1977). "Generalized Eigenvalue Problems for Certain Unsymmetric Band Matrices," *Lin. Alg. and Its Applic. 29*, 139–50.

A. Jennings and W.J. Stewart (1975). "Simultaneous Iteration for the Partial Eigensolution of Real Matrices," *J. Inst. Math. Applic. 15*, 351–62.

L.S. Jennings and M.R. Osborne (1974). "A Direct Error Analysis for Least Squares," *Numer. Math. 22*, 322–32.

P.S. Jenson (1972). "The Solution of Large Symmetric Eigenproblems by Sectioning," *SIAM J. Num. Anal. 9*, 534–45.

E.R. Jessup and D.C. Sorensen (1994). "A Parallel Algorithm for Computing the Singular Value Decomposition of a Matrix," *SIAM J. Matrix Anal. Appl. 15*, 530–548.

Z. Jia (1995). "The Convergence of Generalized Lanczos Methods for Large Unsymmetric Eigenproblems," *SIAM J. Matrix Anal. Applic 16*, 543–562.

J. Johnson and C.L. Phillips (1971). "An Algorithm for the Computation of the Integral of the State Transition Matrix," *IEEE Trans. Auto. Cont. AC-16*, 204–5.

O.G. Johnson, C.A. Micchelli, and G. Paul (1983). "Polynomial Preconditioners for Conjugate Gradient Calculations," *SIAM J. Numer. Anal. 20*, 362–376.

R.J. Johnston (1971). "Gershgorin Theorems for Partitioned Matrices," *Lin. Alg. and Its Applic. 4*, 205–20.

S.L. Johnsson (1985). "Solving Narrow Banded Systems on Ensemble Architectures," *ACM Trans. Math. Soft. 11*, 271–288.

S.L. Johnsson (1986). "Band Matrix System Solvers on Ensemble Architectures," in *Supercomputers: Algorithms, Architectures, and Scientific Computation*, eds. F.A. Matsen and T. Tajima, University of Texas Press, Austin TX., 196–216.

S.L. Johnsson (1987). "Communication Efficient Basic Linear Algebra Computations on Hypercube Multiprocessors," *J. Parallel and Distributed Computing*, No. 4, 133–172.

S.L. Johnsson (1987). "Solving Tridiagonal Systems on Ensemble Architectures," *SIAM J. Sci. and Stat. Comp. 8*, 354–392.

S.L. Johnsson and C.T. Ho (1988). "Matrix Transposition on Boolean n-cube Configured Ensemble Architectures," *SIAM J. Matrix Anal. Appl. 9*, 419–454.

S.L. Johnsson and W. Lichtenstein (1993). "Block Cyclic Dense Linear Algebra," *SIAM J. Sci.Comp. 14*, 1257–1286.

S.L. Johnsson and K. Mathur (1989). "Experience with the Conjugate Gradient Method for Stress Analysis on a Data Parallel Supercomputer," *International Journal on Numerical Methods in Engineering 27*, 523–546.

P. Joly and G. Meurant (1993). "Complex Conjugate Gradient Methods," *Numerical Algorithms 4*, 379–406.

M.T. Jones and M.L. Patrick (1993). "Bunch-Kaufman Factorization for Real Symmetric Indefinite Banded Matrices," *SIAM J. Matrix Anal. Appl. 14*, 553–559.

M.T. Jones and M.L. Patrick (1994). "Factoring Symmetric Indefinite Matrices on High-Performance Architectures," *SIAM J. Matrix Anal. Appl. 15*, 273–283.

T. Jordan (1984). "Conjugate Gradient Preconditioners for Vector and Parallel Processors," in G. Birkoff and A. Schoenstadt (eds), *Proceedings of the Conference on Elliptic Problem Solvers*, Academic Press, NY.

W. Joubert (1992). "Lanczos Methods for the Solution of Nonsymmetric Systems of Linear Equations," *SIAM J. Matrix Anal. Appl. 13*, 926–943.

B. Kågström (1977). "Numerical Computation of Matrix Functions," Department of Information Processing Report UMINF-58.77, University of Ümea, Sweden.

B. Kågström (1977). "Bounds and Perturbation Bounds for the Matrix Exponential," *BIT 17*, 39–57.

B. Kågström (1985). "The Generalized Singular Value Decomposition and the General $A - \lambda B$ Problem," *BIT 24*, 568-583.

B. Kågström (1986). "RGSVD: An Algorithm for Computing the Kronecker Structure and Reducing Subspaces of Singular $A - \lambda B$ Pencils," *SIAM J. Sci. and Stat. Comp. 7*,185–211.

B. Kågström (1994). "A Perturbation Analysis of the Generalized Sylvester Equation $(AR - LB, DR - LE) = (C, F)$," *SIAM J. Matrix Anal. Appl. 15*, 1045–1060.

B. Kågström, P. Ling, and C. Van Loan (1991). "High-Performance Level-3 BLAS: Sample Routines for Double Precision Real Data," in *High Performance Computing II*, M. Durand and F. El Dabaghi (eds), North-Holland, 269–281.

B. Kågström, P. Ling, and C. Van Loan (1995). "GEMM-Based Level-3 BLAS: High-Performance Model Implementations and Performance Evaluation Benchmark," in *Parallel Programming and Applications*, P. Fritzon and L. Finmo (eds), ISO Press, 184–188.

B. Kågström and P. Poromaa (1992). "Distributed and Shared Memory Block Algorithms for the Triangular Sylvester Equation with sep^{-1} Estimators," *SIAM J. Matrix Anal. Appl. 13*, 90–101.

B. Kågström and A. Ruhe (1980). "An Algorithm for Numerical Computation of the Jordan Normal Form of a Complex Matrix," *ACM Trans. Math. Soft. 6*, 398–419.

B. Kågström and A. Ruhe (1980). "Algorithm 560 JNF: An Algorithm for Numerical Computation of the Jordan Normal Form of a Complex Matrix," *ACM Trans. Math. Soft. 6*, 437–43.

B. Kågström and A. Ruhe, eds. (1983). *Matrix Pencils*, Proc. Pite Havsbad, 1982, Lecture Notes in Mathematics 973, Springer-Verlag, New York and Berlin.

B. Kågström and L. Westin (1989). "Generalized Schur Methods with Condition Estimators for Solving the Generalized Sylvester Equation," *IEEE Trans. Auto. Cont.* *AC-34*, 745–751.

W. Kahan (1966). "Numerical Linear Algebra," *Canadian Math. Bull.* *9*, 757–801.

W. Kahan (1975). "Spectra of Nearly Hermitian Matrices," *Proc. Amer. Math. Soc.* *48*, 11–17.

W. Kahan and B.N. Parlett (1976). "How Far Should You Go with the Lanczos Process?" in *Sparse Matrix Computations*, ed. J. Bunch and D. Rose, Academic Press, New York, pp. 131-44.

W. Kahan, B.N. Parlett, and E. Jiang (1982). "Residual Bounds on Approximate Eigensystems of Nonnormal Matrices," *SIAM J. Numer. Anal.* *19*, 470–484.

D. Kahaner, C.B. Moler, and S. Nash (1988). *Numerical Methods and Software*, Prentice-Hall, Englewood Cliffs, NJ.

T. Kailath and J. Chun (1994). "Generalized Displacement Structure for Block-Toeplitz, Toeplitz-Block, and Toeplitz-Derived Matrices," *SIAM J. Matrix Anal. Appl.* *15*, 114–128.

T. Kailath and A.H. Sayed (1995). "Displacement Structure: Theory and Applications," *SIAM Review 37*, 297–386.

C. Kamath and A. Sameh (1989). "A Projection Method for Solving Nonsymmetric Linear Systems on Multiprocessors," *Parallel Computing 9*, 291–312.

S. Kaniel (1966). "Estimates for Some Computational Techniques in Linear Algebra," *Math. Comp. 20*, 369–78.

I.E. Kaporin (1994). "New Convergence Results and Preconditioning Strategies for the Conjugate Gradient Method," *Num. Lin. Alg. Applic. 1*, 179–210.

R.N. Kapur and J.C. Browne (1984). "Techniques for Solving Block Tridiagonal Systems on Reconfigurable Array Computers," *SIAM J. Sci. and Stat. Comp. 5*, 701-719.

I. Karasalo (1974). "A Criterion for Truncation of the QR Decomposition Algorithm for the Singular Linear Least Squares Problem," *BIT 14*, 156–66.

E.M. Kasenally (1995). "GMBACK: A Generalized Minimum Backward Error Algorithm for Nonsymmetric Linear Systems," *SIAM J. Sci. Comp. 16*, 698–719.

T. Kato (1966). *Perturbation Theory for Linear Operators*, Springer-Verlag, New York.

L. Kaufman (1974). "The LZ Algorithm to Solve the Generalized Eigenvalue Problem," *SIAM J. Num. Anal. 11*, 997–1024.

L. Kaufman (1977). "Some Thoughts on the QZ Algorithm for Solving the Generalized Eigenvalue Problem," *ACM Trans. Math. Soft. 3*, 65–75.

L. Kaufman (1979). "Application of Dense Householder Transformations to a Sparse Matrix," *ACM Trans. Math. Soft. 5*, 442–51.

L. Kaufman (1987). "The Generalized Householder Transformation and Sparse Matrices," *Lin. Alg. and Its Applic. 90*, 221–234.

L. Kaufman (1993). "An Algorithm for the Banded Symmetric Generalized Matrix Eigenvalue Problem," *SIAM J. Matrix Anal. Appl. 14*, 372–389.

J. Kautsky and G.H. Golub (1983). "On the Calculation of Jacobi Matrices," *Lin. Alg. and Its Applic. 52/53*, 439–456.

C.S. Kenney and A.J. Laub (1989). "Condition Estimates for Matrix Functions," *SIAM J. Matrix Anal. Appl. 10*, 191–209.

C.S. Kenney and A.J. Laub (1991). "Rational Iterative Methods for the Matrix Sign Function," *SIAM J. Matrix Anal. Appl. 12*, 273–291.

C.S. Kenney and A.J. Laub (1992). "On Scaling Newton's Method for Polar Decomposition and the Matrix Sign Function," *SIAM J. Matrix Anal. Appl. 13*, 688–706.

C.S. Kenney and A.J. Laub (1994). "Small-Sample Statistical Condition Estimates for General Matrix Functions," *SIAM J. Sci. Comp. 15*, 36–61.

D. Kershaw(1982). "Solution of Single Tridiagonal Linear Systems and Vectorization of the ICCG Algorithm on the Cray-1," in G. Roderigue (ed), *Parallel Computation*, Academic Press, NY, 1982.

D.E. Keyes, T.F. Chan, G. Meurant, J.S. Scroggs, and R.G. Voigt (eds) (1992). *Domain Decomposition Methods for Partial Differential Equations*, SIAM Publications, Philadelphia, PA.

A. Kielbasinski (1987). "A Note on Rounding Error Analysis of Cholesky Factorization," *Lin. Alg. and Its Applic. 88/89*, 487–494.

S.K. Kim and A.T. Chronopoulos (1991). "A Class of Lanczos-Like Algorithms Implemented on Parallel Computers," *Parallel Comput. 17*, 763–778.

F. Kittaneh (1995). "Singular Values of Companion Matrices and Bounds on Zeros of Polynomials," *SIAM J. Matrix Anal. Appl. 16*, 333–340.

P.A. Knight (1993). "Error Analysis of Stationary Iteration and Associated Problems," Ph.D. thesis, Department of Mathematics, University of Manchester, England.

P.A. Knight (1995). "Fast Rectangular Matrix Multiplication and the QR Decomposition," *Lin. Alg. and Its Applic. 221*, 69–81.

D. Knuth (1981). *The Art of Computer Programming , vol. 2. Seminumerical Algorithms , 2nd ed.*, Addison-Wesley, Reading, Massachusetts.

E.G. Kogbetliantz (1955). "Solution of Linear Equations by Diagonalization of Coefficient Matrix," *Quart. Appl. Math. 13*, 123–132.

S. Kourouklis and C.C. Paige (1981). "A Constrained Least Squares Approach to the General Gauss-Markov Linear Model," *J. Amer. Stat. Assoc. 76*, 620–25.

V.N. Kublanovskaya (1961). "On Some Algorithms for the Solution of the Complete Eigenvalue Problem," *USSR Comp. Math. Phys. 3*, 637–57.

V.N. Kublanovskaya (1984). "AB Algorithm and Its Modifications for the Spectral Problem of Linear Pencils of Matrices," *Numer. Math. 43*, 329–342.

V.N. Kublanovskaja and V.N. Fadeeva (1964). "Computational Methods for the Solution of a Generalized Eigenvalue Problem," *Amer. Math. Soc. Transl. 2*, 271–90.

J. Kuczyński and H. Woźniakowski (1992). "Estimating the Largest Eigenvalue by the Power and Lanczos Algorithms with a Random Start," *SIAM J. Matrix Anal. Appl. 13*, 1094–1122.

U.W. Kulisch and W.L. Miranker (1986). "The Arithmetic of the Digital Computer," *SIAM Review 28*, 1–40.

V. Kumar, A. Grama, A. Gupta and G. Karypis (1994). *Introduction to Parallel Computing: Design and Analysis of Algorithms*, Benjamin/Cummings, Reading, MA.

H.T. Kung (1982). "Why Systolic Architectures?," *Computer 15*, 37–46.

C.D. La Budde (1964). "Two Classes of Algorithms for Finding the Eigenvalues and Eigenvectors of Real Symmetric Matrices," *J. ACM 11*, 53–58.

S. Lakshmivarahan and S. K. Dhall (1990). *Analysis and Design of Parallel Algorithms: Arithmetic and Matrix Problems*, McGraw-Hill, New York.

J. Lambiotte and R.G. Voigt (1975). "The Solution of Tridiagonal Linear Systems of the CDC-STAR 100 Computer," *ACM Trans. Math. Soft. 1*, 308–29.

P. Lancaster (1970). "Explicit Solution of Linear Matrix Equations," *SIAM Review 12*, 544–66.

P. Lancaster and M. Tismenetsky (1985). *The Theory of Matrices, Second Edition*, Academic Press, New York.

C. Lanczos (1950). "An Iteration Method for the Solution of the Eigenvalue Problem of Linear Differential and Integral Operators," *J. Res. Nat. Bur. Stand. 45*, 255–82.

B. Lang (1996). "Parallel Reduction of Banded Matrices to Bidiagonal Form," *Parallel Computing 22*, 1–18.

J. Larson and A. Sameh (1978). "Efficient Calculation of the Effects of Roundoff Errors," *ACM Trans. Math. Soft. 4*, 228–36.

A. Laub (1981). "Efficient Multivariable Frequency Response Computations," *IEEE Trans. Auto. Cont. AC-26*, 407–8.

A. Laub(1985). "Numerical Linear Algebra Aspects of Control Design Computations," *IEEE Trans. Auto. Cont. AC-30*, 97–108.

C.L. Lawson and R.J. Hanson (1969). "Extensions and Applications of the Householder Algorithm for Solving Linear Least Squares Problems," *Math. Comp. 23*, 787–812.

C.L. Lawson and R.J. Hanson (1974). *Solving Least Squares Problems*, Prentice-Hall, Englewood Cliffs, NJ. Reprinted with a detailed "new developments" appendix in 1996 by SIAM Publications, Philadelphia, PA.

C.L. Lawson, R.J. Hanson, , D.R. Kincaid, and F.T. Krogh (1979). "Basic Linear Algebra Subprograms for FORTRAN Usage," *ACM Trans. Math. Soft. 5*, 308–323.

C.L. Lawson, R.J. Hanson, D.R. Kincaid, and F.T. Krogh (1979). "Algorithm 539, Basic Linear Algebra Subprograms for FORTRAN Usage," *ACM Trans. Math. Soft. 5*, 324–325.

D. Lay (1994). *Linear Algebra and Its Applications*, Addison-Wesley, Reading, MA.

N.J. Lehmann (1963). "Optimale Eigenwerteinschliessungen," *Numer. Math. 5*, 246–72.

R.B, Lehoucq (1995). "Analysis and Implementation of an Implicitly Restarted Arnoldi Iteration," Ph.D. thesis, Rice University, Houston Texas.

R.B. Lehoucq (1996). "Restarting an Arnoldi Reduction," Report MCS-P591-0496, Argonne National Laboratory, Argonne Illinois.

R.B. Lehoucq and D.C. Sorensen (1996). "Deflation Techniques for an Implicitly Restarted Iteration," *SIAM J. Matrix Analysis and Applic*, to appear.

F.T. Leighton (1992). *Introduction to Parallel Algorithms and Architectures*, Morgan Kaufmann, San Mateo, CA.

F. Lemeire (1973). "Bounds for Condition Numbers of Triangular Value of a Matrix," *Lin. Alg. and Its Applic. 11*, 1–2.

S.J. Leon (1980). *Linear Algebra with Applications*. Macmillan, New York.

S.J. Leon (1994). "Maximizing Bilinear Forms Subject to Linear Constraints," *Lin. Alg. and Its Applic. 210*, 49–58.

N. Levinson (1947). "The Weiner RMS Error Criterion in Filter Design and Prediction," *J. Math. Phys. 25*, 261–78.

J. Lewis, ed. (1994). *Proceedings of the Fifth SIAM Conference on Applied Linear Algebra*, SIAM Publications, Philadelphia, PA.

G. Li and T. Coleman (1988). "A Parallel Triangular Solver for a Distributed-Memory Multiprocessor," *SIAM J. Sci. and Stat. Comp. 9*, 485–502.

K. Li and T-Y. Li (1993). "A Homotopy Algorithm for a Symmetric Generalized Eigenproblem," *Numerical Algorithms 4*, 167–195.

K. Li, T-Y. Li, and Z. Zeng (1994). "An Algorithm for the Generalized Symmetric Tridiagonal Eigenvalue Problem," *Numerical Algorithms 8*, 269–291.

R-C. Li (1993). "Bounds on Perturbations of Generalized Singular Values and of Associated Subspaces," *SIAM J. Matrix Anal. Appl. 14*, 195–234.

R-C. Li (1994). "On Eigenvalue Variations of Rayleigh Quotient Matrix Pencils of a Definite Pencil," *Lin. Alg. and Its Applic. 208/209*, 471–483.

R-C. Li (1995). "New Perturbation Bounds for the Unitary Polar Factor," *SIAM J. Matrix Anal. Appl. 16*, 327–332.

R.-C. Li (1996). "Relative Perturbation Theory (I) Eigenvalue and Singular Value Variations," Technical Report UCB//CSD-94-855, Department of EECS, University of California at Berkeley.

R.-C. Li (1996). "Relative Perturbation Theory (II) Eigenspace and Singular Subspace Variations," Technical Report UCB//CSD-94-856, Department of EECS, University of California at Berkeley.

Y. Li (1993). "A Globally Convergent Method for L_p Problems," *SIAM J. Optimization 3*, 609–629.

W-W. Lin and C.W. Chen (1991). "An Acceleration Method for Computing the Generalized Eigenvalue Problem on a Parallel Computer," *Lin.Alg. and Its Applic. 146*, 49–65.

I. Linnik (1961). *Method of Least Squares and Principles of the Theory of Observations*, Pergamon Press, New York.

E. Linzer (1992). "On the Stability of Solution Methods for Band Toeplitz Systems," *Lin.Alg. and Its Applic. 170*, 1–32.

S. Lo, B. Philippe, and A. Sameh (1987). "A Multiprocessor Algorithm for the Symmetric Tridiagonal Eigenvalue Problem," *SIAM J. Sci. and Stat. Comp. 8*, s155-s165.

G. Loizou (1969). "Nonnormality and Jordan Condition Numbers of Matrices," *J. ACM 16*, 580–40.

G. Loizou (1972). "On the Quadratic Convergence of the Jacobi Method for Normal Matrices," *Comp. J. 15*, 274–76.

M. Lotkin (1956). "Characteristic Values of Arbitrary Matrices," *Quart. Appl. Math. 14*, 267–75.

H. Lu (1994). "Fast Solution of Confluent Vandermonde Linear Systems," *SIAM J. Matrix Anal. Appl. 15*, 1277–1289.

H. Lu (1996). "Solution of Vandermonde-like Systems and Confluent Vandermonde-like Systems," *SIAM J. Matrix Anal. Appl. 17*, 127–138.

D. G. Luenberger (1973). *Introduction to Linear and Nonlinear Programming*, Addison-Wesley, New York.

F.T. Luk (1978). "Sparse and Parallel Matrix Computations," PhD Thesis, Report STANS-CS-78-685, Department of Computer Science, Stanford University, Stanford, CA.

F.T. Luk (1980). "Computing the Singular Value Decomposition on the ILLIAC IV," *ACM Trans. Math. Soft. 6*, 524–39.

F.T. Luk (1986). "A Rotation Method for Computing the QR Factorization," *SIAM J. Sci. and Stat. Comp. 7*, 452–459.

F.T. Luk (1986). "A Triangular Processor Array for Computing Singular Values," *Lin. Alg. and Its Applic. 77*, 259–274.

N. Mackey (1995). "Hamilton and Jacobi Meet Again: Quaternions and the Eigenvalue Problem," *SIAM J. Matrix Anal. Appl. 16*, 421–435.

A. Madansky (1959). "The Fitting of Straight Lines When Both Variables Are Subject to Error," *J. Amer. Stat. Assoc. 54*, 173–205.

N. Madsen, G. Roderigue, and J. Karush (1976). "Matrix Multiplication by Diagonals on a Vector Parallel Processor," *Infomation Processing Letters 5*, 41-45.

K.N. Majinder (1979). "Linear Combinations of Hermitian and Real Symmetric Matrices," *Lin. Alg. and Its Applic. 25*, 95–105.

J. Makhoul (1975). "Linear Prediction: A Tutorial Review," *Proc. IEEE 63*(4), 561-80.

M.A. Malcolm and J. Palmer (1974). "A Fast Method for Solving a Class of Tridiagonal Systems of Linear Equations," *Comm. ACM 17*, 14–17.

L. Mansfield (1991). "Damped Jacobi Preconditioning and Coarse Grid Deflation for Conjugate Gradient Iteration on Parallel Computers," *SIAM J. Sci. and Stat. Comp. 12*, 1314–1323.

T.A. Manteuffel (1977). "The Tchebychev Iteration for Nonsymmetric Linear Systems," *Numer. Math. 28*, 307–27.

T.A. Manttueffel (1979). " Shifted Incomplete Cholesky Factorization," in *Sparse Matrix Proceedings*, 1978, ed. I.S. Duff and G.W. Stewart, SIAM Publications, Philadelphia, PA.

M. Marcus (1993). *Matrices and MATLAB: A Tutorial*, Prentice Hall, Upper Saddle River, NJ.

M. Marcus and H. Minc (1964). *A Survey of Matrix Theory and Matrix Inequalities*, Allyn and Bacon, Boston.

J. Markel and A. Gray (1976). *Linear Prediction of Speech*, Springer-Verlag, Berlin and New York.

M. Marrakchi and Y. Robert (1989). "Optimal Algorithms for Gaussian Elimination on an MIMD Computer," *Parallel Computing 12*, 183–194.

R.S. Martin, G. Peters, and J.H. Wilkinson (1970). "The QR Algorithm for Real Hessenberg Matrices," *Numer. Math. 14*, 219–31. See also Wilkinson and Reinsch(1971, pp. 359–71).

R.S. Martin and J.H. Wilkinson (1965). "Symmetric Decomposition of Positive Definite Band Matrices," *Numer. Math. 7*, 355–61.

R.S. Martin and J.H. Wilkinson (1967). "Solution of Symmetric and Unsymmetric Band Equations and the Calculation of Eigenvectors of Band Matrices," *Numer. Math. 9*, 279–301. See also Wilkinson and Reinsch (1971, pp.70–92).

R.S. Martin and J.H. Wilkinson (1968). "Similarity Reduction of a General Matrix to Hessenberg Form," *Numer. Math. 12*, 349–68. See also Wilkinson and Reinsch (1971,pp.339–58).

R.S. Martin and J.H. Wilkinson (1968). "The Modified LR Algorithm for Complex Hessenberg Matrices," *Numer. Math. 12*, 369–76. See also Wilkinson and Reinsch(1971, pp. 396–403).

R.S. Martin and J.H. Wilkinson (1968). "Householder's Tridiagonalization of a Symmetric Matrix," *Numer. Math. 11*, 181-95. See also Wilkinson and Reinsch (1971, pp.212–26).

R.S. Martin and J.H. Wilkinson (1968). "Reduction of a Symmetric Eigenproblem $Ax = \lambda Bx$ and Related Problems to Standard Form," *Numer. Math. 11*, 99–110.

R.S. Martin, G. Peters, and J.H. Wilkinson (1965). "Symmetric Decomposition of a Positive Definite Matrix," *Numer. Math. 7*, 362-83.

R.S. Martin, G. Peters, and J.H. Wilkinson (1966). "Iterative Refinement of the Solution of a Positive Definite System of Equations," *Numer. Math. 8*, 203–16.

R.S. Martin, C. Reinsch, and J.H. Wilkinson (1970). "The QR Algorithm for Band Symmetric Matrices," *Numer. Math. 16*, 85–92. See also See also Wilkinson and Reinsch (1971, pp.266–72).

W.F. Mascarenhas (1994). "A Note on Jacobi Being More Accurate than QR," *SIAM J. Matrix Anal. Appl. 15*, 215–218.

R. Mathias (1992). "Matrices with Positive Definite Hermitian Part: Inequalities and Linear Systems," *SIAM J. Matrix Anal. Appl. 13*, 640–654.

R. Mathias (1992). "Evaluating the Frechet Derivative of the Matrix Exponential," *Numer. Math. 63*, 213–226.

R. Mathias (1993). "Approximation of Matrix-Valued Functions," *SIAM J. Matrix Anal. Appl. 14*, 1061–1063.

R. Mathias (1993). "Perturbation Bounds for the Polar Decomposition," *SIAM J. Matrix Anal. Appl. 14*, 588–597.

R. Mathias (1995). "Accurate Eigensystem Computations by Jacobi Methods," *SIAM J. Matrix Anal. Appl. 16*, 977–1003.

R. Mathias (1995). "The Instability of Parallel Prefix Matrix Multiplication," *SIAM J. Sci. Comp. 16*, 956–973.

R. Mathias and G.W. Stewart (1993). "A Block QR Algorithm and the Singular Value Decomposition," *Lin. Alg. and Its Applic. 182*, 91–100.

K. Mathur and S.L. Johnsson (1994). "Multiplication of Matrices of Arbitrary Shape on a Data Parallel Computer," *Parallel Computing 20*, 919–952.

B. Mattingly, C. Meyer, and J. Ortega (1989). "Orthogonal Reduction on Vector Computers," *SIAM J. Sci. and Stat. Comp. 10*, 372–381.

O. McBryan and E.F. van de Velde (1987). "Hypercube Algorithms and Implementations," *SIAM J. Sci. and Stat. Comp. 8*, s227–s287.

C. McCarthy and G. Strang (1973). "Optimal Conditioning of Matrices," *SIAM J. Num. Anal. 10*, 370–88.

S.F. McCormick (1972). "A General Approach to One-Step Iterative Methods with Application to Eigenvalue Problems," *J. Comput. Sys. Sci. 6*, 354–72.

W.M. McKeeman (1962). "Crout with Equilibration and Iteration," *Comm. ACM. 5*, 553–55.

K. Meerbergen, A. Spence, and D. Roose (1994). "Shift-Invert and Cayley Transforms for the Detection of Rightmost Eigenvalues of Nonsymmetric Matrices," *BIT 34*, 409–423.

V. Mehrmann (1988). "A Symplectic Orthogonal Method for Single Input or Single Output Discrete Time Optimal Quadratic Control Problems," *SIAM J. Matrix Anal. Appl. 9*, 221–247.

V. Mehrmann (1993). "Divide and Conquer Methods for Block Tridiagonal Systems," *Parallel Computing 19*, 257–280.

U. Meier (1985). "A Parallel Partition Method for Solving Banded Systems of Linear Equations," *Parallel Computers 2*, 33–43.

U. Meier and A. Sameh (1988). "The Behavior of Conjugate Gradient Agorithms on a Multivector Processor with a Hierarchical Memory," *J. Comput. Appl. Math. 24,* 13–32.

J.A. Meijerink and H.A. Van der vorst (1977). "An Iterative Solution Method for Linear Equation Systems of Which the Coefficient Matrix is a Symmetric *M*-Matrix," *Math. Comp. 31,* 148–62.

J. Meinguet (1983). "Refined Error Analyses of Cholesky Factorization," *SIAM J. Numer. Anal. 20,* 1243–1250.

R. Melhem(1987). "Toward Efficient Implementation of Preconditioned Conjugate Gradient Methods on Vector Supercomputers," *Int'l J. Supercomputing Applications 1,* 70–98.

M.L. Merriam (1985). "On the Factorization of Block Tridiagonals With Storage Constraints," *SIAM J. Sci. and Stat. Comp. 6,* 182-192.

G. Meurant (1984). "The Block Preconditioned Conjugate Gradient Method on Vector Computers," *BIT 24,* 623–633.

G. Meurant (1989). "Domain Decomposition Methods for Partial Differential Equations on Parallel Computers," to appear *Int'l J. Supercomputing Applications.*

G. Meurant (1992). "A Review on the Inverse of Symmetric Tridiagonal and Block Tridiagonal Matrices," *SIAM J. Matrix Anal. Appl. 13,* 707–728.

C.D. Meyer (1997). *A Course in Applied Linear Algebra,* to be published.

C.D. Meyer and G.W. Stewart (1988). "Derivatives and Perturbations of Eigenvectors," *SIAM J. Num. Anal. 25,* 679–691.

W. Miller (1975). "Computational Complexity and Numerical Stability," *SIAM J. Computing 4,* 97–107.

W. Miller and D. Spooner (1978). "Software for Roundoff Analysis, II," *ACM Trans. Math. Soft. 4,* 369–90.

G. Miminis and C.C. Paige (1982). "An Algorithm for Pole Assignment of Time Invariant Linear Systems," *International J. of Control 35,* 341–354.

L. Mirsky (1960). "Symmetric Gauge Functions and Unitarily Invariant Norms," *Quart. J. Math. 11,* 50-59.

L. Mirsky (1963). *An Introduction to Linear Algebra,* Oxford University Press, Oxford.

J.J. Modi (1988).*Parallel Algorithms and Matrix Computation,* Oxford University Press, Oxford.

J.J. Modi and M.R.B. Clarke (1986). "An Alternative Givens Ordering," *Numer. Math. 43,* 83–90.

J.J. Modi and J.D. Pryce (1985). "Efficient Implementation of Jacobi's Diagonalization Method on the DAP," *Numer. Math. 46,* 443–454.

C.B. Moler (1967). "Iterative Refinement in Floating Point," *J. ACM 14,* 316-71.

C.B. Moler and D. Morrison (1983). "Singular Value Analysis of Cryptograms," *Amer. Math. Monthly 90,* 78–87.

C.B. Moler and G.W. Stewart (1973). "An Algorithm for Generalized Matrix Eigenvalue Problems," *SIAM J. Num. Anal. 10,* 241–56.

C.B. Moler and C.F. Van Loan (1978). "Nineteen Dubious Ways to Compute the Exponential of a Matrix," *SIAM Review 20,* 801–36.

R. Montoye and D. Laurie (1982). "A Practical Algorithm for the Solution of Triangular Systems on a Parallel Processing System," *IEEE Trans. Comp. C-31,* 1076–1082.

M.S. Moonen and B. De Moor, eds. (1995). *SVD and Signal Processing III: Algorithms, Analysis, and Applications.* Elsevier, Amsterdam.

M.S. Moonen, G.H. Golub, and B.L.R. de Moor, eds. (1993). *Linear Algebra for Large Scale and Real-Time Applications,* Kluwer, Dordrecht, The Netherlands.

M.S. Moonen, P. Van Dooren, and J. Vandewalle (1992). "A Singular Value Decomposition Updating Algorithm," *SIAM J. Matrix Anal. Appl. 13,* 1015–1038.

R.B. Morgan (1995). "A Restarted GMRES Method Augmented with Eigenvectors," *SIAM J. Matrix Anal. Applic. 16,* 1154–1171.

R.B. Morgan (1996). "On Restarting the Arnoldi Method for Large Scale Eigenvalue Problems," *Math Comp,* to appear.

M. Mu. (1995). "A New family of Preconditioners for Domain Decomposition," *SIAM J. Sci. Comp. 16*, 289–306.

D. Mueller (1966). "Householder's Method for Complex Matrices and Hermitian Matrices," *Numer. Math. 8*, 72–92.

F.D. Murnaghan and A. Wintner (1931). "A Canonical Form for Real Matrices Under Orthogonal Transformations," *Proc. Nat. Acad. Sci. 17*, 417–20.

N. Nachtigal, S. Reddy, and L. Trefethen (1992). "How Fast Are Nonsymmetric Matrix Iterations," *SIAM J. Matrix Anal. Appl. 13*, 778–795.

N. Nachtigal, L. Reichel, and L. Trefethen (1992). "A Hybrid GMRES Algorithm for Nonsymmetric Linear Systems," *SIAM J. Matrix Anal. Appl. 13*, 796–825.

T. Nanda (1985). "Differential Equations and the QR Algorithm," *SIAM J. Numer. Anal. 22*, 310–321.

J.C. Nash (1975). "A One-Sided Tranformation Method for the Singular Value Decomposition and Algebraic Eigenproblem," *Comp. J. 18*, 74–76.

M.Z. Nashed (1976). *Generalized Inverses and Applications*, Academic Press, New York.

R.A. Nicolaides (1974). "On a Geometrical Aspect of SOR and the Theory of Consistent Ordering for Positive Definite Matrices," *Numer. Math. 12*, 99–104.

W. Niethammer and R.S. Varga (1983). "The Analysis of k-step Iterative Methods for Linear Systems from Summability Theory," *Numer. Math. 41*, 177–206.

B. Noble and J.W. Daniel (1977). *Applied Linear Algebra*, Prentice-Hall, Englewood Cliffs.

Y. Notay (1992). "On the Robustness of Modified Incomplete Factorization Methods," *J. Comput. Math. 40*, 121–141.

C. Oara (1994). "Proper Deflating Subspaces: Properties, Algorithms, and Applications," *Numerical Algorithms 7*, 355–373.

W. Oettli and W. Prager (1964). "Compatibility of Approximate Solutions of Linear Equations with Given Error Bounds for Coefficients and Right Hand Sides," *Numer. Math. 6*, 405–409.

D.P. O'Leary (1980). "Estimating Matrix Condition Numbers," *SIAM J. Sci. Stat. Comp. 1*, 205–9.

D.P. O'Leary (1980). "The Block Conjugate Gradient Algorithm and Related Methods," *Lin. Alg. and Its Applic. 29*, 293–322.

D.P. O'Leary (1987). "Parallel Implementation of the Block Conjugate Gradient Algorithm," *Parallel Computers 5*, 127–140.

D.P. O'Leary (1990). "On Bounds for Scaled Projections and Pseudoinverses," *Lin. Alg. and Its Applic. 132*, 115–117.

D.P. O'Leary and J.A. Simmons (1981). "A Bidiagonalization-Regularization Procedure for Large Scale Discretizations of Ill-Posed Problems," *SIAM J. Sci. and Stat. Comp. 2*, 474–489.

D.P. O'Leary and G.W. Stewart (1985). "Data Flow Algorithms for Parallel Matrix Computations," *Comm. ACM 28*, 841-853.

D.P. O'Leary and G.W. Stewart (1986). "Assignment and Scheduling in Parallel Matrix Factorization," *Lin. Alg. and Its Applic. 77*, 275–300.

S.J. Olszanskyj, J.M. Lebak, and A.W. Bojanczyk (1994). "Rank-k Modification Methods for Recursive Least Squares Problems," *Numerical Algorithms 7*, 325–354.

A.V. Oppenheim (1978). *Applications of Digital Signal Processing* , Prentice-Hall, Englewood Cliffs.

J.M. Ortega (1987). *Matrix Theory: A Second Course*, Plenum Press, New York.

J.M. Ortega (1988). "The *ijk* Forms of Factorization Methods I: Vector Computers," *Parallel Computers 7*, 135–147.

J.M. Ortega (1988). *Introduction to Parallel and Vector Solution of Linear Systems*, Plenum Press, New York.

J.M. Ortega and C.H. Romine (1988). "The *ijk* Forms of Factorization Methods II: Parallel Systems," *Parallel Computing 7*, 149–162.

J.M. Ortega and R.G. Voigt (1985). "Solution of Partial Differential Equations on Vector and Parallel Computers," *SIAM Review 27*, 149–240.

E.E. Osborne (1960). "On Preconditioning of Matrices," *JACM 7*, 338–45.

M.H.C. Paardekooper (1971). "An Eigenvalue Algorithm for Skew Symmetric Matrices," *Numer. Math. 17*, 189–202.

M.H.C. Paardekooper (1991). "A Quadratically Convergent Parallel Jacobi Process for Diagonally Dominant Matrices with Nondistinct Eigenvalues," *Lin.Alg. and Its Applic. 145*, 71–88.

C.C. Paige (1970). "Practical Use of the Symmetric Lanczos Process with Reorthogonalization," *BIT 10*, 183–95.

C.C. Paige (1971). "The Computation of Eigenvalues and Eigenvectors of Very Large Sparse Matrices," Ph.D. thesis, London University.

C.C. Paige (1972). "Computational Variants of the Lanczos Method for the Eigenproblem," *J. Inst. Math. Applic. 10*, 373–81.

C.C. Paige (1973). "An Error Analysis of a Method for Solving Matrix Equations," *Math. Comp. 27*, 355–59.

C.C. Paige (1974). "Bidiagonalization of Matrices and Solution of Linear Equations," *SIAM J. Num. Anal. 11*, 197–209.

C.C. Paige (1974). "Eigenvalues of Perturbed Hermitian Matrices," *Lin. Alg. and Its Applic . 8*, 1–10.

C.C. Paige (1976). "Error Analysis of the Lanczos Algorithm for Tridiagonalizing Symmetric Matrix," *J. Inst. Math. Applic. 18*, 341–49.

C.C. Paige (1979). "Computer Solution and Perturbation Analysis of Generalized Least Squares Problems," *Math. Comp. 33*, 171–84.

C.C. Paige (1979). "Fast Numerically Stable Computations for Generalized Linear Least Squares Problems," *SIAM J. Num. Anal. 16*, 165–71.

C.C. Paige (1980). "Accuracy and Effectiveness of the Lanczos Algorithm for the Symmetric Eigenproblem," *Lin. Alg. and Its Applic. 34*, 235–58.

C.C. Paige (1981). "Properties of Numerical Algorithms Related to Computing Controllability," *IEEE Trans. Auto. Cont. AC-26*, 130–38.

C.C. Paige (1984). "A Note on a Result of Sun J.-Guang: Sensitivity of the CS and GSV Decompositions," *SIAM J. Numer. Anal. 21*, 186–191.

C.C. Paige (1985). "The General Linear Model and the Generalized Singular Value Decomposition," *Lin. Alg. and Its Applic. 70*, 269–284.

C.C. Paige (1986). "Computing the Generalized Singular Value Decomposition," *SIAM J. Sci. and Stat. Comp. 7*, 1126–1146.

C.C. Paige (1990). "Some Aspects of Generalized QR Factorization," in *Reliable Numerical Computations*, M. Cox and S. Hammarling (eds), Clarendon Press, Oxford.

C.C. Paige, B.N. Parlett,and H.A. Van Der Vorst (1995). "Approximate Solutions and Eigenvalue Bounds from Krylov Subspaces," *Numer. Linear Algebra with Applic. 2*, 115–134.

C.C. Paige and M.A. Saunders (1975). "Solution of Sparse Indefinite Systems of Linear Equations," *SIAM J. Num. Anal. 12*, 617–29.

C.C. Paige and M. Saunders (1981). "Toward a Generalized Singular Value Decomposition," *SIAM J. Num. Anal. 18*, 398–405.

C.C. Paige and M.A. Saunders (1982). "LSQR: An Algorithm for Sparse Linear Equations and Sparse Least Squares," *ACM Trans. Math. Soft. 8*, 43–71.

C.C. Paige and M.A. Saunders (1982). "Algorithm 583 LSQR: Sparse Linear Equations and Least Squares Problems," *ACM Trans. Math. Soft. 8*, 195–209.

C.C. Paige and P. Van Dooren (1986). "On the Quadratic Convergence of Kogbetliantz's Algorithm for Computing the Singular Value Decomposition," *Lin. Alg. and Its Applic. 77*, 301–313.

C.C. Paige and C. Van Loan (1981). "A Schur Decomposition for Hamiltonian Matrices," *Lin. Alg. and Its Applic. 41*, 11–32.

C.C. Paige and M. Wei (1993). "Analysis of the Generalized Total Least Squares Problem $AX = B$ when Some of the Columns are Free of Error," *Numer. Math. 65*, 177–202.

C.C. Paige and M. Wei (1994). "History and Generality of the CS Decomposition," *Lin. Alg. and Its Applic. 208/209*, 303–326.

C.-T. Pan (1993). "A Perturbation Analysis of the Problem of Downdating a Cholesky Factorization," *Lin. Alg. and Its Applic. 183*, 103–115.

V. Pan (1984). "How Can We Speed Up Matrix Multiplication?," *SIAM Review 26*, 393–416.

H. Park (1991). "A Parallel Algorithm for the Unbalanced Orthogonal Procrustes Problem," *Parallel Computing 17*, 913–923.

H. Park and L. Eldén (1995). "Downdating the Rank-Revealing URV Decomposition," *SIAM J. Matrix Anal. Appl. 16*, 138–155.

B.N. Parlett (1965). "Convergence of the Q-R Algorithm," *Numer. Math. 7*, 187–93. (Correction in *Numer. Math. 10*, 163–64.)

B.N. Parlett (1966). "Singular and Invariant Matrices Under the QR Algorithm," *Math. Comp. 20*, 611–15.

B.N. Parlett (1967). "Canonical Decomposition of Hessenberg Matrices," *Math. Comp. 21*, 223–27.

B.N. Parlett (1968). "Global Convergence of the Basic QR Algorithm on Hessenberg Matrices," *Math. Comp. 22*, 803–17.

B.N. Parlett (1971). "Analysis of Algorithms for Reflections in Bisectors," *SIAM Review 13*, 197–208.

B.N. Parlett (1974). "The Rayleigh Quotient Iteration and Some Generalizations for Nonnormal Matrices," *Math. Comp. 28*, 679–93.

B.N. Parlett (1976). "A Recurrence Among the Elements of Functions of Triangular Matrices," *Lin. Alg. and Its Applic. 14*, 117–21.

B.N. Parlett (1980). *The Symmetric Eigenvalue Problem*, Prentice-Hall, Englewood Cliffs, NJ.

B.N. Parlett (1980). "A New Look at the Lanczos Algorithm for Solving Symmetric Systems of Linear Equations," *Lin. Alg. and Its Applic. 29*, 323–46.

B.N. Parlett (1992). "Reduction to Tridiagonal Form and Minimal Realizations," *SIAM J. Matrix Anal. Appl. 13*, 567–593.

B.N. Parlett (1995). "The New qd Algorithms," *ACTA Numerica 5*, 459–491.

B.N. Parlett and B. Nour-Omid (1985). "The Use of a Refined Error Bound When Updating Eigenvalues of Tridiagonals," *Lin. Alg. and Its Applic. 68*, 179–220.

B.N. Parlett and W.G. Poole (1973). "A Geometric Theory for the QR, LU, and Power Iterations," *SIAM J. Num. Anal. 10*, 389–412.

B.N. Parlett and J.K. Reid (1970). "On the Solution of a System of Linear Equations Whose Matrix is Symmetric but not Definite," *BIT 10*, 386–97.

B.N. Parlett and J.K. Reid (1981). "Tracking the Progress of the Lanczos Algorithm for Large Symmetric Eigenproblems," *IMA J. Num. Anal. 1*, 135–55.

B.N. Parlett and C. Reinsch (1969). "Balancing a Matrix for Calculation of Eigenvalues and Eigenvectors," *Numer. Math. 13*, 292–304. See also Wilkinson and Reinsch(1971, pp. 315-26).

B.N. Parlett and R. Schreiber (1988). "Block Reflectors: Theory and Computation," *SIAM J. Num. Anal. 25*, 189–205.

B.N. Parlett and D.S. Scott (1979). "The Lanczos Algorithm with Selective Orthogonalization," *Math. Comp. 33*, 217–38.

B.N. Parlett, H. Simon, and L.M. Stringer (1982). "On Estimating the Largest Eigenvalue with the Lanczos Algorithm," *Math. Comp. 38*, 153–166.

B.N. Parlett, D. Taylor, and Z. Liu (1985). "A Look-Ahead Lanczos Algorithm for Unsymmetric Matrices," *Math. Comp. 44*, 105–124.

N. Patel and H. Jordan (1984). "A Parallelized Point Rowwise Successive Over-Relaxation Method on a Multiprocessor," *Parallel Computing 1*, 207–222.

R.V. Patel, A.J. Laub, and P.M. Van Dooren, eds. (1994). *Numerical Linear Algebra Techniques for Systems and Control*, IEEE Press, Piscataway, New Jersey.

D.A. Patterson and J.L. Hennessy (1989). *Computer Architecture: A Quantitative Approach*, Morgan Kaufmann Publishers, Inc., Palo Alto, CA.

M.S. Paterson and L.J. Stockmeyer (1973). "On the Number of Nonscalar Multiplications Necessary to Evaluate Polynomials," *SIAM J. Comp. 2*, 60–66.

K. Pearson (1901). "On Lines and Planes of Closest Fit to Points in Space," *Phil. Mag.* 2, 559–72.

G. Peters and J.H. Wilkinson (1969). "Eigenvalues of $Ax = \lambda Bx$ with Band Symmetric A and B," *Comp. J. 12*, 398-404.

G. Peters and J.H. Wilkinson (1970). "The Least Squares Problem and Pseudo-Inverses," *Comp. J. 13*, 309–16.

G. Peters and J.H. Wilkinson (1970). "$Ax = \lambda Bx$ and the Generalized Eigenproblem," *SIAM J. Num. Anal. 7*, 479–92.

G. Peters and J.H. Wilkinson (1971). "The Calculation of Specified Eigenvectors by Inverse Iteration," in Wilikinson and Reinsch (1971, pp.418–39).

G. Peters and J.H. Wilkinson (1979). "Inverse Iteration, Ill-Conditioned Equations, and Newton's Method," *SIAM Review 21*, 339–60.

D.J. Pierce and R.J. Plemmons (1992). "Fast Adaptive Condition Estimation," *SIAM J. Matrix Anal. Appl. 13*, 274–291.

S. Pissanetsky (1984). *Sparse Matrix Technology*, Academic Press, New York.

R.J. Plemmons (1974). "Linear Least Squares by Elimination and MGS," *J. Assoc. Comp. Mach. 21*, 581–85.

R.J. Plemmons (1986). "A Parallel Block Iterative Scheme Applied to Computations in Structural Analysis," *SIAM J. Alg. and Disc. Methods 7*, 337–347.

R.J. Plemmons and C.D. Meyer, eds. (1993). *Linear Algebra, Markov Chains, and Queuing Models*, Springer-Verlag, New York.

A. Pokrzywa (1986). "On Perturbations and the Equivalence Orbit of a Matrix Pencil," *Lin. Alg. and Applic. 82*, 99–121.

E.L. Poole and J.M. Ortega (1987). "Multicolor ICCG Methods for Vector Computers," *SIAM J. Numer. Anal. 24*, 1394–1418.

D.A. Pope and C. Tompkins (1957). "Maximizing Functions of Rotations: Experiments Concerning Speed of Diagonalization of Symmetric Matrices Using Jacobi's Method," *J. ACM 4*, 459–66.

A. Pothen, S. Jha, and U. Vemapulati (1987). "Orthogonal Factorization on a Distributed Memory Multiprocessor," in *Hypercube Multiprocessors*, ed. M.T. Heath, SIAM Publications, 1987.

M.J.D. Powell and J.K. Reid (1968). "On Applying Householder's Method to Linear Least Squares Problems," *Proc. IFIP Congress*, pp. 122–26.

R. Pratap (1995). *Getting Started with* MATLAB, Saunders College Publishing, Fort Worth, TX.

J.D. Pryce (1984). "A New Measure of Relative Error for Vectors," *SIAM J. Num. Anal. 21*, 202-21.

C. Puglisi (1992). "Modification of the Householder Method Based on the Compact WY Representation," *SIAM J. Sci. and Stat. Comp. 13*, 723–726.

S. Qiao(1986). "Hybrid Algorithm for Fast Toeplitz Orthogonalization," *Numer. Math. 53*, 351–366.

S. Qiao (1988). "Recursive Least Squares Algorithm for Linear Prediction Problems," *SIAM J. Matrix Anal. Appl. 9*, 323–328.

C.M. Rader and A.O. Steinhardt (1988). "Hyperbolic Householder Transforms," *SIAM J. Matrix Anal. Appl. 9*, 269–290.

G. Radicati di Brozolo and Y. Robert (1989). "Parallel Conjugate Gradient-like Algorithms for Solving Sparse Nonsymmetric Linear Systems on a Vector Multiprocessor," *Parallel Computing 11*, 233-240.

P. Raghavan (1995). "Distributed Sparse Gaussian Elimination and Orthogonal Factorization," *SIAM J. Sci. Comp. 16*, 1462–1477.

W. Rath (1982). "Fast Givens Rotations for Orthogonal Similarity," *Numer. Math. 40*, 47–56.

P.A. Regalia and S. Mitra (1989). "Kronecker Products, Unitary Matrices, and Signal Processing Applications," *SIAM Review 31*, 586–613.

L. Reichel (1991). "Fast QR Decomposition of Vandermonde-Like Matrices and Polynomial Least Squares Approximation," *SIAM J. Matrix Anal. Appl. 12*, 552–564.

L. Reichel and L.N. Trefethen (1992). "Eigenvalues and Pseudo-eigenvalues of Toeplitz Matrices," *Lin. Alg. and Its Applic. 162/163/164*, 153–186.

J.K. Reid (1967). "A Note on the Least Squares Solution of a Band System of Linear Equations by Householder Reductions," *Comp. J. 10*, 188–89.

J.K. Reid (1971). "A Note on the Stability of Gaussian Elimination," *J. Inst. Math. Applic. 8*, 374–75.

J.K. Reid (1971). " On the Method of Conjugate Gradients for the Solution of Large Sparse Systems of Linear Equations," in *Large Sparse Sets of Linear Equations* , ed. J.K. Reid, Academic Press, New York, pp. 231–54.

J.K. Reid (1972). "The Use of Conjugate Gradients for Systems of Linear Equations Possessing Property A," *SIAM J. Num. Anal. 9*, 325–32.

C. Reinsch and F.L. Bauer (1968). "Rational QR Transformation with Newton's Shift for Symmetric Tridiagonal Matrices," *Numer. Math. 11*, 264–72. See also Wilkinson and Reinsch (1971, pp.257–65).

J.R. Rice (1966). "A Theory of Condition," *SIAM J. Num. Anal. 3*, 287–310.

J.R. Rice (1966). "Experiments on Gram-Schmidt Orthogonalization," *Math. Comp. 20*, 325–28.

J.R. Rice (1981). *Matrix Computations and Mathematical Software*, Academic Press, New York.

R.F. Rinehart (1955). "The Equivalence of Definitions of a Matric Function," *Amer. Math. Monthly 62*, 395–414.

Y. Robert (1990). *The Impact of Vector and Parallel Architectures on the Gaussian Elimination Algorithm*, Halsted Press, New York.

H.H. Robertson (1977). "The Accuracy of Error Estimates for Systems of Linear Algebraic Equations," *J. Inst. Math. Applic. 20*, 409–14.

G. Rodrigue (1973). "A Gradient Method for the Matrix Eigenvalue Problem $Ax = \lambda Bx$," *Numer. Math. 22*, 1–16.

G. Rodrigue, ed. (1982). *Parallel Computation*, Academic Press, New York.

G. Rodrigue and D. Wolitzer (1984). "Preconditioning by Incomplete Block Cyclic Reduction," *Math. Comp. 42*, 549–566.

C.H. Romine and J.M. Ortega (1988). "Parallel Solution of Triangular Systems of Equations," *Parallel Computing 6*, 109–114.

D.J. Rose (1969). "An Algorithm for Solving a Special Class of Tridiagonal Systems of Linear Equations," *Comm. ACM 12*, 234–36.

D.J. Rose and R. A. Willoughby, eds. (1972). *Sparse Matrices and Their Applications*, Plenum Press, New York, 1972

A. Ruhe (1968). On the Quadratic Convergence of a Generalization of the Jacobi Method to Arbitrary Matrices," *BIT 8*, 210–31.

A. Ruhe (1969). "The Norm of a Matrix After a Similarity Transformation," *BIT 9*, 53–58.

A. Ruhe (1970). "An Algorithm for Numerical Determination of the Structure of a General Matrix," *BIT 10*, 196–216.

A. Ruhe (1970). "Perturbation Bounds for Means of Eigenvalues and Invariant Subspaces," *BIT 10*, 343–54.

A. Ruhe (1970). "Properties of a Matrix with a Very Ill-Conditioned Eigenproblem," *Numer. Math. 15*, 57–60.

A. Ruhe (1972). "On the Quadratic Convergence of the Jacobi Method for Normal Matrices," *BIT 7*, 305–13.

A. Ruhe (1974). "SOR Methods for the Eigenvalue Problem with Large Sparse Matrices," *Math. Comp. 28*, 695–710.

A. Ruhe (1975). "On the Closeness of Eigenvalues and Singular Values for Almost Normal Matrices," *Lin. Alg. and Its Applic. 11*, 87–94.

A. Ruhe (1979). "Implementation Aspects of Band Lanczos Algorithms for Computation of Eigenvalues of Large Sparse Symmetric Matrices," *Math. Comp. 33*, 680–87.

A. Ruhe (1983). "Numerical Aspects of Gram-Schmidt Orthogonalization of Vectors," *Lin. Alg. and Its Applic. 52/53*, 591–601.

A. Ruhe (1984). "Rational Krylov Algorithms for Eigenvalue Computation," *Lin. Alg. and Its Applic. 58*, 391–405.

A. Ruhe (1987). "Closest Normal Matrix Found!," *BIT 27*, 585-598.

A. Ruhe (1994). "Rational Krylov Algorithms for Nonsymmetric Eigenvalue Problems II. Matrix Pairs," *Lin. Alg. and Its Applic. 197*, 283–295.

A. Ruhe (1994). "The Rational Krylov Algorithm for Nonsymmetric Eigenvalue Problems III: Complex Shifts for Real Matrices," *BIT 34*,165–176.

A. Ruhe and T. Wiberg (1972). "The Method of Conjugate Gradients Used in Inverse Iteration," *BIT 12*, 543–54.

H. Rutishauser (1958). "Solution of Eigenvalue Problems with the LR Transformation," *Nat. Bur. Stand. App. Math. Ser. 49*, 47–81.

H. Rutishauser (1966). "Bestimmung der Eigenwerte Orthogonaler Matrizen," *Numer. Math. 9*, 104–108.

H. Rutishauser (1966). "The Jacobi Method for Real Symmetric Matrices," *Numer. Math. 9*, 1–10. See also Wilkinson and Reinsch (1971, pp. 202–11).

H. Rutishauser (1970). "Simultaneous Iteration Method for Symmetric Matrices," *Numer. Math. 16*, 205–23. See also Wilkinson and Reinsch (1971,pp.284–302).

Y. Saad (1980). "On the Rates of Convergence of the Lanczos and the Block Lanczos Methods," *SIAM J. Num. Anal.17*, 687–706.

Y. Saad (1980). "Variations of Arnoldi's Method for Computing Eigenelements of Large Unsymmetric Matrices.," *Lin. Alg. and Its Applic. 34*, 269–295.

Y. Saad (1981). "Krylov Subspace Methods for Solving Large Unsymmetric Linear Systems," *Math. Comp. 37*, 105–126.

Y. Saad (1982). "The Lanczos Biorthogonalization Algorithm and Other Oblique Projection Metods for Solving Large Unsymmetric Systems," *SIAM J. Numer. Anal. 19*, 485–506.

Y. Saad (1984). "Practical Use of Some Krylov Subspace Methods for Solving Indefinite and Nonsymmetric Linear Systems," *SIAM J. Sci. and Stat. Comp. 5*, 203–228.

Y. Saad (1985). "Practical Use of Polynomial Preconditionings for the Conjugate Gradient Method," *SIAM J. Sci. and Stat. Comp. 6*, 865–882.

Y. Saad (1986). "On the Condition Number of Some Gram Matrices Arising from Least Squares Approximation in the Complex Plane," *Numer. Math. 48*, 337–348.

Y. Saad (1987). "On the Lanczos Method for Solving Symmetric Systems with Several Right Hand Sides," *Math. Comp. 48*, 651–662.

Y. Saad (1988). "Preconditioning Techniques for Indefinite and Nonsymmetric Linear Systems," *J. Comput. Appl. Math. 24*, 89–105.

Y. Saad (1989). "Krylov Subspace Methods on Supercomputers," *SIAM J. Sci. and Stat. Comp. 10*, 1200–1322.

Y. Saad (1992). *Numerical Methods for Large Eigenvalue Problems: Theory and Algorithms*, John Wiley and Sons, New York.

Y. Saad (1993). "A Flexible Inner-Outer Preconditioned GMRES Algorithm," *SIAM J. Sci. Comput. 14*, 461–469.

Y. Saad (1996). *Iterative Methods for Sparse Linear Systems*, PWS Publishing Co., Boston.

Y. Saad and M.H. Schultz (1985). "Conjugate Gradient-Like Algorithms for Solving Nonsymmetric Linear Systems," *Math. Comp. 44*, 417–424.

Y. Saad and M.H. Schultz (1986). "GMRES: A Generalized Minimal Residual Algorithm for Solving Nonsymmetric Linear Systems," *SIAM J. Scientific and Stat. Comp. 7*, 856–869.

Y. Saad and M.H. Schultz (1989). "Data Communication in Parallel Architectures," *J. Dist. Parallel Comp. 11*, 131–150.

Y. Saad and M.H. Schultz (1989). "Data Communication in Hypercubes," *J. Dist. Parallel Comp. 6*, 115–135.

A. Sameh (1971). "On Jacobi and Jacobi-like Algorithms for a Parallel Computer," *Math. Comp. 25*, 579–90.

A. Sameh and D. Kuck (1978). "On Stable Parallel Linear System Solvers," *J. Assoc. Comp. Mach. 25*, 81-91.

A. Sameh, J. Lermit and K. Noh (1975). "On the Intermediate Eigenvalues of Symmetric Sparse Matrices," *BIT 12*, 543–54.

M.A. Sanders (1995). "Solution of Sparse Rectangular Systems," *BIT 35*, 588–604.

M.A. Saunders, H.D. Simon, and E.L. Yip (1988). "Two Conjugate Gradient-Type Methods for Unsymmetric Linear Equations," *SIAM J. Num. Anal. 25*, 927–940.

K. Schittkowski and J. Stoer (1979). "A Factorization Method for the Solution of Constrained Linear Least Squares Problems Allowing for Subsequent Data changes," *Numer. Math. 31*, 431–463.

W. Schönauer (1987). *Scientific Computing on Vector Computers*, North Holland, Amsterdam.

P. Schonemann (1966). "A Generalized Solution of the Orthogonal Procrustes Problem," *Psychometrika 31*, 1–10.

A. Schonhage (1964). "On the Quadratic Convergence of the Jacobi Process," *Numer. Math. 6*, 410–12.

A. Schonhage (1979). "Arbitrary Perturbations of Hermitian Matrices," *Lin. Alg. and Its Applic. 24*, 143–49.

R.S. Schreiber (1986). "Solving Eigenvalue and Singular Value Problems on an Undersized Systolic Array," *SIAM J. Sci. and Stat. Comp. 7*, 441–451.

R.S. Schreiber (1988). "Block Algorithms for Parallel Machines," in *Numerical Algorithms for Modern Parallel Computer Architectures*, M.H. Schultz (ed), IMA Volumes in Mathematics and Its Applications, Number 13, Springer-Verlag, Berlin, 197–207.

R.S. Schreiber and B.N. Parlett (1987). "Block Reflectors: Theory and Computation," *SIAM J. Numer. Anal. 25*, 189-205.

R.S. Schreiber and C. Van Loan (1989). "A Storage-Efficient WY Representation for Products of Householder Transformations," *SIAM J. Sci. and Stat. Comp. 10*, 52–57.

M.H. Schultz, ed. (1988). *Numerical Algorithms for Modern Parallel Computer Architectures*, IMA Volumes in Mathematics and Its Applications, Number 13, Springer-Verlag, Berlin.

I. Schur (1909). "On the Characteristic Roots of a Linear Substitution with an Application to the Theory of Integral Equations." *Math. Ann. 66*, 488-510 (German).

H.R. Schwartz (1968). "Tridiagonalization of a Symmetric Band Matrix," *Numer. Math. 12*, 231–41. See also Wilkinson and Reinsch (1971, 273–83).

H.R. Schwartz (1974). "The Method of Coordinate Relaxation for $(A - \lambda B)x = 0$," *Num. Math. 23*, 135–52.

D. Scott (1978). "Analysis of the Symmetric Lanczos Process," Electronic Research Laboratory Technical Report UCB/ERL M78/40, University of California, Berkeley.

D.S. Scott (1979). "Block Lanczos Software for Symmetric Eigenvalue Problems," Report ORNL/CSD-48, Oak Ridge National Laboratory, Union Carbide Corporation, Oak Ridge, Tennessee.

D.S. Scott (1979). "How to Make the Lanczos Algorithm Converge Slowly," *Math. Comp. 33*, 239–47.

D.S. Scott (1984). "Computing a Few Eigenvalues and Eigenvectors of a Symmetric Band Matrix," *SIAM J. Sci. and Stat. Comp. 5*, 658–666.

D.S. Scott (1985). "On the Accuracy of the Gershgorin Circle Theorem for Bounding the Spread of a Real Symmetric Matrix," *Lin. Alg. and Its Applic. 65*, 147–155

D.S. Scott, M.T. Heath, and R.C. Ward (1986). "Parallel Block Jacobi Eigenvalue Algorithms Using Systolic Arrays," *Lin. Alg. and Its Applic. 77*, 345–356.

M.K. Seager (1986). "Parallelizing Conjugate Gradient for the Cray X-MP," *Parallel Computing 3*, 35–47.

J.J. Seaton (1969). "Diagonalization of Complex Symmetric Matrices Using a Modified Jacobi Method," *Comp. J. 12*, 156–57.

S. Serbin (1980). "On Factoring a Class of Complex Symmetric Matrices Without Pivoting," *Math. Comp. 35*, 1231-1234.

S. Serbin and S. Blalock (1979). "An Algorithm for Computing the Matrix Cosine," *SIAM J. Sci. Stat. Comp. 1*, 198–204.

J.W. Sheldon (1955). "On the Numerical Solution of Elliptic Difference Equations," *Math. Tables Aids Comp. 9*, 101–12.

W. Shougen and Z. Shuqin (1991). "An Algorithm for $Ax = \lambda Bx$ with Symmetric and Positive Definite A and B," *SIAM J. Matrix Anal. Appl. 12*, 654–660.

G. Shroff (1991). "A Parallel Algorithm for the Eigenvalues and Eigenvectors of a General Complex Matrix," *Numer. Math. 58*, 779–806.

G. Shroff and C.H. Bischof (1992). "Adaptive Condition Estimation for Rank-One Updates of QR Factorizations," *SIAM J. Matrix Anal. Appl. 13*, 1264–1278.

G. Shroff and R. Schreiber (1989). "On the Convergence of the Cyclic Jacobi Method for Parallel Block Orderings," *SIAM J. Matrix Anal. Appl. 10*, 326–346.

H. Simon (1984). "Analysis of the Symmetric Lanczos Algorithm with Reorthogonalization Methods," *Lin. Alg. and Its Applic. 61*, 101–132.

B. Singer and S. Spilerman (1976). "The Representation of Social Processes by Markov Models," *Amer. J. Sociology 82*, 1–54.

R.D. Skeel (1979). "Scaling for numerical stability in Gaussian Elimination," *J. ACM 26*, 494–526.

R.D. Skeel (1980). "Iterative Refinement Implies Numerical Stability for Gaussian Elimination," *Math. Comp. 35*, 817–832.

R.D. Skeel (1981). "Effect of Equilibration on Residual Size for Partial Pivoting," *SIAM J. Num. Anal. 18*, 449–55.

G.L.G. Sleijpen and D.R. Fokkema (1993). "BICGSTAB(ℓ) for Linear Equations Involving Unsymmetric Matrices with Complex Spectrum," *Electronic Transactions on Numerical Analysis 1*, 11–32.

B.T. Smith, J.M. Boyle, Y. Ikebe, V.C. Klema, and C.B. Moler (1970). *Matrix Eigensystem Routines: EISPACK Guide, 2nd ed.*, Lecture Notes in Computer Science, Volume 6, Springer-Verlag, New York.

R.A. Smith (1967). "The Condition Numbers of the Matrix Eigenvalue Problem," *Numer. Math. 10* 232–240.

F. Smithies (1970). *Integral Equations*, Cambridge University Press, Cambridge.

P. Sonneveld (1989). "CGS, A Fast Lanczos-Type Solver for Nonsymmetric Linear Systems," *SIAM J. Sci. and Stat. Comp. 10*, 36–52.

D.C. Sorensen (1992). "Implicit Application of Polynomial Filters in a k-Step Arnoldi Method," *SIAM J. Matrix Anal. Appl. 13*, 357–385.

D.C. Sorensen (1995). "Implicitly Restarted Arnoldi/Lanczos Methods for Large Scale Eigenvalue Calculations," in *Proceedings of the ICASE/LaRC Workshop on Parallel Numerical Algorithms, May 23–25, 1994*, D.E. Keyes, A. Sameh, and V. Venkatakrishnan (eds), Kluwer.

G.W. Stewart (1969). "Accelerating The Orthogonal Iteration for the Eigenvalues of a Hermitian Matrix," *Numer. Math. 13*, 362–76.

G.W. Stewart (1970). "Incorporating Original Shifts into the QR Algorithm for Symmetric Tridiagonal Matrices," *Comm. ACM 13*, 365–67.

G.W. Stewart (1971). "Error Bounds for Approximate Invariant Subspaces of Closed Linear Operators," *SIAM. J. Num. Anal. 8*, 796–808.

G.W. Stewart (1972). "On the Sensitivity of the Eigenvalue Problem $Ax = \lambda Bx$," *SIAM J. Num. Anal. 9*, 669–86.

G.W. Stewart (1973). "Error and Perturbation Bounds for Subspaces Associated with Certain Eigenvalue Problems," *SIAM Review 15*, 727–64.

G.W. Stewart (1973). *Introduction to Matrix Computations*, Academic Press, New York.

G.W. Stewart (1973). "Conjugate Direction Methods for Solving Systems of Linear Equations," *Numer. Math. 21*, 284–97.

G.W. Stewart (1974). "The Numerical Treatment of Large Eigenvalue Problems," *Proc. IFIP Congress 74*, North-Holland, pp. 666–72.

G.W. Stewart (1975). "The Convergence of the Method of Conjugate Gradients at Isolated Extreme Points in the Spectrum," *Numer. Math. 24*, 85–93.

G.W. Stewart (1975). "Gershgorin Theory for the Generalized Eigenvalue Problem $Ax = \lambda Bx$," *Math. Comp. 29*, 600–606.

G.W. Stewart (1975). "Methods of Simultaneous Iteration for Calculating Eigenvectors of Matrices," in *Topics in Numerical Analysis II* , ed. John J.H. Miller, Academic Press, New York, pp. 185–96.

G.W. Stewart (1976). "The Economical Storage of Plane Rotations," *Numer. Math. 25*, 137–38.

G.W. Stewart (1976). "Simultaneous Iteration for Computing Invariant Subspaces of Non-Hermitian Matrices," *Numer. Math. 25*, 123–36.

G.W. Stewart (1976). "Algorithm 406: HQR3 and EXCHNG: Fortran Subroutines for Calculating and Ordering the Eigenvalues of a Real Upper Hessenberg Matrix," *ACM Trans. Math. Soft. 2*, 275–80.

G.W. Stewart (1976). "A Bibliographical Tour of the Large Sparse Generalized Eigenvalue Problem," in *Sparse Matrix Computations* , ed., J.R. Bunch and D.J. Rose, Academic Press, New York.

G.W. Stewart (1977). "Perturbation Bounds for the QR Factorization of a Matrix," *SIAM J. Num. Anal. 14*, 509–18.

G.W. Stewart (1977). "On the Perturbation of Pseudo-Inverses, Projections and Linear Least Squares Problems," *SIAM Review 19*, 634–662.

G.W. Stewart (1977). "Sensitivity Coefficients for the Effects of Errors in the Independent Variables in a Linear Regression," Technical Report TR-571, Department of Computer Science, University of Maryland, College Park, MD.

G.W. Stewart (1978). "Perturbation Theory for the Generalized Eigenvalue Problem'", in *Recent Advances in Numerical Analysis* , ed. C. de Boor and G.H. Golub, Academic Press, New York.

G.W. Stewart (1979). "A Note on the Perturbation of Singular Values," *Lin. Alg. and Its Applic. 28*, 213–16.

G.W. Stewart (1979). "Perturbation Bounds for the Definite Generalized Eigenvalue Problem," *Lin. Alg. and Its Applic. 23*, 69–86.

G.W. Stewart (1979). "The Effects of Rounding Error on an Algorithm for Downdating a Cholesky Factorization," *J. Inst. Math. Applic. 23*, 203–13.

G.W. Stewart (1980). "The Efficient Generation of Random Orthogonal Matrices with an Application to Condition Estimators," *SIAM J. Num. Anal. 17*, 403–9.

G.W. Stewart (1981). "On the Implicit Deflation of Nearly Singular Systems of Linear Equations," *SIAM J. Sci. and Stat. Comp. 2*, 136-140.

G.W. Stewart (1983). "A Method for Computing the Generalized Singular Value Decomposition," in *Matrix Pencils* , ed. B. Kågström and A. Ruhe, Springer-Verlag, New York, pp. 207–20.

G.W. Stewart (1984). "A Second Order Perturbation Expansion for Small Singular Values," *Lin. Alg. and Its Applic. 56*, 231–236.

G.W. Stewart (1984). "Rank Degeneracy," *SIAM J. Sci. and Stat. Comp. 5*, 403–413.

G.W. Stewart (1984). "On the Asymptotic Behavior of Scaled Singular Value and QR Decompositions," *Math. Comp. 43*, 483–490.

G.W. Stewart (1985). "A Jacobi-Like Algorithm for Computing the Schur Decomposition of a Nonhermitian Matrix," *SIAM J. Sci. and Stat. Comp. 6*, 853–862.

G.W. Stewart (1987). "Collinearity and Least Squares Regression," *Statistical Science 2*, 68–100.

G.W. Stewart (1989). "On Scaled Projections and Pseudoinverses," *Lin. Alg. and Its Applic. 112*, 189–193.

G.W. Stewart (1992). "An Updating Algorithm for Subspace Tracking," *IEEE Trans. Signal Proc. 40*, 1535–1541.

G.W. Stewart (1993). "Updating a Rank-Revealing ULV Decomposition," *SIAM J. Matrix Anal. Appl. 14*, 494–499.

G.W. Stewart (1993). "On the Perturbation of LU Cholesky, and QR Factorizations," *SIAM J. Matrix Anal. Appl. 14*, 1141–1145.

G.W. Stewart (1993). "On the Early History of the Singular Value Decomposition," *SIAM Review 35*, 551–566.

G.W. Stewart (1994). "Perturbation Theory for Rectangular Matrix Pencils," *Lin. Alg. and Applic. 208/209*, 297–301.

G.W. Stewart (1994). "Updating URV Decompositions in Parallel," *Parallel Computing 20*, 151–172.

G.W. Stewart and J.-G. Sun (1990). *Matrix Perturbation Theory*, Academic Press, San Diego.

G.W. Stewart and G. Zheng (1991). "Eigenvalues of Graded Matrices and the Condition Numbers of Multiple Eigenvalues," *Numer. Math. 58*, 703–712.

M. Stewart and P. Van Dooren (1996). "Stability Issues in the Factorization of Structured Matrices," *SIAM J. Matrix Anal. Appl. 18*, to appear.

H.S. Stone (1973). "An Efficient Parallel Algorithm for the Solution of a Tridiagonal Linear System of Equations," *J. ACM 20*, 27–38.

H.S. Stone (1975). "Parallel Tridiagonal Equation Solvers," *ACM Trans. Math. Soft.1*, 289–307.

G. Strang (1988). "A Framework for Equilibrium Equations," *SIAM Review 30*, 283–297.

G. Strang (1993). *Introduction to Linear Algebra*, Wellesley-Cambridge Press, Wellesley MA.

V. Strassen (1969). "Gaussian Elimination is Not Optimal," *Numer. Math. 13*, 354–356.

J.-G. Sun (1982). "A Note on Stewart's Theorem for Definite Matrix Pairs," *Lin. Alg. and Its Applic. 48*, 331–339.

J.-G. Sun (1983). "Perturbation Analysis for the Generalized Singular Value Problem," *SIAM J. Numer. Anal. 20*, 611–625.

J.-G. Sun (1992). "On Condition Numbers of a Nondefective Multiple Eigenvalue," *Numer. Math. 61*, 265–276.

J.-G. Sun (1992). "Rounding Error and Perturbation Bounds for the Cholesky and LDL^T Factorizations," *Lin. Alg. and Its Applic. 173*, 77–97.

J.-G. Sun (1995). "A Note on Backward Error Perturbations for the Hermitian Eigenvalue Problem," *BIT 35*, 385–393.

J.-G. Sun (1995). "On Perturbation Bounds for the QR Factorization," *Lin. Alg. and Its Applic. 215*, 95–112.

X. Sun and C.H. Bischof (1995). "A Basis-Kernel Representation of Orthogonal Matrices," *SIAM J. Matrix Anal. Appl. 16*, 1184–1196.

P.N. Swarztrauber (1979). "A Parallel Algorithm for Solving General Tridiagonal Equations," *Math. Comp. 33*, 185-199.

P.N. Swarztrauber and R.A. Sweet (1973). "The Direct Solution of the Discrete Poisson Equation on a Disk," *SIAM J. Num. Anal. 10*, 900–907.

P.N. Swarztrauber and R.A. Sweet (1989). "Vector and Parallel Methods for the Direct Solution of Poisson's Equation," *J. Comp. Appl. Math. 27*, 241–263.

D.R. Sweet (1991). "Fast Block Toeplitz Orthogonalization," *Numer. Math. 58*, 613–629.

D.R. Sweet (1993). "The Use of Pivoting to Improve the Numerical Performance of Algorithms for Toeplitz Matrices," *SIAM J. Matrix Anal. Appl. 14*, 468–493.

R.A. Sweet (1974). "A Generalized Cyclic Reduction Algorithm," *SIAM J. Num. Anal. 11*, 506–20.

R.A. Sweet (1977). "A Cyclic Reduction Algorithm for Solving Block Tridiagonal Systems of Arbitrary Dimension," *SIAM J. Num. Anal. 14*, 706–20.

H.J. Symm and J.H. Wilkinson (1980). "Realistic Error Bounds for a Simple Eigenvalue and Its Associated Eigenvector," *Numer. Math. 35*, 113–26.

P.T.P. Tang (1994). "Dynamic Condition Estimation and Rayleigh-Ritz Approximation," *SIAM J. Matrix Anal. Appl. 15*, 331–346.

R.A. Tapia and D.L. Whitley (1988). "The Projected Newton Method Has Order $1 + \sqrt{2}$ for the Symmetric Eigenvalue Problem," *SIAM J. Num. Anal. 25*, 1376–1382.

G.L. Thompson and R.L. Weil (1970). "Reducing the Rank of $A - \lambda B$," *Proc. Amer. Math. Sec. 26*, 548–54.

G.L. Thompson and R.L. Weil (1972). "Roots of Matrix Pencils $Ay = \lambda By$: Existence, Calculations, and Relations to Game Theory," *Lin. Alg. and Its Applic. 5*, 207–26.

M.J. Todd (1990). "A Dantzig-Wolfe-like Variant of Karmarker's Interior-Point Linear Programming Algorithm," *Operations Research 38*, 1006–1018.

K.-C. Toh and L.N. Trefethen (1994). "Pseudozeros of Polynomials and Pseudospectra of Companion Matrices," *Numer. Math. 68*, 403–425.

L.N. Trefethen (1992). "Pseudospecta of Matrices," in *Numerical Analysis 1991*, D.F. Griffiths and G.A. Watson (eds), Longman Scientific and Technical, Harlow, Essex, UK, 234–262.

L.N. Trefethen and D. Bau III (1997). *Numerical Linear Algebra*, SIAM Publications, Philadelphia, PA.

L.N. Trefethen and R.S. Schreiber (1990). "Average-Case Stability of Gaussian Elimination," *SIAM J. Matrix Anal. Appl. 11*, 335–360.

L.N. Trefethen, A.E. Trefethen, S.C. Reddy, and T.A. Driscoll (1993). "Hydrodynamic Stability Without Eigenvalues," *Science 261*, 578–584.

W.F. Trench (1964). "An Algorithm for the Inversion of Finite Toeplitz Matrices," *J. SIAM 12*, 515–22.

W.F. Trench (1989). "Numerical Solution of the Eigenvalue Problem for Hermitian Toeplitz Matrices," *SIAM J. Matrix Anal. Appl. 10*, 135–146.

N.K. Tsao (1975). "A Note on Implementing the Householder Transformations." *SIAM J. Num. Anal. 12*, 53–58.

H.W. Turnbull and A.C. Aitken (1961). *An Introduction to the Theory of Canonical Matrices*, Dover Publications, New York, pp. 102–5.

F. Uhlig (1973). "Simultaneous Block Diagonalization of Two Real Symmetric Matrices," *Lin. Alg. and Its Applic. 7*, 281–89.

F. Uhlig (1976). "A Canonical Form for a Pair of Real Symmetric Matrices That Generate a Nonsingular Pencil," *Lin. Alg. and Its Applic. 14*, 189–210.

R. Underwood (1975). "An Iterative Block Lanczos Method for the Solution of Large Sparse Symmetric Eigenproblems," Report STAN-CS-75-495, Department of Computer Science, Stanford University, Stanford, California.

R.J. Vaccaro, ed. (1991). *SVD and Signal Processing II: Algorithms, Analysis, and Applications*. Elsevier, Amsterdam.

R.J. Vaccaro (1994). "A Second-Order Perturbation Expansion for the SVD," *SIAM J. Matrix Anal. Applic. 15*, 661–671.

R.A. Van De Geijn (1993). "Deferred Shifting Schemes for Parallel QR Methods," *SIAM J. Matrix Anal. Appl. 14*, 180–194.

J. Vandergraft (1971). "Generalized Rayleigh Methods with Applications to Finding Eigenvalues of Large Matrices," *Lin. Alg. and Its Applic. 4*, 353–68.

A. Van der Sluis (1969). "Condition Numbers and Equilibration Matrices," *Numer. Math. 14*, 14–23.

A. Van der Sluis (1970). "Condition, Equilibration, and Pivoting in Linear Algebraic Systems," *Numer. Math. 15*, 74–86.

A. Van der Sluis (1975). "Stability of the Solutions of Linear Least Squares Problem," *Numer. Math. 23*, 241–54.

A. Van der Sluis (1975). "Perturbations of Eigenvalues of Non-normal Matrices," *Comm. ACM 18*, 30–36.

A. Van der Sluis and H.A. Van Der Vorst (1986). "The Rate of Convergence of Conjugate Gradients," *Numer. Math. 48*, 543–560.

A. Van der Sluis and G.W. Veltkamp (1979). "Restoring Rank and Consistency by Orthogonal Projection," *Lin. Alg. and Its Applic. 28*, 257–78.

H. Van de Vel (1977). "Numerical Treatment of a Generalized Vandermonde systems of Equations," *Lin. Alg. and Its Applic. 17*, 149–74.

E.F. Van de Velde (1994). *Concurrent Scientific Computing*, Springer-Verlag, New York.

H.A. Van der Vorst (1982). "A Vectorizable Variant of Some ICCG Methods," *SIAM J. Sci. and Stat. Comp. 3*, 350–356.

H.A. Van der Vorst (1982). "A Generalized Lanczos Scheme," *Math. Comp. 39*, 559–562.

H.A. Van der Vorst (1986). "The Performance of Fortran Implementations for Preconditioned Conjugate Gradients on Vector Computers," *Parallel Computing 3*, 49–58.

H.A. Van der Vorst (1986). "An Iterative Solution Method for Solving $f(A)x = b$ Using Krylov Subspace Information Obtained for the Symmetric Positive Definite Matrix A," *J. Comp. and App. Math. 18*, 249–263.

H. Van der Vorst (1987). "Large Tridiagonal and Block Tridiagonal Linear Systems on Vector and Parallel Computers," *Parallel Comput. 5*, 45–54.

H. Van Der Vorst (1989). "High Performance Preconditioning," *SIAM J. Sci. and Stat. Comp. 10*, 1174–1185.

H.A. Van Der Vorst (1992). "BiCGSTAB: A Fast and Smoothly Converging Variant of the Bi-CG for the Solution of Nonsymmetric Linear Systems," *SIAM J. Sci. and Stat. Comp. 13*, 631–644.

P. Van Dooren (1979). "The Computation of Kronecker's Canonical Form of a Singular Pencil," *Lin. Alg. and Its Applic. 27*, 103–40.

P. Van Dooren (1981). "A Generalized Eigenvalue Approach for Solving Riccati Equations," *SIAM J. Sci. and Stat. Comp. 2*, 121–135.

P. Van Dooren (1981). "The Generalized Eigenstructure Problem in Linear System Theory," *IEEE Trans. Auto. Cont. AC-^6*, 111–128.

P. Van Dooren (1982). "Algorithm 590: DSUBSP and EXCHQZ: Fortran Routines for Computing Deflating Subspaces with Specified Spectrum," *ACM Trans. Math. Software 8*, 376–382.

S. Van Huffel (1992). "On the Significance of Nongeneric Total Least Squares Problems," *SIAM J. Matrix Anal. Appl. 13*, 20–35.

S. Van Huffel and H. Park (1994). "Parallel Tri- and Bidiagonalization of Bordered Bidiagonal Matrices," *Parallel Computing 20*, 1107–1128.

S. Van Huffel and J. Vandewalle (1987). "Subset Selection Using the Total Least Squares Approach in Collinearity Problems with Errors in the Variables," *Lin. Alg. and Its Applic. 88/89*, 695–714.

S. Van Huffel and J. Vandewalle (1988). "The Partial Total Least Squares Algorithm," *J. Comp. and App. Math. 21*, 333–342.

S. Van Huffel and J. Vandewalle (1988). "Analysis and Solution of the Nongeneric Total Least Squares Problem," *SIAM J. Matrix Anal. Appl. 9*, 360–372.

S. Van Huffel and J. Vandewalle (1989). "Analysis and Properties of the Generalized Total Least Squares Problem $AX \approx B$ When Some or All Columns in A are Subject to Error," *SIAM J. Matrix Anal. Appl. 10*, 294–315.

S. Van Huffel and J. Vandewalle (1991). *The Total Least Squares Problem: Computational Aspects and Analysis*, SIAM Publications, Philadelphia, PA.

S. Van Huffel, J. Vandewalle, and A. Haegemans (1987). "An Efficient and Reliable Algorithm for Computing the Singular Subspace of a Matrix Associated with its Smallest Singular Values," *J. Comp. and Appl. Math. 19*, 313–330.

S. Van Huffel and H. Zha (1991). "The Restricted Total Least Squares Problem: Formulation, Algorithm, and Properties," *SIAM J. Matrix Anal. Appl. 12*, 292–309.

S. Van Huffel and H. Zha (1993). "An Efficient Total Least Squares Algorithm Based On a Rank-Revealing Two-Sided Orthogonal Decomposition," *Numerical Algorithms 4*, 101–133.

H.P.M. van Kempen (1966). "On Quadratic Convergence of the Special Cyclic Jacobi Method," *Numer. Math. 9*, 19–22.

C.F. Van Loan (1973). "Generalized Singular Values With Algorithms and Applications," Ph.D. thesis, University of Michigan, Ann Arbor.

C.F. Van Loan (1975). "A General Matrix Eigenvalue Algorithm," *SIAM J. Num. Anal. 12*, 819–834.

C.F. Van Loan (1975). "A Study of the Matrix Exponential," Numerical Analysis Report No. 10, Dept. of Maths., University of Manchester, England.

C.F. Van Loan (1976). "Generalizing the Singular Value Decomposition," *SIAM J. Num. Anal. 13*, 76–83.

C.F. Van Loan (1977). "On the Limitation and Application of Padé Approximation to the Matrix Exponential," in *Padé and Rational Approximation*, ed. E.B. Saff and R.S. Varga, Academic Press, New York.

C.F. Van Loan (1977). "The Sensitivity of the Matrix Exponential," *SIAM J. Num. Anal. 14*, 971–81.

C.F. Van Loan (1978). "Computing Integrals Involving the Matrix Exponential," *IEEE Trans. Auto. Cont. AC-23*, 395–404.

C.F. Van Loan (1978). "A Note on the Evaluation of Matrix Polynomials," *IEEE Trans. Auto. Cont. AC-24*, 320–21.

C.F. Van Loan (1982). "Using the Hessenberg Decomposition in Control Theory," in *Algorithms and Theory in Filtering and Control*, D.C. Sorensen and R.J. Wets (eds), Mathematical Programming Study No. 18, North Holland, Amsterdam, pp. 102–11.

C.F. Van Loan (1984). "A Symplectic Method for Approximating All the Eigenvalues of a Hamiltonian Matrix," *Lin. Alg. and Its Applic. 61*, 233–252.

C.F. Van Loan (1985). "How Near is a Stable Matrix to an Unstable Matrix?," *Contemporary Mathematics, Vol. 47*, 465–477.

C.F. Van Loan (1985). "On the Method of Weighting for Equality Constrained Least Squares Problems," *SIAM J. Numer. Anal. 22*, 851–864.

C.F. Van Loan (1985). "Computing the CS and Generalized Singular Value Decomposition," *Numer. Math. 46*, 479–492.

C.F. Van Loan (1987). "On Estimating the Condition of Eigenvalues and Eigenvectors," *Lin. Alg. and Its Applic. 88/89*, 715–732.

C.F. Van Loan (1992). *Computational Frameworks for the Fast Fourier Transform*, SIAM Publications, Philadelphia, PA.

C.F. Van Loan (1997). *Introduction to Scientific Computing: A Matrix-Vector Approach Using Matlab*, Prentice Hall, Upper Saddle River, NJ.

J.M. Varah (1968). "The Calculation of the Eigenvectors of a General Complex Matrix by Inverse Iteration," *Math. Comp. 22*, 785–91.

J.M. Varah (1968). "Rigorous Machine Bounds for the Eigensystem of a General Complex Matrix," *Math. Comp. 22*, 793–801.

J.M. Varah (1970). "Computing Invariant Subspaces of a General Matrix When the Eigensystem is Poorly Determined," *Math. Comp. 24*, 137–49.

J.M. Varah (1972). "On the Solution of Block-Tridiagonal Systems Arising from Certain Finite-Difference Equations," *Math. Comp. 26*, 859–68.

J.M. Varah (1973). "On the Numerical Solution of Ill-Conditioned Linear Systems with Applications to Ill-Posed Problems," *SIAM J. Num. Anal. 10*, 257–67.

J.M. Varah (1979). "On the Separation of Two Matrices," *SIAM J. Num. Anal. 16*, 212–22.

J.M. Varah (1993). "Errors and Perturbations in Vandermonde Systems," *IMA J. Num. Anal. 13*, 1–12.

J.M. Varah (1994). "Backward Error Estimates for Toeplitz Systems," *SIAM J. Matrix Anal. Appl. 15*, 408–417.

R.S. Varga (1961). "On Higher-Order Stable Implicit Methods for Solving Parabolic Partial Differential Equations," *J. Math. Phys. 40*, 220–31.

R.S. Varga (1962). *Matrix Iterative Analysis*, Prentice-Hall, Englewood Cliffs, NJ.

R.S. Varga (1970). "Minimal Gershgorin Sets for Partitioned Matrices," *SIAM J. Num. Anal. 7*, 493–507.

R.S. Varga (1976). "On Diagonal Dominance Arguments for Bounding $\| A^{-1} \|_\infty$," *Lin. Alg. and Its Applic. 14*, 211–17.

S.A. Vavasis (1994). "Stable Numerical Algorithms for Equilibrium Systems," *SIAM J. Matrix Anal. Appl. 15*, 1108–1131.

S.A. Vavasis (1992). "Preconditioning for Boundary Integral Equations," *SIAM J. Matrix Anal. Appl. 13*, 905–925.

K. Veselić (1993). "A Jacobi Eigenreduction Algorithm for Definite Matrix Pairs," *Numer. Math. 64*, 241–268.

K. Veselić and V. Hari (1989). "A Note on a One-Sided Jacobi Algorithm," *Numer. Math. 56*, 627–633.

W.J. Vetter (1975). "Vector Structures and Solutions of Linear Matrix Equations," *Lin. Alg. and Its Applic. 10*, 181–88.

C. Vuik and H.A. van der Vorst (1992). "A Comparison of Some GMRES-like Methods," *Lin. Alg. and Its Applic. 160*, 131–162.

A. Wald (1940). "The Fitting of Straight Lines if Both Variables are Subject to Error," *Annals of Mathematical Statistics 11*, 284–300.

B. Waldén, R. Karlson, J. Sun (1995). "Optimal Backward Perturbation Bounds for the Linear Least Squares Problem," *Numerical Lin. Alg. with Applic. 2*, 271–286.

H.F. Walker (1988). "Implementation of the GMRES Method Using Householder Transformations," *SIAM J. Sci. Stat. Comp. 9*, 152–163.

R.C. Ward (1975). "The Combination Shift QZ Algorithm," *SIAM J. Num. Anal. 12*, 835–853.

R.C. Ward (1977). "Numerical Computation of the Matrix Exponential with Accuracy Estimate," *SIAM J. Num. Anal. 14*, 600–14.

R.C. Ward (1981). "Balancing the Generalized Eigenvalue Problem," *SIAM J. Sci and Stat. Comp. 2*, 141–152.

R.C. Ward and L.J. Gray (1978). "Eigensystem Computation for Skew-Symmetric and A Class of Symmetric Matrices," *ACM Trans. Math. Soft. 4*, 278–85.

D.S. Watkins (1982). "Understanding the QR Algorithm," *SIAM Review 24*, 427–440.

D.S. Watkins (1991). *Fundamentals of Matrix Computations*, John Wiley and Sons, New York.

D.S. Watkins (1993). "Some Perspectives on the Eigenvalue Problem," *SIAM Review 35*, 430–471.

D.S. Watkins and L. Elsner (1991). "Chasing Algorithms for the Eigenvalue Problem," *SIAM J. Matrix Anal. Appl. 12*, 374–384.

D.S. Watkins and L. Elsner (1991). "Convergence of Algorithms of Decomposition Type for the Eigenvalue Problem," *Lin.Alg. and Its Applic. 143*, 19–47.

D.S. Watkins and L. Elsner (1994). "Theory of Decomposition and Bulge-Chasing Algorithms for the Generalized Eigenvalue Problem," *SIAM J. Matrix Anal. Appl. 15*, 943–967.

G.A. Watson (1988). "The Smallest Perturbation of a Submatrix which Lowers the Rank of the Matrix," *IMA J. Numer. Anal. 8*, 295–304.

P.A. Wedin (1972). "Perturbation Bounds in Connection with the Singular Value Decomposition," *BIT 12*, 99–111.

P.A. Wedin (1973). "Perturbation Theory for Pseudo-Inverses," *BIT 13*, 217–32.

P.A. Wedin (1973). "On the Almost Rank-Deficient Case of the Least Squares Problem," *BIT 13*, 344–54.

M. Wei (1992). "Perturbation Theory for the Rank-Deficient Equality Constrained Least Squares Problem," *SIAM J. Num. Anal. 29*, 1462–1481.

M. Wei (1992). "Algebraic Properties of the Rank-Deficient Equality-Constrained and Weighted Least Squares Problems," *Lin. Alg. and Its Applic. 161*, 27–44.

M. Wei (1992). "The Analysis for the Total Least Squares Problem with More than One Solution," *SIAM J. Matrix Anal. Appl. 13*, 746–763.

O. Widlund (1978). "A Lanczos Method for a Class of Nonsymmetric Systems of Linear Equations," *SIAM J. Numer. Anal. 15*, 801–12.

J.H. Wilkinson (1961). "Error Analysis of Direct Methods of Matrix Inversion," *J. ACM 8*, 281–330.

J.H. Wilkinson (1962). "Note on the Quadratic Convergence of the Cyclic Jacobi Process," *Numer. Math. 6*, 296–300.

J.H. Wilkinson (1963). *Rounding Errors in Algebraic Processes*, Prentice-Hall, Englewood Cliffs, NJ.

J.H. Wilkinson (1965). *The Algebraic Eigenvalue Problem,* Clarendon Press, Oxford, England.

J.H. Wilkinson (1965). "Convergence of the LR, QR, and Related Algorithms," *Comp. J. 8,* 77–84.

J.H. Wilkinson (1968). "Global Convergence of Tridiagonal QR Algorithm With Origin Shifts," *Lin. Alg. and Its Applic. I,* 409–20.

J.H. Wilkinson (1968). "Almost Diagonal Matrices with Multiple or Close Eigenvalues," *Lin. Alg. and Its Applic. I,* 1–12.

J.H. Wilkinson (1968). "À Priori Error Analysis of Algebraic Processes," *Proc. International Congress Math.* (Moscow: Izdat. Mir, 1968), pp. 629–39.

J.H. Wilkinson (1971). "Modern Error Analysis," *SIAM Review 13,* 548–68.

J.H. Wilkinson (1972). "Note on Matrices with a Very Ill-Conditioned Eigenproblem," *Numer. Math. 19,* 176–78.

J.H. Wilkinson (1977). "Some Recent Advances in Numerical Linear Algebra," in *The State of the Art in Numerical Analysis,* ed. D.A.H. Jacobs, Academic Press, New York, pp. 1–53.

J.H. Wilkinson (1978). "Linear Differential Equations and Kronecker's Canonical Form," in *Recent Advances in Numerical Analysis* , ed. C. de Boor and G.H. Golub, Academic Press, New York, pp. 231–65.

J.H. Wilkinson (1979). "Kronecker's Canonical Form and the QZ Algorithm," *Lin. Alg. and Its Applic. 28,* 285–303.

J.H. Wilkinson (1984). "On Neighboring Matrices with Quadratic Elementary Divisors," *Numer. Math. 44,* 1–21.

J.H. Wilkinson and C. Reinsch, eds. (1971). *Handbook for Automatic Computation, Vol. 2, Linear Algebra,* Springer-Verlag, New York.

H. Wimmer and A.D. Ziebur (1972). "Solving the Matrix Equations $\Sigma f_p(A)g_p(A) = C$," *SIAM Review 14,* 318–23.

S. Winograd (1968). "A New Algorithm for Inner Product," *IEEE Trans. Comp. C-17,* 693–694.

M. Wolfe (1996). *High Performance Compilers for Parallel Computers,* Addison-Wesley, Reading MA.

A. Wouk, ed. (1986). *New Computing Environments: Parallel, Vector, and Systolic,* SIAM Publications, Philadelphia, PA.

H. Wozniakowski (1978). "Roundoff-Error Analysis of Iterations for Large Linear Systems," *Numer. Math. 30,* 301–314.

H. Wozniakowski (1980). "Roundoff Error Analysis of a New Class of Conjugate Gradient Algorithms," *Lin. Alg. and Its Applic. 29,* 507–29.

A. Wragg (1973). "Computation of the Exponential of a Matrix I: Theoretical Considerations," *J. Inst. Math. Applic. 11,* 369–75.

A. Wragg (1975). "Computation of the Exponential of a Matrix II: Practical Considerations," *J. Inst. Math. Applic. 15,* 273–78.

S.J. Wright (1993). "A Collection of Problems for Which Gaussian Elimination with Partial Pivoting is Unstable," *SIAM J. Sci. and Stat. Comp. 14,* 231–238.

J.M. Yohe (1979). "Software for Interval Arithmetic: A Reasonable Portable Package," *ACM Trans. Math. Soft. 5,* 50–63.

D.M. Young (1970). "Convergence Properties of the Symmetric and Unsymmetric Over-Relaxation Methods," *Math. Comp. 24,* 793–807.

D.M. Young (1971). *Iterative Solution of Large Linear Systems,* Academic Press, New York.

D.M. Young (1972). "Generalization of Property A and Consistent Ordering," *SIAM J. Num. Anal. 9,* 454–63.

D.M. Young and K.C. Jea (1980). "Generalized Conjugate Gradient Acceleration of Nonsymmetrizable Iterative Methods," *Lin. Alg. and Its Applic. 34,* 159–94.

L. Yu. Kolotilina and A. Yu. Yeremin (1993). "Factorized Sparse Approximate Inverse Preconditioning I: Theory," *SIAM J. Matrix Anal. Applic. 14,* 45–58.

L. Yu. Kolotilina and A. Yu. Yeremin (1995). "Factorized Sparse Approximate Inverse Preconditioning II: Solution of 3D FE Systems on Massively Parallel Computers," *Intern. J. High Speed Comput.* 7, 191–215.

H. Zha (1991). "The Restricted Singular Value Decomposition of Matrix Triplets," *SIAM J. Matrix Anal. Appl.* 12, 172–194.

H. Zha (1992). "A Numerical Algorithm for Computing the Restricted Singular Value Decomposition of Matrix Triplets," *Lin.Alg. and Its Applic.* 168, 1–25.

H. Zha (1993). "A Componentwise Perturbation Analysis of the QR Decomposition," *SIAM J. Matrix Anal. Appl.* 4, 1124–1131.

H. Zha and Z. Zhang (1995). "A Note on Constructing a Symmetric Matrix with Specified Diagonal Entries and Eigenvalues," *BIT* 35, 448–451.

H. Zhang and W.F. Moss (1994). "Using Parallel Banded Linear System Solvers in Generalized Eigenvalue Problems," *Parallel Computing* 20, 1089–1106.

Y. Zhang (1993). "A Primal-Dual Interior Point Approach for Computing the L_1 and L_∞ Solutions of Overdetermined Linear Systems," *J. Optimization Theory and Applications* 77, 323–341.

S. Zohar (1969). "Toeplitz Matrix Inversion: The Algorithm of W.F. Trench," *J. ACM* 16, 592–601.

Index